QMW Library
23 1045364 0

D1556702

WITHDRAWN
FROM STOCK
QMUL LIBRARY

ENCYCLOPEDIA OF PHYSICS

EDITOR IN CHIEF

S. FLÜGGE

VOLUME XXXI

CORPUSCLES AND RADIATION IN MATTER I

BY

T. ÅBERG · G. HOWAT · L. KARLSSON
J.A.R. SAMSON · H. SIEGBAHN · A.F. STARACE

EDITOR

W. MEHLHORN

WITH 268 FIGURES

SPRINGER-VERLAG
BERLIN HEIDELBERG NEW YORK
1982

HANDBUCH DER PHYSIK

HERAUSGEGEBEN VON

S. FLÜGGE

BAND XXXI

KORPUSKELN UND STRAHLUNG IN MATERIE I

VON

T. ÅBERG · G. HOWAT · L. KARLSSON
J.A.R. SAMSON · H. SIEGBAHN · A.F. STARACE

BANDHERAUSGEBER

W. MEHLHORN

MIT 268 FIGUREN

SPRINGER-VERLAG
BERLIN HEIDELBERG NEW YORK
1982

214348

Professor Dr. SIEGFRIED FLÜGGE

Physikalisches Institut der Universität, D-7800 Freiburg i. Br.

Professor Dr. WERNER MEHLHORN

Albert-Ludwigs-Universität, Fakultät für Physik
Hermann-Herder-Straße 3, D-7800 Freiburg i. Br.

ISBN 3-540-11313-4 Springer-Verlag Berlin Heidelberg New York
ISBN 0-387-11313-4 Springer-Verlag New York Heidelberg Berlin

Das Werk ist urheberrechtlich geschützt. Die dadurch begründeten Rechte, insbesondere die der Übersetzung, des Nachdruckes, der Entnahme von Abbildungen, der Funksendung, der Wiedergabe auf photomechanischem oder ähnlichem Wege und der Speicherung in Datenverarbeitungsanlagen bleiben, auch bei nur auszugsweiser Verwertung, vorbehalten. Die Vergütungsansprüche des § 54 Abs. 2 UrhG werden durch die „Verwertungsgesellschaft Wort", München, wahrgenommen.

© by Springer-Verlag Berlin Heidelberg 1982.
Library of Congress Catalog Card Number A 56-2942.
Printed in Germany.

Die Wiedergabe von Gebrauchsnamen, Handelsnamen, Warenbezeichnungen usw. in diesem Werk berechtigt auch ohne besondere Kennzeichnung nicht zu der Annahme, daß solche Namen im Sinne der Warenzeichen- und Markenschutz-Gesetzgebung als frei zu betrachten wären und daher von jedermann benutzt werden dürften.

Satz, Druck und Bindearbeiten: Universitätsdruckerei H. Stürtz AG, 8700 Würzburg.
2153/3130-543210

Preface

This volume deals with specific fields of atomic physics: photoionization, photoelectron spectroscopy and the Auger effect. Here a vast amount of experimental and theoretical results has emerged during the last two decades, indeed in the case of photoelectron spectroscopy the field itself has been developed only since the mid-fifties. Experimentally, this was mostly due to the development of high resolution electron spectrometers and the availability or the development of new radiation sources (e.g. synchrotron radiation, resonance radiation of rare gas discharge lamps, monochromatized characteristic X-radiation). Here, not only first-order effects with high precision but, even more important, also second-order effects caused by electron correlation were studied. Parallel to the development of new experimental methods, an important step beyond the independent-particle model was made on the theoretical side through the development of various approaches to treat electron correlation effects seen in photoionization and electron spectroscopy.

Volume 31 is divided into the following chapters: Theory of Photoionization, Atomic Photoionization, Photoelectron Spectroscopy, and Theory of the Auger Effect. A chapter on Auger Electron Spectrometry could not be included since the manuscript was not sent in due time.

Freiburg, September 1982 W. MEHLHORN

Contents

Theory of Atomic Photoionization. By Professor Dr. Anthony F. Starace, Behlen Laboratory of Physics, Department of Physics and Astronomy, The University of Nebraska, Lincoln, NE 68588 (USA). (With 40 Figures) 1

1. Introduction 1

I. General considerations 5

2. Introduction 5
3. Derivation of the general cross-section formula 6
4. The final-state wave function 7
5. Alternative expressions for the transition matrix element 23
6. Cross-section formulae in terms of the final-state channel wave functions 29
7. The angular distribution asymmetry parameter 31
8. The atomic oscillator strength 33
9. Limiting cases and approximations to the general formulae 36

II. The central potential model and its predictions 40

10. The central potential model 40
11. Predictions for the high-energy behavior of the cross section 42
12. Predictions for the near-threshold behavior of the cross section . . . 46
13. Bibliographic note on central potential model calculations 58

III. General methods for including electron correlation 59

a) Whole-space correlation theories 59
14. Introduction 59
15. Configuration interaction in the continuum 60
16. Procedures derived from a variational principle 66
17. Many-body perturbation theory 68
18. The random-phase approximation 75

b) Partitioned-space correlation theories 85
19. Introduction 85
20. Quantum defect theory 86
21. The R matrix theory 94
22. Discrete basis-set methods 100

IV. Specialized topics in the theory of photoionization 101

23. Photoelectron angular distributions 101
24. Multiple photoionization 113

Atomic Photoionization. By Professor Dr. James A.R. Samson, Behlen Laboratory of Physics, Department of Physics and Astronomy, The University of Nebraska, Lincoln, NE 68588 (USA). (With 75 Figures) 123

1. Introduction 123
2. Historical Review 124

I. Absorption of radiation 129
3. General concepts 129
4. Error analysis 134
II. Experimental techniques 136
5. Light sources 136
α) DC glow discharge 137
β) Spark discharge 138
γ) Synchrotron radiation 140
6. Monochromators and spectrographs 143
7. Absorption and photoionization cells 145
α) Double beam 145
β) Split beam 146
γ) Double ion chamber 147
δ) Heat pipes 149
ε) Atomic beams 152
8. Branching-ratio studies 152
α) Mass spectrometers 152
β) Electron energy analyzers 153
γ) Ion chambers 159
δ) Fluorescence spectrometry 159
9. Electron-impact "photoionization" 160
III. Photoionization measurements 163
10. Total photoionization cross sections 164
α) The rare gases 164
β) The alkali metals 169
γ) Cross sections of some miscellaneous atoms 173
δ) Ionic- and excited-state atoms 178
11. Partial photoionization cross sections 182
α) Ionization from specific orbitals 184
β) Multiple ionization 193
γ) Angular distribution of photoelectrons 199
12. Summary of atoms studied and the future 208
General references 212

Photoelectron Spectroscopy. By Dr. HANS SIEGBAHN, Docent of Physics and Dr. LEIF KARLSSON, Docent of Physics, Institute of Physics, University of Uppsala, S-75121 Uppsala 1 (Sweden). (With 139 Figures) 215
I. Introduction 215
1. Scope and abstract 215
2. Historical background 218
II. Experimental procedures and general features of photoelectron spectra . . 224
3. Photon sources 224
α) Introduction 224
β) X-ray sources. Monochromatization 227
γ) UV-source. Filtering and polarization 234
δ) Synchrotron radition 240
4. Electron energy analyzers and detectors 243
α) Introduction 243
β) Dispersive analyzers 244

γ) Nondispersive analyzers 250
δ) Preretardation in dispersive analyzers 252
ε) Electron detection 255
5. Sample preparation procedures 256
α) Vacuum requirements 256
β) Gases 257
γ) Liquids 260
δ) Solids and surfaces 262
6. General features of photoelectron spectra 265
α) Introduction 265
β) Calibration of photoelectron spectra 268
γ) Intensities of photoelectron lines 273
δ) Widths of photoelectron lines 278
7. Theoretical models for photoelectron spectra 283
α) Koopmans' theorem and ΔSCF binding energies 283
β) Correlation in initial and final states 287
γ) Intensities of one-electron and multi-electron transitions 289
δ) Vibrational excitations, Franck-Condon principle 294
III. Atomic photoelectron spectra 296
8. Core electron ionization 296
α) The Ne1s spectrum 296
β) The Xe4s, 4p spectrum 302
9. Valence electron ionization 306
α) Inner valence ionization 306
β) Outer valence ionization 309
10. Tables of electron binding energies and relative multiplet intensities for atoms 315
α) Atomic binding energies 315
β) Relative multiplet intensities for ionization of open shell atoms . . 328
IV. Core level studies in molecules and condensed matter 337
11. Chemical shifts of core electron lines 337
α) Introduction 337
β) Basic model considerations 339
γ) Orbital energy shifts 341
δ) Ground state potential models (GPM) 343
ε) Core shift models including relaxation 345
ζ) Relations to other experimental data. The thermodynamic model . 351
12. Multiple excitations in core ionization of molecules and solids 357
α) Molecular core electron spectra 357
β) Many-electron effects in solid core photoelectron spectra 360
γ) Vibrational excitations in core electron spectra 366
δ) Core line multiplet structure in molecules and solids 376
13. Core electron spectra of molecules on metal surfaces. ESCA diffraction 380
V. Valence level studies in molecules and condensed matter 387
14. General features of molecular valence electron spectra 387
α) Introduction 387
β) Electron binding energies of main bands 389
γ) Intensities of main bands 393
δ) Satellite structure 400
ε) Spin-orbit and Jahn-Teller interactions 403

15. Vibrational and rotational excitations in valence electron spectra . . . 409
α) Vibrational selection rules 409
β) Hot bands 412
γ) Qualitative correspondence between vibrational structure and orbital shapes 412
δ) Calculation of vibrational structure 418
ε) Rotational excitations 422
16. Angular distributions of molecular valence photoelectrons 424
α) General aspects 424
β) Photoelectron energy dependence of the asymmetry parameter . . 426
γ) Vibrational state dependence of the asymmetry parameter 427
δ) Rotational state dependence of the asymmetry parameter 428
ε) Asymmetry parameters for interpretation of valence spectra . . . 430
17. Molecular structure and bonding 431
α) MO methods for characterization of spectra and correlation with chemical reaction parameters 431
β) Biological molecules 436
γ) Molecular conformations, equilibrium systems and chemical reactions 438
δ) Short-lived species 440
ε) Ionically bonded molecules 442
ζ) Negative ions 444
18. Valence electron studies of single crystals and adsorbates 448
α) Selection rules in photoionization from solids 448
β) Two-dimensional band structures 450
γ) Three-dimensional band structures 452
δ) Angular studies of adsorbates 455
19. Density of states structure from photoelectron spectra 458
α) Solid valence photoelectron spectra in the high photon energy limit 458
β) Metal valence bands 459
γ) Alloy valence bands 461
δ) Formation of valence bands in extended systems 464
Appendix: Graph of electron binding energies for the elements 466

Theory of the Auger Effect. By Dr. Teijo Åberg, Associate Professor of Physics, Helsinki Laboratory of Physics, University of Technology, SF-02150 Espoo 15 (Finland), and Dr. George Howat, Computing Officer, Edinburgh Regional Computing Center, University of Edinburgh, Edinburgh EH9 3J2 (Scotland). (With 14 Figures) 469

Introduction 469

I. A nonrelativistic scattering approach to the decay of inner-shell vacancy states 471

1. Boundary conditions 472
2. Nonresonant multichannel scattering 476
3. Resonant scattering including channel interaction 480
4. Autoionization and the Auger effect as resonance scattering. The concept of nonradiative transitions 483
5. Spontaneous photon emission as resonance scattering 489
6. Final-state photon-electron interactions and additivity of nonradiative and radiative widths 492

II. Nonrelativistic theory of nonradiative transitions . . . 495
7. Central field model and classification of nonradiative transitions . . . 496
8. Many-electron theory and frozen-core approximation . . . 501
α) *LS*- coupling . . . 503
β) Intermediate coupling and configuration interaction . . . 506
γ) *jj* coupling . . . 508
δ) Variational principle and Hartree-Fock method . . . 509
ε) Electrons and holes . . . 513
9. Probability of nonradiative transitions . . . 516
α) *LS*- and *jj*-coupling limit . . . 517
β) Intermediate coupling and configuration interaction . . . 520
γ) The open-shell case . . . 521
δ) Relationship between the intensity and probability . . . 523
10. Beyond the independent-electron frozen-core approximation . . . 524
α) Relaxation . . . 524
β) Interchannel interaction . . . 526
γ) Perturbation approaches to the Auger effect . . . 527

III. Relativistic theory of nonradiative transitions . . . 530
11. Basic Hamiltonian and the electron-electron interaction . . . 531
12. A relativistic nonradiative transition probability . . . 536
13. Intermediate coupling . . . 540

IV. Angular distribution of Auger electrons . . . 543
14. General considerations . . . 544
α) Photoionization . . . 547
β) Ionization by high-energy electron impact . . . 548
15. Auger electron angular distributions . . . 549
α) Noncoincidence experiments . . . 549
β) Some features of noncoincidence angular distributions . . . 553
γ) Coincidence experiments . . . 554

V. The calculation of nonradiative transition energies . . . 555
16. General aspects of hole-state and binding energies . . . 556
α) The level shift . . . 556
β) Variational collapse . . . 557
γ) Perturbation analysis of binding energies . . . 560
δ) Relativistic and quantum-electrodynamical effects . . . 562
ε) Correlation . . . 565
17. Semiempirical treatments . . . 566
18. Ab initio calculations . . . 574
α) Self-consistent-field methods . . . 574
β) Correlation effects . . . 578

VI. The calculation of nonradiative transition amplitudes . . . 581
19. Computation of radial wave functions . . . 582
α) The nonrelativistic treatment . . . 582
β) Some approximate potential methods . . . 583
γ) The relativistic treatment . . . 587
δ) Some practical considerations . . . 589
20. On factors affecting the nonradiative transition amplitudes . . . 593
α) Relativistic effects . . . 593
β) The choice of the local potential and kinetic energy . . . 598

γ) The effect of bound orbital basis set and relaxation 600
δ) Correlation effects 604

Appendices
A. Some formulae for the treatment of the continuous spectrum 608
B. Relationship between the configuration interaction and the projection operator approach in the theory of resonances 610
C. A derivation of the Fano profile from scattering theory 613
D. Equivalence of electron and hole wave functions with respect to symmetry 615
E. Conversion and normalization factors 617

General references 619

Subject Index 621

Theory of Atomic Photoionization*

By

A.F. Starace

With 40 Figures

1. Introduction. The theory of atomic photoionization processes has advanced rapidly in response to the challenge presented by new experimental measurements, having accuracies of $\pm 10\%$, that were obtained in the 1960's in the vacuum ultraviolet wavelength range.[1] By the late 1960's it was clear that these new measurements of photoionization cross sections, primarily for the rare gases, could not be completely understood by even the best theoretical independent-particle model calculations. These calculations usually agreed qualitatively with experiment but often differed in magnitude by factors of two or more in the vicinity of the first few ionization thresholds.[2] Such discrepancies signalled the importance of electron correlations within and between the outer subshells of all but the lightest elements. However, of the electron correlations of importance to atomic photoionization the only kinds that were understood at the time were those that induced autoionizing transitions[2-4] or that coupled alternative photoelectron final-state channels.[5,6] The influence on the cross section of those electron correlations responsible for multiple excitations was grossly underestimated, largely because it was not clear in the 1960's how to handle a multitude of such correlations. General theories for calculating the influence of electron correlation on photoionization cross sections were only in their developing stages in the 1960's,[7-9] and detailed

* The Manuscript was completed in December 1978

[1] Samson, J.A.R. (1966): Adv. At. Mol. Phys. *2*, 177

[2] Fano, U., Cooper, J.W. (1968): Rev. Mod. Phys. *40*, 441

[3] Fano, U. (1961): Phys. Rev. *124*, 1866

[4] Fano, U., Cooper, J.W. (1965): Phys. Rev. *137*, A 1364

[5] Seaton, M.J. (1953): Philos. Trans. R. Soc. London Ser. A *245*, 469

[6] Henry, R.J.W., Lipsky, L. (1967): Phys. Rev. *153*, 51

[7] The quantum defect theory was being developed by Seaton, M.J. (1958): Mon. Not. Roy. Astron. Soc. *118*, 504; (1966): Proc. Phys. Soc. *88*, 801. An application of the theory to the calculation of the calcium photoionization cross section was carried out by Moores, D.L. (1966): Proc. Phys. Soc. *88*, 843

[8] The many-body perturbation theory for atoms was being developed by Kelly, H.P. (1968): Adv. Theor. Phys. *2*, 75; (1969): Adv. Chem. Phys. *14*, 129. An application to beryllium photoionization was presented by Kelly, H.P. (1964): Phys. Rev. *136*, B 896. An application to lithium photoionization was presented by Chang, E.S., McDowell, M.R.C. (1968): Phys. Rev. *176*, 126

[9] The random-phase approximation was first applied to atomic photoionization by Altick, P.L., Glassgold, A.E. (1964): Phys. Rev. *133*, A 632; beryllium, magnesium, calcium, and strontium were treated

theoretical studies of multiple-ionization cross sections[10] and of photoelectron angular distributions[11-13] were only just beginning.

By the latter half of the 1970's, several theories of atomic photoionization had been developed which were capable of reproducing the experimental data on total and partial photoionization cross sections for the rare gases to within $\pm 10\%$ in most cases. These theories may be divided into two groups. The multichannel quantum defect theory[7,14,15] and the R-Matrix Theory[16,17] may be described as partitioned-space correlation theories. They treat electron correlations within a sphere of radius r_0 about an atom or ion and assume that for radii greater than r_0 the photoelectron sees a long-range potential, which is usually a pure Coulomb field. Hence the photoelectron's wave function may often be described analytically and in any case may be calculated easily in the region $r > r_0$. These theories are computationally economical when long-range correlation or polarization effects may be ignored for $r > r_0$ (r_0 is typically a few Bohr radii), thereby permitting the partial use of analytic results.

The many-body perturbation theory (MBPT)[18] and the random-phase approximation (RPA)[19,20] on the other hand may be described as whole-space correlation theories, treating the electron correlations they consider over the full range of r. Both of these theories do all calculations completely numerically. The MBPT is a general procedure for treating any kind of weak electron correlation. It is most useful in identifying the relative contributions of alternative kinds of electron correlation; by the third order of perturbation theory, however, the number of terms to be considered becomes very large. Nevertheless, the MBPT may often be employed to treat strong electron correlations by means of either or both of the following two procedures. 1) One may try switching to a different zero-order basis set; often one finds the electron correlations, which are strong in the initial basis set, become weak within a different basis set. 2) Alternatively, one may try to sum the strong electron correlations to all orders of perturbation theory; usually such infinite summations can only be carried out, however, for certain types of electron correlation. These two procedures have greatly extended the range of problems that may be treated by the MBPT. The RPA, in particular, is an example of the use of infinite-order perturbation theory to treat certain kinds of electron correlation. The particular electron correlations considered in the RPA are now known to be the dominant ones for the rare gases and probably for all

---

[10] Byron, F.W., Joachain, C.J. (1967): Phys. Rev. *164*, 1

[11] Cooper, J., Zare, R.N. (1968): J. Chem. Phys. *48*, 942

[12] Cooper, J., Zare, R.N. (1969): Lectures in theoretical physics, Geltman, S., Mahanthappa, K.T., Brittin, W.E. (eds.), New York: Gordon and Breach, Vol. XI-C, pp. 317–37

[13] Cooper, J.W., Manson, S.T. (1969): Phys. Rev. *177*, 157

[14] Dubau, J., Wells, J. (1973): J. Phys. B *6*, 1452

[15] Fano, U. (1975): J. Opt. Soc. Am. *65*, 979

[16] Burke, P.G., Robb, W.D. (1975): Adv. At. Mol. Phys. *11*, 143

[17] Burke, P.G., Taylor, K.T. (1975): J. Phys. B *8*, 2620

[18] Kelly, H.P. (1975): Photoionization and other probes of many-electron interactions, Wuilleumier, F.J. (ed.), New York: Plenum Press, pp. 83–109

[19] Amusia, M.Ya, Cherepkov, N.A. (1975): Case Stud. At. Phys. *5*, 47

[20] Wendin, G. (1971): J. Phys. B *4*, 1080; (1972): ibid., *5*, 110; (1973): ibid., *6*, 42

closed-shell atoms, but their importance for other atoms is only beginning to be investigated.[21,22]

Despite the successes of these four theories for the rare gases and a few other atoms (generally either closed shell or else having low atomic number), theory has yet to develop the same level of understanding for heavier, open-shell atoms. It remains true today that the interaction of radiation with 75% of the atoms in the periodic table is largely unstudied either theoretically or experimentally.[23] This lack of interest in open-shell atoms has been due, on the one hand, to the experimental difficulty of working with atoms that have high melting points or that readily form covalent bonds and, on the other hand, to the theoretical difficulties of working with nonspherically symmetric systems having many final-state channels.

For these reasons, current theoretical knowledge of the interaction of radiation with most elements in the periodic table is still based on calculational surveys[24-27] employing an independent-particle model. Qualitatively, the independent-particle model gives a good overall description of atomic photoionization processes. Quantitatively, as mentioned above, it is not accurate near the outer-subshell ionization thresholds. But it does agree with experiment for high photon energies[28,29], i.e., $\hbar\omega > 1$ keV, where photoionization occurs from inner subshells. It has also been shown to agree with more elaborate calculations for highly stripped positive ions[30], where the nuclear attraction is dominant. Furthermore the recent burgeoning of experimental interest[31] in photoelectron angular distributions is due in large part to theoretical predictions of strong oscillatory behavior in the photoelectron asymmetry parameter as a function of energy that were based on independent-particle model calculations.[32,33]

In addition to these studies of total and partial atomic cross sections and photoelectron angular distributions, some rather specialized areas of atomic photoionization theory are attracting increasing attention. Application of MBPT to the calculation of multiple photoionization cross sections has led to surprising success[34,35] even though the difficult problem of describing the

---

[21a] Starace, A.F., Armstrong, Jr. L. (1976): Phys. Rev. A *13*, 1850

[21b] Starace, A.F., Shahabi, S. (1980): Phys. Scri. *21*, 368; (1982): Phys. Rev. A *25*, 2135

[22] Cherepkov, N.A., Chernysheva, L.V. (1977): Phys. Lett. *60* A, 103

[23] Hudson, R.D., Kieffer, L.J. (1971): At. Data *2*, 205

[24] Burgess, A., Seaton, M.J. (1960): Mon. Not. R. Astron. Soc. *120*, 121

[25] Combet-Farnoux, F. (1967): C.R. Acad. Sci. Paris *264* B, 1728

[26] Manson, S.T., Cooper, J.W. (1968): Phys. Rev. *165*, 126

[27] McGuire, E.J. (1968): Phys. Rev. *175*, 20

[28] Pratt, R.H., Ron, A., Tseng, H.K. (1973): Rev. Mod. Phys. *45*, 273

[29] Cooper, J.W. (1975): Atomic inner-shell processes, Crasemann, B. (ed.), New York: Academic Press, pp. 159–199

[30] Msezane, A., Reilman, R.F., Manson, S.T., Swanson, J.R., Armstrong, Jr., L. (1977): Phys. Rev. A *15*, 668

[31] Samson, J.A.R. (1983): This volume, p. 123

[32] Manson, S.T. (1971): Phys. Rev. Lett. *26*, 219

[33] Kennedy, D.J., Manson, S.T. (1972): Phys. Rev. A *5*, 227

[34] Chang, T.N., Poe, R.T. (1975): Phys. Rev. A *12*, 1432

[35] (a) Carter, S.L., Kelly, H.P. (1976): J. Phys. B *9*, L565; (b) (1977): Phys. Rev. A *16*, 1525

final-state wave function for two or more continuum electrons was avoided. The availability of lasers for experimental studies of negative-ion photodetachment has led to a corresponding increase in activity in the calculation of photodetachment cross sections.[36] Similarly, the use of lasers in experimental studies and the needs of plasma and fusion research have combined with the traditional astrophysical interest in photoionization of positive ions and of excited states of atoms to make these fairly active fields of current research.

In short, the theory of atomic photoionization is now in great ferment as a result of the achievements surveyed above. It is the purpose of this review to present the essence of current theoretical understanding of the photoionization process and consequently the physical basis for the success of the established approximation methods. It is not the purpose of the present review, however, to give an historical account of the subject. Nor is it intended to present a critical survey of all existing theoretical work. Rather, only a few works, chosen primarily for illustrative purposes, will be discussed in detail; other important works will be cited but will only be discussed briefly. Bibliographies of all known theoretical calculations on photoionization to December 1979 have been compiled.[37]

More specifically, this article reviews the nonrelativistic theory of atomic photoionization. That is, we are concerned with photon energies $\hbar\omega \lesssim 10^4$ eV at which the photoelectron's velocity is nonrelativistic and at which retardation effects are negligible. Chapter I presents derivations from first principles of the appropriate formulas for the photoionization cross section and the photoelectron angular distribution in *LS* coupling. Chapter II is devoted to a complete description of the central potential model predictions for photoionization at both low and high (but nonrelativistic) photon energies. Chapter III describes the essential physical content of each of the major approximation methods currently used to include electron correlation effects in the theoretical description of atomic photoionization. The focus in this section is on the accurate calculation of photoionization cross sections near ionization thresholds. Extensive references to the original literature are presented to guide the interested reader to more detailed descriptions of each method. Chapter IV describes progress in understanding some specialized topics in the theory of atomic photoionization: photoelectron angular distributions and multiple (primarily double) photoionization. In order to keep the size of the present review

[36] Massey, H. (1976): Negative Ions, 3rd edn., Cambridge, England: Cambridge University Press

[37a] Kieffer, L.J. (1976): Bibliography of low energy electron and photon cross section data (through December 1974), NBS Special Publication 426, Washington, DC: U.S. Government Printing Office

[37b] Gallagher, J.W., Rumble, Jr., J.R., Beaty, E.C. (1979): Bibliography of low energy electron and photon cross section data (January 1975 through December 1977), NBS Special Publication 426, Supplement 1, Washington, DC: U.S. Government Printing Office

[37c] Gallagher, J.W., Beaty, E.C. (1980): Bibliography of low energy electron and photon cross section data (1978), JILA Information Center Report 18, Boulder, Colorado: University of Colorado

[37d] Gallagher, J.W., Beaty, E.C. (1981): Bibliography of low energy electron and photon cross section data (1979), JILA Information Center Report 21, Boulder, Colorado: University of Colorado

manageable, the following more specialized aspects of the theory of photoionization are omitted: photoionization of ions and of excited atoms, photoionization in the presence of external electric or magnetic fields, multiphoton ionization, and the relativistic theory of photoionization.

A number of other reviews of atomic photoionization complement the present review. The theoretical developments up to the mid-1960's are reviewed by STEWART,[38] BIBERMAN and NORMAN,[39] and FANO and COOPER.[2] Photoionization of one- and two-electron atoms has been reviewed by BETHE and SALPETER[40] and BRANSCOMB.[41] The theoretical treatment of atomic photoionization at very high photon energies is reviewed by PRATT et al.[28] and COOPER.[29] These topics will be touched on only briefly in the present review and the interested reader is referred to these other reviews for more detailed treatments.

I. General considerations

2. Introduction. The earliest theoretical treatments of atomic photoionization date from the early days of quantum mechanics.[42–46] A modern, fairly complete derivation of the formulas for the photoionization cross section is given by SOBEL'MAN.[47] Another useful treatment is that of BETHE and SALPETER.[40] In what follows we indicate the approximations and mathematical procedures used to obtain formulas for the total, partial, and differential photoionization cross sections from first principles. The plan of this section is as follows: In Sect. 3 we examine the coupling of electromagnetic radiation to a many-electron atom and obtain the general formula for the photoionization cross section in the electric dipole approximation. Expressions are then derived in Sect. 4 for the exact final-state wave function in *LS* coupling in terms of the eigenstates of a zero-order model Hamiltonian; both the single-channel and many-channel cases are treated. Alternative forms for the electric dipole transition matrix element are discussed in Sect. 5. We then discuss and present expressions for the total, partial, and differential photoionization cross sections (Sect. 6), the angular distribution asymmetry parameter (Sect. 7), and the atomic oscillator strength (Sect. 8) in terms of the exact many-channel final-state wave function. Finally, in Sect. 9 we examine some useful limits and approximations to the general formulae.

[38] Stewart, A.L. (1967): Adv. At. Mol. Phys. *3*, 1

[39] Biberman, L.M., Norman, G.E. (1967): Usp. Fiz. Nauk *91*, 193 [Sov. Phys. Usp. *10*, 52]

[40] Bethe, H.A., Salpeter, E.E. (1957): Quantum mechanics of one- and two electron atoms, Berlin, Heidelberg, New York: Springer, § IV

[41] Branscomb, L.M. (1969): Physics of the one- and two-electron atoms, Bopp, F., Kleinpoppen, H. (eds.), Amsterdam: North Holland, pp. 669–699

[42] Wentzel, G. (1926): Z. Phys. *40*, 574

[43] Sugiura, M.Y. (1927): J. Phys. (Paris) *8*, 113

[44] Hargreaves, J. (1929): Proc. Cambridge Philos. Soc. *25*, 75

[45] Stobbe, M. (1930): Ann. Phys. (Leipzig) *7*, 661

[46] Hall, H. (1936): Rev. Mod. Phys. *8*, 358

[47] Sobel'man, I.I. (1972): An introduction to the theory of atomic spectra, Oxford: Pergamon Press, Chap. 9 (see especially § 34.2)

Specifically, we examine both the Bates approximation,[48] i.e., the limit of vanishing interchannel interactions, and the central potential approximation.

Regarding units, in Sect. 3 we explicitly indicate all physical constants in order to avoid any confusion concerning the sign of the coupling of the electromagnetic field to the atomic electrons. In later sections, however, we simplify the notation by adopting atomic units, i.e., $|e|=m=\hbar=1$.

3. Derivation of the general cross-section formula. It is assumed that an N-electron atom having nuclear charge Z may be described according to the nonrelativistic, spin-independent Hamiltonian

$$H=\sum_{i=1}^{N}\left(\frac{p_i^2}{2m}-\frac{Ze^2}{r_i}\right)+\sum_{i>j=1}^{N}\frac{e^2}{|\boldsymbol{r}_i-\boldsymbol{r}_j|}, \tag{3.1}$$

where the terms in brackets describe the kinetic and potential energy of each electron in the attractive Coulomb field of the nucleus and the last terms describe the electrostatic interelectron repulsive potential energy. The interaction of this atom with radiation is described by the additional terms obtained upon substituting $\boldsymbol{p}_i+\frac{|e|}{c}\boldsymbol{A}(\boldsymbol{r}_i,t)$ for $\boldsymbol{p}_i$ in (3.1). Thus, the interaction Hamiltonian is

$$H_{\text{int}}=\sum_{i=1}^{N}\left[+\frac{|e|}{2mc}(\boldsymbol{p}_i\cdot\boldsymbol{A}(\boldsymbol{r}_i,t)+\boldsymbol{A}(\boldsymbol{r}_i,t)\cdot\boldsymbol{p}_i)+\frac{e^2}{2mc^2}|\boldsymbol{A}(\boldsymbol{r}_i,t)|^2\right], \tag{3.2}$$

where $\boldsymbol{A}(\boldsymbol{r}_i,t)$ is the vector potential for the incident radiation field.

H_{int} may be further simplified by adopting the following conventions and approximations: 1) If the vector potential is defined in the Coulomb gauge, in which $\boldsymbol{\nabla}\cdot\boldsymbol{A}=0$, then $\boldsymbol{p}_i$ and $\boldsymbol{A}(\boldsymbol{r}_i,t)$ commute and the first two terms in (3.2) may be combined. 2) The field – or more precisely $|e|\boldsymbol{A}/c$ – is treated as a perturbation so that the last term in (3.2) can be dropped, especially since it corresponds to two-photon processes. 3) Through a "happy accident"[49] the correct quantum-mechanical transition rate for photoionization may be calculated by treating the incident radiation classically, provided the proper amplitude for the vector potential is chosen. We choose the following form:[50]

$$\boldsymbol{A}(\boldsymbol{r}_i,t)=\left(\frac{2\pi c^2\hbar}{\omega V}\right)^{1/2}\hat{\boldsymbol{\varepsilon}}\exp\mathrm{i}(\boldsymbol{\kappa}\cdot\boldsymbol{r}_i-\omega t), \tag{3.3}$$

where $\boldsymbol{\kappa}$ and ω are the wave vector and angular frequency of the incident radiation, $\hat{\boldsymbol{\varepsilon}}$ is the polarization direction, and V is the spatial volume. 4) In the electric dipole approximation the factor $\exp(\mathrm{i}\boldsymbol{\kappa}\cdot\boldsymbol{r}_i)$ is replaced by unity. This is appropriate for photoionization of the outer subshell electrons, which require

---

[48] Bates, D.R. (1946): Mon. Not. R. Astron. Soc. *106*, 432

[49] Sakurai, J.J. (1967): Advanced quantum mechanics, Reading, Mass.: Addison-Wesley Publishing Co., p. 37

[50] Ref. [47], p. 292

photons with wavelengths $\lambda \gtrsim 100\,\text{Å}$, since the orbit radii of these outer electrons is only of the order of $1\,\text{Å}$. Therefore substituting (3.3) in (3.2) and invoking the above approximations and conventions, we obtain

$$H_{\text{int}} = +\frac{|e|}{mc}\left(\frac{2\pi c^2 \hbar}{\omega V}\right)^{1/2} \sum_{i=1}^{N} \hat{\boldsymbol{\varepsilon}} \cdot \boldsymbol{p}_i \exp(-\mathrm{i}\omega t). \tag{3.4}$$

If one describes the initial state of the N-electron atom by ψ_0 and the final state – in which the ionized electron has wave vector $\boldsymbol{k}$ – by $\psi_{\boldsymbol{k}}^-$, then standard procedures of time-dependent perturbation theory show that the transition rate in first order is given by[51]

$$dW_{\boldsymbol{k}} = \frac{2\pi}{\hbar}\left| +\frac{|e|}{mc}\left(\frac{2\pi c^2 \hbar}{\omega V}\right)^{1/2} \hat{\boldsymbol{\varepsilon}} \cdot \langle \psi_0 | \sum_{i=1}^{N} \boldsymbol{p}_i | \psi_{\boldsymbol{k}}^- \rangle \right|^2 \delta(E_k + E_{\text{B}} - \hbar\omega)\, k^2\, dk\, d\Omega, \tag{3.5}$$

where E_{B} is the binding energy of the ejected electron, $\mathrm{d}\Omega$ is an element of solid angle about the direction $\hat{\boldsymbol{k}}$, and the final-state wave functions have been normalized according to

$$\int (\psi_{\boldsymbol{k}}^-)^* \psi_{\boldsymbol{k}'}^- \, d^3 \boldsymbol{r} = \delta(\boldsymbol{k} - \boldsymbol{k}'). \tag{3.6}$$

Upon dividing the transition rate by the incident photon current density, c/V and integrating over dk, we obtain the differential photoionization cross section

$$\frac{d\sigma}{d\Omega} = \frac{4\pi^2}{\omega c}\, k \left(\frac{e^2}{m \hbar^2}\right) \left| \hat{\boldsymbol{\varepsilon}} \cdot \langle \psi_i | \sum_{i=1}^{N} \boldsymbol{p}_i | \psi_{\boldsymbol{k}}^- \rangle \right|^2, \tag{3.7}$$

where $\dfrac{\hbar^2 k^2}{2m} + E_{\text{B}} = \hbar\omega$.

4. The final-state wave function. Further reduction of (3.7) requires specification of the initial- and final-state wave functions. Since the interaction H_{int} is only treated in first order, the wave functions are eigenstates of the atomic Hamiltonian H, given by (3.1). Because of the complexity of H for a many-electron atom, these eigenstates must usually be obtained in terms of the eigenstates of a simpler, model Hamiltonian using the theory of configuration interaction for the initial-state wave function and the theory of scattering for the final-state wave function. Specification of the final-state wave function is the more difficult problem partly because of the long-range Coulomb field and partly because of the existence, in general, of many final-state channels. In addition, for ionization processes the final-state wave function satisfies a boundary condition which is different from that applicable to collision processes. In what follows, we examine each of these difficulties in specifying the final-state wave function. We examine first the boundary condition appropriate for ionization processes in the particularly simple one-channel case of an electron moving in

[51] Landau, L.D., Lifshitz, E.M. (1965): Quantum mechanics, 2nd edn., Oxford: Pergamon Press, § 42

a short-range potential. We then indicate the modifications of the single-channel one-electron wave function introduced by the presence of a long-range Coulomb potential. Lastly, we consider the final-state wave function in the many-electron, many-channel case. (As noted in Sect. 2, henceforth for simplicity we shall use atomic units: $\hbar = m = |e| = 1$.)

α) Boundary condition for ionization processes. Consider a system in which the photoelectron moves in a spherically symmetric, short-range potential. That is, we assume the long-range Coulomb potential is not present in the final state, as in the photodetachment of a negative ion. (We also ignore any long-range polarization potential.) The many-electron Hamiltonian H in (3.1) is thus replaced here by the single-electron Hamiltonian H_0,

$$H_0 = \tfrac{1}{2} p^2 + V(r), \tag{4.1}$$

where $V(r)$ decreases asymptotically faster than r^{-2}. The determination of the positive-energy eigenstates of H_0 is the subject of scattering theory, which is treated thoroughly by a number of authors.[52-58] Here we only discuss some of the important results of the scattering theory analysis, without always giving detailed proofs. Our aim is to familiarize the reader with the asymptotic boundary condition imposed on the final-state wave function in photoionization reactions.

The positive-energy eigenstates of the Hamiltonian in (4.1) may be chosen to satisfy a number of alternative boundary conditions at large radial distances r. For a given energy $E = k^2/2$, two such energy eigenstates are the so-called in (i.e., $+$) and out (i.e., $-$) states $\psi_{\boldsymbol{k}}^{\pm}(\boldsymbol{r})$, which have the following asymptotic behaviors:[59]

$$\psi_{\boldsymbol{k}}^{\pm}(\boldsymbol{r}) \xrightarrow[r\to\infty]{} \frac{1}{(2\pi)^{3/2}} [\exp(\mathrm{i}\boldsymbol{k}\cdot\boldsymbol{r}) + f^{\pm}(\boldsymbol{k}', \boldsymbol{k}) \exp(\pm \mathrm{i}kr)/r]. \tag{4.2}$$

[52] Goldberger, M.L., Watson, K.M. (1964): Collision theory, New York: John Wiley and Sons

[53] Joachain, C.J. (1975): Quantum collision theory, Amsterdam: North-Holland Publishing Co.

[54] Mott, N.F., Massey, H.S.W. (1965): The theory of atomic collisions, 3rd edn., London: Oxford University Press

[55] Newton, R.G. (1966): Scattering theory of waves and particles, New York: McGraw-Hill Book Co.

[56] Rodberg, L.S., Thaler, R.M. (1967): Introduction to the quantum theory of scattering, New York: Academic Press

[57] Roman, P. (1965): Advanced quantum theory, Reading, Mass.: Addison-Wesley Publishing Co., Inc.

[58] Wu, T.Y., Ohmura, T. (1962): Quantum theory of scattering, Englewood Cliffs, New Jersey: Prentice-Hall, Inc.

[59] The in- and out-state eigenfunctions are solutions of the following integral equation (see, e.g. Ref. [57], Chap. 4):

$$\psi_{\boldsymbol{k}}^{\pm}(\boldsymbol{r}) = \frac{1}{(2\pi)^{3/2}} \left(\mathrm{e}^{\mathrm{i}\boldsymbol{k}\cdot\boldsymbol{r}} - (2\pi)^{1/2} \int d^3\boldsymbol{r}' \frac{\mathrm{e}^{\pm \mathrm{i}k|\boldsymbol{r}-\boldsymbol{r}'|}}{|\boldsymbol{r}-\boldsymbol{r}'|} V(r') \psi_{\boldsymbol{k}}^{\pm}(\boldsymbol{r}') \right).$$

One easily sees from this equation that for large r these eigenstates have the form of (4.2), in which the scattering amplitude is defined by

$$f^{\pm}(\boldsymbol{k}', \boldsymbol{k}) \equiv -(2\pi)^{1/2} \int d^3\boldsymbol{r}' \mathrm{e}^{\mp \mathrm{i}k\hat{\boldsymbol{r}}\cdot\boldsymbol{r}'} V(r') \psi_{\boldsymbol{k}}^{\pm}(\boldsymbol{r}'),$$

where $\boldsymbol{k}' \equiv k\hat{\boldsymbol{r}}$

The meaning of the vector subscript $\boldsymbol{k}$ on the left of (4.2) will be given shortly; we note here, though, that it does *not* mean that the momentum $\boldsymbol{k}$ is a well-defined quantity for eigenstates of the Hamiltonian in (4.1). On the right of (4.2), the factor $(2\pi)^{-3/2}$ assures that the eigenstates are normalized according to (3.6). The exponential $\exp(\mathrm{i}\boldsymbol{k}\cdot\boldsymbol{r})$ is the wave function for the plane-wave state of momentum $\boldsymbol{k}$. The second term in the parentheses indicates either outgoing [i.e., $\exp(\mathrm{i}kr)/r$] or incoming [i.e., $\exp(-\mathrm{i}kr)/r$] spherical waves having an amplitude $f^{\pm}(\boldsymbol{k}',\boldsymbol{k})$. The function $f^{\pm}(\boldsymbol{k}',\boldsymbol{k})$ is called the scattering amplitude. It represents the probability amplitude for scattering between the momentum states $\boldsymbol{k}$ and $\boldsymbol{k}'$, where for the case of potential scattering $\boldsymbol{k}'$ has the same magnitude as $\boldsymbol{k}$ but is directed along $\hat{\boldsymbol{r}}$, i.e.,

$$\boldsymbol{k}'\equiv k\hat{\boldsymbol{r}}.$$

The notation $f^{\pm}(\boldsymbol{k}',\boldsymbol{k})$ is convenient for later generalization to the many-channel case, but often for potential scattering (in which there is only a single channel) one uses the simpler notation,

$$f^{\pm}(\theta)\equiv f^{\pm}(\boldsymbol{k}',\boldsymbol{k}),$$

where θ is the angle between $\boldsymbol{k}$ and $\hat{\boldsymbol{r}}$ and the dependence of $f^{\pm}(\theta)$ on the magnitude of the momentum, k, is understood. (Obviously, since $V(r)$ is spherically symmetric, the eigenstates $\psi_{\boldsymbol{k}}^{\pm}(\boldsymbol{r})$ and hence the scattering amplitudes must have azimuthal symmetry about the axis along $\boldsymbol{k}$.) We shall use both notations for the scattering amplitudes in this chapter, as convenience dictates. Explicit expressions for $f^{\pm}(\boldsymbol{k}',\boldsymbol{k})$ may be given[59] but are not needed for the discussion which follows.

The physical interpretation of the stationary states $\psi_{\boldsymbol{k}}^{\pm}(\boldsymbol{r})$, the asymptotic form of which is indicated in (4.2), is most easily obtained by constructing a wave packet of such states to represent a localized electron and then examining the time evolution of the wave packet. Since the states $\psi_{\boldsymbol{k}}^{\pm}(\boldsymbol{r})$ are energy eigenstates with energy $E=k^2/2$, their time dependence is $\exp(-\mathrm{i}k^2t/2)$. To represent a wave packet of such states centered about the energy $E_0=k_0^2/2$, we write

$$\Psi^{\pm}(\boldsymbol{k}_0;\boldsymbol{r},t)\equiv\int dk\, C_{k_0}(k)\exp(-\mathrm{i}\tfrac{1}{2}k^2t)\,\psi_{\boldsymbol{k}}^{\pm}(\boldsymbol{r}), \tag{4.3}$$

where the envelope function $C_{k_0}(k)$ is peaked at $k=k_0$ and decreases rapidly as $|k-k_0|$ increases. For comparison, we construct also a wave packet of plane wave states $(2\pi)^{-3/2}\exp(\mathrm{i}\boldsymbol{k}\cdot\boldsymbol{r})$ as follows:

$$\phi(\boldsymbol{k}_0;\boldsymbol{r},t)\equiv\frac{1}{(2\pi)^{3/2}}\int dk\, C_{k_0}(k)\exp(-\mathrm{i}\tfrac{1}{2}k^2t)\exp(\mathrm{i}\boldsymbol{k}\cdot\boldsymbol{r}). \tag{4.4}$$

Note that in the absence of the interaction $V(r)$, (4.3) reduces to (4.4). What is of interest to us is the behavior of the wave packet in (4.3) at time $t=-\infty$, i.e., *before* the electron is incident on the potential $V(r)$, and at time $t=+\infty$, *i.e., after* the electron is incident on the potential $V(r)$. A rather lengthy calculation shows that this behavior is as follows:[60]

$$\Psi^{+}(\boldsymbol{k}_0;\, \boldsymbol{r}, t) \xrightarrow[t\to -\infty]{} \phi(\boldsymbol{k}_0;\, \boldsymbol{r}, t=-\infty), \tag{4.5a}$$

$$\Psi^{+}(\boldsymbol{k}_0;\, \boldsymbol{r}, t) \xrightarrow[t\to +\infty]{} \phi(\boldsymbol{k}_0;\, \boldsymbol{r}, t=+\infty) + \text{a scattered wave}, \tag{4.5b}$$

$$\Psi^{-}(\boldsymbol{k}_0;\, \boldsymbol{r}, t) \xrightarrow[t\to -\infty]{} \phi(\boldsymbol{k}_0;\, \boldsymbol{r}, t=-\infty) + \text{a scattered wave}, \tag{4.5c}$$

$$\Psi^{-}(\boldsymbol{k}_0;\, \boldsymbol{r}, t) \xrightarrow[t\to +\infty]{} \phi(\boldsymbol{k}_0;\, \boldsymbol{r}, t=+\infty). \tag{4.5d}$$

The physical interpretation of the eigenstates $\psi_{\boldsymbol{k}}^{\pm}(\boldsymbol{r})$ is deduced from the time dependence of their corresponding wave packets, shown in (4.5). The wave packet $\Psi^{+}(\boldsymbol{k}_0;\, \boldsymbol{r}, t)$ reduces *before* the scattering occurs (i.e., at $t=-\infty$) to the plane wave packet $\phi(\boldsymbol{k}_0;\, \boldsymbol{r},\, t)$, which has approximately well-defined linear momentum $\boldsymbol{k}_0$ [cf. (4.5a)]. At later times, however, the packet $\Psi^{+}(\boldsymbol{k}_0;\, \boldsymbol{r}, t)$ has contributions from scattered waves and $\boldsymbol{k}_0$ is no longer approximately conserved (i.e., the electron most likely will be scattered in a new direction by the interaction). The wave packet $\Psi^{+}(\boldsymbol{k}_0;\, \boldsymbol{r}, t)$ is thus suitable for describing a collision experiment in which *initially* electrons are *prepared* to have the momentum $\boldsymbol{k}_0$.

The wave packet $\Psi^{-}(\boldsymbol{k}_0;\, \boldsymbol{r}, t)$, on the other hand, reduces to the plane wave packet $\phi(\boldsymbol{k}_0;\, \boldsymbol{r}, t)$ *after* the scattering has occurred [cf. (4.5d)]. Initially, however, other scattered waves contribute to $\Psi^{-}(\boldsymbol{k}_0;\, \boldsymbol{r}, t)$. Thus the wave packet $\Psi^{-}(\boldsymbol{k}_0;\, \boldsymbol{r},\, t)$ is suitable for describing an ionization experiment in which the ionized electrons are *measured after* the ionization has occurred and found to have the linear momentum $\boldsymbol{k}_0$.

While the connection between scattering theory and experiment is described most easily in terms of the behavior of wave packets, it is mathematically simpler (and correct) to use the exact eigenstates $\psi_{\boldsymbol{k}}^{\pm}(\boldsymbol{r})$ rather than linear combinations of such states. Thus the out-states $\psi_{\boldsymbol{k}}^{-}(\boldsymbol{r})$ are the proper ones to use for the final-state electron wave function in a photoionization process. The justification given above is that a wave packet of such states satisfies the boundary condition that at positive infinite times it reduces to what is measured experimentally: a localized wave packet with linear momentum approximately equal to $\boldsymbol{k}$. However, in order to obtain the proper out-state $\psi_{\boldsymbol{k}}^{-}(\boldsymbol{r})$ while avoiding all reference to wave packets, one may simply require the final-state wave function describing an ionized electron to satisfy the asymptotic boundary condition in (4.2) with the minus sign chosen on both sides of the equation. That is, one requires that the final-state wave function satisfies the so-called incoming-wave boundary condition. Namely, at large r the wave function has the form of a plane wave plus *incoming* spherical waves.[61,62]

Since matrix elements of the interaction Hamiltonian in (3.4) are easiest to calculate using eigenstates of angular momentum, we consider now the form of

[60] See Ref. [57], § 4.2

[61] Breit, G., Bethe, H.A. (1954): Phys. Rev. *93*, 888

[62] Altshuler, S. (1956): Nuovo Cimento *III*, 246

the incoming-wave boundary condition when the final-state wave function is expanded in partial waves. The plane wave on the right in (4.2) has the well-known partial wave expansion,

$$e^{i\boldsymbol{k}\cdot\boldsymbol{r}} = \sum_{l=0}^{\infty} i^l (2l+1) P_l(\cos\theta) j_l(kr), \tag{4.6}$$

where θ is the angle between the vectors $\boldsymbol{k}$ and $\boldsymbol{r}$ and $j_l(kr)$ is the spherical Bessel function which is regular at the origin (i.e., $j_l(kr) \xrightarrow[r\to 0]{} 0$). This function has the asymptotic behavior,

$$j_l(kr) \xrightarrow[r\to\infty]{} \frac{\sin(kr - l\pi/2)}{kr} = \frac{e^{i(kr-l\pi/2)} - e^{-i(kr-l\pi/2)}}{2ikr}. \tag{4.7}$$

Substituting (4.6) and (4.7) in (4.2) we see that the plane wave state contributes both outgoing as well as incoming spherical waves to the asymptotic form of the wave function $\psi_{\boldsymbol{k}}^-(\boldsymbol{r})$, whereas the scattering process contributes only incoming spherical waves. This fact will be used later to define a form of the incoming-wave boundary condition in the many-channel case.

Considering now the left side of (4.2), the exact solution $\psi_{\boldsymbol{k}}^-(\boldsymbol{r})$ may also be expanded in partial waves as follows:

$$\psi_{\boldsymbol{k}}^-(\boldsymbol{r}) = \sum_{l=0}^{\infty} a_l P_l(\cos\theta) R_{kl}(r). \tag{4.8}$$

Here the function $R_{kl}(r)$ is an eigenstate of the radial Schrödinger equation corresponding to the Hamiltonian in (4.1). Asymptotically $R_{kl}(r)$ may be written as a linear combination of the regular and irregular spherical Bessel functions, which are solutions of the free-electron radial Schrödinger equation:

$$R_{kl}(r) \xrightarrow[r\to\infty]{} A_l(k)[j_l(kr) - \tan\delta_l n_l(kr)]. \tag{4.9}$$

Equation (4.9) defines the l-th partial wave phase shift δ_l, which is a measure of the asymptotic amplitude of the spherical Bessel function $n_l(kr)$. This function is irregular at the origin, i.e., $n_l(kr) \propto (kr)^{-l}$ for $r \simeq 0$, and thus does not contribute to $R_{kl}(r)$ in the absence of the interaction $V(r)$. The coefficient $A_l(k)$ is a normalization factor. Using the asymptotic behavior of $j_l(kr)$ in (4.7) as well as that of $n_l(kr)$,

$$n_l(kr) \xrightarrow[r\to\infty]{} \frac{-\cos(kr - l\pi/2)}{kr}, \tag{4.10}$$

we find that asymptotically $R_{kl}(r)$ has the form

$$R_{kl}(r) \propto \frac{\sin(kr - l\pi/2 + \delta_l)}{kr}. \tag{4.11}$$

Comparing with the free-electron radial solution in (4.7), we see that in the presence of the interaction the radial solution $R_{kl}(r)$ is shifted by the phase δ_l.

The set of expansion coefficients a_l in (4.8) may now be obtained by comparing the asymptotic forms of the partial wave expansions of both sides of (4.2) (i.e. substitute (4.8) and (4.11) on the left of (4.2), and (4.6) and (4.7) on the right). Since the incoming and outgoing spherical waves are independent functions, we may obtain the coefficients a_l by equating the coefficients of the outgoing spherical waves on both sides of (4.2). We find that the partial wave expansion for the out-state is thus

$$\psi_{\boldsymbol{k}}^{-}(\boldsymbol{r})=\frac{1}{4\pi k^{1/2}}\sum_{l=0}^{\infty} \mathrm{i}^l(2l+1)\exp(-\mathrm{i}\,\delta_l)\,P_l(\cos\theta)\,R_{\varepsilon l}(r). \tag{4.12}$$

Here we have replaced the subscript k in R_k by the corresponding energy $\varepsilon \equiv k^2/2$ to indicate that we have chosen to normalize the radial solutions $R_{\varepsilon l}(r)$ per unit energy, i.e.,

$$\int_0^\infty r^2\,dr\,R_{\varepsilon l}(r)\,R_{\varepsilon' l}(r)=\delta(\varepsilon-\varepsilon'). \tag{4.13}$$

This normalization corresponds to fixing the asymptotic amplitude of $R_{\varepsilon l}(r)$ as follows:[63]

$$R_{\varepsilon l}(r)\xrightarrow[r\to\infty]{}\left(\frac{2}{\pi k}\right)^{1/2}\frac{\sin(kr-l\pi/2+\delta_l)}{r}. \tag{4.14}$$

The partial wave expansion in (4.12) may be given a simple interpretation if we substitute for each Legendre polynomial, the standard expansion

$$P_l(\cos\theta)=\frac{4\pi}{2l+1}\sum_{m=-l}^{+l} Y_{lm}^*(\hat{\boldsymbol{k}})\,Y_{lm}(\hat{\boldsymbol{r}}). \tag{4.15}$$

The out-state $\psi_{\boldsymbol{k}}^{-}(\boldsymbol{r})$ is thus written as,

$$\psi_{\boldsymbol{k}}^{-}(\boldsymbol{r})=\frac{1}{k^{1/2}}\sum_{l=0}^{\infty}\sum_{m=-l}^{+l} \mathrm{i}^l e^{-\mathrm{i}\delta_l} R_{\varepsilon l}(r)\,Y_{lm}(\hat{\boldsymbol{r}})\,Y_{lm}^*(\hat{\boldsymbol{k}}). \tag{4.16}$$

The factors of each term in the summation in (4.16) are interpreted as follows: The energy-normalized single-electron wavefunction corresponding to the angular momentum state (l, m) is $R_{\varepsilon l}(r)\,Y_{lm}(\hat{\boldsymbol{r}})$. The spherical harmonic $Y_{lm}^*(\hat{\boldsymbol{k}})$ gives the probability amplitude for an electron in the angular momentum state (l, m) to have the direction $\hat{\boldsymbol{k}}$. Finally, the factors $\mathrm{i}^l\exp(-\mathrm{i}\delta_l)\,k^{-1/2}$ ensure that the incoming wave boundary condition in (4.2) as well as the overall normalization condition in (3.6) are satisfied.

β) The final-state wave function for the one-channel case. Consider now the situation in which the potential in (4.1) has a long-range Coulomb tail:

$$V(r)\xrightarrow[r\to\infty]{} -\frac{1}{r}. \tag{4.17}$$

[63] Ref. [40], (4.19)

Such a potential might be appropriate for the photoelectron resulting from photoionization of a neutral atom. The purpose of this section is to indicate how the long-range Coulomb potential modifies the single-channel, incoming-wave-normalized final-state wave function $\psi_{\boldsymbol{k}}^{-}(\boldsymbol{r})$ discussed in the previous section.

The major modification is that the photoelectron always sees a Coulomb field asymptotically and hence can never be properly described by a plane wave, no matter how large r is. For this reason the asymptotic form of the out-state $\psi_{\boldsymbol{k}}^{-}(\boldsymbol{r})$ is modified, firstly, by replacing the plane-wave state $(2\pi)^{-3/2}\exp(\mathrm{i}\boldsymbol{k}\cdot\boldsymbol{r})$ on the right in (4.2) by a Coulomb wave $\psi_{\boldsymbol{k}}^{\mathrm{c}}(\boldsymbol{r})$. Furthermore, this Coulomb wave must itself be normalized using the incoming-wave boundary condition so that it has the following asymptotic form:[64]

$$\psi_{\boldsymbol{k}}^{\mathrm{c}}(\boldsymbol{r})\xrightarrow[r\to\infty]{}\frac{1}{(2\pi)^{3/2}}\left\{\exp\mathrm{i}[\boldsymbol{k}\cdot\boldsymbol{r}-\gamma\log(kr+\boldsymbol{k}\cdot\boldsymbol{r})]\left(1+\frac{\mathrm{i}\gamma^2}{kr+\boldsymbol{k}\cdot\boldsymbol{r}}+\ldots\right)\right.$$
$$\left.+f_{\mathrm{c}}(\theta)\frac{\exp-\mathrm{i}(kr-\gamma\log 2kr)}{r}\left(1+\frac{\mathrm{i}(1-\mathrm{i}\gamma)^2}{kr+\boldsymbol{k}\cdot\boldsymbol{r}}+\ldots\right)\right\}, \tag{4.18a}$$

where

$$\gamma\equiv-\frac{(Z-N)}{k}, \tag{4.18b}$$

$$f_{\mathrm{c}}(\theta)\equiv\frac{-\gamma\exp(-2\mathrm{i}\sigma_0)\exp[+\mathrm{i}\gamma\log(\cos^2\theta/2)]}{2k\cos^2\theta/2}, \tag{4.18c}$$

$$\exp(-2\mathrm{i}\sigma_0)\equiv\Gamma(1-i\gamma)/\Gamma(1+\mathrm{i}\gamma), \tag{4.18d}$$

$$\sigma_0\equiv\arg\Gamma(1+\mathrm{i}\gamma). \tag{4.18e}$$

Equation (4.18a) is what corresponds, in the Coulomb case, to the plane wave $(2\pi)^{-3/2}\exp(\mathrm{i}\boldsymbol{k}\cdot\boldsymbol{r})$ in (4.2). In (4.18a) there is a plane wave that is modified by the characteristic Coulomb logarithmic phase factor. In addition, there is a Coulomb scattering amplitude $f_{\mathrm{c}}(\theta)$ giving rise to ingoing spherical waves; an electron in a Coulomb field is thus always scattered.

For an electron moving in an arbitrary potential having a Coulomb tail, then, the generalization of the ingoing-wave boundary condition in (4.2) is

$$\psi_{\boldsymbol{k}}^{-}(\boldsymbol{r})\xrightarrow[r\to\infty]{}\psi_{\boldsymbol{k}}^{\mathrm{c}}(\boldsymbol{r})+\frac{1}{(2\pi)^{3/2}}f^{-}(\boldsymbol{k}',\boldsymbol{k})\frac{\exp-\mathrm{i}(kr-\gamma\log 2kr)}{r}. \tag{4.19}$$

Here $\psi_{\boldsymbol{k}}^{\mathrm{c}}(\boldsymbol{r})$ has the asymptotic form in (4.18a), $f^{-}(\boldsymbol{k}',\boldsymbol{k})$ is the scattering amplitude due to the short-range part of the potential, where $\boldsymbol{k}'\equiv k\hat{\boldsymbol{r}}$, and the incoming spherical waves in (4.19) have the same form as those in (4.18a). The normalization is chosen so that (3.6) holds. We now proceed, as before, to express $\psi_{\boldsymbol{k}}^{-}(\boldsymbol{r})$ in terms of a partial wave expansion (in order to relate $\psi_{\boldsymbol{k}}^{-}(\boldsymbol{r})$ to the usual numerically calculated radial wave functions) and to obtain the

[64] Ref. [53], (6.44) and (6.32)–(6.34). Note that we have converted these equations to those appropriate for an incoming-wave normalized wavefunction by using $\psi_{\boldsymbol{k}}^{-}=(\psi_{-\boldsymbol{k}}^{+})^*$

coefficients of the expansion by comparison, at large r, with the right-hand side of (4.19).

The partial wave expansion of $\psi_{\boldsymbol{k}}^{c}(\boldsymbol{r})$ (in the incoming-wave normalization) is[65]

$$\psi_{\boldsymbol{k}}^{c}(\boldsymbol{r})=\frac{1}{(2\pi)^{3/2}}\frac{1}{kr}\sum_{l=0}^{\infty}(2l+1)\,\mathrm{i}^{l}\exp(-\mathrm{i}\sigma_l)\,F_l(k,r)\,P_l(\cos\theta). \tag{4.20}$$

Here θ is the angle between $\hat{\boldsymbol{r}}$ and $\boldsymbol{k}$, and $F_l(k,r)$ is the spherical Coulomb function which is regular at $r=0$. The asymptotic form of the regular Coulomb function is

$$F_l(k,r)\xrightarrow[r\to\infty]{}\sin(kr-l\pi/2-\gamma\log 2kr+\sigma_l), \tag{4.21}$$

where the Coulomb phase shift is given by

$$\sigma_l=\arg\Gamma(l+1+\mathrm{i}\gamma). \tag{4.22}$$

The final-state wave function $\psi_{\boldsymbol{k}}^{-}(\boldsymbol{r})$, on the other hand, may always be expanded as

$$\psi_{\boldsymbol{k}}^{-}(\boldsymbol{r})=\sum_{l=0}^{\infty}C_l R_{El}(r)\,P_l(\cos\theta), \tag{4.23}$$

where $R_{El}(r)$ has the asymptotic form

$$R_{El}(r)\xrightarrow[r\to\infty]{}\frac{1}{r}\left(\frac{2}{\pi k}\right)^{1/2}\sin(kr-l\pi/2-\gamma\log 2kr+\sigma_l+\delta_l), \tag{4.24}$$

where δ_l is the l-th partial wave phase shift (with respect to Coulomb waves) due to the short-range part of the potential, and where the amplitude $(2/\pi k)^{1/2}$ assures normalization per unit energy,[63]

$$\int_0^{\infty}R_{El}(r)\,R_{E'l}(r)\,r^2\,dr=\delta(E-E'). \tag{4.25}$$

By substituting (4.24) in (4.23) and setting the result equal to the right-hand side of the equation resulting from the substitution of (4.20) and (4.21) into (4.19), one can obtain the coefficients C_l by first expressing the sine functions in (4.21) and (4.24) in terms of positive and negative exponentials and then matching the coefficients of the positive exponentials on both sides of (4.19). The result of this straightforward calculation is

$$C_l=\frac{(2l+1)\,\mathrm{i}^{l}\exp-\mathrm{i}(\sigma_l+\delta_l)}{4\pi k^{1/2}}. \tag{4.26}$$

Substituting (4.26) in (4.23) and using the standard expansion for the Legendre polynomials in (4.15), one obtains finally

$$\psi_{\boldsymbol{k}}^{-}(\boldsymbol{r})=\frac{1}{k^{1/2}}\sum_{l=0}^{\infty}\sum_{m=-l}^{+l}\mathrm{i}^{l}\exp[-\mathrm{i}(\sigma_l+\delta_l)]\;Y_{lm}^{*}(\hat{\boldsymbol{k}})\,R_{El}(r)\,Y_{lm}(\hat{\boldsymbol{r}}). \tag{4.27}$$

[65] Ref. [53], (6.66)

The incoming-wave normalized state in the presence of a long-range Coulomb potential, given by (4.27), is thus of the same form as the corresponding wave function in the absence of the Coulomb potential, given by (4.16). Furthermore, each term in the partial wave expansion in (4.27) has the same interpretation as the terms in (4.16). The major differences between the two expressions are firstly the different radial wave functions [e.g., compare (4.14) and (4.24)] and secondly the presence of the Coulomb phase shift σ_l entering in the normalization factor $\exp(-i\sigma_l)$ for each term in (4.27).

γ) The eigenchannel states for the many-channel case: To generalize (4.27) to the case of photoionization of a many-electron atom, we first obtain the eigenchannel wave functions in the many-electron case. For simplicity we shall assume that one starts with a complete set of one-electron orbitals computed from some model Hamiltonian $H_0(r)$ and forms linear combinations of Slater determinants, having well-defined LS-coupling quantum numbers, to describe the ionic core and an excited continuum electron. We designate these basis states as $|\gamma E\rangle$, where the γ-th channel is identified by the quantum numbers

$$\gamma \equiv \alpha_\gamma L_\gamma S_\gamma l_\gamma L S M_L M_S, \tag{4.28}$$

where $\alpha_\gamma L_\gamma S_\gamma$ identify the state of the ionic core, l_γ is the orbital angular momentum of the photoelectron, and $LSM_L M_S$ refer to the total orbital and spin angular momentum of the ion plus photoelectron system. E is the total energy of the final state,

$$E = E_0 + \omega = E_\gamma + k_\gamma^2/2 + E_0, \tag{4.29}$$

where E_0 is the initial energy of the atom, ω is the photon energy, E_γ is the binding energy for photoionizing an electron and leaving the ion in the γ-th term level, and $k_\gamma^2/2$ is the kinetic energy of the photoelectron corresponding to the γ-th-term level of the ion. We shall assume that the wave function corresponding to the basis state $|\gamma E\rangle$ is real and has the following asymptotic form, which is the many-channel generalization of (4.24) and thus similarly insures normalization per unit energy:

$$\begin{aligned}\langle \boldsymbol{r}_1 s_1, \ldots, \boldsymbol{r}_N s_N | \gamma E\rangle \xrightarrow[r_N \to \infty]{} & \Theta_\gamma(\boldsymbol{r}_1 s_1, \ldots, \hat{\boldsymbol{r}}_N s_N) \left(\frac{2}{\pi k_\gamma}\right)^{1/2} \frac{1}{r_N} \\ & \cdot \sin\left(k_\gamma r_N - 1/2\pi l_\gamma + \frac{1}{k_\gamma}\log 2k_\gamma r_N + \sigma_{l_\gamma} + \delta_\gamma\right).\end{aligned} \tag{4.30}$$

Here

$$\sigma_{l_\gamma} \equiv \arg \Gamma(l_\gamma + 1 - i/k_\gamma), \tag{4.31}$$

δ_γ is the phase shift with respect to Coulomb waves in the γ-th channel arising from the short-range part of the potential in the model Hamiltonian H_0, and Θ_γ is the wave function of the ionic core coupled with the spherical harmonic, $Y_{l_\gamma m_\gamma}(\hat{\boldsymbol{r}}_N)$, and spin function, $\chi(\boldsymbol{s}_N)$, of the photoelectron. Although we have opted for the LS-coupling scheme, which is the most commonly used one, the procedures used in this section are quite general and are not dependent on this choice.

The residual interaction V, defined by

$$V \equiv H - \sum_{i=1}^{N} H_0(\boldsymbol{r}_i), \tag{4.32}$$

where H is given by (3.1), has in general nonzero matrix elements between different basis states $|\gamma E\rangle$. These must be taken into account in any complete treatment of the final-state wave function. The eigenstates of energy E which diagonalize the exact Hamiltonian H are thus, in general, expressed in terms of a linear combination of the basis states $|\gamma E'\rangle$, where E' is not necessarily equal to E. This procedure is similar to ordinary configuration-interaction theory, except that here we deal with states having continuum energies. For ease in numerical calculations, this linear combination is best given in terms of real coefficients, which embody the dynamics of electron correlation effects. Only later, when we wish to compute cross sections, need we apply the incoming-wave boundary condition which introduces complex factors.

The Γ-th eigenstate of the full Hamiltonian H corresponding to the energy E is thus expressed as a linear combination of the basis states $|\gamma E'\rangle$ with real coefficients. The coefficient of each state in the summation is defined, according to scattering theory,[66-68] in terms of matrix elements of the real K matrix as follows:

$$|\Gamma E\rangle = \sum_{\gamma} \left\{ |\gamma E\rangle + \sum_{\gamma'} \mathscr{P} \int dE' \frac{|\gamma' E'\rangle \langle \gamma' E'|K(E)|\gamma E\rangle}{E-E'} \right\} U_{\gamma,\Gamma} \cos \eta_\Gamma. \tag{4.33}$$

In (4.33) the eigenstates $|\Gamma E\rangle$ are labelled by the eigenchannel index Γ and the total energy E. Since each $|\Gamma E\rangle$ is a linear combination of the basis states $|\gamma E'\rangle$, Γ does not correspond to a particular state of the ion and E does not correspond to a particular photoelectron energy. The symbol $\mathscr{P}$ in (4.33) indicates that the Cauchy principal part is to be taken when integrating over the singularity in the denominator. For each particular γ', the integration over total energy E' reduces to a summation over bound excited states and an integration over the photoelectron kinetic energy in channel γ'.

The K matrix in (4.33) is obtained by solving the following integral equation:

$$\begin{aligned} &\langle \gamma' E'|K(E)|\gamma E\rangle \\ &\quad = \langle \gamma' E'|V|\gamma E\rangle + \sum_{\gamma''} \mathscr{P} \int dE'' \frac{\langle \gamma' E'|V|\gamma'' E''\rangle \langle \gamma'' E''|K(E)|\gamma E\rangle}{E-E''}. \end{aligned} \tag{4.34}$$

The orthogonal matrix $U_{\gamma,\Gamma}(E)$, which transforms the channels γ to the eigenchannels Γ at fixed E, as well as the eigenphase shifts η_Γ, which are due to

---

[66] Fano, U., Prats, F. (1963): Proc. Nat. Acad. Sci. (India) A *33*, 553

[67] Ref. [56], Chap. 9

[68] Ref. [53], pp. 419–421

the interactions between states $|\gamma E\rangle$, are obtained by diagonalizing the on-the-energy-shell K matrix as follows:

$$\sum_{\gamma'} \langle \gamma E | K(E) | \gamma' E \rangle U_{\gamma', \Gamma} = -\pi^{-1} \tan \eta_\Gamma U_{\gamma, \Gamma}. \tag{4.35}$$

Here $-\pi^{-1} \tan \eta_\Gamma$ is the Γ-th eigenvalue. The matrix elements of the on-the-energy-shell K matrix, which we shall denote by $\{K(E)\}$, may thus be written as

$$\{K(E)\}_{\gamma\gamma'} \equiv \langle \gamma E | K(E) | \gamma' E \rangle = -\pi^{-1} \sum_{\Gamma} U_{\gamma, \Gamma} \tan \eta_\Gamma \tilde{U}_{\Gamma, \gamma'}, \tag{4.36}$$

where the diagonalizing matrix $U_{\gamma, \Gamma}$ and the eigenphase shifts η_Γ are of course dependent on the energy E. The factors $U_{\gamma, \Gamma} \cos \eta_\Gamma$ in (4.33) serve to normalize the eigenstates $|\Gamma E\rangle$ to a delta function in energy provided the basis states $|\gamma E\rangle$ are so normalized.

We now prove the assertions made above that the states $|\Gamma E\rangle$ are orthonormal energy eigenstates of the full Hamiltonian H. We also examine the asymptotic form of the wave function for the state $|\Gamma E\rangle$. This will be needed when applying the incoming-wave boundary condition to the many-channel final-state wave function.

Eigenstate property. We take the point of view that (4.33) is an ansatz for the supposed eigenstates $|\Gamma E\rangle$ and that the matrix elements of the K matrix appearing in (4.33) are a set of unknown coefficients to be determined. We assume that the states $|\gamma E\rangle$ are energy normalized eigenstates of the model Hamiltonian H_0, i.e.,

$$H_0 |\gamma E\rangle = E |\gamma E\rangle, \tag{4.37a}$$

$$\langle \gamma' E' | \gamma E \rangle = \delta_{\gamma\gamma'} \delta(E' - E). \tag{4.37b}$$

We wish to prove that the states $|\Gamma E\rangle$ in (4.33) are eigenstates of the full Hamiltonian, $H \equiv H_0 + V$, i.e.,

$$(H_0 + V) |\Gamma E\rangle = E |\Gamma E\rangle. \tag{4.38}$$

The proof of (4.38) is straightforward. One substitutes (4.33) for $|\Gamma E\rangle$ in (4.38) and then operates from the left on both sides of (4.38) with the basis state $\langle \gamma' E'|$. The resulting equation is then simplified using the properties of the basis states specified in (4.37). One finds that equality between the two sides of the equation holds provided the matrix elements $\langle \gamma' E' | K(E) | \gamma E \rangle$ satisfy (4.34).

Orthonormality. The proof of the orthonormality of the states $|\Gamma E\rangle$ is not trivial and rather lengthy. We content ourselves here with indicating the crucial arguments of the proof and leave the more straightforward details to the interested reader. Using (4.33) for the eigenstates $|\Gamma E\rangle$, the Hermitian property of the K matrix [cf. (4.34)], and the properties of the basis states in (4.37), we find that the overlap amplitude $\langle \Gamma' E' | \Gamma E \rangle$ may be written as follows:

$$\langle \Gamma' E' | \Gamma E \rangle = \sum_{\gamma', \gamma} \cos \eta_{\Gamma'} \tilde{U}_{\Gamma', \gamma'} \Big\{ \delta_{\gamma\gamma'} \delta(E - E') + \langle \gamma' E' | K(E') | \gamma E \rangle \frac{1}{E' - E}$$
$$+ \frac{1}{E - E'} \langle \gamma' E' | K(E) | \gamma E \rangle + \sum_{\gamma''} \mathscr{P} \int dE'' \langle \gamma' E' | K(E') | \gamma'' E'' \rangle$$
$$\cdot \frac{1}{E' - E''} \frac{1}{E - E''} \langle \gamma'' E'' | K(E) | \gamma E \rangle \Big\} U_{\gamma, \Gamma} \cos \eta_{\Gamma}. \tag{4.39}$$

The product of energy denominators in the integral appearing in (4.39) is now replaced using the following mathematical identity:[69]

$$\frac{1}{E' - E''} \cdot \frac{1}{E - E''} = \frac{1}{E' - E} \left(\frac{1}{E - E''} - \frac{1}{E' - E''} \right) + \pi^2 \delta(E' - E) \delta(E'' - E'). \tag{4.40}$$

Substituting (4.40) in (4.39), one finds that the overlap amplitude may be written as

$$\langle \Gamma' E' | \Gamma E \rangle = \delta(E' - E) A_{\Gamma', \Gamma}(E) + B(\Gamma' E', \Gamma E), \tag{4.41a}$$

where

$$A_{\Gamma', \Gamma}(E) \equiv \sum_{\gamma', \gamma} \cos \eta_{\Gamma'} \tilde{U}_{\Gamma', \gamma'} \{ \delta_{\gamma', \gamma}$$
$$+ \pi^2 \sum_{\gamma''} \langle \gamma' E | K(E) | \gamma'' E \rangle \langle \gamma'' E | K(E) | \gamma E \rangle \} U_{\gamma, \Gamma} \cos \eta_{\Gamma}, \tag{4.41b}$$

and

$$B(\Gamma' E', \Gamma E) \equiv \frac{1}{E' - E} \sum_{\gamma', \gamma} \cos \eta_{\Gamma'} \tilde{U}_{\Gamma', \gamma'} \Big\{ \langle \gamma' E' | K(E') | \gamma E \rangle - \langle \gamma' E' | K(E) | \gamma E \rangle$$
$$+ \sum_{\gamma''} \int dE'' \langle \gamma' E' | K(E') | \gamma'' E'' \rangle \frac{1}{E - E''} \langle \gamma'' E'' | K(E) | \gamma E \rangle \tag{4.41c}$$
$$- \sum_{\gamma''} \int dE'' \langle \gamma' E' | K(E') | \gamma'' E'' \rangle \frac{1}{E' - E''} \langle \gamma'' E'' | K(E) | \gamma E \rangle \Big\} U_{\gamma, \Gamma} \cos \eta_{\Gamma}.$$

The expressions given above for $A_{\Gamma', \Gamma}(E)$ and $B(\Gamma' E', \Gamma E)$ may be reduced to obtain

$$A_{\Gamma', \Gamma}(E) = \delta_{\Gamma', \Gamma}, \tag{4.42a}$$

$$B(\Gamma' E', \Gamma E) = 0. \tag{4.42b}$$

To prove (4.42a) one uses the definition of the unitary matrix $U_{\gamma, \Gamma}$ given by (4.35). To prove (4.42b) one uses (4.34) to substitute for $\langle \gamma'' E'' | K(E) | \gamma E \rangle$ in the second integral in (4.41c) and the Hermititan conjugate of (4.34) to substitute for $\langle \gamma' E' | K(E') | \gamma'' E'' \rangle$ in the first integral in (4.41c). The results, in (4.42), when substituted in (4.41a) give the desired orthonormality relation for the eigenstates $|\Gamma E\rangle$, i.e.,

$$\langle \Gamma' E' | \Gamma E \rangle = \delta_{\Gamma', \Gamma} \delta(E' - E). \tag{4.43}$$

[69] Ref. [3], Appendix A

Asymptotic behavior. We examine here the coordinate-space representation of (4.33) in the limit that the radial coordinate of the N-th electron, r_N, becomes large. Making use of the asymptotic behavior of the basis-state wave functions in (4.30), we find

$$\langle \boldsymbol{r}_1, \ldots, \boldsymbol{r}_N | \Gamma E \rangle \xrightarrow[r_N \to \infty]{} \sum_{\gamma} \theta_{\gamma} \left(\frac{2}{\pi k_{\gamma}}\right)^{1/2} \frac{1}{r_N} \sin\left[k_{\gamma} r_N + \alpha(k_{\gamma}, r_N)\right] U_{\gamma,\Gamma} \cos\eta_{\Gamma}$$
$$+ \sum_{\gamma', \gamma} \theta_{\gamma'} \mathscr{P} \int dE' \left(\frac{2}{\pi k_{\gamma'}}\right)^{1/2} \frac{1}{r_N} \frac{\sin\left[k_{\gamma'} r_N + \alpha(k_{\gamma'}, r_N)\right]}{E - E'}$$
$$\cdot \langle \gamma' E' | K(E) | \gamma E \rangle U_{\gamma,\Gamma} \cos\eta_{\Gamma}, \tag{4.44a}$$

where we have used the following abbreviations:

$$\theta_{\gamma} \equiv \theta_{\gamma}(\boldsymbol{r}_1 \boldsymbol{s}_1, \ldots, \hat{\boldsymbol{r}}_N \boldsymbol{s}_N), \tag{4.44b}$$

$$\alpha(k_{\gamma}, r_N) \equiv -\pi l_{\gamma}/2 + \frac{1}{k_{\gamma}} \log 2k_{\gamma} r_N + \sigma_{l_{\gamma}} + \delta_{\gamma}. \tag{4.44c}$$

The crucial step needed for the further reduction of (4.44a) is to prove

$$\lim_{r_N \to \infty} \frac{\sin\left[k_{\gamma'} r_N + \alpha(k_{\gamma'}, r_N)\right]}{E - E'} = -\pi \frac{\delta(k_{\gamma'} - q)}{k_{\gamma'}} \cos\left[k_{\gamma'} r_N + \alpha(k_{\gamma'}, r_N)\right], \tag{4.45}$$

where q is defined as the momentum of the N-th electron in channel γ' corresponding to an energy E for the N-electron system, i.e.,

$$E = I_{\gamma'} + q^2/2, \tag{4.46}$$

where $I_{\gamma'}$ is the binding energy of the N-th electron in channel γ'. The energy E' of course satisfies a similar relation,

$$E' = I_{\gamma'} + k_{\gamma'}^2/2. \tag{4.47}$$

To prove (4.45) we change variables by defining $k_{\gamma'}$ in terms of the difference x between $k_{\gamma'}$ and q, i.e.,

$$k_{\gamma'} \equiv q + x, \tag{4.48}$$

and use (4.46) and (4.47) to replace the energies E and E'. The left-hand side of (4.45) may then be written as

$$\lim_{r_N \to \infty} \frac{\sin\left[k_{\gamma'} r_N + \alpha(k_{\gamma'}, r_N)\right]}{E - E'}$$
$$= - \lim_{r_N \to \infty} \frac{\{\sin(x r_N) \cos\left[q r_N + \alpha(k_{\gamma'}, r_N)\right] + \cos(x r_N) \sin\left[q r_N + \alpha(k_{\gamma'}, r_N)\right]\}}{x(q + x/2)}. \tag{4.49}$$

The second term in the numerator on the right of (4.49) vanishes in the limit of large r_N due to the rapid oscillations of the trigonometric functions. The first term, however, must be examined more carefully. We note that

$$\frac{\sin x r_N}{x} = \frac{1}{2} \int_{-r_N}^{r_N} dy\, e^{ixy} \xrightarrow[r_N \to \infty]{} \pi \delta(x). \tag{4.50}$$

Using (4.50) to reduce the right side of (4.49), we obtain the desired result in (4.45).

Use of (4.45) on the right in (4.44a) permits the integration over energy to be carried out trivially. Noting then that equality of $k_{\gamma'}$ and q implies also equality of E' and E, we may use (4.35) to make the following replacement on the right of (4.44a):

$$\sum_{\gamma} \langle \gamma' E | K(E) | \gamma E \rangle\, U_{\gamma,\Gamma} \cos \eta_\Gamma = -\pi^{-1} \sin \eta_\Gamma\, U_{\gamma',\Gamma}. \tag{4.51}$$

The result of these operations on the second summation in (4.44a) may then be combined (after changing the summation index from γ' to γ) with the first summation in (4.44a) to give finally

$$\langle \boldsymbol{r}_1, \ldots, \boldsymbol{r}_N | \Gamma E \rangle \xrightarrow[r_N \to \infty]{} \sum_{\gamma} \theta_\gamma \left(\frac{2}{\pi k_\gamma} \right)^{1/2} \frac{1}{r_N} \sin [k_\gamma r_N + \alpha(k_\gamma, r_N) + \eta_\Gamma]\, U_{\gamma,\Gamma}. \tag{4.52}$$

Equation (4.52) indicates that the eigenchannel wave function is represented asymptotically as a linear combination of channel wave functions and that the radial wavefunction of the N-th electron in each channel is phase shifted by the eigenphase η_Γ.

δ) Incoming-wave-normalized final-state wave functions for the many-channel case. Having established the properties of the eigenchannel states $|\Gamma E\rangle$, we are now ready to obtain the incoming-wave-normalized final-state wave function for the many-channel case. In fact, we shall obtain two such final-state wave functions, corresponding to two alternative final-state measurements. In one type of measurement one determines the *angular* momentum of the photoelectron, its coupling to the ion's angular momentum, and their *mutual* orientation. In another type of measurement one determines the *linear* momentum of the electron, the angular momentum of the ion, and their *separate* orientations. More specifically in what follows we shall first determine the incoming-wave-normalized channel wave function $\psi^-_{\gamma E}$, which describes the final state in an experiment that measures the quantum numbers γ in (4.28). Then we shall obtain the incoming-wave-normalized momentum wave function $\psi^-_{\gamma(\mathrm{i})\boldsymbol{k}_\gamma}$, where the quantum numbers $\gamma(\mathrm{i})$ refer only to the state of the ion,

$$\gamma(\mathrm{i}) \equiv \alpha_\gamma L_\gamma S_\gamma M_{L_\gamma} M_{S_\gamma}, \tag{4.53}$$

and $\boldsymbol{k}_\gamma$ represents the electron's momentum and implicitly the electron's spin orientation. (Note that the energy E of the system described by $\psi^-_{\gamma(\mathrm{i})\boldsymbol{k}_\gamma}$ is determined by (4.29).)

The incoming-wave-normalized channel wave function corresponding to a particular asymptotic channel γ may be expressed as a linear combination of the eigenchannel states in (4.33):

$$\psi_{\gamma E}^{-}(\boldsymbol{r}_1 s_1, \ldots, \boldsymbol{r}_N s_N) = \sum_{\Gamma} \langle \boldsymbol{r}_1 s_1 \ldots \boldsymbol{r}_N s_N | \Gamma E \rangle \, C_{\Gamma\gamma}(E). \tag{4.54}$$

$\psi_{\gamma E}^{-}$ differs from the basis-state wave functions $\langle \boldsymbol{r}_1 s_1 \ldots \boldsymbol{r}_N s_N | \gamma E \rangle$ in that it takes into account the final-state interactions between the channels γ. The coefficients $C_{\Gamma\gamma}(E)$ are obtained by requiring that as $r_N \to \infty$, $\psi_{\gamma E}^{-}$ contains outgoing spherical waves only in channel γ (i.e., $\psi_{\gamma E}^{-}$ satisfies the incoming-wave boundary condition). Substituting the asymptotic form of the eigenchannel wave functions $\langle \boldsymbol{r}_1 s_1 \ldots \boldsymbol{r}_N s_N | \Gamma E \rangle$ given by (4.52) into (4.54), expanding the sine in terms of outgoing and incoming spherical waves, and requiring all outgoing waves in channels other than γ to vanish determines the coefficients $C_{\Gamma\gamma}(E)$ in the expansion (4.54). The channel wave function is thus

$$\psi_{\gamma E}^{-}(\boldsymbol{r}_1 s_1, \ldots, \boldsymbol{r}_N s_N) = \sum_{\Gamma} \langle \boldsymbol{r}_1 s_1 \ldots \boldsymbol{r}_N s_N | \Gamma E \rangle \, \mathrm{e}^{-\mathrm{i}\eta_\Gamma} \, \tilde{U}_{\Gamma,\gamma} \, \mathrm{e}^{-\mathrm{i}\delta_\gamma}. \tag{4.55}$$

The phase $\exp(-\mathrm{i}\delta_\gamma)$ in (4.55) insures that the outgoing Coulomb spherical wave in channel γ is not phase shifted. Substituting (4.52) into (4.55), one sees that the asymptotic form of the final-state channel wave function is

$$\begin{aligned}
\psi_{\gamma E}^{-}(\boldsymbol{r}_1 s_1, \ldots, \boldsymbol{r}_N s_N) \xrightarrow[r_N \to \infty]{} & \theta_\gamma(\boldsymbol{r}_1 s_1, \ldots, \hat{\boldsymbol{r}}_N s_N) \frac{1}{\mathrm{i}(2\pi k_\gamma)^{1/2}} \frac{1}{r_N} \\
& \cdot \exp \mathrm{i}\left(k_\gamma r_N - \tfrac{1}{2}\pi l_\gamma + \frac{1}{k_\gamma} \log 2k_\gamma r_N + \sigma_{l_\gamma}\right) \\
& - \sum_{\gamma'} \theta_{\gamma'}(\boldsymbol{r}_1 s_1, \ldots, \hat{\boldsymbol{r}}_N s_N) \frac{1}{\mathrm{i}(2\pi k_{\gamma'})^{1/2}} \frac{1}{r_N} \\
& \cdot \exp -\mathrm{i}\left(k_{\gamma'} r_N - \tfrac{1}{2}\pi l_{\gamma'} + \frac{1}{k_{\gamma'}} \log 2k_{\gamma'} r_N + \sigma_{l_{\gamma'}}\right) \mathrm{e}^{-\mathrm{i}\delta_{\gamma'}} S^{\dagger}_{\gamma'\gamma} \, \mathrm{e}^{-\mathrm{i}\delta_\gamma}.
\end{aligned} \tag{4.56}$$

In (4.56) we have defined the matrix $S_{\gamma\gamma'}$ to be

$$S_{\gamma\gamma'} = \sum_{\Gamma} U_{\gamma,\Gamma} \, \mathrm{e}^{2\mathrm{i}\eta_\Gamma} \, \tilde{U}_{\Gamma,\gamma'}. \tag{4.57}$$

Using the definition of the on-the-energy-shell K matrix in (4.36), one may verify that the S matrix is related to $\{K(E)\}$ as follows:

$$S_{\gamma\gamma'} = \sum_{\gamma''} (1 - \mathrm{i}\pi\{K(E)\})_{\gamma,\gamma''} (1 + \mathrm{i}\pi\{K(E)\})^{-1}_{\gamma'',\gamma'}. \tag{4.58}$$

Equation (4.58) shows that $S_{\gamma\gamma'}$ is the scattering matrix of collision theory. The meaning of the scattering matrix is that $S_{\gamma\gamma'}$ equals the probability amplitude for scattering from the basis state $|\gamma' E\rangle$ to the basis state $|\gamma E\rangle$.[70] An expression identical to (4.56) has been presented by HENRY and LIPSKY,[6] although these authors include the phase shifts δ_γ of the basis-state wave functions in their definition of the S matrix.[71]

We proceed now to obtain the incoming-wave-normalized momentum wave function, which is needed to compute photoelectron angular distributions. This wave function is determined by imposing the many-channel generalization of the incoming-wave boundary condition in (4.19). That is, we seek a wave function, normalized according to (3.6), that represents asymptotically a definite LS-coupling state of the ion plus a photoelectron having wave vector k and a definite spin orientation. Such a generalization of (4.19) may be written as

$$\begin{aligned}\psi^-_{\gamma(\mathrm{i})\boldsymbol{k}_\gamma}(\boldsymbol{r}_1 s_1, \ldots, \boldsymbol{r}_N s_N) \xrightarrow[r_N\to\infty]{} & \Phi_{\gamma(\mathrm{i})}(\boldsymbol{r}_1 s_1, \ldots, \boldsymbol{r}_{N-1} s_{N-1})\, \chi(s_N)\, \psi^{\mathrm{c}}_{\boldsymbol{k}_\gamma}(\boldsymbol{r}_N) \\ & + \frac{1}{(2\pi)^{3/2}} \sum_{\gamma'} \Phi_{\gamma'}(\boldsymbol{r}_1 s_1, \ldots, \boldsymbol{r}_{N-1} s_{N-1}) \frac{f(\boldsymbol{k}_{\gamma'}, \boldsymbol{k}_\gamma)}{r_N} \chi(s_N) \\ & \cdot \exp -\mathrm{i}\left(k_{\gamma'} r_N + \frac{1}{k_{\gamma'}} \log 2k_{\gamma'} r_N\right). \end{aligned} \tag{4.59}$$

Here $\Phi_{\gamma(\mathrm{i})}$ is the wave function for the ion, where $\gamma(\mathrm{i})$ specifies the quantum numbers in (4.53); $\chi(s_N)$ is the spin funcion for the photoelectron; $\psi^{\mathrm{c}}_{\boldsymbol{k}_\gamma}(\boldsymbol{r}_N)$ is the Coulomb wavefunction defined in (4.20) for an electron having wave vector $\boldsymbol{k}_\gamma$; and $f(\boldsymbol{k}_{\gamma'}, \boldsymbol{k}_\gamma)$ is the many-channel generalization of the scattering amplitude, where $\boldsymbol{k}_{\gamma'} \equiv k_{\gamma'} \hat{r}_N$. Now $\psi^-_{\gamma(\mathrm{i})\boldsymbol{k}_\gamma}$ may be written in terms of the channel functions $\psi^-_{\gamma E}$ as follows:

$$\begin{aligned}\psi^-_{\gamma(\mathrm{i})\boldsymbol{k}_\gamma}(\boldsymbol{r}_1 s_1, \ldots, \boldsymbol{r}_N s_N) = & \sum_{l_\gamma m_\gamma} C_{l_\gamma m_\gamma} Y^*_{l_\gamma m_\gamma}(\hat{\boldsymbol{k}}_\gamma) \\ & \cdot \sum_{\substack{LM_L \\ SM_S}} C(L_\gamma l_\gamma L; M_{L_\gamma} m_\gamma M_L)\, C(S_\gamma \tfrac{1}{2} S; M_{S_\gamma} m_{1/2} M_S) \\ & \cdot \psi^-_{\gamma E}(\boldsymbol{r}_1 s_1, \ldots, \boldsymbol{r}_N s_N). \end{aligned} \tag{4.60}$$

In (4.60) the Clebsch-Gordan coefficients serve to uncouple the spin and orbital angular momenta of the photoelectron from that of the ion, the spherical harmonic serves to project the angular momentum states $l_\gamma m_\gamma$ of the photoelectron onto the direction $\hat{\boldsymbol{k}}_\gamma$, and the wave function subscripts $\gamma(\mathrm{i})$ and γ on the left- and right-hand sides of (4.60) specify the quantum numbers in (4.53) and (4.28), respectively. (Note also that in (4.60) we have, for simplicity,

[70] Ref. [53], § 14.2

[71] Strictly speaking, the S matrix we have defined in (4.57) is a reduced S matrix since the basis states $|\gamma E\rangle$ already include part of the short-range interaction giving rise to the phase shifts δ_γ. From the point of view of a basis set of pure Coulomb waves, the matrix

$$\mathrm{e}^{-\mathrm{i}\delta_{\gamma'}} S^\dagger_{\gamma'\gamma}\, \mathrm{e}^{-\mathrm{i}\delta_\gamma},$$

which appears in (4.56), is the Hermitian conjugate of the full S matrix computed the wave function corresponding to a photoelectron whose z component of spin is $m_{1/2}$ on the same z axis used to describe the orientation of the ion; and arbitrary state of the photoelectron's spin may be obtained by suitably combining wave functions $\psi^-_{\gamma(i)\boldsymbol{k}_\gamma}$ having different values of $m_{1/2}$.) One determines the coefficients $C_{l_\gamma m_\gamma}$ in (4.60) by requiring the asymptotic form of (4.60) to be the same as the asymptotic form in (4.59). Therefore, substituting (4.56) in (4.60), substituting (4.15), (4.20), and (4.21) in (4.59), and comparing the results, one obtains

$$C_{l_\gamma m_\gamma} = \frac{\mathrm{i}^{l_\gamma} \exp(-\mathrm{i}\sigma_{l_\gamma})}{k_\gamma^{1/2}}. \tag{4.61}$$

Thus, the many-channel generalization of (4.27) for the incoming-wave-normalized final-state wave function is

$$\begin{aligned}\psi^-_{\gamma(i)\boldsymbol{k}_\gamma}(\boldsymbol{r}_1 s_1, \ldots, \boldsymbol{r}_N s_N) = &\sum_{l_\gamma m_\gamma} \frac{\mathrm{i}^{l_\gamma} \exp(-\mathrm{i}\sigma_{l_\gamma})}{k_\gamma^{1/2}} Y^*_{l_\gamma m_\gamma}(\hat{\boldsymbol{k}}_\gamma)\\ &\cdot \sum_{\substack{LM_L\\ SM_S}} C(L_\gamma l_\gamma L; M_{L_\gamma} m_\gamma M_L)\, C(S_\gamma \tfrac{1}{2} S; M_{S_\gamma} m_{1/2} M_S)\\ &\cdot \psi^-_{\gamma E}(\boldsymbol{r}_1 s_1, \ldots, \boldsymbol{r}_N s_N),\end{aligned} \tag{4.62}$$

where the energy-normalized channel function $\psi^-_{\gamma E}$ is given by (4.55) and has the asymptotic form in (4.56). An equation identical to (4.62) has been given by BURKE,[72] except that BURKE's wave function is energy normalized whereas (4.62) is normalized in momentum space according to (3.6).

5. Alternative expressions for the transition matrix element. The general formula for the differential photoionization cross section has been given in terms of the matrix element of the momentum operator, $\sum_{i=1}^{N} \boldsymbol{p}_i$. Alternative expressions for this matrix element may be obtained by considering the following commutation relations involving the exact Hamiltonian in (3.1):

$$\sum_{i=1}^{N} \boldsymbol{p}_i = -\mathrm{i}\left[\sum_{i=1}^{N} \boldsymbol{r}_i, H\right], \tag{5.1a}$$

$$\left[\sum_{i=1}^{N} \boldsymbol{p}_i, H\right] = -\mathrm{i} \sum_{i=1}^{N} \frac{Z\boldsymbol{r}_i}{r_i^3}. \tag{5.1b}$$

Taking matrix elements of (5.1) between N-electron energy eigenstates $\langle E_1|$ and $|E_2\rangle$ having total energies E_1 and E_2, we find

$$\langle E_1|\sum_{i=1}^{N} \boldsymbol{p}_i|E_2\rangle = -\mathrm{i}(E_2 - E_1)\langle E_1|\sum_{i=1}^{N} \boldsymbol{r}_i|E_2\rangle, \tag{5.2a}$$

[72] Burke, P.G. (1976) in: Atomic processes and applications, Burke, P.G., Moiseiwitsch, B.L. (eds.), Amsterdam: North-Holland, Chap. 7, Eq. (18)

$$\langle E_1|\sum_{i=1}^{N} \boldsymbol{p}_i|E_2\rangle = \frac{-\mathrm{i}}{(E_2-E_1)}\langle E_1|\sum_{i=1}^{N} \frac{Z\boldsymbol{r}_i}{r_i^3}|E_2\rangle. \tag{5.2b}$$

The matrix elements of $\sum_{i=1}^{N} \boldsymbol{p}_i$, $\sum_{i=1}^{N} \boldsymbol{r}_i$, and $\sum_{i=1}^{N} Z\boldsymbol{r}_i/r_i^3$ in (5.2) are known as the "velocity," "length," and "acceleration" forms of the electric dipole matrix element.

As first discussed by CHANDRASEKHAR,[73] equality in (5.2) holds only for exact eigenstates of H. For approximate wave functions the expressions on the right and left in (5.2) will differ in general. The length form tends to weight the large r part of the wave functions, the acceleration form the small r part of the wave functions, and the velocity form the intermediate part of the wave functions. Hence for wave functions determined variationally to give good energy expectation values, the intermediate range of r is probably best determined, and hence the velocity form for the electric dipole matrix element would seem to give the best results. Other such qualitative criteria are discussed in the reviews of STEWART[38] and of CROSSLEY.[74]

However, for a certain class of approximate wave functions, namely, those obtained as the exact solutions of a model Hamiltonian H_{mod}, justification may be given for using the length formula exclusively.[75] Consider the model Hamiltonian $H_{\mathrm{mod}} = \sum_{i=1}^{N} \frac{1}{2}\boldsymbol{p}_i^2 + V$, where V is a potential chosen to approximate that in (3.1) as well as possible. The relations in (5.1) hold in general when H_{mod} is substituted for H only if V is a local potential. In many cases however V is a nonlocal potential. Hence we must reexamine (5.1) for the case of a nonlocal potential.

The effect of operating with a general potential operator V on an eigenstate $|\psi\rangle$ is represented by $V|\psi\rangle$ or in coordinate representation by $\langle \boldsymbol{r}|V|\psi\rangle$. This latter quantity may be expanded by inserting a complete set of coordinate states between V and $|\psi\rangle$ and integrating to obtain

$$\langle \boldsymbol{r}|V|\psi\rangle = \int \langle \boldsymbol{r}|V|\boldsymbol{r}'\rangle \langle \boldsymbol{r}'|\psi\rangle \, d\boldsymbol{r}' = \int \langle \boldsymbol{r}|V|\boldsymbol{r}'\rangle \, \psi(\boldsymbol{r}') \, d\boldsymbol{r}', \tag{5.3}$$

where we have replaced $\langle \boldsymbol{r}'|\psi\rangle$ by its usual notation, $\psi(\boldsymbol{r}')$. The right-hand side of (5.3) describes the interaction of a nonlocal potential $\langle \boldsymbol{r}|V|\boldsymbol{r}'\rangle \equiv V(\boldsymbol{r}, \boldsymbol{r}')$ and a coordinate wave function $\psi(\boldsymbol{r}')$. A local potential is diagonal in coordinate representation, having matrix element

$$\langle \boldsymbol{r}|V|\boldsymbol{r}'\rangle = V(\boldsymbol{r})\,\delta(\boldsymbol{r}-\boldsymbol{r}'), \tag{5.4}$$

which when substituted in (5.3) gives the familiar product $V(\boldsymbol{r})\,\psi(\boldsymbol{r})$.

Substituting H_{mod} for H in (5.1a) gives

$$-\mathrm{i}\left[\sum_{i=1}^{N} \boldsymbol{r}_i, H_{\mathrm{mod}}\right] = \sum_{i=1}^{N} \{\boldsymbol{p}_i - \mathrm{i}\,[\boldsymbol{r}_i, V(\boldsymbol{r}_i, \boldsymbol{r}_i')]\}, \tag{5.5}$$

[73] Chandrasekhar, S. (1945): Astrophys. J. *102*, 223
[74] Crossley, R.J.S. (1969): Adv. At. Mol. Phys. *5*, 237
[75] Starace, A.F. (1971): Phys. Rev. A *3*, 1242; (1973): Phys. Rev. A *8*, 1141

and a calculation similar to that in (5.3) shows that the commutator of the position operator with the nonlocal potential gives

$$[\boldsymbol{r}, V(\boldsymbol{r}, \boldsymbol{r}')] = (\boldsymbol{r} - \boldsymbol{r}')\, V(\boldsymbol{r}, \boldsymbol{r}'). \tag{5.6}$$

(Note that if the potential is local, then (5.4) applies and the commutator in (5.6) vanishes.) Taking matrix elements of (5.5) between N-electron eigenstates $\langle E_1|$ and $|E_2\rangle$ of H_{mod} gives

$$-\mathrm{i}(E_2 - E_1)\langle E_1| \sum_{i=1}^{N} \boldsymbol{r}_i |E_2\rangle = \langle E_1| \sum_{i=1}^{N} \boldsymbol{p}_i |E_2\rangle - \mathrm{i}\langle E_1| \sum_{i=1}^{N} [\boldsymbol{r}_i, V(\boldsymbol{r}_i, \boldsymbol{r}_i')]\,|E_2\rangle. \tag{5.7}$$

Comparison of (5.7) with (5.2a) shows that the nonlocality of V introduces the nonzero commutator $[\boldsymbol{r}, V(\boldsymbol{r}, \boldsymbol{r}')]$, which causes the length and velocity forms of the electric dipole moment to differ. Similar considerations show that the velocity and acceleration forms of the electric dipole matrix element are no longer equal. Hence all three expressions for the transition matrix element differ from one another for a model Hamiltonian containing a nonlocal potential. However, if the model Hamiltonian has only a local potential, then all three expressions are equal. Thus it is important to realize that equality or inequality of the three forms of the matrix element says little in itself about how accurate the results will be but serves mainly as a measure of the nonlocality of the model potential.

Although in the presence of nonlocal potentials the length, velocity, and acceleration forms for the dipole matrix element differ, another property of nonlocal potentials enables one to single out the length formula for use in calculating the transition matrix element. This property is that a nonlocal potential may always be expressed as a local momentum-dependent potential in the following manner. On the right-hand side of (5.3) we express $\psi(\boldsymbol{r}')$ in terms of its value at $\boldsymbol{r}$ by means of the translation operator,[76]

$$\psi(\boldsymbol{r}') = \exp[\mathrm{i}(\boldsymbol{r}' - \boldsymbol{r}) \cdot \boldsymbol{p}]\, \psi(\boldsymbol{r}), \tag{5.8}$$

where $\boldsymbol{p} = -\mathrm{i}\nabla_r$. Substitution of (5.8) on the right in (5.3) gives

$$\int \langle \boldsymbol{r}|V|\boldsymbol{r}'\rangle\, \psi(\boldsymbol{r}')\, d\boldsymbol{r}' = \{\int \langle \boldsymbol{r}|V|\boldsymbol{r}'\rangle \exp[\mathrm{i}(\boldsymbol{r}' - \boldsymbol{r}) \cdot \boldsymbol{p}]\, d\boldsymbol{r}'\}\psi(\boldsymbol{r}) \equiv V(\boldsymbol{r}, \boldsymbol{p})\, \psi(\boldsymbol{r}), \tag{5.9}$$

where the local momentum-dependent potential $V(\boldsymbol{r}, \boldsymbol{p})$ is defined by the expression in brackets.

The model Hamiltonian now has two terms dependent on momentum: the kinetic-energy operator and the potential operator. Hence in considering the interaction of electrons described by this model Hamiltonian with electromagnetic radiation, the standard substitution $\boldsymbol{p} \to \boldsymbol{p} + \boldsymbol{A}/c$ must be made in the

[76] Expansion of the exponential in a power series shows (5.8) to be a Taylor series expansion of $\psi(\boldsymbol{r}')$ about its value at $\psi(\boldsymbol{r})$. See Landau, L., Lifshitz, E.M. (1965): Quantum mechanics. 2nd edn., Reading, Mass.: Addison-Wesley, p. 45

potential also, provided one wishes the Schrödinger equation to be gauge invariant.[77] That is, the additional electromagnetic interaction induced by the nonlocal potential may be found from $V(\boldsymbol{r}, \boldsymbol{p}+\boldsymbol{A}/c)$, which is given by

$$V(\boldsymbol{r}, \boldsymbol{p}+\boldsymbol{A}/c)=\int\langle\boldsymbol{r}|V|\boldsymbol{r}'\rangle \exp[\mathrm{i}(\boldsymbol{r}'-\boldsymbol{r})\cdot(\boldsymbol{p}+\boldsymbol{A}/c)]\,d\boldsymbol{r}'. \tag{5.10}$$

In the electric dipole approximation the vector A may be regarded as constant over the region of integration in (5.10). Thus, expanding the right side of (5.10) to first order in $\boldsymbol{A}/c$, we obtain

$$V(\boldsymbol{r}, \boldsymbol{p}+\boldsymbol{A}/c)=\int\langle\boldsymbol{r}|V|\boldsymbol{r}'\rangle\{1-\mathrm{i}(\boldsymbol{r}-\boldsymbol{r}')\cdot\boldsymbol{A}/c+\ldots\}\exp[\mathrm{i}(\boldsymbol{r}'-\boldsymbol{r})\cdot\boldsymbol{p}]\,d\boldsymbol{r}'. \tag{5.11}$$

Comparison of (5.11) with (5.6) shows that the additional interaction with the electromagnetic field is proportional to the commutator of the position operator $\boldsymbol{r}$ with the nonlocal potential. Upon rewriting (5.11) in nonlocal form and combining the interaction that is first order in $\boldsymbol{A}/c$ with the usual one arising from the kinetic-energy operator [cf. (3.2)], we find that in the presence of a nonlocal potential the coupling of an electron to the electromagnetic field is

$$H_{\mathrm{int}}=\sum_{i=1}^{N}\{\boldsymbol{p}_i-\mathrm{i}[\boldsymbol{r}_i, V(\boldsymbol{r}_i, \boldsymbol{r}_i')]\}\cdot\frac{\boldsymbol{A}}{c}. \tag{5.12}$$

Taking matrix elements of (5.12) between exact eigenstates of H_{mod} gives

$$\langle E_1|H_{\mathrm{int}}|E_2\rangle=-\mathrm{i}\omega\langle E_1|\sum_{i=1}^{N}\boldsymbol{r}_i|E_2\rangle, \tag{5.13}$$

where we have used (5.5) and (5.7) and set the photon energy ω equal to (E_2-E_1). Equation (5.13) shows that the length formula is the one that is consistent with gauge invariance of the Schrödinger equation when one uses a

[77] According to Maxwell's equations the electric and magnetic field vectors are unchanged under the following gauge transformations of the vector (A) and scalar (ϕ) potentials,

$$\boldsymbol{A}'=\boldsymbol{A}+\nabla\chi, \quad \phi'=\phi-\frac{1}{c}\frac{\partial\chi}{\partial t},$$

where χ is an arbitrary scalar function. (If one chooses $\nabla\cdot\boldsymbol{A}'=\nabla\cdot\boldsymbol{A}=0$ then χ must satisfy $\nabla^2\chi=0$ but is otherwise arbitrary.) It is well known (e.g., see Schiff, L.I. (1968): Quantum mechanics, 3rd edn., New York: McGraw-Hill Book Co., pp. 398–9) that for local potentials the form of the Schrödinger equation in the presence of electromagnetic fields is invariant under the above transformations of the fields provided one simultaneously transforms the wavefunction according to

$$\psi'=\psi\exp(-\mathrm{i}\chi/c).$$

The invariance of the form of the Schrödinger equation under the above three simultaneous transformations is called the "gauge invariance of the Schrödinger equation." R.G. Sachs and N. Austern (Phys. Rev. *81*, 705 (1951)) and A.M. Korolev (Yadern. Fiz. *6*, 353 (1967) [Sov. J. Nucl. Phys. *6*, 257 (1968)]) have shown that this gauge invariance holds also when there are nonlocal potentials provided the replacement $\boldsymbol{p}\to\boldsymbol{p}+\boldsymbol{A}/c$ is made in the momentum-dependent form of the nonlocal potentials as well as in the kinetic-energy operator model Hamiltonian containing a nonlocal potential. While the Schrödinger equation involving the exact Hamiltonian *must* be gauge invariant, there has been some debate over whether one should require the Schrödinger equation involving H_{mod} to be gauge invariant.[78] Not doing so of course requires one to consider the length, velocity, and acceleration formulas on an equal footing unless qualitative arguments indicate otherwise.

The best-known model Hamiltonian employing a nonlocal potential is the Hartree-Fock Hamiltonian, which includes the Fock exchange potential,

$$\langle \boldsymbol{r}| V_{\text{Fock}} |\boldsymbol{r}'\rangle = -\sum_{i=1}^{N} \frac{\psi_i(\boldsymbol{r})\,\psi_i^*(\boldsymbol{r}')}{|\boldsymbol{r}-\boldsymbol{r}'|}, \tag{5.14}$$

where the $\psi_i(\boldsymbol{r})$ are the N Hartree-Fock orbitals of the ground-state atom or ion. Related to Hartree-Fock calculations are perturbation theory calculations that start from a Hartree-Fock basis set. These latter calculations begin with a nonlocal model potential, the nonlocal effects of which on the transition matrix element are removed to n-th order in the perturbation if the transition matrix element is improved to the n-th order in the perturbation. Only in infinite order of course will the velocity and acceleration matrix elements finally equal the length matrix elements.

A less obvious model Hamiltonian employing a nonlocal potential is that used in configuration-interaction calculations. Here the model Hamiltonian consists firstly of a zero-order Hamiltonian H_0, which is used to generate a complete set of N-electron wavefunctions. The full perturbation is thus given by

$$V \equiv H - H_0, \tag{5.15}$$

where H is the exact Hamiltonian in (3.1). Usually in configuration-interaction calculations, however, only a part of the full perturbation (5.15) is considered, namely, that portion of V having nonzero matrix elements between the configurations to be mixed. More formally, one may define a projection operator P,

$$P \equiv \sum_{i \in P} |i\rangle\langle i|, \tag{5.16}$$

where each state $|i\rangle$ represents one of the N-electron configurations to be mixed. The portion of V under consideration is then given by PVP. The model Hamiltonian in configuration-interaction calculations is thus

$$H_{\text{mod}} = H_0 + PVP, \tag{5.17}$$

and this Hamiltonian is usually diagonalized exactly. Now one may see easily that PVP is nonlocal by considering its coordinate space representation,

$$\begin{aligned}\langle \boldsymbol{r}_1, \ldots, \boldsymbol{r}_N | PVP | \boldsymbol{r}_1', \ldots, \boldsymbol{r}_N' \rangle &= \sum_{i,j \in P} \langle \boldsymbol{r}_1, \ldots, \boldsymbol{r}_N | i\rangle \langle i| V |j\rangle \langle j | \boldsymbol{r}_1', \ldots, \boldsymbol{r}_N'\rangle \\ &= \sum_{i,j \in P} \psi_i^*(\boldsymbol{r}_1, \ldots, \boldsymbol{r}_N) \langle i| V |j\rangle\, \psi_j(\boldsymbol{r}_1', \ldots, \boldsymbol{r}_N').\end{aligned} \tag{5.18}$$

[78] Grant, I.P., Starace, A.F. (1975): J. Phys. B *8*, 1999

Equation (5.18) shows that even if the residual interaction V is local, PVP is not local since the set of states $i,j \in P$ are not complete. If the states $i,j \in P$ were complete, then the projection operator P would equal the unit operator 1 and (5.18) would reduce to $\langle \boldsymbol{r}_1, \ldots, \boldsymbol{r}_N | V | \boldsymbol{r}'_1, \ldots, \boldsymbol{r}'_N \rangle$, which would be local if V is local. Thus gauge invariance of the Schrödinger equation corresponding to the nonlocal configuration-interaction Hamiltonian in (5.17) requires the use of the length formula for electric dipole transition matrix elements.

Rather than asking which formula for the electric dipole transition matrix element is to be used in a particular approximate calculation, one may ask alternatively which approximation procedures preserve the equality of the various formulas for the electric dipole matrix element. At the present time this very interesting question cannot be answered in general. From the discussion above, it is obvious that any approximation procedure employing a local potential will preserve the desired equality. Consider now the more general approximation of improving both initial and final states by configuraion interaction. In particular, we imagine that one diagonalizes the residual interaction V [cf. (5.15)] within the subspace of states defined by a projection operator $P = P_{\mathrm{i}} + P_{\mathrm{f}}$, where P_{i} is a projector operator that singles out the configurations that are mixed in the initial state and where P_{f} singles out the configurations that are mixed in the final state. Since electric dipole transitions only connect initial and final states of opposite parity, the configurations in P_{i} cannot interact with those in P_{f} so that we have

$$PVP = P_{\mathrm{i}} V P_{\mathrm{i}} + P_{\mathrm{f}} V P_{\mathrm{f}}. \tag{5.19}$$

As discussed above, PVP is nonlocal in coordinate representation. Hence as shown in (5.7), its nonzero commutator with the position operator induces an inequality between the length and velocity forms of the electric dipole transition matrix element. However, if we calculate the matrix element of this commutator between the initial state $\langle \psi_0 |$ and the final state $| \psi_E \rangle$, we find the following interesting result:

$$\langle \psi_0 | [\sum_j \boldsymbol{r}_j, PVP] | \psi_E \rangle = \langle \psi_0 | \sum_j \boldsymbol{r}_j P_{\mathrm{f}} V P_{\mathrm{f}} | \psi_E \rangle - \langle \psi_0 | P_{\mathrm{i}} V P_{\mathrm{i}} \sum_j \boldsymbol{r}_j | \psi_E \rangle. \tag{5.20}$$

[To obtain (5.20) we have used the fact that the initial and final configurations have opposite parity so that $\langle \psi_0 | P_{\mathrm{f}} = P_{\mathrm{i}} | \psi_E \rangle = 0$.] Equation (5.20) suggests that in those cases where the interaction has the same sign in both initial and final states, then simultaneous improvement of both the initial and final states will lead to cancellations between the matrix elements on the right in (5.20). Thus the effects of nonlocal potentials may be minimized by judicious simultaneous improvement of both initial and final states rather than improvement of only one. In fact, the only nontrivial approximation method that preserves strict equality of the length and velocity formulas, namely, the random-phase approximation,[19] does simultaneously account for very similar correlations in *both* initial *and* final states (cf. Sect. 18). Unfortunately, it is not clear at present how to retain this equality in any approximation procedure that goes beyond the random-phase approximation.

6. Cross-section formulae in terms of the final-state channel wave functions. The energy-normalized channel wave functions $\psi_{\gamma E}^{-}$ defined by (4.55) are the wave functions usually computed numerically. Expressions for the differential, partial, and total photoionization cross sections may be obtained in terms of these wave functions as follows. Substituting (4.62) in (3.7), summing over final magnetic quantum numbers, and averaging over initial magnetic quantum numbers gives

$$\frac{d\sigma_{\gamma(\mathrm{i})}}{d\Omega}=\frac{4\pi^2}{\omega c}[L_0]^{-1}[S_0]^{-1}\sum_{\substack{M_{L_0}M_{S_0}\\ M_{L_\gamma}M_{S_\gamma}m_{1/2}}}\left|\sum_{\substack{l_\gamma m_\gamma\\ LM_LSM_S}}\mathrm{i}^{l_\gamma}\exp(-\mathrm{i}\sigma_{l_\gamma})\,Y^*_{l_\gamma m_\gamma}(\Omega)\right.$$
$$\left.\cdot\, C(L_\gamma l_\gamma L;M_{L_\gamma}m_\gamma M_L)\,C(S_\gamma \tfrac{1}{2}S;M_{S_\gamma}m_{1/2}M_S)\,\hat{\boldsymbol{\varepsilon}}\cdot\langle\psi_0|\sum_{j=1}^{N}\boldsymbol{p}_j|\psi_{\gamma E}^{-}\rangle\right|^2, \tag{6.1}$$

where the subscript $\gamma(\mathrm{i})$ on the left now specifies only the quantum numbers $\alpha_\gamma L_\gamma S_\gamma$ of the ionic state [cf. (4.53)], the symbol $[X]$ is defined to be $2X+1$, and $\hat{\boldsymbol{k}}$ has been replaced by Ω. The differential cross section in terms of the length formula is obtained from (6.1) by replacing the transition operator $\sum_{i=1}^{N}\boldsymbol{p}$ by $\omega\sum_{i=1}^{N}\boldsymbol{r}_i$ [cf. 5.2a)]. Standard techniques of the theory of angular distributions, reviewed by Blatt and Biedenharn,[79] may be used to reduce (6.1) to the form,

$$\frac{d\sigma_{\gamma(\mathrm{i})}}{d\Omega}=\frac{\sigma_{\gamma(\mathrm{i})}}{4\pi}[1+\beta P_2(\cos\theta)], \tag{6.2}$$

where $\sigma_{\gamma(\mathrm{i})}$ is the partial photoionization cross section for leaving the ion in the state $\gamma(\mathrm{i})$. β is the asymmetry parameter, $P_2(\cos\theta)\equiv\frac{3}{2}\cos^2\theta-\frac{1}{2}$, and θ is the direction of the outgoing photoelectron as measured from the polarization vector $\hat{\boldsymbol{\varepsilon}}$ of the incident light. The form of (6.2) follows from symmetry principles, as shown by Yang,[80] provided only that the target atom is randomly oriented, the incident light is linearly polarized, and the electric dipole approximation holds (cf. Sect. 3). Only slight modifications of (6.2) are necessary when the incident light is unpolarized,[81] circularly polarized,[82, 83] partially polarized,[84] or elliptically polarized.[85, 86] The requirement that $d\sigma_{\gamma(\mathrm{i})}/d\Omega$ is positive restricts β to the range $-1\leqq\beta\leqq2$. Alternatively, (6.1) may be reduced using the angular momentum transfer formulation for the angular distribution of Dill and Fano.[87-89] This formulation is discussed in Sect. 7 below.

---

[79] Blatt, J.M., Biedenharn, L.C. (1952): Rev. Mod. Phys. *24*, 258
[80] Yang, C.N. (1948): Phys. Rev. *74*, 764
[81] Cooper, J.W., Manson, S.T. (1969): Phys. Rev. *177*, 157
[82] Peshkin, M. (1970): Adv. Chem. Phys. *18*, 1
[83] Jacobs, V.L. (1972): J. Phys. B *5*, 2257
[84] Samson, J.A.R. (1969): J. Opt. Soc. Am. *59*, 356; (1970): Phil. Trans. Roy. Soc. A *268*, 141
[85] Schmidt, V. (1973): Phys. Lett. *45* A, 63
[86] Samson, J.A.R., Starace, A.F. (1975): J. Phys. B *8*, 1806; (1979): J. Phys. B *12*, 3993
[87] Fano, U., Dill, D. (1972): Phys. Rev. A *6*, 185
[88] Dill, D., Fano, U. (1972): Phys. Rev. Lett. *29*, 1203
[89] Dill, D. (1973): Phys. Rev. A *7*, 1976

The partial cross section $\sigma_{\gamma(i)}$ is obtained by integrating (6.1) over angles Ω and using the orthonormality of the spherical harmonics to give

$$\sigma_{\gamma(i)}=\frac{4\pi^2}{\omega c}[L_0]^{-1}[S_0]^{-1}\sum_{\substack{l_\gamma LM_L SM_S\\ M_{L_0}M_{S_0}}}\left|\hat{\boldsymbol{\varepsilon}}\cdot\langle\psi_0|\sum_{j=1}^{N}\boldsymbol{p}_j|\psi_{\gamma E}^{-}\rangle\right|^2 \tag{6.3a}$$

$$=\frac{4\pi^2\omega}{c}[L_0]^{-1}[S_0]^{-1}\sum_{\substack{l_\gamma LM_L SM_S\\ M_{L_0}M_{S_0}}}\left|\hat{\boldsymbol{\varepsilon}}\cdot\langle\psi_0|\sum_{j=1}^{N}\boldsymbol{r}_j|\psi_{\gamma E}^{-}\rangle\right|^2. \tag{6.3b}$$

The total cross section is obtained by summing (6.3) over all ionic states accessible at the photon energy ω. Using (4.55) for the channel wave functions, we find

$$\sigma_{\text{TOTAL}}=\sum_{\alpha_\gamma L_\gamma S_\gamma}\sigma_{\gamma(i)}=\frac{4\pi^2}{\omega c}[L_0]^{-1}[S_0]^{-1}\sum_{\Gamma M_{L_0}M_{S_0}}\left|\hat{\boldsymbol{\varepsilon}}\cdot\langle\psi_0|\sum_{j=1}^{N}\boldsymbol{p}_j|\Gamma E\rangle\right|^2 \tag{6.4a}$$

$$=\frac{4\pi^2\omega}{c}[L_0]^{-1}[S_0]^{-1}\sum_{\Gamma M_{L_0}M_{S_0}}\left|\hat{\boldsymbol{\varepsilon}}\cdot\langle\psi_0|\sum_{j=1}^{N}\boldsymbol{r}_j|\Gamma E\rangle\right|^2. \tag{6.4b}$$

In (6.4), the transition matrix element is expressed in terms of the eigenchannel states $|\Gamma E\rangle$ defined by (4.33).

Finally, we note that the partial and total photoionization cross-section formulas in (6.3) and (6.4) may be simplified further by performing analytically the summations over magnetic quantum numbers. Using the length formula for the electric dipole transition operator, for example, one proceeds by writing its scalar product with the polarization vector $\hat{\boldsymbol{\varepsilon}}$ in spherical tensor form:

$$\sum_{i=1}^{N}\hat{\boldsymbol{\varepsilon}}\cdot\boldsymbol{r}_i=\sum_{i=1}^{N}\sum_{q=-1}^{+1}\varepsilon_q^{*}r_{iq}^{(1)}. \tag{6.5}$$

Application of the well-known Wigner-Eckart theorem to matrix elements of the tensor operators $r_{iq}^{(1)}$ serves to separate these matrix elements into a geometrical factor (which describes the dependence on the orientation of the initial and final states) and a reduced matrix element (which describes the dependence on the other, dynamical parameters of the initial and final states). This reduction is rather lengthy, but straightforward, and is described in detail in works concerned with spectroscopic theory.[90, 91] For the length form, in the LS-coupling scheme the partial and total cross sections [(6.3b) and (6.4b)] become

$$\sigma_{\gamma(i)}=\frac{4\pi^2}{3c}\omega[L_0]^{-1}[S_0]^{-1}\sum_{l_\gamma LS}[S]\left|\langle\psi_0\|\sum_{j=1}^{N}r_j^{(1)}\|\psi_{\gamma E}^{-}\rangle\right|^2, \tag{6.6a}$$

$$\sigma_{\text{TOTAL}}=\frac{4\pi^2}{3c}\omega[L_0]^{-1}[S_0]^{-1}\sum_{\Gamma}[S]\left|\langle\psi_0\|\sum_{j=1}^{N}r_j^{(1)}\|\Gamma E\rangle\right|^2. \tag{6.6b}$$

---

[90] Shore, B.W., Menzel, D.H. (1965): Astrophys. J. Suppl. Ser. No. 106, *12*, 187

[91] Ref. [47], § 31

In (6.6) the double bars in the matrix elements indicate that these are reduced matrix elements, the factor 1/3 arises from replacing the component of $\boldsymbol{r}$ along $\hat{\varepsilon}$ by the tensor $r^{(1)}$, the factor $[S]$ arises from the summations over magnetic quantum numbers, and now, obviously, γ and Γ no longer specify any magnetic quantum numbers. Equation (6.6) is given in atomic units; we note however that in cgs units the constant factor

$$\frac{4\pi^2}{3c} = 2.68909\,\text{Mb}, \tag{6.7}$$

where Mb stands for the megabarn (which equals $10^{-18}\,\text{cm}^2$). Expressions similar to (6.6) hold for the velocity form. Explicit, but approximate, formulas for the reduced matrix elements appearing in (6.6) are given in Sect. 9.

7. The angular distribution asymmetry parameter. Although formulas for the angular distribution asymmetry parameter β in LS coupling have been given by LIPSKY[92] and by JACOBS and BURKE,[83,93] the recent angular momentum transfer formulation for β of DILL and FANO[87-89] is presented here because of two advantages the theory possesses. Firstly, it represents β as an incoherent sum of contributions characteristic of a given value of the angular momentum transfer j_t, defined below. This decomposition of the problem into smaller pieces simplifies the theoretical analysis in many cases. Secondly, the angular momentum transfer formulation permits a transparent reduction after application of certain approximations to the simpler formulation for β of COOPER and ZARE[12] (cf. Sect. 9).

Consider the photoionization process

$$\text{A}(\alpha_0 L_0 S_0 \pi_0) + \omega(j_\text{p}=1, \pi_\text{p}=-1) \rightarrow \text{A}^+(\alpha_\gamma L_\gamma S_\gamma \pi_\gamma) + \text{e}^-[l_\gamma, s=\tfrac{1}{2}, \pi_\text{e}=(-1)^{l_\gamma}], \tag{7.1}$$

where A is an arbitrary atom. In LS coupling, the angular momentum transfer j_t is defined as

$$\boldsymbol{j}_\text{t} = \boldsymbol{j}_\text{p} - \boldsymbol{l}_\gamma = \boldsymbol{L}_\gamma - \boldsymbol{L}_0. \tag{7.2}$$

Thus j_t equals the orbital angular momentum transferred between the atom and the ion. The angular momentum and parity-conservation equations, i.e.,

$$\boldsymbol{L}_0 + \boldsymbol{j}_\text{p} = \boldsymbol{L}_\gamma + \boldsymbol{l}_\gamma \tag{7.3}$$

$$\pi_0 \pi_\text{p} = \pi_\gamma \pi_\text{e} \quad \text{or} \quad -\pi_0 = \pi_\gamma (-1)^{l_\gamma}, \tag{7.4}$$

combine with the two relations in (7.2) to determine the allowed values of j_t for a given photoionization process.

For a given value of j_t, the first relation in (7.2) shows that the photoelectron orbital angular momentum l_γ can assume the values $j_t \pm 1$ and j_t since the

[92] Lipsky, L. (1967): Fifth international conference on the physics of electronic and atomic collisions: Abstracts of papers, Leningrad: Nauka, pp. 617–18

[93] Jacobs, V.L., Burke, P.G. (1972): J. Phys. B 5, L 67

photon angular momentum j_p is unity. But the photoelectron has a well-defined parity so that for a given value of j_t *either* $l_\gamma = j_t \pm 1$ *or* $l_\gamma = j_t$. In these two cases, (7.4) becomes

$$\pi_0 \pi_\gamma = (-1)^{j_t} \quad \text{or} \quad \pi_0 \pi_\gamma = -(-1)^{j_t}, \tag{7.5}$$

where have used the fact that the photon parity π_p is odd. DILL and FANO[88] label the former case "parity favored" and the latter "parity unfavored" for the following reason. In the former case, the angular distribution asymmetry parameter for the value j_t [i.e., $\beta_{\text{fav}}(j_t)$] can assume a normal range of values between -1 and $+2$ due to interference between the two transition amplitudes corresponding to the two allowed values of $l_\gamma = j_t \pm 1$. In the latter case, however, l_γ is restricted to the single value $l_\gamma = j_t$ and hence $\beta_{\text{unf}}(j_t)$ has a fixed value, $\beta_{\text{unf}}(j_t) = -1$. From (6.2) one sees that if $\beta = -1$, then the differential cross section varies as $\sin^2\theta$. This implies that the photoelectron angular distribution vanishes along the electric vector of the incident light and peaks at 90° from the electric vector, a result completely at odds with independent-particle models.

In the *LS*-coupling form of the angular momentum transfer analysis, one proceeds by first determining the allowed values of j_t from (7.2)–(7.4). For each allowed value one then determines whether the transition is parity favored or unfavored according to (7.5). For each allowed value of the photoelectron orbital angular momentum l_γ and for each of the corresponding allowed values of j_t one then calculates the following reduced scattering amplitudes:

$$\bar{S}_{l_\gamma}(j_t) \equiv \left(\frac{4}{3}\frac{\pi}{c}\omega\right)^{1/2} \mathrm{i}^{l_\gamma} \exp(-\mathrm{i}\sigma_{l_\gamma}) \cdot \sum_L [L]^{1/2} \begin{Bmatrix} L_\gamma & l_\gamma & L \\ 1 & L_0 & j_t \end{Bmatrix} \langle \alpha_0 L_0 S_0 \| \sum_{j=1}^{N} r_j^{(1)} \| \psi_{\gamma E}^- \rangle. \tag{7.6}$$

In (7.6) ω is the photon energy, the initial state is indicated simply by the quantum numbers $\alpha_0 L_0 S_0$, the energy-normalized channel wave function $\psi_{\gamma E}^-$ is defined by (4.55), and the phase factors in front of the summation symbol come from the incoming-wave normalization [cf. (4.62)]. The reduced scattering amplitudes $\bar{S}_{l_\gamma}(j_t)$ are simply those linear combinations of the reduced electric dipole transition matrix elements which recouple the angular momenta of the system to particular values of the angular momentum transfer j_t. The advantage of introducing the amplitudes $\bar{S}_{l_\gamma}(j_t)$ is the relatively simple form one obtains for the angular distribution asymmetry parameter.

According to the analysis of DILL and FANO,[87-89] the asymmetry parameter β appropriate for substitution in (6.2) is given by the following weighted average,

$$\beta = \frac{\sum_{j_t} \sigma(j_t)\,\beta(j_t)}{\sum_{j_t} \sigma(j_t)}, \tag{7.7}$$

where the sums extend over all allowed values of j_t and where $\beta(j_t)$ and $\sigma(j_t)$ are the asymmetry parameter and partial photoionization cross section characteristic of a given value of j_t. $\beta(j_t)$ and $\sigma(j_t)$ are defined in terms of the scattering amplitudes $\bar{S}_{l_\gamma}(j_t)$ as follows:

$$\beta_{\text{fav}}(j_t)=\frac{(j_t+2)|\bar{S}_+(j_t)|^2+(j_t-1)|\bar{S}_-(j_t)|^2-3[j_t(j_t+1)]^{1/2}\{\bar{S}_+(j_t)\bar{S}_-^\dagger(j_t)+\text{c.c.}\}}{(2j_t+1)\{|\bar{S}_+(j_t)|^2+|\bar{S}_-(j_t)|^2\}} \tag{7.8a}$$

$$\beta_{\text{unf}}(j_t)=-1, \tag{7.8b}$$

$$\sigma_{\text{fav}}(j_t)=\pi\frac{[j_t]}{[L_0]}\{|\bar{S}_+(j_t)|^2+|\bar{S}_-(j_t)|^2\}, \tag{7.8c}$$

$$\sigma_{\text{unf}}(j_t)=\pi\frac{[j_t]}{[L_0]}|\bar{S}_0(j_t)|^2. \tag{7.8d}$$

In (7.8) "c.c." denotes complex conjugate; the subscripts $+$, $-$, and 0 on the amplitudes $\bar{S}_{l_\gamma}(j_t)$ denote whether $l_\gamma=j_t\pm1$ or $l_\gamma=j_t$, respectively; and the subscripts fav and unf indicate whether j_t is favored or unfavored. The denominator in (7.7) is the partial photoionization cross section $\sigma_{\gamma(i)}$ [cf. (6.6a)] for leaving the ion in the state $\gamma(i)$, i.e.,

$$\sigma_{\gamma(i)}=\sum_{j_t}\sigma(j_t)\equiv\sum_{j_t}^{\text{fav}}\sigma_{\text{fav}}(j_t)+\sum_{j_t}^{\text{unf}}\sigma_{\text{unf}}(j_t). \tag{7.9}$$

(This is easily verified by carrying out the summation over j_t in (7.9), using (7.6) and (7.8), and comparing the result with (6.6a), provided one sets $S=S_0$ in the latter equation.)

8. The atomic oscillator strength

α) Definition. At each energy E we may define a continuum oscillator strength by

$$\frac{df}{dE}\equiv2\omega[L_0]^{-1}[S_0]^{-1}\sum_{\Gamma M_{L_0}M_{S_0}}\left|\hat{\varepsilon}\cdot\langle\psi_0|\sum_{j=1}^{N}\boldsymbol{r}_j|\Gamma E\rangle\right|^2 \tag{8.1a}$$

$$=\frac{2\omega}{3}[L_0]^{-1}[S_0]^{-1}\sum_{\Gamma}[S]\left|\langle\psi_0\|\sum_{j=1}^{N}r_j^{(1)}\|\Gamma E\rangle\right|^2. \tag{8.1b}$$

In (8.1) the summation $[L_0]^{-1}[S_0]^{-1}\sum_{M_{L_0}M_{S_0}}$ serves to average over the orientation of the initial state $\langle\psi_0|$, and Γ refers to all quantum numbers specifying the LS-coupling eigenchannel states (4.33). Equation (8.1b) is obtained from (8.1a) by carrying out the summation over both initial and final magnetic quantum numbers [and hence Γ includes final magnetic quantum numbers in (8.1a) but does not in (8.1b)]. Note that since at each energy E the summations in (8.1) are carried out over all other final-state quantum numbers, one is free to represent the final state using any energy eigenstate of the full Hamiltonian.

In particular we might have used the channel eigenfunctions $\psi_{\gamma E}^{-}$ [defined by (4.55)] and summed over the quantum numbers γ. Also, one might define each term in the summation in (8.1) as the continuum oscillator strength for a particular transition, but usually such a specific notation is only needed when defining oscillator strengths for discrete transitions.

Comparison of (8.1) with (6.6) shows that the total photoionization cross section at energy E is simply proportional to the continuum oscillator strength,

$$\sigma_{\mathrm{TOT}} = \frac{2\pi^2}{c} \frac{df}{dE}. \tag{8.2}$$

Equations (8.1) and (8.2) are both given in atomic units (a.u.); in ordinary units (8.2) becomes

$$\sigma_{\mathrm{TOT}} = 4.03364 \text{ Mb} \cdot \text{a.u.} \frac{df}{dE}, \tag{8.3}$$

where the energy unit a.u. in the numerator is cancelled by df/dE, which has the dimensions of an inverse energy in a.u., where 1 a.u. = 27.2108 eV.

β) The Thomas-Reiche-Kuhn sum rule. The n-th spectral moment of the oscillator strength distribution is defined by

$$S(n) \equiv \int dE (E - E_0)^n \frac{df}{dE}, \tag{8.4}$$

where n may be a positive or negative integer and where the integration over energy implicitly includes a summation over all discrete energy levels. The spectral moments may often be related to theoretical or experimental properties of the initial state of the atom.[2,38] In the theory of atomic photoionization, however, the most useful sum rule is that for $n=0$, which is known as the Thomas-Reiche-Kuhn sum rule:[94,95]

$$S(0) = \int dE \frac{df}{dE} = N. \tag{8.5}$$

In (8.5), N is the total number of electrons in the initial state.

The proof of (8.5) is as follows. Substituting (8.1a) in (8.5) and using the fact that $\omega \equiv E - E_0$, one obtains

$$S(0) = [L_0]^{-1} [S_0]^{-1} \sum_{M_{L_0} M_{S_0}} \sum_{\Gamma} 2 \int dE (E - E_0) \cdot \langle \psi_0 | \hat{\boldsymbol{\varepsilon}} \cdot \sum_{i=1}^{N} \boldsymbol{r}_i | \Gamma E \rangle \langle \Gamma E | \hat{\boldsymbol{\varepsilon}} \cdot \sum_{j=1}^{N} \boldsymbol{r}_j | \psi_0 \rangle. \tag{8.6}$$

---

[94] Reiche, F., Thomas, W. (1925): Naturwissenschaften *13*, 627

[95] Kuhn, W. (1925): Z. Phys. *33*, 408

Use of the velocity formula for the transition matrix elements, (5.2a), and the completeness of the states $|\Gamma E\rangle$ in (8.6) gives

$$S(0)=[L_0]^{-1}[S_0]^{-1}\sum_{M_{L_0}M_{S_0}} \mathrm{i}\langle\psi_0|\left[\hat{\boldsymbol{\varepsilon}}\cdot\sum_{i=1}^{N}\boldsymbol{p}_i\,,\ \hat{\boldsymbol{\varepsilon}}\cdot\sum_{j=1}^{N}\boldsymbol{r}_j\right]|\psi_0\rangle. \tag{8.7}$$

Using now the commutation relation,

$$[\hat{\boldsymbol{\varepsilon}}\cdot\boldsymbol{p}_i,\hat{\boldsymbol{\varepsilon}}\cdot\boldsymbol{r}_j]=-\mathrm{i}\delta_{ij}, \tag{8.8}$$

as well as the fact that

$$\frac{1}{[L_0][S_0]}\sum_{M_{L_0}M_{S_0}}\langle\psi_0|\psi_0\rangle=1, \tag{8.9}$$

we obtain the result in (8.5).

The meaning of (8.5) is that the number of electrons in the atom is equal to the oscillator strength integrated over all photon energies, starting from the photon energy needed to excite the outermost atomic electrons into the nearest unoccupied discrete levels and continuing up to infinitely high photon energies. Note that (8.5) does *not* hold for each atomic subshell, but only for the atom as a whole. Furthermore, we note that in proving the sum rule we used the exact equality in (5.2a) for the length and velocity forms of the transition matrix element. This equality holds in general only for exact eigenstates of the full Hamiltonian, as discussed in Sect. 5. In approximate calculations, therefore, the integral of the calculated oscillator strength does not in general equal N. However, as noted before, the random-phase approximation (see Sect. 18) does give strict equality between length and velocity matrix elements, and hence the oscillator strength calculated in the random phase approximation satisfies the sum rule in (8.5).[19]

γ) Relation to the generalized oscillator strength. In the theory of electron impact excitation or ionization of atoms a familiar quantity is the generalized oscillator strength, defined as[96,97]

$$\frac{df}{dE}(\boldsymbol{K})=\frac{1}{[L_0][S_0]}\sum_{\Gamma M_{L_0}M_{S_0}}[S]2(E-E_0)\left|\frac{1}{K}\langle\psi_0|\sum_{j=1}^{N}\exp(\mathrm{i}\boldsymbol{K}\cdot\boldsymbol{r}_j)|\Gamma E\rangle\right|^2. \tag{8.10}$$

The transition matrix element in (8.10) is the first Born approximation to the interaction between a fast incident electron and the atom.[96,97] The momentum transfer $\boldsymbol{K}$ is defined by

$$\boldsymbol{K}\equiv\boldsymbol{k}-\boldsymbol{k}', \tag{8.11}$$

where $\boldsymbol{k}$ is the incident electron's momentum before collision, and $\boldsymbol{k}'$ is its momentum after collision. For any $\boldsymbol{K}$ one may show that the generalized oscillator strength $df/dE(\boldsymbol{K})$ satisfies the same sum rule that the optical oscillator strength satisfies in (8.5).[96,97]

---

[96] Bethe, H. (1930): Ann. Phys. (Leipzig) *5*, 325

[97] Inokuti, M. (1971): Rev. Mod. Phys. *43*, 297

Of relevance to atomic photoionization is the fact that the generalized oscillator strength in (8.10) reduces to the optical oscillator strength in (8.1) in the limit of vanishing momentum transfer, i.e.,

$$\lim_{K\to 0} \frac{df}{dE}(\boldsymbol{K}) = \frac{df}{dE}. \tag{8.12}$$

The proof of (8.12) follows straightforwardly upon expanding the exponential in (8.10) and noting the orthogonality of the ground and excited states. Experimentally, the limit of vanishing momentum transfer can be attained by using very large incident electron energies and observing the scattered electrons in the forward direction.[31] Equation (8.12) thus serves as a common boundary between electron-impact excitation and ionization on the one hand and photoexcitation and photoionization on the other.

9. Limiting cases and approximations to the general formulae. Expressions in terms of radial matrix elements for the partial and total cross sections in (6.6), the reduced scattering amplitudes in (7.6) (from which the asymmetry parameter β may be calculated), and the atomic oscillator strength in (8.1) may be obtained by application of standard techniques of Racah algebra.[47,90] In what follows we first present expressions for the reduced electric dipole matrix element in terms of radial integrals for the limiting case of vanishing final-state interactions. These reduced electric dipole matrix elements may be substituted in the general formulas in Sects. 6–8 to obtain the cross sections, asymmetry parameters, and oscillator strengths in this limit. We then examine the further reduction of the formulas for the partial cross section and for the asymmetry parameter in the central potential approximation.

α) The limit of vanishing final-state interactions. Consider the form of the final-state wave functions (obtained in Sect. 4) in the approximation of BATES,[48] who assumed LS coupling and ignored final-state interactions. With no final-state interactions the K matrix vanishes, the transformation matrix $U_{\gamma,\Gamma}$ becomes the unit matrix, and the eigenchannel phase shift η_Γ vanishes. Thus (4.33) becomes

$$|\Gamma E\rangle \xrightarrow[\text{interactions}]{\text{no final-state}} |\gamma E\rangle, \tag{9.1}$$

by which we mean that the eigenchannel states become the basis states. Similarly, (4.55) becomes

$$\psi_{\gamma E}^{-}(\boldsymbol{r}_1 s_1, \ldots, \boldsymbol{r}_N s_N) \xrightarrow[\text{interactions}]{\text{no final-state}} \langle \boldsymbol{r}_1 s_1, \ldots, \boldsymbol{r}_N s_N | \gamma E\rangle \, e^{-i\delta_\gamma}, \tag{9.2}$$

i.e., the channel wave functions also reduce to the basis wave functions but have in addition the phase factor $\exp(-i\delta_\gamma)$, which ensures incoming-wave normalization for the channel wave function. Substituting (9.2) in (6.6a) and (7.6), we find that the partial photoionization cross section and the reduced scattering amplitude (from which the asymmetry parameter may be calculated) for the ionic state $L_\gamma S_\gamma$ are given by

$$\sigma_{L_\gamma S_\gamma} = \frac{4\pi^2}{3c}\,\omega [L_0]^{-1}[S_0]^{-1}\sum_{lLS}[S]\left|\langle L_0 S_0\| \sum_{j=1}^{N} r_j^{(1)} \|(L_\gamma S_\gamma)\,\varepsilon l LS\rangle\right|^2 \tag{9.3}$$

$$\bar{S}_l(j_t) = \left(\frac{4\pi}{3c}\,\omega\right)^{1/2} \mathrm{i}^l \exp(-\mathrm{i}\sigma_l)\sum_L [L]^{1/2}\begin{Bmatrix} L_\gamma & l & L \\ 1 & L_0 & j_t \end{Bmatrix}$$
$$\cdot\langle L_0 S_0\| \sum_{j=1}^{N} r_j^{(1)} \|(L_\gamma S_\gamma)\,\varepsilon l LS\rangle \exp(-\mathrm{i}\,\delta_l^{L_\gamma S_\gamma L}). \tag{9.4}$$

In (9.3) and (9.4) we have made the following replacement:

$$|\gamma E\rangle \equiv |(L_\gamma S_\gamma)\,\varepsilon l LS\rangle, \tag{9.5}$$

where L_γ and S_γ refer to the ionic term level, l is the photoelectron orbital angular momentum, L and S are the total orbital and spin angular momenta of the ion-electron system, and the total energy E on the left is replaced on the right by the kinetic energy ε of the photoelectron. We have suppressed, for simplicity, explicit notation of any other quantum numbers needed to specify uniquely the initial, ionic, or final states. The reduced matrix element appearing in (9.3) and (9.4) must be calculated separately for each pair of initial and final configurations using either the graphical techniques of Briggs,[98] or the algebraic techniques of Fano and Racah[99] as done, for example, in[47, 90]. These reduced matrix elements are presented for three very general cases.

1) *Photoionization of the open subshell of an arbitrary open-shell atom having configuration (closed shells) nl_0^q (closed shells)*

$$\langle nl_0^q L_0 S_0\| \sum_{j=1}^{N} r_j^{(1)} \| nl_0^{q-1}(L_\gamma S_\gamma)\,\varepsilon l LS\rangle$$
$$= \delta_{S_0 S}\, q^{1/2} (l_0^q L_0 S_0\{|l_0^{q-1} L_\gamma S_\gamma)\,\{S_0 S_\gamma 1/2\}(-1)^{l_0+L_\gamma+L_0}$$
$$\cdot [L_0]^{1/2}[L]^{1/2}\begin{Bmatrix} L_0 & L_\gamma & l_0 \\ l & 1 & L \end{Bmatrix}\langle nl_0\| r^{(1)} \|\varepsilon l\rangle. \tag{9.6}$$

2) *Photoionization of an inner closed subshell $nl_0^{4l_0+2}$ of an arbitrary open-shell atom having configuration (closed shells) $nl_0^{4l_0+2} n'l'^q$ (closed shells)*

$$\langle nl_0^{4l_0+2} n'l'^q L_0 S_0\| \sum_{j=1}^{N} r_j^{(1)} \| nl_0^{4l_0+1} n'l'^q (L_\gamma S_\gamma)\,\varepsilon l LS\rangle$$
$$= \delta_{SS_0}(-1)^{q+l_0+L_0+L_\gamma}\{S_0 S_\gamma 1/2\}\,[L_\gamma]^{1/2}[L]^{1/2}[S_\gamma]^{1/2}[S_0]^{-1/2}$$
$$\cdot\begin{Bmatrix} L_0 & L_\gamma & l_0 \\ l & 1 & L \end{Bmatrix}\langle nl_0\| r^{(1)} \|\varepsilon l\rangle. \tag{9.7}$$

---

[98] Briggs, J.S. (1971): Rev. Mod. Phys. *43*, 189

[99] Fano, U., Racah, G. (1959): Irreducible tensorial sets, New York: Academic Press

3) Photoionization of an outer closed subshell $nl_0^{4l_0+2}$ of an arbitrary open-shell atom having configuration (closed shells) $n'l'^q nl_0^{4l_0+2}$ (closed shells)

$$\langle n'l'^q nl_0^{4l_0+2} L_0 S_0 \| \sum_{j=1}^{N} r_j^{(1)} \| n'l'^q nl_0^{4l_0+1}(L_\gamma S_\gamma)\varepsilon l LS\rangle$$
$$=(-1)^{q+l_0+L_0+L_\gamma+S_\gamma-1/2-S_0}$$
$$\cdot\langle nl_0^{4l_0+2} n'l'^q L_0 S_0 \| \sum_{j=1}^{N} r_j^{(1)} \| nl_0^{4l_0+1} n'l'^q(L_\gamma S_\gamma)\,\varepsilon l LS\rangle. \tag{9.8}$$

In (9.6) and (9.7) the symbol $[X]$ is defined to be $2X+1$; the symbol $\{abc\}$ indicates the triangular delta function, which is unity if

$$a+b \geqq c \geqq |a-b| \tag{9.9a}$$

and

$$a+b+c=\text{integer}, \tag{9.9b}$$

and is zero otherwise; the symbols $(l_0^q L_0 S_0\{|l_0^{q-1} L_\gamma S_\gamma)$ are coefficients of fractional parentage, which are tabulated by SHORE and MENZEL;[100] the $6-j$ symbols are tabulated by ROTENBERG et al.;[101] and the reduced one-electron matrix element is given by

$$\langle nl_0 \| r^{(1)} \| \varepsilon l\rangle = (-1)^{l_0-l_>} l_>^{1/2} \int_0^\infty P_{nl_0}^{L_0 S_0}(r)\, r\, P_{\varepsilon l}^{L_\gamma S_\gamma L}(r)\, dr. \tag{9.10}$$

In (9.10) $l_>$ is the greater of l_0 and l, where l can only assume the values $l_0 \pm 1$. Thus $l_>$ may be written as

$$l_> = \frac{l_0+l+1}{2}. \tag{9.11}$$

The one-electron radial wave functions are $r^{-1}P_{nl_0}(r)$ and $r^{-1}P_{\varepsilon l}(r)$, where the superscripts in (9.10) indicate the dependence of these radial wave functions on the term levels of the initial and final states. The one-electron continuum wave function $r^{-1}P_{\varepsilon l}(r)$ is normalized per unit energy in atomic units according to (4.25). Its asymptotic form is given by

$$P_{\varepsilon l}^{L_\gamma S_\gamma L}(r) \xrightarrow[r\to\infty]{} \left(\frac{2}{\pi k_\gamma}\right)^{1/2} \sin\left(k_\gamma r - \tfrac{1}{2} l\pi + \frac{1}{k_\gamma}\log 2k_\gamma r + \sigma_l + \delta_l^{L_\gamma S_\gamma L}\right), \tag{9.12}$$

where ε is $k_\gamma^2/2$, the photoelectron's kinetic energy, and $\delta_l^{L_\gamma S_\gamma L}$ is the phase shift (with respect to Coulomb waves) of the photoelectron in the channel $L_\gamma S_\gamma l L$.

Note that if the velocity form of the electric dipole transition operator were used in place of the length form in (9.3) and (9.4), then the photon energy ω in

---

[100] Shore, B.W., Menzel, D.H. (1968): Principles of atomic spectra, New York: Wiley, pp. 385–90

[101] Rotenberg, M., Bivens, R., Metropolis, N., Wooten, Jr., J.K. (1959): The $3j$- and $6j$-Symbols, Cambridge, Mass.: The Technology Press, M.I.T.

these equations would appear in the denominator instead of in the numerator, and in (9.6) and (9.7) the reduced one-electron matrix element $\langle nl_0 \| r^{(1)} \| \varepsilon l \rangle$ would be replaced by the one-electron reduced matrix element of the gradient operator,[102]

$$\langle nl_0 \| \nabla^{(1)} \| \varepsilon l \rangle$$
$$= (-1)^{l_0 - l_>} l_>^{1/2} \int_0^\infty dr\, P_{nl_0}^{L_0 S_0}(r) \left\{ \frac{l(l+1) - l_0(l_0+1)}{2r} + \frac{d}{dr} \right\} P_{\varepsilon l}^{L_\gamma S_\gamma L}(r), \quad (9.13)$$

where $l_>$ is given by (9.11).

When calculating the asymmetry parameter β from the reduced scattering amplitudes in (9.4), one finds that many of the factors included in $\bar{S}_l(j_t)$ cancel in the numerator and denominator of the expression for β, (7.7). Due to this cancellation, it is possible to represent $\bar{S}_l(j_t)$ in (9.4) by a single formula for all three of the cases considered above, the reduced dipole matrix elements of which have been given in (9.6)–(9.8). That is,

$$\bar{S}_l(j_t) \propto \mathrm{i}^{+l} \exp(-\mathrm{i}\sigma_l)(-1)^{l_0 - l_>} l_>^{1/2}$$
$$\cdot \sum_L \exp(-\mathrm{i}\,\delta_{\varepsilon l}^{L_\gamma S_\gamma L}) R_{\varepsilon l}^{L_\gamma S_\gamma L} [L] \begin{Bmatrix} L_0 & L_\gamma & j_t \\ l & 1 & L \end{Bmatrix} \begin{Bmatrix} L_0 & L_\gamma & l_0 \\ l & 1 & L \end{Bmatrix}. \quad (9.14)$$

In (9.14) the constant of proportionality that has been omitted depends only on properties of the initial state and the ionic state. Also, we have represented the radial integral by

$$R_{\varepsilon l}^{L_\gamma S_\gamma L} \equiv \int_0^\infty P_{nl_0}^{L_0 S_0}(r)\, r\, P_{\varepsilon l}^{L_\gamma S_\gamma L}(r)\, dr \quad (9.15)$$

so that its dependence on the quantum numbers L_γ, S_γ, and L is explicitly indicated. As before, $l_>$ is defined by (9.11). The LS-coupling form of the reduced scattering amplitude, (9.14), has been derived by DILL et al.[103,104]

β) The central potential approximation. In the central potential approximation, the radial wave function of the photoelectron no longer depends on the coupling between the photoelectron and the ion, i.e., it no longer depends on the quantum numbers L_γ, S_γ, and L. Similarly, for very light atoms the interaction between the photoelectron and the ion is approximately isotropic. In either case we have

$$\exp(-\mathrm{i}\delta_{\varepsilon l}^{L_\gamma S_\gamma L}) R_{\varepsilon l}^{L_\gamma S_\gamma L} \xrightarrow[\text{interactions}]{\text{isotropic}} \exp(-\mathrm{i}\,\delta_{\varepsilon l}) R_{\varepsilon l}. \quad (9.16)$$

The approximation in (9.16) permits analytic summations over term level quantum numbers that greatly simplify expressions for cross sections and

[102] Ref. 101, pp. 6–7

[103] Dill, D., Manson, S.T., Starace, A.F. (1974): Phys. Rev. Lett. *32*, 971

[104] Dill, D., Starace, A.F., Manson, S.T. (1975): Phys. Rev. A *11*, 1596

angular distribution asymmetry parameters. In what follows we examine some consequences of (9.16).

Consider first the partial cross section for photoionizing an electron from the subshell nl_0^q of an arbitrary atom, where q is the occupation number of the subshell. Substituting (9.6) in (9.3) and using (9.10), (9.11), and (9.16), we can perform the summations over L and S analytically. Summing also over all ionic term levels and using the normalization property of the coefficients of fractional parentage, we find

$$\sum_{L_\gamma S_\gamma} \sigma_{L_\gamma S_\gamma} = \frac{4\pi^2}{3c} \omega q \sum_l \frac{l_0+l+1}{2[l_0]} (R_{\varepsilon l})^2. \tag{9.17}$$

In (9.17) the summation over photoelectron orbital angular momenta l is carried out for the two values $l = l_0 \pm 1$, and the radial integral $R_{\varepsilon l}$ is given by (9.15) with the dependencies of the initial and final radial wave functions on the term-level structure omitted.

Consider next the angular distribution asymmetry parameter β. Substituting (9.16) in (9.14) for the reduced scattering amplitude permits us to sum over L analytically to obtain

$$\bar{S}_l(j_t) \propto \delta(j_t, l_0)\, \mathrm{i}^{+l} \exp(-\mathrm{i}\sigma_l)\,(-1)^{l_0-l_>}\, l_>^{1/2} [l_0]^{-1} \exp(-\mathrm{i}\delta_{\varepsilon l})\, R_{\varepsilon l}. \tag{9.18}$$

Note that the delta function in (9.18) restricts j_t to the *single* value $j_t = l_0$. Substitution of (9.18) in (7.7) and (7.8) and use of (9.11) gives the Cooper-Zare formula for β,[12]

$$\beta_{CZ} = \frac{l_0(l_0-1) R_{l_0-1}^2 + (l_0+1)(l_0+2) R_{l_0+1}^2 - 6 l_0(l_0+1) R_{l_0-1} R_{l_0+1} \cos\Delta}{(2l_0+1)[l_0 R_{l_0-1}^2 + (l_0+1) R_{l_0+1}^2]}, \tag{9.19a}$$

where

$$\Delta \equiv \sigma_{l_0+1} + \delta_{l_0+1} - \sigma_{l_0-1} - \delta_{l_0-1}. \tag{9.19b}$$

In a special case, (9.19) holds more generally than just in the central potential approximation. In photoionization of atoms having $L_0 = 0$ (e.g., rare gas atoms) in which the resulting ion has $L_\gamma = l_0$, the exact LS-coupling amplitude (7.6) reduces to a form similar to (9.18), although the definitions of the dipole amplitudes and phase shifts are different. Thus a β-parameter formula similar to (9.19) is valid.

II. The central potential model and its predictions

10. The central potential model. The formal development of the previous section resulted in the description of the photoionization cross section in terms of dynamical quantities: the radial dipole matrix elements and the photoelectron phase shifts. Approximate calculational procedures are needed to obtain these dynamical quantities. In this section we examine the simplest approximation procedure, the central potential model of the atom, and discuss its predictions for these dynamical quantities. These predictions are quantitatively accurate for inner-shell electrons at photon energies well beyond their threshold and qualitatively useful for lower photon energies or for outer-subshell electrons.

The physical content of the central potential model is that the exact Hamiltonian in (3.1) is approximated by

$$H_{\mathrm{CP}}=\sum_{i=1}^{N}\left\{\frac{p_i^2}{2m}+V(r_i)\right\}, \tag{10.1}$$

where $V(r_i)$ is the potential experienced by the i-th electron. The potential $V(r)$ must approximate the nuclear attraction and the inter-electron repulsion in (3.1) as well as possible and, in particular, should have the following limiting behavior for a neutral atom:

$$V(r)\xrightarrow[r\to 0]{}-\frac{Z}{r} \quad\text{and}\quad V(r)\xrightarrow[r\to\infty]{}-\frac{1}{r}. \tag{10.2}$$

The Hamiltonian in (10.1) is separable in spherical coordinates and its eigenstates can be written as Slater determinants of one-electron orbitals of the form $r^{-1}P_{nl}(r)\,Y_{lm}(\Omega)$ for bound orbitals and of the form $r^{-1}P_{\varepsilon l}(r)\,Y_{lm}(\Omega)$ for continuum orbitals. The one-electron radial wave functions are obtained as solutions of

$$\frac{d^2P_{\varepsilon l}(r)}{dr^2}+2\left[\varepsilon-V(r)-\frac{l(l+1)}{2r^2}\right]P_{\varepsilon l}(r)=0. \tag{10.3}$$

A similar equation holds for discrete orbitals $P_{nl}(r)$. All of the radial wave functions satisfy the boundary condition $P_{\varepsilon l}(0)=0$.

A realistic central potential that has been widely used is that of Herman and Skillman,[105] which approximates exchange interactions between electrons by the average exchange potential appropriate to a free electron gas. This potential is obtained by solving the Hartree-Fock-Slater equations[106] self-consistently for the ground state of the neutral atom. Since the resulting potential goes to zero at large r, a Latter-type correction[107] to the potential is made by joining it to the Coulomb potential $-r^{-1}$ at $r=r_0$, where r_0 is the radius at which the self-consistent atomic potential and the Coulomb potential are equal.

The behavior of the Herman-Skillman central potential as a function of atomic number Z and coordinate r is presented in Fig. 1. As discussed by Rau and Fano,[108] Fig. 1 shows periodicities and trends due to inner-atomic-shell

---

[105] Herman, F., Skillman, S. (1963): Atomic structure calculations, Englewood Cliffs, New Jersey: Prentice-Hall

[106] Slater, J.C. (1951): Phys. Rev. *81*, 385

[107] Latter, R. (1955): Phys. Rev. *99*, 510

[108] Rau, A.R.P., Fano, U. (1968): Phys. Rev. *167*, 7

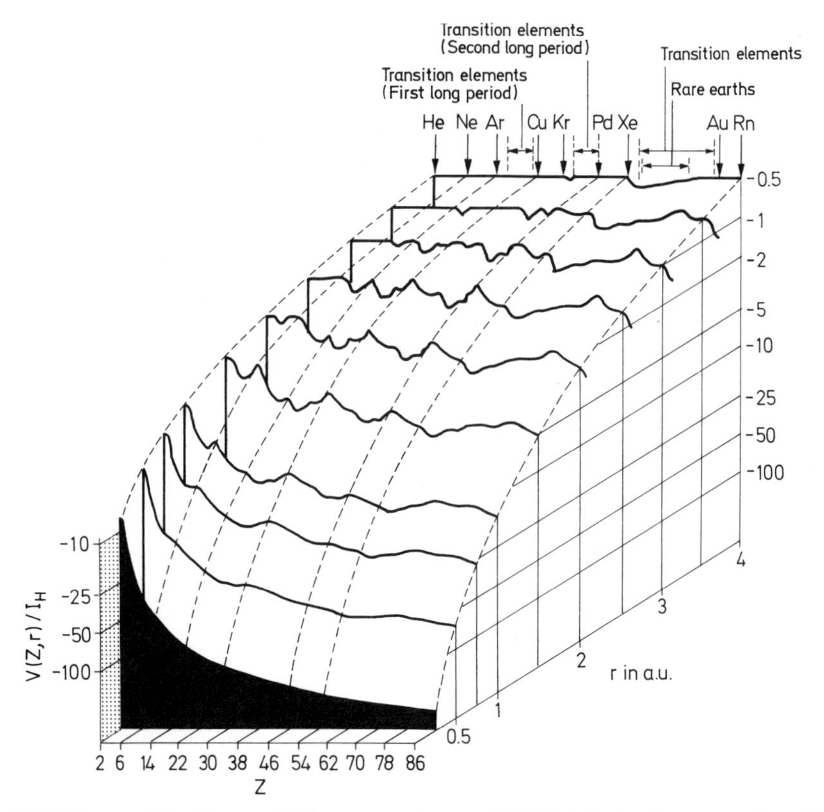

Fig. 1. Relief map of the Herman-Skillman central potential $V(Z,r)$ (in Rydbergs on a logarithmic scale) as a function of atomic number Z and coordinate r. (From [108])

structure that are not observable in the periodic table of the elements, which is based on *chemical* properties. Chemical properties of the elements, of course, depend on the outer electron structure and hence are cyclic in the filling of p subshells. Hence chemical properties such as atomic ionization potentials have maxima at the rare gases. This is so because filling of d and f subshells occurs at radii somewhat inside the atom (i.e., there are always s or p subshell electrons at larger radii). This inner-shell structure is clearly seen, however, in Fig. 1 where the potential has sharp peaks at Cu, Pd, and Au for radii in the range 1.5 a.u. $\leqq r \leqq$ 3.0 a.u. These elements, of course, correspond to filling of the $3d$, $4d$, and $5d$ subshells.

11. Predictions for the high-energy behavior of the cross section. As seen in Fig. 1, the central potential for an atom having more than one electron is distinctly nonhydrogenic. However, for high photon energies (i.e., $\omega \gtrsim 1$ keV) one can justify a hydrogenic treatment. For hydrogen the photoionization cross section is known to have a maximum at threshold and to decrease monotonically with increasing photon energy (cf. Fig. 21). Motivation for a hydrogenic treatment at high energies stems from the fact that sharp onsets at threshold followed by monotonic decreases above threshold are precisely the behavior seen in x-ray photoabsorption cross-section measurements.

A simple hydrogenic approximation at high energies may be justified intuitively as follows.[109] Firstly, at high photon energies the inner electrons are far more likely to be ionized than the outer electrons. A free electron cannot be photoionized due to kinematical considerations, and conversely, at a given photon energy the most strongly bound electrons will have the largest photoionization cross sections. Secondly, the wave function $P_{nl}(r)$ for an inner electron is concentrated in a very small range of r. The integrand of the radial dipole matrix element will therefore be negligible except in the small range of r where $P_{nl}(r)$ is greatest. Hence the final-state wave function $P_{\varepsilon l'}(r)$ only needs to be known in the same small region of r. Thirdly, in the small region of r occupied by the bound nl electrons the potential may be approximated by the screened Coulomb potential,

$$V_{nl}(r) = -\frac{(Z - s_{nl})}{r} + V_{nl}^0. \tag{11.1}$$

Here s_{nl} is the "inner-screening" parameter, which accounts for the screening of the nuclear charge by the other atomic electrons, and V_{nl}^0 is the "outer-screening" parameter, which accounts for the lowering of the nl electrons' binding energy due to the repulsion between the outer electrons and the photo-electron as the latter leaves the atom. The inner-screening constants may be obtained from the semiempirical estimates of SLATER[110] or the computed values of FROESE,[111] while the outer-screening parameters are usually determined so that the binding energy of the bound electron in the potential (11.1) equals the experimental binding energy $I_{nl}^{\exp}$. Lastly, both the bound wave function $P_{nl}(r)$ and the continuum wave function $P_{\varepsilon l}(r)$ are computed in the potential $V_{nl}(r)$ in (11.1) and used to calculate the cross section. As expected, one finds that the cross section is maximum at threshold and monotonically decreases above threshold.

The calculation of the continuum electron's wave function $P_{\varepsilon l}(r)$ in the potential (11.1), where $\varepsilon = \omega - I_{\exp} - V_{nl}^0$, seems like a drastic approximation, since in reality the photoelectron does not see the potential V_{nl}^0 at large r but sees zero potential. For elements having large nuclear charge Z and for subshells $n < 3$, however, the simple hydrogenic model described above predicts that the energy variation of the photoionization cross section is mainly dependent on the photon energy ω and is only weakly dependent on the outer-screening parameter V_{nl}^0.[109] The main function of the outer-screening parameter is to ensure the correct onset of photoionization. These predictions of the simple hydrogenic model are clearly confirmed by the accurate photoionization cross-section calculations of BOTTO et al.[112] for the ions of Fe. Figure 2

---

[109] Ref. [40], Sects. 69–71

[110] Slater, J.C. (1930): Phys. Rev. *36*, 57

[111] Froese, C. (1963): Can. J. Phys. *41*, 50

[112] Botto, D.J., McEnnan, J., Pratt, R.H. (1978): Phys. Rev. A *18*, 580

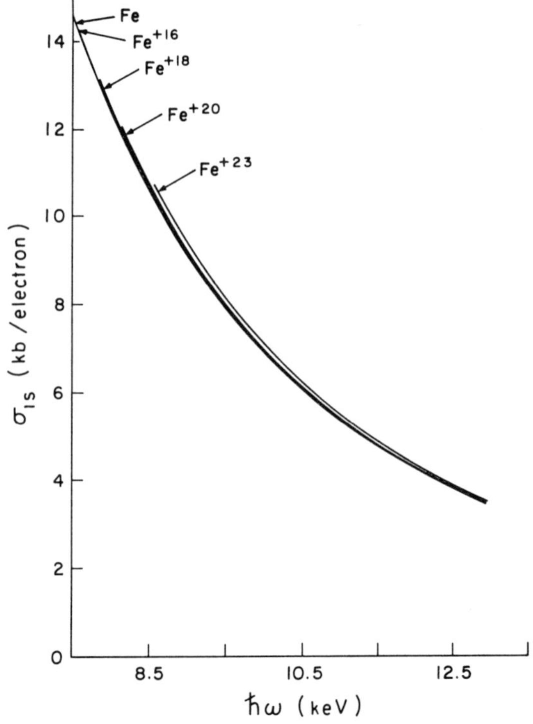

Fig. 2. K-shell photoionization cross sections of Fe, Fe^{+16}, Fe^{+18}, Fe^{+20}, and Fe^{+23} vs photon energy. (From[112])

shows the K-shell cross sections for Fe, Fe^{+16}, Fe^{+18}, Fe^{+20}, and Fe^{+23}. One sees that changes in the outer screening only affect the binding energy of the $1s$ electron and hence the onset of photoionization, as predicted by (11.1). The shape of the initial and final wave functions is not affected (in their region of overlap) by changes of outer screening, and hence for a given photon energy the value of the photoionization cross section is nearly the same for all stages of ionization, i.e., it is nearly independent of V_{nl}^0.

To improve upon the above simple hydrogenic model one must consider the radial dependence of the screening parameters. Alternatively, and more straightforwardly, one may solve (10.3) numerically for the initial and final wave functions using either the Herman-Skillman potential or any similar self-consistent field potential. Typical results obtained using the Herman-Skillman potential are compared with experiment in Table 1. One sees that theory and experiment generally agree to within the experimental error. Other such calculations have been reviewed by PRATT, RON, and TSENG,[113] and the results have been tabulated by STORM and ISRAEL[114] and by SCOFIELD.[115] A better understanding of the photoionization process at high energies, however, can be obtained using a more analytical approach.

[113] Pratt, R.H., Ron, A., Tseng, H.K. (1973): Rev. Mod. Phys. *45*, 273
[114] Storm, E., Israel, H.I. (1970): Nucl. Data Tables A 7, 565
[115] Scofield, J.H. (1973): Lawrence Livermore Laboratory Report No. UCRL-51326

Table 1. Comparison of nonrelativistic central potential model calculations with experimental measurements of the photoionization cross sections for the K shell of neutral atoms

Atomic number Z	Photon energy ω (keV)	K Shell binding energy I_{1s} (keV)	ω/I_{1s}	Theoretical cross section[a] σ_{Th} (barns)	Experimental cross section σ_{Exp} (barns)	σ_{Exp}/σ_{Th}
40	145	17.6	8.2	32.0	32 ± 2[b]	1.00 ± 0.06
47	145	24.7	5.9	62.1	68 ± 4[b]	1.10 ± 0.06
50	145	28.2	5.1	79.6	83 ± 5[b]	1.04 ± 0.06
73	145	62.2	2.3	342	348 ± 21[b]	1.02 ± 0.06
82	145	79.3	1.8	522	537 ± 32[b]	1.03 ± 0.06
47	74.409	24.7	3.0	445	428 ± 42[c]	0.96 ± 0.09
50	74.409	28.2	2.6	561	509 ± 41[c]	0.91 ± 0.07
74	74.409	64.0	1.2	2,238	$2{,}123\pm217$[c]	0.95 ± 0.10
40	36.818	17.6	2.1	1,799	$1{,}726\pm143$[c]	0.96 ± 0.08
47	36.818	24.7	1.5	3,171	$2{,}846\pm239$[c]	0.90 ± 0.08
50	36.818	28.2	1.3	3,910	$3{,}600\pm302$[c]	0.92 ± 0.08

[a] Calculated by Dr. H.K. Tseng (private communication) using the electric dipole approximation and the Herman-Skillman central potential (Ref. 105)
[b] Gowda, R., Sanjeevaiah, B. (1973): Phys. Rev. A**8**, 2425
[c] Allawadhi, K.L., Sood, B.S. (1975): Phys. Rev. A**11**, 1928

Oh et al.[116] have presented analytic nonrelativistic expressions for K- and L-shell photoionization cross sections that agree with direct numerical calculations to within $\pm1\%$ over an energy range from just above threshold to as high as 100 keV above in the case of s subshells. (A nonrelativistic dipole-approximation treatment appears to be valid to such high photon energies due to cancellations between relativistic and multipole effects.[116]) The formulas are obtained using the analytic perturbation theory for screened Coulomb potentials of McEnnan et al.[117] In this theory the atomic potential is expanded as follows:

$$V(r) = -\frac{Z\alpha}{r}[1 + V_1(\lambda r) + V_2(\lambda r)^2 + V_3(\lambda r)^3 + \ldots], \tag{11.2}$$

where α is the fine-structure constant, λ is the perturbation parameter ($\lambda \simeq \alpha Z^{1/3}$), and the V_n are coefficients of order unity that have alternating sign and determine the radial variation of the nuclear screening. (Note that (11.2) is written in natural units, which are the units adopted in [117]. In these units $\hbar = m_e = c = 1$, $\alpha = e^2$, r is measured in units of electron Compton wavelengths, and energies are measured in units of electron rest mass.) The coefficients V_n are determined by fitting to a realistic atomic potential. When the expansion in (11.2) is substituted in (10.3), both bound and continuum wave functions may be obtained as power-series expansions in λ. For sufficiently high photon energies both bound and continuum wave functions reach their asymptotic

[116] Oh. S.D.. McEnnan, J., Pratt, R.H. (1976): Phys. Rev. A *14*, 1428
[117] McEnnan, J., Kissel, L., Pratt, R.H. (1976): Phys. Rev. A *13*, 532; *14*, 521

forms for values of $r \lesssim 1/\lambda$; hence it is found that even the normalizations of the wave functions can be obtained as power series in λ.[117]

The results of OH et al.[116] are expressed in terms of the point-Coulomb potential results [i.e., the results obtained for the lowest order term in (11.2), $V(r) = -Z\alpha/r$] times correction factors due to screening of the nuclear charge and other effects. In the limit of high photon energy (which is estimated to begin at about twice the K-shell binding energy) the K- and L-shell photoionization cross sections may be written as

$$\sigma_{nl} = \sigma_{nl}^{\mathrm{C}} \left(\frac{N_{nl}}{N_{nl}^{\mathrm{C}}}\right)^2 [1 + O(\lambda^2/\omega)]. \tag{11.3}$$

Here σ_{nl}^{C} and N_{nl}^{C} are the point-Coulomb photoionization cross section and normalization for an nl electron moving in the field of the unscreened nucleus. N_{nl} is the normalization of the nl-electron wave function expressed as a power series in λ. The brackets indicate corrections due to changes in the shape of the bound and continuum wave functions and changes in the continuum normalization due to terms of order λ and higher in the potential in (11.2). These corrections vanish as λ^2/ω, where ω is the photon frequency.

The results of OH et al.[116] in (11.3) thus reduce, in the limit of high energies, to the normalization screening theory of PRATT and TSENG,[118] which predicts that the only effect of electron screening on the high energy photoionization cross section is a change in the normalization of the bound electron wave function. For high, but still nonrelativistic, photon energies, i.e., $I_{nl} \ll \omega \ll mc^2$, the energy dependence of the cross section in (11.3) is[109]

$$\sigma_{nl} \sim \omega^{-l-7/2}. \tag{11.4}$$

12. Predictions for the near-threshold behavior of the cross section. For low photon energies, $10\,\mathrm{eV} \lesssim \omega \lesssim 1{,}000\,\mathrm{eV}$, particularly in the vicinity of outer-electron ionization thresholds, the energy dependence of the photoionization cross section is often distinctly nonhydrogenic. Instead of a *maximum at threshold* and a *monotonic decrease* above threshold, one often finds an energy dependence such as that in Fig. 3, which shows the cross section[119] for the process,

$$\mathrm{Xe}\,4d^{10}\,5s^2\,5p^6 + \omega \rightarrow \mathrm{Xe}^+\,4d^9\,5s^2\,5p^6 + \mathrm{e}^-. \tag{12.1}$$

One sees that at threshold the cross section is practically zero and only reaches a *delayed maximum* approximately 3 Ry above threshold. Furthermore, far from decreasing monotonically at energies above the maximum, the cross section in Fig. 3 has a *minimum* at 10 Ry above threshold as well as a *second maximum* at 25 Ry above threshold, after which it then decreases monotonically. These low-

[118] Pratt, R.H., Tseng, H.K. (1972): Phys. Rev. A *5*, 1063

[119] Kennedy, D.J., Manson, S.T. (1972): Phys. Rev. A *5*, 227. Fig. 3 shows the calculated cross section in Hartree-Fock approximation; the experimental results differ as to the position of the maxima and minima but are otherwise similar

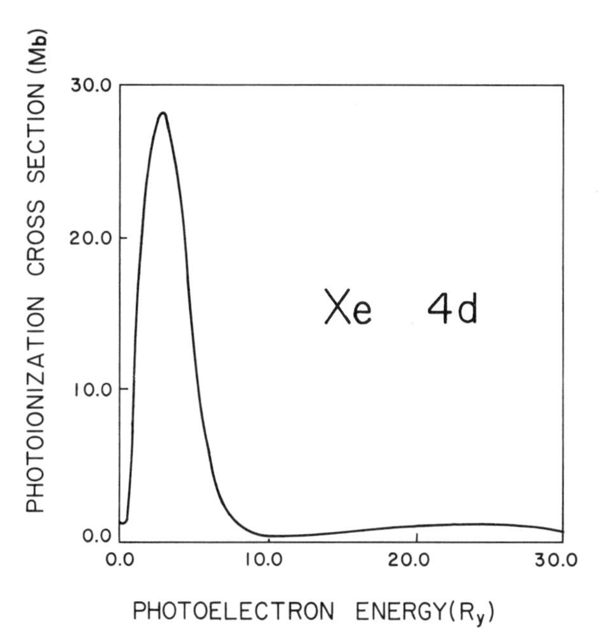

Fig. 3. Theoretical photoionization cross section for the $4d$ subshell in Xe vs photoelectron energy (Hartree-Fock-length results from [119])

energy near-threshold effects of a delayed maximum and of a cross-section minimum and second maximum may be understood qualitatively in terms of a central potential model of the atom. In addition, certain features of rare-earth and transition-metal photoabsorption spectra may be understood on the basis of an atomic central potential model. All of these nonhydrogenic behaviors may be classified as due to either an effective potential barrier or to a zero in the radial dipole matrix element. We consider each of these in turn.

α) Potential barrier effects. The radial equation (10.3) for the excited-electron-orbital wave function is a one-dimensional equation having the effective potential,

$$V_{\mathrm{eff}}(r) \equiv V(r) + \frac{l(l+1)}{2r^2}. \tag{12.2}$$

For Coulomb potentials $V(r) = -Z/r$, and (12.2) is always a single-well potential having a repulsive centrifugal barrier near the origin and an attractive Coulomb field at larger r. However, realistic atomic potentials $V(r)$ are highly non-Coulombic for ranges of r corresponding to the orbits of outer-shell electrons. For orbital angular momenta $l \geqq 2$, $V_{\mathrm{eff}}(r)$ for certain atoms is in fact a two-well potential,[120] as shown in Fig. 4, which employs the Herman-Skillman[105] atomic potentials $V(r)$. Thus, Fig. 4 shows that for $l=2$, Ar, Cu, and Kr have potential barriers [i.e., $V_{\mathrm{eff}}(r) > 0$] for $1\,\mathrm{a.u.} \lesssim r \lesssim 3\,\mathrm{a.u.}$, while for $l=3$, Xe, La, Eu, and Au have potential barriers for $1\,\mathrm{a.u.} \lesssim r \lesssim 4\,\mathrm{a.u.}$ These

[120] Goeppert-Mayer, M. (1941): Phys. Rev. *60*, 184

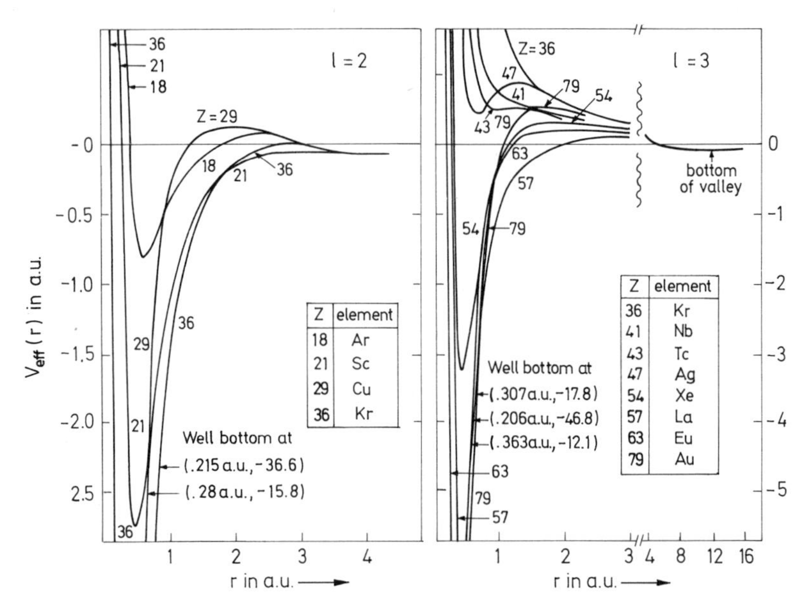

Fig. 4. Effective potential $V_{\text{eff}}(r)$ plotted vs coordinate r for $l=2$ and $l=3$ electrons. $V_{\text{eff}}(r)$ is the sum of the Herman-Skillman atomic potential $V(r)$ and the centrifugal potential $l(l+1)/(2r^2)$. (From[108])

potential barriers in effect raise the ionization threshold for photoelectrons having orbital angular momenta $l \geqq 2$ for the atoms indicated. More precisely, the final-state photoelectron radial wave functions have very low amplitude at small r (i.e., inside the barrier) until the photoelectron has a kinetic energy comparable to the height of the barrier.

In Fig. 5 we show three $l=3$ radial wave functions for xenon computed according to (10.3), where $V(r)$ is the Herman-Skillman[105] atomic potential for xenon. Note that the discrete wave function $4f$ has been normalized per unit energy so that it may be compared with the continuum wavefunctions (cf. Sect. 20). The barrier region is indicated. One sees that both the $4f$ and the $\varepsilon = 0.0$ wave functions have very low amplitudes inside the barrier region. On the other hand the $\varepsilon = 0.3$ a.u. wave function, which describes a photoelectron having kinetic energy comparable to the height of the barrier for Xe in Fig. 4, has substantial amplitude inside the barrier region. Alternatively, one may interpret Fig. 5 in terms of a two-well potential: the $4d$ wave function is an "eigenfunction" of the inner well, the $4f$ and $\varepsilon = 0.0$ wave functions are "eigenfunctions" of the outer well, and the $\varepsilon = 0.3$ a.u. wave function is an "eigenfunction" of both wells. In general, the continuum electron's amplitude in the inner well reaches a maximum at kinetic energies comparable to the barrier height. Subsequently, the amplitude in the inner well decreases with increasing energy.

The consequences of these potential barrier effects on the Xe $4d$-subshell photoionization cross section have been discussed by COOPER.[121] Since the $4d$

[121] Cooper, J.W. (1964): Phys. Rev. Lett. *13*, 762

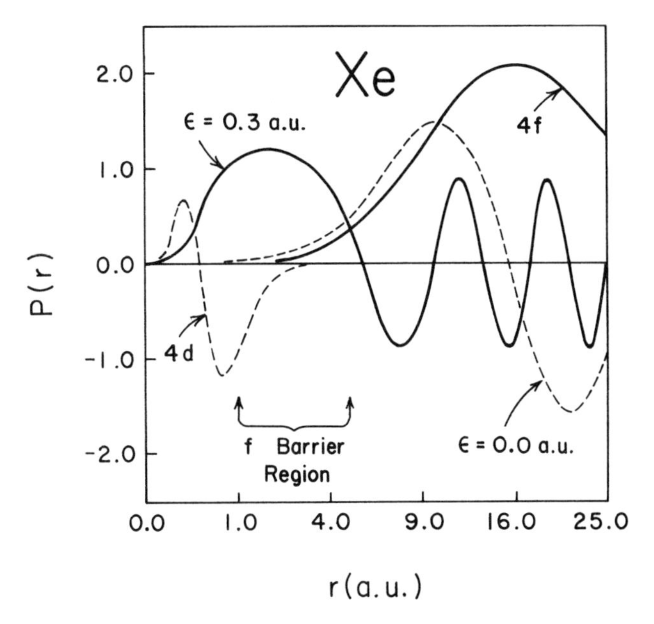

Fig. 5. Xenon $l=3$ orbitals: $4f$ (energy normalized), $\varepsilon=0.0$ a.u., and $\varepsilon=0.3$ a.u. radial wave functions computed using the Herman-Skillman atomic potential. The xenon $4d$ orbital, normalized to unity, is also shown. For the energy-normalized $4f$ wave function, as well as the continuum wave functions, the ordinate values are in units of $(\text{a.u.})^{-1/2}$; for the $4d$ wave function, the ordinate values are dimensionless

wave function in xenon has maximum amplitude at $r=0.73$ a.u.,[105] the radial dipole matrix element for the one-electron transition $4d \rightarrow \varepsilon f$ will be small at threshold (i.e., $\varepsilon=0.0$) and largest above threshold at kinetic energies comparable to the barrier height. Comparing these central potential model predictions with a more accurate (intrachannel) calculation,[122] in Fig. 6 we see that the height of the repulsive potential barrier in the central potential model is too small. The experimentally measured maximum[123] for xenon $4d$ subshell photoionization, also shown in Fig. 6, occurs at about 30 eV above threshold, thus lying between the central potential model and intrachannel calculation predictions, although the energy dependence of the experimental cross section is closer to that of the intrachannel calculation (cf. Sect. 15).

Note that in the above example we have ignored the transition $4d \rightarrow \varepsilon p$. Electric dipole selection rules only allow the photoelectron to have angular momenta $l_0 \pm 1$, where l_0 is the initial angular momentum of the bound electron. Furthermore, it is well known[124] that the transitions $l_0 \rightarrow l_0+1$ usually have much larger cross sections (often by an order of magnitude) than the transitions $l_0 \rightarrow l_0-1$. At threshold, however, in the case where there is a potential barrier in the l_0+1 channel, the opposite may be the case. This is also the case in Fig. 6 where the nonzero cross section at threshold in each of

[122] Starace, A.F. (1970): Phys. Rev. A *2*, 118

[123] Haensel, R., Keitel, G., Schreiber, P., Kunz, C. (1969): Phys. Rev. *188*, 1375

[124] Ref. [40], pp. 256–7

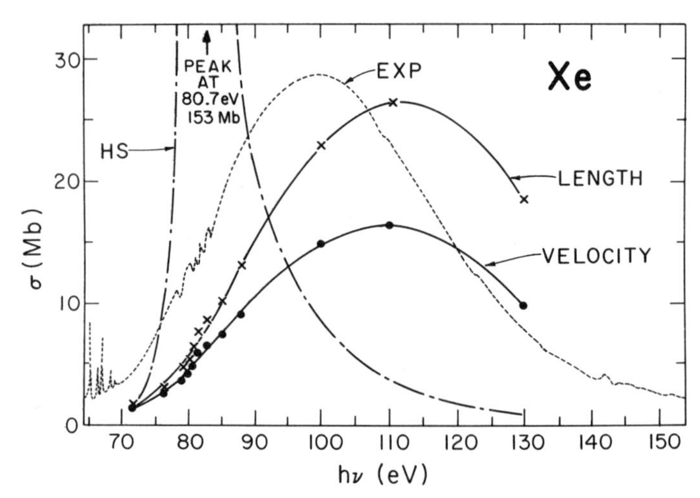

Fig. 6. Photoionization cross section for the $4d$ subshell of xenon. Dotted line is the experimental result of HAENSEL et al.,[123] dot-dash line is the Herman-Skillman central potential model calculation, and the solid lines give length and velocity results of an intrachannel calculation (cf. Sect. 15). (From [122])

the theoretical calculations may be attributed[121, 122] to the $4d \rightarrow \varepsilon p$ channel, the $4d \rightarrow \varepsilon f$ channel having negligible cross section.

The central potential model thus explains qualitatively the phenomenon of a delayed maximum in the photoionization cross section in terms of an effective potential barrier. Survey calculations have been carried out on properties of the final-state wave function, using a central potential model, which further demonstrate the effects of a potential barrier. MANSON[125] surveyed the behavior of photoelectron phase shifts (with respect to Coulomb waves) as functions of energy and of atomic number Z. In Fig. 7, the zero-energy phase shifts for $l=0, 1, 2$, and 3 are shown as a function of atomic number Z. For $l=3$, one sees that at low Z the phase shift is zero but increases at $Z=56$ by π. GOEPPERT-MAYER[120] and GRIFFEN et al.,[126] using different central potentials, showed that between $Z=56$ and $Z=57$ the $4f$ orbital moves from the outer potential well to the inner potential well. This move of the bound orbital is mimicked by the first loop of the $\varepsilon=0$ orbital, which also moves into the inner potential well and thereby acquires a phase shift of π. The second loop is kept out of the inner-well region until $Z \approx 88$ at which point it moves into the inner well (along with the $5f$ bound orbital) and contributes an additional phase shift of π. The behavior of the $\varepsilon=0$ phase shift for $l<3$ becomes less steplike as l decreases due to the decreasing significance of the potential barrier. Shell structure is still observed, however, even for $l=0$. Additional survey studies of the phase shifts as well as of the amplitudes of continuum-electron radial wave functions in a Herman-Skillman[105] potential have been presented by FANO et al.[127]

[125] Manson, S.T. (1969): Phys. Rev. *182*, 97

[126] Griffen, D.C., Andrew, K.L., Cowan, R.D. (1969): Phys. Rev. *177*, 62

[127] Fano, U., Theodosiou, C.E., Dehmer, J.L. (1976): Rev. Mod. Phys. *48*, 49

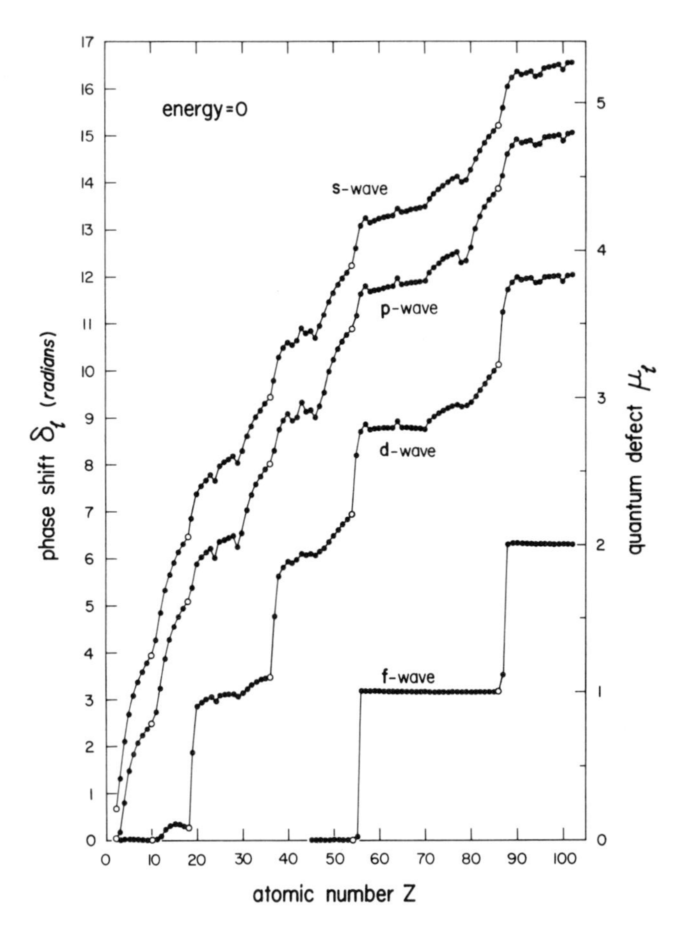

Fig. 7. Zero-energy phase shifts for $l=0, 1, 2,$ and 3 vs atomic number Z. Vertical scale on left is in radians and on right in units of π. (From [125, 127])

Despite the fact that the zero-energy phase shifts for $l=3$ increase by π whenever one of the bound orbitals moves from the outer well to the inner well, the zero-energy wave function amplitudes in the inner-well region are generally very small. The amplitude for the bound orbital on the other hand is usually large in the inner well. Figure 8 shows the $4f$ and $5f$ orbitals in cerium, which has the ground-state configuration Ce $1s^2 \ldots 4d^{10}\, 4f^2\, 5s^2\, 5p^6\, 6s^2$. The region of the potential barrier is indicated, and one clearly sees that the $4f$ orbital is an eigenstate of the inner well, whereas the $5f$ orbital, as well as all other bound f orbitals and low-energy continuum f orbitals, are eigenstates of the outer well. Note particularly that even though the $5f$ orbital has a "loop" in the inner-well region, the amplitude of this first loop is very small. The contrast between the $4f$ orbital in Xe and the $4f$ orbital in Ce is shown in Fig. 9. Whereas the $4f$ orbital in Xe is bound in the outer well, the $4f$ orbital in Ce is bound in the inner well. This difference in behavior stems from the increase in the nuclear charge in Ce, relative to Xe, which causes the inner well

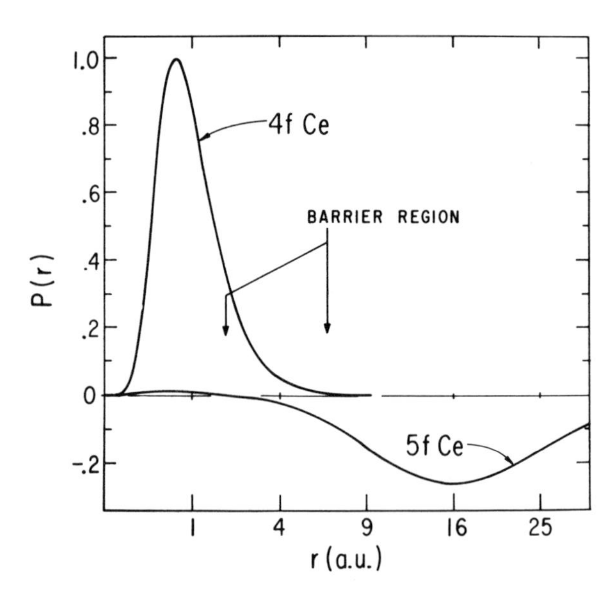

Fig. 8. The $4f$ and $5f$ radial wave functions for cerium calculated in the Herman-Skillman potential. (From [133])

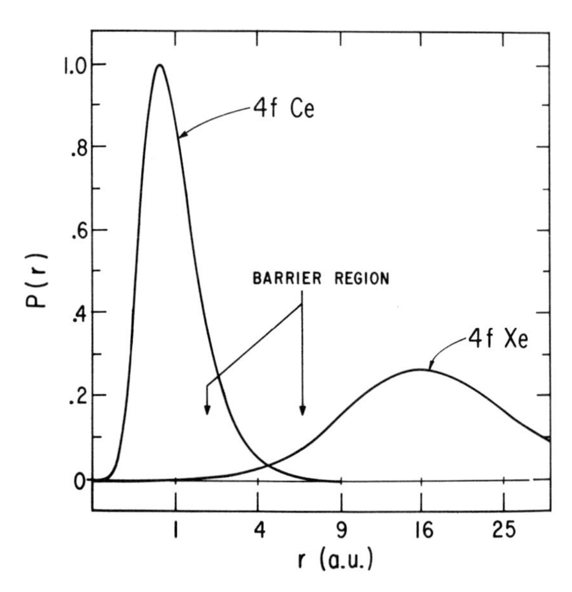

Fig. 9. The $4f$ radial wave functions for xenon and cerium calculated in the Herman-Skillman potential. (From [133])

in the $l=3$ effective potential for Ce to be deeper than in Xe and able to accommodate a bound f orbital.

The effect of an unoccupied bound orbital located in the inner well on the electric dipole radial matrix element is illustrated in Fig. 10. The left-hand side

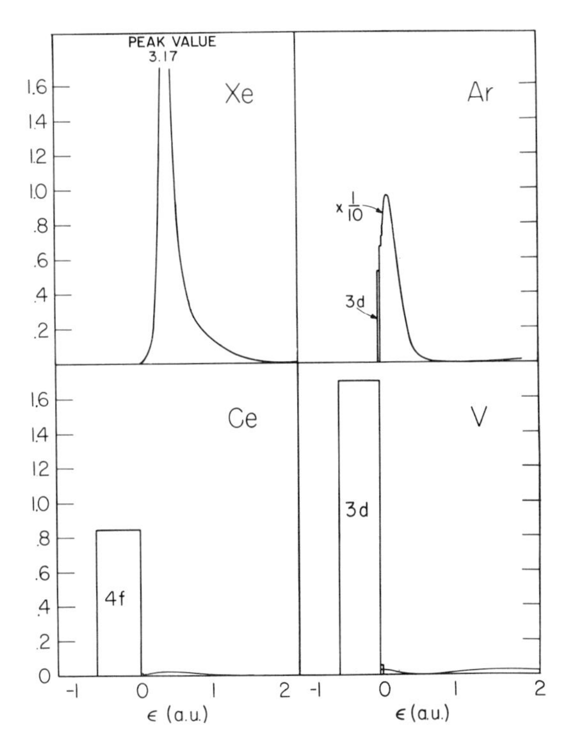

Fig. 10. Squared radial dipole matrix elements $\left[\int_0^\infty P_{nl}(r)\, r\, P_{\varepsilon l'}(r)\, dr\right]^2$ plotted as a function of ε using Herman-Skillman wave functions. For Ar and V, $nl=3p$ and $l'=d$; for Xe and Ce, $nl=4d$ and $l'=f$. (From [132])

of Fig. 10 gives the squared radial dipole matrix elements for $4d \to \varepsilon f$ transitions in Xe and in Ce; the right-hand side of Fig. 10 gives the squared radial dipole matrix elements for $3p \to \varepsilon d$ transitions in Ar and V. [Discrete final-state wave functions for *nf* or *nd* orbitals have been normalized per unit energy (cf. Sect. 20) so that dipole matrix elements to discrete final states may be plotted on the same graph with dipole matrix elements to continuum final states.] Whereas in Xe the dipole matrix element is maximum at $\varepsilon \approx 0.3$ a.u. *above* threshold due to the effective potential barrier, in Ce the dipole matrix element is maximum for the *discrete* transition $4d \to 4f$, which steals all of the transition strength, leaving practically none for transitions to the continuum. Although the effective potential barrier for $l=2$ electrons is not as strong, for $3p \to \varepsilon d$ transitions Ar and V behave similarly: in Ar nearly all the transition strength is in the continuum, peaking above threshold, while in V, where the $3d$ orbital is bound in the inner potential well, all of the strength is in the $3p \to 3d$ transition.

It is to be emphasized again that one does not expect the central potential model predictions in Fig. 10 for the squared dipole matrix element (and hence for the photoionization cross section) to be quantitatively accurate, but one does expect them to be qualitatively correct. In fact the above central potential model predictions for the rare earth atom Ce may be used to interpret the extensive experimental data on photoabsorption by the $4d$ subshell in the rare-earth metals ($57 \leqq Z \leqq 71$).[128-131] These metallic data show a) a *structured* peak *above* threshold, b) a decrease in the intensity of this peak as Z increases until it vanishes at $Z=71$, and c) many weak absorption lines in the vicinity of threshold, both below and above threshold. The peak above threshold at first sight appears to be a delayed maximum, just as in xenon $4d$-subshell photoabsorption, but the structure in the main peak as well as the weak structure in the vicinity of threshold are unusual as is the decrease in intensity in the main peak with increasing Z.

These spectra have been explained, largely on the basis of Fig. 10, as due to the following primary transition,[132-138]

$$4d^{10}\,4f^{N}+\omega \rightarrow 4d^{9}\,4f^{N+1}, \tag{12.3}$$

where N varies between 0 and 14 as Z varies between 57 and 71. In other words (12.3) implies that all of the transition strength is taken up by the $4d \rightarrow 4f$ transition in the rare earths. This interpretation of the experimental data explains all of the observed structure as due to the various term levels of the configuraton $4d^9\,4f^{N+1}$. These term levels are, of course, degenerate in the central potential model. But by introducing electrostatic and spin-orbit interactions within the configuration $4d^9\,4f^{N+1}$ this degeneracy is removed and one finds[132, 135] the term levels spread over an energy range of 20 eV or more, some of them appearing above threshold. These last term levels are generally the optically allowed term levels, which thus give the strong, structured peaks observed above threshold. The optically forbidden peaks (in *LS* coupling) are predicted to be weak and spread both below and above threshold, again as observed experimentally. The decrease in intensity of the main peak as Z increases is explained as due to the filling of the $4f$ subshell. The peaks above threshold eventually autoionize to the continuum (cf. Sect. 15):

$$4d^9\,4f^{N+1} \xrightarrow[\text{autoionization}]{} 4d^9\,4f^{N}+e^{-}. \tag{12.4}$$

---

[128] Zimkina, T.M., Fomichev, V.A., Gribovskii, S.A., Zhukova, I.I. (1967): Fiz. Tverd. Tela *9*, 1447 [Sov. Phys.-Sol. State *9*, 1128]

[129] Fomichev, V.A., Zimkina, T.M., Gribovskii, S.A. Zhukova, I.I. (1967): Fiz. Tverd. Tela *9*, 1490 [Sov. Phys.-Sol. State *9*, 1163]

[130] Haensel, R., Rabe, P., Sonntag, B. (1970): Solid State Commun. *8*, 1845

[131] Zimkina, T.M., Gribovskii, S.A. (1971): J. Phys. Colloq. C 4, *32*, 282

[132] Dehmer, J.L., Starace, A.F., Fano, U., Sugar, J., Cooper, J.W. (1971): Phys. Rev. Letters *26*, 1521

[133] Dehmer, J.L., Thesis, Ph.D. (1971): Department of Chemistry, The University of Chicago

[134] Starace, A.F. (1972): Phys. Rev. B *5*, 1773

[135] Sugar, J. (1972): Phys. Rev. B *5*, 1785

[136] Dehmer, J.L., Starace, A.F. (1972): Phys. Rev. B *5*, 1792

[137] Starace, A.F. (1974): J. Phys. B *7*, 14

[138] Dehmer, J.L. (1974): Physica Fennica *9S*, 60

The $4d$-subshell cross sections for many of the lanthanide vapors have recently been measured[139] and found to be in good agreement with the above theoretical picture.

Existing data[140a] on $3p$-subshell absorption in the transition metals are very similar to the data on $4d$-subshell absorption in the rare earths.[128–131] This suggests that the transition-metal data may be explained as due to the primary transition

$$3p^6\,3d^N + \omega \rightarrow 3p^5\,3d^{N+1}, \tag{12.5}$$

in accordance with Fig. 10. Recent experimental measurements[140b] as well as more detailed theoretical calculations[140c] of the $3p$-subshell absorption spectra of transition-metal vapors are also consistent with this interpretation.

The central potential model thus gives a good qualitative understanding for observed delayed maxima in photoabsorption by the rare gases and for certain structured absorption peaks observed in rare-earth and transition-metal photoabsorption. In studying an arbitrary element it is often helpful to see whether the central potential model predicts a potential barrier for the transition of interest, and, if so, whether there are any unoccupied, bound levels located in the inner potential well. Quantitatively, these predictions should only be taken as a guide since electron correlations (or even only a Hartree-Fock calculation) may substantially alter the central potential model's predictions.

β) Effects of zeros in the dipole matrix element. The sensitivity of the radial dipole integral to cancellations between positive and negative portions of the integrand has been discussed by Bates.[48] In particular, the existence of a minimum in the cross section for photoionization of the outer s electron in the alkalis has long been known to be due to a change in sign of the dipole integral with increasing photoelectron kinetic energy.[48, 141] The photoionization cross section never goes precisely to zero but has a nonzero minimum due to the existence of alternative photoelectron channels. Thus, for the alkalis, Seaton[141] has shown that spin-orbit interactions introduce a slight difference between the final state $p_{3/2}$ and $p_{1/2}$ continuum wave functions that causes the zeroes in dipole matrix elements for these two alternative channels to occur at slightly different energies.

Based on his survey calculations of photoionization from outer atomic subshells, Cooper[142] gave the following rule: *The photoionization cross sections for subshells with nodeless wave functions (i.e., $1s$, $2p$, $3d$, and $4f$) do not have*

---

[139] Mansfield, M.W.D., Connerade, J.P. (1976): Proc. R. Soc. A *352*, 125; Wolff, H.W., Bruhn, R., Radler, K., Sonntag, B. (1976): Phys. Letters *59*A, 67; Radtke, E.-R. (1979): J. Phys. B *12*, L71; (1979): ibid, *12*, L77

[140a] Sonntag, B., Haensel, R., Kunz, C. (1969): Solid State Commun. *7*, 597

[140b] Connerade, J.P., Mansfield, M.W.D., Martin, M.A.P. (1976): Proc. R. Soc. A *350*, 405; Sonntag, B. (1978): J. Phys. (Paris) *39*, Colloq. C4-9; Bruhn, R., Sonntag, B., Wolff, H.W. (1978): Phys. Lett. *69* A, 9; (1979): J. Phys. B *12*, 203

[140c] Dietz, R.E., McRae, E.G., Yafet, Y., Caldwell, C.W. (1974): Phys. Rev. Letters *33*, 1372; Davis, L.C., Feldkamp, L.A. (1976): Solid State Commun. *19*, 413

[141] Seaton, M.J. (1951): Proc. R. Soc. A *208*, 418

[142] Cooper, J.W. (1962): Phys. Rev. *128*, 681

minima, while the cross sections for all other subshells may. In their review of photoionization, FANO and COOPER[143] gave a more refined rule: *The radial dipole matrix element for the transition $nl \to n'l'$ (where n' extends over all discrete excited states $n' > n$ as well as the continuum) will change sign as a function of n' when l'-states with $n' = n$ exist but are not occupied in the ground state of the atom.* Exceptions to this rule occur occasionally when the subshell nl' is just filled.[143] However, the rule excludes the $1s$, $2p$, $3d$, and $4f$ subshells in agreement with Cooper's rule, and it limits the occurrence of sign changes to the dominant transition $l \to l+1$ rather than $l \to l-1$, since the $n(l-1)$ subshell is always occupied in the ground state of the atom. Note, however, that the Fano-Cooper rule includes changes of sign in the dipole matrix element in the discrete range of excited states. Such a change of sign would not be observed in the photoionization cross section. However, the energy location of the "Cooper" zero in the dipole matrix element is highly sensitive to electron correlations. Thus while the central potential model may predict a zero in the discrete energy range, electron correlations may shift this zero into the continuum (and, vice versa, a continuum zero located very near threshold might be shifted into the discrete).

The physical basis for the above rules comes partly from the known results for atomic hydrogen and partly from calculational evidence based on the central potential model. It is known that for hydrogen wave functions the radial dipole matrix element is always positive except when $n' = n$.[143] As discussed in Sect. 11, at high photon energies the photoionization cross section for an arbitrary atom becomes hydrogenic, and hence the radial dipole matrix element must be positive. From central potential model calculations on photoionization of atoms in their ground state, it is found that for any particular channel the radial dipole matrix element either changes sign once or not at all. Combining these facts, one concludes that if the radial dipole matrix element is to have a zero, it must be *negative* in the region of threshold. The rules of COOPER[142] and of FANO and COOPER[143] thus amount to predictions for the occurrence of a negative radial dipole matrix element for low-energy final-state wave functions.

Consider, for example, the $p \to d$ transitions in neon and argon. The Ne $2p$ and Ar $3p$ wave functions as well as the $\varepsilon = 0.0d$ wave functions for each element are shown in Fig. 11. Since $l = 2$ orbitals do not exist for $n = 2$, the Ne $2p$ wave function is nodeless and positive. The dipole matrix element for $2p \to \varepsilon d$ is obviously positive for $\varepsilon = 0$ and hence the channel $2p \to \varepsilon d$ is predicted to have no minimum. In argon, however, the $3p$ orbital has a major, negative loop. The dipole matrix element for the channel $3p \to \varepsilon d$ is negative for $\varepsilon = 0$ and hence the channel $3p \to \varepsilon d$ is predicted to have a zero for some $\varepsilon > 0$. In a similar way, one may show that the radial dipole matrix elements for $4p \to \varepsilon d$ in Kr and $5p \to \varepsilon d$ in Xe are negative at threshold and hence these channels are also predicted to have a Cooper minimum.

The minimum at $\varepsilon \approx 10$ Ry in the Xe $4d$-subshell photoionization cross section shown in Fig. 3 may now be interpreted as a Cooper minimum on the

[143] Ref. [2], Sect. 4

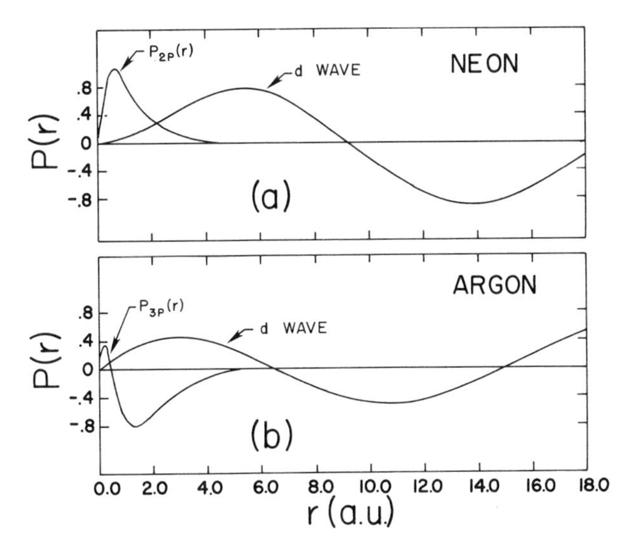

Fig. 11. (**a**) The neon $2p$ and εd radial wave functions, where $\varepsilon = 0.0$. (**b**) The argon $3p$ and εd radial wave functions, where $\varepsilon = 0.0$. The discrete wave functions are normalized to unity and the continuum wave functions are normalized per unit energy (Ry). (From [142])

basis of the above rules or else by reference to Fig. 5, which shows that the radial dipole matrix element for $4d \rightarrow \varepsilon f$ transitions is negative for $\varepsilon = 0.0$. The cross section has a minimum and not a zero due to contributions to the cross section from $4d \rightarrow \varepsilon p$ transitions. At energies above the Cooper zero the cross section increases again due to increasing positive values of the $4d \rightarrow \varepsilon f$ matrix element. At still higher energies the cross section decreases monotonically in accordance with the hydrogenic model. Hence the Cooper zero and the monotonically decreasing behavior in the $4d \rightarrow \varepsilon f$ channel bracket the second maximum in the $4d$-subshell cross section shown in Fig. 3.

For photoionization from excited-state orbitals, the above rules no longer hold since excited-state orbitals are generally quite diffuse. Msezane and Manson [144] have shown that the cross section for photoionization of the excited $5d$ orbital in Cs has the energy dependence displayed in Fig. 12. The gross shape of the cross section is due to the $5d \rightarrow \varepsilon f$ channel, which is found to have *two* minima. At threshold, the radial dipole matrix element for the $5d \rightarrow \varepsilon f$ transition is positive. However the εf orbital moves in rapidly with increasing ε causing the dipole matrix element to pass through a zero at $\varepsilon \approx 0.17$ Ry. The cross section appears to show an autoionizing "window" resonance (cf. Sect. 15), but in actuality this sharp drop in the cross section is due to the rapid change in sign of the dipole matrix element from positive to negative. The "normal" Cooper minimum due to the change of the matrix element from negative to positive occurs at much higher energies, near $\varepsilon \approx 7.0$ Ry, shown in the inset. Msezane and Manson [144] further find that the

[144] Msezane, A., Manson, S.T. (1975): Phys. Rev. Letters *35*, 364

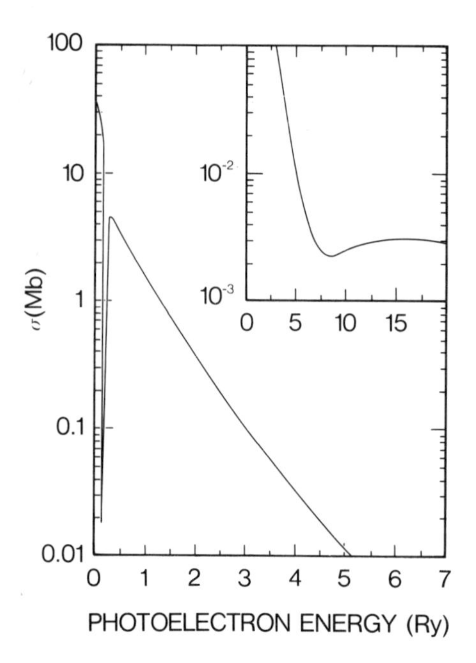

Fig. 12. Photoionization cross section for the excited $5d$ orbital of Cs. Inset shows the high-energy behavior of the cross section. (From [144])

radial dipole integral for the weak transition $5d \rightarrow \varepsilon p$ also has a zero near $\varepsilon \approx 0.07$ Ry, but this minimum is not noticeable on the scale used in Fig. 12.

The central potential model thus permits the prediction of whether or not there will be a minimum in a subshell photoionization cross section. For photoionization of unexcited atoms one may either do a calculation to see whether the radial dipole matrix element for the channel of interest is negative at threshold or else one may use the usually reliable rules of COOPER [142] and of FANO and COOPER.[143] Note that for any particular subshell nl, a Cooper minimum in the photoionization cross section will generally move in toward the threshold with increasing atomic number Z and may even pass below threshold into the discrete energy region. For photoionization from excited orbitals the initial orbital is so diffuse that there are at present no rules for predicting the occurrence of cross-section minima. The work of MSEZANE and MANSON,[144] however, indicates the utility of a central potential model calculation for distinguishing between cross-section minima and resonance features.

13. Bibliographic note on central potential model calculations. In view of the usefulness of central potential model predictions, we summarize the major calculational surveys of the photoionization cross section and related quantities. For high photon energies, $\omega > 1$ keV, there are the numerical tabulations of STORM and ISRAEL [114] and of SCOFIELD,[115] which are based on the Herman-Skillman [105] potential primarily. In the photon energy range from threshold to

100 eV above, COOPER[142] has carried out Hartree central potential calculations for the rare gases and for Na, Cu^+, and Ag^+. (All of the other central potential model calculations mentioned below use the Herman-Skillman[105] potential.) Photoionization cross sections for the $n=3$ to $n=6$ subshells in the photon energy range from threshold to ≈ 2 keV have been calculated for numerous elements by COMBET-FARNOUX and HENO,[145] by COMBET-FARNOUX,[146-148] and by MANSON and COOPER.[149] Furthermore, MCGUIRE[150] has calculated the photoionization cross sections for all subshells of all elements from He to Xe as a function of photon wavelength in the range from threshold to $\lambda=0.2$ Å. Closely related to the photoabsorption spectrum of an atom, as discussed in Sect. 8, are the atomic-oscillator-strength distribution and its moments. The moments $S(n)$ [defined by (8.4)] have been calculated by DEHMER et al.[151] for $-6 \leqq n \leqq 1$ for all elements from He to Ar. These moments provide information on ground-state properties of the atom (cf. Sect. 8). Lastly, central potential model calculations of the photoelectron angular distribution asymmetry parameter are surveyed in Sect. 23.

III. General methods for including electron correlation

a) Whole-space correlation theories

14. Introduction. The plethora of methods for obtaining quantitatively accurate photoionization cross sections may be conveniently grouped into two classes: a) those which treat interelectron correlations numerically over the entire region of coordinate space, considered in this section, and b) those which explicitly recognize that most electron correlations are important only in a finite region of coordinate space surrounding the atom, considered in Sects. 19-22. In the former class, the configuration interaction and the variational methods focus on calculating improved wave functions, which are then used to calculate dipole matrix elements; the many-body perturbation and random-phase approximation (RPA) methods focus on improving the dipole matrix element directly. While the RPA is considered last in our account of whole-space correlation theories, historically it was the first to provide accurate photoionization cross sections near threshold for the heavier rare gases, thus presenting a challenge to the practitioners of competing approximation methods to do at least as well. In some cases this challenge has now been met, as described below. We use the photoionization of the outer $3p$ subshell in

---

[145] Combet-Farnoux, F., Heno, Y. (1967): C.R. Acad. Sci. Paris *264* B, 138
[146] Combet-Farnoux, F. (1967): C.R. Acad. Sci. Paris *264* B, 1728
[147] Combet-Farnoux, F. (1969): J. Phys. (Paris) *30*, 521
[148] Combet-Farnoux, F., Lamoureux, M. (1976) J. Phys. B *9*, 897
[149] Manson, S.T., Cooper, J.W. (1968): Phys. Rev. *165*, 126
[150] McGuire, E.J. (1968): Phys. Rev. *175*, 20
[151] Dehmer, J.L., Inokuti, M., Saxon, R.P. (1975): Phys. Rev. A *12*, 102

argon to demonstrate the capabilities of each method. Of course, it is precisely in the outer subshells of the heavier atoms (i.e., $Z>10$) that electron correlations are strongest. This is due, on the one hand, to the reduced influence of the nuclear attraction with increasing orbital size (which renders a central field approximation to the electrostatic potential "seen" by a slow photoelectron increasingly unrealistic) and, on the other hand, to the closer proximity of the atomic and ionic ground-state configurations to unoccupied excited-state configurations (which renders the single-configuration approximation for the atom and for the ion increasingly unrealistic). The essential features of the different approximation methods that account for these electron correlations are discussed below.

15. Configuration interaction in the continuum. The most straightforward method for including electron correlation effects in the final-state wave function is to extend the configuration-interaction procedures commonly used for bound-state problems to the continuum. Thus, one may write the "exact" final-state eigenchannel wave function having total energy E as[152]

$$\Psi_{\Gamma E}(\boldsymbol{r}_1 s_1 \dots \boldsymbol{r}_N s_N) = \sum_{i=1}^{n} a_i(E)\, \psi_i(\boldsymbol{r}_1 s_1 \dots \boldsymbol{r}_N s_N) + \sum_{\gamma=1}^{m} \int dE'\, b_{\gamma E'}(E)\, \psi_{\gamma E'}(\boldsymbol{r}_1 s_1 \dots \boldsymbol{r}_N s_N). \tag{15.1}$$

In (15.1) each ψ_i represents the wave function for an excited discrete-energy configuration, while each $\psi_{\gamma E'}$ represents a channel wave function, where γ specifies the state of the ionic core and its coupling to the photoelectron and where E' is the total energy of the configuration. For each ionic state γ, the integral over total energy E' represents both a sum over the discrete energy levels of channel γ as well as an integration over the continuum energies of the photoelectron in channel γ. The eigenchannel index Γ on the "exact" wave

[152] As with all configuration-interaction treatments, one must choose the configurations included in one's calculation sensibly since one cannot treat an infinite number of configurations. We note here that the form of (15.1) implies such a choice. Thus the separation of the linear combination on the right of (15.1) into discrete configurations, represented by ψ_i, and configurations belonging to channels, represented by $\psi_{\gamma E}$, implies neglect of configurations having energies much greater than the energy E of interest in (15.1). (Among these neglected configurations are those belonging to the channels defined by successive promotion of a *single* electron from the discrete configuration ψ_i to higher and higher energies.) In a formally exact treatment, the right-hand side of (15.1) would consist only of an infinite (rather than finite) sum over channel indices γ as well as an integration over all energies for each channel γ. (The discrete energy members of the infinite set of channels γ would then comprise all possible discrete configurations.) Most of the infinite set of channels γ would be unimportant for describing states of energy E. The more practical "exact" eigenchannel wave function we have written in (15.1) consists then of a finite sum over those channels γ, the ionization threshold of which is either below the energy E or else not too far above the energy E. The wave functions $\psi_{\gamma E'}$ (as a function of energy E') represent both discrete and continuum states belonging to the channel γ. *Additional* discrete configurations that are deemed to be important in characterizing states of energy E are then included separately, with wave function ψ_i

function has the same dimension as γ (i.e., there are as many channels Γ as there are channels γ), but Γ does not correspond to any particular ionic core state. Equation (15.1) thus expresses $\Psi_{\Gamma E}$ as a superposition of "configurations" ψ_i and $\psi_{\gamma E'}$, which are usually obtained as the solutions of a zero-order Hamiltonian.

The goal of the configuration-interaction theory is to obtain the unknown coefficients $a_i(E)$ and $b_{\gamma E'}(E)$ by diagonalizing the submatrix of the Hamiltonian formed by the configurations included in (15.1). (An appropriate linear combination of the eigenchannel wave functions may then be formed to satisfy the asymptotic final-state boundary conditions, as discussed in Sect. 4.) In bra and ket notation the equations to be solved are

$$\langle i|H|\Gamma E\rangle = E\langle i|\Gamma E\rangle \qquad (\text{for } 1 \leqq i \leqq n) \tag{15.2a}$$

and

$$\langle \gamma E'|H|\Gamma E\rangle = E\langle \gamma E'|\Gamma E\rangle \qquad (\text{for } 1 \leqq \gamma \leqq m), \tag{15.2b}$$

where we have used the fact that $|\Gamma E\rangle$ is an eigenstate of H. One may distinguish several different specialized cases of these equations.

α) Interchannel interaction calculations. In the absence of the discrete configurations ψ_i in (15.1) (i.e., in the absence of discrete configurations other than the discrete energy states belonging to the channels γ), $\Psi_{\Gamma E}$ is a superposition of channel wave functions $\psi_{\gamma E'}$. The matrix elements on the left in (15.2b) reduce, after substitution of (15.1) for $|\Gamma E\rangle$, to a linear combination of matrix elements between the channel basis states $|\gamma E'\rangle$. When $\gamma \neq \gamma'$, one refers to $\langle \gamma E'|H|\gamma' E''\rangle$ as an "interchannel" matrix element, while when $\gamma = \gamma'$, one refers to $\langle \gamma E'|H|\gamma E''\rangle$ as an "intrachannel" matrix element.[153] If the channel states $|\gamma E\rangle$ are normalized per unit energy, then the matrix elements $\langle \gamma E'|H|\gamma' E''\rangle$ are dimensionless numbers, the magnitude of which when compared to unity, serves as an index of the interaction strength.[153] Usually the magnitudes of the interchannel matrix elements are weak, while those of the intrachannel matrix elements depend on the choice of basis states $|\gamma E\rangle$. The solution for the eigenchannel state $|\Gamma E\rangle$ has been given by (4.33) in terms of the K matrix of scattering theory. [The linear combination of these eigenchannel states, which asymptotically represents an outgoing electron in channel γ, given by (4.55), is equivalent to the result of a close-coupling calculation (cf. Sect. 16).]

β) Intrachannel interaction calculations. In the absence of both the discrete configurations ψ_i and interchannel interactions (but including still those discrete energy states belonging to the channel γ), (15.1) reduces to

$$\Psi_{\gamma E} = \int dE'\, b_{\gamma E'}(E)\, \psi_{\gamma E'}. \tag{15.3}$$

Solving for the coefficients $b_{\gamma E'}(E)$ is called an intrachannel calculation, which yields a channel wave function $\Psi_{\gamma E}$ that is equivalent to that obtained from a

[153] Ref. [2], § 5, § 6, and § 8

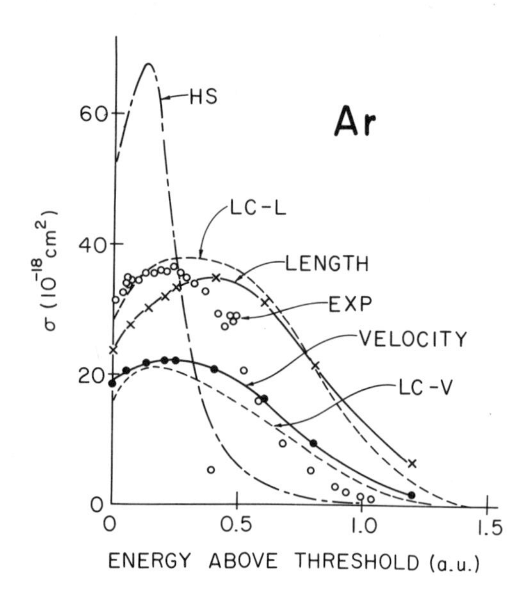

Fig. 13. Photoionization cross sections for the $3p$ subshell of argon. Dot-dash lines: Herman-Skillman central potential model calculation; solid lines: intrachannel calculation of STARACE[122]; dashed lines: close-coupling calculation of LIPSKY and COOPER[174]; open circles: experimental results of SAMSON[1]. (From [122])

Hartree-Fock calculation (cf. Sect. 16).[153] Intrachannel calculations[122] have been carried out for the $4d \to \varepsilon f$ channel in Xe and for the $3p \to \varepsilon d$ channel in Ar starting from a basis of channel functions $\psi_{\gamma E}$ constructed from one-electron orbitals calculated in the Herman-Skillman[105] central potential. The initial state was a Slater determinant of Herman-Skillman orbitals, and the cross sections for the weak transitions $4d \to \varepsilon p$ in Xe and $3p \to \varepsilon s$ in Ar were calculated in the central potential model approximation. The length and velocity results for Xe and Ar are shown in Fig. 6 and Fig. 13, respectively. In each case there is a great improvement in the cross section over that predicted by the Herman-Skillman central potential model due to inclusion of the intrachannel matrix elements

$$\langle \mathrm{Xe}^+\, 4d^9\, 5s^2\, 5p^6\, \varepsilon f\, (^1P_1)|H|\mathrm{Xe}^+\, 4d^9\, 5s^2\, 5p^6\, \varepsilon' f\, (^1P_1)\rangle \tag{15.4a}$$

and

$$\langle \mathrm{Ar}^+\, 3p^5\, \varepsilon d(^1P_1)|H|\mathrm{Ar}^+\, 3p^5\, \varepsilon' d(^1P_1)\rangle. \tag{15.4b}$$

These matrix elements, the main contribution of which comes from the $R^1(4d\varepsilon f, \varepsilon' f 4d)$ and $R^1(3p\varepsilon d, \varepsilon' d\, 3p)$ Slater integrals, respectively,[153] are large and repulsive, thus broadening the central potential model cross section and shifting it to higher energy. In comparison with experiment, however, the intrachannel interaction overcorrects the central potential model calculation. This remaining discrepancy with experiment was a great puzzlement until the random-phase-approximation calculations indicated its cause (cf. Sect. 18).

γ) Autoionization. When some of the discrete configuratons ψ_i lie above the threshold for ionization of one or more photoelectron channels γ, interelectron correlations can cause a transition of an excited electron from the discrete state ψ_i to an unbound continuum state. One says that the state ψ_i decays by "autoionization". A detailed treatment of autoionization by configuration interaction has been given by FANO,[3] and specific applications to rare gas photoionization have been presented by FANO and COOPER.[2, 4]

The treatment of the simplest case of a single discrete state ψ_i interacting with a single continuum channel illustrates the method. Equation (15.1) becomes

$$\Psi_E = a(E)\,\psi_i + \int dE'\, b_{E'}(E)\,\psi_{E'}, \tag{15.5}$$

where we have dropped the channel indices of γ and Γ for this one-channel case. Substituting Ψ_E for the ket $|\Gamma E\rangle$ and $\psi^*_{E'}$ for the bra $\langle\gamma E'|$ in (15.2) and using (15.5), we obtain

$$a(E)\,E_i + \int dE'\, b_{E'}(E)\, V^*_{E'} = E\,a(E), \tag{15.6a}$$

$$a(E)\,V_{E'} + E'\,b_{E'}(E) = E\,b_{E'}(E), \tag{15.6b}$$

where we have defined

$$\langle i|H|i\rangle \equiv E_i \tag{15.7a}$$

and

$$\langle \psi_E|H|i\rangle \equiv V_E. \tag{15.7b}$$

In obtaining (15.6) it has been assumed that an intrachannel calculation has already been carried out within the single continuum channel so that

$$\langle \psi_E|H|\psi_{E'}\rangle = E\,\delta(E-E').$$

Equation (15.6b) shows that $b_{E'}(E) = a(E)\,V_{E'}/(E-E')$ except at $E=E'$. Detailed consideration of this singularity shows that the solution of (15.6) is[3]

$$b_{E'}(E) = \left\{\frac{\mathscr{P}}{E-E'} + \frac{\delta(E-E')}{|V_E|^2}\left[E - E_i - \mathscr{P}\int \frac{dE''\,|V_{E''}|^2}{E-E''}\right]\right\} V_{E'}\,a(E), \tag{15.8}$$

where the symbol $\mathscr{P}$ indicates that the Cauchy principal value is to be taken in any integration over the singularities. Normalization of the eigenfunction Ψ_E in (15.5) after substitution of (15.8) determines the coefficients $a(E)$ up to a phase:[3]

$$|a(E)|^2 = \frac{|V_E|^2}{\left[E - E_i - \mathscr{P}\int \frac{dE'\,|V_{E'}|^2}{E-E'}\right]^2 + \pi^2 |V_E|^4}. \tag{15.9}$$

Substitution of (15.8) and (15.9) in (15.5) finally determines the eigenfunction Ψ_E:

$$\Psi_E = \{[E - E_i(E)]^2 + \pi^2 |V_E|^4\}^{-1/2}\,\{\Phi_i V^*_E + [E - E_i(E)]\,\psi_E\}. \tag{15.10}$$

In (15.10) we have defined a modified discrete-state energy and a modified discrete-state wave function as follows:

$$E_i(E) \equiv E_i + \mathscr{P} \int \frac{dE' |V_{E'}|^2}{E - E'}, \tag{15.11a}$$

$$\Phi_i \equiv \psi_i + \mathscr{P} \int \frac{dE' \psi_{E'} V_{E'}}{E - E'}. \tag{15.11b}$$

In each case, the modification is energy dependent and arises from the interaction between the discrete state and the continuum channel.

The expected resonance behavior is clearly demonstrated by the final-state wave function in (15.10). The wave function has an overall resonance denominator which has a minimum when $E = E_i(E)$. In addition the coefficient of ψ_E in (15.10) changes sign as E passes through $E_i(E)$ so that Φ_i and ψ_E interfere constructively on one side of the resonance and destructively on the other.

Taking electric dipole matrix elements between the eigenstate Ψ_E in (15.10) and the atomic ground state ψ_0 and comparing them to the dipole matrix element that would obtain in the absence of the discrete state ψ_i one finds [3]

$$\frac{|\langle \psi_0 | \sum_i \boldsymbol{r}_i | \Psi_E \rangle|^2}{|\langle \psi_0 | \sum_i \boldsymbol{r}_i | \psi_E \rangle|^2} = \frac{(q + \varepsilon)^2}{1 + \varepsilon^2}, \tag{15.12}$$

where the Fano "q" parameter and the reduced energy ε are defined as

$$q \equiv \frac{\langle \psi_0 | \sum_i \boldsymbol{r}_i | \Phi_i \rangle}{\pi \langle \psi_0 | \sum_i \boldsymbol{r}_i | \psi_E \rangle V_E}, \tag{15.13a}$$

$$\varepsilon \equiv \frac{E - E_i(E)}{\pi |V_E|^2}. \tag{15.13b}$$

Usually the halfwidth of the resonance, $\pi |V_E|^2$, is only of the order of 10^{-2} eV so that the matrix elements defining the q parameter are effectively constant over the energy range of the resonance. Figure 14 illustrates the resonance line profile implied by (15.12) for various values of q. The ordinate value unity corresponds to the "shape" that would apply in the absence of the resonance. For $|q| > 1$ one observes primarily a resonance peak, whereas for $|q| < 1$ one observes primarily a resonance "window." For any value of q there is a zero minimum on one side of the resonance due to interference between the transition matrix elements to the discrete state Φ_i and the continuum states ψ_E [cf. (15.10)]. The line profile is asymmetric about the resonance energy $\varepsilon = 0$ (unless $q = 0$ or $|q| \to \infty$) and the profile is reversed when $q \to -q$.

There have been several extensions of the above treatment that attempt to obtain parameterized formulas for resonance behavior. FANO [3] obtained formulas for the line profile when several discrete states interact with a single

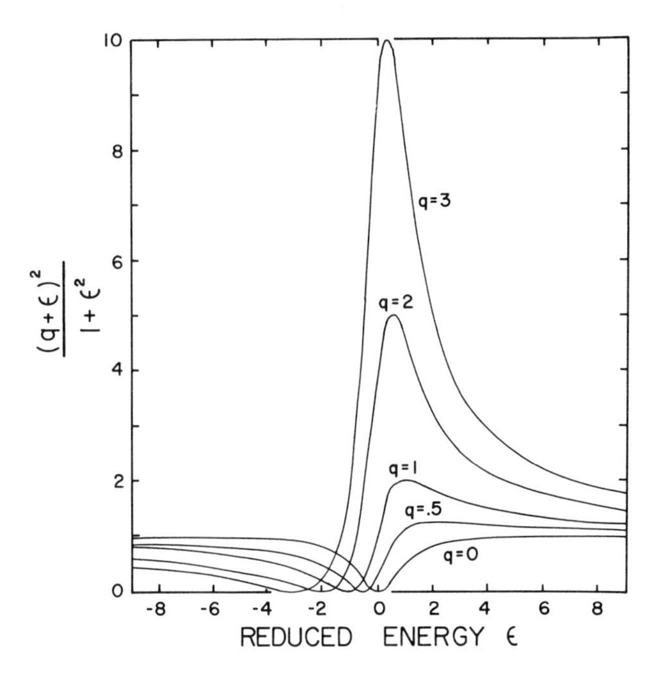

Fig. 14. Resonance line profiles for different values of q. (From [3])

continuum and also when a single discrete state interacts with many continuum channels. SHORE[154] and MIES[155] present profile formulas applicable to the case when two or more discrete states are close together and have overlapping profiles. Recently, parameterized formulas have been given for the behavior of photoelectron angular distributions[156] and photoelectron branching ratios[157] within a resonance. It should be noted that although most autoionizing resonances are weak and narrow, on occasion some are both strong and wide, influencing the cross section over a range of a few eV.[158] In such cases the "parameter" q is usually energy dependent.

The above work has been concerned primarily with parameterized formulas for resonance behavior. ALTICK and co-workers[159–161] have used the configuration interaction method to obtain ab initio photoionization cross sections. They did calculations for He, Be, and Mg by treating the Rydberg-series levels converging to the higher threshold photoelectron channels as separate

---

[154] Shore, B.W. (1968): Phys. Rev. *171*, 43

[155] Mies, F.H. (1968): Phys. Rev. *175*, 164

[156] Kabachnik, N.M., Sazhina, I.P. (1976): J. Phys. B *9*, 1681

[157] Starace, A.F. (1977): Phys. Rev. A *16*, 231

[158] A striking example is the $3s3p^2\,{}^2D$ resonance in Al, which increases the threshold photoionization cross section for Al by a factor of three. See Lin, C.D. (1974): Astrophys. J. *187*, 385 and Weiss, A.W. (1974): Phys. Rev. A *9*, 1524

[159] Altick, P.L., Moore, E.N. (1966): Phys. Rev. *147*, 59

[160] Altick, P.L. (1968): Phys. Rev. *169*, 21

[161] Bates, G.N., Altick, P.L. (1973): J. Phys. B *6*, 653

discrete states ψ_i interacting with the single photoelectron channel associated with the lowest ionization threshold. For example, ALTICK and MOORE[159] treated the interaction of the lowest-lying excited configurations in He (i.e., $2snp$, $2pnd$, and $2pns$) with the principal final-state channel, $1s\varepsilon p(^1P_1)$. The initial-state wave function was obtained from a highly accurate variational calculation. Be and Mg were treated in a similar way.[160,161] In all three cases agreement with experiment and with alternative methods of calculation is excellent.

16. Procedures derived from a variational principle. Rather than expressing the final-state many-channel wave function as a superposition of basis-state functions, as in the configuration-interaction method, one may solve for the continuum electron's radial wave function directly by solving a differential equation. Computationally this method is simpler and more accurate. However, since all of the final-state interactions are incorporated into the differential equation, it is not as easy as in the configuration interaction method to assess their relative importance.

α) The close-coupling method. The final-state wave function representing a photoelectron in channel γ may be written as

$$\psi_{\gamma E}^{-}(\boldsymbol{r}_1 s_1, \ldots, \boldsymbol{r}_N s_N) = \frac{1}{r_N} \mathscr{A} \sum_{\gamma'} \Theta_{\gamma'}(\boldsymbol{r}_1 s_1, \ldots, \hat{\boldsymbol{r}}_N s_N) F_{\gamma'\gamma}(r_N), \tag{16.1}$$

where $\Theta_{\gamma'}$ represents an antisymmetrized ionic-core wave function coupled to the angular and spin parts of the photoelectron wave function, $F_{\gamma'\gamma}(r_N)$ represents the radial part of the photoelectron wavefunction, and $\mathscr{A}$ represents the operator which antisymmetrizes the complete wave function. The expansion implied by (16.1) is infinite, extending over the complete set of ionic eigenstates. In practice, of course, only a few ionic states are included and their wave functions are approximated by the appropriate Hartree-Fock solutions. The coupled differential equations for the radial functions $F_{\gamma'\gamma}(r_N)$ are often obtained[162] by substituting (16.1) into the KOHN[163] variational principle. However, there are at least two other equivalent methods of deriving the equations,[5,164] one of which is to require the following stationary condition to hold:[164]

$$\langle \delta\psi_{\gamma E}^{-} | H - E | \psi_{\gamma E}^{-} \rangle = 0. \tag{16.2}$$

Here the variation in the wave function $\psi_{\gamma E}^{-}$ is due to variation in the photoelectron's radial wave function. The reduction of the equations implied by (16.2) to radial differential equations for the functions $F_{\gamma'\gamma}(r_N)$ is cumbersome. This reduction is discussed by SEATON,[5] by BURKE and SEATON,[164] and by SMITH et al.[165] Usually the radial functions are required to be orthogonal to

[162] Smith, K. (1971): The calculation of atomic collision processes, New York: Wiley, § 2.5
[163] Kohn, W. (1948): Phys. Rev. *74*, 1763
[164] Burke, P.G., Seaton, M.J. (1971): Methods Comput. Phys. *10*, 1
[165] Smith, K., Henry, R.J.W., Burke, P.G. (1966): Phys. Rev. *147*, 21

all ionic bound orbitals. The various numerical procedures used to solve the coupled differential equations are reviewed by BURKE and SEATON.[164]

The method outlined above for obtaining the final-state channel functions in (16.1) is known as the close-coupling method. It is to be compared with (4.55) and (4.33), which together represent $\psi_{\gamma E}^-$ as a superposition of basis states according to the configuration-interaction method. In the vocabulary of Sect. 15, the close-coupling equations account for both inter- and intra-channel interactions in the final state. Discussion of the close-coupling method as applied specifically to photoionization has been given by HENRY and LIPSKY.[6]

Extensions of the ordinary close-coupling method have been developed to meet two different shortcomings. Firstly, because of the truncation of the summation in (16.1), the short-range correlations at small radial distances are not always well accounted for by the ordinary close-coupling method. These short-range correlations may be taken into account by including appropriate additional N-electron functions on the right in (16.1), at least for small radial distances. The R matrix method, discussed in Sect. 21, is the most suitable means for accomplishing this. Secondly, the ionic-core wave functions in (16.1) are held fixed. Hence the modification of the core wave functions by the photoelectron and the resulting long-range polarization interaction between the ion and the photoelectron are not taken into account by the ordinary close-coupling method. DAMBURG and KARULE,[166] however, have proposed a modification of the close-coupling method – related to earlier work of TEMKIN[167] – which accounts for the long-range dipole and quadrupole polarization interactions.

The close-coupling method has been used to calculate the photoionization cross section for numerous atoms.[6, 168–174] As an example, we consider the close-coupling calculation of LIPSKY and COOPER[174] for the photoionization cross section of Ar presented in Fig. 13. The following three channels were considered:

$$\mathrm{Ar}\,3s^2\,3p^6\,(^1S_0)+\omega \to \mathrm{Ar}^+\,3s^2\,3p^5\,(^2P)\,\varepsilon d(^1P_1), \qquad (16.3\mathrm{a})$$

$$\to \mathrm{Ar}^+\,3s^2\,3p^5\,(^2P)\,\varepsilon s(^1P_1), \qquad (16.3\mathrm{b})$$

$$\to \mathrm{Ar}^+\,3s\,3p^6\,(^2S)\,\varepsilon p(^1P_1). \qquad (16.3\mathrm{c})$$

The autoionizing resonances due to the discrete np levels of channel (16.3c) are not shown in Fig. 13. One sees that the interchannel interactions included in the close-coupling calculation cause only a slight change from the intrachannel

[166] Damburg, R., Karule, E. (1967): Proc. Phys. Soc. *90*, 637

[167] Temkin, A. (1957): Phys. Rev. *107*, 1004; (1959): Phys. Rev. *116*, 358; Temkin, A., Lamkin, J.C. (1961): Phys. Rev. *121*, 788

[168] Burke, P.G., McVicar, D.D. (1965): Proc. Phys. Soc. *86*, 989

[169] Bell, K.L., Kingston, A.E. (1967): Proc. Phys. Soc. *90*, 31

[170] Henry, R.J.W. (1967): Planet. Space Sci. *15*, 1747

[171] Conneely, M.J., Smith, K., Lipsky, L. (1970): J. Phys. B *3*, 493

[172] Chapman, R.D., Henry, R.J. (1972): Astrophys. J. *173*, 243

[173] Combet-Farnoux, F. (1976): Photoionization and other probes of many-electron interactions, Wuilleumier, F.J. (ed.), New York: Plenum, p. 407

[174] Lipsky, L., Cooper, J.W. (unpublished). Results presented in Fig. 22 of Ref. [2]

calculation,[122] which is also shown in Fig. 13. Most of the difference between the two calculations can be traced to the use of different Ar$3p$ orbitals to compute the dipole matrix element.[33] These two calculations demonstrate the weakness of final-state interchannel interactions for the rare gases and strongly hint that the remaining discrepancy with experiment is due to the neglect of initial-state correlations (cf. Sects. 17 and 18).

β) The Hartree-Fock method. In the approximation that only a single channel γ is considered in the expansion (16.1), then $F_{\gamma\gamma}(r_N)$ becomes the continuum Hartree-Fock radial wave function. The radial Hartree-Fock equations resulting from reduction of the variational equation (16.2) have been presented by Dalgarno et al.[175] for atoms with ground-state configurations $1s^2 2s^2 2p^q$ $(1 \leqq q \leqq 6)$. Solution of the Hartree-Fock differential equation is equivalent to performing an intrachannel calculation in the configuration-interaction method. Kennedy and Manson[33] have performed an extensive survey of photoionization cross sections and photoelectron angular distributions for the rare gases in the Hartree-Fock approximation. The various kinds of continuum Hartree-Fock wave functions are discussed in the next section.

17. Many-body perturbation theory. The methods discussed thus far have been concerned primarily with improving the final-state wave function by diagonalizing the submatrix of the Hamiltonian corresponding to intra- and inter-channel correlations. Photoionization cross sections are obtained by calculating electric dipole matrix elements between these final-state wave functions and the best available ground-state wave functions. The many-body perturbation theory (MBPT) focuses instead on the dipole matrix element directly and considers simultaneously both initial- and final-state correlation effects. An advantageous feature of the MBPT is that each term in the perturbation expansion can be related to a particular kind of physical interaction. Thus the relative importance of the alternative interactions is easily assessed (provided that the perturbation expansion converges rapidly).

The definition of the perturbation depends on the choice of the single-particle basis states, the wavefunctions of which are obtained as the eigenfunctions of the following single-particle Hamiltonian:

$$H_0 \phi_n \equiv \left(-\frac{1}{2} \nabla^2 - \frac{Z}{r} + V \right) \phi_n = \varepsilon_n \phi_n. \tag{17.1}$$

Here V is some suitable potential meant to approximate the electrostatic interactions with the other electrons. The N-electron Hamiltonian is thus

$$H = \sum_{i=1}^{N} H_0(i) + H' + H_{\text{int}}, \tag{17.2}$$

[175] Dalgarno, A., Henry, R.J.W., Stewart, A.L. (1964): Planet. Space Sci. *12*, 235

where H_{int}, given by (3.4), describes the interaction of the atom with electromagnetic radiation and H' defines the perturbation due to electron correlation,

$$H' \equiv \sum_{i>j}^{N} \frac{1}{|\boldsymbol{r}_i - \boldsymbol{r}_j|} - \sum_{i=1}^{N} V(i). \tag{17.3}$$

The MBPT method for calculating photoionization cross sections consists in treating the interaction with radiation, H_{int}, to first order only, as is usual, and treating the perturbation H' due to electron correlations to higher orders.

Of the many perturbation theories suitable for N-particle systems,[176] among them the algebraic theories of Rayleigh-Schrödinger and of Brillouin-Wigner, most practitioners of MBPT employ the diagrammatic perturbation theory of BRUECKNER[177] and GOLDSTONE.[178] The latter theory has the advantage that certain groups of terms that appear in Rayleigh-Schrödinger perturbation theory – the so-called unliked clusters – are excluded from consideration since they may be shown to cancel.[176–178] There is also the additional advantage that one may think in terms of physical processes, each represented by a diagram, and rely upon a detailed set of rules to reduce any diagram to an algebraic expression. The first applications of the Brueckner-Goldstone MBPT to the calculation of atomic processes were made by KELLY.[179]

One of the principal findings of KELLY's early work was the great influence of the choice of single-particle basis set on the convergence of the perturbation expansion.[179a, b, c] Given that the best approximation for the wave function of the ground state of an N-electron atom is the Hartree-Fock approximation,[180, 181] the problem is to choose a potential in which to calculate excited one-electron orbitals. Using the one-electron orbitals ϕ_i $(1 \leqq i \leqq N)$ generated by a Hartree-Fock calculation for the atomic ground state, KELLY initially defined what is now known as the V^N potential,[179a] which is given in terms of its matrix elements as follows:

$$\langle m|V^N|n\rangle \equiv \sum_{i=1}^{N} \{\langle mi|v|ni\rangle - \langle mi|v|in\rangle\}. \tag{17.4}$$

Here the matrix element of electrostatic interaction has been defined as

$$\langle ab|v|cd\rangle \equiv \iint \phi_a^*(\boldsymbol{r})\,\phi_b^*(\boldsymbol{r}')\,\frac{1}{|\boldsymbol{r}-\boldsymbol{r}'|}\,\phi_c(\boldsymbol{r})\,\phi_d(\boldsymbol{r}')\,d\boldsymbol{r}\,d\boldsymbol{r}'. \tag{17.5}$$

---

176 Kumar, K. (1962): Perturbation theory and the nuclear many-body problem, Amsterdam: North-Holland

177 Bruekner, K.A. (1955): Phys. Rev. *100*, 36; (1959): The many-body problem, les houches – Session 1958, New York: John Wiley, pp. 47–242

178 Goldstone, J. (1957): Proc. R. Soc. A *239*, 267

179 (a) Kelly, H.P. (1963): Phys. Rev. *131*, 684; (b) (1964): Phys. Rev. *136*, B896; (c) (1966): Phys. Rev. *144*, 39; (d) (1968): Adv. Theor. Phys. *2*, 75; (e) (1969): Adv. Chem. Phys. *14*, 129

180 Slater, J.C. (1960): Quantum theory of atomic structure, Vol. II, New York: McGraw-Hill, Chap. 17

181 Froese-Fischer, C. (1977): The Hartree-Fock method for atoms, New York: John Wiley

When the V^N potential defined by (17.4) is substituted in the single-electron Schrödinger equation (17.1), the orbitals ϕ_n for $1 \leqq n \leqq N$ are once again the N Hartree-Fock orbitals for the ground state while the orbitals ϕ_n for $n > N$ are excited-state Hartree-Fock orbitals. KELLY found that the excited-state orbitals all had continuum energies due to the fact that asymptotically the N lowest-occupied orbitals completely screen the attractive nuclear charge.[179a] In other words, use of V^N corresponds to having an electron in the field of a neutral atom. Understandably then, Kelly found that use of this basis set to calculate properties of an excited electron in the field of a singly charged ion by MBPT gave very slow convergence.

Much faster convergence was obtained through use of the V^{N-1} potential,[179b,c] defined as

$$\langle m|V^{N-1}|n\rangle \equiv \sum_{i=1}^{N-1} \{\langle mi|v|ni\rangle - \langle mi|v|in\rangle\}. \tag{17.6}$$

Excited one-electron orbitals computed from the V^{N-1} potential see a singly charged ion asymptotically since one of the bound occupied orbitals (usually the one from which the excited electron came) is excluded from the summation in (17.6). One problem with the V^{N-1} potential defined in (17.6), however, is that it does not regenerate the N ground-state Hartree-Fock orbitals. This problem can be remedied by introducing projection operators to insure that the N lowest orbitals obtained from (17.1) are those of the ground-state Hartree-Fock solution and that the excited orbitals see a singly charged ion asymptotically.[182-184]

So far the discussion has been concerned with choosing a *single* basis set of one-electron orbitals, usually in an $m_l m_s$ scheme, for use in the MBPT. It is now generally accepted, however, that for faster convergence and for greater flexibility it is best to use *multiple* basis sets.[185-187] Thus, for example, in a photoionization calculation one might use a restricted Hartree-Fock basis for the initial ground state of the atom and a different restricted Hartree-Fock basis for the continuum orbitals obtained from the equation resulting from the variational principle (16.2), thereby taking into account all intrachannel interactions. (By "restricted Hartree-Fock basis" we mean that the N-electron wave functions are represented by a linear combination of Slater determinants having well-defined orbital and spin angular momenta.[181]) The potential in which the continuum orbitals are calculated is called a $V^{N-1}(LS)$ potential. In effect the use of multiple basis sets is equivalent to summing certain classes of diagrams, which would appear in a single basis-set calculation, to infinite order. Extra care is required in a multiple basis-set calculation to avoid double counting.

---

[182] Pu, R.T., Chang, E.S. (1966): Phys. Rev. *151*, 31

[183] Silverstone, H.J., Yin, M.L. (1968): J. Chem. Phys. *49*, 2026

[184] Huzinaga, S., Arnau, C. (1970): Phys. Rev. A *1*, 1285

[185] Amus'ya, M.Ya., Cherepkov, N.A., Chernysheva, L.V. (1971): Zh. Eksp. Teor. Fiz. *60*, 160 [Sov. Phys. - JETP *33*, 90]

[186] Ishihara, T., Poe, R.T. (1972): Phys. Rev. A *6*, 111

[187] Kelly, H.P., Simons, R.L. (1973): Phys. Rev. Lett. *30*, 529

A different numerical problem in the MBPT that arises in the calculation of perturbation terms of second order or higher is the summation over a complete set of intermediate states. Usually this summation is carried out by computing a large number of intermediate-state wave functions and numerically integrating the relevant matrix elements up to some high-energy cutoff. MORRISON[188] and CHANG and POE[189] have shown, however, that the summation over intermediate states usually amounts to the calculation of an effective charge density in a small region of coordinate space about the atom. They obtain this effective charge density by solving an appropriate differential equation thereby both saving time and improving the numerical accuracy.

Detailed applications of the MBPT for the calculation of autoionization line profiles and of atomic photoionization cross sections have been made for many atoms.[179b, 187, 189–200] Many of these calculations have been reviewed by KELLY.[18] Here we review some of the principal results, as illustrated by photoionization of the $3p$-subshell of argon,

$$\mathrm{Ar}\,3p^6(^1S)+\omega \rightarrow \mathrm{Ar}^+\,3p^5\,\varepsilon l(^1P_1), \tag{17.7}$$

where $l=0$ or 2.

The lowest-order Brueckner-Goldstone MBPT diagrams contributing to the electric dipole matrix element for the reaction (17.7) are shown in Fig. 15.[187] The dotted lines connecting one small dot and one large dot indicate the one-body interaction H_{int} of the electrons with the electromagnetic field. The dotted lines connecting two small dots indicate the two-body electrostatic interaction H'; both direct and exchange terms are implied. The sequence of events is from bottom to top. Lines with arrows pointing upward represent occupied excited-state orbitals or "particle states," and lines with arrows pointing downward represent unoccupied unexcited-state orbitals or "hole states". Figure 15a represents the lowest-order transition: an electric dipole transition of an electron from the ground-state atomic orbital p to the excited or continuum orbital k. Figure 15b represents an electric dipole excitation of the ground state q orbital to the state k'; the electron in k' is then scattered by H' into the state k and the ion core undergoes a rearrangement whereby the vacancy in orbital q is replaced by a vacancy in orbital p. Figure 15c represents the lowest-order ground-state correlation term contributing to reaction (17.7): the electrostatic interaction H' excites two electrons from orbitals p and

---

[188] Morrison, J.C. (1973): J. Phys. B *6*, 2205
[189] Chang, T.N., Poe, R.T. (1975): Phys. Rev. A *11*, 191
[190] Chang, E.S., McDowell, M.R.C. (1968): Phys. Rev. *176*, 126
[191] Wendin, G. (1970): J. Phys. B *3*, 455; (1970): J. Phys. B *3*, 466
[192] Kelly, H.P., Ron, A. (1972): Phys. Rev. A *5*, 168
[193] Ishihara, T., Poe, R.T. (1972): Phys. Rev. A *6*, 116
[194] Kelly, H.P. (1972): Phys. Rev. A *6*, 1048
[195] Fliflet, A.W., Kelly, H.P. (1974): Phys. Rev. A *10*, 508
[196] Chang, T.N. (1975): J. Phys. B *8*, 743
[197] Chang, J.J., Kelly, H.P. (1975): Phys. Rev. A *12*, 92
[198] Carter, S.L., Kelly, H.P. (1975): J. Phys. B *8*, L 467
[199] Fliflet, A.W., Kelly, H.P. (1976): Phys. Rev. A *13*, 312
[200] Carter, S.L., Kelly, H.P. (1976): Phys. Rev. A *13*, 1388

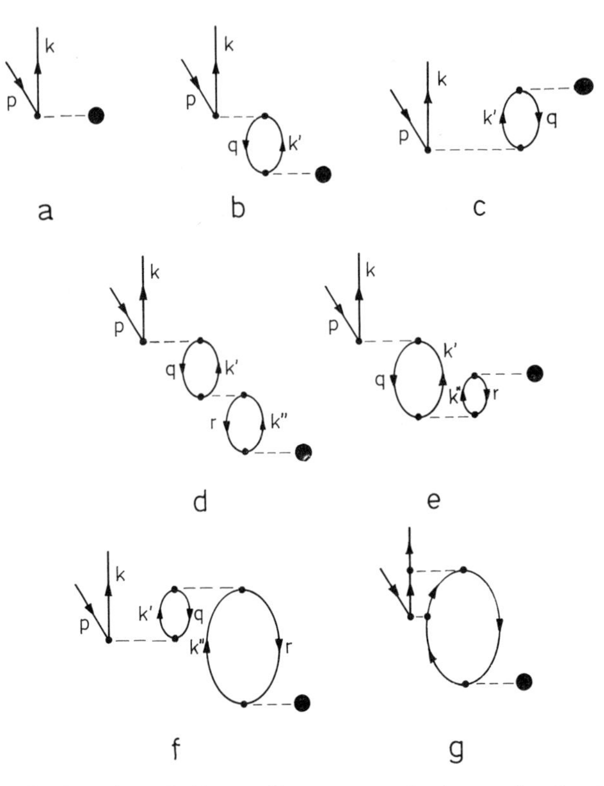

Fig. 15a–g. Low-order Brueckner-Goldstone diagrams contributing to the electric dipole matrix element. (From [187])

q into the excited orbitals k und k'; the electron in k' is subsequently deexcited back to the orbital q under the influence of the electromagnetic field.

The diagrams in Fig. 15 are those that appear when one uses a single one-electron orbital basis set. One may estimate how fast the perturbation expansion is likely to converge by computing the lowest-order diagram in Fig. 15a (which is simply the ordinary electric dipole matrix element) for various possible basis sets, as done by AMUSIA and CHEREPKOV,[19] and seeing how close the corresponding photoionization cross sections are to experiment. Figure 16 shows the cross sections, using both the length and velocity formulas, obtained with one-electron orbitals calculated in the V^N, V^{N-1}, and $V^{N-1}(LS)$ potentials described above. For comparison, the Herman-Skillman[105] central potential model cross section and the experimental cross section[1] are also shown. The short dashes indicate σ^N while the long dashes indicate σ^{N-1}, each calculated with orbitals defined in an $m_l m_s$ scheme. The Coulomb charge seen by the orbitals calculated in the V^{N-1} potential results in σ^{N-1} being pulled in toward lower energies as compared with σ^N. However, σ^{N-1} is in very poor agreement with experiment, and in fact it is found that the perturbation expansion in the V^{N-1} basis diverges![187] The reason for the divergence is the large magnitude of the repulsive intrachannel interactions (15.4b) between the $l=2$ electrons. If

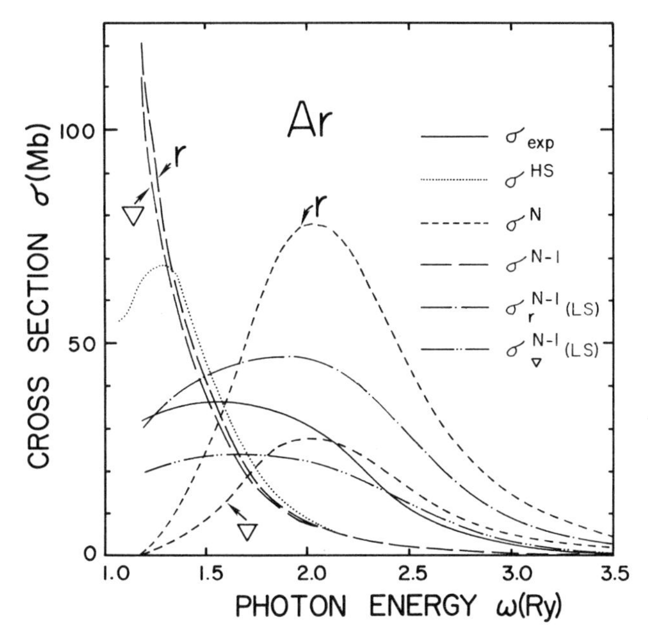

Fig. 16. Photoionization cross sections for the $3p$ subshell of argon using various types of one-electron orbitals. σ_{exp} is the experimental result from [1]; σ^{HS} is the result using Herman-Skillman central potential model wave functions; σ^{N}, σ^{N-1}, and $\sigma^{N-1}(LS)$ indicate the results using wave functions calculated in the V^{N}, V^{N-1}, and $V^{N-1}(LS)$ Hartree-Fock potentials, respectively; the subscripts r and ∇ indicate use of the length and velocity formula, respectively. (From [19])

the $l=2$ orbitals are instead calculated in the $V^{N-1}(LS)$ potential obtained by requiring (15.4b) to be stationary with respect to changes in the $l=2$ orbitals on the left of (15.4b), then one obtains the cross sections $\sigma^{N-1}(LS)$ in Fig. 16. One sees that the repulsive intrachannel interactions push $\sigma^{N-1}(LS)$ to higher energies, as compared with σ^{N-1}, giving much better agreement with experiment.

Use of excited orbitals calculated in the $V^{N-1}(LS)$ potential amounts to accounting for the intrachannel interactions (15.4b). In diagrammatic language, one says that use of this basis set amounts to summing the infinite sequence of Figs. 15a, b, d, etc. (in which the vacancy remains the same, i.e., $p=q=r$) so that only the lowest-order diagram in Fig. 15a need be calculated. Similarly the infinite sequence of Figs. 15c, e, etc., (in which the vacancy seen by the scattered particle remains the same, i.e., $p=q$ in Fig. 15e) is summed and may be represented by its lowest order diagram in Fig. 15c. Alternatively, one may say that in the $V^{N-1}(LS)$ basis there are no intrachannel scattering interactions so that Figs. 15b, d, e do not appear in a MBPT expansion as long as the ionic-core vacancy remains fixed.

The significance of the MBPT calculation of Kelly and Simons [187] is their finding that in the $V^{N-1}(LS)$ basis one may obtain excellent agreement with experiment using only a low-order perturbation expansion. Their results are shown in Fig. 17 together with the zero-order cross sections and experiment.

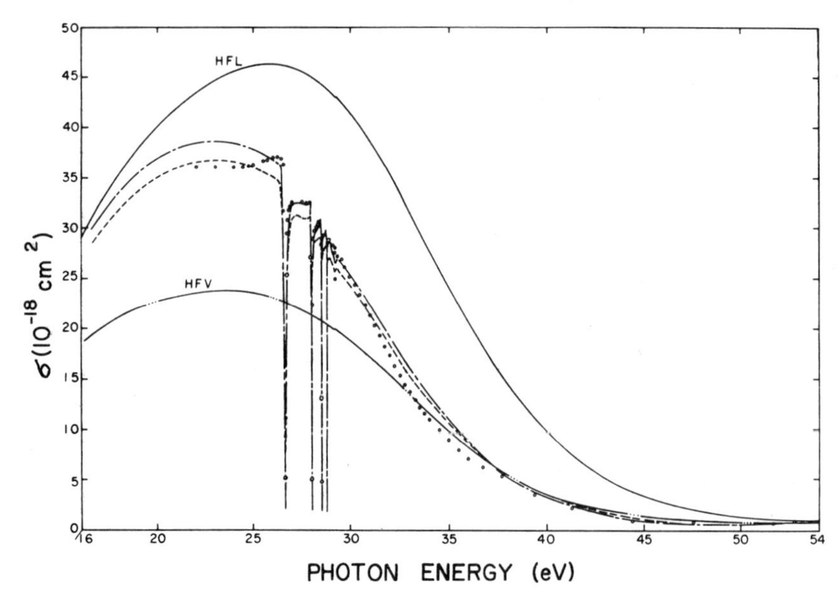

Fig. 17. Photoionization cross section for the $3p$ and $3s$ subshells of argon. HFL and HFV indicate the length and velocity results obtained using Hartree-Fock orbitals calculated in a $V^{N-1}(LS)$ potential. The dot-dash and dashed lines represent the length and velocity results of the MBPT calculation of KELLY and SIMONS[187]. Only the four lowest $3s \rightarrow np$ resonances are shown; the series converges to the $3s$ threshold at 29.24 eV. Experimental results are those of SAMSON[1] above 37 eV and of R.P. MADDEN et al. [Phys. Rev. *177*, 136 (1969)] below 37 eV. (From [187])

The difference between experiment and the zero-order cross sections (the latter being nearly identical to the cross sections obtained from the intrachannel[122] and close-coupling[174] calculations shown in Fig. 13) is due to the double excitations in the ground-state configuration and in the ionic-core configuration that are represented by Figs. 15c, f, respectively. The first accurate calculation of the Ar $3p$-subshell photoionization cross section was by AMUSIA et al.[201] using the random-phase approximation (cf. Sect. 18), which sums interactions of the type shown in Fig. 15a–f to all orders of perturbation theory. The work of KELLY and SIMONS[187] shows that a low-order perturbation treatment suffices provided the proper basis set is chosen.

Prior to the work of AMUSIA et al.,[201] the importance of the virtual double excitations indicated by Figs. 15c, f was underestimated due to the large energy denominators introduced by such highly excited intermediate states. The summation over intermediate states k', however, results in a sizable contribution to the dipole matrix element. One may understand better the importance of Figs. 15c, f perhaps from the point of view of a single one-electron basis set in which all the diagrams in Fig. 15 contribute. One notices that topologically Fig. 15c is very similar to Fig. 15b, and Fig. 15f is very similar to Fig. 15d. In Sect. 15 it was mentioned that the intrachannel matrix elements (15.4b), the effects of which are represented by the Figs. 15b, d, are large primarily because

[201] Amusia, M.Ya., Cherepkov, N.A., Chernysheva, L.V., Sheftel, S.I. (1969): Abstracts of papers, VI I.C.P.E.A.C., Cambridge, Mass.: M.I.T. Press, p. 132; (1970): Phys. Lett. *31* A, 553

of the large Slater integrals $R^1(3p\varepsilon d, \varepsilon' d3p)$. These same large integrals, in the form $R^1(3p3p, \varepsilon d\varepsilon' d) = R^1(3p\varepsilon d, \varepsilon' d3p)$, are the primary contributors to Figs. 15c, f. When the excited orbitals are calculated in the $V^{N-1}(LS)$ basis, Figs. 15c, d, e vanish and Figs. 15c, f become smaller, but the latter are still very important for obtaining good agreement with experiment.

18. The random-phase approximation. The current theoretical understanding of rare gas and other closed-shell-atom photoabsorption spectra is drawn from the first successful calculations of such spectra, which used the random-phase approximation (RPA). These calculations reduced the discrepancies between theory and experiment from factors of two or so to less than 10% in many cases. As compared to existing intrachannel and close-coupling calculations, the first RPA calculations demonstrated the important influence on closed-shell photoabsorption spectra of certain virtual double excitations in the ground and ionic states.

The RPA was borrowed from nuclear physics in an attempt to take electron correlations into account for heavier atoms, which could not be treated by the older "best wave function" methods. The initial application of the RPA to atomic photoionization was in 1964 by ALTICK and GLASSGOLD,[202] who calculated oscillator strengths and photoionization cross sections for the outer *s* subshells of Be, Mg, Ca, and Sr. They demonstrated the importance of the correlations included in the RPA and obtained improved agreement with experiment. But due to neglect of interchannel interactions, discussed below, and use of only a Hartree basis set of single-particle orbitals, the calculated results still differed substantially from experiment. In 1969, using a Hartree-Fock basis set and also neglecting interchannel interactions, AMUSIA et al.[201] applied the RPA to the calculation of $3p$-subshell photoionization of Ar and obtained agreement with experiment to within $\approx 10\%$. (It is now known that interchannel interactions are much more important for *s* subshells than for *p*, *d*, and *f* subshells, which partially explains the much better initial results of AMUSIA et al.[201] as compared to those of ALTICK and GLASSGOLD.[202]) This work marked the beginning of an extensive series of very accurate calculations by AMUSIA and co-workers,[19, 185, 201, 203, 204] by WENDIN,[205–208] and by others,[209–212] who employed the RPA to calculate photoionization cross sections for primarily closed-shell atoms.

---

[202] Altick, P.L., Glassgold, A.E. (1964): Phys. Rev. *133*, A632

[203] Amusia, M.Ya., Ivanov, V.K., Cherepkov, N.A., Chernysheva, L.V. (1974): Zh. Eksp. Teor. Fiz. *66*, 1537 [Sov. Phys. - J.E.T.P. *39*, 752 (1975)]

[204] Amusia, M.Ya., Cherepkov, N.A., Živanović, Dj., Radojević, V. (1976): Phys. Rev. A *13*, 1466

[205] Wendin, G. (1971): J. Phys. B *4*, 1080

[206] Wendin, G. (1972): J. Phys. B *5*, 110

[207] Wendin, G. (1973): J. Phys. B *6*, 42

[208] Wendin, G. (1974): Vacuum ultraviolet radiation physics, Koch, E.E., Haensel, R., Kunz, C. (eds.), Braunschweig: Vieweg, p. 225

[209] Lin, C.D. (1974): Phys. Rev. A *9*, 171

[210] Lin, C.D. (1974): Phys. Rev. A *9*, 181

[211] Fliflet, A.W., Chase, R.L., Kelly, H.P. (1974): J. Phys. B *7*, L443

[212] Chang, T.N. (1977): Phys. Rev. A *15*, 2392

α) Definition of the RPA. Despite its successes, the application of the RPA to atomic photoionization poses a number of difficulties. The first is to define what is meant by the RPA. The RPA was originally introduced to describe the collective behavior of an electron gas.[213] The term "random phase" was used to describe the approximation of neglecting certain matrix elements which, in the limit of a very large number of electrons, cancel each other due to their random phases.[213] (Certain exchange matrix elements are also neglected in treating an infinite electron gas; although these exchange matrix elements are always included in RPA calculations for finite systems, some authors[19] prefer the more explicit name "random-phase approximation with exchange (RPAE).") There are several equivalent ways of deriving the RPA. As applied to finite systems these are: 1) the equations-of-motion method,[202, 214–217] 2) the time-dependent Hartree-Fock method,[19, 218, 219] 3) the Green's-function method,[220, 221] and 4) the diagrammatic perturbation theory method.[19, 206] The latter method is perhaps the easiest in which to define the RPA. Specifically, the RPA for calculating electric dipole matrix elements consists in summing the class of diagrams known as the "bubble" diagrams to all orders of perturbation theory and neglecting all other diagrams. The "bubble" diagrams are those which at any vertex have only a single "particle" line and a single "hole" line. The lowest-order RPA diagrams are shown in Fig. 15a–f. Fig. 15g is not an RPA diagram since at two of the vertices a particle state is scattered into another particle state rather than into a hole state. Thus, in the diagrammatic perturbation theory method, the RPA image of the atom is that of a bubbling sea of electrons: at any given instant there may be many electrons excited out of their ground-state orbitals (leaving a corresponding number of vacancies or holes behind), but they never stay excited long enough to interact with each other since they quickly fall back into their own vacancy. In the more standard language of configuration interaction, the RPA as applied to photoionization treats not only final-state intra- and inter-channel interactions but also some of the effects of virtually-excited pairs of electrons in the ground and ionic configurations.

β) Difficulties and advantages of the RPA. Other difficulties with the RPA relate to its inclusion of certain kinds of electron correlation (i.e., the "bubble" diagrams) to infinite order while ignoring other interactions which appear in the second and higher orders. There is no rigorous theoretical justification for this procedure, although as the number of electrons in a given atomic subshell

---

[213] Bohm, D., Pines, D. (1951): Phys. Rev. *82*, 625

[214] Rowe, D.J. (1968): Rev. Mod. Phys. *40*, 153

[215] Shibuya, T.I., McKoy, V. (1970): Phys. Rev. A *2*, 2208

[216] Brown, G.E. (1971): Unified theory of nuclear models and forces, 3rd edn., Amsterdam: North-Holland, Chap. V

[217] Fetter, A.L., Walecka, J.D. (1971): Quantum theory of many-particle systems, New York: McGraw-Hill, § 59

[218] Dalgarno, A., Victor, G.A. (1966): Proc. R. Soc. A *291*, 291

[219] Jamieson, M.J. (1973): Wave mechanics: The first fifty years, Price, W.C., Chissick, S.S., Ravensdale, T. (eds.), New York: Wiley, p. 133

[220] Csanak, G., Taylor, H.S., Yaris, R. (1971): Adv. At. Mol. Phys. *7*, 287

[221] Schneider, B. (1973): Phys. Rev. A *7*, 557

increases one comes closer to the infinite electron gas case. Furthermore, while perturbation theory expansions always satisfy the Pauli principle in each order of the perturbation, not every diagram satisfies the Pauli principle. Thus, for a given diagram intermediate states are usually summed over all excited single-particle states, even those which are already occupied. This causes no problem because there exist other diagrams which will cancel the given diagram whenever a Pauli principle violation occurs. Since the RPA includes only certain diagrams and not others, however, this cancellation does not occur and Pauli principle violations arise starting with the second order of perturbation theory. AMUSIA and CHEREPKOV[19] estimate that these violations impose a limit on the accuracy of the RPA of $\approx 10\%$. Given these difficulties, perhaps the most convincing justification of the RPA procedure so far is that the high-order electron-correlation effects that it includes are not very important: i.e., as shown by the MBPT (cf. Sect. 17), it is only necessary to take into account the RPA type of correlations to at most second order, provided an appropriate basis set is used.

Despite its difficulties, the RPA procedure of summing a certain class of diagrams to all orders has nevertheless several advantages. Firstly, it is in some respects computationally simpler to calculate the sum of an infinite number of related diagrams than to calculate separately even only the first few of them, as in MBPT. Secondly, one may show rigorously that the length and velocity formulas for the electric dipole matrix element are equal in the RPA[19] (cf. Sect. 5). Thirdly, like the MBPT, the RPA focuses attention on the electric dipole matrix element directly, treating initial- and final-state correlations simultaneously. Lastly, should the MBPT calculation for some atom fail to converge, then methods of taking interactions into account to infinite order, such as the RPA, must be used.

γ) Configuration interaction approaches to the RPA. Recently CHANG and FANO[222] have shown that the RPA follows straightforwardly from a new configuration-interaction approach of calculating transition matrix elements directly. The method consists firstly in defining a transition operator consisting of the outer product of two N-electron configuration wave functions, one representing the initial state $\langle i|$ and one representing the final state $|f\rangle$,

$$\langle \boldsymbol{r}_1 s_1, \ldots, \boldsymbol{r}_N s_N | f \rangle \langle i | \boldsymbol{r}'_1 s'_1, \ldots, \boldsymbol{r}'_N s'_N \rangle. \tag{18.1}$$

This transition operator satisfies the following equation of motion obtained by taking its commutator with the exact Hamiltonian:

$$H|f\rangle\langle i| - |f\rangle\langle i|H = (E_f - E_i)|f\rangle\langle i|. \tag{18.2}$$

The main task of the method is the construction of the single-particle transition matrix, $\langle \boldsymbol{r}_1 | \Gamma | \boldsymbol{r}'_1 \rangle$, which is defined as follows:

$$\langle \boldsymbol{r}_1 | \Gamma | \boldsymbol{r}'_1 \rangle \equiv \left(\prod_{j=2}^{N} \iint d\boldsymbol{r}_j \, d\boldsymbol{r}'_j \, \delta(\boldsymbol{r}_j - \boldsymbol{r}'_j) \right) \langle \boldsymbol{r}_1 \ldots \boldsymbol{r}_N | f \rangle \langle i | \boldsymbol{r}'_1 \ldots \boldsymbol{r}'_N \rangle, \tag{18.3}$$

[222] Chang, T.N., Fano, U. (1976): Phys. Rev. A *13*, 263; (1976): ibid, *13*, 282

where we have omitted specification of the spin variables for simplicity. This matrix is constructed using the solutions of the one-dimensional equation of motion obtained by integrating (18.2) over coordinates $2 \leqq j \leqq N$. Once the single-particle transition matrix is obtained, the electric dipole matrix element is simply

$$\hat{\boldsymbol{\varepsilon}} \cdot \langle i | \sum_{j=1}^{N} \boldsymbol{r}_j | f \rangle = \hat{\boldsymbol{\varepsilon}} \cdot \iint d\boldsymbol{r}_1 \, d\boldsymbol{r}_1' \, \boldsymbol{r}_1 \, \delta(\boldsymbol{r}_1 - \boldsymbol{r}_1') \langle \boldsymbol{r}_1 | \Gamma | \boldsymbol{r}_1' \rangle. \tag{18.4}$$

Some examples will clarify the procedure. Consider $3p \to \varepsilon d$ transitions in Ar [cf. (17.7)]. The simplest choice for the transition operator is

$$|f\rangle\langle i| \equiv |3p^5 \psi_{\varepsilon d} \, {}^1P_1\rangle \langle 3p^6 \, {}^1S_0|, \tag{18.5}$$

where atomic Hartree-Fock orbitals are used to describe the N-lowest-bound orbitals of neutral argon. The symbol $\psi_{\varepsilon d}$ represents the as yet undetermined one-electron orbital for the photoelectron. After a considerable amount of Racah algebra one finds that (18.3) yields the following form for the single-particle transition matrix:[222]

$$\langle \boldsymbol{r}_1 | \Gamma | \boldsymbol{r}_1' \rangle = \psi_{\varepsilon d}(r_1) \langle l_f = 2, \hat{\boldsymbol{r}}_1 | w_{00}^{[0,1]} | l_i = 1, \hat{\boldsymbol{r}}_1' \rangle \, \chi_{3p}(\boldsymbol{r}_1'), \tag{18.6}$$

where $\psi_{\varepsilon d}(r_1)$ is the undetermined photoelectron wavefunction, $\chi_{3p}(r_1')$ is the Hartree-Fock wave function for the $3p$ orbital, and the matrix element of the double tensor operator w results from the integration over angular and spin variables. Upon integrating the equation of motion (18.2) over all variables except the first, one obtains the Hartree-Fock equation for $\psi_{\varepsilon d}(r_1)$ with the $V^{N-1}(LS)$ potential (cf. Sect. 17). Solving for $\psi_{\varepsilon d}(r_1)$ then completes the construction of $\langle \boldsymbol{r}_1 | \Gamma | \boldsymbol{r}_1' \rangle$, and the dipole matrix element (18.4) becomes

$$\hat{\boldsymbol{\varepsilon}} \cdot \langle 3p^6 \, {}^1S_0 | \sum_{j=1}^{N} \boldsymbol{r}_j | 3p^5 \psi_{\varepsilon d} \, {}^1P_1 \rangle = (\tfrac{4}{3})^{\frac{1}{2}} \langle \chi_{3p} | r | \psi_{\varepsilon d} \rangle. \tag{18.7}$$

A better approximation for the transition operator is

$$|f\rangle\langle i| \equiv |3p^5 \psi_{\varepsilon d} \, {}^1P_1\rangle \{\langle 3p^6 \, {}^1S_0| + \langle 3p^4 \phi_a \phi_b \, {}^1S_0|\}, \tag{18.8}$$

where, again, the bound occupied orbitals of neutral argon are known from a standard Hartree-Fock calculation, and $\psi_{\varepsilon d}$, ϕ_a, and ϕ_b are excited d-orbital wave functions that remain to be determined. The admixture coefficient of the doubly excited configuration has been absorbed in its normalization. Substituting (18.8) in (18.3) and performing the integrations gives

$$\begin{aligned} \langle \boldsymbol{r}_1 | \Gamma | \boldsymbol{r}_1' \rangle = {} & \psi_{\varepsilon d}(r_1) \langle l_f = 2, \hat{\boldsymbol{r}}_1 | w_{00}^{[0,1]} | l_i = 1, \hat{\boldsymbol{r}}_1' \rangle \, \chi_{3p}(r_1') \\ & + \chi_{3p}(r_1) \langle l_f = 1, \hat{\boldsymbol{r}}_1 | w_{00}^{[0,1]} | l_i = 2, \hat{\boldsymbol{r}}_1' \rangle \, \phi(r_1'), \end{aligned} \tag{18.9}$$

where $\phi(r)$ is a linear combination of the two d orbitals ϕ_a and ϕ_b:

$$\phi(r)=\sum_{LS}(-1)^L\begin{Bmatrix}2 & 2 & L\\ 1 & 1 & 1\end{Bmatrix}(3p^6\,{}^1S\{|3p^4(^{2S+1}L)\,\phi_a\phi_b)$$

$$\cdot[\langle\psi_{\varepsilon d}|\phi_b\rangle\,\phi_a(r)+\langle\psi_{\varepsilon d}|\phi_a\rangle\,\phi_b(r)]. \qquad (18.10)$$

The orbital wave functions $\psi_{\varepsilon d}(r_1)$ and $\phi(r_1')$ are determined by solving the equation of motion (18.2) after integrating over the coordinates for particles $2\leqq j\leqq N$. The equations that result are a coupled pair of differential equations which, after dropping some terms deemed to be small, becomes equivalent to the time-dependent Hartree-Fock equations.[222] Having found $\psi_{\varepsilon d}$ and ϕ, the single-particle transition matrix is now defined by (18.9) and the electric dipole matrix element may be found from (18.4) to be

$$(\tfrac{4}{3})^{\frac{1}{2}}[\langle\chi_{3p}|r|\psi_{\varepsilon d}\rangle-\langle\phi|r|\chi_{3p}\rangle], \qquad (18.11)$$

which may be shown to be equivalent to the RPA result.[222]

The Chang-Fano procedure[222] thus interprets the RPA corrections due to excitation of many pairs of electrons in the ground and ionic configurations in terms of a much simpler configuration interaction. Namely, Chang and Fano have found that in their representation one need only consider a *single* pair of excited electrons in the ground state. Furthermore, only the average amplitude of this pair, defined by (18.10), enters the dipole matrix element (18.11) and the equation of motion. The final-state effects of these virtual double excitations are embodied in the final-state wave function $\psi_{\varepsilon d}$, the differential equation of which is coupled with that for ϕ. In short, Chang and Fano have shown that if one uses their coupled-differential-equation procedure (instead of using a complete single-particle basis set to calculate corrections to transition matrix elements by means of the MBPT or the RPA), then one may give a very simple interpretation of the improvements incorporated in the dipole matrix element. The improvements are those resulting from interaction of the configurations chosen to represent the transition operator (18.1).

Chang[212] has calculated the outer-subshell photoionization cross section for Ne and Ar. His results for $3p$-subshell photoionization of argon are shown in Fig. 18 in three levels of approximation. Curves I correspond to choosing the transition operator in the form (18.5), which is equivalent to an intrachannel calculation or else a Hartree-Fock calculation in which the continuum orbitals are calculated in the $V^{N-1}(LS)$ potential. Curves II result from the choice of (18.8) for the transition operator and use of (18.11) for the dipole matrix element. However, in curves II the pair of differential equations for $\psi_{\varepsilon d}$ and for ϕ were uncoupled so that the double excitations did not affect the orbital $\psi_{\varepsilon d}$. This is equivalent to including the double excitations only in the initial state. Curves III represent the results obtained by solving the coupled differential equations for $\psi_{\varepsilon d}$ and ϕ. These latter curves agree with the results of calculations using MBPT[187] and the ordinary RPA.[19]

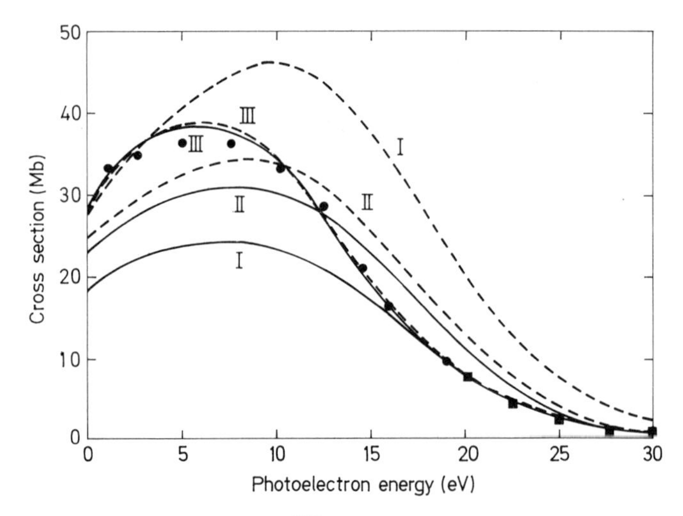

Fig. 18. Theoretical calculations of Chang[212] for the photoionization cross section of the $3p$-subshell of argon. Dashed and solid lines give length and velocity results respectively in three levels of approximation, discussed in the text. The experimentally measured values of the argon cross section are indicated by the solid circles (SAMSON[1]) and by the solid squares (SAMSON, unpublished). (From[212])

SWANSON and ARMSTRONG[223,224] have recently developed a simpler alternative configuration-interaction approach that agrees with the CHANG and FANO interpretation of rare-gas photoionization processes as well as with the findings of MBPT and RPA calculations for the rare gases. Their procedure is to calculate initial- and final-state wave functions that explicitly account for double excitations in the ground and ionic configurations. Ordinarily this would be a monumental task since, for example, to represent pair excitations only to d orbitals in the ground state of argon one must write

$$|i\rangle = a|3p^6\,{}^1S_0\rangle + \sum_{\substack{nn'\\LS}} b_{nn'}^{LS} |(3p^4)\,LS(nd\,n'd)\,LS,\,{}^1S_0\rangle, \tag{18.12}$$

where the sums over n and n' extend over both the discrete and the continuum. FROESE-FISCHER,[225] however, has shown that the double summation in (18.12) is related by an orthogonal transformation to a rapidly convergent single sum over natural orbitals $\bar{n}d$, where each $\bar{n}d$ orbital is a linear combination extending over all energies of the original nd orbitals. In terms of these natural orbitals the ground state of argon may be represented as

$$|i\rangle = a|3p^6\,{}^1S_0\rangle + \sum_{\bar{n}LS} c_{\bar{n}}^{LS} |(3p^4)\,LS(\bar{n}d^2)\,LS,\,{}^1S_0\rangle. \tag{18.13}$$

---

[223] Swanson, J.R., Armstrong, Jr., L. (1977): Phys. Rev. A *15*, 661
[224] Swanson, J.R., Armstrong, Jr., L. (1977): Phys. Rev. A *16*, 1117
[225] Froese-Fischer, C. (1973): J. Comput. Phys. *13*, 502

SWANSON and ARMSTRONG calculated all of the orbitals as well as the coefficients a and $c_{\bar{n}}^{LS}$ from a multiconfiguration Hartree-Fock procedure[181] and find that $a \approx 0.97$, $c_{\bar{3}}^{LS} \approx 0.1$, and $c_{\bar{4}}^{LS} \approx 0.01$.[223] Thus they find that it is sufficient to include only the $\bar{3}d$-natural orbital in the sum in (18.13), and moreover, the same $\bar{3}d$-orbital is used for all three allowed term levels LS.

SWANSON and ARMSTRONG calculated the final-state wave function in two ways. First, they assumed a configuration $|3p^5 \varepsilon l\, {}^1P_1\rangle$ for the final state, where the continuum l orbitals were calculated in the $V^{N-1}(LS)$ potential.[223] Their results for the $3p$-subshell cross section in argon are equivalent to curve II of CHANG[212] shown in Fig. 18. In a second calculation, SWANSON and ARMSTRONG expanded the ionic state in a form analogous to (18.13) for the ground state. The same orbitals obtained for the ground state were used in the ionic-state expansion; only the coefficients in the ionic-state expansion thus had to be determined. The photoelectron orbital wave function was then calculated by solving a differential equation containing the usual $V^{N-1}(LS)$ potential plus additional terms arising from the double excitations in the ionic states. The resulting cross sections for argon are very close to curves III of CHANG[212] shown in Fig. 18. Thus the calculations of SWANSON and ARMSTRONG confirm the interpretation of CHANG and FANO[222] that in a suitable basis the RPA may be said to go beyond a Hartree-Fock calculation by including the effects on the dipole matrix element of a single pair of excited electrons in the ground and ionic states. (Note that whether or not one must *explicitly* consider double excitations in the ionic state depends on whether one solves for the final-state and initial-state wave function separately, as in [223, 224], or simultaneously as in [222].)

δ) RPA Predictions of interchannel interaction effects. Some of the most striking predictions of the RPA have concerned the effects of interchannel interactions, particularly on s-subshell photoionization cross sections. Consider for example the following reaction in argon:

$$\mathrm{Ar}\, 3s^2 3p^6 ({}^1S_0) + \omega \rightarrow \mathrm{Ar}^+\, 3s\, 3p^6 \varepsilon p ({}^1P_1). \tag{18.14}$$

AMUSIA et al.[226] and, independently, LIN[210] have found that the process (18.14) is strongly affected by interaction with the outer-subshell channel $3p \rightarrow \varepsilon d$. In diagrammatic language, the interaction in lowest order is represented by Figs. 15b, e, where q represents a $3p$ vacancy and k an excited p orbital. In each case the photoelectron starts out in the $3p \rightarrow \varepsilon d$ channel but is subsequently scattered by electrostatic interaction into the $3s \rightarrow \varepsilon p$ channel. Since the direct $3s \rightarrow \varepsilon p$ dipole matrix element is weak, these higher-order interactions strongly modify the $3s$-subshell cross section.

The calculations of AMUSIA et al.[19, 226] are shown in Fig. 19. The curves marked L and V show the single-channel Hartree-Fock results using the length and velocity formulas, while the curves a and b show the RPA interchannel interaction calculations using the experimental and Hartree-Fock $3s$-orbital

[226] Amusia, M.Ya., Ivanov, V.K., Cherepkov, N.A., Chernysheva, L.V. (1972): Phys. Lett. *40* A, 361

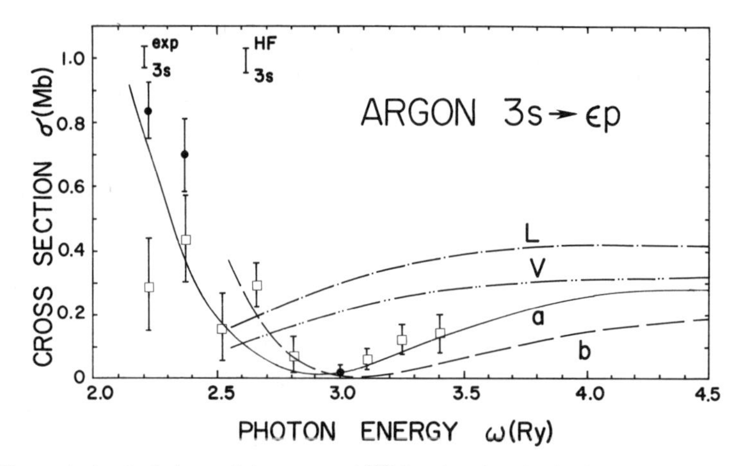

Fig. 19. Theoretical calculations of AMUSIA et al.[226] for the photoionization cross section of the 3s subshell of argon, showing the influence of interchannel interactions. See text for description of curves. (From [19])

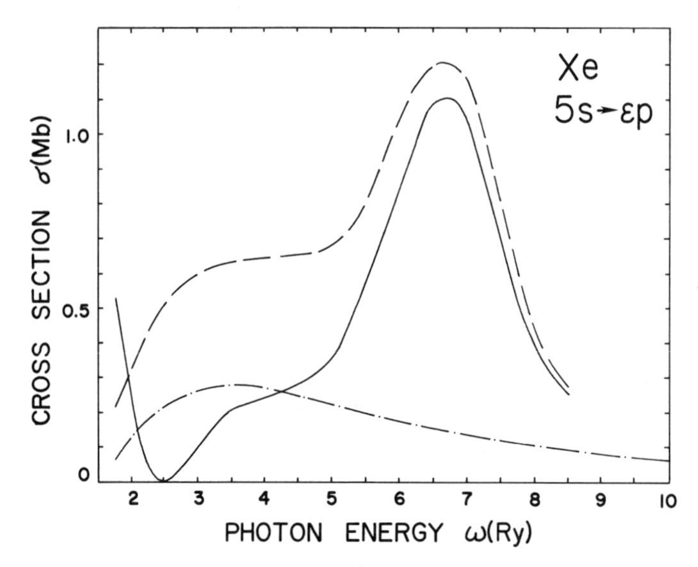

Fig. 20. Theoretical calculations of AMUSIA et al.[19] for the photoionization cross section of the 5s subshell of xenon, showing the influence of interchannel interactions. See text for description of curves. (From [19])

ionization thresholds, respectively. The effect of the interchannel interactions is seen to be an interference leading to a zero in the 3s-subshell cross section. [This zero is not predicted by the rules for the occurrence of a Cooper minimum since those rules are based on a single-channel model (cf. Sect. 12).] The original predictions of Amusia et al.[226] have now been verified by experimental measurements,[227,228] also shown in Fig. 19.

[227] Lynch, M.J., Gardner, A.B., Codling, K., Marr, G.V. (1973): Phys. Lett. *43* A, 237
[228] Samson, J.A.R., Gardner, J.L. (1974): Phys. Rev. Lett. *33*, 671

Even more striking is the analogous $5s \to \varepsilon p$ cross section in xenon,

$$\mathrm{Xe}\,4d^{10}5s^2 5p^6(^1S_0)+\omega \to \mathrm{Xe}^+\,4d^{10}5s5p^6(^2S)\,\varepsilon p(^1P_1). \tag{18.15}$$

The $5s$-subshell photoionization cross section is shown in Fig. 20 in three different approximations.[19] The dot-dash line represents the direct Hartree-Fock length result using continuum orbitals calculated in a $V^{N-1}(LS)$ potential. The dashed line represents an RPA calculation including interchannel interaction with the $4d \to \varepsilon f$ channel. One sees that the large delayed maximum in the $4d$-subshell cross section (cf. Figs. 3 and 6) is mirrored in the $5s$-subshell cross section. (This has been qualitatively verified by experiment, but the experimental $5s$-subshell cross section in the neighborhood of the $4d$-subshell threshold is more than a factor of two below the theoretical predictions.[229]) The solid line represents an RPA calculation including interchannel interaction with both the $4d \to \varepsilon f$ and the $5p \to \varepsilon d$ channels. One sees that interchannel interaction with the outer $5p$ subshell produces interference leading to a zero in the $5s$-subshell cross section.

Amusia et al. have carried out s-subshell cross sections for a number of other atoms.[19] They that the ns^2-subshell cross sections for Be, Mg, Ca, Zn, and Ba as well as the ns-subshell cross sections for the alkalis are strongly influenced by interchannel interactions. However, agreement between the RPA calculations and experiment is not as good as for the rare gases due to correlations not included in the RPA, such as those due to polarization of the ionic core.

ε) Extensions of the RPA. The sucess of the RPA in calculating accurate photoionization cross sections for the outer subshells of the rare gases has stimulated interest in using the RPA for inner subshells and for other atoms. Many of these newer applications have required extensions of the RPA.

In calculating inner-shell ionization cross sections, relaxation of the ion core strongly influences the ionization threshold energy as well as the charge field seen by the photoelectron. Extensions of the RPA to include relaxation effects have been discussed by Amusia[19,230] and by Wendin.[231] Often these relaxation effects can be taken into account by using experimental ionization energies and by calculating continuum orbitals in the $V^{N-1}(LS)$ potential constructed from the single-particle orbital wave functions *of the ion.*[230]

As originally developed, the RPA was applicable only to systems having spherical symmetric ground states, which restricted the use of the RPA primarily to closed-shell atoms and to other atoms, such as the alkalis, having zero total orbital angular momentum. The theoretical difficulty with non-spherically symmetric systems is their degeneracy in an $m_l m_s$ scheme. However, this problem can be removed by dealing with states and operators having well-

---

[229] West, J.B., Woodruff, P.R., Codling, K., Houlgate, R.G. (1976): J. Phys. B 9, 407; Adam, M.Y., Wuilleumier, F., Sandner, N., Krummacher, S., Schmidt, V., Mehlhorn, W. (1978): Jpn. J. Appl. Phys. *17*, Supplement 17-2, 170

[230] Amusia, M.Ya. (1977): Atomic Physics, Vol. 5, Marrus, R., Prior, M., Shugart, H. (eds.), New York: Plenum, p. 537

[231] Wendin, G. (1976): Photoionization and other probes of many-electron interactions, Wuilleumier, F.J. (ed.), New York: Plenum, p. 61

defined orbital angular momentum. General theories for an open-shell RPA have been developed by ROWE and coworkers[232-234], by ARMSTRONG,[235] and by STARACE and SHAHABI.[21b] The first application of the open-shell RPA of ARMSTRONG[235] for calculating the photoionization cross section of a non-spherically symmetric atom was the calculation by STARACE and ARMSTRONG[21a] of $3p$-subshell ionization of atomic chlorine. In their calculation only intrachannel interactions were calculated, using the RPA. Virtual double excitations in the ground state of Cl were found to have a small effect within each channel and the effect on the $3p$-subshell ionization cross section was even smaller due to cancellations between different channels.[21a] CHEREPKOV and CHERNYSHEVA,[22] using a different open-shell RPA, confirmed the small magnitude of the effect of double excitations within each channel, but found, however, that interchannel interactions are extremely large in Cl. The interchannel interactions are so large, in fact, that, as compared to the intrachannel calculation of [21a], they reduce the magnitude of the cross section just above threshold by a factor of two.[22] No experimental data exist for atomic Cl.

STARACE and SHAHABI[21b] have developed an RPA for transition matrix elements[222] in a way that is independent of whether one is dealing with a closed- or an open-shell atom. The calculation of the first-order transition matrix uses a graphical method,[21b] which greatly simplifies the treatment of antisymmetrization and of angular momentum algebra. Following CHANG and FANO,[222] it is assumed that the ground state may contain a single pair of virtually excited orbitals. Two approximations are made in evaluating the interactions of these virtually excited orbitals with the residual ion: firstly, it is assumed that there is no exchange of orbital or spin angular momentum, and, secondly, that the Pauli Principle is ignored. These approximations define the open-shell RPA. It is found that the first-order transition matrix for an atom having N final-state channels is determined by a set of N final-state radial functions and N' initial-state radial functions which satisfy $(N+N')$ coupled differential equations. (The relation of N' to N depends on the particular atom studied.) These $(N+N')$ coupled differential equations reduce to familiar forms in the following limiting cases: (1) When initial-state correlations are ignored, one obtains the N-coupled differential equations of the close-coupling approximation[165]. (2) When the atom has only closed shells, $N=N'$ and one obtains the $2N$-coupled differential equations of the CHANG-FANO[222] version of the RPA.

Although the effects of relativistic interactions are outside the scope of this review, we note that a relativistic RPA for closed-shell atoms has been developed by JOHNSON and LIN.[236] Their Hamiltonian consists of a sum of one-electron Dirac Hamiltonians plus the electrostatic interactions between elec-

---

[232] Rowe, D.J., Wong, S.S.M. (1969): Phys. Lett. *30* B, 147

[233] Rowe, D.J. (1972): Dynamic structure of nuclear states, Rowe, D.J., et al. (eds.), Toronto: University of Toronto Press, p. 101

[234] Rowe, D.J., Ngo-Trong, C. (1975): Rev. Mod. Phys. *47*, 471

[235] Armstrong, Jr., L. (1974): J. Phys. B7, 2320

[236] Johnson, W.R., Lin, C.D. (1976): Phys. Rev. A *14*, 565; (1979): ibid, *20*, 964

trons. The Breit interaction is treated as a perturbation. The first applications were to the calculation of discrete oscillator strengths[237a] and to the calculation of the photoionization cross sections for He and Be.[237b] More recently, extensive calculations of the photoionization cross sections and photoelectron angular distributions[238a] as well as of the photoelectron spin polarization[238b] for the heavier rare gas atoms have been carried out. The results are generally in excellent agreement with experiment, even in the case of effects that are sensitive to very weak relativistic interactions.

b) Partitioned-space correlation theories

19. Introduction. Most of the "whole-space" correlation theories described in the previous section and the "partitioned-space" correlation theories considered in this section have been developed to treat primarily short-range electron correlations. The difference between the two groups is that the "partitioned-space" theories explicitly recognize the fact and incorporate it into their formulation whereas the "whole-space" theories do not and hence must always perform numerical calculations over the entire region of coordinate space. Truly long-range correlations, such as polarization interactions, are difficult to treat in all methods except by the introduction of ad hoc potentials.

Of the three "partitioned-space" correlation theories considered in this section, the quantum defect theory (QDT) and the R matrix theory are closely related. Each assumes that beyond a radius r_0, which is of the order of a few Bohr radii, the photoelectron sees only a long-range potential, which is usually assumed to be Coulombic. Hence the final-state wave function for $r \geqq r_0$ can be expressed analytically in terms of a few scattering parameters or else easily calculated numerically. In either theory these parameters may be obtained either semiempirically or else by an ab initio calculation. The two theories are complementary: the QDT studies the region $r > r_0$ and determines the scattering parameters of the final-state wave function by application of boundary conditions at $r = \infty$ while the R matrix theory studies the region $r < r_0$ and determines the scattering parameters by application of boundary conditions at $r = r_0$. The QDT has been used to calculate very detailed cross sections in the vicinity of ionization thresholds, particularly in the region of autoionizing resonances. The R matrix theory has been used to calculate photoionization cross sections over a larger energy range with results that are competitive with those of the best alternative calculational methods.

In contrast, the discrete basis-set theories go to the extreme of ignoring the asymptotic form of the photoelectron wave function altogether. Instead the photoelectron wave function for small r is approximated by an expansion in a basis set of discrete (i.e., finite-range) wave functions. However, applications of

---

[237a] Lin, C.D., Johnson, W.R., Dalgarno, A. (1977): Phys. Rev. A *15*, 154; Lin, C.D., Johnson, W.R. (1977): Phys. Rev. A *15*, 1046

[237b] Johnson, W.R., Lin, C.D. (1977): J. Phys. B *10*, L331

[238a] Johnson, W.R., Cheng, K.T. (1978): Phys. Rev. Lett. *40*, 1167; (1979): Phys. Rev. A *20*, 978

[238b] Cheng, K.T., Huang, K.-N., Johnson, W.R. (1980): J. Phys. B *13*, L45

these discrete basis-set methods to atomic photoionization have been limited to only a few light elements.

20. Quantum defect theory. The quantum defect theory (QDT) uses the analytically known properties of excited electrons moving in a pure Coulomb field to describe atomic photoabsorption and electron-ion scattering processes in terms of a few parameters. These parameters are usually nearly independent of energy in the threshold energy region (i.e., within a few eV of the atomic ionization threshold) and smoothly varying functions of energy above threshold. Hence the determination of these parameters in a restricted energy region suffices to predict the variation over a much wider energy region of numerous atomic properties, such as photoionization cross sections, autoionization line profiles, photoelectron angular distributions, discrete line strengths, etc., many of which are strongly energy dependent and difficult to calculate by other methods. Yet all these phenomena depend on only a few essential parameters which represent the proper interface between theory and experiment. The determination of these parameters by theoretical or experimental means thus might be considered preferable to the calculation or measurement of the various phenomena dependent on these parameters.

α) Single-channel QDT. The essential features of the QDT are exemplified by the single-channel theory, which has been presented by HAM[239] and by SEATON.[240] The theory would apply, for example, to the photoabsorption spectrum of the outer s electron in any of the alkali atoms. It is assumed that the excited electron sees a pure Coulomb field for $r \geqq r_0$, where r_0 is roughly the ionic radius. The energy ε of the excited electron is measured relative to the ionization threshold in terms of the parameter ν, where

$$\varepsilon = -0.5\,\nu^{-2}. \tag{20.1}$$

Below threshold ν is real and represents an effective quantum number; above threshold ν is imaginary, i.e., $\nu = \mathrm{i}/k$, where k is the photoelectron's momentum, and thus ε in (20.1) simply gives the photoelectron's kinetic energy, $k^2/2$.

The radial Schrödinger equation for $r \geqq r_0$ has two solutions for each value of orbital angular momentum l. We may specify a particular pair of solutions by requiring them to have the following asymptotic forms for photo-electron energies above threshold:[241, 242]

$$\mathscr{f}(\nu, r) \xrightarrow[r\to\infty]{} \left(\frac{2}{\pi k}\right)^{1/2} \sin(kr+\theta), \tag{20.2a}$$

$$\mathscr{g}(\nu, r) \xrightarrow[r\to\infty]{} -\left(\frac{2}{\pi k}\right)^{1/2} \cos(kr+\theta), \tag{20.2b}$$

---

[239] Ham, F.S. (1955): Solid State Phys. *1*, 127
[240] Seaton, M.J. (1958): Mon. Not. R. Astron. Soc. *118*, 504
[241] Eissner, W., Nussbaumer, H., Saraph, H.E., Seaton, M.J. (1969): J. Phys. B *2*, 341
[242] Fano, U. (1970): Phys. Rev. A *2*, 353; (1977): ibid., *15*, 817

where

$$\theta = -\tfrac{1}{2}l\pi + \frac{1}{k}\ln(2kr) + \arg\Gamma(l+1-\mathrm{i}/k). \tag{20.2c}$$

For energies below threshold these functions have the following asymptotic behavior:[241,242]

$$f(v,r) \xrightarrow[r\to\infty]{} u(v,r)\sin\pi v - v(v,r)\cos\pi v, \tag{20.3a}$$

$$g(v,r) \xrightarrow[r\to\infty]{} -u(v,r)\cos\pi v - v(v,r)\sin\pi v, \tag{20.3b}$$

where $u(v,r)$ and $v(v,r)$ are, respectively, exponentially increasing and decreasing functions of r. Near the origin, f is regular and g is irregular with the following dependence on r:

$$f(v,r) \propto r^{l+1} \qquad (r\to 0), \tag{20.4a}$$

$$g(v,r) \propto r^{-l} \qquad (r\to 0). \tag{20.4b}$$

Note that in general the coefficients of proportionality in (20.4) are energy dependent. However, near threshold (i.e., for $\varepsilon \simeq 0$) these proportionality coefficients become nearly independent of energy thus indicating that in the energy region of the Rydberg levels just below threshold as well as in the energy region just above threshold the pair of Coulomb functions are nearly independent of energy for small r.

A general solution of the radial Schrödinger equation for $r \geqq r_0$ is a linear combination of $f(v,r)$ and $g(v,r)$ with coefficients to be determined by application of boundary conditions at infinity and at r_0. This general solution may be written

$$\psi(v,r) = N_v\{f(v,r)\cos\delta - g(v,r)\sin\delta\} \qquad \text{for } r \geqq r_0. \tag{20.5}$$

In (20.5) N_v is a normalization factor which is determined by the behavior of $\psi(v,r)$ at large r and hence is sensitively dependent on energy. On the other hand, δ is the relative phase with which the regular and irregular Coulomb functions are superposed at $r = r_0$. Its value is determined by the behavior of the wave function for values of $r < r_0$. Owing to the deep potential seen by the photoelectron for $r < r_0$, the behavior of the wave function in this core region is quite insensitive to the photoelectron's kinetic energy, apart from the overall normalization factor. Hence δ is also expected to be insensitive to the kinetic energy, especially near threshold where f and g are nearly independent of energy for small r.

Table 2 gives an explicit demonstration of this insensitivity. Radial wave functions for the first ten p orbitals of uranium were calculated in a central potential and the positions of the first maximum, first node, and first minimum of each wave function are tabulated. These positions are seen to coincide for orbitals $6p$ through $11p$, which span an energy range of 0.95 a.u. Furthermore, the ratio of the amplitudes of each wave function at its first maximum and first minimum is seen to be the same for the orbitals $7p$ to $11p$. In taking this ratio the normalization factor [cf. (20.5)] cancels, showing the insensitivity of the

Table 2. Behavior of uranium np radial wave functions[a] $r^{-1}P_{np}(r)$ for small radial distances r

Orbital	Energy (a.u.)[a]	r_{max} [b]	r_{node} [b]	r_{min} [b]	$P(r_{max}) \div P(r_{min})$
2p	−624.781	0.0456	–	–	–
3p	−153.292	0.0347	0.0707	0.1490	−0.593
4p	− 36.910	0.0334	0.0666	0.1218	−0.692
5p	− 7.422	0.0330	0.0657	0.1181	−0.714
6p	− 0.959	0.0329	0.0655	0.1173	−0.717
7p	− 0.104	0.0329	0.0655	0.1173	−0.718
8p	− 0.048	0.0329	0.0655	0.1173	−0.718
9p	− 0.028	0.0329	0.0655	0.1173	−0.718
10p	− 0.018	0.0329	0.0655	0.1173	−0.718
11p	− 0.013	0.0329	0.0655	0.1173	−0.718

[a] Wave functions and energies computed using the central potential of Herman, F., Skillman, S. (1963): Atomic structure calculations, Englewood Cliffs, New Jersey: Prentice-Hall

[b] r_{max}, r_{node}, and r_{min} are, respectively, the positions (in a.u.) of the first maximum, first node, and first minimum of the radial wave function $P_{np}(r)$

unnormalized wave function to changes in energy. Of course, r_0 for most applications of QDT will be of the order of several Bohr radii, instead of the much smaller range of r shown in Table 2. In practice, one usually finds that δ, which depends on the wave function for $r \leqq r_0$, is either constant or weakly dependent on energy so that its energy dependence may be approximated by a linear function of energy over an energy range of a few eV about the ionization threshold.

The parameter δ is determined either theoretically or experimentally from its relationship to other physical quantities. For energies below threshold, the wave function $\psi(v,r)$ must be finite at large r. Substituting the asymptotic forms of the Coulomb functions from (20.3) in (20.5) gives

$$\psi(v,r) \xrightarrow[r\to\infty]{} N_v\{u(v,r)\sin\pi(v+\mu) - v(v,r)\cos\pi(v+\mu)\}, \tag{20.6a}$$

where we have defined

$$\mu \equiv \delta/\pi. \tag{20.6b}$$

Requiring the coefficient of the exponentially increasing function $u(v,r)$ to be zero implies that $v+\mu=n$, where n is an integer. Substituting for μ in the expression for the energy, (20.1), one obtains

$$\varepsilon = -0.5\,v^{-2} = -0.5(n-\mu)^{-2}. \tag{20.7}$$

Thus μ is seen to be the quantum defect of spectroscopy, and it is clear that μ can be obtained by fitting (20.7) to experimentally determined energy-level data. For energies above threshold, on the other hand, substitution of (20.2) in (20.5) gives

$$\psi(k,r) \xrightarrow[r\to\infty]{} N_k\left(\frac{2}{\pi k}\right)^{1/2}\sin(kr+\theta+\delta), \tag{20.8}$$

which indicates that δ is the phase shift of the photoelectron with respect to a Coulomb wave. Hence the quantum defect μ may be obtained theoretically by calculating the phase shift δ for electron-ion scattering near threshold and using (20.6b). SEATON[240, 243] first showed this connection between discrete energy-level data (which determine μ) and the scattering phase shift δ.

A major application of the single-channel QDT is the "Coulomb approximation" for calculating photoionization cross sections. The method is based on the observation of BATES and DAMGAARD[244] that electric dipole matrix elements, using the length formula, are very insensitive to the form of either the initial- or final-orbital wave function in the non-Coulomb region $r \leqq r_0$. Hence, for discrete transitions, both initial- and final-orbital wave functions are chosen to be bounded Coulomb wave functions having quantum defects determined from experimental energies. Ad hoc forms for the wave functions may be chosen for small radii in order to insure that the proper boundary conditions are satisfied at $r=0$. BURGESS and SEATON[245] extended the method to bound-free transitions by employing continuum Coulomb wave-functions having phase shifts determined (via the quantum defect) from experimental energy-level data. A large number of photoionization cross sections have been calculated for the region from threshold to as much as 0.5 a.u. above, primarily for the alkalis and for light elements.[245, 246] The single-channel QDT gives good results for alkali-like systems but of course is too simple for more complex atoms.

Another application of the single-channel QDT serves to relate the oscillator strength for discrete transitions just below an ionization threshold to the associated photoionization cross section just above threshold. This relationship is based on the observation that near threshold, electric dipole matrix elements depend on the final electron's energy primarily through the normalization factor N_ν in the final electron's wave function [cf. (20.5)]. It may be shown[247, 248] that N_ν is proportional to $\nu^{-3/2}=(n-\mu)^{-3/2}$ for discrete (i.e., negative) electron energies and that N_ν is independent of energy for positive electron energies (assuming normalization per unit energy). This implies that multiplication of discrete dipole matrix elements by $N_\nu^{-1}=(n-\mu)^{+3/2}$ will produce renormalized matrix elements that 1) vary slowly from one discrete level to another and 2) join smoothly onto the continuum dipole matrix elements.[248] (This renormalization serves to give the discrete final-state wave functions continuum-type normalizations.)

This renormalization procedure is illustrated in Fig. 21 for H and Li discrete and continuous oscillator strengths.[248] The discrete oscillator strength for transition to the n-th Rydberg level in the single-channel case is defined as

$$f_n=\tfrac{2}{3}(E_n-E_0)\left|\langle\psi_0|\sum_{i=1}^{N} \boldsymbol{r}_i|\psi_n\rangle\right|^2; \tag{20.9}$$

[243] Seaton, M.J. (1955): C.R. Acad. Sci. Paris *240*, 1317
[244] Bates, D.R., Damgaard, A. (1949): Philos. Trans. R. Soc. A *242*, 101
[245] Burgess, A., Seaton, M.J. (1960): Mon. Not. R. Astron. Soc. *120*, 121
[246] Peach, G. (1967): Mem. R. Astron. Soc. *71*, 13; (1967): ibid, *71*, 29; (1970): ibid., *73*, 1
[247] Dehmer, J.L., Fano, U. (1970): Phys. Rev. A *2*, 304
[248] Ref. [2], § 2

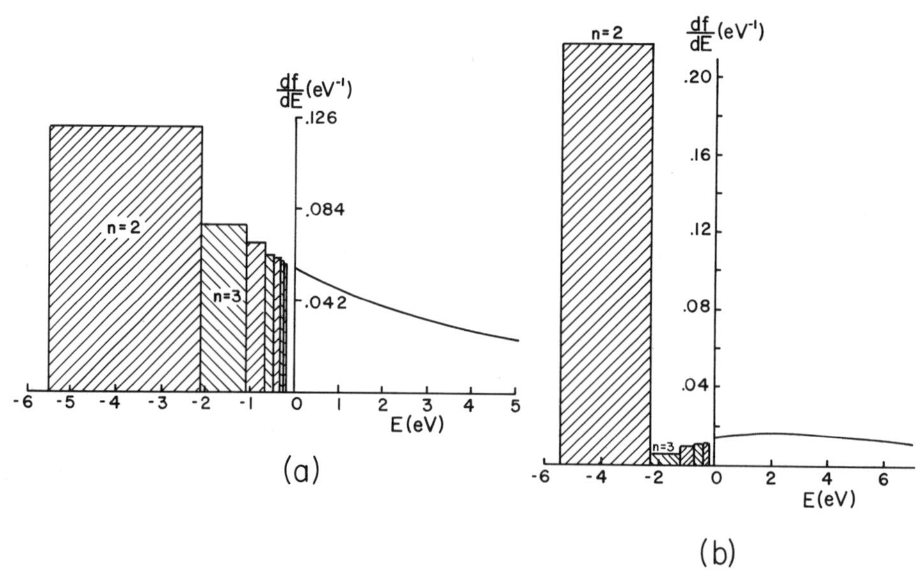

Fig. 21a, b. Oscillator strength distribution in the discrete spectra and part of the continuous spectra of (**a**) H (theory) and (**b**) Li (experiment). (From [2])

a similar definition holds for the continuum oscillator strength, df/dE, for transition to a continuum state of energy E (cf. Sect. 8). The area of each rectangle in Fig. 21 corresponds to the value of f_n, the height of each rectangle equals $(n-\mu)^3 f_n$, and the base is equal to $d\varepsilon/dn=(n-\mu)^{-3}$ [cf. (20.7)]. When plotted in this way, Fig. 21 shows that the discrete oscillator strength joins smoothly onto the continuum oscillator strength at threshold. The deviation from constancy near theshold is due to the residual weak-energy dependence of the radial dipole matrix element and, in the case of Li, of the quantum defect μ. Knowledge of discrete oscillator strengths as well as of the quantum defect thus enables one, by extrapolation, to predict the energy dependence near threshold of the photoionization cross section (which is simply related to the continuum oscillator strength).

β) Multichannel QDT. In the more typical case where the final state has many alternative photoelectron channels, the QDT represents the final-state wave function by the close-coupling expansion [cf. (16.1) for definitions]:

$$\psi_{jE}^{-}(\boldsymbol{r}_1 s_1 \ldots \boldsymbol{r}_N s_N)=\mathscr{A}\sum_i \Theta_i(\boldsymbol{r}_1 s_1 \ldots \hat{\boldsymbol{r}}_N s_N)\frac{F_{ij}(E, r_N)}{r_N}. \tag{20.10}$$

The radial wave function $F_{ij}(E,r)$ for the photoelectron in channel i is represented, for $r \geqq r_0$, by a linear combination of regular and irregular Coulomb functions [cf. (20.5)],

$$F_{ij}(E,r)=f(v_i,r)\,A_{ij}(E)-g(v_i,r)\,B_{ij}(E), \tag{20.11}$$

where each ν_i is related to the kinetic energy of the photoelectron in channel i by a formula analogous to (20.1). The object of the multichannel QDT, of course, is to obtain the coefficients A_{ij} and B_{ij} either by ab inito calculations or by semiempirical means.

SEATON and coworkers[241, 249–254] have developed methods for obtaining the following reaction matrix,*

$$K_{ij}(E) = \sum_k B_{ik}(E) A_{kj}^{-1}(E), \tag{20.12}$$

as well as other matrices simply related to K_{ij}. One method is to obtain K_{ij} semiempirically from experimental energy-level data below all ionization thresholds and then to extrapolate K_{ij} to energies above the ionization thresholds.[251] More typical, however, is the method used by DUBAU and WELLS,[253] who calculate the functions $F_{ij}(E, r)$ above all ionization thresholds by solving the close-coupling equations (cf. Sect. 16). Comparison at large r of the numerically determined wave functions $F_{ij}(E, r)$ with the corresponding analytic expressions in (20.11) allows one to obtain the matrix elements K_{ij}. These matrix elements are then extrapolated to lower energies, assuming a linear dependence on energy. Knowledge of K_{ij} at lower energies permits the extrapolation of the dipole matrix elements and hence the computation of the photoionization cross section in the autoionization region between different ionization thresholds.

The results of this method[253] as applied to photoionization of beryllium are shown in Fig. 22. Three channels were considered,

$$\begin{aligned} \mathrm{Be}\,1s^2\,2s^2(^1S_0) + \omega &\to \mathrm{Be}^+\,1s^2\,2s(^2S)\,\varepsilon p(^1P_1) \\ &\to \mathrm{Be}^+\,1s^2\,2p(^2P)\,\varepsilon d(^1P_1) \\ &\to \mathrm{Be}^+\,1s^2\,2p(^2P)\,\varepsilon s(^1P_1), \end{aligned} \tag{20.13}$$

where the first channel belongs to the lower "$2s$" threshold and the other two channels belong to the upper "$2p$" threshold. The solid line in Fig. 22 shows the close-coupling calculation, while the dashed line shows the multichannel QDT calculation obtained by extrapolation from the close-coupling results above the "$2p$" ionization threshold. The two calculations are in very close agreement in the energy region from the second ionization threshold down to the spurious "$2p\,2d$" resonance. However, due to the strong oscillations in the cross section, the close-coupling calculation must be carried out over a closely-

* Note that SEATON and coworkers write the reaction or K matrix as R_{ij}. We prefer to use K_{ij}, which is consistent with our notation in Sect. 4 and which avoids confusion with the R matrix discussed in Sect. 21. Note also that the reaction matrix in (20.12) is defined with respect to Coulomb waves, whereas the reduced reaction matrix in Sect. 4 is defined with respect to a set of basis states (cf. footnote 71).

[249] Seaton, M.J. (1966): Proc. Phys. Soc. *88*, 801; (1966): ibid, *88*, 815

[250] Bely, O. (1966): Proc. Phys. Soc. *88*, 833

[251] Moores, D.L. (1966): Proc. Phys. Soc. *88*, 843

[252] Seaton, M.J. (1970): Comments At. Mol. Phys. *2*, 37

[253] Dubau, J., Wells, J. (1973): J. Phys. B *6*, 1452

[254] Dubau, J. (1976): Electron and photon interactions with atoms, Kleinpoppen, H., McDowell, M.R.C. (eds.), New York: Plenum, p. 99

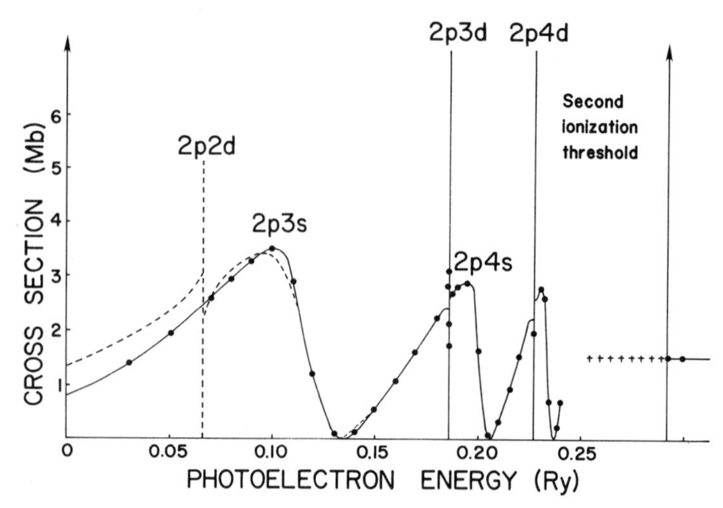

Fig. 22. Theoretical calculations of DUBAU and WELLS[253] for photoionization of beryllium. *Solid circles:* results of a close-coupling calculation, the trend of which is indicated by the solid line. *Dashed line:* resuts of a quantum defect theory extrapolation of the close-coupling results from above the "$2p$" ionization threshold. Note that the "$2p2d$" resonance is spurious and may be eliminated by alternative QDT extrapolation techniques. *Pluses:* cross section averaged over autoionizing resonances. (From [253])

spaced energy mesh, shown by the dots. The multichannel QDT result, on the other hand, is obtained simply by extrapolation of slowly varying parameters. The pluses in Fig. 22 show the cross section averaged over the autoionizing resonances. This averaged cross section is seen to join smoothly onto the photoionization cross section above the second ionization threshold, in agreement with the theorems of BAZ[255] and GAILITIS.[256]

FANO, LU, and coworkers[242, 257–264] have developed an alternative, semi-empirical procedure for determining the multichannel radial wave function in (20.11). Rather than calculating matrix elements of K_{ij}, these authors calculate the parameters characterizing the diagonal representation of K_{ij},

$$K_{ij} = \sum_{\alpha} U_{i\alpha} \tan \pi \mu_{\alpha} U^{\dagger}_{\alpha j}. \tag{20.14}$$

In (20.14) the μ_{α} represent eigenchannel quantum defects and the unitary matrix $U_{i\alpha}$ transforms from the asymptotic channels i to the eigenchannels α. The matrix $U_{i\alpha}$ is called the "frame transformation matrix" since it transforms

[255] Baz, A.I. (1959): Zh. Eksp. Teor. Fiz. *36*, 1762 [Sov. Phys. JETP *9*, 1256 (1959)]
[256] Gailitis, M. (1963): Zh. Eksp. Teor. Fiz. *44*, 1974 [Sov. Phys. JETP *17*, 1328 (1963)]
[257] Lu, K.T., Fano, U. (1970): Phys. Rev. A *2*, 81
[258] Lu, K.T. (1971): Phys. Rev. A *4*, 579
[259] Starace, A.F. (1973): J. Phys. B *6*, 76
[260] Dill, D. (1973): Phys. Rev. A *7*, 1976
[261] Lee, C.M., Lu, K.T. (1973): Phys. Rev. A *8*, 1241
[262] Lu, K.T. (1974): J. Opt. Soc. Am. *64*, 706
[263] Fano, U. (1975): J. Opt. Soc. Am. *65*, 979
[264] Starace, A.F. (1976): Photoionization and other probes of many electron interactions, Wuilleumier, F.J. (ed.), New York: Plenum, p. 395

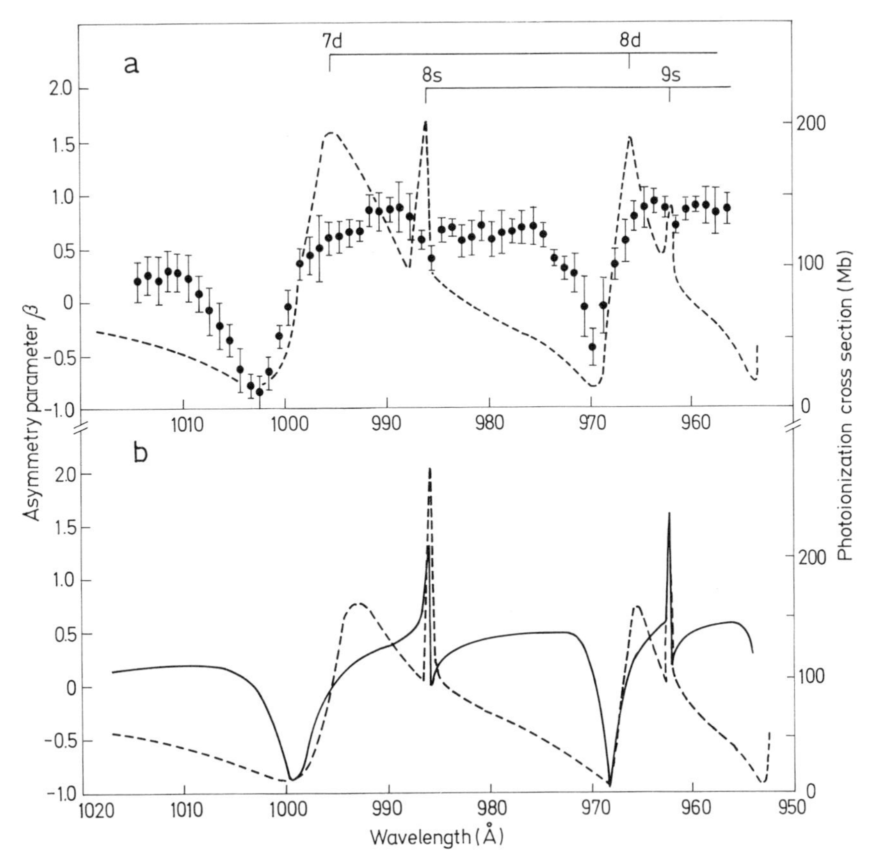

Fig. 23a, b. Wavelength dependence of the asymmetry parameter β for electrons photoionized from the $5p$ subshell of xenon in the autoionization region between the $^2P_{3/2}$ and $^2P_{1/2}$ thresholds for producing $Xe^+ 5p^5(^2P_{3/2,1/2})$. (**a**) Experimental values (Samson and Gardner[267]) for β (solid circles) and for the photoionization cross section (dashed line). (**b**) Theoretical results (Dill[260]) using multichannel QDT for β (solid line) and for the photoionization cross section (dashed line). (From [267])

the states α, which describe the ion-electron system in the strong-interaction frame appropriate for small r, to the states i, which describe the system in the weak-interaction frame appropriate for large r. The eigenchannel quantum defects μ_α are easily determined from a Lu-Fano plot[257] of experimental energy-level data as are certain combinations of matrix elements $U_{i\alpha}$. Complete determination the matrix $U_{i\alpha}$ requires fitting to additional experimental data. The closeness of $U_{i\alpha}$ to the $jj-LS$ transformation matrix, in the case of atoms, is frequently very helpful in carrying through the fitting procedure. Although initially developed for spectra converging to two ionization thresholds, the Lu-Fano plot may be used for spectra having any number of ionization thresholds. For spectra having N ionization thresholds the Lu-Fano plot will be N dimensional. The case of three thresholds has recently been treated.[265] In

[265] Armstrong, J.A., Esherick, P., Wynne, J.J. (1977): Phys. Rev. A *15*, 180

addition to determining the QDT parameters, these plots have been of great help to spectroscopists in classifying new energy-level data.[265, 266]

A striking success of the above multichannel QDT was the prediction of DILL[260] for the photoelectron angular distribution asymmetry parameter β between the $^2P_{3/2}$ and $^2P_{1/2}$ ionization thresholds of xenon (in which the ion is in the states Xe^+ $5p^5(^2P_{3/2,1/2})$). Figure 23b shows the predictions of DILL[260] for the photoionization cross section and for the β parameter in the autoionizing region between these thresholds. These predictions were made by extrapolation of the multichannel QDT parameters (which were obtained by a fitting to experimental energy-level data in the discrete part of the spectrum) to the energy region above the first ionization threshold. Figure 23a shows the experimental results of SAMSON and GARDNER,[267] which are in beautiful general agreement with the theoretical predictions. A slight discrepancy in the positions of the cross section minima near $\lambda = 1{,}000$ Å and $\lambda = 970$ Å and a more serious discrepancy in the magnitude of the photoionization cross section are probably due to residual errors in the theoretical determination of the multichannel QDT parameters.

Ab initio calculation of the multichannel QDT parameters μ_α and $U_{i\alpha}$ have been carried out by FANO and LEE,[268, 269] using an R matrix method (cf. Sect. 21). More recently, FANO[270] has shown the connection between the analytic expressions of the multichannel QDT for the final-state wave function and the corresponding expressions of continuum configuration-interaction calculations. Such ab initio methods for computing the QDT parameters are essential for studying atoms for which experimental energy-level data are sparse and for which close-coupling calculations cannot be relied upon.

21. The *R* matrix theory. The R matrix theory was originally developed by WIGNER and EISENBUD[271, 272] for application to nuclear reactions. Detailed reviews of these applications to nuclear physics have been given by LANE and THOMAS[273] and by BREIT.[274] More recently BURKE and coworkers[275–283] and

[266] Brown, C.M., Tilford, S.G., Ginter, M.L. (1975): J. Opt. Soc. Am. *65*, 385
[267] Samson, J.A.R., Gardner, J.L. (1973): Phys. Rev. Lett. *31*, 1327
[268] Fano, U., Lee, C.M. (1973): Phys. Rev. Lett. *31*, 1573
[269] Lee, C.M. (1974): Phys. Rev. A *10*, 584
[270] Fano, U. (1978): Phys. Rev. A *17*, 93
[271] Wigner, E.P. (1946): Phys. Rev. *70*, 15; (1946): ibid., *70*, 606
[272] Wigner, E.P., Eisenbud, L. (1947): Phys. Rev. *72*, 29
[273] Lane, A.M., Thomas, R.G. (1958): Rev. Mod. Phys. *30*, 257
[274] Breit, G. (1959): Handbuch der Physik *XLI/1*, 1
[275] Burke, P.G. (1973): Comments At. Mol. Phys. *4*, 157
[276] Burke, P.G. (1974): Comput. Phys. Commun. *6*, 288
[277] Burke, P.G., Robb, W.D. (1975): Adv. At. Mol. Phys. *11*, 143
[278] Burke, P.G., Taylor, K.T. (1975): J. Phys. B *8*, 2620
[279] LeDourneuf, M., Lan, V.K., Burke, P.G., Taylor, K.T. (1975): J. Phys. B *8*, 2640
[280] Taylor, K.T., Burke, P.G. (1976): J. Phys. B *9*, L353
[281] LeDourneuf, M., Lan, V.K., Hibbert, A. (1976): J. Phys. B *9*, L359
[282] Burke, P.G. (1977): Atomic physics. Vol. 5, Marrus, R., Prior, M. Shugart, H. (eds.), New York: Plenum, p. 293
[283a] Combet Farnoux, F., Lamoureux, M., Taylor, K.T. (1978): J. Phys. B *11*, 2855
[283b] Combet Farnoux, F. (1980): Phys. Rev. A *21*, 1975

FANO and LEE[268, 269] have developed R matrix theories for application to atomic reactions, in particular, to atomic photoionization.[269, 278–281, 283] Critical reviews and comparisons of various forms of the R matrix theory as used in atomic physics are given elsewhere; [277, 282, 284] here we are concerned with the essential features common to all forms of the R matrix theory.

As in the QDT, the R-matrix theory divides configuration space into two regions: an inner region, $0 \leqq r \leqq r_0$, where electron correlations are strong and difficult to treat; and an outer region, $r_0 \leqq r \leqq \infty$, where the form of the final-state wave function is either known analytically or else is easily calculable by solving the close-coupling equations. The boundary radius r_0 is chosen to be large enough so that all ionic-core wave functions are effectively zero for $r > r_0$. Hence one may assume that for $r > r_0$ the photoelectron sees a pure Coulomb field and perhaps some additional long-range potentials. In the pure Coulomb field case the form of the final-state wave function is known analytically for $r > r_0$, and in any case the final-state wave function is easily obtained numerically by solving the close-coupling equations since exchange integrals involving the photoelectron and the ionic-core orbitals are effectively zero for $r > r_0$.

The R matrix theory obtains the final-state wave function in the inner region, $0 \leqq r \leqq r_0$, by expanding it in terms of an infinite complete set of *discrete* radial functions spanning the inner region. The major approximation is the necessity of truncating the infinite summation. A major advantage of the R matrix theory is that the expansion coefficients needed to define the final-state wave function in the inner region are obtained by standard bound-state procedures. Both the R matrix theory developed by BURKE and coworkers[275–283] and the closely related eigenchannel method developed by FANO and LEE[268, 269] permit the ab initio calculation of the scattering parameters which characterize the final-state wave function in the region outside r_0. Each of these theories has been used to calculate photoionization cross sections for atoms as heavy as argon with accuracies generally comparable to those obtained by MBPT and RPA calculations.

α) Single-channel R matrix theory. The essential features of the R matrix theory may be illustrated by consideration of the single-channel case. For simplicity we assume that an electron moves in a central potential $V(r)$ having a Coulomb tail, i.e.,

$$V(r) = -1/r \qquad \text{for } r \geqq r_0. \tag{21.1}$$

If the electron has orbital angular momentum l and energy ε, its reduced radial wave function satisfies (10.3), i.e.,

$$u_\varepsilon''(r) + 2\left[\varepsilon - V(r) - \frac{l(l+1)}{2r^2}\right] u_\varepsilon(r) = 0. \tag{21.2}$$

In the external region $r \geqq r_0$, the wave function $u_\varepsilon(r)$ is known analytically (cf. Sect. 20) to have the form of (20.5), i.e.,

$$u_\varepsilon(r) = N_\varepsilon \{\mathscr{f}(\varepsilon, r) \cos\delta - \mathscr{g}(\varepsilon, r) \sin\delta\} \qquad \text{for } r \geqq r_0. \tag{21.3}$$

[284] "The R-matrix method" (1978): Electronic and atomic collisions: Invited papers and progress reports, Watel, G. (ed.), Amsterdam: North-Holland Publishing Co., pp. 201–280

Our aim is to determine $u_\varepsilon(r)$ in the region $0 \leqq r \leqq r_0$ and at the same time determine the phase shift δ.

In the internal region, $u_\varepsilon(r)$ may be written as the following expansion:

$$u_\varepsilon(r) = \sum_{\lambda=1}^{\infty} c_\lambda(\varepsilon)\, v_\lambda(r) \qquad (0 \leqq r \leqq r_0), \tag{21.4}$$

where each $v_\lambda(r)$ is an eigenfunction of the Schrödinger equation for $0 \leqq r \leqq r_0$, i.e.,

$$v''_\lambda(r) + 2\left[\varepsilon_\lambda - V(r) - \frac{l(l+1)}{2r^2}\right] v_\lambda(r) = 0 \qquad (0 \leqq r \leqq r_0), \tag{21.5}$$

and satisfies the boundary conditions

$$v_\lambda(0) = 0 \tag{21.6a}$$

and

$$\left(\frac{v'_\lambda}{v_\lambda}\right)_{r=r_0} = \alpha. \tag{21.6b}$$

Over the interval $0 \leqq r \leqq r_0$, the functions $v_\lambda(r)$ are orthonormal:

$$\int_0^{r_0} v_\lambda(r)\, v_{\lambda'}(r)\, dr = \delta_{\lambda\lambda'}. \tag{21.7}$$

The expansion coefficients $c_\lambda(\varepsilon)$ are determined as follows: multiply (21.2) from the left by $v_\lambda(r)$ and (21.5) from the left by $u_\varepsilon(r)$; integrate each equation over the interval $0 \leqq r \leqq r_0$ and subtract the results; simplify the result by application of Green's Theorem, by use of the boundary conditions in (21.6), and also by use of the boundary condition $u_\varepsilon(0) = 0$ to obtain

$$c_\lambda(\varepsilon) = \frac{v_\lambda(r_0)}{2(\varepsilon_\lambda - \varepsilon)} \{u'_\varepsilon(r_0) - \alpha u_\varepsilon(r_0)\}. \tag{21.8}$$

The wave function in the internal region $0 \leqq r \leqq r_0$ is thus given by (21.4) and (21.8) as

$$u_\varepsilon(r) = \sum_{\lambda=1}^{\infty} \frac{v_\lambda(r_0)\, v_\lambda(r)}{2(\varepsilon_\lambda - \varepsilon)} \{u'_\varepsilon(r_0) - \alpha u_\varepsilon(r_0)\}. \tag{21.9}$$

In the external region $r_0 \leqq r \leqq \infty$ the wave function is given by (21.3). At the boundary $r = r_0$ (21.9) becomes

$$u_\varepsilon(r_0) = R\, \{u'_\varepsilon(r_0) - \alpha u_\varepsilon(r_0)\}, \tag{21.10}$$

where the R matrix is defined as

$$R \equiv \sum_{\lambda=1}^{\infty} \frac{[v_\lambda(r_0)]^2}{2(\varepsilon_\lambda - \varepsilon)}, \tag{21.11}$$

and is a one-by-one matrix in this single-channel case. The phase shift δ is determined by substituting in (21.10) the values of $u_\varepsilon(r_0)$ and $u'_\varepsilon(r_0)$ obtained from (21.3). One finds that the normalization factor N_ε cancels from both sides of (21.10) and δ is given by

$$\tan\delta = \frac{+f(\varepsilon, r_0) - R(f'(\varepsilon, r_0) - \alpha f(\varepsilon, r_0))}{g(\varepsilon, r_0) - R(g'(\varepsilon, r_0) - \alpha g(\varepsilon, r_0))}. \tag{21.12}$$

Hence the wavefunction $u_\varepsilon(r)$ is now completeley determined. Note further that $\tan\delta$ is the reaction or K matrix in the one-channel case; also, for energies in the vicinity of threshold (i.e., $\varepsilon \lesssim 0$), the quantum defect of the Rydberg levels is given by $\mu = \delta/\pi$ (cf. Sect. 20). Hence the R matrix determined in the inner region $r < r_0$ permits the determination of the K matrix, the phase shift, and the quantum defect through the boundary condition at r_0 given by (21.10) or equivalently (21.12).

The main task of the R matrix theory is thus the calculation of the discrete radial functions $v_\lambda(r)$. In the simple case considered above, these are the solutions of a central potential problem. In most cases, however, the exact Hamiltonian H for $0 \leqq r \leqq r_0$ is too difficult to solve directly and instead each $v_\lambda(r)$ is expressed as a finite expansion over the first N solutions of a simpler, model Hamiltonian H_0 for the inner region. The coefficients of this expansion are then obtained by diagonalizing the exact Hamiltonian H within the $N \times N$ space of the first N solutions of H_0. In this way one obtains the first N eigenfunctions $v_\lambda(r)$. The rest of the eigenfunctions $v_\lambda(r)$ are then usually approximated by the corresponding solutions of the model Hamiltonian H_0. Suitable choices for N and for H_0 are crucial to obtaining reasonable results by the R Matrix method.

There are two major methods of handling the boundary condition (21.6b) on the logarithmic derivative of the eigenfunctions $v_\lambda(r)$ at $r = r_0$. In the usual R matrix theory, used by BURKE and coworkers,[275–283] the value α is fixed once and for all at the beginning of the calculation. Hence the $v_\lambda(r)$ are determined only once, and from then on the calculation of each $u_\varepsilon(r)$ requires only the simple determination of the expansion coefficients $c_\lambda(\varepsilon)$. The only problem with this procedure is that since α is fixed, the derivative of $u_\varepsilon(r)$ is usually discontinuous at $r = r_0$. For photoionization calculations this causes no problem since in most cases the entire contribution to the radial dipole integral comes from the interior region $0 \leqq r \leqq r_0$. However, lack of continuity of $u'_\varepsilon(r)$ may cause the R-matrix calculation to be slowly convergent as N is increased.[277]

An alternative procedure, known as the "eigenchannel method," ensures that $u'_\varepsilon(r)$ is continuous at $r = r_0$ at the cost of a greater amount of computational labor. The eigenchannel method was first developed for applicaton in nuclear physics[285, 286] and has been applied in atomic physics by FANO and LEE.[268, 269] The difference between the eigenchannel method and the R-matrix

---

[285] Danos, M., Greiner, W. (1966): Phys. Rev. *146*, 708

[286] Barrett, R.F., Biedenharn, L.C., Danos, M., Delsanto, P.P., Greiner, W., Wahsweiler, H.G. (1973): Rev. Mod. Phys. *45*, 44

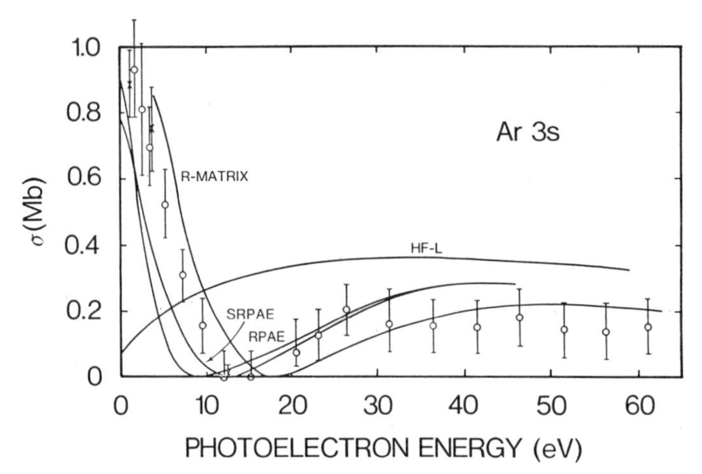

Fig. 24. Photoionization cross section for the $3s$ subshell of argon. R-MATRIX: R-matrix (length) calculation of BURKE and TAYLOR[278]; RPAE: RPA calculation of AMUSIA et al. (1972): [Phys. Letters *40A*, 361]; SRPAE: simplified RPA calculation of LIN (1974): [Phys. Rev. A *9*, 181]; HF-L: Hartree-Fock (length) calculation of KENNEDY and MANSON (1972): [Phys. Rev. A *5*, 227]; X - experimental data of SAMSON and GARDNER (1974): [Phys. Rev. Letters *33*, 671]; O - experimental data of HOULGATE et al. (1976): [J. Elec. Spec. Rel. Phen. *9*, 205]. (From HOULGATE et al., loc. cit.)

method described above is that, in the former, the logarithmic derivative α in (21.6b) is a function of energy ε. The value of $\alpha(\varepsilon)$ is determined iteratively *at each* ε to ensure that $u'_\varepsilon(r)$ is continuous at $r=r_0$. With this boundary condition the infinite summation in (21.9) reduces to a single term. That is, $u_\varepsilon(r)$ is itself the eigenstate of the exact Hamiltonian in the inner region, with eigenvalue ε. Thus in this method there is no need to compute an R-matrix, the phase shift δ being determined directly by matching logarithmic derivatives at $r=r_0$. The procedure necessary to determine $u_\varepsilon(r)$ in the inner region requires a diagonalization of the exact Hamiltonian, within the space spanned by the zero-order basis set, once per iteration. This extra labor produces an "eigenchannel" wave function $u_\varepsilon(r)$ with a continuous derivative at $r=r_0$, which may be important, for example, in photodetachment calculations where the radial dipole integral will have significant contribution from the region $r>r_0$.[282]

β) Results of multichannel R matrix theory calculations. The multichannel R matrix theory proceeds in a similar manner to the single-channel theory described above. For computing radial dipole matrix elements, the final-state radial wave functions only have to be known in the inner region $0 \leqq r \leqq r_0$. However, in order to determine the coefficients in the expansion for the radial wave functions in the inner region, these radial wave functions must be known in the outer region [cf. (21.8)]. BURKE and coworkers[275-283] obtain these radial wave functions in the outer region by solving the close-coupling equations numerically and obtain the reaction matrix K_{ij} (cf. Sect. 20) by matching the two expressions for the radial wave functions at $r=r_0$. FANO and LEE[268, 269] express the radial wave functions for $r>r_0$ analytically in terms of the parameters μ_α and $U_{i\alpha}$ characterizing the diagonal representation of the reaction matrix [cf. (20.14)]. These parameters are then determined by matching the

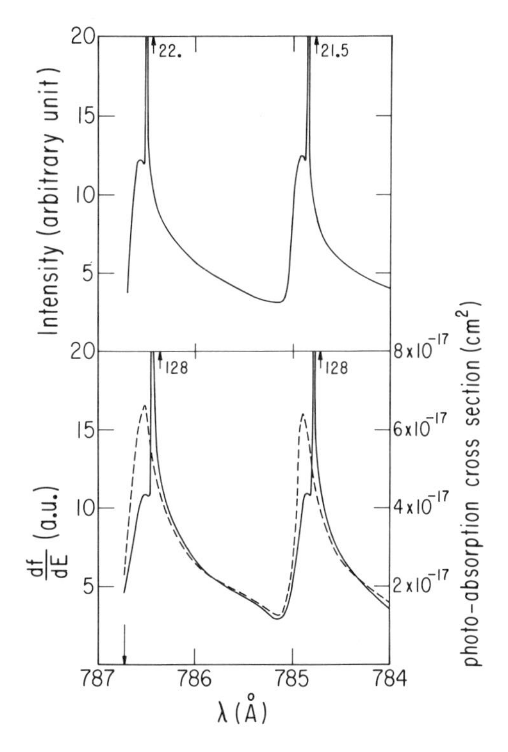

Fig. 25. Intensity profiles of autoionization lines in the region between the $^2P_{3/2}$ and $^2P_{1/2}$ ionization thresholds for producing Ar^+ $3p^5(^2P_{3/2,\,1/2})$. Upper figure: ——— high resolution, relative measurements of YOSHINO (1970): [J. Opt. Soc. Am. *60*, 1220]; Lower figure: - - - - low resolution, absolute measurements from HUDSON and KIEFFER (1971): [At. Data *2*, 205]: ——— ab initio calculation of LEE[269] using the eigenchannel method. (From [269])

analytic expression for the wave function in the outer region to the numerically determined wave function in the inner region at $r = r_0$.

The application of the ordinary R matrix theory to atomic photoionization is described in detail by BURKE and TAYLOR.[278] Both initial- and final-state wave functions are computed in the inner region by means of R matrix theory expansions over discrete eigenstates having the proper LS-coupling symmetry. Photoionization cross sections have been calculated for Ne and Ar,[278] for Al,[279] for C and O,[280] for N,[281] for K^{++},[283a] and for Ni.[283b] Figure 24 shows the photoionization cross section for the $3s$ subshell in Ar calculated by BURKE and TAYLOR[278] (cf. Fig. 19). In comparison with the RPA calculations, the R matrix calculation seems to give slightly worse agreement with experiment at threshold but significantly better agreement with experiment above the zero in the cross section.

The eigenchannel method has been applied by LEE[269] to obtain the multichannel QDT parameters for the Ar $3p$-subshell photoabsorption spectrum. Using these parameters, LEE computed the photoionization cross section for the $3p$ subshell of Ar in the energy range between the $^2P_{3/2}$ and $^2P_{1/2}$ ionization thresholds. Figure 25 shows LEE's result for the autoionization line profiles in this energy region. It is clear from the figure that there is excellent agreement with experiment despite the rapid variations in the cross section over a very small energy range.

22. Discrete basis-set methods. While the R matrix theory expresses the continuum electron's wave function as an expansion over discrete eigenstates of the Hamiltonian in the interior region $0 \leqq r \leqq r_0$, some recent theories of photoionization have simply expressed the continuum wave function as an expansion over an arbitrary discrete basis set of functions for all r. Hence these theories have no need to compute continuum wave functions. While many of these theories have been developed for molecular photoionization, for which continuum wave functions are extremely difficult to calculate, a number of calculations have been performed for a few very light atoms and ions. However, since the applications of these methods to atomic photoionization have been so limited, we restrict our discussion to a bibliographic survey.

The procedure common to most of these methods is, firstly, the calculation of one or more physical observables within a complete set of discrete functions instead of the usual complete set of bound and continuum functions and, secondly, the extraction by a suitable mathematical procedure of the photoionization cross section. These methods are somewhat reminiscent of the QDT procedure of obtaining the photoionization cross section at the ionization threshold by extrapolation of the (suitably renormalized) discrete oscillator strength (cf. Sect. 20). Thus, LANGHOFF and coworkers[287-289] and others[290] use discrete functions to calculate the energy moments of the oscillator strength distribution, from which a histogram approximation to the photoionization cross section can be obtained ("moment theory method"). BROAD and REINHARDT[291] and MCKOY and coworkers,[292-294] on the other hand, construct the complex frequency polarizability $\alpha(\omega)$ using a basis of discrete functions. The photoionization cross section is then extracted either by using the analytic properties of $\alpha(\omega)$ as a function of complex frequency ("analytic continuation method")[291, 292] or else by use of a coordinate transformation on the coordinates used to calculate the dipole matrix elements in $\alpha(\omega)$ ("coordinate rotation method").[293, 294]

Some other procedures compute dipole matrix elements directly. Thus, DALGARNO and coworkers,[295, 296] construct improved dipole matrix elements using a discrete basis representation for the standing-wave Green's function.

---

[287] Langhoff, P.W., Corcoran, C.T. (1974): J. Chem. Phys. *61*, 146

[288] Langhoff, P.W., Sims, J., Corcoran, C.T. (1974): Phys. Rev. A *10*, 829

[289] Langhoff, P.W., Corcoran, C.T., Sims, J.S., Weinhold, F., Glover, R.M. (1976): Phys. Rev. A *14*, 1042

[290] Broad, J.T., Reinhardt, W.P. (1976): Chem. Phys. Lett. *37*, 212

[291] Broad, J.T., Reinhardt, W.P. (1974): J. Chem. Phys. *60*, 2182

[292] Rescigno, T.N., McCurdy, C.W., McKoy, V. (1974): Phys. Rev. A *9*, 2409

[293] Rescigno, T.N., McKoy, V. (1975): Phys. Rev. A *12*, 522

[294] Rescigno, T.N., McCurdy, C.W., McKoy, V. (1976): J. Chem. Phys. *64*, 477

[295] Dalgarno, A., Doyle, H., Oppenheimer, M. (1972): Phys. Rev. Lett. *29*, 1051

[296] Doyle, H., Oppenheimer, M., Dalgarno, A. (1975): Phys. Rev. A *11*, 909

More recently, BROAD and REINHARDT[297] have developed procedures for solving the close-coupling equations entirely within a basis of nonorthogonal Laguerre functions ("J-matrix method").

Applications of the above methods to the calculation of total photoionization cross sections of atoms and ions have been limited to the following very light species: H, H^-, He, and Li. NESBET,[298] however, has made a detailed study of the moment-theory method of LANGHOFF and coworkers[287–289] with a view toward extending the method to the calculation of photoionization cross sections for complex atoms. His tentative conclusion, based on a calculation of the boron photoionization cross section, is that the difficulties inherent in complex spectra, due to their many channels, can be overcome.[298] Thus the discrete basis-set methods are presently in an early stage of development and cannot yet be considered on a par with the other theoretical methods for treating electron correlations that have been discussed in Chap. III. In particular these methods provide as yet no way of obtaining the photoelectron phase shifts needed to obtain partial photoionization cross sections and photoelectron angular distributions (cf. Chap. I).

IV. Specialized topics in the theory of photoionization

23. Photoelectron angular distributions. A large number of recent experimental studies[31] of photoelectron angular distributions as a function of photon energy have stimulated the theoretical study of the asymmetry parameter β (cf. Sect. 7). Knowledge of β is crucial for the proper interpretation of all experimental measurements of photoionization processes, unless the measurements are carried out at the "magic angle", $\theta = 54° 44'$, in which case (6.2) shows that the differential photoionization cross section becomes proportional to the total cross section.[299, 300] Knowledge of β is important also for the interpretation of natural phenomena, e.g., the interaction of solar radiation with the earth's upper atmosphere. For these reasons we present below a survey of extant theoretical calculations of the photoelectron angular distribution asymmetry parameter β. Furthermore, we interpret the results obtained using the general theoretical formulas presented in Sects. 7 and 9.

α) Central potential model calculations. MANSON and coworkers[13, 301–306] have carried out an extensive series of calculations for β using the Herman-

[297] Broad, J.T., Reinhardt, W.P. (1976): J. Phys. B *9*, 1491; (1976): Phys. Rev. A *14*, 2159

[298] Nesbet, R.K. (1976): Phys. Rev. A *14*, 1065

[299] Samson, J.A.R. (1969): J. Opt. Soc. Am. *59*, 356

[300] Samson, J.A.R., Gardner, J.L. (1972): J. Opt. Soc. Am. *62*, 856

[301] Manson, S.T., Cooper, J.W. (1970): Phys. Rev. A *2*, 2170

[302] Manson, S.T. (1971): Phys. Rev. Lett. *26*, 219

[303] Manson, S.T. (1973): J. Electron. Spectrosc. Relat. Phenom. *1*, 413; (1973): ibid., *2*, 206; (1973): ibid., *2*, 482

[304] Manson, S.T. (1973): Chem. Phys. Lett. *19*, 76

[305] Manson, S.T., Kennedy, D.J., Starace, A.F., Dill, D. (1974): Planet. Space Sci. *22*, 1535

[306] Reilman, R.F., Msezane, A., Manson, S.T. (1976): J. Electron. Spectrosc. Relat. Phenom. *8*, 389

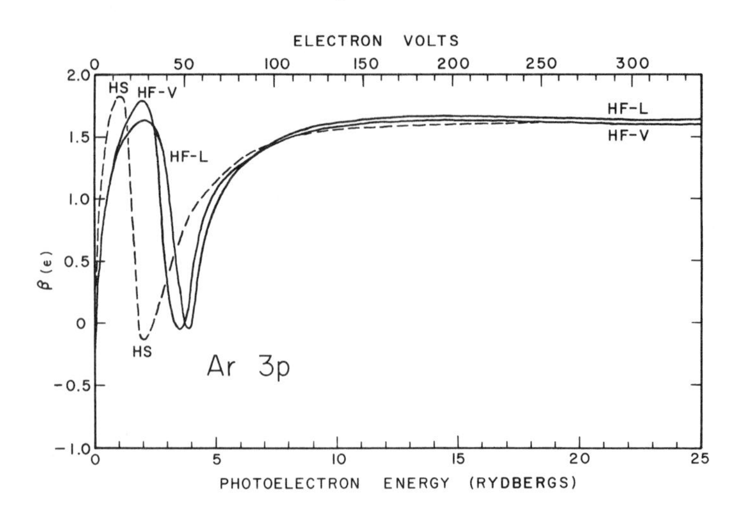

Fig. 26. Asymmetry parameter $\beta(\varepsilon)$ for electrons ejected from the $3p$ subshell of argon using both Herman-Skillman (HS) central potential model wave functions and Hartree-Fock (HF) wave functions. L and V refer to use of dipole length and velocity formulas. (From [303])

Skillman[105] central potential model wave functions. The Cooper-Zare formula for β, (9.19), is used in these calculations since in the central potential model the photoelectron's orbital wave function and phase shift are independent of the photoelectron's orbital and spin angular momentum coupling with the residual ion. Hence for each subshell nl, one associates a single asymmetry parameter β_{nl} for the photoelectrons ionized from that subshell. Calculations have been carried out for the $5d$ subshell of mercury,[304] for the $2p$ subshells of atomic oxygen, nitrogen, and carbon,[305] and for various subshells of the rare gases.[13, 301, 302] However, the most extensive work is a survey by MANSON of β_{np} for all shells of all atoms for photon energies from threshold to 18 Ry above.[303] More recently this work has been extended by tabulating β_{np} for all shells of all atoms at three commonly used x-ray photon energies.[306]

The most significant finding of these calculations is the rapid variation of β as a function of photon energy near the ionization threshold, particularly at energies corresponding to the Cooper minimum (cf. Sect. 12) in the associated photoionization cross section.[13, 301-303] Figure 26 shows β_{3p} for argon in both Herman-Skillman (HS) and Hartree-Fock (HF) approximations. As is typical, at high energies ($\simeq 7$ Ry in this case) both HS and HF approximations give nearly identical results. However, in the region of threshold, where β_{3p} varies rapidly, the two approximations give results which are shifted with respect to one another but which qualitatively have the same energy dependence. The rapid rise of β_{3p} to a maximum value of $\simeq 1.8$ just above threshold may be shown to be due to a rapid increase in the Coulomb phase-shift difference, $\sigma_d - \sigma_s$, in (9.19).[303] The subsequent drop in β_{3p} to negative values is associated with the Cooper minimum in the $3p$-subshell photoionization cross section.[301, 303] If we write (9.19) for β_{3p} in the following form,

$$\beta_{3p}(\varepsilon)=\frac{2\gamma(\varepsilon)\{\gamma(\varepsilon)-2\cos(\sigma_d+\delta_d-\sigma_s-\delta_s)\}}{1+2[\gamma(\varepsilon)]^2}, \tag{23.1a}$$

where $\gamma(\varepsilon)$ is the ratio of the radial dipole matrix elements,

$$\gamma(\varepsilon)\equiv R_{\varepsilon d}/R_{\varepsilon s}, \tag{23.1b}$$

then when $R_{\varepsilon d}$ passes through zero as ε increases, β_{3p} becomes zero also. Hence the asymmetry parameter is extremely sensitive to the exact energy location of the Cooper zero in the radial dipole matrix element.

β) More accurate calculations of β for the outer p subshell of the rare gases. A number of more accurate theoretical calculations of β have been carried out for the rare gases. Kennedy and Manson[33, 307] have performed HF calculations for many p and d subshells of the rare gases. Amusia et al.[308] calculated β for the outer p subshells of the rare gases using the RPA (cf. Sect. 18). The configuration interaction approaches to the RPA of Chang[212] and of Swanson and Armstrong[224] give results for the outer p subshells of the rare-gases that are equivalent to those of Amusia et al.[308] Substantially equivalent results for the outer p-subshell β parameters for neon and argon have also been obtained by Taylor[309] using the R matrix theory (cf. Sect. 21). The HF calculations employ the Cooper-Zare formula for β, (9.19), since the rare gases have a spherically symmetric ground state (cf. Sect. 9). The RPA calculations as well as the R matrix calculations employ more general formulas for β, equivalent to (7.6)–(7.8), since they take into account interchannel interactions and ground-state correlations that are ignored in the HF calculations. The HF results of Kennedy and Manson[33] for β_{3p} of argon are compared to the RPA results of Amusia et al.[308] and to experiment[310, 311] in Fig. 27. Near threshold and at high energies the HF and RPA calculations agree; but in the vicinity of the Cooper minimum the RPA calculation gives much better agreement with experiment.

The effect of interchannel interactions on the asymmetry parameter can be substantial. Figure 28 shows the energy dependence of β_{5p} for Xe with and without interchannel interactions between the $5p$ and $4d$ subshells.[312] The RPA calculation without interchannel interaction shows a single minimum in β_{5p} at $\simeq 6$ Ry due to the Cooper zero in the $5p\rightarrow\varepsilon d$ radial dipole matrix element. The RPA interchannel calculation for β_{5p}, however, shows two minima in β_{5p}: the first is due to the ordinary Cooper zero, which is shifted, however, to lower energy; the second is due to a minimum in the *effective* radial dipole matrix

[307] Manson, S.T., Kennedy, D.J. (1970): Chem. Phys. Lett. *7*, 387
[308] Amusia, M.Ya., Cherepkov, N.A., Chernysheva, L.V. (1972): Phys. Lett. *40* A, 15
[309] Taylor, K.T. (1977): J. Phys. B *10*, L699
[310] Dehmer, J.L., Chupka, W.A., Berkowitz, J., Jivery, W.T. (1975): Phys. Rev. A *12*, 1966
[311] Houlgate, R.G., West, J.B., Codling, K., Marr, G.V. (1976): J. Electron. Spectrosc. Rel. Phenom. *9*, 205
[312] Amusia, M.Ya., Ivanov, V.K. (1976): Phys. Lett. *59* A, 194

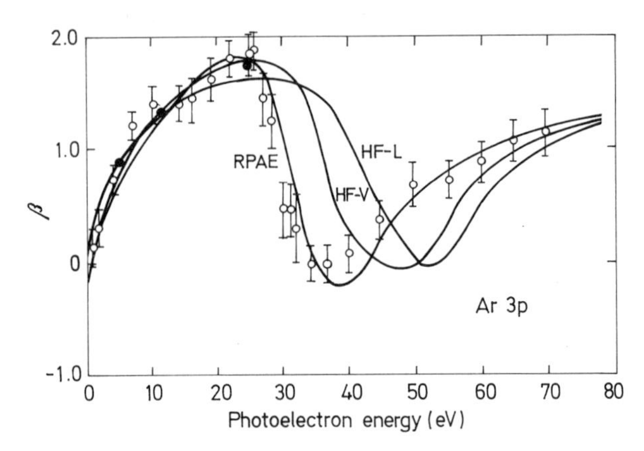

Fig. 27. Asymmetry parameter β for photoelectrons from the $3p$ subshell of argon. See text for description of curves. (From [311])

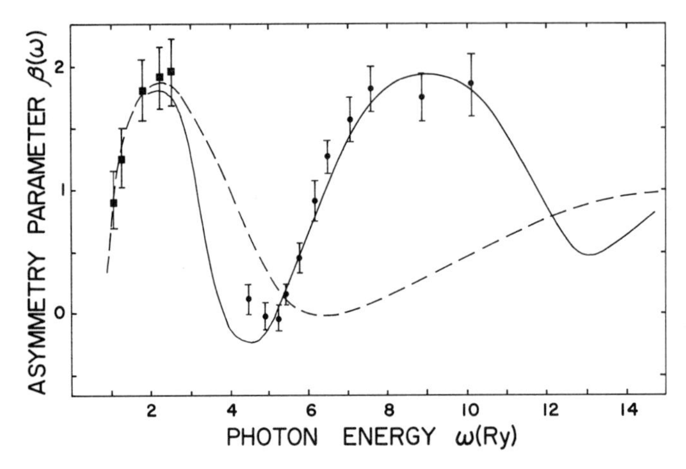

Fig. 28. Asymmetry parameter β for photoelectrons from the $5p$ subshell of xenon. Solid (dashed) line shows the RPA calculation of AMUSIA and IVANOV [312] with (without) interchannel coupling between the $5p$ and $4d$ subshells. Solid squares are the experimental results of LYNCH et al.[313]; solid circles are those of TOROP et al.[314]. (From [312])

element for the $5p \rightarrow \varepsilon d$ transition arising from cancellations between the $5p \rightarrow \varepsilon d$ and $4d \rightarrow \varepsilon f$ transitions.[312] The interchannel RPA calculation is seen to be in excellent agreement with recent experimental measurements.[313, 314]

γ) More accurate calculations of β for the outer p-subshell of open-shell atoms. For open-shell atoms, when one goes beyond the central potential approximation, the interaction between the photoelectron and the residual ion is dependent on the ionic term level as well as on the orbital and spin angular momentum coupling of the ion-electron system. These "anisotropic" (i.e., term-

[313] Lynch, M.J., Codling, K., Gardner, A.B. (1973): Phys. Lett. *43* A, 213
[314] Torop, L., Morton, J., West, J.B. (1976): J. Phys. B *9*, 2035

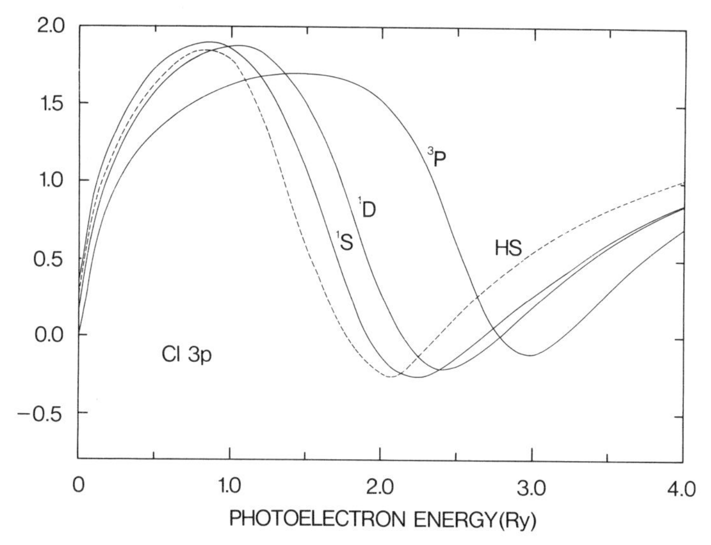

Fig. 29. Energy dependence of the asymmetry parameters $\beta(^2P \to {}^3P, {}^1D, {}^1S)$ for the photoionization reactions $\mathrm{Cl}\,3p^5(^2P) + \omega \to \mathrm{Cl}^+\,3p^4(^3P, {}^1D, {}^1S) + e^-$. The length formula was used with HF wavefunctions calculated in a $V^{N-1}(LS)$ potential. HS denotes the β parameter calculated using the term-independent Herman-Skillman central potential model wave functions and phase shifts

dependent) interactions have significant effects on the photoelectron angular distribution.[103, 104] In particular, one can no longer associate a single asymmetry parameter β_{nl} with a given subshell but must associate as many asymmetry parameters $\beta_{nl}(L_\gamma S_\gamma)$ as there are ionic-core term levels $L_\gamma S_\gamma$, where $\beta_{nl}(L_\gamma S_\gamma)$ is given by (7.6)–(7.8). Calculations have been performed for $2p$-subshell ionization of atomic oxygen using HF[315] and close-coupling[316] wave functions. However, the anisotropic interactions are so weak in light elements (i.e., elements with $Z \leqq 10$) that the more accurate calculations nearly coincide with the HS central potential model calculations except in the region from threshold to $\simeq 0.5$ Ry above.[315]

The angular distributions of photoelectrons ionized from the $3p$ subshells of sulfur[103, 104] and of chlorine[317] show much stronger effects of anisotropic interactions. Figure 29 shows the three asymmetry parameters β for photoionization of atomic chlorine (corresponding to the three ionic term levels ^3P, ^1D, ^1S of $\mathrm{Cl}^+\,3p^4$) calculated in the approximation of vanishing interchannel interactions using HF wave functions [cf. Sect. 9, especially (9.14)]. For comparison, β_{3p} computed in the central potential approximation [cf. Sect. 9, especially (9.19)] using HS wave functions is shown also. One sees that there are large differences among the three asymmetry parameters corresponding to the three photoelectron energy groups produced at a given photon energy, particularly in the vicinity of the Cooper minimum. An intrachannel RPA calculation[21a] for

[315] Starace, A.F., Manson, S.T., Kennedy, D.J. (1974): Phys. Rev. A *9*, 2453
[316] Smith, E.R. (1976): Phys. Rev. A *13*, 1058
[317] Manson, S.T., Starace, A.F., Dill, D. (1974): Bull. Am. Phys. Soc. *19*, 1203

atomic chlorine resulted in β parameters only slightly changed from those in Fig. 29.

δ) Photoelectron angular distributions of s electrons. One of the most striking predictions of the Cooper-Zare formula for β, (9.19), is that photoionization of an electron having orbital angular momentum $l=0$ leads to a pure $\cos^2\theta$ photoelectron angular distribution, i.e., $\beta=2$, irrespective of the photon's energy. Classically, this result is intuitively obvious: since the initial state of the electron is spherically symmetric, the photoelectron angular distribution is centered about the electric vector of the incident light. Quantum mechanically, neglecting retardation and relativistic effects, this result holds exactly for photoionization of atomic hydrogen.[40] For more complicated atoms, this result follows from the central potential model approximation [in which β is given by (9.19)] since in the final state there is only a *single* outgoing p wave. In general, however, one expects that whenever there are *multiple* final-state channels, β will be a function of photon energy and will vary between -1 and $+2$.

Indeed, STARACE et al.[318] have shown that even in a nonrelativistic LS-coupling approximation, β may be *expected* to differ from 2 for photoionization of $l=0$ electrons in *open-shell atoms.* Consider photoionization of the $3s$ subshell in atomic chlorine,

$$\mathrm{Cl}\,3s^2 3p^5(^2P)+\omega \rightarrow \mathrm{Cl}^+\,3s3p^5(^{1,3}P)\,\varepsilon p(^2D, {}^2P, {}^2S). \tag{23.2}$$

One sees that for each ionic term there are three final-state channels. Thus (7.6)–(7.8) show that β will in general be a function of photon energy due to interferences between these three final-state channels.[318] Figure 30 shows the calculated HF results for $\beta(^1P)$ and $\beta(^3P)$ which indicate a large deviation of β from 2 near threshold. (Note that in these calculations (9.14) is substituted for (7.6) since interchannel interactions are ignored; for this reason Fig. 30 does not indicate any resonances in the asymmetry parameter due to autoionizing levels.) It is instructive to compare these results in Cl with the corresponding nonrelativistic results for photoionization of the $3s$ subshell in Ar,

$$\mathrm{Ar}\,3s^2 3p^6(^1S^e)+\omega \rightarrow \mathrm{Ar}^+\,3s3p^6(^2S^e)\,\varepsilon p(^1P^o). \tag{23.3}$$

Comparing (23.3) with the general photoionization reaction in (7.1), we see that $L_0=L_\gamma=0$, $L=1$, and $\pi_0=\pi_\gamma=+1$ so that according to (7.2), the only allowed value of the angular momentum transfer is $j_t=0$. Furthermore, (7.5) indicates that $j_t=0$ is a parity-favored transition. Hence (7.7) and (7.8a) imply that

$$\beta_{3s}^{\mathrm{Ar}}=+2. \tag{23.4}$$

Thus since, nonrelativistically, only a single final-state channel is allowed in reaction (23.3), the asymmetry parameter for the $3s$ subshell of Ar is a constant, independent of photon energy due to the lack of interference between

[318] Starace, A.F., Rast, R.H., Manson, S.T. (1977): Phys. Rev. Letters *38*, 1522

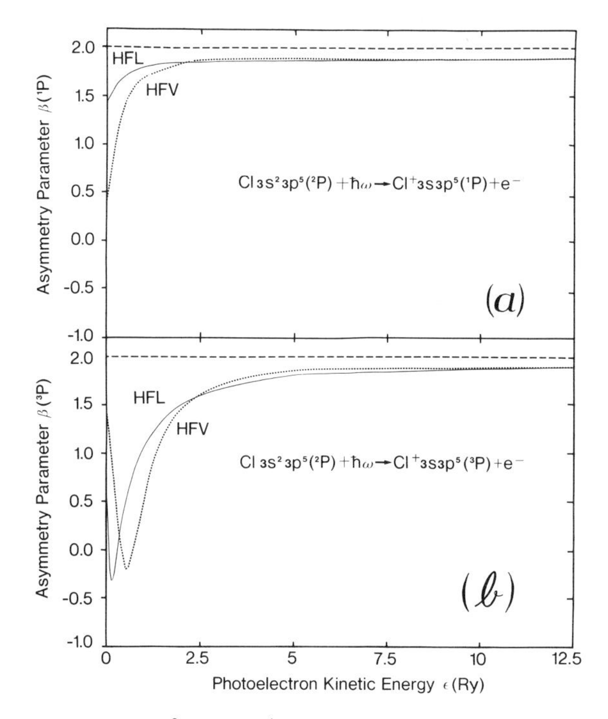

Fig. 30. Asymmetry parameters $\beta(^3P)$ and $\beta(^1P)$ for photoelectrons ionized from the $3s$ subshell of chlorine. Solid (dotted) lines represent results using the length (velocity) formula and HF wave functions calculated in the $V^{N-1}(LS)$ potential. Dashed lines indicate the value $\beta=2$ that obtains in central potential model calculations. (From [318])

competing final-state channels. These two results for Cl$(Z=17)$ and Ar$(Z=18)$ demonstrate the significance of an open shell in an atom even though the subshell being ionized (i.e., $3s^2$) is closed: the open shell permits several final-state channels, the nonrelativistic anisotropic interactions of which with the residual ion result in significant departures of β from the constant value 2.

Recently, Chang and Taylor[319] have shown that for open-shell atoms anisotropic electron-ion interactions may produce classically forbidden photoelectron angular distributions even when only a single final-state channel is allowed. They calculated the angular distribution of photoelectrons ionized from the $2s$ subshell of the ^3P ground state of carbon. Consider the reaction which leaves the ion in the ^2S term level:

$$\mathrm{C}\,2s^2 2p^2\,(^3P^e)+\omega \to \mathrm{C}^+\,2s2p^2(^2S^e)\,\varepsilon p(^3P^o). \tag{23.5}$$

Reaction (23.5) shows that only a single final-state channel is allowed nonrelativistically. Comparing (23.5) with the general photoionization reaction in (7.1), we see that $L_0=L=1$, $L_\gamma=0$, and $\pi_0=\pi_\gamma=+1$ so that according to (7.2), the only allowed value of the angular momentum transfer is $j_t=1$. Further-

[319] Chang, E.S., Taylor, K.T. (1978): J. Phys. B *11*, L 507

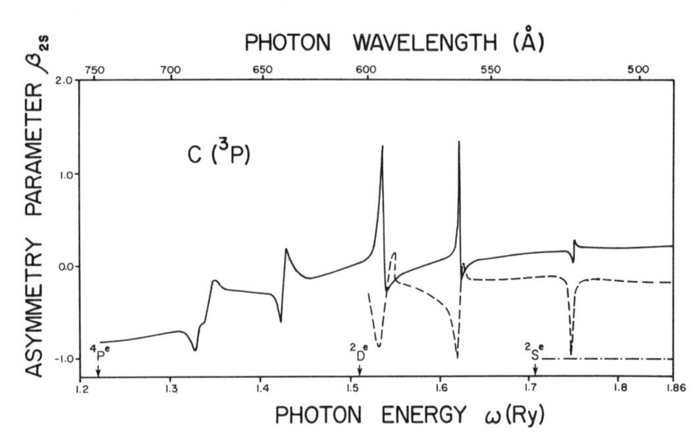

Fig. 31. Energy dependence of the asymmetry parameters $\beta(^3P \rightarrow ^4P, ^2D, ^2S)$ for the photoionization reactions $C\,2s^2\,2p^2(^3P) + \omega \rightarrow C^+\,2s\,2p^2(^4P, ^2D, ^2S) + e^-$ calculated using the R-matrix method. Solid line indicates $\beta(^4P)$, dashed line $\beta(^2D)$, and dot-dash line $\beta(^2S)$. Ionization thresholds are indicated by the arrows. (From [319])

more, (7.5) indicates that $j_t = 1$ is a parity-unfavored transition. Hence (7.7) and (7.8b) imply that the asymmetry parameter β for reaction (23.5) is -1, independent of photon energy, i.e., the angular distribution is $\sin^2\theta$ at all energies. The parity-unfavored transition is made possible by the anisotropic interaction of the photoelectron with the $2p$ subshell during its escape from the atom: initially the two $2p$ electrons are coupled to ^3P, but in the ion they are coupled to ^1S. Such interactions are ignored in the central potential approximation. The asymmetry parameter for reaction (23.5) as well as those for transitions to the ^4P and ^2D states of the ion are shown in Fig. 31. The calculations were carried out using the R matrix method.[319] Resonances in the asymmetry parameters for the ^4P and ^2D ionic states occur due to autoionizing resonances converging to higher ionic thresholds. Note also that the asymmetry parameters for the ^4P and ^2D states of the ion are energy dependent due to the existence of multiple final-state channels, as for the Cl example discussed above.

The experimental evidence on the angular distribution of photoelectrons having initial angular momentum $l = 0$ is sparse, and that which exists is for closed-shell atoms only. Theoretically, in the nonrelativistic approximation, an s subshell of a closed-shell atom should have an asymmetry parameter $\beta = 2$ since there is only a single final-state channel and since the transition is parity favored (cf. the analysis for the $3s$ subshell of argon above). In fact, the value $\beta = 2$ for the $1s$ subshell of He is consistent with recent experimental measurements.[310,320] Deviations from the value $\beta = 2$, however, have been found experimentally for the $6s$ subshell in mercury[321] and the $5s$ subshell in xenon.[322,323] Such results can only be explained theoretically by introducing

[320] Watson, W.S., Stewart, D.T. (1974): J. Phys. B *7*, L466

[321] Niehaus, A., Ruf, M.W. (1972): Z. Phys. *252*, 84

[322] Dehmer, J.L., Dill, D. (1976): Phys. Rev. Lett. *37*, 1049

[323] White, M.G., Southworth, S.H., Kobrin, P., Poliakoff, E.D., Rosenberg, R.A., Shirley, D.A. (1979): Phys. Rev. Lett. *43*, 1661

relativistic interactions, which serve to make the dynamics of the photoionization process dependent either upon the coupling of the photoelectron's spin with the spin of the ion (in an LS-coupling scheme) or upon the coupling of the photoelectron's spin with its own orbital angular momentum (in a jj-coupling scheme). Thus, for example, the experimental observations in mercury[321] and in xenon[322, 323] may be understood (in LS-coupling) as due to the interference between the allowed $^1P_1$ final-state channel and the spin-orbit-populated $^3P_1$ final-state channel, as suggested by DILL.[260, 322] Significantly, the experimental measurements in mercury[321] are in an autoionizing region, while the experimental measurements in xenon[322, 323] are near a cross-section minimum; the enhancement of spin-orbit effects in narrow energy regions (i.e., near resonances or near cross-section minima) is well known.[324] Although the relativistic treatment of atomic photoionization is beyond the scope of this article, we indicate below how relativistic (mainly spin-orbit) interactions influence the angular distribution of s-subshell electrons in closed-shell atoms as well as the angular distribution of the outer s electron in the alkalis.

Consider the photoionization of the $5s$ electron in xenon, i.e.,

$$\mathrm{Xe}\, 5s^2 5p^6\, (^1S_0) + \omega \xrightarrow[LS\text{-coupling}]{} \mathrm{Xe}^+\, 5s5p^6 (^2S)\, \varepsilon p(^{1,3}P_1) \tag{23.6a}$$

$$\xrightarrow[jj\text{-coupling}]{} \mathrm{Xe}^+\, 5s5p^6 (^2S_{1/2})\, \varepsilon p_{3/2,\,1/2} (J=1). \tag{23.6b}$$

The possible final states are indicated on the right of (23.6) in both LS coupling and jj coupling. In actuality, the true final state can only be described as having some intermediate coupling, but the essential feature is that in *any* coupling scheme there are *two* final-state channels. If one designates $A(^1P)$ and $A(^3P)$ as the amplitudes for the transitions in (23.6a) and $A(3/2)$ and $A(1/2)$ as the amplitudes for the transitions in (23.6b), then the asymmetry parameters for reactions (23.6a) and (23.6b) are given by[238a]

$$\beta_{5s} \underset{LS\text{-coupling}}{=} 2 - \frac{3|A(^3P)|^2}{[|A(^3P)|^2 + |A(^1P)|^2]} \tag{23.7a}$$

$$\beta_{5s} \underset{jj\text{-coupling}}{=} 2 - \frac{2|A(1/2) - A(3/2|^2}{[|A(1/2)|^2 + 2A(3/2)|^2]}. \tag{23.7b}$$

Usually, $A(^3P) \ll A(^1P)$ and $A(1/2) \simeq A(3/2)$ so that $\beta_{5s} \simeq 2$. In xenon, however, the $5s$-subshell photoionization cross section has a minimum just above threshold (cf. Fig. 20). Near this minimum the singlet amplitude is so small that it is comparable in magnitude to the triplet amplitude according to calculations of JOHNSON and CHENG[238a] using a relativistic RPA. The results of JOHNSON and CHENG[238a] for β_{5s} are shown in Fig. 32 in two approximations: one which includes interchannel interactions between the $5s$ and $5p$ subshell channels and one which includes interchannel interactions between the $4d$, $5s$, and $5p$ subshell channels. The dashed curve represents a single channel Dirac-

---

[324] Fano, U. (1970): Comments At. Mol. Phys. *2*, 30

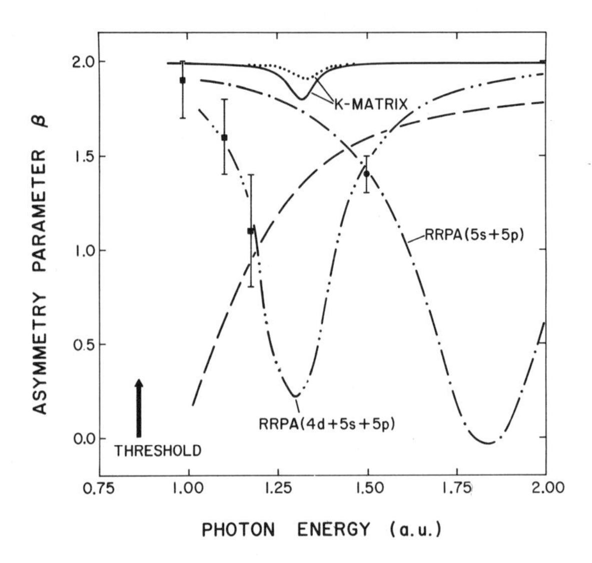

Fig. 32. Photoelectron angular distribution asymmetry parameter β for the $5s$ subshell in Xe: RRPA, relativistic RPA calculations of JOHNSON and CHENG[238a] including interchannel correlations between the $5s+5p$ and the $4d+5s+5p$ subshells; ---, Dirac-Fock calculation of ONG and MANSON;[325] K-matrix: calculations of HUANG and STARACE[327] including final-state spin orbit and $5s+5p$ interchannel correlations in dipole length (········) and velocity (——) approximation; ■, experimental results of WHITE et al.;[323] ●, experimental result of DEHMER and DILL.[322] (From [327])

Fock calculation of ONG and MANSON.[325] Results similar to ONG and MANSON's[325] had been obtained earlier by WALKER and WABER[326] in the Dirac-Slater approximation. All of these fully relativistic calculations exhibit strong interference between the forbidden ^3P channel and the allowed ^1P channel causing β_{5s} to vary dramatically with energy. Not surprisingly, the largest variations in β occur near the minimum in the $5s$-subshell cross section. One sees that interchannel interactions shift the position of the minimum in β_{5s}, indicating the sensitivity of the calculations to various electron correlation effects. Curiously, each of the fully relativistic results passes close to the first experimentally measured point.[322] Only the relativistic RPA calculation including interchannel interactions among the $4d$, $5s$, and $5p$ subshells, however, reproduces the more recent experimental data at lower energies.[323] Figure 32 shows also that for photon energies above the minimum in the $5s$-subshell cross section (cf. Fig. 20), the asymmetry parameter approaches the nonrelativistic value 2. Similar relativistic RPA calculations[237b] for He and Be found no deviation of β from the value 2, indicating the weakness of spin-orbit interactions for these light elements.

The K-matrix calculation[327] shown in Fig. 32 starts from a nonrelativistic basis of HF wavefunctions and only treats spin orbit interactions in the final state (as well as interchannel interactions between the $5s$ and $5p$ subshells).

[325] Ong, W., Manson, S.T. (1978): J. Phys. B *11*, L 65

[326] Walker, T.E.H., Waber, J.T. (1974): J. Phys. B 7, 674

[327] Huang, K.-N., Starace, A.F. (1980): Phys. Rev. A *21*, 697

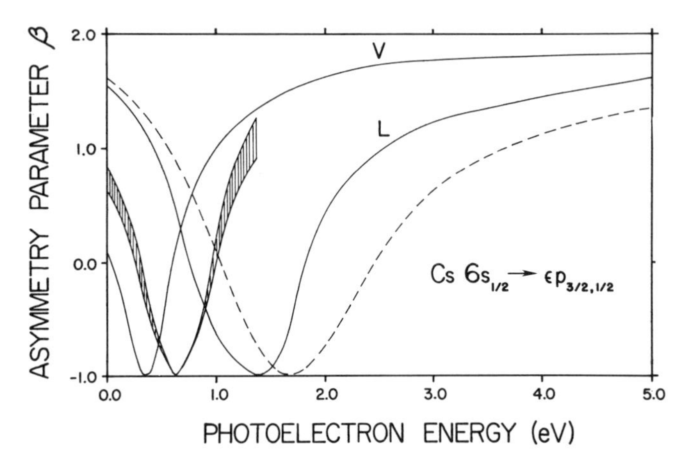

Fig. 33. Energy dependence of the asymmetry parameter for the $6s$ electron in cesium. Dashed line: Dirac-Fock calculation of ONG and MANSON[331]; hatched line: semiempirical calculation of MARR[329] based upon data of BAUM et al.[330]; solid line: dipole length (L) and velocity (V) results of HUANG and STARACE[332] including final-state spin-orbit interactions. (From[332])

The small deviation in β from the nonrelativistic value 2 that is obtained in this calculation is due to the weakness of the ^3P dipole amplitudes that are obtained. Comparison with the other calculations emphasizes the importance of using relativistic core wavefunctions when calculating forbidden dipole amplitudes.

Relativistic (mainly spin-orbit) interactions are also expected theoretically[328] to cause the asymmetry parameter for the outer s electron in the alkalis to deviate from 2. Consider the photoionization of the outer electron of cesium,

$$\mathrm{Cs}\,6s(^2S_0)+\omega\rightarrow \mathrm{Cs}^+(^1S_0)\,\varepsilon p(^2P_{3/2,\,1/2}). \tag{23.8}$$

Once again there are two final state channels. The asymmetry parameter is given by (23.7b), where $A(1/2)$ [resp., $A(3/2)$] is the amplitude for transition to the $^2P_{1/2}$ (resp. $^2P_{3/2}$) final state. As first noted by SEATON,[141] the nonzero minima in the alkali photoionization cross sections just above threshold are due to the occurrence of zeros in the amplitudes $A(1/2)$ and $A(3/2)$ at slightly *different* photon energies. The effect of these zeros on the asymmetry parameter β is to make β vary rapidly with photon energy near the cross section minimum. In particular, according to (23.7b), $\beta_{5s}=+1$ when $A(1/2)=0$, and $\beta_{5s}=0$ when $A(3/2)=0$.[328] Three recent calculations for reaction (23.8) in cesium are shown in Fig. 33. The hatched curve represents the semiempirical calculation of MARR,[329] which is based on experimental spin-polarization data.[330] The dashed curve represents the Dirac-Fock dipole-velocity calculations of ONG and MANSON.[331] The solid curves are the results of HUANG and STARACE,[332] who treat the final state spin-orbit interactions by collision theory

[328] Walker, T.E.H., Waber, J.T. (1973): Phys. Rev. Lett. *30*, 307; (1973): J. Phys. B. *6*, 1165
[329] Marr, G.V. (1974): J. Phys. B 7, L 47
[330] Baum, G., Lubell, M.S., Raith, W. (1972): Phys. Rev. A *5*, 1073
[331] Ong, W., Manson, S.T. (1978): Phys. Lett. *66* A, 17
[332] Huang, K.-N., Starace, A.F. (1979): Phys. Rev. A *19*, 2335

using the K-matrix and starting from a basis set of nonrelativistic HF wave functions. The K-matrix calculation is much more successful for β_{6s} in Cs than it was for β_{5s} in Xe. In Cs the deviation of β from the relativistic value 2 is due to interferences arising from a fine-structure splitting of an allowed dipole amplitude [cf. (23.8)]. Thus in Cs a forbidden dipole amplitude does not have to be calculated, and treatment of only final state spin orbit interactions within a nonrelativistic set of basis functions appears not to be a bad approximation. Finally, once again we see that as the photon energy increases (i.e., as one moves away from the cross-section minimum) the asymmetry parameter tends toward the nonrelativistic value 2.

ε) Behavior of the asymmetry parameter in the vicinity of a resonance. The variation with photon energy of the asymmetry parameter β depends on the variation with energy of the dipole amplitudes for each allowed final-state channel. In the vicinity of a resonance, the dipole amplitudes for the various final-state channels are affected differently by the resonant state leading to sharp variations in the β parameter. The behavior of dipole amplitudes in the vicinity of an isolated resonance has been treated in general by STARACE[157] and in the limit of vanishing interchannel interactions by KABACHNIK and SAZHINA.[156] These latter authors use their results to parameterize β in the vicinity of an isolated resonance as follows:[156]

$$\beta(\varepsilon)=\beta_0\left\{\frac{\varepsilon^2+C_1\varepsilon+C_2}{\varepsilon^2+\left(\frac{2q\sigma_a}{\sigma_a+\sigma_b}\right)\varepsilon+\left(\frac{q^2\sigma_a+\sigma_b}{\sigma_a+\sigma_b}\right)}\right\}. \tag{23.9}$$

Here β_0 is the value of the asymmetry parameter away from the resonance, ε is the reduced energy given by (15.13b) (for the single-channel case) in terms of the resonance energy and the resonance half width, C_1 and C_2 are two parameters to be determined by fitting to experimental data, and the remaining parameters are defined by fitting the experimental cross-section data to the FANO formula,[3]

$$\sigma(\varepsilon)=\sigma_a\frac{(q+\varepsilon)^2}{1+\varepsilon^2}+\sigma_b, \tag{23.10}$$

where q is the profile index [defined for the single-channel case by (15.13a)], σ_b is the minimum value of the cross section in the neighborhood of the resonance, and $(\sigma_a+\sigma_b)$ is the value of the cross section away from the resonance. Since the quadratic polynomials in ε in the numerator and denominator of (23.9) have their zeros at different values of ε (in general) one expects $\beta(\varepsilon)$ to vary rapidly as a function of ε. Formulas similar to (23.9) for the spin polarization[156] and for the partial cross sections and their branching ratios[157] in the vicinity of an isolated resonance have also been obtained.

A number of calculations of the asymmetry parameter in the vicinity of a resonance have also been carried out. DILL[260] has used the multichannel quantum defect theory (QDT) to calculate the variation of β_{5p} in xenon in the region between the $Xe^+(^2P_{3/2,\,1/2})$ thresholds. In this QDT calculation the resonances are not treated as isolated resonances but as members of Rydberg series. The results are shown in Fig. 23 and are seen to be in excellent agreement with the experimental measurements of SAMSON and GARDNER.[267] The R matrix theory has also been used to calculate asymmetry parameters for carbon[319] and for neon and argon[333] in the vicinity of resonances. The results for carbon[319] are shown in Fig. 31. There are no experimental data for β within resonances in carbon, neon, or argon.

24. Multiple photoionization. Great advances have been made recently in understanding the electron correlations responsible for double photoionization at low photon energies. It was CARLSON[334] who showed that the measured double photoionization cross sections for He, Ne, and Ar at low photon energies were several times larger than predicted by "sudden" approximation calculations. The "sudden" or "shake-off" approximation applies only at high photon energies; in particular, at energies capable of producing a deep inner-shell vacancy or vacancies.[335,336] In the "sudden" approximation the photoelectron is assumed to leave the atom very rapidly so that no relaxation of the ion occurs until after the photoelectron is gone. Relaxation of the ion then occurs, i.e., the ionic orbitals adjust to the vacancy left by the photoelectron. As the ionic charge contracts, one additional electron may be squeezed out of the singly charged ion resulting in a doubly charged ion. The probability amplitude for ejecting the second electron is calculated simply as the overlap of unrelaxed and relaxed electron radial wave functions, as discussed in detail by ÅBERG.[335,336] This picture of double photoionization is however inaccurate near the threshold for double ionization since at low photon energies one cannot neglect the strong correlations between the photoelectron and the ion during the escape process. In this section we survey theoretical calculations of double photoionization cross sections at these low photon energies where the "sudden" approximation is inadequate. We also summarize theoretical work on the threshold energy dependence of multiple photoionization cross sections.

α) Double photoionization of He. Helium is a small enough atom so that calculations based on a "best wave function" approach have achieved good agreement with experiment. BYRON and JOACHAIN[10] carried out the first sophisticated calculation. They used a partial wave expansion for the correlated initial-state wave function; the final-state wave function was a symmetrized product of uncorrelated Coulomb wave functions for an ionic charge $Z=2$. Using the velocity form of the dipole matrix element, the calculated cross section agreed with the measurements of CARLSON[334] to within experimental error. Their calculation demonstrated the importance of ground-state correlations in the double ionization process: upon replacing their correlated ground-state wave function by Hartree-Fock wave functions, the calculated

---

[333] Taylor, K.T. (1977): J. Phys. B *10*, L699

[334] Carlson, T.A. (1967): Phys. Rev. *156*, 142

[335] Åberg, T. (1969): Ann. Acad. Sci. Fenn. Ser. A VI, *308*, 1

[336] Åberg, T. (1976): Photoionization and other probes of many-electron interactions, Wuilleumier, F. (ed.), New York: Plenum Press, pp. 49–59

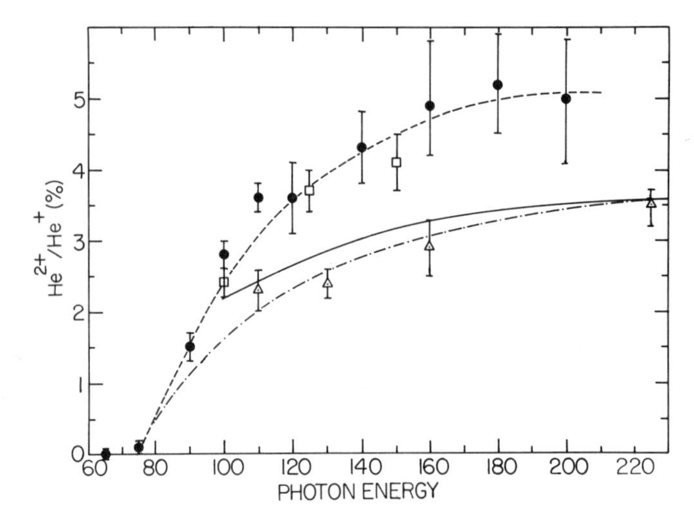

Fig. 34. Ratio of the double photoionization to single photoionization cross sections for He as a function of photon energy. *Dash-dot line:* Theoretical (dipole velocity) results of BYRON and JOACHAIN[10]; *Solid line:* Theoretical (dipole velocity) results of BROWN[337]; *Triangles:* Experimental measurements of CARLSON[334]; *Squares:* Experimental measurements of SCHMIDT et al.[338]; *Solid circles:* Experimental measurements of WIGHT and VAN DER WIEL[339]. Dashed line indicates the trend of the more recent experimental measurements

double to single ionization ratio, σ^{++}/σ^{+}, became smaller than experiment by a factor of 6! A similar "best wave function" calculation by BROWN,[337] using a six-parameter Hylleraas ground-state wave function, also gave good agreement with experiment. Both theoretical calculations, however, are 35 % lower than more recent experimental measurements,[338, 339] as shown in Fig. 34.

A number of other theoretical studies of He double ionization have been carried out which have not yet been compared with experiment. ÅBERG has used the sudden approximation to predict a constant ratio of double to single ionization of 1.66 % at very high photon energies.[340] AMUSIA et al.[341] have also investigated He double photoionization in the limit of very high photon energies using first-order perturbation theory and Coulomb orbital wave functions. Lastly, TALUKHAR and CHATTERJI[342] have used MBPT to predict the photoelectron angular distribution resulting from double ionization of He.

The helium atom is the prototype system for studying the correlations of two electrons, and double photoionization is only one kind of two electron transition process dependent on such correlations. Other two-electron processes that have been studied in He are single photoionization accompanied by

[337] Brown, R.L. (1970): Phys. Rev. A *1*, 586

[338] Schmidt, V., Sandner, N., Kuntzemüller, H., Dhez, P., Wuilleumier, F., Kallne, E. (1976): Phys. Rev. A *13*, 1748

[339] Wight, G.R., Van der Wiel, M.J. (1976): J. Phys. B *9*, 1319

[340] Åberg, T. (1970): Phys. Rev. A *2*, 1726

[341] Amusia, M.Ya., Drukarev, E.G., Gorshkov, V.G., Kazachkov, M.P. (1975): J. Phys. B *8*, 1248

[342] Talukhar, B., Chatterji, M. (1975): Phys. Rev. A *11*, 2214

excitation,

$$\mathrm{He} + \omega \to (\mathrm{He}^+)^* + \mathrm{e}^-, \tag{24.1a}$$

and double photoexcitation,

$$\mathrm{He} + \omega \to \mathrm{He}^{**}. \tag{24.1b}$$

With regard to processes of the type (24.1a), a number of authors[343-349] have calculated the cross sections for leaving the ion with an electron in either the $2s$ or $2p$ level. With regard to processes of the type (24.1b) we note the hyperspherical coordinate approach of MACEK[350] and of FANO and LIN.[351-354] In this approach the six electron coordinates $\boldsymbol{r}_1$ and $\boldsymbol{r}_2$ are replaced by R, α, $\hat{r}_1$, and $\hat{r}_2$, where

$$R = (r_1^2 + r_2^2)^{1/2}$$
$$\alpha = \arctan(r_2/r_1).$$

For low excitation energies, the Schrödinger equation in hyperspherical coordinates is found to be quasi-separable in the radial and angular coordinates implying the usefulness of an adiabatic approximation to the two-electron wave function. This separable – or adiabatic – approximation has been very successful in classifying and in understanding the physical properties of doubly excited states of He, H^-, and of heavier atoms and ions.[355, 356]

Two recent applications of the hyperspherical approach to atomic photoionization have shown its utility for describing two electron wave functions even for low energies in the continuum. The first calculation[357] is for photoionization of He,

$$\mathrm{He}(^1S) + \omega \to \mathrm{He}^+\ 1s(^2S) + \mathrm{e}^-(^1P). \tag{24.2}$$

Separable approximation hyperspherical wavefunctions are used to describe the initial ^1S ground state and the final ^1P excited state. The calculated cross section, shown in Fig. 35, agrees with experiment[358] to within 1% at thresh-

343 Dalgarno, A., Stewart, A.L. (1960): Proc. Phys. Soc. *76*, 49
344 Salpeter, E.E., Zaidi, M.H. (1962): Phys. Rev. *125*, 248
345 Brown, R.L. (1970): Phys. Rev. A *1*, 341
346 Brown, R.L., Gould, R.J. (1970): Phys. Rev. D *1*, 2252
347 Jacobs, V.L. (1971): Phys. Rev. A *3*, 289
348 Jacobs, V.L., Burke, P.G. (1972): J. Phys. B *5*, L 67
349 Hyman, H.A., Jacobs, V.L., Burke, P.G. (1972): J. Phys. B *5*, 2282
350 Macek, J. (1968): J. Phys. B *1*, 831
351 Lin, C.D. (1974): Phys. Rev. A *10*, 1986
352 Fano, U., Lin, C.D. (1975): Atomic physics, Vol. 4, zu Putlitz, G., Weber, E.W., Winnaker, A. (eds.), New York: Plenum Press, pp. 47–70
353 Fano, U. (1976): Photoionization and other probes of many-electron interactions, Wuilleumier, F.J. (ed.), New York: Plenum Press, pp. 11–30
354 Fano, U. (1976): Physics Today *29*, No. 9, pp. 32–41
355 Starace, A.F. (1982): XII International conference on the physics of electronic and atomic collisions, Invited Papers, Datz, S. (ed.), Amsterdam: North-Holland, pp. 431–446
356 Fano, U. (1981): Physica Scripta *24*, 656
357 Miller, D.L., Starace, A.F. (1980): J. Phys. B *13*, L525
358 Samson, J.A.R. (1976): Phys. Reports *28*C, 303

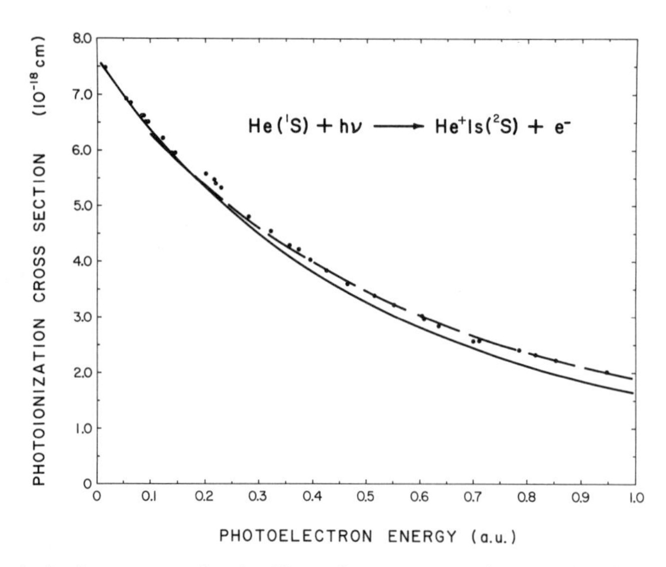

Fig. 35. Photoionization cross section for He. *Full curve:* separable approximation (single channel) hyperspherical calculation of MILLER and STARACE [357]; *Dots:* Experimental results of SAMSON [358]; *Dashed curve:* $1s-2\bar{s}-2\bar{p}$ (four channel) close-coupling calculation of JACOBS [347]. (From [357])

old, lies within the experimental error bars of $\pm 3\%$ up to 0.4 a.u. above threshold, but is systematically lower than experiment at higher energies due to the limitations of the separable approximation. Of the many other theoretical calculations, the one with the best overall agreement with experiment is also shown: The four channel ($1s-2\bar{s}-2\bar{p}$) close-coupling calculation of JACOBS.[347] In comparison with the close-coupling results, the hyperspherical results are in better agreement with experiment below $\varepsilon=0.2$ a.u. and are systematically lower above $\varepsilon=0.2$ a.u. The second calculation by GREENE[359] extends the hyperspherical coordinate approach to an atom with more than two electrons and in addition goes beyond the separable approximation. The Be photoionization cross section in the vicinity of the $2s$ and $2p$ thresholds is calculated. The outer two electrons of Be are described in hyperspherical coordinates while the effect of the inner two electrons is described by a potential. The coupling between the lowest hyperspherical adiabatic channels is taken into account. The results, shown in Fig. 36, are in reasonable agreement with the QDT and close-coupling calculations of DUBAU and WELLS[253], shown in Fig. 22.

β) Many-body perturbation theory calculations of double photoionization. The first application of MBPT to double photoionization was that of CHANG et al.,[360] who analyzed the various electron correlations contributing to double ionization of the $2p$ subshell of neon at a photon energy of 278 eV. Such an analysis is not possible when a "best wave function" approach is used, as in

[359] Greene, C.H. (1981): Phys. Rev. A *23*, 661
[360] Chang, T.N., Ishihara, T., Poe, R.T. (1971): Phys. Rev. Lett. *27*, 839

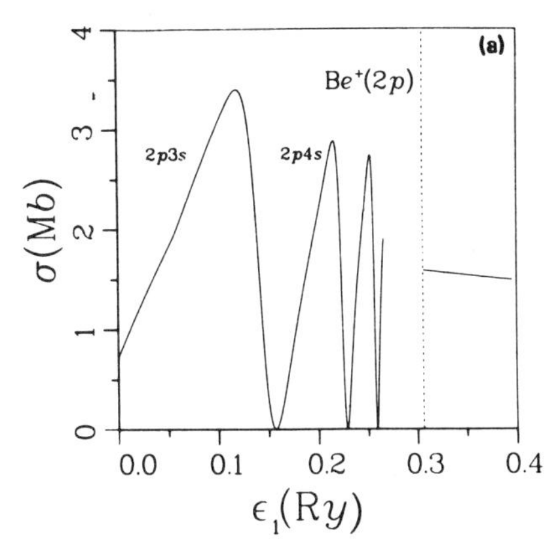

Fig. 36. Hyperspherical coordinate calculation of GREENE[359] for the Be photoionization cross section plotted vs. photoelectron energy ε_1. (From[359])

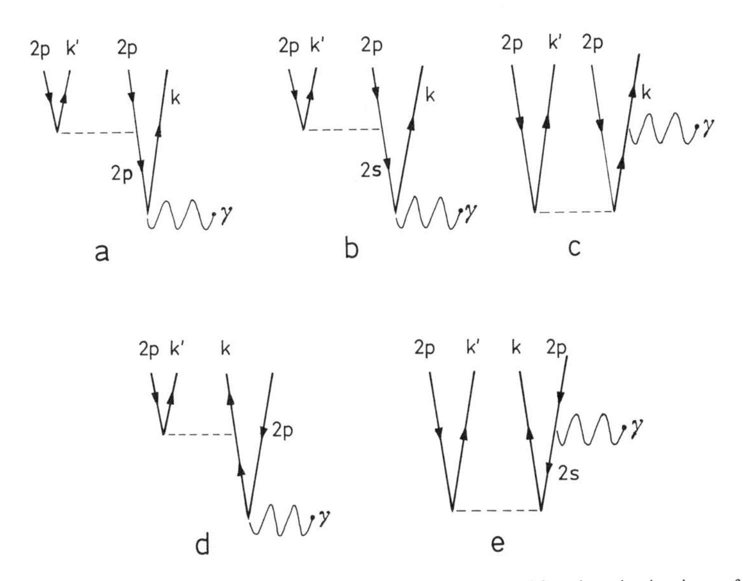

Fig. 37a–e. Lowest-order Brueckner-Goldstone diagrams for double photoionization of the $2p$-subshell of neon. (From[360])

the calculations for He described above. Figure 37 shows the lowest order MBPT diagrams in which two $2p$ vacancies are produced in neon. Figure 37a describes photoionization of a $2p$ electron to the continuum orbital k followed by a "core rearrangement" in which a second $2p$ electron is promoted to the continuum via the electron-electron electrostatic interaction. This type of electron correlation is approximately included in a shake off theory calculation. Figure 37b describes a "virtual Auger" process in which photoionization of a $2s$ electron is followed by a rearrangement in which one $2p$ electron is promoted to the continuum and another falls into the $2s$ vacancy. In a similar

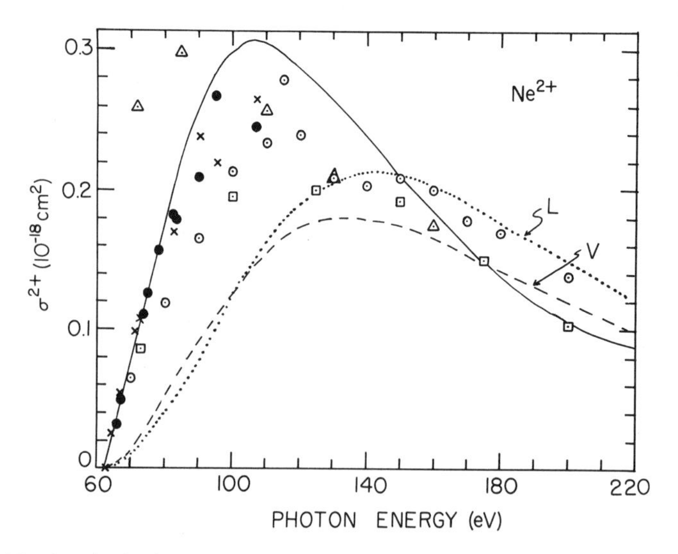

Fig. 38. Double photoionization cross section of neon as a function of photon energy. *Solid line:* theoretical calculations of CHANG and POE[34] using the dipole velocity formula; *dashed line* and *dotted line:* theoretical calculations of CARTER and KELLY[35b] using the dipole velocity (V) and dipole length (L) formulas; *triangles:* experimental measurements of CARLSON[334]; *solid circles:* experimental measurements of SAMSON and HADDAD[361]; *crosses:* experimental measurements of SAMSON and KEMENY (unpublished); *squares:* experimental measurements of SCHMIDT et al.[338]; *open circles:* experimental measurements of WIGHT and VAN DER WIEL[339]

manner, Figs. 37c–e are seen to describe the creation of two $2p$ vacancies by means of "initial-state correlation involving two electrons," "electron scattering," and "initial-state correlation involving three electrons." CHANG et al. found the double to single ionization cross-section ratio, $Ne^{++}\,2p^4$: $Ne^{+}\,2p^5$, to be $(11.1 \pm 0.4)\,\%$ at a photon energy of 278 eV. This result is in excellent agreement with CARLSON's[334] measured value of $(11 \pm 1)\,\%$. The "core rearrangement" process (Fig. 37a) was found to have contributed $\simeq 47\,\%$ of the calculated $Ne^{++}\,2p^4$ cross section; the "virtual Auger" process (Fig. 37b) $\simeq 22\,\%$; and the "two-electron ground-state correlation" process (Fig. 37c) $\simeq 31\,\%$. The processes represented by Figs. 37d and 37e were found to give negligible contribution.

More recently, CHANG and POE[34] have extended their MBPT calculation to obtain the energy dependence of the neon double photoionization cross section over the energy range from threshold to 220 eV above threshold. Excellent agreement with experiment[334, 338, 339, 361] is achieved near threshold and at high energies, but for photon energies of approximately 100 eV the calculated results are significantly higher than the experimental measurements, as shown in Fig. 38. Regarding the distribution of energy between the two ionized electrons, it is found that at low photon energies all energy distributions are nearly equally probable whereas at high photon energies it is much more probable for one electron to take nearly all the available kinetic energy

[361] Samson, J.A.R., Haddad, G.N. (1974): Phys. Rev. Lett. *33*, 875

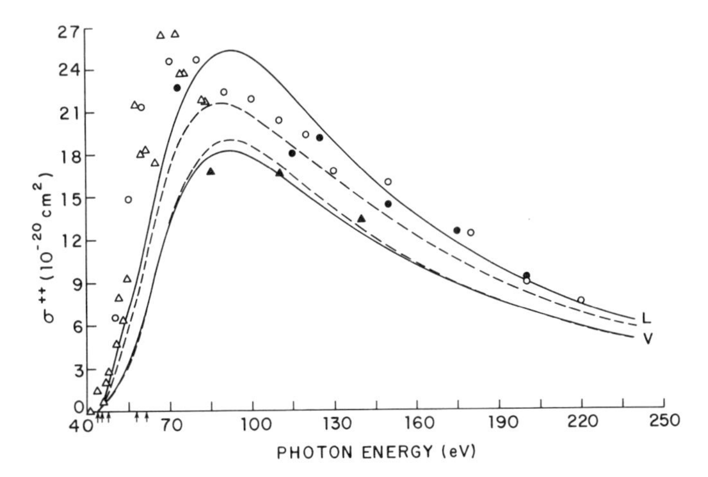

Fig. 39. Double photoionization cross section for argon. Solid lines are results obtained by including electron correlations only to lowest order. Dashed lines are results including some important second-order correlations. L and V indicate results using length and velocity formulas for the electric dipole matrix element. *Full triangles*, experimental data of CARLSON[334]; *open triangles*, SAMSON and HADDAD[361]; *full circles*, SCHMIDT et al.[338]; *open circles*, WIGHT and VAN DER WIEL[339]. The arrows mark, respectively, the experimental excitation thresholds for $Ar^{++}\,3s^2\,3p^4\,(^3P, {}^1D, {}^1S)$ and $Ar^{++}\,3s\,3p^5\,(^3P, {}^1P)$. (From [35a])

and the other to have very little. CARTER and KELLY[35b] have also carried out a MBPT calculation for double photoionization of neon. As shown in Fig. 38, their results agree very well with experimental measurements for photon energies above 120 eV. For lower photon energies, however, their results disagree with both experiment and with the calculation of CHANG and POE.[34]

The double photoionization cross section for argon over an energy range from threshold to $\simeq 200$ eV above has been calculated by CARTER and KELLY.[35a] Their calculated cross section is shown in Fig. 39, the solid lines showing the result of including all first-order correlations and the dashed lines showing the result of including also some second-order correlations. Overall, the theoretical results are in reasonable agreement with experiment. However, the calculation that includes some second-order interactions is slightly below the experimental results[334, 338, 339, 361] in the region of the cross-section maximum. This appears to indicate that other second-order and higher interactions are necessary to achieve convergence to the experimental cross section.

In addition to the theoretical studies reported thus far for the rare gases, there are calculations of the carbon[362] and beryllium[363] double photoionization cross sections. The work of CARTER and KELLY[362] on carbon, using MBPT, is the first study of double photoionization for an open-shell atom. The work of WINKLER[363] on beryllium, also using MBPT, is the first calculation to include polarization of the ion core by the photoelectrons. The results for Be are shown in Fig. 40, where the double photoionization cross section due to

[362] Carter, S.L., Kelly, H.P. (1976): J. Phys. B *9*, 1887

[363] Winkler, P. (1977): J. Phys. B *10*, L 693

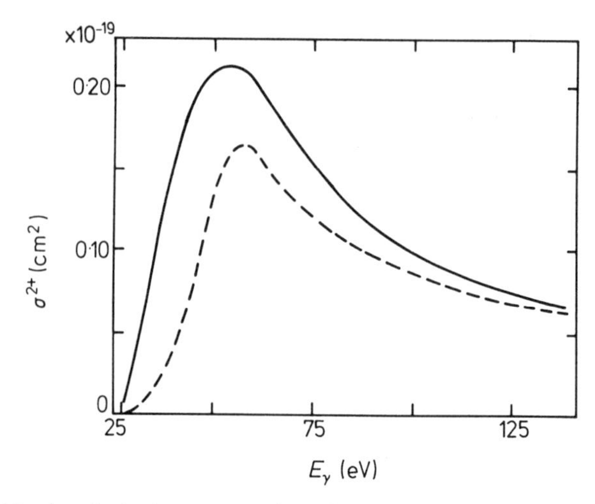

Fig. 40. Double photoionization cross section of beryllium as a function of photon energy. *Solid line:* both final-state and initial-state correlations included. *Dashed line:* only initial-state correlations included. (From [363])

initial-state correlation effects is compared with that due to both initial- and final state correlation effects. One sees that the final-state correlations are most significant near threshold but diminish in importance relative to initial state correlations at high photon energies.

All of the MBPT calculations described above, except that for Be, employ excited one-electron orbitals calculated in a V^{N-1} Hartree-Fock potential so that asymptotically the electrons see a *singly charged* ion. In the Be calculation[363] a V^{N-2} Hartree-Fock potential was used so that asymptotically the electrons would see a *doubly charged* ion. (In actuality, the electrons see a net charge between +1 and +2, depending upon their mutual screening.) A comparison of results using these two basis sets has been given by CARTER and KELLY[35b] for neon in the lowest order of perturbation theory. As might be expected, they find that the double photoionization cross section calculated using the V^{N-2} basis set is shifted to lower energies as compared with that calculated using the V^{N-1} basis set. In view of the importance that the choice of basis has in MBPT calculations for single ionization cross sections (cf. Sect. 17) it would be of great interest to see further comparisons of the results for double ionization using V^{N-1} and V^{N-2} basis sets. In infinite order either basis set would in principle give the same result.

γ) Threshold law for multiple ionization. Detailed approximate consideration of the wave function for two excited electrons having a very small positive kinetic energy enables one to predict the energy dependence near threshold of the cross section for processes in which there are two continuum electrons in the final state. The energy dependence of the cross section, in other words, is independent of the particular reaction process.[364] By such

[364] Wigner, E.P. (1948): Phys. Rev. *73*, 1002

considerations, WANNIER,[365] RAU,[366] and PETERKOP[367] showed that the threshold energy dependence of electron impact ionization cross sections for $L=0$ final states is,

$$\sigma \propto E^{1/2\mu - 1/4}, \tag{24.3a}$$

where

$$\mu = \frac{1}{2}\left(\frac{100Z-9}{4Z-1}\right)^{1/2}. \tag{24.3b}$$

For electron ionization of neutral atoms the net ionic charge in the final state is $Z=1$, so that $\sigma \propto E^{1.127}$. This energy dependence was subsequently confirmed experimentally by CVEJANOVIĆ and READ.[368]

Generalizations of the above threshold law are of interest for application to multiple photoionization. ROTH[369] has found that for $L=1$ final states the same threshold law holds. Hence for double photoionization of the rare gases, in which $L=1$ and $Z=2$ in the final state, (24.3) predicts that $\sigma \propto E^{1.056}$. The range of validity of this result is expected to be $0 \leqq E \leqq 1.3\,\text{eV}$.[369] KLAR and SCHLECHT[370] have analyzed the general case of threshold multiple ionization and present results for various final-state spin and orbital angular momenta for both double and triple ionization.

Acknowledgements. I wish to acknowledge the Alfred P. Sloan Foundation and the U.S. Department of Energy under Contract No. EY-76-S-02-2892 for their support of my research on the theory of atomic photoionization. I am most grateful to Professor Ugo Fano, to Dr. Mitio Inokuti, and to Professor Werner Mehlhorn for their critical reading of the manuscript and for their numerous suggestions for its improvement. I am also grateful to Professor James Samson for numerous discussions during the course of writing this article and to Mr. Steven Alston for verifying many of the formulas in Chap. I. Numerous other colleagues have helped by providing figures and tabular material for use in this article. Lastly, I wish to thank especially Mrs. Cathleen Oslzly for her competent and rapid typing of several versions of the manuscript.

---

[365] Wannier, G. (1953): Phys. Rev. *90*, 817
[366] Rau, A.R.P. (1971): Phys. Rev. A *4*, 207
[367] Peterkop, R. (1971): J. Phys. B *4*, 513
[368] Cvejanović, S., Read, F.H. (1974): J. Phys. B *7*, 1841
[369] Roth, T.A. (1972): Phys. Rev. A *5*, 476
[370] Klar, H., Schlecht, W. (1976): J. Phys. B *9*, 1699

Atomic Photoionization *

By

J. A. R. Samson

With 75 Figures

1. Introduction. The process of ejecting an electron from an atom or molecule was termed "electrification of a gas" prior to the discovery of the photoelectric effect in 1887. In the early 1900's it was called "photoelectric ionization of a gas". However, it soon took on its permanent name of "photoionization".

The field of photoionization is, of course, part of the more general area of study involving the question, "how does radiant energy interact with atoms and molecules?" Photoionization has a very precise onset. It starts at the photon energy necessary to eject the least tightly bound electron. In most cases this is a step function and occurs between 4.3 and 24.6 eV for all elements and molecules. With the exception of a very few atoms the photoionization thresholds lie in the vacuum ultraviolet spectral region. In recent years a wealth of experimental data have been and are still being accumulated. This has provided an incentive for theorists to explain the interaction process. It has become clear that the interaction is not simply between the incident photon and the ejected electron but includes an interaction between the ejected electron and many of the remaining bound electrons (correlation effects). In the last few years theorists have been making dramatic steps forward in the explanation and prediction of photoionization absorption processes [*1*]. Photoionization studies thus provide data to test the various theoretical approximations. However, they also provide the necessary basic data for many fields such as atomic and molecular physics, aeronomy, astrophysics, and plasma physics. Although plasma physics now includes the high-temperature devices of fusion physics, it is interesting to note that it was the complex problem of electrical breakdown in gas discharges that contributed to the early interest in photoionization (it was suspected of acting as a precurser to the spread of the general discharge). Photoionization has now been shown to be an important mechanism in sparks, in coronas, and in lightning discharges. The advent of the "space age" also has been a major stimulus to the recent advances in photoionization studies. With precise measurements of the solar flux in the

* The manuscript was completed in December 1978.

vacuum ultraviolet and knowledge of the number densities and nature of the atoms and molecules in planetary atmospheres, there has been a demand for accurate data on photoionization cross sections.

When we look at the overall photoionization process, we are struck with how much detail (specific ionization processes) that must be integrated to give a total probability for photoionization. High-lying superexcited states can be populated and these may or may not autoionize, correlation effects can cause absorption maxima (and minima) in the ionization continuum, multiple ionization and excitation can take place, and electrons can be ejected from various orbits. To study the photoionization process in detail we must look at the positive ions and ion fragments that are produced, the electrons and their angular distribution, the absorbed radiation, and the emitted fluorescence radiation. Each technique provides valuable information. Although this review is concerned primarily with atomic photoionization, the discussions in Chaps. 2 and 3, concerning the absorption of radiation and the experimental techniques used to measure photoionization cross sections, electron energies, etc., apply equally well to the study of molecules. Because of the additional absorption channels in molecules, e.g., dissociative ionization, vibrational, and rotational excitation, their absorption spectra are more complicated than those for atoms. To completely understand the specific absorption processes in a molecule it is necessary to include a study of the kinetic energies of the ion fragments and make more use of coincidence techniques between the emitted ion fragments (or electrons) and any fluorescent radiation. Also, because neutral channels compete with continuum ionization, the ionization efficiency of a molecule is not always 100%. Thus, this becomes an important measurement and requires knowledge of how to measure the absolute intensity of radiation in the vacuum uv spectral region. This technique is described in Sect. 7.3.

In the following sections, after a brief historical review, we will discuss in detail the experimental techniques available to us in studying photoionization processes and will present the results of typical atoms and discuss some of the problems still unsolved. In general, the discussion will be concerned primarily with the photoionization of atoms in the spectral range from the ionization threshold to about 100 Å.

2. Historical Review. In 1887 HERTZ[1] made the first observation of the photoelectric effect. Although this observation was of ultraviolet light interacting with a solid, nevertheless it was the first observation of the ionization of matter and the precursor to the photoelectric effect in gases, now called photoionization. It was during this period of interest in the photoelectric effect that ROENTGEN[2] in 1895 published his discovery of x rays. Almost immediately ROENTGEN observed that "x rays discharged electrified bodies, whether the electrification is positive or negative." In 1896 VILLARI[3] and J.J. THOMSON[4]

[1] Hertz, H. (1887): Wien. Ann. *31*, 983

[2] Roentgen, W.C. (1895): Sitzungsber. der Würzburger Physik-Medic. Gesellsch.; (1898): reprinted in: Ann. Phys. *64*, 1; translated by Stanton, A. (1896): Science *3*, 227 and 726

[3] Villari (1896): C. R. *123*, 418 and 446

[4] Thomson, J.J. (1896): Electrician Feb. 4

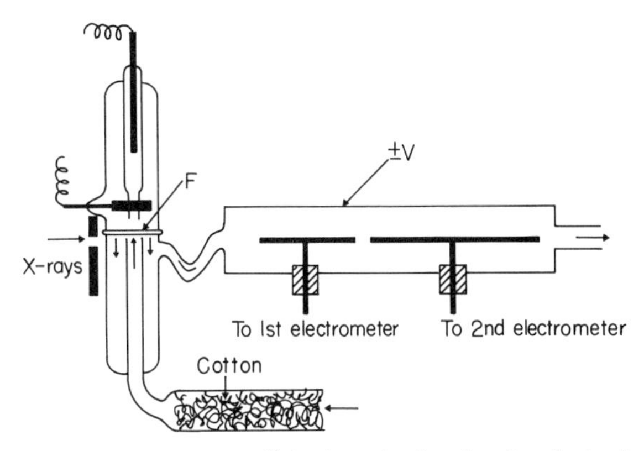

Fig. 1. Apparatus used by HUGHES[10] for investigating the photoionization of air

showed that the discharging effect depended upon the pressure and nature of the surrounding gas, and thus, the first observation of the photoionization of a gas was observed. Soon after, the ionization chamber was to replace the photographic plate as a quantitative detector of x rays. The process of ionization in a gas produced by x rays was not pursued at this time. The electron had not been discovered (THOMSON, 1897),[5] monochromatic x rays were not available (BRAGG, 1912),[6] and the concept of energy levels had not yet arrived (BOHR, 1913).[7] However, continuing interest in the photoelectric effect prompted C. T. R. WILSON[8] in 1899 to direct intense ultraviolet light into his newly developed cloud chamber. Apparent ionization was observed. This was the start of the first systematic investigation of the ionization of air by ultraviolet light.

LENARD[9] pursued this line of investigation and was able to show in many cases that the ionization observed was caused by impurities in the gas, possibly by the photoelectric effect on dust particles or water droplets. They were further plagued with scattered light striking the ion collector electrodes and thus causing an apparent ion current. By 1910 HUGHES[10] reported a "definitive" experiment for true ionization of air using ultraviolet light from a hydrogen discharge with fluorite and quartz windows. His experimental arrangement is shown in Fig. 1. Ionization occurred only with the fluorite window. Thus, the ionizing band of radiation must have occurred between 1,250 and 1,450 Å. To test whether the ionization was produced by the photoelectric effect on dust particles or truly from the air molecules Hughes measured the mobility of the ions produced, both by ultraviolet light and by x rays. The mobilities were identical. In light of our present day knowledge none of the major constituents of air can be ionized by radiation of wavelength longer

[5] Thomson, J.J. (1897): Philos. Mag. *44*, 293
[6] Bragg, W.L. (1912): Proc. Cambridge Philos. Soc. *17*, 43
[7] Bohr, N. (1913): Philos. Mag. *6*, 1, 476, and 857
[8] Wilson, C.T.R. (1899): Philos. Trans. R. Soc. *192*, 403
[9] Lenard, P. (1900): Ann. Phys. *1*, 486; (1900): *3*, 298
[10] Hughes, A.L. (1910): Proc. Cambridge Philos. Soc. *15*, 483

Table 1. Ionization thresholds for some permanent gases[a]

Gas	Ionization potential/eV	Threshold wavelength/Å
NO	9.266	1,338.0
NO_2	9.80	1,265.1
NH_3	10.19	1,216.7
O_2	12.071[b]	1,027.1
O_3	12.3	1,008.0
H_2O	12.61	983.6
CO	14.013	884.8
CO_2	13.769	900.4
H_2	15.427	803.7
N_2	15.576	796.0
Xe	12.130	1,022.1
Kr	13.999	885.6
Ar	15.759	786.7
Ne	21.564	574.9
He	24.586	504.3

[a] Franklin, J.L., Dillard, J.G., Rosenstock, H.M., Herron, J.T., Draxl, K. (1969): NSRDS-NBS 26 (Superintendent of Documents, Washington, DC 20402)
[b] Samson, J.A.R., Gardner, J.L. (1975): Can. J. Phys. *53*, 1948

than 1,050 Å! Trace impurities of NO and NO_2 could account for the observed ionization as they have thresholds at 1,338.0 Å and 1,265.1 Å, respectively. Other possibilities include the contact of highly excited molecules with the positively charged fluorite window causing the molecules to lose an electron. Radiation between 1,250 and 1,300 Å does eject photoelectrons from fluorite [2] and thus the window would be charged on the surface facing the excited molecules. However, this is only speculation. We do know that the observed ionization was not caused by a primary photoionization process in O_2, N_2, or CO_2, for example. The ionization thresholds for some of the permanent gases are listed in Table 1 to give an idea of the energy range required for direct photoionization.

It is interesting to note the comments of HUGHES and DUBRIDGE[11] many years later in 1932 that "although experiments on the photoionization of air and other permanent gases have been carried out intermittently ever since the photoelectric effect was first discovered, it must be admitted that the state of our knowledge regarding the ionization of these gases is extremely unsatisfactory." And so it was to remain until about 1950 when the techniques of vacuum uv spectroscopy had advanced sufficiently to work in the windowless region below the fluorite cutoff. Of course, ever since the publication of BOHR's theory of atomic structure (1913), it had been recognized that beyond each absorption series there would be a continuous absorption band arising from the removal of an electron from the atom with kinetic energy. However,

[11] Hughes, A.L., DuBridge, L.A. (1932): Photoelectric Phenomena. New York: McGraw-Hill Book Co. p. 273

photoionization studies continued to be concentrated in the x-ray region where they were used to determine the energy levels of atoms.

By 1923 and 1924 evidence had accumulated that Cs and K vapors were being ionized by radiation in the range 2,800–3,220 Å.[12–14] The first unambiguous results of direct ionization by photons in the ultraviolet were achieved in May, 1925 by FOOTE and MOHLER.[15] They reported on the "photoelectric ionization of caesium vapor" between 2,500 and 3,500 Å by using an ingenious experimental technique to overcome the problem of scattered light ejecting photoelectrons from the ion-collecting electrodes. The principle of the method involved thermionic emission from a tungsten filament that was space-charge limited. A platinum cylinder surrounded the filament and was held slightly positive with respect to the filament (~ 1 V). If an ion appeared by any method, it tended to neutralize the space charge and a resulting large increase in the filament current was observed. By this method relative photoionization cross sections were obtained.[16] Soon afterwards, in 1929, a simple ionization chamber was used to obtain absolute photoionization cross sections for Cs and Rb between 2,200 and 3,200 Å.[17] These were the first aboslute cross sections to be measured. Concurrent with this research, absorption spectroscopy of alkali vapors was carried out by several groups using photographic plates as detectors.[18–20] These early measurements confirmed the presence of continuous absorption of radiation at wavelengths shorter than the series limit.

During 1922 and 1923 COMPTON[21] and WILSON[22] observed that the most probable direction for the ejection of photoelectrons by x rays in a cloud chamber was in the direction at right angles to the incident radiation. BUBB[23] refined the experiments by observing the effect with polarized x rays and confirmed the suggestion that most photoelectrons were ejected in the direction of the electric vector. However, in many experiments there was a decided forward component in the direction of the incident photons. These results stimulated interest in the angular distributions of photoelectrons. During this period, theorists[24, 25] showed that for nonrelativistic electrons and polarized light the number of electrons $dN/d\Omega$ ejected per unit of solid angle Ω in the direction θ was

$$\frac{dN}{d\Omega} \propto a + b\cos^2\theta, \tag{2.1}$$

---

[12] Williamson, R.C. (1923): Phys. Rev. *21*, 107
[13] Kunz, J., Williams, E.H. (1923): Phys. Rev. *22*, 456
[14] Samuel, R. (1924): Z. Phys. *29*, 209
[15] Foote, P.D., Mohler, F.L. (1925): Phys. Rev. *26*, 195
[16] Mohler, F.L., Foote, P.D., Chenault, R.L. (1926): Phys. Rev. *27*, 37
[17] Mohler, F.L., Boeckner, C. (1929): J. Res. Nat. Bur. Stand. *3*, 303
[18] Harrison, G.R. (1924): Phys. Rev. *24*, 466
[19] Ditchburn, R.W. (1928): Proc. R. Soc. *117*, 486
[20] Trumpy, B. (1928): Z. Phys. *47*, 804
[21] Compton, A.H. (1922): Bull. Nat. Res. Council No. 20, p. 25
[22] Wilson, C.T.R. (1923): Proc. R. Soc. *104*, 1
[23] Bubb, F.W. (1924): Phys. Rev. *23*, 137
[24] Auger, P., Perrin, F. (1927): J. Phys. *8*, 93
[25] Wentzel, G. (1930): Z. Phys. *40*, 574 (1926); *41*, 828

where θ is the angle between the electric vector and the direction of the electron or, more generally (for s electrons only),

$$\frac{dN}{d\Omega} \propto a + b\left(1 + 4\frac{v}{c}\cos\alpha\right)\cos^2\theta, \tag{2.2}$$

where α is the angle between the photon direction and photoelectron direction, v is the velocity of the photoelectrons, and c is the velocity of light. The parameters a and b define the asymmetry of the distribution. For example, a represents the spherically symmetrical component of the distribution. The experiments with hard x rays verified the forward component predicted by (2.2). The cloud chamber experiments of AUGER[26] also revealed tracks caused by lower-velocity electrons ejected simultaneously with the photoelectrons, that is, double ionization. These electrons tended to have a symmetrical distribution. This period ended in 1931 with the only measurement made with ultraviolet radiation, the work of LAWRENCE[27] and CHAFFEE.[28] The lowest photon energy previously used was about 15,000 eV. They used 5 eV! In their experiment they ionized potassium vapor with polarized radiation of 2,400 Å and found the electron distribution to be proportional to $\cos^2\theta$ (i.e., $a=0$).

This decade of investigation apparently satisfied the interest in the phenomenon of angular distribution of photoelectrons for the next 35 years! There are probably many reasons for this, but certainly one reason must have been the inaccessibility of the vacuum ultraviolet spectral region (namely, 2–2,000 Å).

Between 1930 and 1950 there was very little activity in photoionization. Emphasis was primarily on photographic absorption spectra to study atomic and molecular energy levels.[29] The discovery by HOPFIELD[30] in 1930 of an emission continuum between 600 and 1,000 Å in a helium discharge allowed the observation of numerous absorption lines within the ionization continuum. The identification of the Rydberg series in many gases gave precise ionization potentials. The main exception to this line of research was the work by DITCHBURN and his group who measured the absolute cross sections of alkali vapors down to 1,600 Å[31–35]. At shorter wavelengths SCHNEIDER[36] measured the absorption cross section of air between 380 and 1,600 Å, and TERENIN[37] discovered that thallium and lead halide vapors, among others, could be photoionized at wavelengths shorter than 2,140 Å.

---

[26] Auger, P. (1925): C. R. *180*, 65; (1925): J. Phys. *6*, 205; (1926): C. R. *182*, 773

[27] Lawrence, E.O., Chaffee, M.A. (1930): Phys. Rev. *36*, 1099

[28] Chaffee, M.A. (1931): Phys. Rev. *37*, 1233

[29] For example, see Tanaka, Y., Takamine, T. (1942): Tokyo: Sci. Pap. Inst. Phys. Chem. Res. *39*, 427 and 437; (1942): Worley, R.E. (1943): Phys. Rev. *64*, 207; Price, W.C., Collins, G. (1935): Phys. Rev. *48*, 714

[30] Hopfield, J.J. (1930): Phys. Rev. *35*, 1133; (1930): *36*, 784; (1930): Astrophys. J. *72*, 133

[31] Ditchburn, R.W., Harding, J. (1936): Proc. R. Soc. A **157**, 66

[32] Ditchburn, R.W. (1937): Z. Phys. *107*, 719

[33] Ditchburn, R.W., Gilmour, J.C. (1941): Rev. Mod. Phys. *13*, 310

[34] Ditchburn, R.W., Tunstead, J., Yates, J.G. (1943): Proc. R. Soc. A *181*, 386

[35] Ditchburn, R.W., Jutsum, P.J., Marr, G.V. (1953): London: Proc. R. Soc. *219*, 89

[36] Schneider, E.G. (1940): J. Opt. Soc. Am. *30*, 128

[37] Terenin, A.N. (1930): Phys. Rev. *36*, 147; (1931): Usp. Fiz. Nauk. *11*, 276

The 1950's saw a major increase in photoionization research. WEISSLER and his group[38] were opening up the spectral range into the "windowless" region, that is, for wavelengths less than about 1,050 Å; WATANABE and colleagues[39] were initiating studies of the absolute ionization efficiencies of molecules in the region 1,050–2,000 Å; spectroscopic studies of Rydberg series and autoionizing lines were being conducted by PRICE,[40] TANAKA,[41] GARTON,[42] and others; and TERENIN and VILESSOV[43] were making thorough investigations of the photoionization of organic compounds. The introduction of photoelectron spectroscopy by VILESSOV[44] in 1961 and independently by TURNER[45] in 1962 has led to an explosion in this field. The expansion of vacuum ultraviolet radiation physics and particularly photoionization research has continued to expand. The first review article to appear on photoionization during this "modern" period was that by WEISSLER in 1956 [*3*]. Subsequently, reviews have been written by VILESSOV [*4*], SAMSON [*5*, *6*], MARR [*7*], and AMUSIA and CHEREPKOV [*8*]. FRANKLIN et al. [*9*] have compiled an extensive table of ionization and appearance potentials of gaseous ions. In photoelectron spectroscopy, books and conference proceedings are appearing almost daily! References [*10–21*] cover most of the progress through 1978. The published proceedings of the 4th International Conference on VUV Radiation Physics, 1974 [*22*], and of the NATO Advanced Study Institute, Carry-le-Rouet, 1975 [*23*], also discuss photoionization at some length. A bibliography of original reports of studies of photoionization and photoabsorption is available for the period 1921–1977 [*24*].

I. Absorption of radiation

3. General concepts

α) Bouguer-Lambert-Beer law.★ The discovery that light was absorbed exponentially was first enunciated by Pierre BOUGUER (1698–1758)[46] in his *Essai*

★ I would like to express my thanks to John H. Hubbell of the National Bureau of Standards for his kindness in obtaining copies of the original works of these authors.

[38] Weissler, G.L., Po Lee (1952): J. Opt. Soc. Am. *42*, 200; Po Lee (1955): J. Opt. Soc. Am. *45*, 703; Wainfan, N., Walker, W.C., Weissler, G.L. (1953): J. Appl. Phys. *24*, 1318; Weissler, G.L., Samson, J.A.R., Ogawa, M., Cook, G.R. (1959): J. Opt. Soc. Am. *49*, 338

[39] Watanabe, K., Marmo, F.F., Inn, E.C.Y. (1953): Phys. Rev. *91*, 1155; *90*, 155. – Watanabe, K. (1953): J. Chem. Phys. *22*, 1564; (1957): *26*, 542

[40] Price, W.C., Bralsford, R., Harris, P.V., Ridley, R.C. (1959): Mol. Spectrosc., p. 54, New York: Pergamon Press; Price, W.C. (1959): Adv. Spectrosc., Vol. 1, p. 56, New York: Interscience

[41] Tanaka, Y., Jursa, A.S., LeBlanc, F. (1957): J. Chem. Phys. *26*, 862; (1960): *32*, 1199

[42] Garton, W.R.S., Codling, K. (1960): Proc. Phys. Soc. *75*, 87; Garton, W.R.S., Wilson, M. (1966): Astrophys. J. *145*, 333

[43] Terenin, A.N., Vilessov, F.I. (1964): Adv. in Photochemistry, Vol. 2, p. 385, New York: Interscience; (1964): Vilessov, F.I., Terenin, A.N. (1957): Dokl. Akad. Nauk. SSSR *115*, 744

[44] Vilessov, F.I., Kurbatov, B.L., Terenin, A.N. (1961): Dokl. Akad. Nauk. SSSR *138*, 1329; (1961): *140*, 797 [English transl.: Sov. Phys.-Dokl. *6*, 490 (1961) and *6*, 883 (1962)]

[45] Turner, D.W. (1962): J. Chem. Phys. *37*, 3007

[46] Bouguer, P. (1729): Essai d'optique sur la graduation de la lumiere. This work was reprinted in 1921 by Gauthier-Villars, Eds., Librakries du Bureau des Longitudes, de l'Ecole Polytechnique, Quais des Grands-Augustine, 55

d'Optique sur la Graduation de la Lumiere, which was published in 1729. Bouguer, at the age of 15, occupied the Chair at the College of Jesuite de Vannes, which was left vacant upon his father's death. Apparently, he was a very able Professor even at that young age! It was his interest in navigation that led to his consideration of the attenuation of starlight in the Earth's atmosphere. His experiments led to his enunciation of the law of absorption: "We can therefore establish as a principle that when light travels through varying thicknesses in the same body, there is always the same relation between the difference of the logarithms of the two ordinates or quantities of light *QB* and *RF* and the thickness *BF* which is between the two, as between the difference of the logarithms of all other ordinates or quantities of light and the corresponding thickness".

Bouguer's work is generally little known and credit for describing the law of absorption is frequently given to LAMBERT (1728–1777).[46a] LAMBERT was interested primarily in optics and studied the absorption of light in great detail. He describes the absorption noting that if a *parallel* beam of *monochromatic* radiation passed through a *homogeneous* absorbing medium, the intensity of the radiation I was reduced by the same fractional amount $-\Delta I/I$ for each element of path length Δx traversed. That is,

$$-\Delta I = I\mu\Delta x, \tag{3.1}$$

where the proportionality constant μ depends only on the properties of the medium at that particular monochromatic wavelength. Integrating (3.1) we obtain the familar expression

$$I = I_0 \exp(-\mu x), \tag{3.2}$$

where I_0 is the intensity of the incident beam, I is the intensity transmitted through a distance x of the medium, and μ is referred to as the linear absorption coefficient.

Over a century ago BEER[47] found that the absorption coefficient was proportional to the concentration of the absorbing medium. This resulted in expressing the value of μ in (3.2) at some standard concentration. For gases this is taken at the standard conditions of temperature and pressure (STP), namely, at 273 K and 1 atmosphere (760 Torr) pressure. Thus, when μ is given at STP, the distance x is related to the actual measured distance l of the absorption cell by

$$x = \frac{P(\text{torr})}{760} \frac{273}{T(\text{K})} l. \tag{3.3}$$

From Beer's law we can also write that

$$\mu = \sigma_a n_0, \tag{3.4}$$

[46a] Lambert, J.H. (1760): Photometria sive de mensura et gradibus luminis, colorum et umbrae, Augsburg. A German translation appeared as "Lambert's Photometrie" translated by E. Andig, (1892) Leipzig: Verlag von Wilhelm Engelmann

[47] Beer, A.V. (1852): Ann. Phys. Chem. *86*, 78

where the concentration or density of the gas n_0 is given by Loschmidt's number ($n_0=2.687\cdot 10^{19}$ atoms or molecules cm^{-3}) when μ is measured at STP. The constant of proportionality σ_a has the units of an area and is, therefore, referred to as the total absorption cross section. It is more customary to express the Lambert-Beer law in terms of σ_a, thus

$$I=I_0\exp(-\sigma_a nl), \tag{3.5}$$

where n is the actual number density of the gas and is determined from the relation

$$n=n_0\left(\frac{P(\text{torr})}{760}\cdot\frac{273}{T(\text{K})}\right). \tag{3.6}$$

σ_a is usually expressed in cm^2 or in megabarns ($1\,\text{Mb}=10^{-18}\,cm^2$) and does in general obey both Lambert and Beer's law provided the atoms and molecules are sufficiently separated so that the intermolecular forces are negligible. This is usually the case for the pressures encountered in absorption spectroscopy of gases. However, an important exception occurs if σ_a varies within the band width $\Delta\lambda$ of the "monochromatic" radiation used. Then the value of the cross section will vary with gas pressure and with $\Delta\lambda$. This occurs when the radiation is absorbed by discrete energy levels. The problem has been discussed in detail in connection with infrared absorption studies.[48] However, it has often been the practice in vacuum uv spectroscopy to measure $\ln(I_0/I)$ vs n. If a straight line is obtained and passes through the origin one concludes that Beer's Law holds and that σ_a is proportional to the slope of the straight line. HUDSON[49] points out, however, that this is not necessarily true. He shows that when $\Delta\lambda$ is equal to ΔL (the halfwidth of a Lorentzian profile) the ratio of the average cross section $\bar{\sigma}_a$, taken from the slope of a Beer's law plot, to the true value of σ_a at the center of the band width is equal to 0.68. The results for several bandwidths are shown in Fig. 2. It can also be seen from the figure that even a very small bandwidth, relative to the absorption line profile, still does not give the true peak-absorption cross section. The effect of a finite bandwidth is to reduce the peak values of the cross section and to raise the values measured in a minimum. Care must, therefore, be taken in regions of structure.

In an absorption process we are interested in what fraction of the absorbed photons produces ionization. In the total absorption process some of the absorbed photons ΔI_n will be lost in excitation or dissociative processes; some, ΔI_i, will cause ionization, and a certain number ΔI_s will be scattered out of the incident beam. If ΔI represents the total number of photons absorbed from the incident beam, then $\Delta I=\Delta I_n+\Delta I_i+\Delta I_s$, and thus

$$\sigma_a=\left(\frac{\Delta I_n}{\Delta I}\right)\sigma_a+\left(\frac{\Delta I_i}{\Delta I}\right)\sigma_a+\left(\frac{\Delta I_s}{\Delta I}\right)\sigma_a=\sigma_n+\sigma_i+\sigma_s, \tag{3.7}$$

---

[48] Nielsen, J.R., Thornton, V., Dale, E.B. (1944): Rev. Mod. Phys. *16*, 307

[49] Hudson, R.H., Carter, V.L. (1968): J. Opt. Soc. Am. *58*, 227

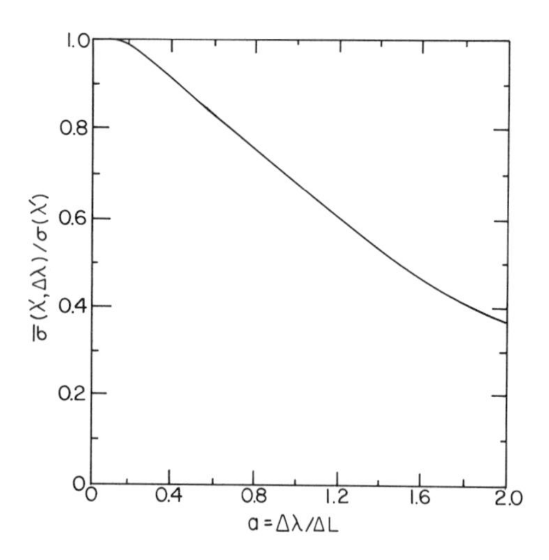

Fig. 2. Ratio of the cross section obtained from a Beer's law plot to the true cross section at the center of an absorption line as a function of the monochromator band pass ($\Delta\lambda$) divided by the half width ΔL of the absorption line. (From HUDSON, CARTER [49])

where σ_n, σ_i, and σ_s represent, respectively, the cross sections for producing neutral products, ionized products, and scattering. The fractions $\Delta I_i/\Delta I$, etc. are the *yields* or *efficiencies* for that process, usually denoted by γ_i, etc. Because our discussion is only concerned with ionization we will drop the subscript i and use γ as the ionization efficiency. Thus the photoionization cross section is given by

$$\sigma_i = \gamma \sigma_a. \tag{3.8}$$

It is important at this point to emphasize the definition of the photoionization efficiency. Neglecting multiphoton absorption processes as found in high-powered laser beams, we note that when a photon is absorbed in an ionizing process, we equate it to the production of a single ion. Thus ΔI_i is equal to the number of ions produced and γ is the number of ions produced per photon absorbed. In the vacuum uv the scattering cross section is usually considered to be negligible although no measurements of σ_s have been reported at wavelengths shorter than 1,200 Å. For atoms, σ_n is usually zero for photon energies greater than the ionization threshold. The only primary absorption mechanism, other than ionization, is absorption by discrete energy levels that lie in the ionization continuum. In nearly all cases autoionization occurs and the end result is ionization. Thus, γ can be considered to be unity and $\sigma_a = \sigma_i$ for atoms. For molecules, especially near the ionization threshold, this is not true. Many of the superexcited levels simply decay by fluorescence or cross over into a dissociative state, while others will autoionize.

We have defined γ as the number of ions produced per photon absorbed. The ion may be singly, doubly, or multiply charged, but it is still a single ion and σ_i refers to the total ionization cross section which cannot exceed σ_a, that is, $\gamma \leqq 1$. The process of one photon producing a multiply charged ion is but one channel in photoionization. Considering all possible channels (e.g., ejection of electrons from different orbitals, simultaneous ionization and excitation, dissociative ionization, etc.), we must set up a similar relation as given by (3.7). This results in the following,

$$\sigma_j = (N_j / \sum_j N_j)\, \sigma_i, \tag{3.9}$$

where N_j represents the number of ions produced by a specific mechanism and $\sum_j N_j$ represents the total number of ions produced. This ratio is the yield for the specific process but is often called the *branching ratio*. The cross section for the specific process is σ_j and is usually referred to as the partial photoionization cross section. Equation (3.9) applies to atoms and molecules.

When the product $(\sigma_a n l)$ is $\ll 1$, the photoionization cross section can be obtained by measuring the number of ions produced per incident photon as follows. From the definition of yield

$$\gamma = \frac{N_i}{\Delta I},$$

where N_i is the number of ions produced and ΔI is the number of photons absorbed. When $\sigma_a n l$ is $\ll 1$ then $\Delta I = I_0 \sigma_a n l$. Thus,

$$\gamma \sigma_a = \sigma_i = \frac{N_i}{I_0 n l}. \tag{3.10}$$

If the product nl remains constant the relative photoionization cross section can be obtained simply by measuring the number of ions produced per *incident* photon. If the number of ions (N_i) refers to the number of doubly charged ions, then (3.10) would give the cross section for producing doubly charged ions, etc. In the wavelength range where only single ionization can occur, N_i is equal to i/e, where i is the ion current in amperes and e is the electronic charge. Otherwise the ratio i/e is equal to the number of charges produced.

The measurement of an absorption cross section involves simply a measurement of a ratio I_0/I. The absolute intensities are not required. However, to measure the photoionization cross section in absolute units it is necessary to know the intensity of the incident flux if $\gamma \neq 1$ (that is, for molecules) or if (3.10) is to be used. The problem of absolute intensity measurements will be discussed later in Sect. 7.

The absorption of radiation by matter is often characterized by the concept of an optical oscillator strength f_o or by an optical oscillator strength density df_o/dE for absorption into the ionization continuum, where E is the photon energy. The oscillator strength was a term that originated from classical mechanics. It was assumed that an electron would undergo forced oscillations about a fixed point in the presence of an electromagnetic field and thereby absorb energy from the field. However, this classical model was incapable of calculating the absorption cross section and one must resort to a quantum-mechanical treatment to determine either σ_a or df_o/dE [1]. These quantities are related by the following expression regardless of whether a classical or quantum-mechanical caculation is used[49a]:

$$\sigma_a(E) = (\pi e^2 h/mc)(df_o/dE) = 8.067 \cdot 10^{-18} (df_o/dE[\mathrm{Ry}])\ \mathrm{cm}^2 \tag{3.11}$$

when the photon energy is measured in Rydbergs (1 Ry = 13.6058 eV), or

$$\sigma_a(E) = 109.8 \cdot 10^{-18} (df_o/dE[\mathrm{eV}])\ \mathrm{cm}^2$$

when the photon energy is measured in electron volts (eV).

Thus, measurements of the total absorption cross section as a function of photon energy give the total oscillator strength, that is,

$$\begin{aligned} f_o(\text{total}) &= \sum f_o(\text{line strength}) + \int_{E_i}^{\infty} (df_o/dE)\, dE \\ &= f_o(\text{total line strengths}) + (1/109.8) \int_{E_i}^{\infty} \sigma_a[\mathrm{Mb}]\, dE\, [\mathrm{eV}], \end{aligned} \tag{3.12}$$

where σ_a is in megabarns [Mb]. In (3.12) the summation is over all discrete transitions and the lower limit of the integration E_i is the ionization threshold energy. A very important f-sum rule is the *Thomas-Reiche-Kuhn* rule for the sum of the oscillator strengths for all transitions, which states that the sum must equal the total number of electrons Z in the system.[49a] That is,

$$f_0(\text{total}) = Z. \tag{3.13}$$

Thus, by subtracting (3.13) and (3.12) and measuring the total continuum absorption cross section, it is possible to obtain the total discrete oscillator strengths [5].

4. Error analysis. Consider (3.5) written in the form

$$\sigma = \frac{1}{nl} \ln (I_0/I),$$

then for random errors the *possible error* $\Delta\sigma/\sigma$ is obtained from

$$\Delta\sigma = \left|\left(\frac{\partial\sigma}{\partial n}\right)\right| \Delta n + \left|\left(\frac{\partial\sigma}{\partial l}\right)\right| \Delta l + \left|\left(\frac{\partial\sigma}{\partial I_0}\right)\right| \Delta I_0 + \left|\left(\frac{\partial\sigma}{\partial I}\right)\right| \Delta I.$$

[49a] Fano, U., Cooper, J.W. (1968): Rev. Mod. Phys. *40*, 441

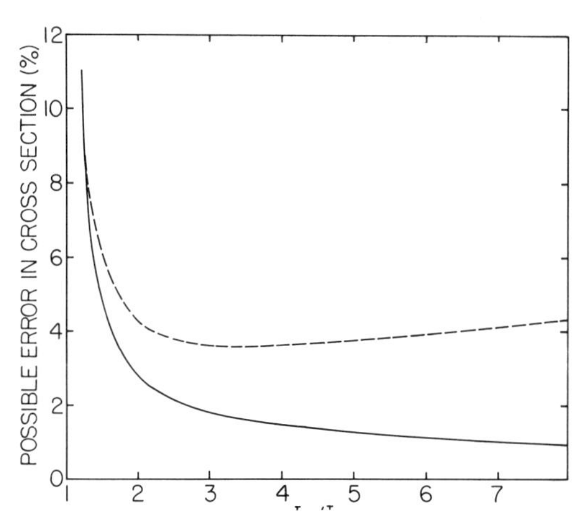

Fig. 3. Variation in the error of the cross section as a function of (I_0/I) assuming an error of 1 % in $(\Delta I_0/I_0)$. The dashed line is obtained if we assume $\Delta I_0 = \Delta I$, whereas the solid curve is obtained if $(\Delta I_0/I_0) = (\Delta I/I) = 1\,\%$

That is,

$$\frac{\Delta\sigma}{\sigma} = \frac{\Delta n}{n} + \frac{\Delta l}{l} + \frac{1}{\ln(R)}\left(\frac{\Delta I_0}{I_0} + \frac{\Delta I}{I}\right), \tag{4.1}$$

where $R = I_0/I$ and the ΔI's refer to the uncertainties in measuring the intensity of the radiation.

For the moment let us consider the errors in the number density and length measurements to be negligible in (4.1), then

$$\frac{\Delta\sigma}{\sigma} = \frac{1}{\ln(R)}\left(\frac{\Delta I_0}{I_0} + \frac{\Delta I}{I}\right). \tag{4.2}$$

In the past it has been common practice to state that the error given by (4.2) will be a minimum when the ratio I_0/I is about 3, and therefore experimental measurements should be taken at pressures that give this ratio. However, the origin of this concept is based on the assumption that the absolute errors $\Delta I_0 = \Delta I$, which is not true in general. If we assume that $\Delta I_0 = \Delta I$, then (4.2) becomes

$$\frac{\Delta\sigma}{\sigma} = \frac{\Delta I_0}{I_0}\,\frac{(1+R)}{\ln(R)}. \tag{4.3}$$

If we differentiate (4.3) with respect to R to find the minimum value of the error we find that when $\Delta I_0/I_0$ is 1 %, the minimum error is $\Delta\sigma/\sigma = 3.59\,\%$ and $R = 3.59$. The variation of the possible error as a function of R is shown in Fig. 3 for the condition given by (4.3) with $\Delta I_0/I_0 = 1\,\%$ and $\Delta I_0 = \Delta I$. Also shown in the figure are the results with both errors chosen equal to 1 %, that is, $\Delta I_0/I_0 = \Delta I/I = 1\,\%$. Thus, it is clear that a high value of R is desirable, consistent with the ability to maintain the 1 % error in I. In practice, one should consider the individual errors either in (4.1) or in the percent *standard deviation* equation as follows:

$$\frac{\Delta\sigma}{\sigma}=\pm\left\{\left(\frac{\Delta n}{n}\right)^2+\left(\frac{\Delta l}{l}\right)^2+\left[\left(\frac{\Delta I_0}{I}\right)^2\Big/\ln^2(R)\right]\left[\left(\frac{\Delta I}{I}\right)^2\Big/\ln^2(R)\right]\right\}^{1/2}. \quad (4.4)$$

$A\pm1\%$ error in each of the variables in (4.4) yields a total standard deviation of $\pm1.63\%$ for $R=3$ and $\pm1.17\%$ for $R=10$.

The error in measuring the length of an absorption cell is usually negligible for the permanent gases and can be set equal to zero. For metal vapors this is not true (see Sect. 7.4). Systematic errors must be considered in a windowless cell or with very thin films for windows, which may bulge slightly owing to the pressure differentials. But with care these errors can be reduced to a negligible amount. The problem of measuring the number densities is much more serious. Although accurate mercury McLeod gauges are available, it is difficult to measure pressures more accurately than about 2%, and few measurements in the past have considered the effect of "mercury streaming" on the observed pressure. This effect occurs as a steady flow of mercury leaves the gauge reservoir and flows to the cold trap protecting the absorption cell. This impedes the flow of gas to the gauge. It is particularly effective at low gas pressures. By far the most accurate method for measuring gas pressures is by use of the capacitance monometer. These are usually rated at better than 1% absolute accuracy. Thus, it appears possible that all random errors can be reduced to 1% or less, making it possible to measure cross sections of the permanent gases to an accuracy approaching $\pm1\%$. No such data have been reported to date. However, accuracies of $\pm3\%$ have been achieved [*6*]. For nonpermanent gases such as vapors, excited atoms, positive, and negative ions, the problem of measuring the number density is a serious one. However, for vapors this problem may be reduced, if not eliminated, by the use of heat-pipe ovens, which are discussed in Sect. 7.

II. Experimental techniques

5. Light sources. The choice of light source to be used in studying photoionization processes depends upon the nature of the measurement, on the target atom, and on the accuracy of the desired measurement. For atoms whose ionization thresholds lie between 4 and 6 eV conventional Hg−Xe arcs are useful, but for studies at shorter wavelengths more specialized sources are required. There is one exception. The synchrotron radiation emitted from electron storage rings or synchrotrons provides a continuum of radiation from the infrared to the x-ray region. However, as noted above the source must be tailored to suit the needs of the measurement. For example, to study the photoionization of atoms in excited states requires a source of great intensity. The number densities of excited species available in an experiment are typically about $10^9\,cm^{-3}$, whereas with a permanent gas the densities used are of the order of $10^{14}\,cm^{-3}$. Thus, a laser must be used at present to study excited species because the flux from the present synchrotron sources is not quite sufficient. But there are more specific reasons for a particular choice of source. The two classes of light sources are discrete and continuum sources. In principle, the continuum source is the ideal source. However, when the radiation is dispersed by a monochromator it contains a certain percentage of scattered radiation of all wavelengths and a certain amount of second- and higher-order wavelengths. Filters can be used to minimize these effects but usually at the expense of accuracy. However, to study discrete atomic and molecular structure the continuum source is indispensable. But to obtain continuum photoionization cross sections, the highest degree of accuracy is obtained with a discrete line emission source, particularly if each line is sufficiently isolated from its neighbor. Then the accuracy of the wavelength identification is usually in the second or third decimal place. The bandwidth is of the order of 0.01 Å regardless of the monochromator setting. But perhaps the most important advantage is that the scattered background level is easily identified. The most useful sources are described briefly below. For more details the reader is referred to [*2*].

α) DC glow discharge. This is a high-voltage dc discharge in a low-pressure gas. The gas pressure is in the vicinity of 1 Torr and a starting voltage of 2–3 kV is required to strike the discharge. Once the glow discharge has been established the voltage across the lamp drops to about 300 to 500 V depending upon the gas used, its pressure, and the length of the discharge. The discharge current ranges from a few milliamperes to about 1 A. The intensity of the light emitted increases if the discharge is confined to a capillary of about 1 to 3 mm in diameter. The source should be windowless and have a steady flow of gas passing through the capillary.

The electrons in a glow discharge seldom get sufficient energy to excite levels in the ions. Consequently, the radiation emitted is primarily that of neutral excited atoms and in particular the atomic resonance lines. Figure 4 shows the He I spectrum for a discharge in pure helium. The ratio of the intensities of the first to second members of the series, I(584 Å)/I(537 Å), varies with pressure. The ratio is about 70:1 at high pressures and drops to about 50:1 at the lowest pressures. At very low pressures the ion spectrum begins to appear. For helium, the main emission from the ions is the He II resonance line at 303.782 Å. Figure 5 shows its relative intensity as a function of He pressure with respect to the He I 584.334 Å line. Because the major voltage drop in a dc discharge occurs at the cathode, this is the region where the ion spectrum is most intense. Thus, hollow-cathode sources should produce more intense ion spectra.[50] However, the criteria is a low pressure to enable the electrons to reach sufficiently high energies for excitation of the ions. This can be achieved by applying magnetic fields such as in the duoplasmatron light source.[51] Table 2 lists the first resonance lines of the rare gas atoms and ions.

---

[50] Paresce, F., Kumar, S., Bowyer, C.S. (1971): Appl. Opt. *10*, 1904

[51] Samson, J.A.R., Liebl, H. (1962): Rev. Sci. Instrum. *33*, 1340

Fig. 4. He I spectrum for a glow discharge in helium. The ratio of the intensities of the 584 Å line to the 537 Å line is about 70:1 depending upon the gas pressure

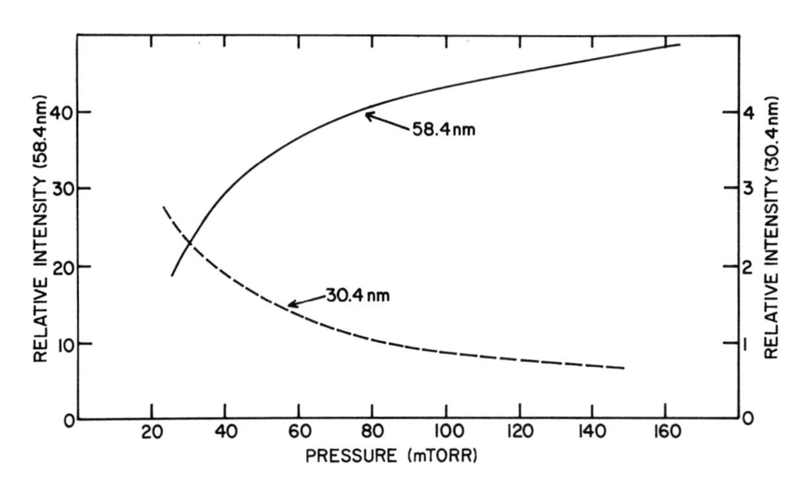

Fig. 5. Relative intensity of the He I 584 Å and He II 304 Å lines as a function of pressure within the glow discharge

When hydrogen is used in a glow discharge, a profuse many-line spectrum is produced between 850 and 1,670 Å. Beyond 1,670 Å a weaker continuum exists, continuing into the visible region of the spectrum.

β) Spark discharge. In this type of source a capacitor is stored with a given amount of energy at some appropriate voltage and then discharged rapidly in

Table 2. The first resonance lines of the rare gas atoms and ions

Gas	First resonance line		Gas	First resonance line	
	(Å)	(eV)		(Å)	(eV)
He I	584.334	21.217	Ar I	1,048.219	11.828
He II	303.782	40.812		1,066.659	11.623
Ne I	735.895	16.848	Ar II	919.782	13.479
	743.718	16.670		932.053	13.302
Ne II	460.725	26.910	Kr I	1,164.867	10.643
	462.388	26.813		1,235.838	10.032
			Xe I	1,295.586	9.569
				1,469.610	8.436

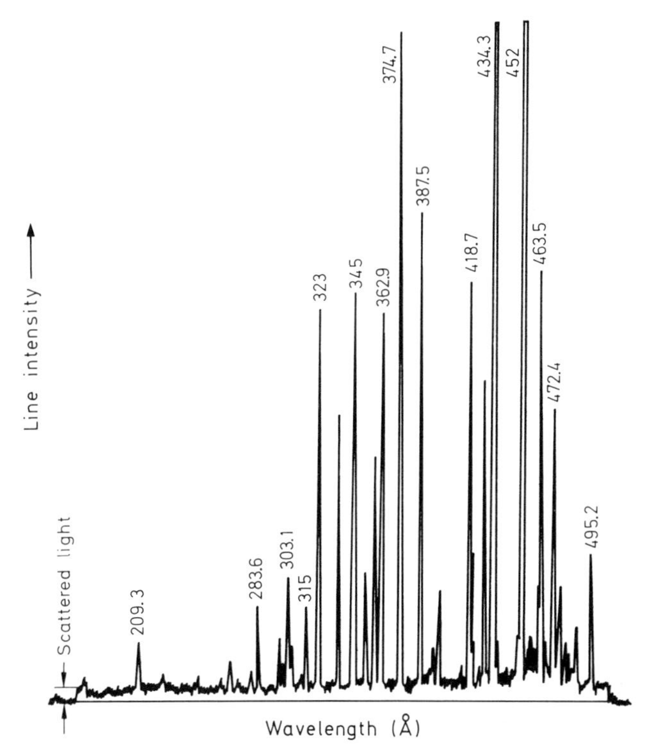

Fig. 6. Emission spectrum of a spark discharge source used with N_2 and measured with a 1/2 m Seya manochromator. The source can produce lines down to at least 90 Å

a low-pressure gas. Again a capillary is used, about 2–3 mm in diameter, to confine the discharge. Typical parameters are a low-inductance capacitor (0.25 μF) charged to about 6 kV and fired at repetition rates of 50 to 100 pulses per second. Radiation from about 90 to 1,000 Å is emitted from highly ionized atoms (four to six times ionized). The spectrum is characteristic of the gas used. Figure 6 shows the emission lines produced when nitrogen is used in the source and observed with a normal incidence monochromator. The figure also illustrates how easy it is to identify the scattered background radiation.

If the voltage and capacitance is increased in the spark discharge a continuum can be produced, the Lyman continuum.[52] The discharge needs a small amount of gas to start, but the nature of the continuum is independent of the gas used. Helium is used to allow the continuum to extend to about 500 Å. An improved version of the Lyman source has been described by GARTON.[53]

The continuum from the spark source can be extended into the x-ray region as shown by BALLOFFET et al.[54] They operated their source in high vacuum and use a uranium electrode. A sliding spark trigger pulse ablates some of the insulating material to allow a discharge to occur. The continuum comes from a narrow sheath surrounding the uranium electrode.

The continuum spark sources are particularly suitable for flash absorption spectroscopy of transient species. However, their low repetition rate and non-uniformity from spark to spark make them less desirable for accurate measurements of photoionization cross sections. The line emission spectrum from the spark source with a gas is usually very stable and reproducible.

The Hopfield continuum[30, 54a] is produced by a mildly condensed spark discharge ($C \approx 0.002\,\mu F$) in high-pressure helium ($P \approx 50$ Torr). Typically, a voltage of about 10 kV is used with a repetition rate of 5 kHz. A useful continuum is produced between 600 and 1,000 Å.

γ) Synchrotron radiation. Synchrotron radiation is produced whenever high-energy electrons are constrained to move in a segment of a circle. In a conventional synchrotron, the electrons are radiating at various rates as their energy is increased from near zero to a maximum in the sinusoidally varying field. In a storage ring, the electrons are brought up to a fixed energy and then allowed to circulate for several hours. Because of their relativistic energies, the radiation pattern of the orbiting electrons is extremely directional. The radiation is emitted in the same direction as the instantaneous velocity vector of the electrons and is contained in a cone of half angle ψ, which is of the order of 1 mrad but varies with the electron energy and the wavelength of the observed radiation. As the electrons circulate, this cone of light sweeps a path in a horizontal plane, as shown in Fig. 7. The horizontal angle is limited only by the diameter of the vacuum pipe that transmits the radiation from the storage ring to the monochromator. The intensity of the radiation entering the monochromator depends on whether the incident beam is focused onto the monochromator entrance slit and whether the monochromator is orientated so that the entrance slit geometry is matched to the focused beam. This usually means that the conventional vertical slit (with horizontal dispersion) is rotated 90° to give a horizontal rectangular slit to match the shape of the incoming beam.

---

[52] Lyman, T. (1924): Astrophys. J. *60*, 1; (1926): Science *64*, 89

[53] Garton, W.R.S. (1959): J. Sci. Instrum. *36*, 11; (1953): *30*, 119

[54] Balloffet, G., Romand, J., Vodar, B (1969): C. R. *252*, 4139. See also Damany, H., Roncin, J-Y., Damany-Astoin, N. (1960): Appl. Opt. *5*, 297

[54a] Huffman, R.E., Tanaka, Y., Larrabee, J.C. (1963): Appl. Opt. *2*, 617

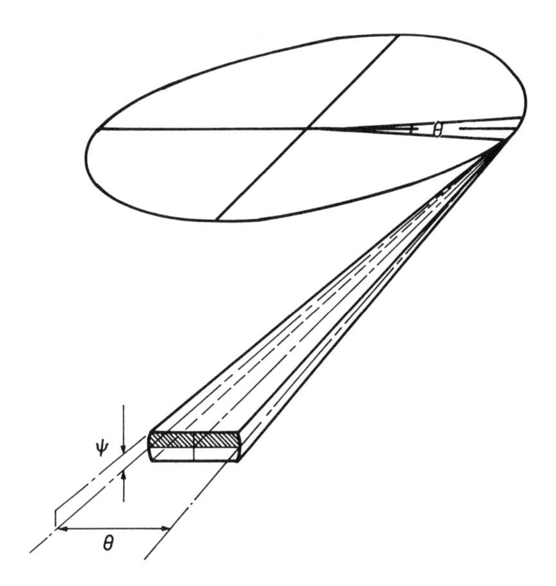

Fig. 7. The directional cone of light produced by a relativistic electron traveling in a circular path

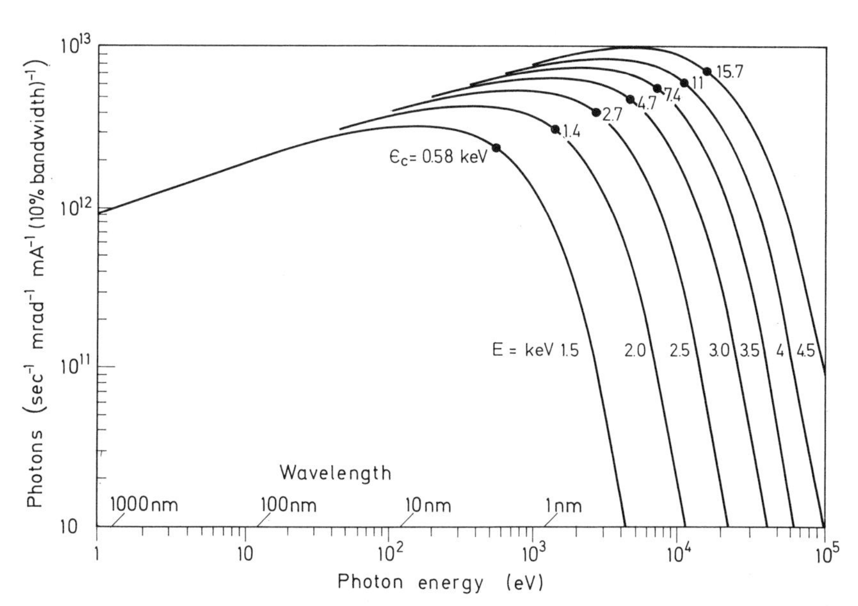

Fig. 8. Calculated spectral distribution as a function of electron energy of the SPEAR synchrotron radiation (radius = 12.7 m) at a constant 10% bandwidth (i.e., $\Delta\lambda/\lambda = 0.1$). The characteristic wavelength λ_c expressed in keV, is shown for each curve as ε_c

Figure 8 shows the calculated intensity of the synchrotron radiation as a function of wavelength and photon energy emitted from SPEAR, the storage ring at Stanford University. A family of curves are shown for energies of 1.5–

4.5 GeV and for a radius $R=12.7$ m. The characteristic energy ε_c is shown for each curve. This is a useful parameter used in the analysis of the spectral distribution of synchrotron radiation and is characteristic of a given machine and its operating parameters. Its magnitude is always close to the peak value of the spectral distribution and is given by

$$\varepsilon_c(\text{keV}) = 2.22\, E^3(\text{GeV})/R(\text{m}), \tag{5.1}$$

where E is the energy of the stored electrons in GeV and R is the radius of the ring in meters. Because the operating currents vary, the plots are given for a current of 1 mA. Typical current operation of the SPEAR machine is 50 mA, in which case the flux shown in Fig. 8 would be multiplied by 50. The curves assume all the radiation emitted in the vertical plane (~ 1 or 2 mrad) is collected and gives the flux per mrad in the horizontal plane. This can be of the order of 1–30 mrad. It is interesting to note that for work in the far ultraviolet, there is little advantage, in terms of flux output, to increasing the energy of the machine from 1.5 to 4.5 GeV. In fact, it is more troublesome because of the unwanted scattered short wavelengths.

In addition to its continuum nature, synchrotron radiation exhibits several other features not found in conventional laboratory light sources. For instance, it is found that the radiation observed in the plane of the orbiting electrons is essentially 100% plane polarized with the electric vector in the plane of the orbit. As observations are made off the plane, the radiation becomes elliptically polarized. Another unusual feature is the extremely short duration of each light pulse. Examples from the SPEAR storage ring show that the duration of a pulse, at present, is 1 ns, with durations of 0.15 ns planned. The pulses arrive spaced 781 ns apart. In TANTALUS (University of Wisconsin) the pulse length is variable from 1 to 6 ns, and the pulses are separated by 31 ns.

As a far uv light source, the storage ring is much more attractive than the synchrotron because it is possible to work in the area of the experiments without radiation hazards and the position of the electron beam is much more stable. In addition, it is possible to monitor the precise beam current and hence any variation in the light intensity. The beam current in the storage ring will be slowly depleted because of electron collisions with residual gas atoms. Typical half-lives for the beams in Tantalus are about 2 hours for an 80-mA beam and 20 hours for a 1-mA beam.

The major problem with synchrotron radiation is accounting for the higher-order spectra and scattered radiation. Monochromators tend to be specially designed for synchrotrons and storage rings to minimize these problems. Several recent reviews have appeared describing the characteristics of synchrotron radiation.[55–58a]

---

[55] Godwin, R.P. (1969): Springer Tracts Mod. Phys. *51*, 1

[56] Codling, K. (1974): Phys. Scr. *9*, 247

[57] Codling, K., Madden, R.P. (1965): J. Appl. Phys. *36*, 380

[58] Kunz, C. (1974): In: Vacuum ultraviolet radiation physics, Koch, E.E., Haensel, R., Kunz, C. (eds.), p. 753

[58a] Marr, G.V., Munro, I.H., Sharp, J.C.C. (1972): Synchrotron radiation: A bibliography, Daresbury Nuclear Physics Laboratory Report DNPL/R24

6. Monochromators and spectrographs. The spectrograph with photographic detection is still one of the most useful devices to observe discrete absorption structure in the ionization continuum. It enables a rapid survey to be made of the absorption properties of gases and provides the most precise determination of the energies of the discrete lines. Figure 9 shows a particularly beautiful photographic absorption spectrum of Kr [*2*]. However, in order to obtain detailed information on the absolute photoionization cross sections and partial cross sections a scanning vacuum uv monochromator is essential. Many types exist [*2*],[59] some have high radiation throughputs with poor resolution and others sacrifice intensity for high resolution. Each has its advantage depending upon the problem being investigated. But there are some important aspects that should be considered. Where one cannot tolerate much scattered radiation, a holographic grating should be used. This type of diffraction grating produces the lowest amount of scattering because it does not have a mechanically ruled surface. However, at present they are not quite as efficient as the blazed gratings produced on a standard ruling machine. For grazing-incidence monochromators a toroidal grating should be used to reduce astigmatism, thus greatly increasing the efficiency of the monochromator.

Most monochromators produce partially or elliptically polarized light to some degree. Thus, it is important to know the polarization characteristics of

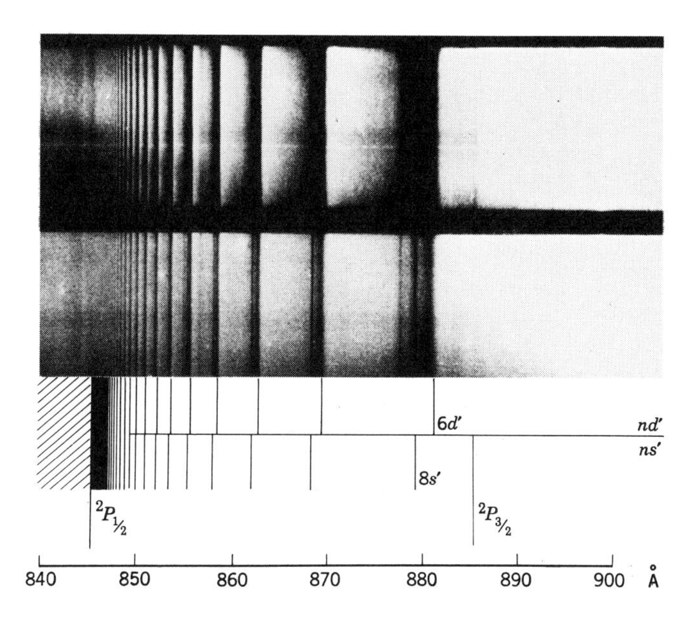

Fig. 9. Absorption spectrum of Kr showing the discrete autoionizing structure between the $^2P_{1/2}$ and $^2P_{3/2}$ states of the ion. (From SAMSON [*2*])

[59] For monochromators used with synchrotron radiation see, West, J.B., Codling, K., Marr, G.V. (1974): J. Phys. E *7*, 137; Dietrich, H., Kunz, C. (1972): Rev. Sci. Instrum. *43*, 434; Kunz, C., Haensel, R., Sonntag, B. (1968): Opt. Soc. Am. *58*, 1415; Codling, K., Mitchell, P. (1970): J. Phys. E *3*, 685; Jaeglé, P., Dhez, P., Wuilleumier, F. (1977): Rev. Sci. Instrum. *48*, 978

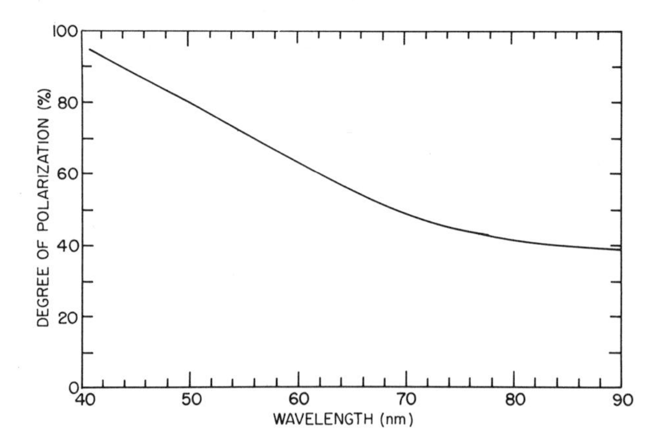

Fig. 10. Degree of polarization produced by a Seya monochromator ($\alpha = 35°$) using unpolarized incident radiation. The grating was gold coated but had some carbon contamination. (From SAMSON [2])

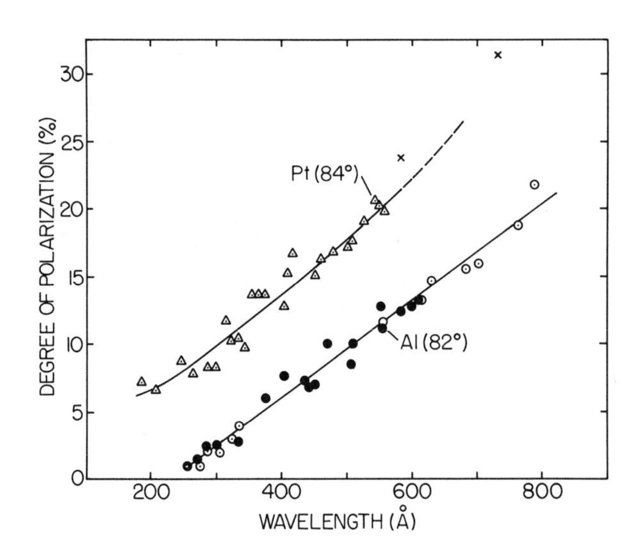

Fig. 11. Degree of polarization produced by grazing-incidence monochromators with Pt ($\alpha = 84°$) and Al ($\alpha = 82°$) coatings using unpolarized incident radiation. All grazing incidence monochromators exhibit a decrease in polarization towards shorter wavelengths

the instrument. This can be important for some experiments (e.g., measuring the angular distribution of photoelectrons) and also to enable the monochromator to be used in its most efficient configuration. The maximum degree of polarization from a monochromator is usually in the direction parallel to the grating rulings and thus parallel to the exit slit. Thus, when coupled to a source of synchrotron radiation, which has its maximum degree of polarization in the horizontal plane, the monochromator should be mounted in the vertical plane with the length of the entrance slit (and the grating rulings) oriented in the horizontal plane. Examples of the degree of polarization produced by a Seya monochromator and grazing-incidence monochromator are shown in Figs. 10 and 11.[60,61] At all wavelengths the Seya instrument produces a high degree of polarization which generally increases as the wavelength scanned decreases. For grazing angles of incidence the opposite is true and reflects the fact that the optical constants n and k of the diffracting surface tend to decrease with wavelength below about 800 Å.

7. Absorption and photoionization cells. To measure total absorption cross sections requires a measurement of the ratio of the light incident upon a sample to that transmitted by the sample. To do this, several techniques have been developed to meet the specialized needs of various experiments. An obvious approach, and one that has been used quite extensively, is shown in Fig. 12. It consists of a simple absorption cell of length l with a window or narrow slit facing the incident radiation. The end of the cell is defined by a window coated with sodium salicylate and followed by a photomultiplier (PM). With the cell empty the signal I_0 entering the cell is recorded. The cell is then filled with a gas of number density n and the PM reading taken again to record the intensity I that is transmitted through the length l of the gas. The total cross section is then given by

$$\sigma = (1/nl) \ln (I_0/I).$$

The accuracy of this method depends strongly upon the light source intensity remaining constant at each wavelength during the entire measurement. Few sources can remain stable and be reproducable to $\pm 1\%$ over this time period. Thus, techniques to measure I_0 and I over very small time intervals are used (double-beam method) or they are measured simultaneously (split-beam and double-ion-chamber methods). These techniques are described below.

α) Double beam. This method involves a mechanical oscillation of a mirror or detector, which actually intercepts the entire incident beam. One method is shown in Fig. 13, where a wedge-shaped mirror is forced to oscillate about the incident beam.[62] The intercepted beam is focused onto the entrance slit (or window) of a gas-filled cell during one-half of the cycle and then focused onto an empty reference cell during the next half cycle. The signals from two photomultipliers are adjusted to be identical when both cells are empty. A similar technique, which applies only to a windowless system, uses a single cell and periodically interrupts the incident beam with a chopped blade oriented at 45° to the beam.[63] The blade is coated with sodium salicylate and the fluorescent

---

[60] Samson, J.A.R. (1978): Nuclear Instrum. Methods *152*, 225

[61] Arakawa, E.T., Williams, M.W., Samson, J.A.R. (1978): Appl. Opt. *17*, 2502

[62] This is the principle of the McPherson double-beam intrument (McPherson Instrument Co., Acton, MA). A similar device is reported by Schmitt, R.G., Brehm, R.K. (1966): Appl. Opt. *5*, 1111

[63] Onaka, R., Ejiri, A. (1963): Appl. Opt. *2*, 321

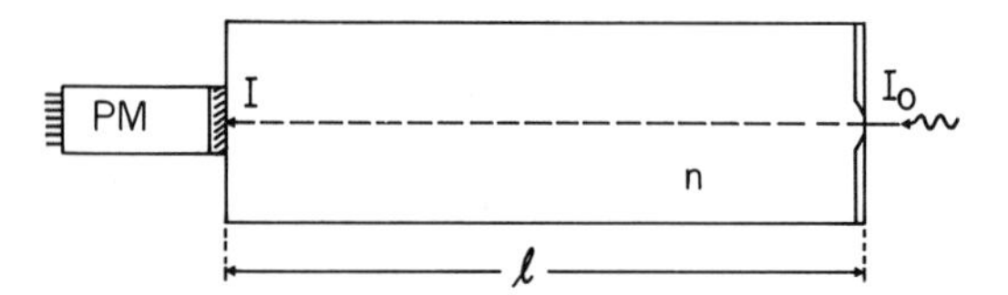

Fig. 12. Simple absorption cell where measurements must be made of the incident light intensity with and without gas in the cell. I_0 and I are, respectively, the incident and transmitted light intensity, n is the gas number density, l is the length of the cell, and PM is a photomultiplier

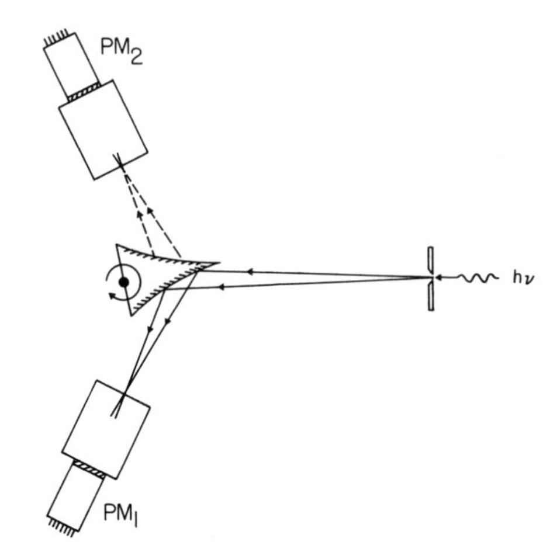

Fig. 13. Dual-beam absorption measurements utilizing a gas cell, an evacuated reference cell, and an oscillating mirror. $PM_{1,2}$ are photomultiplier detectors

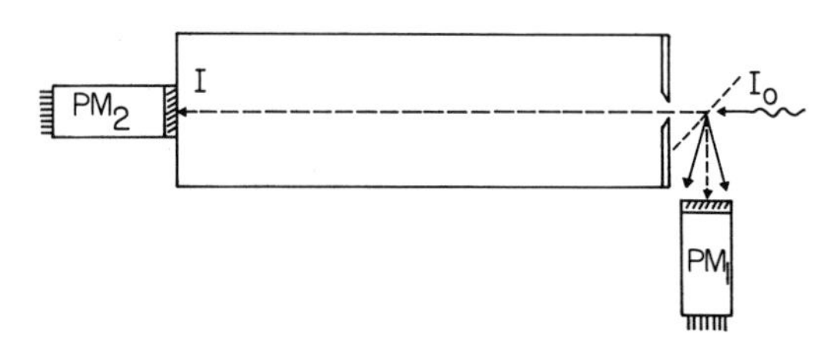

Fig. 14. Split-beam absorption measurements utilizing a mesh coated with sodium salicylate or a mirror intercepting a fraction of the incident radiation at the entrance to the absorption cell

radiation is observed with a photomultiplier. Again, the two multipliers must be adjusted to give identical signals when there is not gas in the cell.

β) Split beam. No moving parts are involved with this method. However, the technique is similar to the double-beam method. In the split-beam method a portion of the incident beam is reflected into PM_1 (Fig. 14) giving a signal proportional to I_0. The remainder of the incident beam is transmitted to PM_2. The photomultiplier tubes must be balanced when there is no gas in the cell.

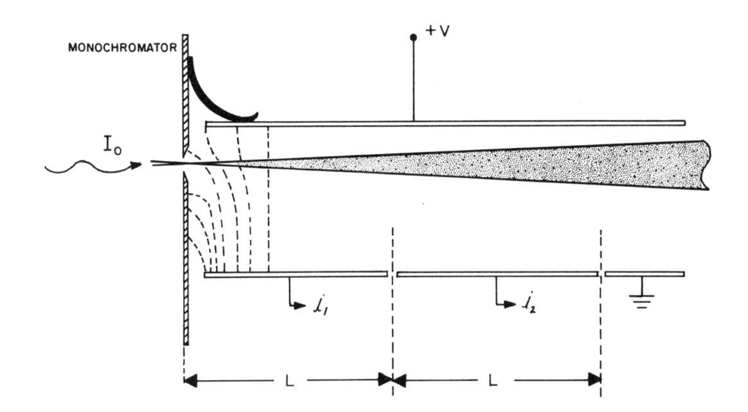

Fig. 15. Double ion chamber with parallel plates and guard ring. The diagram shows the monochromator exit slit acting simultaneously as the entrance slit for the ion chamber, in which case it must be held at the same potential as the repellor electrode. However, for absorption measurements the two slits may be separated with the monochromator exit slit at ground potential. The drawing is not to scale. (From SAMSON[69])

The splitting of the beam can be achieved either with a reflecting mesh or with a mirror intercepting part of the beam. Because of the poor reflectance of most materials in the vacuum uv, it is often preferable to coat the mesh directly with sodium salicylate and focus the fluorescent radiation into a PM tube either with a lens or a light pipe. This method or variations thereof have been reported in the literature, especially for use with metal vapors and solids.[64–67]

γ) Double ion chamber. The double ionization chamber can measure both total absorption cross sections (σ_a) and ionization yields (γ). Thus, it can provide absolute photoionization cross sections ($\sigma_i = \gamma\sigma_a$). The principle of the method is described below.[68–70]

In one configuration, the double ion chamber consists of two parallel electrodes. One continuous electrode is held at a positive voltage and another one consists of three parts, namely, two ion collector electrodes and a guard electrode to maintain a uniform field. The separation of the parallel plates should be small compared to the widths of the plates, and the length of the guard electrode should be about three times the separation distance between the parallel plates. This configuration is shown in Fig. 15. The length of the collector electrodes L are identical and the electric-field distribution is constant at the junction between each electrode. That is, the ions formed above each

[64] Hudson, R.D., Carter, V.L. (1965): Phys. Rev. *137* A, 1648
[65] Codling, K., Hamley, J.R., West, J.B. (1977): J. Phys. B *10*, 2797
[66] Boursey, E., Roncin, J-Y., Damany, H., Damany, N. (1971): Rev. Sci. Instrum. *42*, 526
[67] Castex, M-C., Romand, J., Vodar, B. (1968): Rev. Sci. Instrum. *39*, 331
[68] Samson, J.A.R. (1976): U.S. Patent No. 3,952,197
[69] Samson, J.A.R. (1964): J. Opt. Soc. Am. *54*, 6
[70] Samson, J.A.R., Haddad, G.N. (1974): J. Opt. Soc. Am. *64*, 47

electrode must be collected by that electrode. Under these conditions, we obtain the following expression for the number of photons absorbed (ΔI) above each electrode:

$$\Delta I_1 = I_0[1-\exp(-\sigma_a nL)] \tag{7.1}$$

and

$$\Delta I_2 = I_0 \exp(-\sigma_a nL)[1-\exp(-\sigma_a nL)]. \tag{7.2}$$

We define the yield Y for producing charges as follows:

$$Y = (\text{Number of charges produced})/(\text{Number of photons absorbed}).$$

Thus,

$$\Delta I_1 = (i_1/e)/Y \tag{7.3}$$

and

$$\Delta I_2 = (i_2/e)/Y, \tag{7.4}$$

where i_1 and i_2 are the ion currents in amperes and e is the electronic charge. Substituting (7.3) and (7.4) into (7.1) and (7.2), respectively, and taking the ratio of the resulting equations, we find that

$$\sigma_a = (1/nL)\ln(i_1/i_2), \tag{7.5}$$

This result holds even for photon energies that are capable of producing multiply charged ions. The ratio (i_1/i_2) remains constant even if the ejected photoelectrons have sufficient energy to cause secondary ionization. In this case it is preferable to design a symmetric ion chamber as shown in Fig. 16. This design is superior to the parallel-plate ion chamber, because the photoelectrons are given less acceleration by the ion-retarding field and consequently produce less secondary ionization. Further, the design does not depend on the equipotential lines being parallel to the collector plates except in the vicinity of the center line. It also prevents stray fields from penetrating into either region above the collector electrodes and perturbing the system.

There are several major advantages in using the double ion chamber for absorption measurements.

a) Gas escaping from the exit aperture of a windowless absorption cell into the monochromator will absorb some of the incident radiation, thereby giving a false reading of I in the conventional method. With the double ion chamber only the photons absorbed within the chamber are recorded.

b) Any fluorescence produced within the double ion chamber will not affect its operation.

c) Scattered radiation from the monochromator will not be detected except for that fraction with wavelengths shorter than the ionization threshold of the gas. Thus, helium is insensitive to scattered radiation of wavelength greater than 504.3 Å.

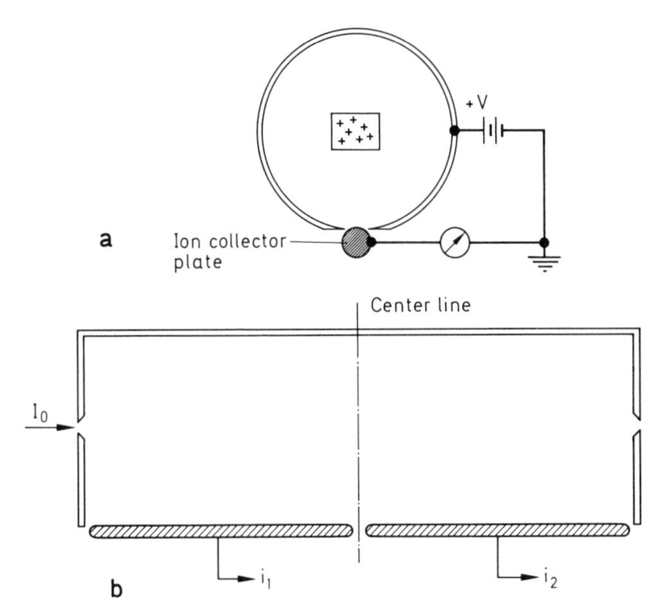

Fig. 16a, b. Double ion chamber with cylindrical cross section. (**a**) Formation of the photoionized products in a weak electric field although the potential V is sufficient to retard electrons from reaching the ion collector electrode. This design minimizes the energy gain that the electrons can achieve from the electric field. (**b**) Symmetrical arrangement of the ion chamber

To obtain the ionization efficiency of a gas, the absolute intensity of the incident radiation must be known. This is obtained from the double ion chamber as follows.

Equating (7.1) and (7.3) we find that

$$I_0 Y = (i_1/e)/[1-\exp(-\sigma_a nL)]. \tag{7.6}$$

Substituting (7.5) into (7.6) gives

$$I_0 Y = (i_1^2/e)/(i_1 - i_2). \tag{7.7}$$

A rare gas is used to determine I_0. At wavelengths longer than the double ionization threshold, $Y=\gamma=1$. For shorter wavelengths we must know Y. These values are now available (see Sect. 11β). The efficiency of the unknown gas (molecular) is then compared with that of the rare gas, again using (7.7). Thus, γ is found.

δ) Heat pipes. To determine the absorption cross section of the majority of the elements it is necessary first to vaporize the element by the use of some type of furnace. Several major problems present themselves. The most difficult problem is in determining the number density of the atoms in the vapor. Tables of vapor pressure versus temperature are not always very precise. Further, the temperature must be constant over the region of the measurement.

Vapor pressure data are given by HONIG,[71] NESMEYANOV,[72] and by the JANAF tables.[73] Variations in the quoted vapor pressures at a given temperature lie between 10 and 15%. DITCHBURN and GILMOUR[33] used the expression

$$\log_{10} P(\text{mm}) = -\frac{A}{T} - B \log_{10} T - C \tag{7.8}$$

to obtain the vapor pressure P(mm) in millimeters of Hg, where A, B, and C are constants they tabulated, and T is the absolute temperature.

A more recent development in furnaces suitable for absolute cross section measurements is the *heat-pipe oven.*[74–76a] The heat pipe consists of a metal tube, the inner wall of which is covered with mesh. When a vapor condenses out as a liquid onto the mesh the liquid is transposed along the entire length of the mesh by capillary action similar to the action of a wick. Thus, by heating the central portion of the tube and vaporizing a material it will diffuse outwards in both directions until condensing on cooler surfaces. The wick returns the liquid to the center of the tube where it is reevaporated. This process of heat transfer rapidly produces a long segment of tube that is at a constant temperature. By flowing a gas such as helium through the heat pipe it is possible to get a direct measure of the vapor pressure. The gas and vapor actually separate because the vapor stream drives the inert gas to the end of the heat pipe. At that point, dependent upon the heat input and the pressure of the inert gas, the vapor will liquify.

In a steady-state condition the heat input is sufficient to maintain a column of vapor of a particular length such that the temperature produces a vapor pressure identical to the pressure of the inert gas. As the heat input is increased (at constant inert-gas pressure) the vapor column expands until the heat losses are equal to the heat input. Alternatively, if the gas pressure is increased (at constant heat input) the length of the column will shrink so that the available heat is used to raise the temperature of the vapor until the vapor pressure again equals that of the inert gas. A schematic arrangement of the heat-pipe oven constructed by VIDAL and COOPER[75] is shown in Fig. 17 and the temperature profile of the vapor as a function of position in the tube is shown in Fig. 18 for two different heat input levels, The helium pressure remained constant. The increase in length of the constant temperature portion of the inner curve is given by twice $l/2$ as shown in the figure. The heat pipe has recently been used by CODLING et al.[65] to measure the cross sections of Na in the region 45–250 eV and by EDERER et al.[77] to study Li.

A serious problem with any type of furnace is the tendency of vapors to form molecular species. If only the atomic and diatomic species are present

[71] Honig, R.E. (1962): R.C.A. Rev. *23*, 567

[72] Nesmeyanov, A.N. (1963): In: Vapor pressure of the chemical elements, Gary, R. (ed.). New York: Elsevier

[73] JANAF Thermochemical Tables (1962): The Dow Chemical Company, Midlands, Michigan

[74] Grover, G.M., Cotter, T.P., Erikson, G.F. (1964): J. Appl. Phys. *35*, 1990

[75] Vidal, C.R., Cooper, J. (1969): J. Appl. Phys. *40*, 3370

[76] Dunn, P., Reay, D.A. (1975): Heat Piping Oxford: Pergamon Press

[76a] Connerade, J.P. (1978): Nucl. Instrum. Methods *152*, 271

[77] Ederer, D.L., Lucatorto, T., Madden, R.P. (1970): Phys. Rev. Lett. *25*, 1537

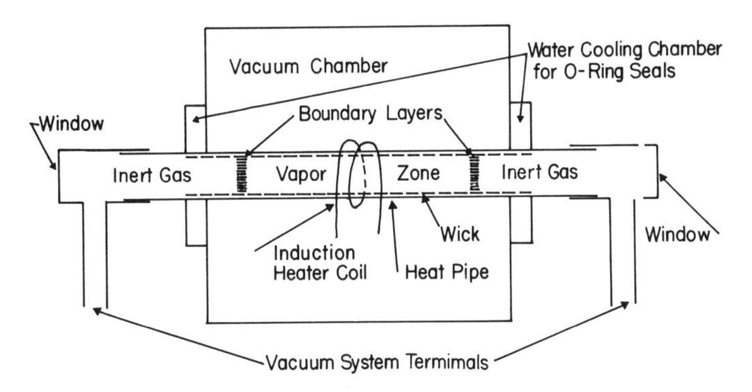

Fig. 17. Schematic arrangement of the heat-pipe oven. (From VIDAL, COOPER[75])

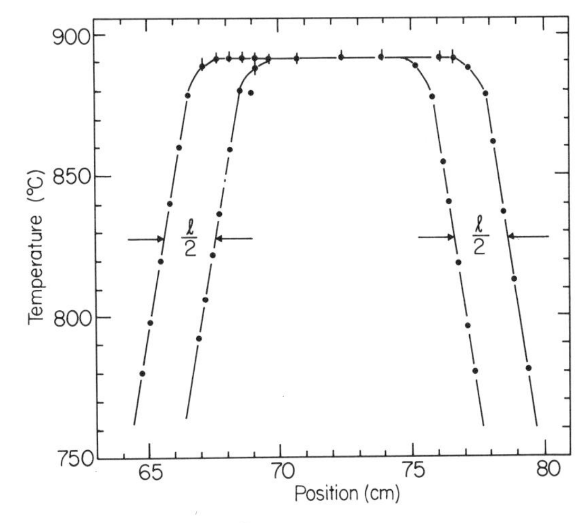

Fig. 18. Temperature profiles measured at two different powers with a heat-pipe oven operating with lithium and using helium as the containing gas at a pressure of 10.9 Torr. (From VIDAL, COOPER[75])

then the following equation must be solved:

$$\ln[I_0(\lambda)/I(\lambda)] = C_{at}\sigma_{at}(\lambda)\,L + C_m\sigma_m(\lambda)\,L, \tag{7.9}$$

where $I_0(\lambda)$ and $I(\lambda)$ are, respectively, the flux incident on and transmitted through the column of vapor of length L, C_{at} and C_m are the concentrations, and σ_{at} and σ_m are the absorption cross section of the atomic and molecular species, respectively. The values of C_{at} and C_m can be calculated using the vapor-pressure data given in the JANAF tables.[73] This, of course, presents another source of error. It is desirable to measure the cross sections as a function of wavelength at a very high vapor temperature to reduce the number of molecular species. When this is impractical the relative cross sections should be measured by use of an atomic beam, which would allow much higher temperatures. This result can then be compared to the absolute values given by the heat-pipe oven to determine the influence of molecular species.

ε) Atomic beams. In some cases it may be more practical to measure relative cross sections of vaporized materials. This can be done by the use of atomic beams emerging from a furnace. A simple measurement of ions per photon gives the relative photoionization cross section, see (3.10). Atomic-beam ovens of various descriptions have been reported for many applications. In fact, one of the earliest measurements of photoionization (Williamson, 1923)[12] used an atomic-beam oven to study photoionization of potassium. More recently, Cairns et al.[78] have used atomic beams to study the relative cross sections of Cd, Zn, and Hg. A brief review of furnaces has been given by Connerade.[76a]

8. Branching-ratio studies. When an atom is ionized, the ion may be singly or multiply charged. Although the ions will have a kinetic energy appropriate to their ambient temperature the electrons will have varying kinetic energies dependent upon the orbital from which they are ejected. Excitation may accompany the ionization process with subsequent decay through fluorescent emission. For molecules, further channels are available such as dissociative ionization. In this case the fragment ions generally have varying amounts of kinetic energy.

The partial photoionization cross sections are given by (3.9),

$$\sigma_j = (N_j / \sum_j N_j)\, \sigma_i .$$

Thus, the quantity to be measured is the ratio of the number of ions of a particular nature N_j to the total number of ions formed. This is the branching ratio. To study these specific processes requires specialized apparatus designed to observe the particular phenomena of interst. Some examples are given below.

α) Mass spectrometers are required to identify fragment ions from molecular photoionization and to identify multiply charged ions. They also help to distinguish the ion under investigation from an impurity background gas. Terenin and Popov[79] in 1932 were the first to apply mass-spectrometric analysis to identify the products of photoionization of molecules using an undispersed source of radiation. This work was expanded in the late 1950's by groups working with undispersed[80-83] and dispersed[84,85] radiation.

---

[78] Cairns, R.B., Harrison, H., Schoen, R.I. (1972): In: Advances in atomic and molecular physics, Bates, D.R., Esterman, I. (eds.), Vol. 8, pp. 131–162. New York: Academic Press

[79] Terenin, A.N., Popov, B. (1932): Phys. Z. Sowjetunion *2*, 299

[80] Lossing, F.P., Tanaka, I. (1956): J. Chem. Phys. *25*, 1031

[81] Herzog, R.F., Marmo, F.F. (1957): J. Chem. Phys. *27*, 1202

[82] Schönheit, E. (1957): Z. Phys. *149*, 153

[83] Vilessov, R.I., Kurbatov, B.L., Terenin, A.N. (1958): Dokl. Akad. Nauk. SSSR *122*, 94

[84] Hurzeler, H., Inghram, M.G., Morrison, J.D. (1957): J. Chem. Phys. *27*, 313 (1958): *28*, 76

[85] Weissler, G.L., Samson, J.A.R., Ogawa, M., Cook, G.R. (1959): J. Opt. Soc. Am. *49*, 338

There are a great variety of mass spectrometers that can be used in photoionization studies. However, to obtain accurate branching ratios the mass spectrometer should not discriminate against fragment ions formed with kinetic energy neither should the residence time of the ions be so long that there would be any chance of a molecular ion dissociating in flight. These problems do not occur for atomic photoionization. However, a problem common to both atoms and molecules is the response of the detector to ions of different charge and different energy. There is no problem when the detector is a Faraday cage. However, usually an electron multiplier is required to observe the small signals. In this case discrimination effects can be observed.[86,87] When the multiplier is operated in the current mode, the output signal increases with ion energy. At a constant ion energy the signal decreases with increasing mass of the ion but increases as the charge on the ion increases. This makes it difficult to obtain accurate branching ratios. However, SCHRAM et al.[86] have shown that the output signal caused by a particular mass increases linearly with velocity and independently of the charge of the ion. In the pulse-counting mode there is less dicrimination. BURROUS et al.[88] show that for channel electron multipliers there is an almost constant counting efficiency for rare-gas ions with energies in excess of 4 keV. In either of the two modes of operation care must be taken to check for any discrimination.

Relative values of the partial cross section as a function of wavelength can be obtained from a mass spectrometer by measuring the ratio of ions produced per photon incident.[88a] The problem with this measurement is that we must use a photon detector whose response to radiation as a function of wavelength is either constant or has been measured. A photodiode with a simple metal photocathode is usually quite suitable [2] but it must be calibrated. This is best done by use of the double-ion-chamber technique described in Sect. 7γ. A typical efficiency curve for an aluminum photocathode is shown in Fig. 19, where the photoelectric or yield is defined as the number of electrons ejected per photon incident upon the photocathode.

β) Electron energy analyzers. By studying the kinetic energy of the ejected photoelectrons, detailed knowledge can be obtained regarding the binding energies of electrons in various orbits and of the cross sections for ejecting electrons from these specific orbits. The electron energies are given by the Einstein equation

$$E_j = h\nu - I_j, \tag{8.1}$$

where E_j is the kinetic energy of the electron ejected from the jth orbit, $h\nu$ is the photon energy, and I_j is the ionization potential or binding energy of the

[86] Schram, B.L., Boerboom, A.J.H., Kleine, W., Kistemaker, J. (1966): Physica *32*, 749

[87] Inghram, M.G., Hayden, R.J. (1954): Nucl. Sci. Ser., Report No. 14. Washington: National Research Council

[88] Burrous, C.N., Lieber, A.J., Zaviantseff, V.T. (1967): Rev. Sci. Instrum. *38*, 1477

[88a] There have been numerous studies of photoionization-mass spectrometry of molecules. A few of the more recent references follow: Kronebusch, P.L., Berkowitz, J. (1976): Int. J. Mass. Spectrom. Ion Phys. *22*, 283; Chupka, W., Dehmer, P., Jivery, W.T. (1975): J. Chem. Phys. *63*, 3929; McCulloch, K.E. (1973): J. Chem. Phys. *59*, 4250; Dibeler, V.H., Walker, J.A. (1967): J. Opt. Soc. Am. *57*, 1007

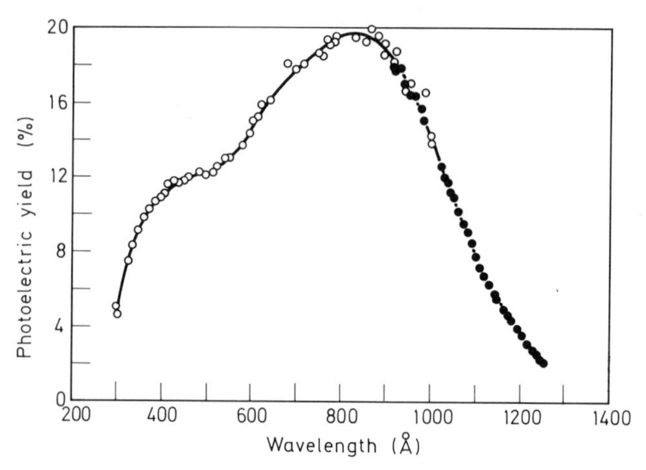

Fig. 19. Photoelectric yield of an aluminum photocathode as a function of wavelength

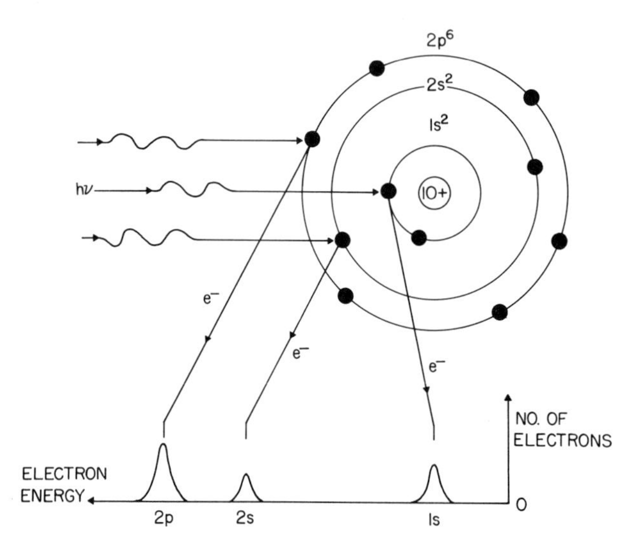

Fig. 20. Neon atom irradiated with a monoenergetic photon beam. Photoelectrons are illustrated as being ejected from the various orbitals

electron in the jth orbit. This field is called *photoelectron spectroscopy* [*10–21*]. The technique is illustrated in Fig. 20, which shows atomic Ne irradiated with a monoenergetic photon beam of sufficient energy to eject a K-shell electron. There is a finite probability that a photon will eject an electron from the $2p$ shell of one atom while another photon may interact with a different atom ejecting an electron from the $2s$ shell, etc. This results in groups of electrons ejected with discrete energies as given by (8.1). The figure does not show the effects of multiple ionization of Ne nor the effects of Auger electrons. Electrons from multiple ionization will have a continuum of energies over the allowed range, whereas Auger electrons appear at fixed energies regardless of the

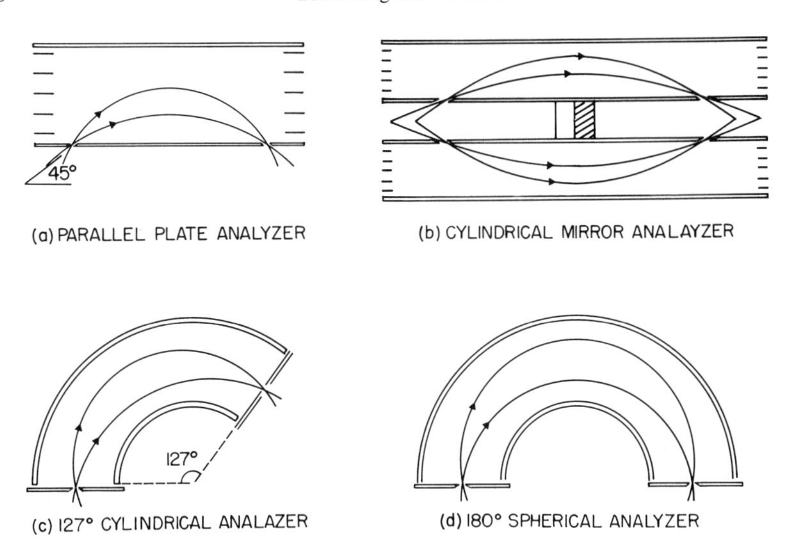

Fig. 21a–d. Typical types of electron energy analyzers in present use in photoelectron spectroscopy

change in photon energy beyond the corresponding inner-shell ionization energy.

The main types of electron energy analyzers used today in photoelectron spectroscopy are shown in Fig. 21. These and other varieties of analyzers are described in detail in [*10*]–[*21*].

If all of the electrons produced were detected and analyzed, then the number of observed electrons ejected from each electronic state would be equal to the number of ions produced in that state. Normally, analyzers sample only those electrons ejected within a given solid angle and in a specific direction. If the angular distribution of the electrons is isotropic, or at least constant for all electronic states, then true branching ratios can be obtained. However, the angular distribution of the electrons depends on their initial kinetic energy, on the energy level from which they are ejected (that is, on the angular momentum of the electron), and on the degree of polarization of the incident radiation. Thus, sampling of the electrons at a fixed angle will not give, in general, a true representation of the branching ratios. It has been shown theoretically[89, 90] that for dipole transitions the number of electrons $dN_j/d\Omega$ ejected per unit solid angle in a specific direction by plane-polarized radiation is given by

$$\frac{dN_j}{d\Omega} \propto (\sigma_j/4\pi)[1+\beta P_2(\cos\theta)], \tag{8.2}$$

where σ_j is the partial photoionization cross section, β is an asymmetry parameter that depends on the photon energy and can take on values ranging from -1 to $+2$, θ is the angle between the electric vector and the direction of

[89] Yang, C.N. (1948): Phys. Rev. *74*, 764

[90] Cooper, J., Zare, R.N. (1968): J. Chem. Phys. *48*, 942

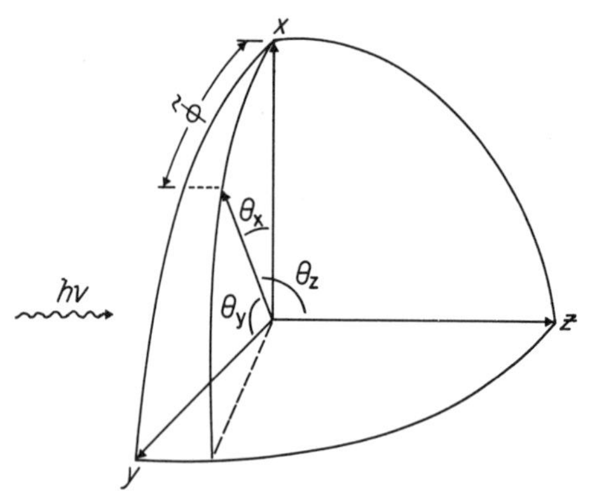

Fig. 22. Coordinate system showing the direction of ejection of a photoelectron with respect to the x, y, and z axis. The photon beam is in the direction of the z axis

the ejected electron, and the Legendre polynomial $P_2(\cos\theta) = 3/2\cos^2\theta - 1/2$. Equation (8.2) has been generalized to include the case of partially plane-polarized or elliptically polarized radiation[91-94] and takes the form

$$\frac{dN_j}{d\Omega} \propto (\sigma_j/4\pi)\{1 + 3/4\beta[(1+P)\cos^2\theta_x + (1-P)\cos^2\theta_y - 2/3]\}, \qquad (8.3)$$

when the photon direction is along the z axis, as shown in Fig. 22, and the angles θ_x and θ_y refer to the direction of the photoelectron with respect to the x and y axes, respectively. Either the x or y axis must be oriented parallel to the direction of maximum polarization or in the direction of the major axis of the ellipse. P is the degree of polarization of the incident radiation and can be defined, for either partially plane-polarized radiation or elliptically polarized radiation, by the relation $P = |(I_x - I_y)/(I_x + I_y)|$. The general expression in (8.3) is necessary because most vacuum uv monochromators produce partially (or elliptically) polarized radiation and synchrotron light sources are all elliptically polarized except in the plane of the orbiting electrons. When $\theta_x = \theta_y = 54°44'$, the expression containing β and P in (8.3) vanishes and the number of electrons observed is independent of their angular distribution. This is also true for analyzers accepting a conical shell centered on 54° 44'.[92,94] Thus all analyzers should observe electrons at this "magic angle" in order to measure true branching ratios. If angles other than 54° 44' are used, then we must have a knowledge of β and P and use (8.3). RON et al.[95] have pointed out that if multipole effects become important, as they do for photon energies in the keV range, then (8.2) is not strictly correct and there would be no magic angle.

---

[91] Samson, J.A.R. (1969): J. Opt. Soc. Am. *59*, 356
[92] Samson, J.A.R., Gardner, J.L. (1972): J. Opt. Soc. Am. *62*, 856
[93] Samson, J.A.R., Starace, A.F. (1975): J. Phys. B*8*, 1806
[94] Schmidt, V. (1973): Phys. Lett. *45* A, 63
[95] Ron, A., Pratt, R.H., Tseng, H.K. (1977): Chem. Phys. Lett. *47*, 377

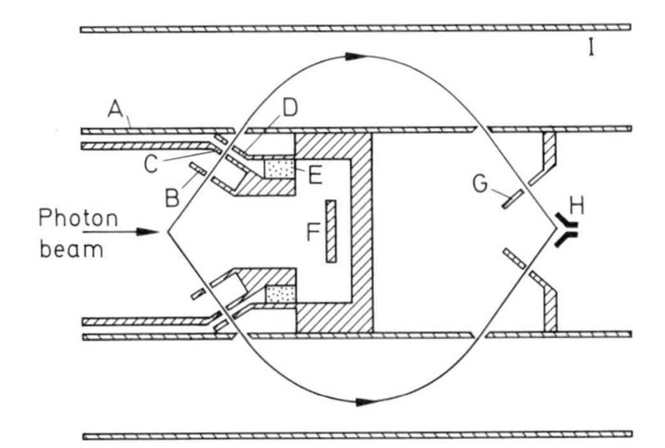

Fig. 23. Cylindrical-mirror electron energy analyzer with retarding/accelerating lens system BCD. The photodiode F monitors the incident-beam intensity. H is a channeltron electron multiplier. A and I represent the inner and outer cylinders of the analyzer

They give the more general expression (for unpolarized radiation),

$$\frac{dN_j}{d\Omega} \propto (\sigma_j/4\pi) \sum_{n=0}^{\infty} B_n P_n(\cos\theta_z), \tag{8.4}$$

with $B_0 \equiv 1$. When the photon energies are less than one or two hundred electron volts, B_1, B_3, B_4, etc., are usually small and (8.4) tends to that obtained by the dipole approximation.

Another important aspect of obtaining true branching ratios is to use a calibrated electron energy analyzer. All analyzers discriminate between electrons of low and high energy. Thus it is imperative the relative luminosity of an instrument be known as a function of electron energy. The luminosity is defined as the ratio of the total number of electrons detected to the total number that are produced. Contact potentials, stray magnetic fields, and fringing electric fields all contribute to the loss of the number of electrons that are detected. An absolute method to calibrate an analyzer is to measure the ratio of the photoelectron signal from a rare gas to the incident light intensity. This should be proportional to the known total photoionization cross sections. Any deviations constitute the necessary correction factors.[96–98]

The problem of calibrating an analyzer can be facilitated if it is equipped with an electrostatic lens system that is capable of accelerating or retarding the electrons such that they are always analyzed at a fixed energy.[97–99] A typical retarding/accelerating lens system for a cylindrical-mirror analyzer is shown in Fig. 23. A photodiode is located at position F to monitor the absolute flux entering the analyzer. The analyzer can be operated in either of two modes. When using the retarding/accelerating lens, the electrons are always analyzed at a constant energy, that is, at a fixed band-pass energy. In the other mode,

---

[96] Gardner, J.L., Samson, J.A.R. (1973): J. Electron. Spectrosc. *2*, 267
[97] Gardner, J.L., Samson, J.A.R. (1975): J. Electron. Spectrosc. *6*, 53
[98] Gardner, J.L., Samson, J.A.R. (1976): J. Electron. Spectrosc. *8*, 469
[99] Poole, R.T., Leckey, R.C.G., Liesegang, J., Jenkin, J.G. (1973): J. Phys. E. *6*, 226

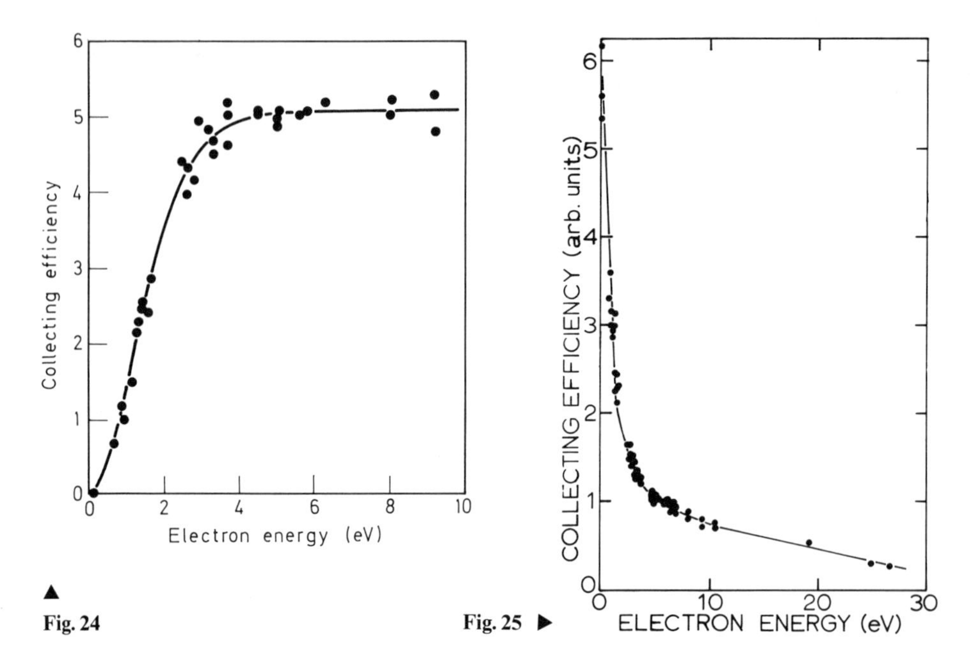

▲ Fig. 24

Fig. 25 ▶

Fig. 24. Collecting efficiency curve of the mirror analyzer without the use of the lens. In this mode of operation the analyzer discriminates against low-energy electrons. (From GARDNER, SAMSON[97])

Fig. 25. Collecting efficiency curve of the mirror analyzer with the use of a retarding/accelerating lens (pass energy = 3V). In this mode of operation the analyzer is extremely efficient for collecting low-energy electrons. (From GARDNER, SAMSON[97])

without the lens, the electrons are analyzed at different energies. The luminosity of the analyzer is quite different for each mode of operation. The collecting efficiency curve without the lens is shown in Fig. 24. In this mode an analyzer tends to discriminate strongly against the low-energy electrons. The reverse is true when a lens system is used, as shown in Fig. 25. By accelerating the low-energy electrons up to the constant pass energy, the fringing field of the lens efficiently collects more low-energy electrons. Thus, for this particular instrument a discrimination of about 5:1 is experienced between 1 and 10 eV electrons.

The techniques of photoelectron spectroscopy are also used to measure the angular distribution of the ejected photoelectrons. The analyzer [usually of the types shown in Fig. 21b, d] is adjusted to accept a small solid angle and then rotated either in a plane containing the incident ionizing radiation, or else in a plane at right angles to it.[100] An alternate method is to polarize the incident radiation then with a fixed energy analyzer rotate the polarizer and hence the plane of polarization.[101,102]

---

[100] See, for example, Carlson, T.A., Jones, A.E. (1971): J. Chem. Phys. *55*, 4913; Houlgate, R.G., West, J.B., Codling, K., Marr, G.V. (1974): J. Phys. B. *7*, L470

[101] Hancock, W.H., Samson, J.A.R. (1976): J. Electron. Spectrosc. *9*, 211

[102] Chaffee, M.A. (1931): Phys. Rev. *37*, 1233

γ) Ion chambers. The ionization chamber when used with a rare gas can measure the absolute intensity of the incident radiation. Then when filled with a molecular gas, because the light intensity is known, the ionization yield of the gas can be measured. Near the ionization threshold it is likely to be <1. At photon energies beyond the double ionization threshold the yield, measured in units of charge per incident photon, will be greater than unity.[103–105] Although such multiple ionization data may be obtained with a mass spectrometer, the ionization chamber has certain advantages. For example, there is no problem in distinguishing between N^+ and N_2^{2+} nor is there a problem if the doubly ionized molecule dissociates, because the ion chamber measures the total charge produced.

To determine multiple ionization cross sections of atoms, the ion chamber has the advantage that it does not dicriminate in collecting ions of various charges. On the other hand, mass spectroscopy is more sensitive and with care or in conjunction with ion chamber measurements good branching-ratio data should be obtainable.

δ) Fluorescence spectrometry. This technique has barely been used to study atomic photoionization in the vacuum uv region. It is a well known and used technique in the x-ray region where the efficiency of fluorescence versus Auger electron emission has been studied. The availability of dedicated sychrotron radiation sources is really necessary for a quantitative study of fluorescence from ionized species, especially when resonances are involved. Usually, a high lying resonance (above the first ionization potential of the atom) will decay via an autoionizing process. For those atoms that do not decay by this process it is desirable to determine the efficiency of autoionization. This can be done by measuring the ionization efficiency of the particular resonance. However, if the efficiency is not quite 100%, it is difficult to say with certainty that no fluorescence exists. That is, we are looking for a small effect on top of a large background. Thus, fluorescence should be investigated and with a tunable light source. Examples of states not fully autoionizing are the $3d'(^3P^0)$, $3s''(^3P^0)$, and $2s\,2p^5\,^3P^0$ levels of atomic oxygen.[106] Other areas of fluorescence in atomic photoionization include double electron excitations (where one electron is bound and the other is ejected into the ionization continuum) and inner shell vacancies with subsequent cascading from the outer shells. Results for He will be discussed in Sect. II α.

With molecular photoionization, fluorescence spectroscopy is especially valuable. This method has been used particularly by JUDGE and his colleagues to obtain quantitative fluorescence data.[107] The methods for studying atomic and molecular fluorescence are identical. A typical arrangement, as used by JUDGE, is shown in Fig. 26. In this arrangement a Czerny-Turner monochro-

---

[103] Samson, J.A.R., Haddad, G.N. (1974): Phys. Rev. Lett. *33*, 875

[104] Cole, B.E., Dexter, R.N. (1978): J. Phys. B *11*, 1011

[105] Samson, J.A.R., Kemeny, P.C., Haddad, G.N. (1977): Chem. Phys. Lett. *51*, 75 (1977)

[106] Samson, J.A.R., Petrosky, V.E. (1974): J. Electron. Spectrosc. *3*, 461

[107] Judge, D.L., Bloom, G.S., Morse, A.L. (1969): Can. J. Phys. *47*, 489; Judge, D.L., Lee, L.C. (1972): J. Chem. Phys. *57*, 455; Lee, L.C., Carlson, R.W., Judge, D.L. (1976): J. Phys. B *9*, 855; and references therein.

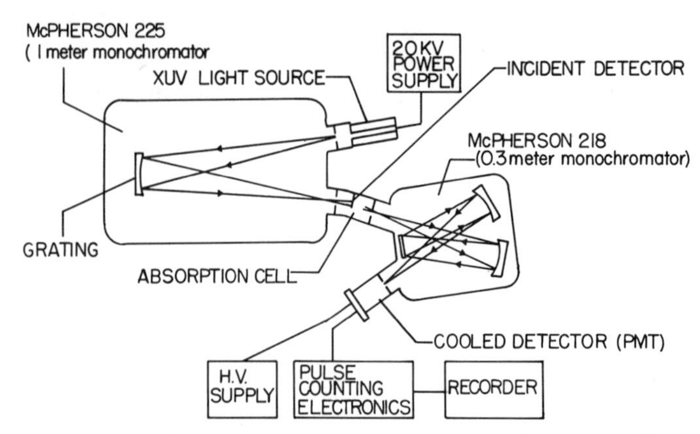

Fig. 26. Fluorescence measuring system. (From JUDGE, LEE[107])

mator is used to observe and disperse the fluorescent radiation. This is a particularly useful instrument because of its large light gathering power. However, its useful short wavelength limit is about 1,100 or 1,200 Å. For shorter wavelengths a high-efficiency normal or grazing-incidence vacuum uv monochromator is required. In some cases a simple narrow-band filter can be used.[108]

9. Electron-impact "photoionization". When electrons collide with atoms and molecules they excite both optical and nonoptical transitions. The possibility of obtaining pure optical information from electron-atom collisions, that is, dipole oscillator strengths for specific transitions, was pointed out by BETHE in 1930.[109] In the collision process electrons can transfer energy to the atom in a manner similar to photons, provided their velocities are much greater than those of the atomic-orbital electrons and provided the interaction distance is large. Then the atom experiences the passage of the electron as a uniform electric field that is pulsed in time (essentially, a δ function). The Fourier transform of such a function will be a spectrum of constant intensity over a large range of frequencies of the electric field. This simulates a continuum or "white light" spectrum. The range of constant intensity will increase as the velocity of the electrons increases.

The differential scattering cross section $d\sigma(K,E)/d\Omega$ is given by[109, 110]

$$\frac{d\sigma(K,E)}{d\Omega}=\frac{2}{E}(k_n/k_0)\frac{1}{K^2}\frac{df(K,E)}{dE}, \tag{9.1}$$

where E is the energy lost by the electron in the collision, k_0 and k_n are the magnitudes of the incident and scattered momenta, and K is the magnitude of

[108] Lee, L.C., Carlson, R.W., Judge, D.L., Ogawa, M. (1974): J. Chem. Phys. *61*, 3261
[109] Bethe, H. (1930): Ann. Phys. *5*, 325
[110] Inokuti, M. (1971): Rev. Mod. Phys. *43*, 297

the momentum transferred. Note, that all quantities in (9.1) are in atomic units, which are used throughout this section. In these units, $e=\hbar=m=1$ and 1 a.u. of energy equals 27.2 eV or 2 Ry.

The generalized oscillator strength $df(K,E)/dE$ is defined as

$$\frac{df(K,E)}{dE}=2E\left|\frac{1}{K}\langle\psi_f|e^{iKr}|\psi_i\rangle\right|^2, \qquad (9.2)$$

where ψ_f and ψ_i are the final and initial state wavefunctions, respectively, and $\boldsymbol{r}$ is the coordinate of the atomic electron under consideration. In (9.2) one assumes the atom is spherical or oriented at random and one presumes the usual average over the magnetic quantum number of ψ_i and the same over the magnetic quantum number of ψ_f. Then, $df(K,E)/dE$ is a function of the scalar variable K^2. When the operator $\exp(i\boldsymbol{K}\cdot\boldsymbol{r})$ is expanded in a power series of K, (9.2) becomes

$$\frac{df(K,E)}{dE}=\frac{df_o}{dE}+K^2\left(\frac{df_o}{dE}\right)'+\frac{K^4}{2!}\left(\frac{df_o}{dE}\right)''+\ldots, \qquad (9.3)$$

where

$$\frac{df_o}{dE}=\frac{df}{dE}(0,E)=2E\langle\psi_f|r\cos(K,r)|\psi_i\rangle$$

is the optical-oscillator-strength density, and $(df_o/dE)'$ and $(df_o/dE)''$ are the first and second derivatives with respect to K^2 of the generalized oscillator strength evaluated at $K^2=0$. In the limit $K\to 0$ the higher terms in the expansion vanish and the generalized-oscillator-strength density becomes equal to the optical-oscillator-strength density (which in turn is proportional to the total absorption cross section σ_a, see (3.11)). When $K\to 0$ this implies very little momentum transfer and hence a large impact distance. It also implies a very small scattering angle. However, this depends on the energy of the incident electrons. For small-angle scattering, K^2 is given by

$$K^2=(E^2/2E_0)+2E_0\theta^2, \qquad (9.4)$$

where E_0 is the incident energy of the electron, E is the energy loss, and θ is the angle of scattering. Thus, K^2 is finite at any E_0 even for $\theta=0$, but it can be made to be small by increasing E_0 and using $\theta\sim 0$. Therefore, using fast electrons it is possible to make quantitative "optical" measurements in which the energy loss E simulates the photon energy.

Measurements of the generalized oscillator strengths and their application to obtain optical oscillator strengths have been pioneered by LASSETTRE and coworkers.[111] They measured $df(K,E)/dE$ over a range of K^2 values and extrapolated $df(K,E)/dE$ to $K^2=0$. Most of this work was below the ionization limit of the atoms and molecules studied.

VAN DER WIEL[112–114] has exploited the dipole-approximation method by

[111] Lassettre, E.N., Skerbele, A. (1974): Inelastic electron scattering, in: Molecular physics, 2nd edn., Williams, D. (ed.), Vol. 3B, pp. 868–951. New York: Academic Press

[112] Van der Wiel, M.J. (1970): Physica *49*, 411

[113] Van der Wiel, M.J., Wiebes, G.: (1971): Physica *54*, 411; (1971): *53*, 225

[114] El-Sherbini, Th.M., Van der Wiel, M.J. (1972): Physica *62*, 119

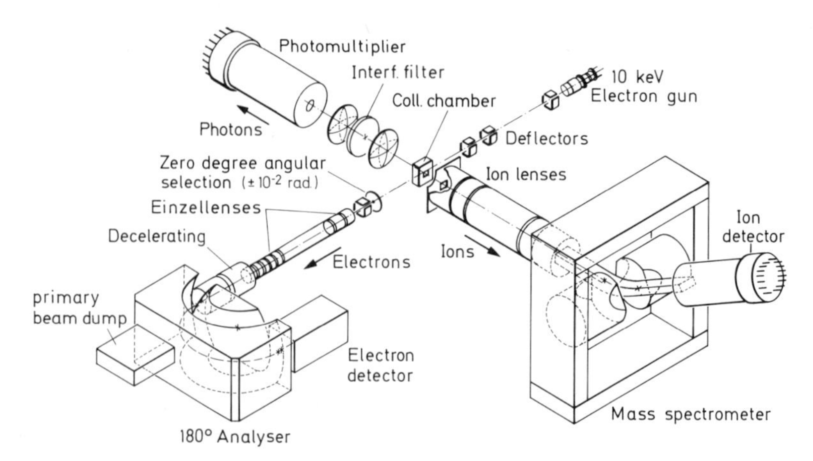

Fig. 27. Electron-impact simulation of a continuous photon source. Electron energy analyzer measures energy loss of scattered electrons, equivalent to the "photon" energy. Ions produced can be identified by the mass spectrometer and any fluorescence produced can be monitored by the photomultiplier-filter system. (From BACKX et al.[120])

using $\theta \sim 0$ (forward scattering) and incident electron energies in the range 3–10 kV. This idea of electron-impact simulation of a continuous photon source for analysis of ionization products was applied by VAN DER WIEL and colleagues for the first time in an electron-ion coincidence experiment.[112–114] This experiment is the analogue of photoionization mass spectrometry when the ions are mass analyzed and detected in delayed coincidence with the scattered electrons. Figure 27 illustrates what measurements are made. The primary electrons are fired into the collision cell and enter, along with the forward-scattered electrons, an electron energy analyzer. The primary beam is barely deflected and enters a beam dump, whereas the forward-scattered electrons are energy analyzed to give E, the energy-loss spectrum. When the signal from the analyzer is detected in coincidence with the mass spectrometer signal, photoionization mass spectrometry is simulated. The figure also illustrates that fluorescent photons may be detected. If they are detected in coincidence with the forward-scattered electrons, photofluorescence is simulated. If they are detected in triple coincidence with the ions, fluorescence from specific states of ionization can be determined.[115] When the mass spectrometer is replaced with an electron energy analyzer to detect the ejected electrons and these are measured in coincidence with the scattered electrons, photoelectron spectrometry is simulated.[116–118] The apparatus as used by BRION and colleagues is shown in Fig. 28. The energy analyzer that observes the ejected electrons is allowed to rotate about the collision chamber. This permits measurements of β, the anisotropy parameter. HAMNETT et al.[119] showed that if the analyzer was

[115] Backx, C., Dlewer, M., Van der Wiel, M.J. (1973): Chem. Phys. Lett. *20*, 100

[116] Van der Wiel, M.J., Brion, C.E. (1972–1973): J. Electron Spectrosc. *1*, 309

[117] Brion, C.E. (1975): Radiat. Res. *64*, 37

[118] Tan, K.H., Brion, C.E. (1978): J. Electron. Spectrosc. *13*, 77

[119] Hamnett, A., Stoll, W., Branton, G., Brion, C.E., Van der Wiel, M.J. (1976): J. Phys. B *9*, 945

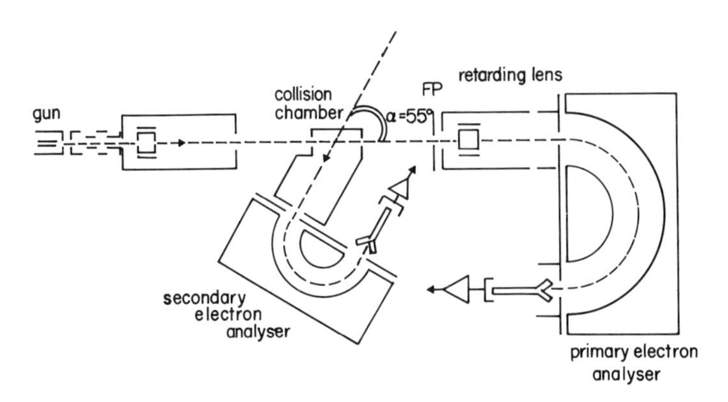

Fig. 28. "Photoelectron Spectroscopy" apparatus by electron-impact technique. (From BRION[117])

set at 54° 44′ with respect to the incident beam the results were independent of the parameter β just as in the case of photon-impact experiments. Finally, if instead of energy analyzing the ejected electrons they were simply all collected and measured in coincidence with the scattered signal, then the total photoionization cross section would be simulated. Thus, we see that this method can provide exactly the same optical information as photon-impact experiments. To obtain absolute cross sections the data must either be normalized to photon-impact results or else data must be taken over a sufficiently large energy-loss range to enable application of the oscillator sum rule, namely, that the total oscillator strength must equal the number of electrons in the atom. Care must be exercised because it is not always clear what are the contributions of the higher terms in the expansion given by (9.3). Although K^2 may be small $(df_0/dE)'$ may be sufficiently large to cause error if that term is neglected. These experiments are now performed at two different angles of scattering (still in the small-angle scattering mode).[120] Neglecting terms of K^4 and higher it is possible to obtain both df_0/dE and $(df_0/dE)'$. This has resulted in more precise data. The technique appears to be extremely powerful and should provide valuable data on "photoionization". Some recent reviews on the technique have been provided by VAN DER WIEL,[121] BRION,[117] and BACKX.[122]

III. Photoionization measurements

Measurements of the photoionization of atoms should include atoms in excited and ionic states (positive and negative) as well as ground-state neutral

[120] Backx, C., Tol, R.R., Wight, G.R., Van der Wiel, M.J. (1975): J. Phys. B *8*, 2050

[121] Van der Wiel, M.J. (1973): The physics of electronic and atomic collisions, 8th ICPEAC Conference, Cobic, B.C., Kurepa, M.V. (eds.), pp. 417–442. Belgrade: Institute of Physics

[122] Backx, C., Van der Wiel, M.J. (1974): Coincidence measurements with electron impact excitation, in: Vacuum ultraviolet radiation physics, Koch, E.E., Haensel, R., Kunz, C. (eds.), pp. 137–153. Oxford-Braunschweig: Pergamon/Vieweg

atoms. In addition, different classes of atoms should be studied, for example, closed-shell versus open-shell atoms. Specific processes should be studied, such as probing autoionizing resonances, electron angular distributions, spin polarization of the photoelectrons, alignment of the resulting ions' total angular momentum, and fluorescence. Most of these projects have scarcely been touched, with the exception of photoionization of the neutral rare gases. Below we will discuss the present status of photionization measurements.

10. Total photoionization cross sections

α) The rare gases. Total photoionization cross sections of the rare gases have been measured from their respective ionization thresholds down into the x-ray region. The accuracy of the measurements vary from about 3% to 10% over the entire wavelength range studied. The results of these measurements have been tabulated and reviewed by SAMSON [5] and by WEST and MARR.[123] At present, measurements are being repeated in the author's laboratory in an effort to obtain results accurate to 1–2%. This level of accuracy will be needed as refinements in theory continue to bring the measured and calculated values of the cross sections closer together. The present status of the cross-section measurements of the rare gases is summarized in Fig. 29. The absolute absorption cross sections are plotted as a function of the incident photon energy, starting at the $^2P_{1/2}$ threshold of the ion. As mentioned earlier (Sect. 3), if we neglect scattering, the total absorption cross section is identical to the total photoionization cross section provided a single photon produces a single ion.

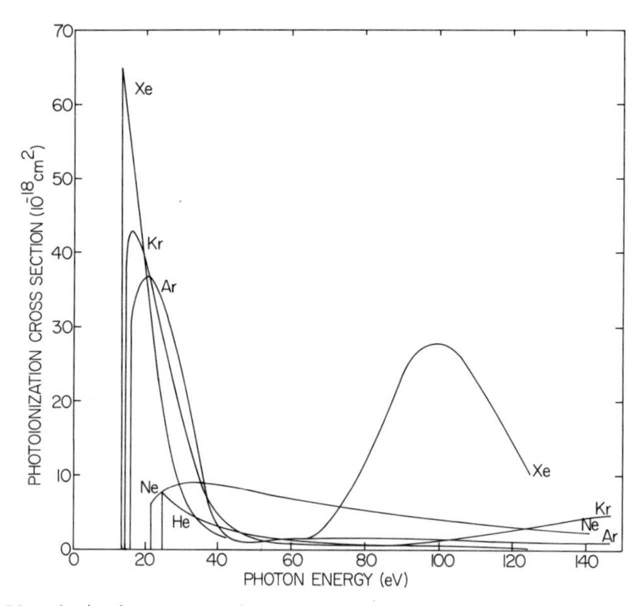

Fig. 29. Photoionization cross sections of the rare gases as a function of photon energy

[123] West, J.B., Marr, G.V. (1976): Proc. R. Soc. London A*349*, 397

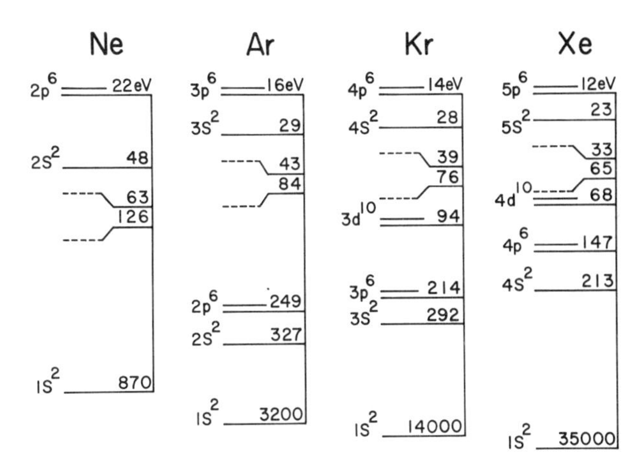

Fig. 30. Energy-level diagram for Ne, Ar, Kr, and Xe. Dashed lines indicate double and triple ionization thresholds

The ion may be multiply charged but it is still a single ion. Thus, the *total photoionization* cross section is the sum of the partial cross sections that give singly, doubly, etc., charged ions. A molecule, of course, has a dissociative absorption channel that can produce neutral fragments with kinetic energy (and hence no ion is observed) or a doubly charged ion can dissociate producing two singly charged ions. Thus, we are not free to equate absorption with the production of a single ion in the case of molecular photoionization.

We can see from Fig. 29 that most of the oscillator strength for ejecting the outer *s*- and *p*-shell electrons from Ar, Kr, and Xe occurs within the first 25 to 30 eV above the ionization threshold. The remaining oscillator strength does not decrease monotonically for higher photon energies because the oscillator strength transfers from one channel to another. We will illustrate this result later when we discuss partial cross sections for ejecting 5*s*- and 5*p*-shell electrons from xenon. It is interesting to note that at the 4*d*-shell threshold (68 eV) there is no sudden step in the cross section. In fact, the cross section starts to increase before this threshold is reached and continues to increase smoothly beyond the threshold. The delayed onset is caused by an effective potential barrier (see discussion by STARACE [*1*]). A similar effect occurs in Kr. In this case the 3*d* threshold lies at 94 eV but the maximum cross section for *d*-shell ejection is not reached until 150 eV (82.7 Å).

The continuum photoionization cross sections are not without structure, although for simplicity we have shown them this way in Fig. 29. It is to be expected that resonances will occur at energies corresponding to the excitation of inner-shell electrons. These excited levels lie above the first ionization potential of the atom and may interact with the ionization continuum causing autoionizing transitions. The electron is ejected with the same kinetic energy as an electron ejected directly into the continuum. Figure 30 shows an energy-level diagram for Ne, Ar, Kr, and Xe. The electron binding energy for each shell is given in electron volts. Preceding the threshold for ejection of these

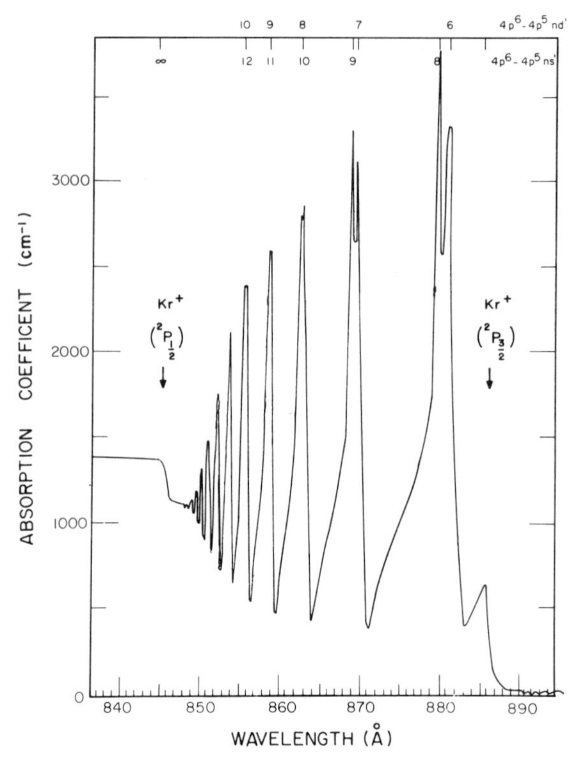

Fig. 31. Absorption coefficients of Kr as a function of wavelength in the autoionizing region between $^2P_{1/2}$ and $^2P_{3/2}$. The step in the curve at the $^2P_{1/2}$ threshold is an instrument artifact and is caused by the finite bandwidth of the monochromator, which was about 0.5 Å. (HUFFMAN, TANAKA, LARRABEE, private communication)

electrons discrete autoionizing structure occurs in the continuum photoionization cross section. Depending on the nature of the interaction between the discrete and continuum states, the cross section may show an increase, a decrease, or an asymmetric shape in the cross section. Such structure was first observed by BEUTLER[124] in his studies of the rare gases between the $^2P_{3/2}$ and $^2P_{1/2}$ states of the ion. Examples of these autoionizing states are shown in Figs. 31 and 32 for Kr and Xe, respectively (see also Fig. 9). The step observed at the $^2P_{1/2}$ threshold is an instrument artifact. This occurs because the bandwidth of the monochromator was much greater than the width of the resonances near the series limit. This has the effect of reducing the value of the peak-absorption cross section of a discrete line as discussed in Sect. 3α. Quantum theory predicts a smooth transition of the cross section between the discrete levels and the continuum.[49a] Thus, a step in the cross section is expected at the $^2P_{1/2}$ threshold when a large bandpass is used. The height of the step decreases with higher wavelength resolution (smaller bandpass) and lower gas pressure.

[124] Beutler, H. (1935): Z. Phys. *93*, 177

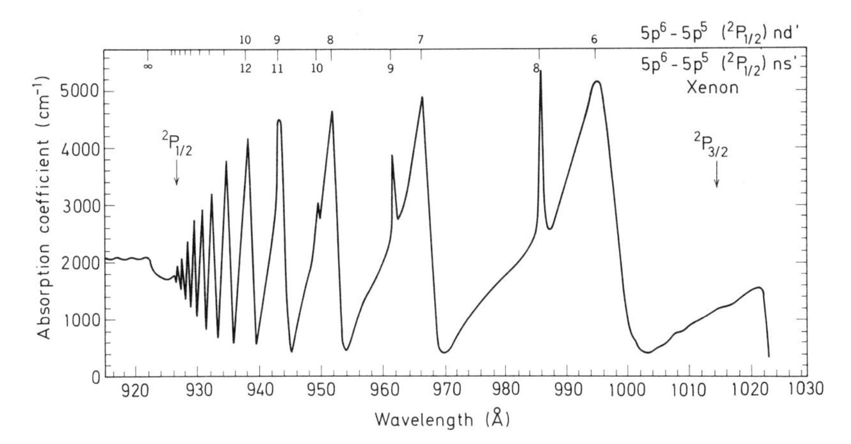

Fig. 32. Absorption coefficients of Xe as a function of wavelength in the autoionizing region between $^2P_{1/2}$ and $^2P_{3/2}$. The step in the curve at the $^2P_{1/2}$ threshold is an instrument artifact and is caused by the finite bandwidth of the monochromator, which was about 0.5 Å. (From HUFFMAN, TANAKA, LARRABEE[155])

FANO[125,126] has shown that the profile of an absorption line in the ionization continuum can be represented by the formula

$$\sigma(E)=\sigma_a[(q+\varepsilon)^2/(1+\varepsilon^2)]+\sigma_b, \tag{10.1}$$

where ε represents the energy difference between the incident photon energy (E) and the idealized resonance energy (E_r) and is expressed in units of the half-width ($\Gamma/2$) of the resonance line. That is,

$$\varepsilon=2(E-E_r)/\Gamma \tag{10.2}$$

$\sigma(E)$ represents the absorption cross section, whereas σ_a and σ_b represent the portions of the continuum cross section that, respectively, do and do not interact with the discrete autoionizing state. The parameter q is a numerical index that characterizes the line profile. The cross section that would be observed in the absence of the autoionizing state is $\sigma_c=\sigma_a+\sigma_b$. Thus,

$$\sigma(E)/\sigma_c(E)=1+\rho^2(q^2-1+2q\varepsilon)/(1+\varepsilon^2), \tag{10.3}$$

where $\rho^2=\sigma_a/(\sigma_a+\sigma_b)$. Figure 33 shows typical line profiles for excitation of s-shell electrons that have been observed and that differ quite dramatically from one another.[49a] A photograph of the original absorption spectrum taken by MADDEN and CODLING[127] of the rare gases, in the region of s-shell excitation, is shown in Fig. 34. Figure 35 shows the first measurements of absolute cross

---

[125] Fano, U. (1961): Phys. Rev. *124*, 1866
[126] Fano, U., Cooper, J.W. (1965): Phys. Rev. *137*, A1364
[127] Madden, R.P., Codling, K. (1963): Phys. Rev. Lett. *10*, 516

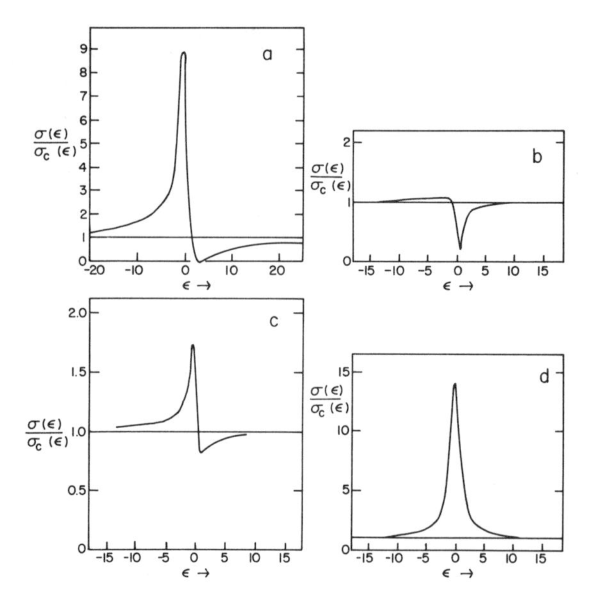

Fig. 33a–d. Profiles of autoionization lines in the rare gases. (**a**) He $2s\,2p(^1P^0)$, $q=-2.8$ and $\rho^2=1$. (**b**) Ar $3s\,3p^6\,4p(^1P^0)$, $q=-0.22$ and $\rho^2=0.86$. (**c**) Ne $2p^4(^3P)\,3s\,3p(^1P^0)$, $q=-2.0$ and $\rho^2=0.17$. (**d**) Xe $4d^9\,5s^2\,5p^6\,6p(^1P^0)$, $q\sim200$ and $\rho^2\sim0.0003$. (From FANO, COOPER[49a])

sections within these resonances for Kr.[128] Where measurements exist, the ionization efficiency within the resonances are found to be unity.[129] Thus, the total absorption cross sections can be equated to the total photoionization cross sections in these cases. The decrease in the absorption cross section at these resonances can be visualized as a collective oscillation of the electrons in both the *s*- and *p*-shells, analogous to the effect in solids where a metal becomes transmitting at the plasma frequency (i.e., collective oscillations of the free electrons).[130] For many experiments it is desirable to know the position and nature of these autoionizing structures. Most of the lines observed from the ionization threshold of the rare gases to about 150 eV have been tabulated by CODLING and MADDEN.[131–136] EDERER[137] has discussed the cross-section profiles of many lines in Kr and Xe.

The accurate measurements of the photoionization cross sections of the rare gases have provided a stimulus to theoretical work to understand the photoionization process. The early independent-particle model calculations could

[128] Samson, J.A.R. (1963): Phys. Rev. *132*, 2122
[129] Samson, J.A.R. (1964): J. Opt. Soc. Am. *54*, 6
[130] Bohm, D., Pines, D. (1951): Phys. Rev. *82*, 625
[131] Codling, K., Madden, R.P. (1964): Phys. Rev. Lett. *12*, 106
[132] Madden, R.P., Codling, K. (1965): Astrophys. J. *141*, 364
[133] Codling, K., Madden, R.P. (1965): Appl. Opt. *4*, 1431
[134] Codling, K., Madden, R.P. (1967): Phys. Rev. *155*, 26
[135] Madden, R.P., Ederer, D.L., Codling, K. (1969): Phys. Rev. *177*, 136
[136] Codling, K., Madden, R.P. (1972): J. Res. Nat. Bur. (Stand.) 76 A, 1
[137] Ederer, D.L. (1971): Phys. Rev. A *4*, 2263

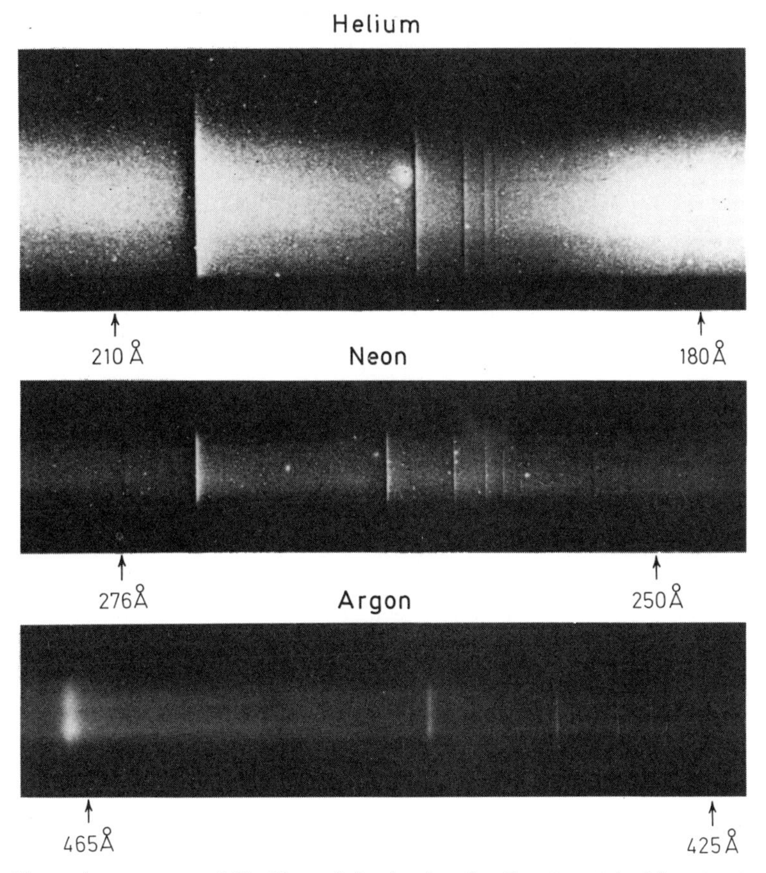

Fig. 34. Absorption spectrum of He, Ne, and Ar showing the discrete autoionizing structure in the photoionization continuum. He and Ne show examples of asymmetric resonances, whereas Ar illustrates window-type resonances. (From MADDEN, CODLING[127])

not reproduce the experimental cross sections of the rare gases heavier than Ne. With the realization that electron correlations were very important, work started on many-body type calculations[138] [*1*]. As a result, theory is in good agreement (within $\pm 10\%$) with the experimental cross sections of the rare gases. This actually represents agreement with the special case of closed-shell atoms. It is not known whether the same theoretical approximations are appropriate for open-shell atoms. Thus, there is a need for good experimental measurements on open-shell atoms to guide theoretical calculations in this area.

β) The alkali metals. With the exception of the rare gases the only other group of atoms that have been extensively studied are the alkali metals. Because of their low-ionization-threshold potentials (2,300–3,200 Å) these were the first atoms to be studied.[15] However, experimentally, it has been difficult to

[138] Ya Amusia, M., Cherepkov, N.A. (1975): Case Stud. At. Phys. 5, 47

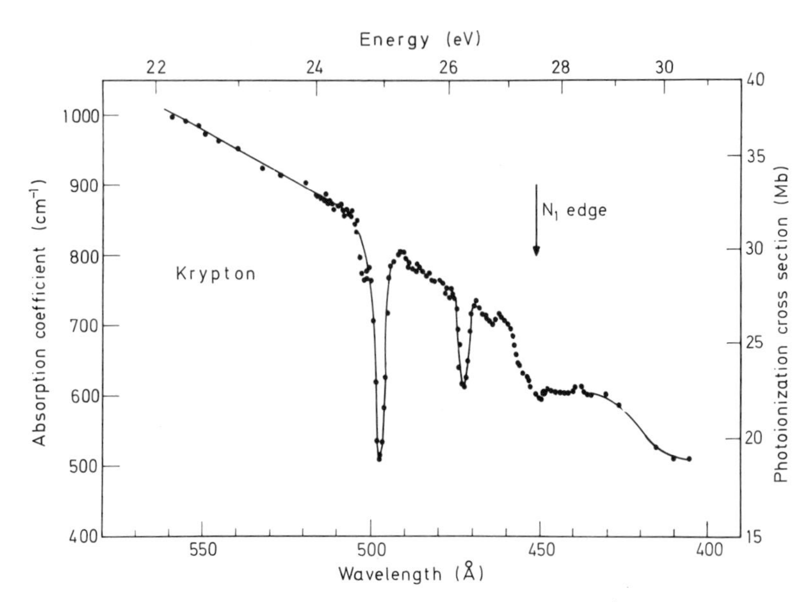

Fig. 35. Photoionization cross sections for Kr between 400 and 460 Å illustrating window-type resonances. (From SAMSON[128])

obtain accurate cross sections because of the problems of measuring accurately the number density of the absorbing atoms and in correcting for the presence of molecular species. The use of heat pipes (Sect. 7δ) should improve the accuracy of number-density measurements. Experimental values of the cross sections of the alkali metals, between 600 and 3,200 Å, are illustrated in Fig. 36.[139, 140] However, the recent measurements of the Cs cross sections near threshold by COOK et al.[141] are smaller by about a factor of two than the Cs data shown in Fig. 36. The new data of COOK et al. is in agreement with the calculations of WEISHEIT[142] and NORCROSS[143] but disagree with the RPAE calculations of AMUSIA and CHEREPKOV.[138] The experimental results of Na have been difficult to duplicate by theory.[144-147] The cross sections calculated by CHANG[144] are shown in Fig. 36. All of the theoretical results agree in their general shape. However, none reproduce the broad peak at 600 Å. It is not clear whether this is an experimental artifact or not. Thus, it is necessary to remeasure the cross sections in this wavelength region. Probably the best method will be to measure relative cross sections by studying the number of

[139] Marr, G.V., Creek, D.M. (1968): Proc. R. Soc. A *304*, 233
[140] Hudson, R.D., Kieffer, L.J. (1971): At. Data *2*, 205
[141] Cook, T.B., Dunning, F.B., Foltz, G.W., Stebbings, R.F. (1977): Phys. Rev. A *15*, 1526
[142] Weisheit, J.C. (1972): Phys. Rev. A *5*, 1621
[143] Norcross, D.W. (1973): Phys. Rev. A *7*, 606
[144] Chang, T.N. (1974): J. Phys. B *8*, 743
[145] Chang, J.J., Kelly, H.P. (1975): Phys. Rev. A *12*, 92
[146] Ya Amusia, M., Cherepkov, N.A., Pavlin, I., Radojevic, V., Zivanovic, Dj. (1977): J. Phys. B *10*, 1413
[147] Laughlin, C. (1978): J. Phys. B *11*, 1399

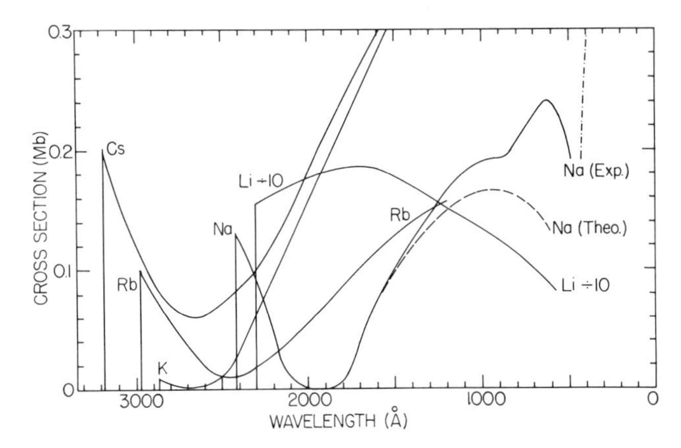

Fig. 36. Photoionization cross sections of the alkali metals as a function of wavelength. Dashed curve represents the calculations by CHANG[144] for Na

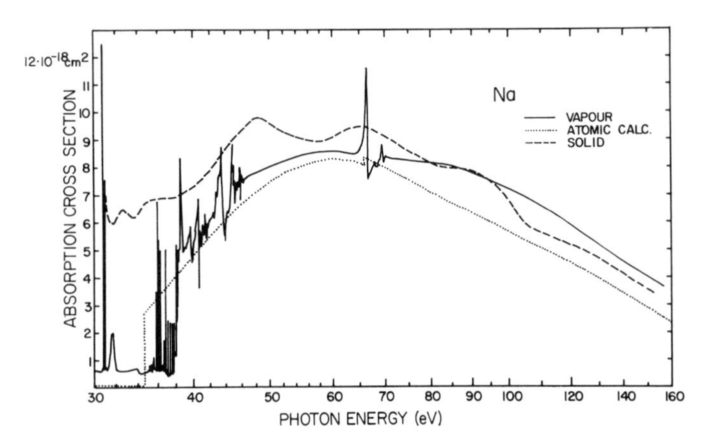

Fig. 37. Relative absorption cross sections of solid and atomic Na normalized to the theoretical resuts of McGUIRE. (From WOLFF et al.[148])

ions produced in a mass spectrometer per incident photon. This removes the problem of impurities or the presence of dimers, which have troubled earlier measurements. With the exception of a small energy interval between 25 and 30 eV, Na has been studied from threshold to a photon energy of 250 eV. This extended region has been measured by WOLFF et al.[148] and CODLING et al.[149] by the use of synchrotron radiation. The results of WOLFF et al. shown in Fig. 37, are given for the solid and vapor phase and compared with theory.[148] The peak cross section of the continuum in this region is about thirty to eighty times larger than that found at threshold (Fig. 36). This is caused by the higher oscillator strength for ejection of the $2p$ electrons. The resonant structure caused by excitation of the $2p^6$ electrons can be seen in the region between 35

[148] Wolff, H.W., Radler, K., Sonntag, B., Haensel, R. (1972): Z. Phys. *257*, 353

[149] Codling, K., Hamley, J.R., West, J.B. (1977): J. Phys. B *10*, 2797

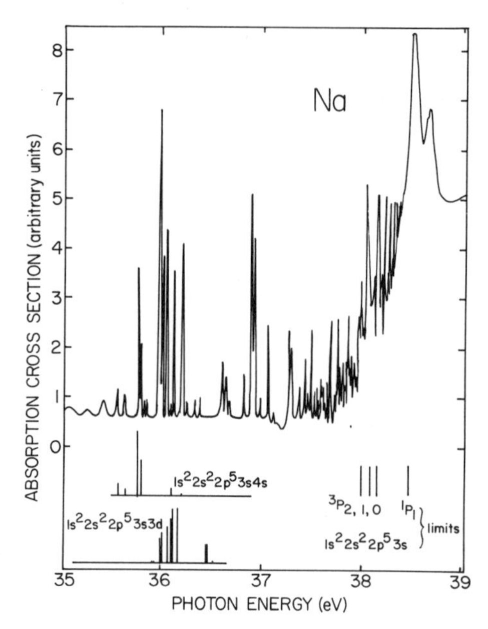

Fig. 38. Relative absorption cross section of atomic Na between 35 and 39 eV showing the autoionizing structure produced by excitation of the $2p^6$ electrons. (From WOLFF et al.[148])

and 39 eV. A more detailed curve of this region is shown in Fig. 38. The structure is presumably caused by the following transitions,

$$2p^6\,3s \rightarrow 2p^5\,3s\,4s$$
$$\rightarrow 2p^5\,3s\,3d.$$

Transitions from the $2s$ level in Na give rise to the strong asymmetric resonances at 66.6 eV and 69.6 eV. WOLFF et al.[148] have made a fit of the line profiles to 10.1 and obtained q values of -2 and -1.8, respectively.

The cross sections for Cs have also been extended by measurements using synchrotron radiation.[150] The results for excitation and ionization of a $4d$ electron are shown in Fig. 39 and are compared with the calculations of AMUSIA.[151] The gross features of the spectrum are similar to the Xe $4d$ absorption (Fig. 29) and again represents an example of a delayed transition from the $4d$ shell.[152,153]

Although no cross-section measurements exist between 600 and 1,000 Å for Cs, the absorption spectrum has been observed by BEUTLER and

---

[150] Petersen, H., Radler, K., Sonntag, B., Haensel, R. (1975): J. Phys. B *8*, 31

[151] Ya Amusia, M. (1974) in: Vacuum ultraviolet radiation physics, Koch, E.E., Haensel, R., Kunz, C. (eds.), p. 205. Oxford-Braunschweig: Pergamon/Vieweg

[152] Ederer, D.L. (1964): Phys. Rev. Lett. *13*, 760

[153] Haensel, R., Keitel, G., Schreiber, P., Kunz, C. (1969): Phys. Rev. *188*, 1375

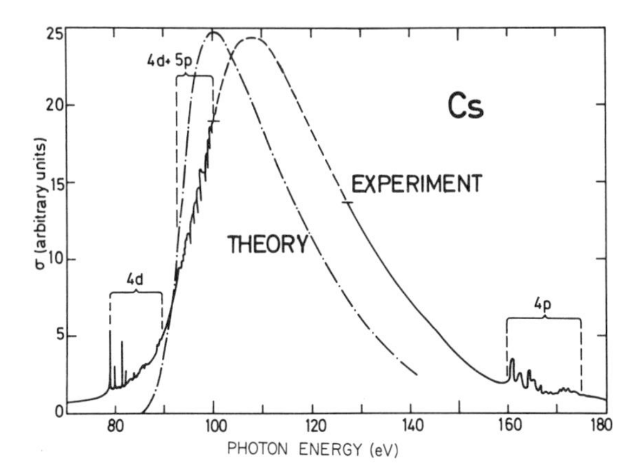

Fig. 39. Relative absorption cross section of Cs vapor as a function of photon energy. For comparison, AMUSIA's theoretical results[151] are included. (From PETERSON et al.[150])

GUGGENHEIMER.[153a] They observed and cataloged numerous absorption lines (presumably autoionizing) of the type $sp^5(^{1,3}P_{1,2})nd, ms$ leading to the Cs II $^3P_{1,2}$ limits at 17.21 and 17.27 eV and to the $^{1,3}P_{0,1}$ limits at 19.07 and 19.13 eV.

γ) Cross sections of some miscellaneous atoms. The absorption of radiation by nearly half of the elements in the periodic table have been studied to some extent. However, absolute absorption cross sections of only a few elements have been measured and then only over limited wavelength regions. Nevertheless, a general picture of the photoionization process is beginning to emerge. For example, the giant resonances caused by the delayed onset of a potential barrier have been observed frequently (e.g., Cs, Xe, Ba, and the rare earths[154]), the picturesque autoionization series shown in Fig. 32 for Xe is common to all the rare gases and Hg[155–157], and the asymmetric resonances shown in Figs. 33 and 34 are found in virtually all atomic species and even in some molecules. Many atoms with outer *s*-shell electrons appear to have hydrogen- and helium-like cross sections in the vicinity of their ionization thresholds, that is, a step function at threshold and an immediate monotonic decrease beyond threshold [5]. Actually, this decrease is produced because of correlation effects between electrons within the same shell and with electrons in other shells. It is the purpose of this section to present the results for a few atoms to illustrate the appearance of the photoionization cross sections in the vicinity of their ionization thresholds.

[153a] Beutler, H., Guggenheimer, K. (1934): Z. Phys. *88*, 25
[154] Mansfield, M.W.D., Connerade, J.P. (1976): Proc. R. Soc. London A *352*, 125
[155] Huffman, R.E., Tanaka, Y., Larrabee, J.C. (1963): J. Chem. Phys. *39*, 902
[156] Brehm, B. (1966): Z. Naturforsch, *21* a, 196
[157] Garton, W.R.S., Connerade, J.P. (1969): Astrophys. J. *155*, 667

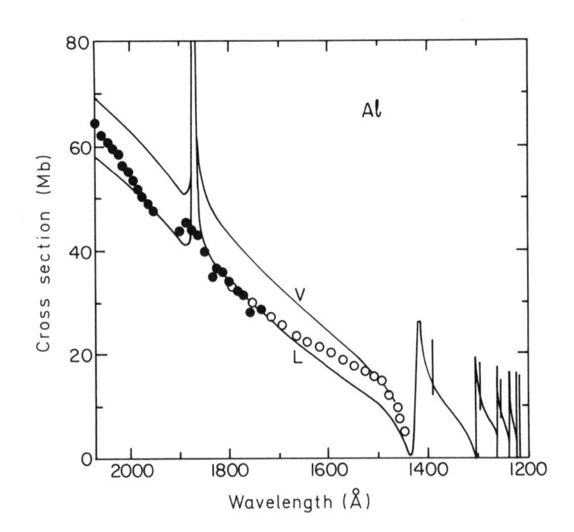

Fig. 40. Photoionization cross section of aluminum vapor as a function of wavelength. *Experimental:* Solid circles, KOHL, PARKINSON[160]; open circles, ROIG[161]. *Theoretical:* solid line, length (*L*) and velocity (*V*) approximations, DOURNEUF et al.[163]

Al: (Ne) $3s^2 3p^1$: Ionization threshold = 5.986 eV or 2,071.3 Å

Absolute cross sections for Al vapor and its series (Ga, In, and Tl) have been measured near the ionization threshold by several groups.[158–162] Figure 40 illustrates the results for Al as obtained by KOHL and PARKINSON[160] and by ROIG.[161] The theoretical results of DOURNEUF et al.[163] are also shown for the dipole velocity (*V*) and length (*L*) approximations. ROIG's data is relative and is placed on an absolute scale with reference to the results of KOHL and PARKINSON, whose data in turn are relative to the oscillator strength of the $3p\,^2P^0 \rightarrow 4s\,^2S^0$ transition of neutral aluminum that was observed at the same time. Both sets of data used photographic absorption techniques.

Ba: (Xe) $6s^2$: Ionization threshold = 5.211 eV or 2,379.1 Å

The absolute cross section of Ba has been measured by HUDSON et al.[164] from the ionization threshold down to 1,700 Å. Their results at threshold and betwen 1,700 and 2,100 Å are shown in Fig. 41. Photographic absorption spectra have been taken in the spectral regions 2,100–2,379.8 Å[164a], 1,560–

[158] Kozlov, M.G., Nikonova, E.I., Startsev, G.P. (1966): Opt. Spectrosc. *21*, 298; (1969): *27*, 383 [(1969): Opt. Spektrosk. *21*, 532; (1969): *27*, 704]

[159] Kozlov, M.G., Startsev, G.P. (1968): Opt. Spectrosc. *24*, 3 [(1967): Opt. Spektrosc. *24*, 8]

[160] Kohl, J.L., Parkinson, W.H. (1973): Astrophys. J. *184*, 641

[161] Roig, R.A. (1975): J. Phys. B *8*, 2939

[162] Marr, G.V., Heppinstall, R. (1966): Proc. Phys. Soc. *87*, 547; (1966): *87*, 293

[163] Le Dourneuf, M., Lan, Vo. Ky., Burke, P.G., Taylor, K.T. (1975): J. Phys. B *8*, 2640

[164] Hudson, R.D., Carter, V.L., Young, P.A. (1970): Phys. Rev. A *2*, 643

[164a] Garton, W.R.S., Codling, K. (1959): Proc. Phys. Soc. (London) *75*, 87; Garton, W.R.S., Parkinson, W.H., Reeves, E.M. (1962): Proc. Phys. Soc. (London), *80*, 860

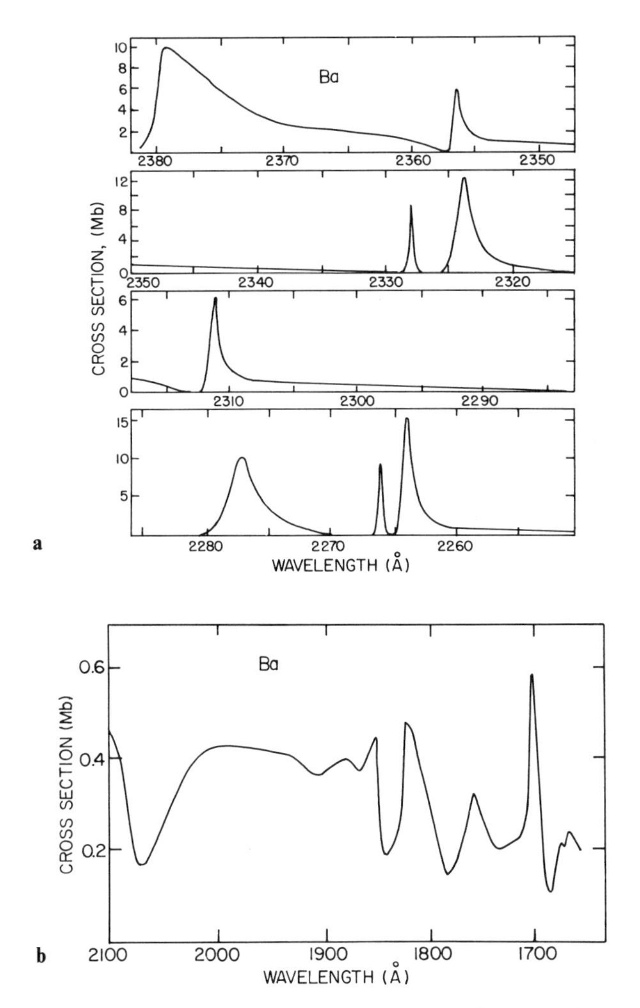

Fig. 41a, b. Photoionization cross section of Ba vapor as a function of wavelength. (**a**) The autoionizing series $6s^2(^1S_0)$-$5dnp(^1P, ^3P, ^3D)$ and $6s^2(^1S_0)$-$5dnf(^1P, ^3P)$ and (**b**) a series suggested to be the $6pns$ transitions. (From HUDSON et al.[164])

1,770 Å [164b], 10–200 Å [165, 166], and 455–912 Å [167]. These regions are all dominated by discrete structure. Of particular interest is the chance coincidence of the 584 Å He I resonance line with an unidentified autoionizing peak. BREHM and colleagues[168, 169] and HOTOP and MAHR[169a] have observed anomolous

[164b] Brown, C.M., Ginter, M.L. (1978): J. Opt. Soc. Am. *68*, 817

[165] Connerade, J.P., Mansfield, M.W.D. (1974): Proc. R. Soc. London A *341*, 267

[166] Ederer, D.L., Lucatorto, T.B., Saloman, E.B., Madden, R.P., Sugar, J. (1975): J. Phys. B *8*, L 21

[167] Connerade, J.P., Tracy, D.H. (1977): J. Phys. B *10*, L 235

[168] Brehm, B., Bucher, A. (1974): Intern. J. Mass Spectrom. Ion Phys. *15*, 463

[169] Brehm, B., Höfler, K. (1975): Intern. J. Mass Spectrom. Ion Phys. *17*, 371

[169a] Hotop, H., Mahr, D. (1975): J. Phys. B *8*, L 301

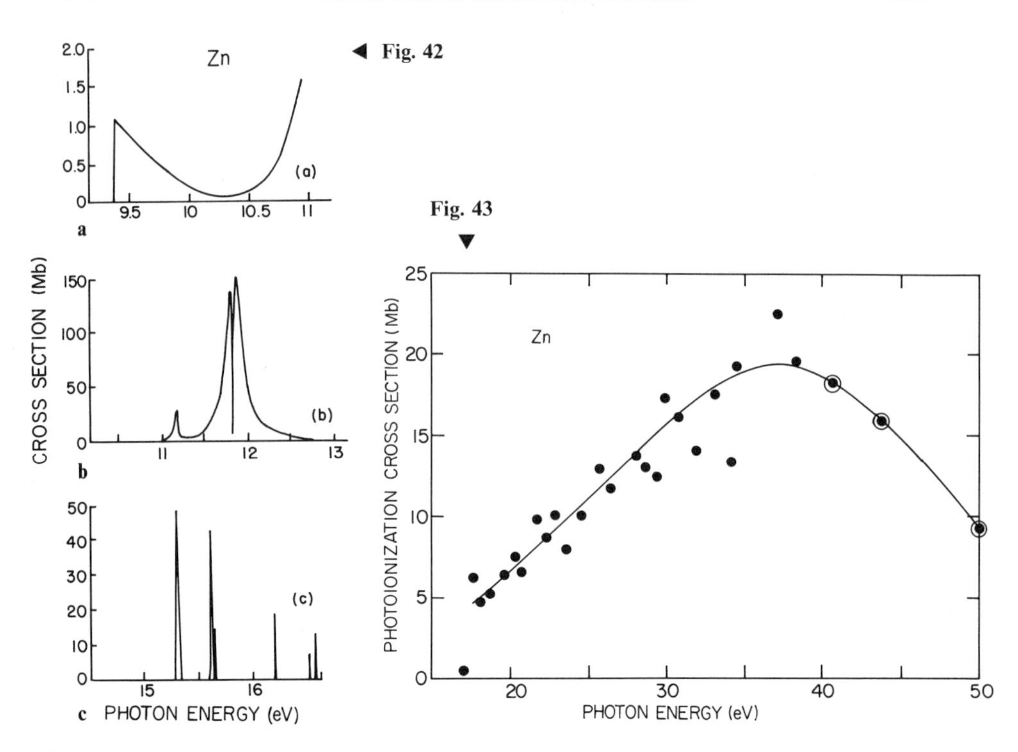

Fig. 42a–c. Photoionization cross section of zinc vapor as a function of photon energy from the ionization threshold to 16.5 eV. (From MARR, AUSTIN[170])

Fig. 43. Photoionization cross section of zinc vapor as a function of photon energy between 15 and 50 eV. There may be some question about the accuracy of the circled data points beyond 40 eV. (From HARRISON et al.[171])

photoelectron spectra using the 584 Å line, whereas a more normal spectrum is obtained with the Ne I line at 736 Å. They observed a ratio for Ba^{2+}/Ba^{+} of 2.4 at 584 Å compared with a ratio of 0.25 at 736 Å.

Zn: (Ar) $3d^{10}4s^2$*: Ionization threshold = 9.394 eV or 1,319.8 Å*

MARR and AUSTIN[170] have measured the absolute absorption cross sections of Zn vapor from threshold to 748 Å. Their results are shown in Fig. 42. At threshold Zn shows He-like behavior with a characteristic decrease at higher photon energies. The pure continuum cross section is interrupted by the autoionizing Rydberg series $3d^{10}4s^2 \rightarrow 3d^9 4s^2(np)$, which dominates the cross-section curve in this energy region. Beyond 17.17 eV the $3d^{10}$ channel opens up and the continuum cross section increases. HARRISON et al.[171] have measured the relative cross sections in this region by use of a mass spectrometer and a sodium salicylate coated detector. Their results are shown in Fig. 43 but have

---

[170] Marr, G.V., Austin, J.M. (1969): J. Phys. B *2*, 107

[171] Harrison, H., Schoen, R.I., Cairns, R.B. (1969): J. Chem. Phys. *50*, 3930

been put on an absolute scale by normalization to the data of MARR and AUSTIN. The normalization is only approximate ($\pm 10\%$) and is intended to give an idea of the magnitude of the $3d^{10}\,4s^2 \rightarrow 3d^9\,4s^2(\varepsilon p)$ continuum absorption.

Calculations of the continuum cross sections have been made by FLIFLET and KELLY from 20 eV to 1 keV.[172] There is general agreement, within a factor of two, between their results and those shown in Fig. 43. However, the theoretical cross section reaches a maximum at about 60 eV, whereas the experimental curve reaches a maximum at 37 eV. There may be some question about the accuracy of the experimental data between 40 and 50 eV owing to the relative efficiency of sodium salicylate that was used in determining the relative cross sections.

Cd: (Kr) $4d^{10}\,5s^2$*: Ionization threshold = 8.994 eV or 1,378.6 Å*

Cadmium, like zinc, again shows He-like threshold. Figure 44 shows the threshold data of ROSS and MARR.[173] Beyond 11 eV, the $4d^{10}\,5s^2 \rightarrow 4d^9\,5s^2(np)$ autoionizing series is the main absorption feature, shown in Fig. 45a, b from the data of MARR and AUSTIN.[174] Figure 45c shows the start of the $4d^9(\varepsilon p)$ continuum, and Fig. 46 shows the continuation of this continuum as obtained by CODLING et al.[175] by use of synchrotron radiation. Figure 46 also shows the experimental data of CAIRNS et al.[176] and the theoretical results of CARTER and KELLY.[177]

Hg: (Xe) $4f^{14}\,5d^{10}\,6s^2$*: Ionization threshold = 10.438 eV or 1,187.9 Å*

The continuum ionization of Hg at threshold, caused by the ejection of the outer $6s$ electron, cannot be observed because it overlaps with the discrete autoionizing series $5d^{10}\,6s^2 \rightarrow 5d^9\,6s^2(np)$. The threshold behavior for the $6s\,\varepsilon p$ continuum would be expected to be similar to that of Zn and Cd in the absence of the resonance line. However, the interaction between this continuum and the first member of the above series ($n=6$) completely distorts the expected form of the $6s\varepsilon p$ continuum. BREHM[156] has obtained relative cross sections between 14.5 eV and threshold. These data are shown in Fig. 47 and have been put on an absolute basis by comparison with the absolute measurements of the oscillator strength of the 1,126.6-Å autoionizing line (i.e., $n=6$) as obtained by LINCKE and STREDELE.[178] These autoionizing lines have, of course, been observed in photographic absorption measurements by BEUTLER[179] in 1933 and more recently by GARTON and CONNERADE.[157]

The $5d^9(\varepsilon p)$ continuum cross sections are shown in Fig. 48. This curve represents the smoothed relative cross-section data of CAIRNS et al.[180] that has

[172] Fliflet, A.W., Kelly, H.P. (1976): Phys. Rev. A *13*, 312
[173] Ross, K.J., Marr, G.V. (1965): Proc. Phys. Soc. (London) *85*, 193
[174] Marr, G.V., Austin, J.M. (1969): Proc. R. Soc. A *310*, 137
[175] Codling, K., Hamley, J.R., West, J.B. (1978): J. Phys. B *11*, 1713
[176] Cairns, R.B., Harrison, H., Schoen, R.I. (1969): J. Chem. Phys. *51*, 5440
[177] Carter, S.L., Kelly, H.P. (1976): J. Phys. B *9*, L565
[178] Lincke, R., Stredele, B. (1970): Z. Phys. *238*, 164
[179] Beutler, H. (1933): Z. Phys. *86*, 710
[180] Cairns, R.B., Harrison, H., Schoen, R.I. (1970): J. Chem. Phys. *53*, 96

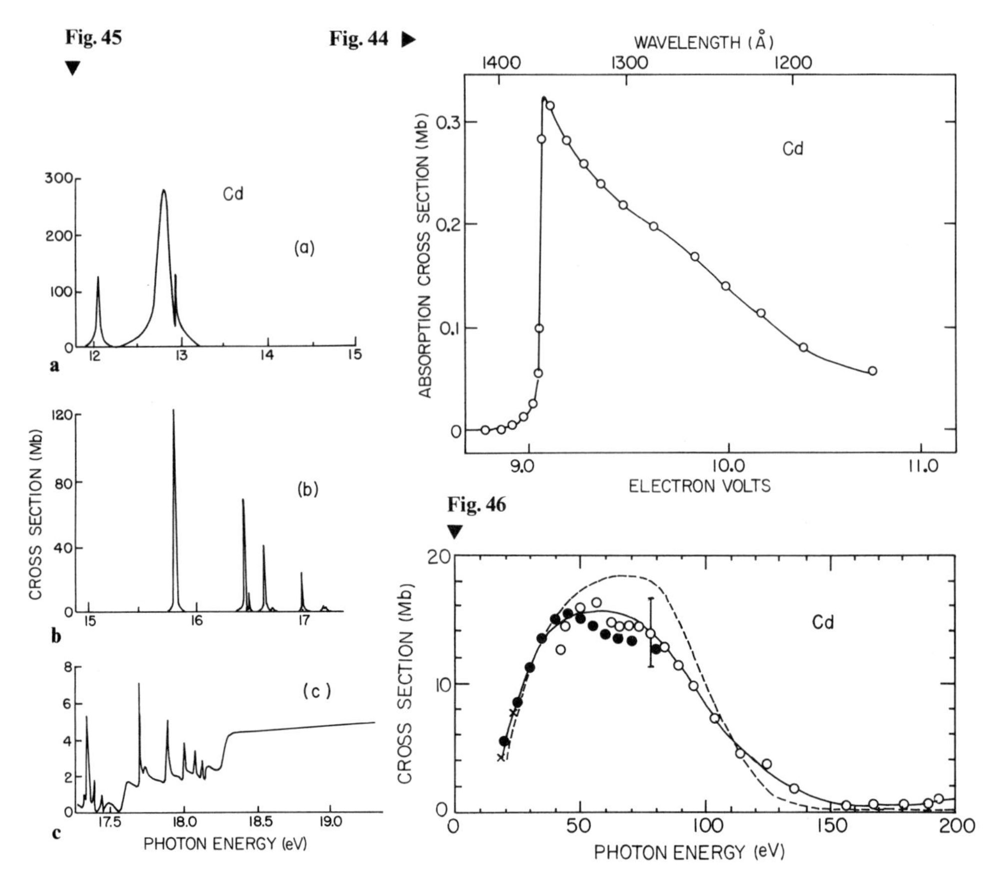

Fig. 44. Photoionization cross section of Cd vapor from threshold to 11 eV. (From ROSS, MARR[173])

Fig. 45a–c. Photoionization cross section of Cd vapor from 12 to 19.5 eV. (From MARR, AUSTIN[174])

Fig. 46. Photoionization cross section of Cd vapor from 20 to 200 eV. (From CODLIG et al.[175])

been normalized to the data of LINCKE and STREDELE[178] by DEHMER and BERKOWITZ[181]. The partial cross sections leading to production of the $6s(^2S_{1/2})$ and $5d^9(^2D_{5/2,\,3/2})$ ionization continuum are also shown.

δ) Ionic- and excited-state atoms. The more specialized photoionization studies of excited- and ionic-state atoms are more difficult and few measurements exist. The only absolute cross-section measurement of a positive ion is that by CAIRNS and WEISSLER[182] who obtained the cross section of Xe^+ at the single wavelength of 555 Å and quoted a value of approximately $1.6 \times 10^{-18}\,\mathrm{cm}^2$. Photographic absorption spectra of several positive ions have

---

[181] Dehmer, J.L., Berkowitz, J. (1974): Phys. Rev. A *10*, 484

[182] Cairns, R.B., Weissler, G.L. (1962): Bull. Am. Phys. Soc. 7, 129

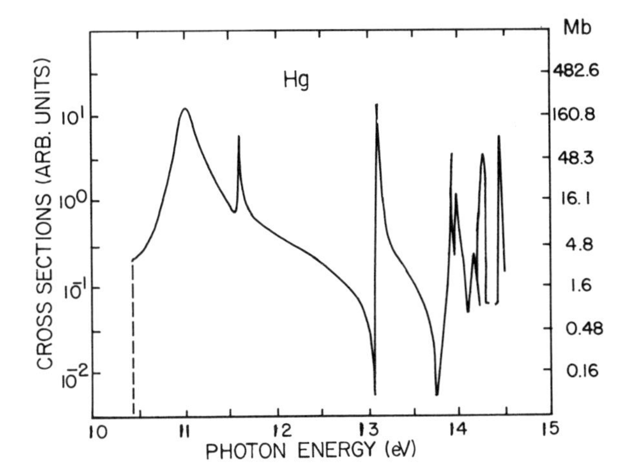

Fig. 47. Photoionization cross section of Hg vapor from 10 to 15 eV showing the $5d^{10}6s^2$–$5d^9 6s^2(np)$ autoionizing series. The absolute scale (in megabarns) is shown on the right-hand side of the figure. (From BREHM[156])

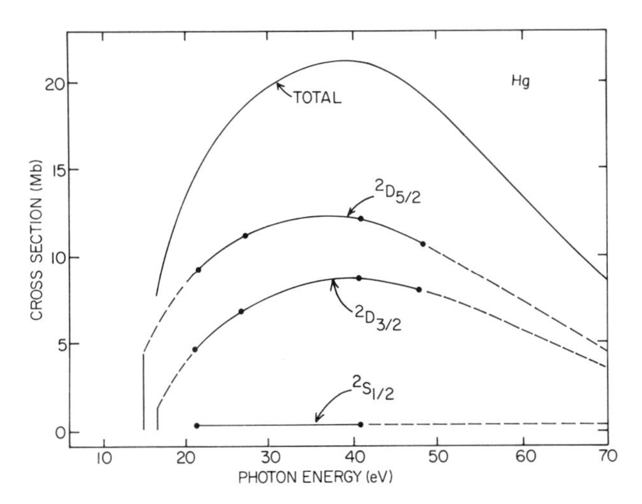

Fig. 48. Photoionization cross sections of Hg vapor from 15 to 70 eV showing the partial cross sections for ejection of a $5d$ electron. (From DEHMER, BERKOWITZ[181])

been reported[182a], namely, Be^+, Na^+, and Li^+. No quantitative data on absorption cross sections are available. However, the absorption spectra show numerous autoionizing lines, which have been identified and classified.

The ionization threshold for most positive ions lies above 15 eV, and therefore the vacuum uv spectral region must be used. In contrast, most negative ions can be ionized by photons of energy less than 2 or 3 eV. This is

[182a] Mehlman, G., Estava, J.M. (1974): Astrophys. J. *188*, 191, (Be^+); Lucatorto, T.B., McIlrath, T.J. (1976): Phys. Rev. Lett. *37*, 428, (Na^+); Carroll, P.K., Kennedy, E.T. (1977): Phys. Rev. Lett. *38*, 1068, (Li^+); McIlrath, T.J., Lucatorto, T.B. (1977): Phys. Rev. Lett. *38*, 1390, (Li^+)

Table 3. Electron affinities for atoms in the ground state in electron volts[a]

Atom	Electron affinity	Atom	Electron affinity	Atom	Electron affinity	Atom	Electron affinity
1 H	0.754	20 Ca	<0	39 Y	0.03	58 rare earths 71	<0.5
2 He	<0	21 Sc	<0	40 Zr	0.53		
3 Li	0.620	22 Ti	0.22	41 Nb	1.03		
4 Be	<0	23 V	0.52	42 Mo	1.02		
5 B	0.281	24 Cr	0.66	43 Tc	0.75	72 Hf	<0
6 C	1.268	25 Mn	<0	44 Ru	1.13	73 Ta	0.64
7 N	−0.078	26 Fe	0.25	45 Rh	1.23	74 W	0.64
8 O	1.462	27 Co	0.72	46 Pd	0.63	75 Re	0.15
9 F	3.399	28 Ni	1.15	47 Ag	1.303	76 Os	1.13
10 Ne	<0	29 Cu	1.226	48 Cd	≃0	77 Ir	1.62
11 Na	0.546	30 Zn	≃0	49 In	0.23	78 Pt	2.128
12 Mg	<0	31 Ga	0.30	50 Sn	1.25	79 Au	2.309
13 Al	0.463	32 Ge	1.21	51 Sb	1.05	80 Hg	<0
14 Si	1.385	33 As	0.80	52 Te	1.971	81 Tl	0.32
15 P	0.743	34 Se	2.021	53 I	3.061	82 Pb	1.12
16 S	2.077	35 Br	3.364	54 Xe	<0	83 Bi	1.12
17 Cl	3.615	36 Kr	<0	55 Cs	0.472	84 Po	1.93
18 Ar	<0	37 Rb	0.486	56 Ba	<0	85 At	2.82
19 K	0.501	38 Sr	<0	57 La	0.53	86 Rn	<0

[a] Hotop, H., Lineberger, W.C. (1975): J. Phys. Chem. Ref. Data *4*, 539

particularly advantageous because of the availability of lasers in this spectral region. Consequently, considerable progress has been made in the study of negative ions near their ionization threshold. However, no measurements have been made in the vacuum uv region.

Photoionization of negative ions is generally called *photodetachment*. Photodetachment studies were first introduced by BRANSCOMB and FITE[183] in 1954 by the use of carbon-arc light sources. These sources have since been replaced with tunable lasers. The impetus to study negative ions was to determine the electron affinities of the various atoms, particularly, for application to the upper atmosphere.[184] The status of the electron affinities (or binding energies) has been reviewed by HOTOP and LINEBERGER.[185] Table 3 lists the recommended values of electron affinities for most of the atoms in the periodic table. Only about thirty-seven of the atoms have been studied by photoionization (or photodetachment), whereas the remainder of the atomic electron affinities have been determined by semiempirical extrapolations. No error limits are quoted in Table 3 and the reader is referred to the original review[185] for detailed discussions. Attention is now being directed to measurements of the photodetachment cross sections as a function of wavelength. Autoionizing structure near the ionizing threshold has been observed with extremely good resolution. Figure 49 illustrates an example of this for Cs^-.[186]

[183] Branscomb, L.M., Fite, W.L. (1954): Phys. Rev. *93*, *651* A
[184] Branscomb, L.M. (1964): Ann. Geophys. *20*, 88
[185] Hotop, H., Lineberger, W.C. (1975): J. Phys. Chem Ref. Data *4*, 539
[186] Patterson, T.A., Hotop, H., Kasdan, A., Norcross, D.W., Lineberger, W.C. (1974): Phys. Rev. Lett. *32*, 189

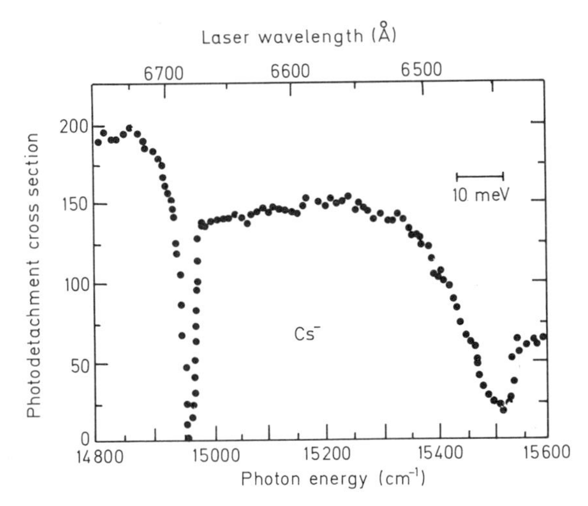

Fig. 49. Photodetachment cross section of Cs^- between 6,400 and 6,750 Å. (From PATTERSON et al.[186])

The first resonance has a width (FWHM) of about 1 mV. PATTERSON et al.[186] have tentatively assigned these resonances to the double electron excited states $Cs^-(6p7s)$. Photodetachment measurements, as a function of wavelength or by the use of photoelectron spectroscopy, can also provide the values of any fine-structure splitting in the negative ion.[187]

Photoionization of excited-state neutral atoms is another area where lasers must be used. Again, no measurements have been made in the vacuum uv. STEBBINGS et al.[188–190] have measured the photoionization cross sections for several rare-gas metastable atoms in the spectral region between 2,300 and 4,400 Å. Their results for $He(2\,^{1,3}S)$ atoms are shown in Fig. 50 along with some theoretical results.[191–194] GALLAGHER and YORK[195] have studied metastable Ba, and BRADLEY et al.[196] have investigated autoionization of excited Ba and Mg. The only nonlaser studies have been those by NYGAARD et al.,[197] who photoionized the excited $6\,^2P$ states of Cs by use of a Hg–Xe high-power lamp. The number densities of excited states, positive ions, and negative ions that can be produced easily are, at present, too small for studies

[187] Hotop, H., Patterson, T.A., Lineberger, W.C. (1973): Phys. Rev. A *8*, 762

[188] Stebbings, R.F., Dunning, F.B., Tittel, F.K., Rundel, R.D. (1973): Phys. Rev. Lett. *30*, 815

[189] Dunning, F.B., Stebbings, R.F. (1974): Phys. Rev. A *9*, 2378

[190] Stebbings, R.F., Dunning, F.B., Rundel, R.D. (1976) in: Proceedings of the Fourth International Conference on Atomic Physics, Putlitz, G. zu, Weber, E.W., Winnacker, A. (eds.), p. 713. New York: Plenum Press

[191] Burgess, A., Seaton, M.J. (1960): Mon. Not. R. Astr. Soc. *120*, 121

[192] Norcross, D.W. (1971): J. Phys. B *4*, 652

[193] Jacobs, V. (1971): Phys. Rev. A *4*, 939

[194] Dalgarno, A., Doyle, H., Oppenheimer, M. (1973): Phys. Rev. Lett. *29*, 1051

[195] Gallagher, A.C., York, G. (1974): Rev. Sci. Instrum. *45*, 662

[196] Bradley, D.J., Ewart, P., Nicholas, J.V., Shaw, J.R.D. (1973): J. Phys. B *6*, 1594; (1973): Phys. Rev. Lett. *31*, 263

[197] Nygaard, K.J., Hebner, R.E., Jr., Jones, J.D., Corbin, R.J. (1975): Phys. Rev. A *12*, 1440

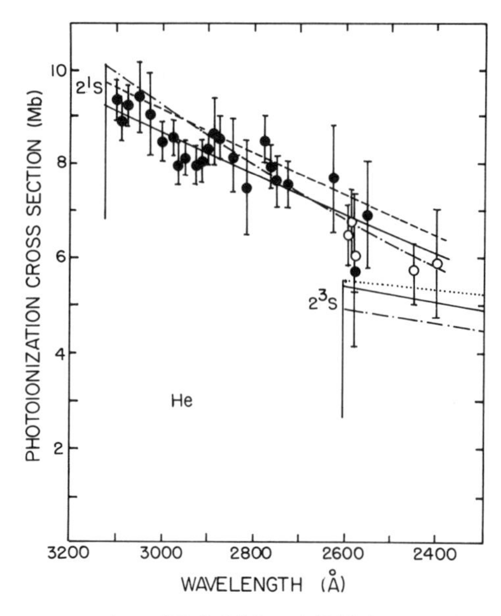

Fig. 50. Photoionization cross section of He* ($2\,^1S$) and ($2\,^3S$) between 2,300 and 3,200 Å. Solid and open circles are experimental points. Theoretical data are given by the curves: ----- BURGESS, SEATON[191]; -·-·-·-· NORCROSS[192]; ——— JACOBS[193]; ······ DALGARNO et al.[194] (From STEBBINGS et al.[190])

in the vacuum uv. However, the availability of high-intensity synchrotron light sources has increased the probability that measurements in the vacuum uv may soon be started on these exotic species.

The development of photoelectron spectroscopy and the improved technology in the vacuum uv region has given impetus to investigations of specific photoionization processes, for example, studies of the angular distribution of photoelectrons and multiple photoionization. In most cases measurements of this nature provide a more sensitive test of the validity of a given theoretical approach. A considerable amount of work is currently being performed in efforts to measure the absolute partial cross sections for specific photoionization processes. Measurements of this nature are described in the following section.

11. Partial photoionization cross sections. Historically, research usually proceeds from the simple to the more difficult measurements. This has been especially true of studies of photoabsorption. Total photoabsorption cross sections σ_a require a simple measurement of the ratio of the intensity of the light incident to that transmitted by a gas. To obtain the probability that an absorbed photon would produce an ion required the knowledge of how to measure the absolute intensity of the incident radiation. With this obstacle overcome[69, 70] (Sect. 7.3) the photoionization efficiency γ could be measured. γ is defined as the number of positive ions formed (N_i) divided by the number of photons absorbed by the gas. Thus, the total photoionization cross section σ_i is given by

$$\sigma_i(\text{total}) = \gamma \sigma_a. \tag{11.1}$$

The value of γ must be $\leqq 1$. For atoms we can expect it to be unity except in regions of discrete structure. Here we must determine if the structure completely autoionizes or only partially does so with some emission of fluorescent ratiation. There have been few direct measurements of this type.[106]

Once the probability has been measured that an absorbed photon will produce ionization, we must next determine the specific process that produced ionization. From what orbital was the electron ejected, in what direction, how many electrons, etc.? The cross sections for these specific processes are called partial cross sections and can be represented by σ_j, where the subscript j indicates the specific orbital of the electron involved in the ionization process. The partial cross sections are determined from the relation

$$\sigma_j = (N_j / \sum_j N_j)\, \sigma_i(\text{total}), \tag{11.2}$$

where N_j represents the number of ions formed in the specific state j and $\sum_j N_j$ represents the total number of ions formed in all states. The ratio of the two quantities is often called the branching ratio or abundance. The N_j can be equated to the number of electrons produced in the specific process *provided* the photon energy is less than the double ionization threshold. In the case of double ionization, although only a single ion is produced, two electrons are created. Thus, in using the technique of photoelectron spectroscopy N_j for electrons would be twice the value found by measuring the number of ions by mass spectroscopy.

Care must be taken in determining the partial cross sections. There is danger in measuring the same event twice. As an example, the ejection of a $4d$ electron from Xe is a primary process and photoelectron spectroscopy can give the branching ratio for this process. However, the atom subsequently undergoes an Auger transition ejecting another electron and leaving the residual ion doubly charged. The Auger electron cannot be included if a true branching ratio for $4d$ ejection is to be obtained. The branching ratio, as obtained by mass spectrometry, contains a measure of the secondary processes and will not necessarily equal the true production of d-shell electrons. A comparison between the two measurements would provide information on the degree of the Auger process.

By the use of the techniques described in Sects. 8 and 9 it is possible to obtain a complete picture of the absorption process. The end result of such measurements should yield absolute cross sections for each absorption process. An example is shown in Fig. 51 for Ne. Here the partition of the total photoionization cross section is shown for single ionization (from each of the $1s$, $2s$, and $2p$ shells), multiple ionization, and simultaneous excitation and ionization.[198]

---

[198] Wuilleumier, F., Krause, M.O. (1974): Phys. Rev. A *10*, 242

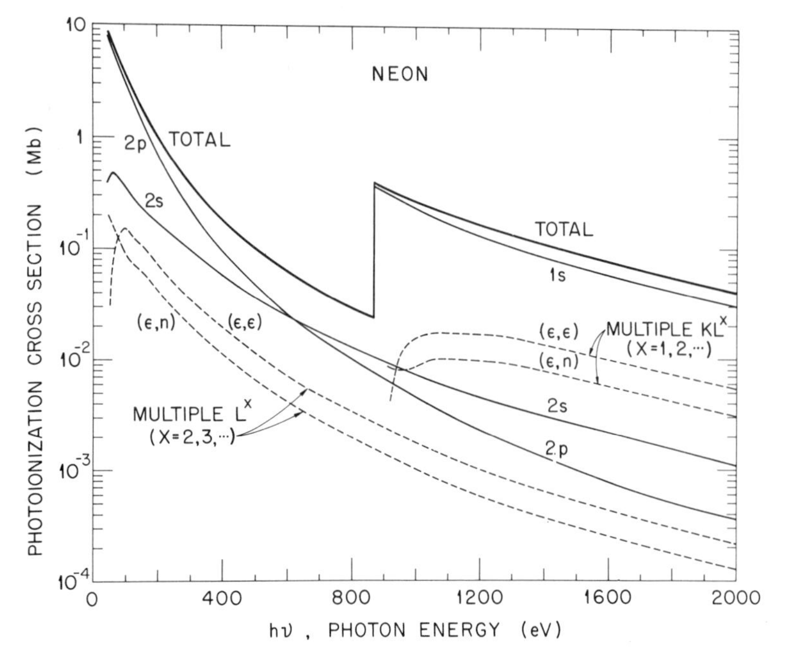

Fig. 51. Partition of total photoionization cross section of neon into its components of single ionization in $2p$, $2s$, and $1s$ subshells, multiple ionization, and simultaneous excitation and ionization in various subshells. Symbols $(\varepsilon, \varepsilon)$ represent multiple ionization from either the L shell or where one electron is ejected from the K shell and the remainder from the L shell (KL^x). Symbols (ε, n) represent ionization of one electron and simultaneous excitation of another. (From WUILLEUMIER, KRAUSE[198])

α) Ionization from specific orbitals. Photoelectron spectroscopy is the ideal technique to measure the probability for ejecting either a single electron from a specific orbit or observing a simultaneous excitation and ionization event. There is a considerable amount of work currently being performed in this area. Much of the work has been performed at a few discrete wavelengths (e.g., He 584 Å, He 304 Å, and Ne 736 Å), however, there are also measurements utilizing synchrotron radiation. Some specific examples of the type of results that are available are given below, but no attempt is made to refer to the large volume of material becoming available.

Some of the earliest studies that utilized photoelectron spectroscopy were concerned with measuring the ratio for producing the rare-gas ions in their $^2P_{1/2,\,3/2}$ states. From their statistical weights a ratio of 2 was expected. However, most measurements have shown the ratio to be nonstatistical.[199–202] This

[199] Samson, J.A.R., Cairns, R.B. (1968): Phys. Rev. *173*, 80

[200] Samson, J.A.R., Gardner, J.L., Starace, A.F. (1975): Phys. Rev. A *12*, 1459

[201] Comes, F.J., Salzer, H.G. (1964): Z. Naturforsch. A *19*, 1230

[202] Wuilleumier, F., Adam, M.Y., Dhez, P., Sander, N., Schmidt, V., Mehlhorn, W. (1977): Phys. Rev. A *16*, 646

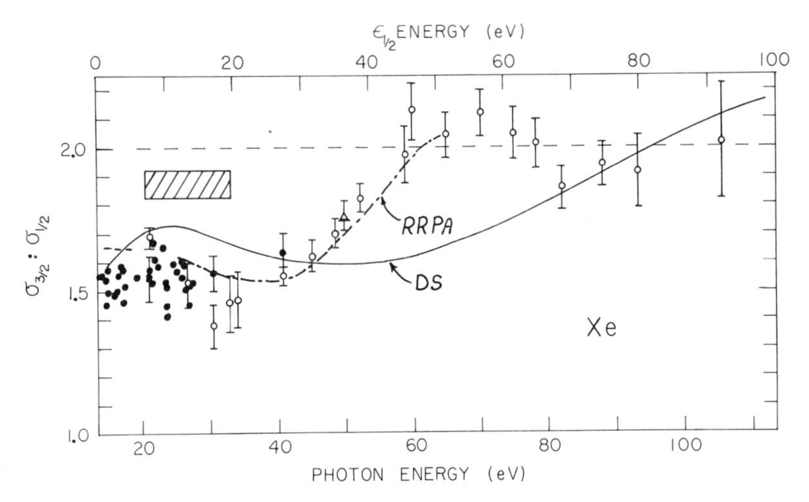

Fig. 52. $^2P_{3/2}$:$^2P_{1/2}$ branching ratio for Xe as a function of photon energy. *Theoretical data:* ----- statistical weight value; ———— *DS*, Ong, Manson[206]; —·—·—· RRPA, Johnson, Cheng[207]. *Experimental data:* ○ Wuilleumier et al.[202]; ● Samson et al.[199, 200]; △ Dehmer as given in[202]. Shaded area represents region of numerous autoionizing lines

effect has also been observed in Cd, Hg, and Cs.[203–205] Although these ratios appeared to be nearly constant as a function of photon energy, it has recently been shown for Cs and Xe that if the ratios are studied over a sufficiently extended photon-energy range, the ratios will tend toward the statistical values.[202, 204] The experimental values of the ratios found in Xe, for the photon-energy range 13.4–110 eV, are shown in Fig. 52. There is considerable scatter in the data between threshold (13.436 eV) and 33 eV. This is possibly caused by the presence of numerous autoionizing resonances in this region (shown by the shaded rectangle).[136] In fact, Wuilleumier et al.[202] have shown that a pressure effect occurs ($\simeq 10\%$) at photon energies of 21.2 and 30.55 eV but not at 40.81 eV. Their data, plotted at these two photon energies, represent their lowest pressure measurements. Error bars are placed on the closed-circle data at the above energies to indicate the spread to be expected. These error bars are typical of all the closed-circle data points. The rise in the ratio to the statistical value for higher photon energies occurs just in the region of the Cooper minimum, between 40 and 60 eV. The theoretical results of Ong and Manson[206] are shown by the solid curve in Fig. 52. They used a Dirac-Fock formulation that includes the effect of spin-orbit coupling and treats exchange exactly. However, the effects of interchannel coupling between the $5p$ and $4d$ channels were not included in the calculation. The dash-dot curve represents the relativistic RPA calculations of Johnson and Cheng,[207] which includes

[203] Walker, T.E.H., Berkowitz, J., Dehmer, J.L., Waber, J.T. (1973): Phys. Rev. Letters *31*, 678
[204] Rowe, J.E., Margaritondo, G. (1976): Phys. Rev. Lett. 57 A, 314
[205] Frost, D.C., McDowell, C.A., Vroom, D.A. (1967): Chem. Phys. Lett. *1*, 93
[206] Ong, W., Manson, S.T. (1978): J. Phys. B *11*, L163
[207] Johnson, W.R., Cheng, K.T. (1979): Phys. Rev. A *20*, 978

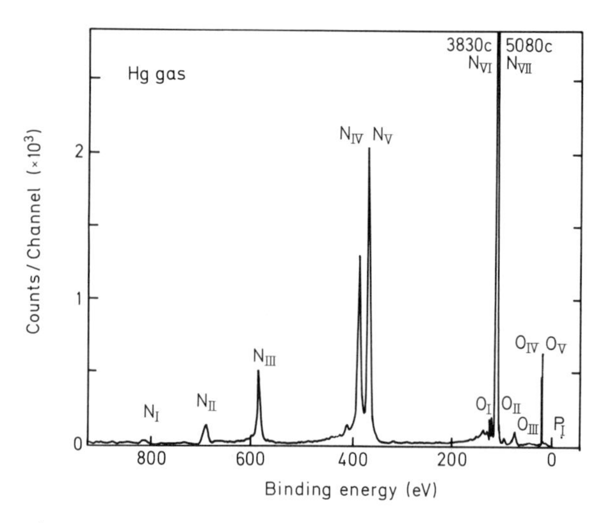

Fig. 53. Photoelectron spectrum of Hg vapor excited by monochromatized Al K_α x rays =1,486.6 eV. (From SVENSSON et al.[210])

interchannel coupling. The agreement with the experimental data is vastly improved.

The $^2P_{1/2,\,3/2}$ ratio is found to vary dramatically within autoionizing resonances.[200, 208] For example, in the vicinity of the Xe $5s\,5p^6\,6p(^1P_1)$ window resonance at 20.95 eV, the ratio is observed to change from 1.54 to 3.0. The calculations of STARACE[209] predict that the ratio will increase to about 9.0 at the minimum of the resonance, and in fact, variations in the ratio within resonances are expected as a general phenomenon.

The photoelectron spectrum of Hg excited by the Al K_α x ray (1,486.6 eV) is shown in Fig. 53.[210] All electronic levels with principal quantum numbers n =4, 5, and 6 are reached. If the transmission of the analyzer is calibrated, then the ratio of the intensity of any line to that of the sum of the intensities of all lines gives the branching ratio. An expanded version of the spectrum showing the valence electrons more clearly is shown in the 584 Å spectrum in Figs. 54 and 55.[211] When spectra, like those shown in Figs. 53–55, are taken as a function of wavelength, then by the use of (11.2) the partial cross sections σ_j are obtained. The results for Hg, using synchrotron radiation as an ionizing source, are shown in Fig. 56.[212] In Fig. 55 we see another example (for Hg) of the nonstatistical branching ratio of the $^2P_{1/2,\,3/2}$ states. The observed ratio is about 0.28 rather than the statistical weight value of 2.0. From Mansfield's

---

[208] Kemeny, P.C., Samson, J.A.R., Starace, A.F. (1977): J. Phys. B *10*, L201

[209] Starace, A.F. (1977): Phys. Rev. A *16*, 231

[210] Svensson, S., Martensson, N., Basiliev, E., Malmqvist, P.A., Gelius, U., Siegbahn, K. (1976): J. Electron. Spectrosc. *9*, 51

[211] Mahr, D. (1976): Diplomarbeit, University of Freiburg, Germany, as communicated by Hotop, H.

[212] Shannon, S.P., Codling, K. (1978): J. Phys. B *11*, 1193

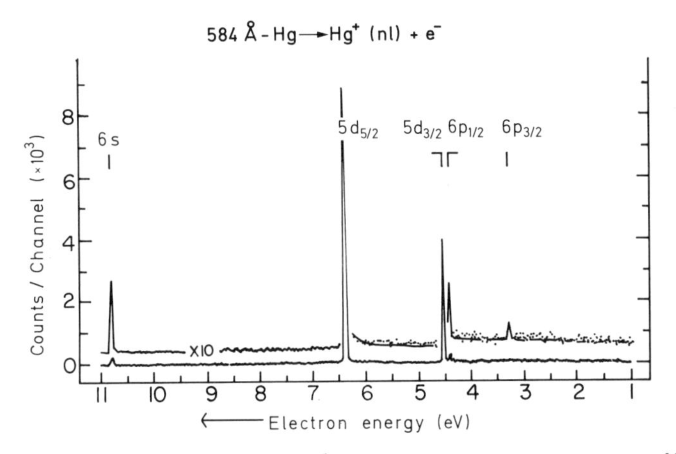

Fig. 54. Photoelectron spectrum (584 Å) of Hg vapor. (From MAHR, HOTOP[211])

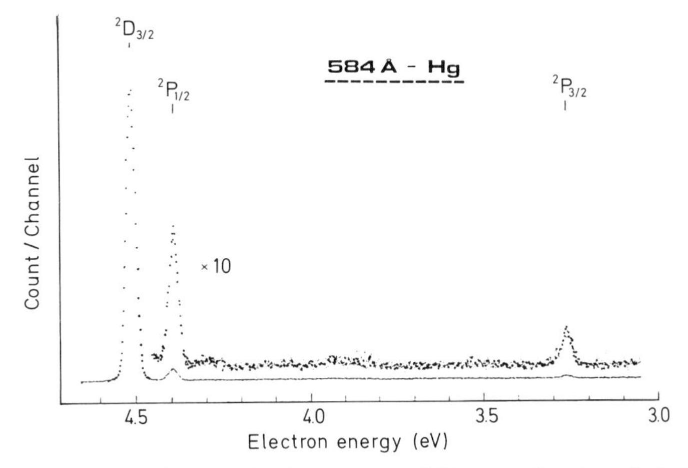

Fig. 55. Portion of the 584-Å photoelectric spectrum of Hg as a function of electron energy illustrating the relative strengths of the $^2P_{1/2}$ and $^2P_{3/2}$ lines. Data are not corrected for the analyzer transmission. (From MAHR, HOTOP[211])

absorption spectrum of Hg[213] the 584 Å line appears to coincide with a broad absorption feature and this may perturb the ratio. More data in the vicinity of the 584-Å line will be necessary to clarify this result.

A study of the partial cross sections can provide a very sensitive test for the various theoretical approximations used in calcuating photoionization cross sections. A very interesting case is that of the 5s electron in Xe. The one-electron central-field approximation predicts a low value for the partial cross section at threshold. The cross section then increases slowly with photon energy beyond the threshold to a maximum at a photon energy of about 40 eV. The experimental results show the opposite effect. The cross section is a maximum at threshold and falls rapidly with increasing photon energy to

[213] Mansfield, M.W.D. (1973): Astrophys. J. *180*, 1011

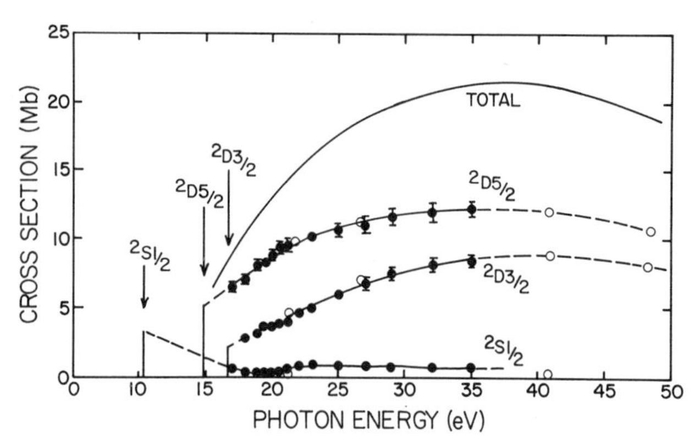

Fig. 56. Partial cross sections for ejection of $6s$ and $5d$ electrons from Hg. The open circles represent the data from Fig. 48. (From SHANNON, CODLING[212])

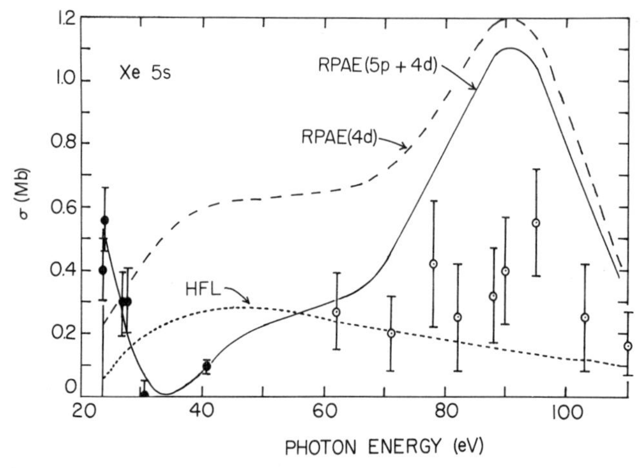

Fig. 57. Partial photoionization cross section for the $5s$ electrons in Xe as a function of photon energy. *Theoretical curves:* The dashed curve represents the RPAE calculation taking into account correlations between the $5s$ and $4d$ shells, whereas the solid curve represents correlations between the $5s$ and $(4d+5p)$ shells. Dotted curve (Hartree-Fock length approximation) does not take correlations into account (AMUSIA, CHEREPKOV[138]). *Experimental data:* ● SAMSON, GARDNER[214]; ⊙ WEST et al.[215]

nearly zero at 37 eV. AMUSIA[151, 138] has shown that this phenomenon is caused by the interaction between the outgoing electron and the other electrons in the $5p$, $5s$, and $4d$ shells. It is not sufficient to consider interactions simply between the s and d electrons, although this step does qualitatively give the correct behavior for the photoionization of the $5s$ electron in the vicinity of the d-shell ionization threshold. We must also consider the s- and p-electron interaction. The theoretical results are shown in Fig. 57 along with the experimental results.[214, 215] This correlation effect is very strong in all the outer s

[214] Samson, J.A.R., Gardner, J.L. (1974): Phys. Rev. Lett. *33*, 671

[215] West, J.B., Woodruff, P.R., Codling, K., Houlgate, R.G. (1976): J. Phys. B *9*, 407

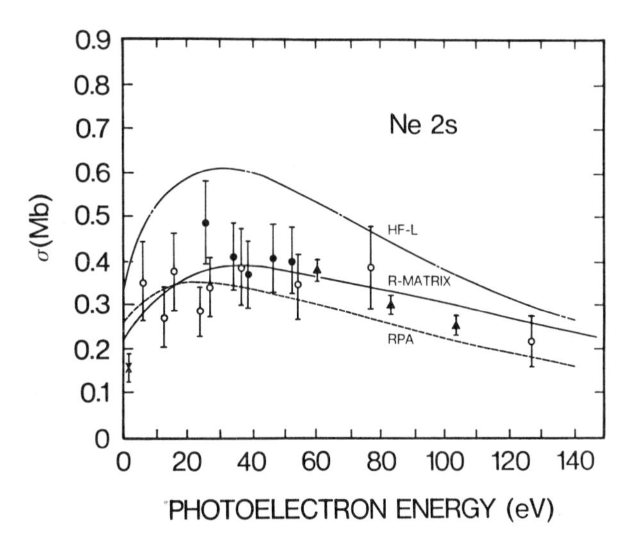

Fig. 58. Partial photoionization cross section for the 2*s* electrons in neon as a function of the photoelectron energy. *Theoretical curves:* *R* matrix, BURKE, TAYLOR[219]; RPA, AMUSIA[151]; HFL, KENNEDY, MANSON[220]. *Experimental data:* ● and ○ CODLING et al.[216]; × SAMSON, GARDNER[214]; ▲ WUILLEUMIER, KRAUSE[198]. (From CODLING et al.[216])

subshells of the rare gases.[118,214–221] Figures 58–60 show the experimental and theoretical data for photoionization of the Ne(2*s*), Ar(3*s*), and Kr(4*s*) subshells. In each case, the inclusion of electron correlation effects are very dramatic.

Observations of simultaneous ionization of one electron and excitation of a second electron from within the same orbit are becoming more frequent.[222–224] This is another manifestation of electron correlation effects. The probability for this type of double electron excitation has been calculated for He by several authors with conflicting results.[225–230] The calculated ratio $He^+(n=2)/He^+(n=1)$ by JACOBS and BURKE[227] and SALPETER and ZAIDI[225]

[216] Codling, K., Houlgate, R.G., West, J.B., Woodruff, P.R. (1976): J. Phys. B *9*, L83
[217] Houlgate, R.G., West, J.B., Codling, K., Marr, G.V. (1976): J. Electron. Spectrosc. *9*, 205
[218] Samson, J.A.R., Cairns, R.B. (1968): Phys. Rev. *173*, 80
[219] Burke, P.G., Taylor, K.T. (1975): J. Phys. B *8*, 2620
[220] Kennedy, D.J., Manson, S.T. (1972): Phys. Rev. A *5*, 227
[221] Lin, C.D. (1974): Phys. Rev. A *9*, 181
[222] Samson, J.A.R. (1969): Phys. Rev. Lett. *22*, 693
[222a] Woodruff, P.R., Samson, J.A.R. (1980): Phys. Rev. Lett. *45*, 110
[222b] Woodruff, P.R., Samson, J.A.R. (1982): Phys. Rev. A *25*, 848
[223] Krause, M.O., Wuilleumier, F. (1972): J. Phys. B *5*, L143
[224] Adam, M.Y., Wuilleumier, F., Sander, N., Schmidt, V., Wendin, G. (1978): J. Phys. *39*, 129
[225] Salpeter, E.E., Zaidi, M.H. (1962): Phys. Rev. *125*, 248
[226] Jacobs, V.L. (1971): Phys. Rev. *3* A, 289
[227] Jacobs, V.L., Burke, P.G. (1972): J. Phys. B *5*, L67
[228] Hyman, H.A., Jacobs, V.L., Burke, P.G. (1972): J. Phys. B *5*, 2282
[229] Brown, R.L., Gould, R.J. (1970): Phys. Rev. D *1*, 2252
[230] Dalgarno, A., Stewart, A.L. (1960): Proc. Phys. Soc. *76*, 49

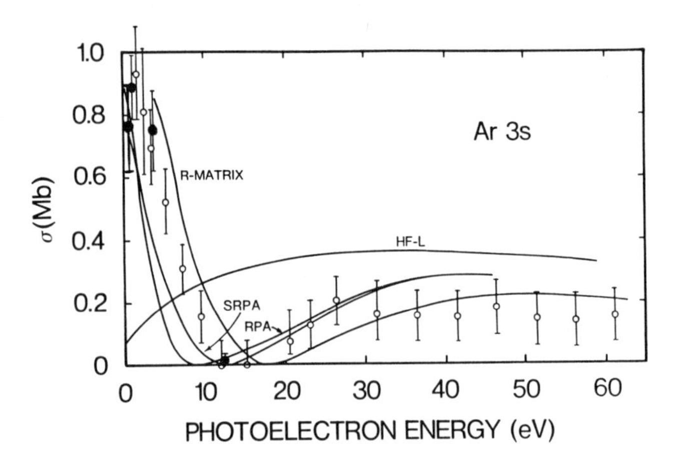

Fig. 59. Partial photoionization cross section for the $3s$ electrons in argon as a function of the photoelectron energy. *Theoretical curves:* R matrix, BURKE, TAYLOR[219]; HF-L, KENNEDY, MANSON[220]; RPA, AMUSIA[151]; SRPA, LIN[221]. *Experimental data:* ○ HOULGATE et al.[217]; ● SAMSON, GARDNER[214]. (From HOULGATE et al.[217])

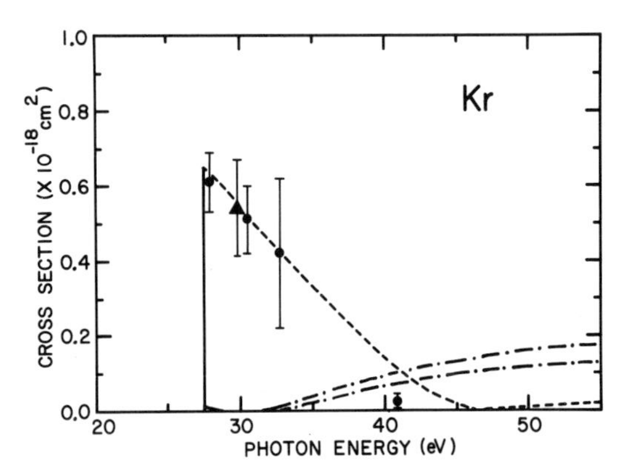

Fig. 60. Partial photoionization cross section for the $4s$ electrons in krypton as a function of the photoelectron energy. *Theoretical curves:* ---- AMUSIA[151]; -·-·-·, upper curve is dipole-length approximation and lower curve is the velocity approximation, KENNEDY, MANSON[220]. *Experimental data:* ● and ▲ SAMSON, GARDNER[214]. (From SAMSON, GARDNER[214])

are shown in Fig. 61 along with the experimental values.[222, 223] The results of KRAUSE and WUILLEUMIER[223] tend to support the more sophisticated calculations of JACOBS and BURKE.[227] However, recent measurements by WUILLEUMIER et al.[230a] with photon energies between 65 and 100 eV disagree with both calculations near threshold. The He^+ $(n=2)$ state can be de-excited with subsequent emission of fluorescent radiation at 304 Å. This provides an independent technique to measure the $n=2/n=1$ ratio. These measurements

[230a] Wuilleumier, F., Adam, M.Y., Sander, N., Schmidt, V., Mehlhorn, W. (1977): Abstracts of papers, X. Int. Conf. on the physics of electronic and atomic collisions, Paris, p. 1170; and Wuilleumier, F., Adam, M.Y., Sander, N., Schmid (1980): Phys. Lett. *41*, L375

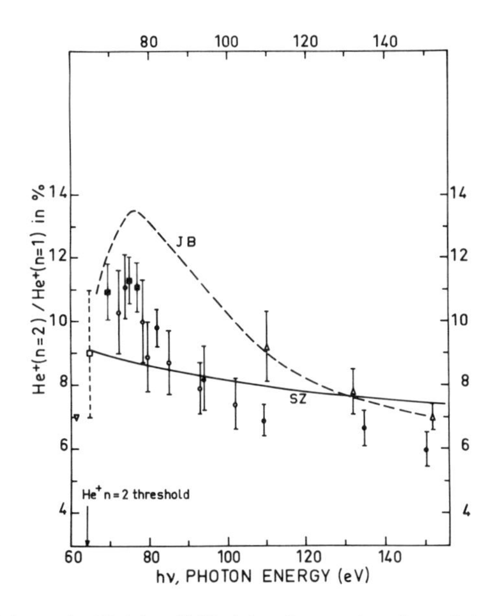

Fig. 61. Variation of the ratio He^+ $(n=2)/He^+$ $(n=1)$ as a function of the photon energy. Experimental data points: ○, ●, ■ WUILLEUMIER et al.[231]; △ KRAUSE and WUILLEUMIER[223]; □ SAMSON[222]. Theoretical data: ——— SALPETER, ZAIDI[225]; ----- JACOBS, BURKE[227]

have recently been made by WOODRUFF and SAMSON.[222a,b] The results are in excellent agreement with those of WUILLEUMIER et al.[231] and with the recent calculations of CHANG.[230a]

Simultaneous photoionization and excitation of helium can proceed via the transitions $1s^2$–$2s\varepsilon p$ and $1s^2$–$2p\varepsilon s$. Measurement of the ratio $\sigma(2p)/\sigma(2s)$, that is, the ratio of the probabilities for producing He^+ in its $2p\,^2P_{1/2,\,3/2}$ to $2s\,^2S_{1/2}$ states, would be a sensitive test of theory. At present, calculations using Coulomb wave functions obtain very small values of $\sigma(2p)/\sigma(2s)$. Values range from zero[225–230] to one percent[229]. However, when close coupling is included between the degenerate final states a ratio of about 3:1 is obtained at the threshold.[226,228] In most of these calculations accurate correlated wave functions were used. However, the inclusion of close coupling causes theory to predict more $2p\,^2P\,He^+$ ions. Photoelectron spectroscopy cannot distinguish between the εp and εs electrons because of the near degeneracy between the $2p\,^2P$ and $2s\,^2S$ states of He^+. The angular distribution of the two groups of electrons will be different. Thus, it may be possible to measure the combined angular distributions and compare this result with theoretical predictions. The results of KRAUSE and WUILLEUMIER shown in Fig. 61, show qualitatively that there is an asymmetry in the total angular distribution. A more direct approach would be to study the expected fluorescence from the $n=2$ state of He^+. The $2s\,^2S$ state is metastable and should not fluoresce because of its long lifetime. However, the $sp\,^2P$–$1s\,^2S$ transition gives the resonance line of He II at 303.782 Å. A measure of the relative amount of fluorescence per incident

[231] Chang, T.N. (1980): J. Phys. B *13*, L551

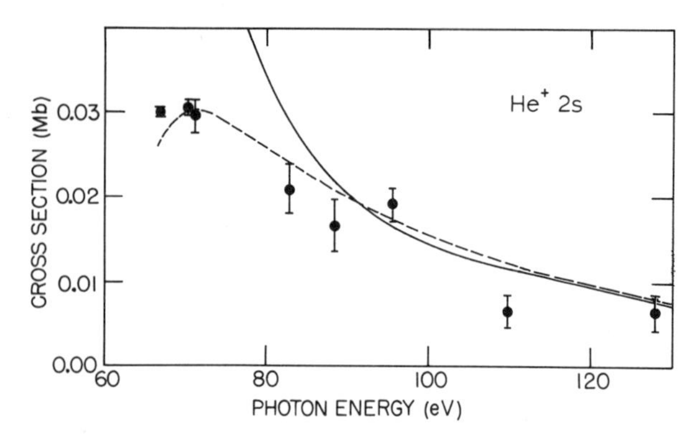

Fig. 62. The partial cross section for producing $He^{+}(2s)$ from neutral helium. ● WOODRUFF, SAMSON[222b]; ———— CHANG[231]; ----- JACOBS, BURKE[227]

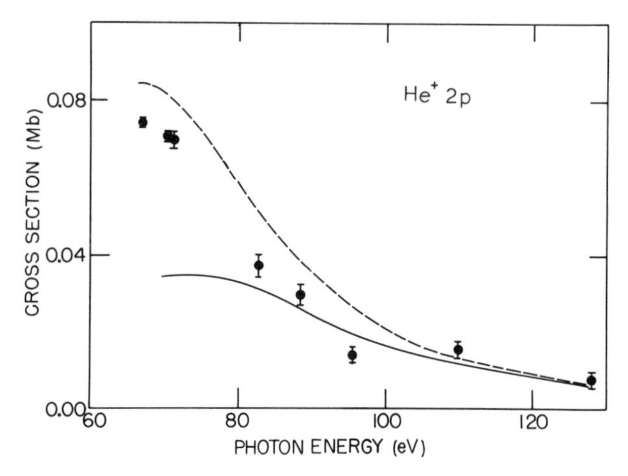

Fig. 63. The partial cross section for producing $He^{+}(2p)$ from neutral helium. ● WOODRUFF, SAMSON[222b]; ———— CHANG[231]; ----- JACOBS, BURKE[227]

photon would give the relative cross section $\sigma(2p)$. Then applying an electric or magnetic field to the interaction region a mixing of the ^2S and ^2P states should take place. If mixing is complete, a relative cross section $\sigma(2p+2s)$ should be obtained and hence $\sigma(2p)/\sigma(2s)$ can be determined. These experiments have just been completed in the author's laboratory. The measured photoionization cross sections for $He(1s^2) \to He^{+}(2s)$, $He^{+}(2p)$ are shown in Figs. 62 und 63. The error bars represent random errors from counting statistics. The data are compared with the calculated cross sections of JACOBS and BURKE[227] and CHANG.[231] For the $2s$ cross section (Fig. 62) above 85 eV, both theories lies approximately 10% higher than the data. The data for the $2p$ cross section (Fig. 63) in the same energy region agrees reasonably well with both theories. At low energies the data clearly favor the close coupling calculation of JACOBS and BURKE. This is a little unexpected in view of the excellent agreement between experiment and calculation of CHANG for the total cross section for the $N=2$ level and serves to underline the value of

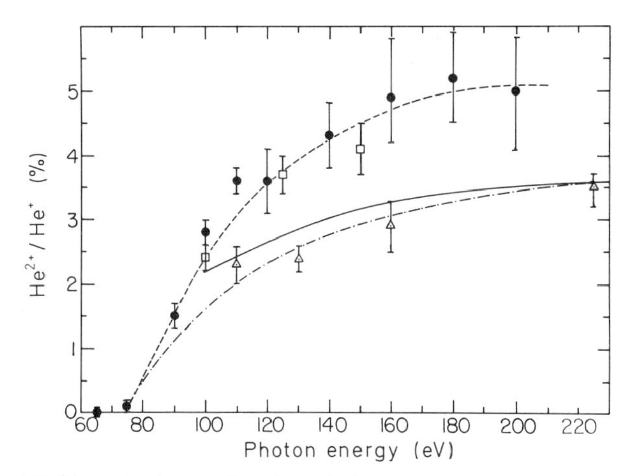

Fig. 64. Ratio of He^{2+} to He^{+} as a function of photon energy. *Experimental data:* ● WIGHT, VAN DER WIEL[234]; ⊡ SCHMIDT et al.[235]; ⚠ CARLSON[232]. The dashed line represents an estimate of the true ratio. *Theoretical data:* ——— BROWN[240]; –·–·–· BYRON, JOACHAIN[239]

measurements of the partial cross sections in comparing photoionization theories.

The extension of simultaneous ionization and excitation leads to double ionization, which is discussed in the next section.

β) Multiple ionization. Multiple ionization of the rare gases by single photons has been studied in the range from their double ionization thresholds to about 250 eV.[103, 113, 114, 232–237] Other atoms that have been studied are Cd,[176] Zn,[176] Hg,[180] Yb,[238, 238a] Ba,[238a] Rb & Cs,[238b] Tl & Pb,[238c] Ca & Sr[238d] and Sm, Eu, & Tm.[238e] Double ionization of Ba has been observed at 584 Å in photoelectron studies.[168, 169, 169a]

Studies of double ionization have shown the importance of considering electron correlation effects in any calculation of a photon-atom interaction. For example, BYRON and JOACHAIN[239] have shown that the ratio He^{2+}/He^{+} remains less than 1% when a Hartree-Fock ground-state wave function is used to describe the effect of initial-state interactions. From Fig. 64 it can be seen that the ratio from experimental data rises to between 3 and 5%. However, by

[232] Carlson, T.A. (1967): Phys. Rev. *156*, 142

[233] Cairns, R.B., Harrison, H., Schoen, R.I. (1969): Phys. Rev. *183*, 53

[234] Wight, G.R., Van der Wiel, M.J. (1976): J. Phys. B *9*, 1319

[235] Schmidt, V., Sander, N., Kuntzemuller, H., Dhez, P., Wuilleumier, F., Kallne, E. (1976): Phys. Rev. A *13*, 1748

[236] Van der Wiel, M.J., Chang, T.N. (1978): J. Phys. B *11*, L125

[237] Holland, D.M.P., Codling, K., West, J.B., Marr, G.V. (1979): J. Phys. B *12*, 2465

[238] Parr, A.C., Ingram, M.G. (1970): J. Chem. Phys. *52*, 4916

[238a] Holland, D.M.P., Codling, K., Chamberlain, R.N. (1981): J. Phys. B *14*, 839

[238b] Holland, D.M.P., Codling, K. (1981): J. Electron Spectrosc. Relat. Phenom. *23*, 275

[238c] Holland, D.M.P., Codling, K. (1980): J. Phys. B *13*, L745

[238d] Holland, D.M.P., Codling, K. (1981): J. Phys. B *14*, 2345

[238e] Holland, D.M.P., Codling, K. (1981): J. Phys. B *14*, L359

[239] Byron, F.W., Joachain, C.J. (1967): Phys. Rev. *164*, 1

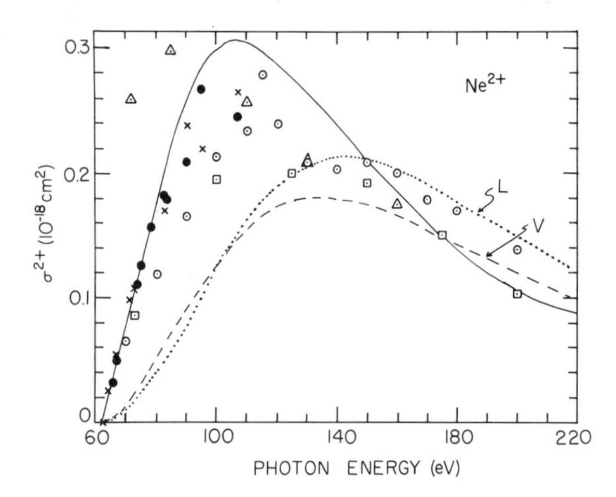

Fig. 65. Double ionization cross section of Ne as a function of photon energy. *Experimental data:* ● SAMSON, HADDAD[103]; × SAMSON, KEMENY, unpublished data; ⊙ WIGHT and VAN DER WIEL[234]; ⊡ SCHMIDT et al.[235]; △ CARLSON[232]. *Theoretical data:* ———— CHANG, POE[241]; ······· dipole-length (L) and - - - - - dipole-velocity (V) approximation, CARTER, KELLY[242]

the use of fully correlated initial-state wave functions much better agreement with experiment is achieved. Figure 64 contains BYRON and JOACHAIN's more sophisticated calculations along with those of BROWN,[240] who also included the effects of correlation in the initial state but used uncorrelated Coulomb waves for the free electrons. Although the theoretical curves are in good agreement with the experimental results of CARLSON,[232] there is a significant difference compared to the other experimental results[234, 235] shown in Fig. 62. Over most of the photon energy range the discrepancy is about 50 %. Better agreement between the experimental and theoretical results have been obtained for Ne and Ar[177, 241, 242] (Figs. 65 and 66). Figure 65 illustrates a variety of experimental data for the ratio Ne^{2+}/Ne^{+} along with the calculations of CHANG and POE[241] and CARTER and KELLY.[242] It should be pointed out that the absolute values of the experimental cross sections depend directly on the values of the total absorption cross section chosen (see Sect. 11). The experimental data shown in Figs. 65 and 66 have been normalized to the total cross sections of [5] and to new unpublished data.[243] The calculations of CHANG and POE used many-body perturbation theory, which allowed a quantitative study of the relative importance of various contributing effects. In addition to ground-state correlations, other factors such as core rearrangement, virtual Auger transitions, and inelastic internal collisions were included in the calculations. Figure 65 shows only the sum of those factors. However, the calcuations indicate that near the double ionization threshold internal collisions play a dominant role (they are about three times more effective than the sum of all other contributions considered). Calculations by CARTER and KELLY

[240] Brown, R.L. (1970): Phys. Rev. *1*, 586
[241] Chang, T.N., Poe, R.T. (1975): Phys. Rev. A *12*, 1432
[242] Carter, S.L., Kelly, H.P. (1977): Phys. Rev. A *16*, 1525
[243] Samson, J.A.R., Haddad, G.N. to be published

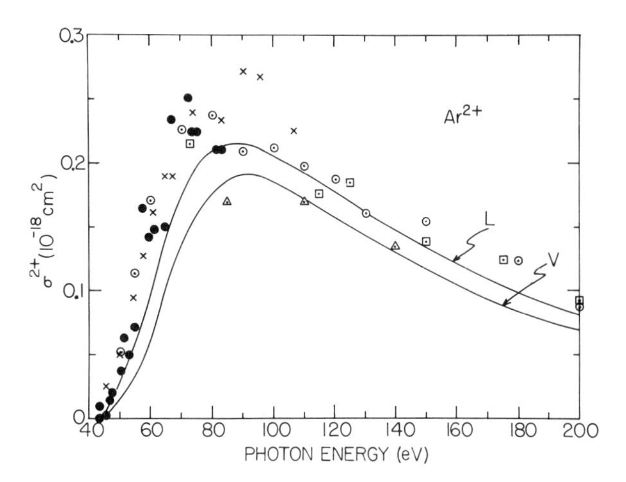

Fig. 66. Double ionization cross section of Ar as a function of photon energy. *Experimental data:* ● SAMSON, HADDAD[103]; × SAMSON, KEMENY, unpublished data; ⊙ WIGHT, VAN DER WIEL[234]; ⊡ SCHMIDT[235]; △ CARLSON[232]. The dashed line represents an estimate of the true ratio. *Theoretical data:* ——— dipole-length (L) and velocity (V) approximation, CARTER, KELLY[242]

reproduce the general shape of the curves in Fig. 65 with excellent agreement for photon energies greater than 140 eV but with poor agreement near threshold. Their calculations also used many-body perturbation theory and their approach considered the production of doubly charged ions via the two processes

$$\mathrm{Ne}\,2s^2\,2p^6\,(^1S) \rightarrow \mathrm{Ne}^{2+}\,2s^2\,2p^4(^3P, {}^1D, {}^1S)$$
$$\rightarrow \mathrm{Ne}^{2+}\,2s\,2p^5(^3P, {}^1P).$$

Their calculations showed that the most probable transition resulted in the ejection of two electrons from the $2p$ shell (82 % at a photon energy of 130.6 eV). The results of their calculation of double ionization in Ar[242] are shown in Fig. 66 and are compared to the experimental data. Their results include second-order correlations and are shown in the dipole-length (L) and velocity (V) approximations. The dashed curve indicates the average of the experimental data. As in the case of neon, the most probable transition results in the ejection of two electrons from the outer $3p$ shell (79 % at a photon energy of 90 eV). In this energy range there is no complication with Auger processes producing double ionization. Thus, for He, Ne, and Ar double ionization is a primary absorption process. CARLSON[232] has observed a small amount of triply charged Ne, about 3–4 % of the Ne^{2+} abundance. However, no Ar^{3+} has been observed in this energy range.[237]

The measurement of direct double ionization in Kr and Xe cannot be carried out over such a large energy range as that covered for He, Ne, and Ar. The reason for this is the proximity of the d shell to the double ionization threshold (see Fig. 30). When a d-shell electron is ejected, Auger processes become possible giving rise to double and triple ionization. However, this provides us with an ideal example of the necessity to study partial photoionization cross sections and their value in studying correlation effects.

Consider the total absorption cross section for Xe in the range 65–130 eV as shown in Fig. 29. This large increase in cross section is caused by the availability of the $4d$-shell electrons to participate in the photoionization process. Correlation effects between electrons within a given shell (intrashell correlations) and in other shells (intershell correlations) will modify the course of the partial cross sections for $5s$- and $5p$-shell photoionization (see Fig. 57). We will see that it also influences the direct double ionization cross sections.

The various possible channels by which photons can be absorbed by Xe in this energy region are given below.

$$4d^{10}\,5s^2\,5p^6 \rightarrow 4d^{10}\,5s^2\,5p^5 + e \qquad \sigma^+(p) \tag{11.3a}$$

$$4d^{10}\,5s^1\,5p^6 + e \qquad \sigma^+(s) \tag{11.3b}$$

$$4d^9\,5s^2\,5p^6 + e \qquad \sigma^+(d) \tag{11.3c}$$

$$4d^{10}\,5s^2\,5p^4\,ns, d + e \qquad \sigma^{+*} \tag{11.3d}$$

$$\left.\begin{array}{l} 4d^{10}\,5s^2\,5p^4 + 2e \\ 4d^{10}\,5s^1\,5p^5 + 2e \\ 4d^{10}\,5s^0\,5p^6 + 2e \end{array}\right\} \qquad \sigma^{2+}\ \text{(direct)}. \tag{11.3e}$$

These are the primary absorption processes. The partial cross section for each process is denoted by σ^+ for single ionization, and σ^{2+} for double ionization. The letters in parenthesis denote the specific orbital electrons involved. σ^{+*} denotes the family of possible simultaneous ionization and excitation processes. σ^{2+} (direct) is the cross section for producing double ionization directly without going through an Auger process. In the above we are assuming σ^{3+} (direct) is very much smaller and is $\simeq$ zero.

Processes (11.3a–d) produce discrete photoelectron energies and can be measured by photoelectron spectroscopy. Direct double ionization produces two electrons with a continuum of energies from zero to the maximum allowed for by single ionization. This is difficult to observe and to account for above a scattered-electron background in photoelectron spectroscopy. Mass analysis is generally the best method to measure the ratio of double to singly charged ions. However, we must distinguish between doubly charged ions caused by the secondary process of Auger transitions and those caused by direct double ionization. The Auger process proceeds as follows:

$$4d^{10}\,5s^2\,5p^6 \rightarrow 4d^9\,5s^2\,5p^6 + e_d,$$

followed by the Auger transitions

$$\left.\begin{array}{l} \rightarrow 4d^{10}\,5s\,5p^5 + e_A \\ 4d^{10}\,5s^2\,5p^4 + e_A \end{array}\right. \qquad \sigma^{2+}(\mathrm{A}) \tag{11.4a}$$

$$\left.\begin{array}{l} 4d^{10}\,5s\,5p^4 + 2e_A \\ 4d^{10}\,5s^2\,5p^3 + 2e_A, \end{array}\right. \qquad \sigma^{3+}(\mathrm{A}) \tag{11.4b}$$

where the subscripts d and A denote the d shell and Auger electrons, respectively. The Auger electrons from process (11.4a) will be monoenergetic and thus can be observed by photoelectron spectroscopy. In (11.4b) the electrons may have a continuum of energies. The energies of the pertinent Auger electrons have been tabulated by SIEGBAHN et al. [*10*].

The partial cross sections are determined from the relation given by (11.2) and must add up to the total cross section as follows,

$$\sigma(\text{total}) = \sigma^{+}(s) + \sigma^{+}(p) + \sigma^{+}(d) + \sigma^{+*} + \sigma^{2+}(\text{direct}). \tag{11.5}$$

Information about the Auger transitions are contained in $\sigma^{+}(d)$. Provided only Auger transitions are involved in filling the d-shell vacancy (i.e., no fluorescence occurs), then

$$\sigma^{+}(d) = \sigma^{2+}(\text{A}) + \sigma^{3+}(\text{A}). \tag{11.6}$$

However, photoionization mass spectrometry determines the total charge production without distinction between direct and Auger transitions, thus

$$\sigma(\text{total}) = \sigma^{+}(\text{total}) + \sigma^{2+}(\text{total}) + \sigma^{3+}(\text{total}), \tag{11.7}$$

where

$$\sigma^{2+}(\text{total}) = \sigma^{2+}(\text{direct}) + \sigma^{2+}(\text{A}), \tag{11.8}$$

and

$$\sigma^{3+}(\text{total}) = \sigma^{3+}(\text{direct}) + \sigma^{3+}(\text{A}). \tag{11.9}$$

From (11.6, 8, 9)

$$\sigma^{2+}(\text{direct}) = \sigma^{2+}(\text{total}) + \sigma^{3+}(\text{total}) - \sigma^{+}(d), \tag{11.10}$$

provided we assume that $\sigma^{3+}(\text{total}) = \sigma^{3+}(\text{A})$. Thus, the two techniques of photoelectron spectroscopy and mass spectrometry are required to completely detail the photoionization process, including the determination of $\sigma^{2+}(\text{direct})$.

The partial cross sections $\sigma^{+}(s)$, $\sigma^{+}(p)$, and $\sigma^{+}(d)$ in the energy range 60–140 eV have been determined by WEST et al.[215] using the technique of photoelectron spectroscopy and estimating the effect of $\sigma^{2+}(\text{direct})$ from the results of [114]. Their results are shown in Fig. 67 and include the theoretical results of AMUSIA et al.[244, 151] The major contribution to the total cross section is the ejection of d-shell electrons. However, the interesting point is that the probability for ejecting s- and p-shell electrons is influenced by the d shell as evidenced by the increase in the s and p partial cross sections in this region. A more complete theoretical curve for $\sigma^{+}(s)$ is shown in Fig. 57, illustrating the necessity for considering electron correlations.

Figure 68 shows the experimental data for threshold double ionization in Xe.[103, 114, 233, 236, 237] The solid curve indicates the detailed data of EL-SHERBINI and VAN DER WIEL[114] using the technique of electron-impact "photo-

---

[244] Ya Amusia, M., Ivanov, V.K., Cherepkov, N.A., Chernysheva, L.V. (1974): Sov. Phys.-J.E.T.P. *39*, 752

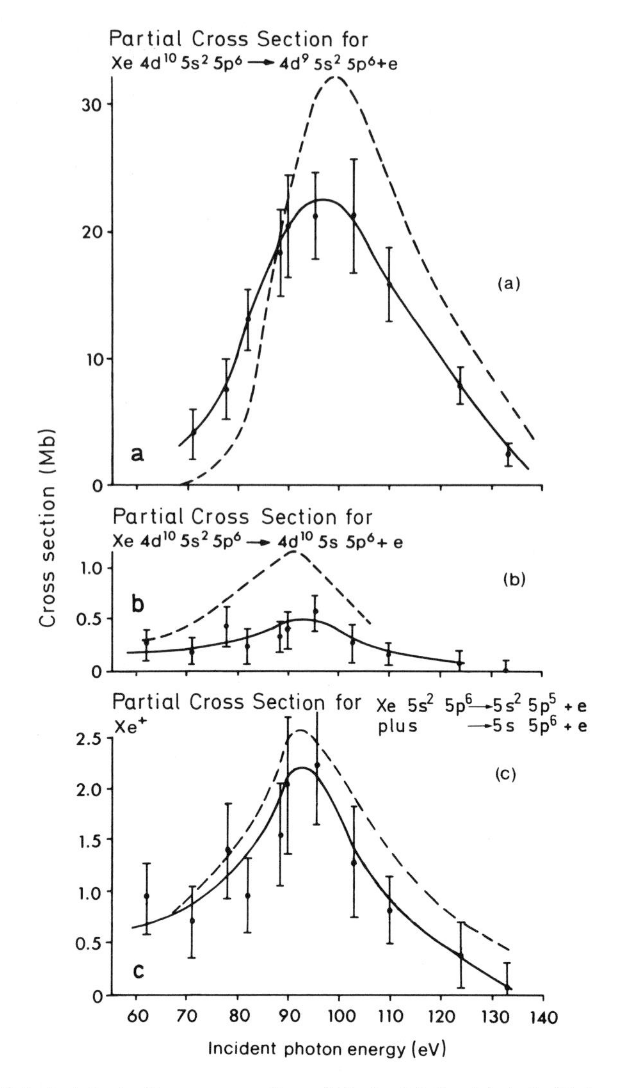

Fig. 67a–c. Partial photoionization cross section of Xe for (**a**) the 4*d* shell, (**b**) the 5*s* shell, and (**c**) the $5p+5s$ shell. The dashed curves represent the data of AMUSIA et al.[151, 244]. (From WEST et al.[215])

ionization," whereas the dashed curve represents VAN DER WIEL's estimate for direct double ionization [σ^{2+} (direct)]. The total cross section for producing double ionization (solid line) continues to increase with photon energy and in fact follows the cross-section curve for producing *d*-shell electrons. From (11.10) and [114, 215] an estimate of σ^{2+}(direct) at 100 eV gives a value of about 8 Mb. Regardless of the precise value, the result is again evidence of correlation effects transferring a fraction of the 4*d* oscillator strength to the $n=5$ shell of Xe.

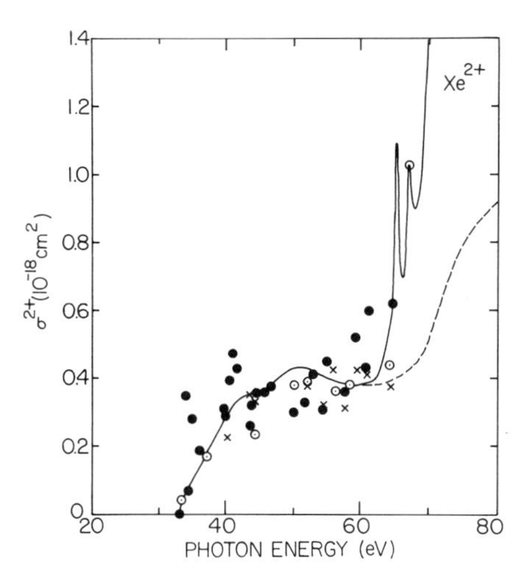

Fig. 68. Threshold double ionization of Xe as a function of photon energy. Solid line, EL-SHERBINI, VAN DER WIEL[114]; dashed line, estimate for *direct* double ionization, VAN DER WIEL, CHANG[236]; ● SAMSON, HADDAD[103]; × SAMSON, KEMENY, unpublished data; ⊙ CAIRNS et al.[233]

There is a wealth of information stored in the total photoionization cross sections. This information can be revealed by studying the partial cross sections using many of the techniques discussed in Sect. II.

γ) Angular distribution of photoelectrons. The angular distribution of photoelectrons is given, in the dipole approximation for linearly polarized radiation, by the relation[24, 89, 90, 245]

$$\frac{d\sigma_j}{d\Omega}=(\sigma_j/4\pi)\,[1+\beta P_2(\cos\theta_x)], \tag{11.11}$$

where P_2 is the Legendre polynomial of the second order, σ_j is the photoionization cross section for channel j, θ_x is the angle between the electric vector of the polarized radiation and the ejected photoelectron (see Fig. 22), Ω is the solid angle of acceptance, and β is the asymmetry parameter and can have a magnitude between $+2$ and -1.

When higher multipoles are considered (11.11) becomes for unpolarized radiation[95, 246]

$$\frac{d\sigma_j}{d\Omega}=(\sigma_j/4\pi)\sum_{n=1}^{\infty}[1+B_n P_n(\cos\theta_z)], \tag{11.12}$$

where the B_n terms replace the asymmetry term β. For photon energies in the vacuum ultraviolet and the lighter elements, all the B_n terms except B_2 tend to be very small and can be neglected. Thus, (11.12) becomes identical to the dipole approximation. There is no clear transition energy to decide when

[245] Cooper, J., Zare, R.N. (1969): In: Lectures in theoretical physics, Geltman, S., Mahanthappa, K., Brittin, N. (eds.), Vol. 11C, p. 317–337. New York: Gordon & Breach

[246] Tseng, H.K., Pratt, R.H., Yu, S., Ron, A. (1978): Phys. Rev. A*17*, 1061

(11.12) will give a better description of the angular distribution so care must be taken.

Equation (11.11) can be generalized to give the differential cross section when the incident radiation has some arbitrary degree of polarization P, regardless of whether P describes partially plane-polarized light or elliptically polarized light.[91-94] Thus, (11.11) becomes

$$\frac{d\sigma_j}{d\Omega}=(\sigma_j/4\pi)\{1-\tfrac{1}{2}\beta[P_2(\cos\theta_z)-\tfrac{3}{2}P(\cos^2\theta_x-\cos^2\theta_y)]\}, \qquad (11.13)$$

where the x and y axes are oriented along the major and minor axes of polarization. We note that for the low pressures used in electron-angular-distribution measurements the total number of electrons ejected is given by $N_j = Inl\sigma_j$, thus

$$\frac{dN_j}{d\Omega}=Inl\frac{d\sigma_j}{d\Omega}, \qquad (11.14)$$

where $dN_j/d\Omega$ represents the number of electrons ejected into a unit of solid angle, I represents the number of photons incident on the target, n is the number density of the target, and l is the ionizing pathlength observed. Thus,

$$\frac{dN_j}{d\Omega}=Inl(\sigma_j/4\pi)\{1+\tfrac{3}{4}\beta[(1+P)\cos^2\theta_x+(1-P)\cos^2\theta_y-\tfrac{2}{3}]\}. \qquad (11.15)$$

Equation (11.15) shows that β can be determined by measuring the number of electrons ejected at two different angles. However, it is advisable to use as many angles as possible. Three basic methods for varying the angle of observation have been used: an electron energy analyzer is rotated in a plane containing the incident photon beam (and hence the volume observed varies with angle), the analyzer is rotated in a plane at right angles to a linearly or partially polarized photon beam, or a polarizer is used to rotate the plane of polarization of the incident beam and the analyzer remains in a fixed position. The latter method is the most accurate technique. Any variation in contact potentials or stray magnetic fields experienced when the analyzer is rotated is eliminated in this mode of operation. However, some loss of light intensity is experienced when using a vacuum uv polarizer.[101]

The angular distribution of photoelectrons from atoms has been studied by numerous investigators using discrete resonance lines.[101, 247-253] The most

[247] Berkowitz, J., Ehrhardt, H. (1966): Phys. Lett. *21*, 531

[248] Niehaus, A., Ruf, M.W. (1972): Z. Phys. *252*, 84

[249] Morgenstern, R., Niehaus, A., Ruf, M.W. (1970): Chem. Phys. Lett. *4*, 635; (1971): Electronic and atomic collisions, Branscomb, L.M. et al. (eds.) (Abstracts VIIth Int. Conf., Amsterdam, July 1971) p. 167. Amsterdam: North-Holland Pub. Co.

[250] Dehmer, J.L., Chupka, W.A., Berkowitz, J., Jivery, W.T. (1975): Phys. Rev. A *12*, 1966

[251] Carlson, T.A., Jones, A.E. (1971): J. Chem. Phys. *55*, 4913

[252] Karlsson, L., Mattson, L., Jadrny, R., Siegbahn, K., Thimm, K. (1976): Phys. Lett. *58* A, 381

[253] Harrison, H. (1970): J. Chem. Phys. *52*, 901

Table 4. The angular distribution of photoelectrons ejected by 584-Å radiation from Ar, Kr, and Xe

Atom and Orbital	HANCOCK, SAMSON[a]	MORGENSTERN et al.[b]	DEHMER et al.[c]	CARLSON, JONAS[d]	KARLSSON et al.[e]
Ar $3p_{3/2}$	0.95 ± 0.02	0.95 ± 0.02	0.89 ± 0.04	0.85 ± 0.05	–
$3p_{1/2}$	–	0.95 ± 0.02	–	0.85 ± 0.05	–
Kr $4p_{3/2}$	–	1.37 ± 0.02	1.29 ± 0.05	1.20 ± 0.05	1.30 ± 0.05
$4p_{1/2}$	–	1.37 ± 0.03	1.25 ± 0.05	1.20 ± 0.05	1.23 ± 0.05
Xe $5p_{3/2}$	1.78 ± 0.04	1.71 ± 0.02	1.77 ± 0.05	1.45 ± 0.05	–
$5p_{1/2}$	1.64 ± 0.04	1.64 ± 0.06	1.63 ± 0.05	1.35 ± 0.05	–

[a] Hancock, W.H., Samson, J.A.R. (1976): J. Electron. Spectrosc. *9*, 211

[b] Morgenstern, R., Niehaus, A., Ruf, M.W. (1971): In: Electronic and atomic collisions, Branscomb, L.M., et al. (eds.) (Abstracts VIIth Int. Conf. Amsterdam, The Netherlands, July 1971), p. 167. Amsterdam: North-Holland

[c] Dehmer, J.L., Chupka, W.A., Berkowitz, J., Jivery, W.T. (1975): Phys. Rev. A *12*, 1966

[d] Carlson, T.A., Jones, A.E. (1971): J. Chem. Phys. *55*, 4913

[e] Karlsson, L., Mattson, L., Jadrny, R., Siegbahn, K., Thimm, K. (1976): Phys. Lett. *58A*, 381

commonly used line is the He I resonance line at 584 Å. Table 4 lists the various results of the β parameter for Ar, Kr, and Xe, as obtained by different groups. However, for practical applications and theoretical understanding it is desirable to have data as a function of photon energy. These data have now been obtained for the outer p electrons of Ne, Ar, Kr, and Xe from threshold to about 80 eV by the use of synchrotron radiation[216, 217, 254–256] and by the electron-impact "photoionization" method[257, 258]. The β parameter exhibits rapid fluctuations in the vicinity of the ionization threshold as can be seen from Figs. 69–72. Theoretical analysis ascribes these fluctuations, for example, to phase differences and interference between the various channels that are available to the outgoing waves. These aspects are discussed in detail in the review by STARACE [*1*]. The solid-line curves shown in Figs. 69–72 are the theoretical values calculated by AMUSIA and colleagues using the RPAE method.[259, 260] Of particular interest is the example of Xe. When electron correlations between the $5p$ and $4d$ shells are ignored, there is no agreement with the experimental data for photon energies greater than 40–50 eV. However, as soon as this interaction is considered almost perfect agreement is obtained (see STARACE, Fig. 28 [*1*]). Thus, the angular distributions of photoelectrons are shown to be sensitive indicators regarding the dynamical interaction between electrons of different shells.

In the nonrelativistic treatment of the dipole approximation of angular distributions, $\beta = 2$ is predicted for all s-shell electrons. WATSON and

[254] Miller, D.L., Dow, J.D., Houlgate, R.G., Marr, G.V., West, J.B. (1977): J. Phys. B *10*, 3205

[255] Torop, L., Morton, J., West, J.B. (1976): J. Phys. B *9*, 2035

[256] Watson, W.S., Stewart, D.T. (1974): J. Phys. B *7*, L466

[257] Van der Wiel, M.J., Brion, C.E. (1972/1973): J. Electron. Spectrosc. *1*, 439

[258] Branton, G.R., Brion, C.E. (1974): J. Electron Spectrosc. *3*, 123

[259] Ya Amusia, M., Cherepkov, N.A., Chernysheva, L.V. (1972): Phys. Lett. *40* A, 15

[260] Ya Amusia, M., Ivanov, V.K. (1976): Phys. Lett. *59* A, 194

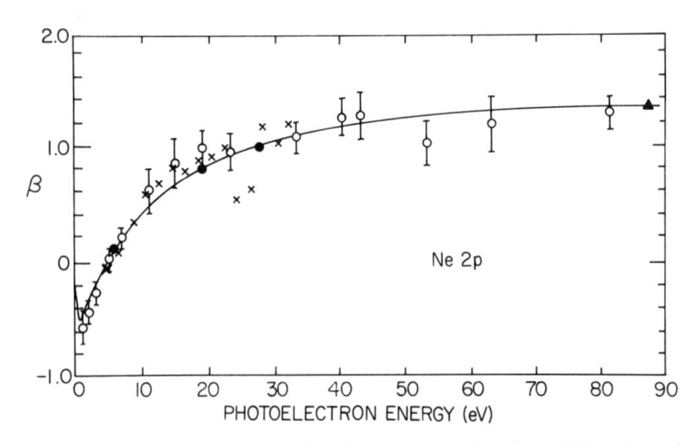

Fig. 69. The asymmetry parameter β for Ne $2p$ electrons as a function of the photoelectron energy. *Experimental data:* ○ CODLING et al.[216]; ● DEHMER et al.[250]; ▲ WUILLEUMIER, KRAUSE[198]; × VAN DER WIEL, BRION[257]. The two low data points at about 25 eV coincide with autoionizing resonance lines. *Theoretical data:* ——— AMUSIA et al.[259]

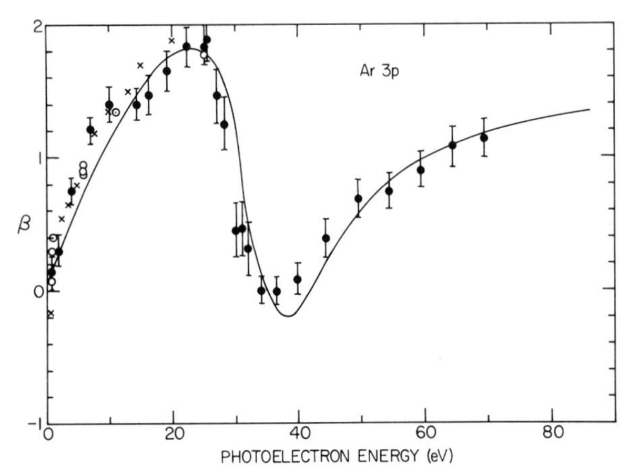

Fig. 70. The asymmetry parameter β for Ar $3p$ electrons as a function of the photoelectron energy. *Experimental data:* ● HOULGATE et al.[217]; × WATSON, STEWART[256]; ⊙ data from Table 4 and references therein. *Theoretical data:* ——— AMUSIA et al.[259]

STEWART[256] and DEHMER et al.[250] have shown this to be true for He. However, it appears that He and other light elements may be special cases. WALKER and WABER[261] have predicted on the basis of their calculations using relativistic wave functions that β for s-shell electrons may deviate from $\beta=2$. This is characterized by the appearance of spin-orbit interactions allowing $s\to p_{1/2}$ and $s\to p_{3/2}$ transitions. The change in the β parameter is caused by the interference between these outgoing waves. Within the dipole approximation they find that the relativistic correction for s orbitals is given by

$$\beta=[2\gamma^2+4\gamma\cos(\delta_{1/2}-\delta_{3/2})]/1+2\gamma^2, \tag{11.16}$$

[261] Walker, T.E.H., Waber, J.T. (1973): Phys. Rev. Lett. *30*, 307; (1973): J. Phys. B *6*, 1165

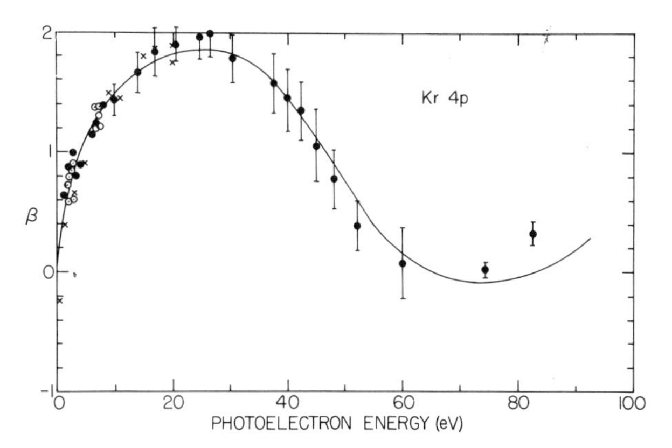

Fig. 71. The asymmetry parameter β for Kr $4p$ electrons as a function of the photoelectron energy. *Experimental data:* ● MILLER et al.[254]; × WATSON, STEWART[256]; ⊙ data from Table 4 and references therein. *Theoretical data:* ——— AMUSIA et al.[259]

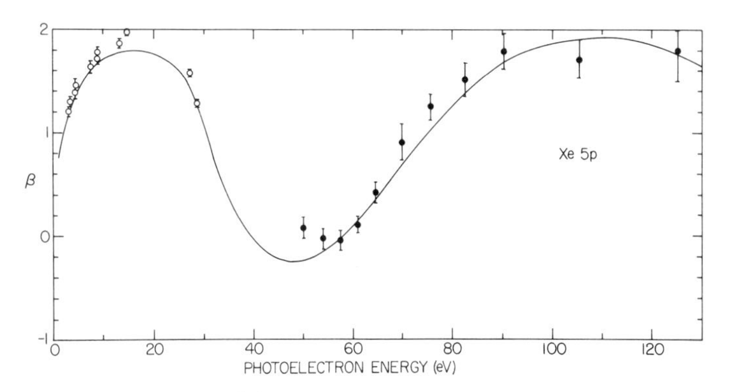

Fig. 72. The asymmetry parameter β for Xe $5p$ electrons as a function of the photoelectron energy. *Experimental data:* ● TOROP et al.[255]; ○ DEHMER et al.[250]. *Theoretical data:* ——— AMUSIA, IVANOV[260]

where $\gamma = R_{3/2}/R_{1/2}$ and $R_{1/2,\,3/2}$ are the radial matrix elements for the $s \to P_{1/2}$ and $s \to P_{3/2}$ transitions, respectively and $\delta_{1/2,\,3/2}$ are the respective phase shifts. We can see from (11.16) that when $R_{1/2} = R_{3/2}$, then $\beta = 2$ as predicted from the nonrelativistic dipole theory. When $R_{3/2} = 0$, $\beta = 0$ and when $R_{1/2} = 0$, $\beta = 1$. In general, the two matrix elements will go to zero for different photon energies causing variations in β (that is, in the "Cooper minima").

A measurement of the angular distribution of the Xe $5s$ electron at 304 Å by DEHMER and DILL[262] gave a value of $\beta = 1.4 \pm 0.1$. The value calculated by WALKER and WABER[263] was $\beta = 1.7$. ONG and MANSON[264] predict a value between 1.4 and 1.6. However, both calculations neglect interchannel electron correlations. Interchannel correlations have been shown by AMUSIA to be very important in calculations of the Xe $5s$ photoionization cross section (see

---

[262] Dehmer, J.L., Dill, D. (1976): Phys. Rev. Lett. *37*, 1049

[263] Walker, T.E.H., Waber, J.T. (1974): J. Phys. B *7*, 674

[264] Ong, W., Manson, S.T. (1978): J. Phys. B *11*, L 65

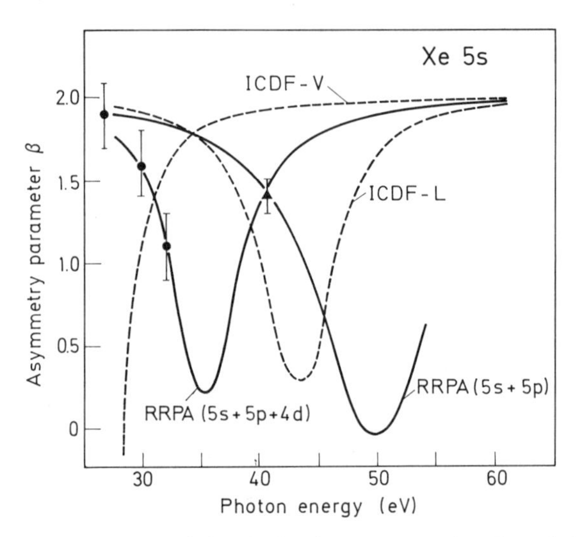

Fig. 73. The asymmetry parameter β for Xe $5s$ electrons as a function of the photon energy. Theory: The intermediate coupling Dirac-Fock (ICDF) length (L) and velocity (V) calculations (dashed lines) are compared with the relativistic random-phase approximations (RRPA) calculations (solid lines), which onsider electron correlations between $(5s+5p)$ electrons and $(5s+5p+4d)$ electron, from JOHNSON and CHENG[265]. The square experimental datum point is that of DEHMER and DILL[262]. The circular data points are these of WHITE et al.[266b)].

Sect. 11.1). Recently, JOHNSON and CHENG[265, 207] have calculated β for the $5s \rightarrow \varepsilon p$ transition in Xe using a relativistic random phase approximation (RRPA) and have shown a wide variation in the predicted β values caused by the different approximations used, and in particular between the RRPA calculations that consider only $(5s+5p)$ electron correlations and those that include $(5s+5p+4d)$ correlations (see Fig. 73). From past experience with Xe we would expect the RRPA $(5s+5p+4d)$ curve to be the more realistic. A similar calculation has been made by CHEREPKOV[266] yielding almost identical results, except that the minimum of the β curve goes all the way to -1. The K matrix calculation of HUANG and STARACE[266a] also shows a variation of β within the Cooper minimum. However, their calculation starts from a non-relativistic basis of Hartree-Fock wave functions and then treats the spin-orbit interaction only in the final state. In all approximations that include relativistic effects β varies rapidly between 0 and 2 within the Cooper minimum. Ironically, the single datum point of DEHMER and DILL's occurs at the intersection of the two RRPA curves shown in Fig. 73. This emphasizes the need for data at many photon energies. These data are now available. WHITE et al.[266b] have measured β for several photon energies near threshold and obtain excellent agreement with the RRPA $(5s+5p+4d)$ curve as shown in Fig. 73.

---

[265] Johnson, W.R., Cheng, K.T. (1978): Phys. Rev. Lett. *40*, 1167

[266] Cherepkov, N.A. (1978): Phys. Lett. *66* A, 204

[266a] Huang, K.-N., Starace, A.F. (1980): Phys. Rev. A*21*, 697

[266b] White, M.G., Southworth, S.H., Korbin, P., Poliakoff, E.D., Rosenberg, R.A., Shirley, D.A. (1979): Phys. Rev. Lett. *43*, 1661

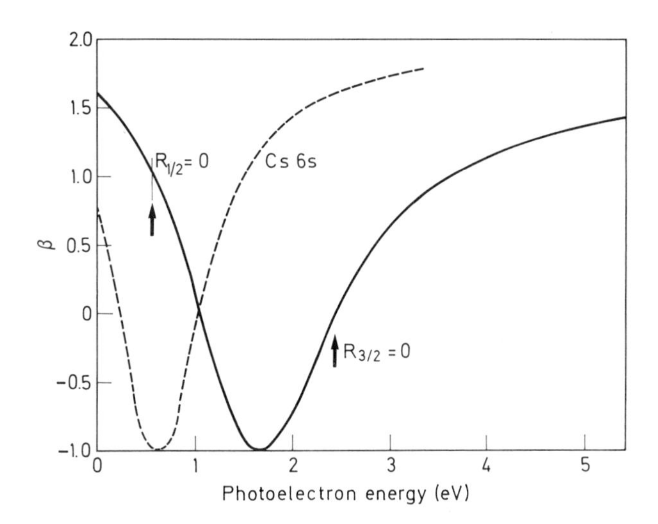

Fig. 74. The asymmetry parameter β for Cs 6*s* electrons as a function of the photo-electron energy. Theoretical curve (solid line): The deviation from the expected value $\beta=2$ is obtained by considering spin-orbit interactions. The vertical arrows indicate the photon energies where $R_{3/2}$ and $R_{1/2}$ go to zero (ONG, MANSON[267]). Semiempirical curve (dashed line) obtained from spin polarization measurements (MARR[258])

A variation in β for the outer *s* electrons in the alkali metals has also been predicted.[261] The strongest variation is expected in the Cooper minima where the radial matrix elements tend to zero. Interchannel coupling is expected to be small and the deviation of β from $\beta=2$ should be caused by the spin-orbit interaction. ONG and MANSON[267] have calculated β for the 6*s* electron of Cs. Their results are shown in Fig. 74. The vertical arrows indicate the photon energies where $R_{3/2}$ and $R_{1/2}$ go to zero. As can be seen from the figure, very dramatic changes in β are expected. Also shown in Fig. 74 (dashed curve) are the results calculated by MARR[268] using the experimental results obtained for the degree of spin polarization of photoelectrons ejected from Cs. There is excellent agreement in the shape of the two curves. However, there is a considerable displacement in their relative positions. The only direct measurement of β that has been made for the alkali metals is that by CHAFFEE in 1931 for potassium.[28] His measurement was made at about 2,400 Å, which is on the high-energy side of the Cooper minimum cross section (see Fig. 36). The data, although quoted to have a cosine square distribution, appear to have a β value of 1.5.

Another example of an outer *s* shell β anomaly is that of the Hg $6s^2$ electrons. NIEHAUS and RUF[248] have obtained a β value of 1.68 ± 0.1 by use of 584-Å radiation. Again, the explanation of this deviation from $\beta=2$ appears to be the spin-orbit interaction. Although, as WALKER and WABER[261] point out, when the dipole matrix elements tend to zero, it may be necessary to include

[267] Ong, W., Manson, S.T. (1978): Phys. Lett. *66* A, 17
[268] Marr, G.V. (1974): J. Phys. B 7, L47
[269] Baum, G., Lubell, M.S., Raith, W. (1972): Phys. Rev. A 5, 1073

the higher-order terms in the multipole expansion. This is also the point made by RON et al.[95] in their consideration of the Hg $6s^2$ electrons.

STARACE et al.[270] have predicted that even in the nonrelativistic case, the photoelectron angular distribution of s-electrons in open-shell atoms (other than $l=0$ for the outer shell) will fluctuate with the energy of the incident radiation because the various possible couplings of the continuum p waves to the ion core give rise to several distinct final states. Calculations for chlorine have been reported that show dramatic variations of β near threshold. However, no measurements have been made so far on any open-shell atom (other than those with outer s shells).

Thus, we can expect, in general, that whenever an s-shell electron has the opportunity to be ejected from the atom via more than one channel, the β value will deviate from $\beta=2$.

So far our discussion has involved the angular distribution of continuum photoelectrons. However, as we have seen, the continuum-photoionization cross section of atoms is overlaid by numerous autoionizing resonances, particularly in the ionization threshold region. DILL[271] predicted that the asymmetry parameter β would vary rapidly between $+2$ and -1 within the autoionizing resonances of Xe between the $^2P_{3/2}$ and $^2P_{1/2}$ thresholds of the ion. Subsequent measurements by SAMSON, GARDNER[272] verified his prediction. These results are shown in Fig. 75. This rapid variation in β within an autoionizing resonance appears to be a general phenomena.[273] BALASHOV et al.[274] have calculated the β parameter for Ne in the vicinity of the first autoionizing state, $2s\,2p^6\,3p\,^1P$, and found that β made a complete oscillation between 2 and 0 within the width of the resonance. Recent calculations within resonances have been made by KABACHNIK and SAZHINA[275] and by TAYLOR[276] confirming this behavior. KABACHNIK and SAZHINA[275] have parametrized the variation of β yielding the result that for an isolated resonance β is given by the following relation:

$$\beta=\frac{-2(X\varepsilon^2+Y\varepsilon+Z)}{(A\varepsilon^2+B\varepsilon+C)}, \tag{11.17}$$

where ε represents the energy difference between the incident photon energy and the idealized resonance energy [see (10.2)],

$$A=(\sigma_a+\sigma_b)/4\pi, \quad B=(2q\sigma_a)/4\pi, \quad \text{and} \quad C=(\sigma_a q^2+\sigma_b)/4\pi.$$

X, Y, and Z are considered as parameters the values of which may be chosen to fit the shape of the observed or calculated variation. When only one decay

[270] Starace, A.F., Rast, R.H., Manson, S.T. (1977): Phys. Rev. Lett. *38*, 1522

[271] Dill, D. (1973): Phys. Rev. A 7, 1976

[272] Samson, J.A.R., Gardner, J.L. (1973): Phys. Rev. Lett. *31*, 1327

[273] Dill, D. (1972): Phys. Rev. A 6, 160

[274] Balashov, V.V., Kabachnick, N.M., Sazhina, I.P. (1973): Proc. 8th Int. Conf. on Phys. of Electronic and At. Collisions, Belgrade, pp. 527–8, Belgrade: Institute of Physics; (1973): Vestnik MGU *14*, 733

[275] Kabachnik, N.M., Sazhina, I.P. (1976): J. Phys. B 9, 1681

[276] Taylor, K.T. (1977): J. Phys. B *10*, L699

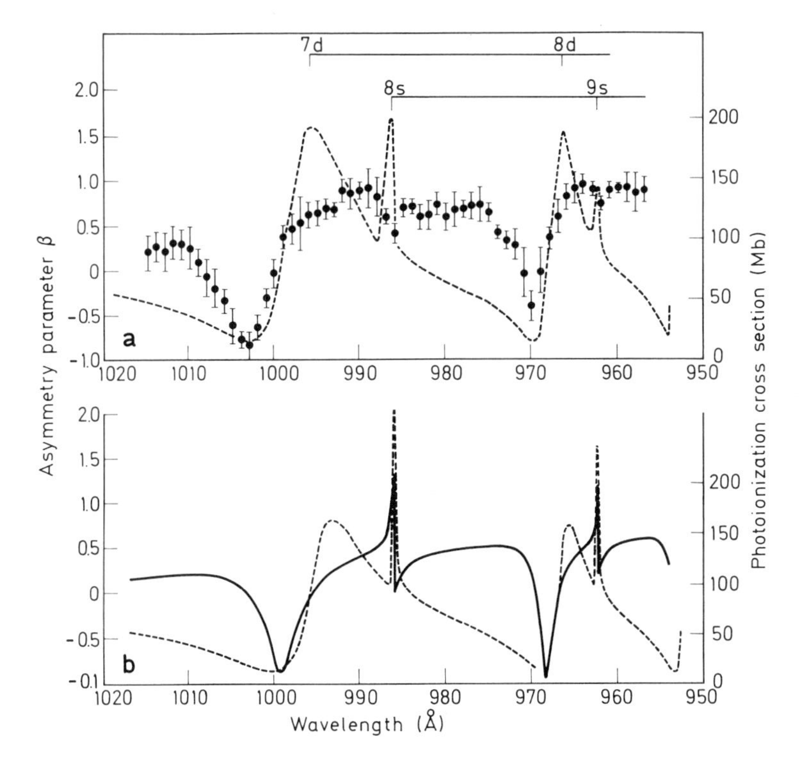

Fig. 75a, b. The asymmetry parameter β for Xe $5p$ electrons within the autoionizing resonances between the $^2P_{3/2}$ and $^2P_{1/2}$ thresholds. (**a**) *Experimental data:* SAMSON, GARDNER[272], (**b**) *Theoretical data:* DILL[271]. (From SAMSON, GARDNER[272])

channel exists, the value of β is the same as the continuum β value. Also, the value of β for one member of a Rydberg series is the same for all members of the series.

Experimentally, there have been no direct measurements of β within resonances lying above the $^2P_{1/2}$ ion threshold for the rare gases. However, measurements of the $^2P_{3/2}$:$^2P_{1/2}$ partial-cross-section ratios have been made and were found to deviate dramatically within these resonances relative to the almost constant value for the ratio in regions free from the autoionizing structure.[218, 277, 278] STARACE[279] has predicted that the cross-section ratio should vary within a resonance and this has been verified.[278] However, in other measurements that were sensitive to the angular distributions of the photoelectrons, it was shown that the variation in the ratios could be explained by assuming values for β, within the resonance, that were different from the neighboring continuum.[277]

---

[277] Samson, J.A.R., Gardner, J.L., Starace, A.F. (1975): Phys. Rev. *12*, 1459

[278] Kemeny, P.C., Samson, J.A.R., Starace, A.F. (1977): J. Phys. B *10*, L201

[279] Starace, A.F. (1977): Phys. Rev. A *16*, 231

12. Summary of atoms studied and the future. Photoionization cross sections and absorption spectra have been measured for 52 of the elements over various wavelength ranges. These elements and the wavelength ranges are given in Table 5. Some of the studies consist purely of photographic absorption spectra with no quantitative data on cross sections. These are indicated with an asterisk. The other measurements contain cross section data of varying degrees of completeness. The rare gases have been studied extensively, to the point where the absolute experimental photoionization cross sections are known to an accuracy of $\pm 3\%$ to $\pm 10\%$ over the range from 100 Å to their ionization thresholds. The autoionizing structure of these gases also has been studied extensively. Less complete information is available for most of the other elements listed in Table 5. From the theoretical aspects tremendous strides have been made in recent years, particularly with the rare gases where an accurate comparison with experimental data is possible. The various forms of the many-body theory are providing data in very good agreement with experiment. In fact, the point has been reached where it is now necessary to achieve as high an accuracy as possible in cross-section measurements in order to distinguish between the correctness of the various approximations. In the case of He most of the recent sophisticated calculations agree with the experimental data. For the heavier rare gases the spread in the various theoretical calculations is larger.

Partial-cross-section measurements as a function of wavelength have been made only for the rare gases, Cd, and Hg. The measurements include the two outermost subshells of the atoms.

The angular distribution of photoelectrons ejected from the outermost shells has been measured only for the rare gases, Zn, Cd, Hg, and O. The rare gases have been studied over a large photon energy range, whereas the other atoms have been studied only at a few resonance lines (e.g., 584, 304, 736, and 1,048 Å).

Multiple-ionization studies have been made as a function of wavelength in the rare gases, Zn, Cd, Hg, and Yb. The remaining studies have been for Ba and Eu using resonance lines at 584 Å and 736 Å.

A new area of photoionization has opened up recently, namely, the study of spin-polarized electrons produced by the photoionization of the alkali atoms when the ionizing radiation is circularly polarized. This field has attracted some attention since the phenomenon was first predicted by FANO[280] in 1969. Experimental verification followed soon after.[281-284] The progress to date has been surveyed by LUBELL[285] and KESSLER.[286] Studies have been confined mainly to Cs and Rb and in the spectral range near their ionization thresholds

---

[280] Fano, U. (1969): Phys. Rev. *178*, 131

[281] Lubell, M.S., Raith, W. (1969): Phys. Rev. Lett. *23*, 211

[282] Baum, G., Lubell, M.S., Raith, W. (1970): Phys. Rev. Lett. *25*, 267

[283] Kessler, J., Lorenz, J. (1970): Phys. Rev. Lett. *24*, 87

[284] Heinzmann, U., Kessler, J., Lorenz, J. (1970): Z. Phys. *240*, 42

[285] Lubell, M.S. (1977): In: At. Phys. 5, Marrus, R., Prior, M., Shugart, H. (eds.), p. 325. New York: Plenum Press

[286] Kessler, J. (1976): Polarized electrons, pp. 123–146. Berlin-Heidelberg-New York: Springer

Table 5. Atomic gases and vapors for which absorption spectra and absolute absorption cross sections are available. The numbers in parenthesis respresent the wavelength range in Å. The asterisk after the parenthesis indicates only photographic absorption spectra are available[a]

Closed shells		Open shells	
s^2, s^2p^6		s^2p	
He	(1.5–524)	B	(950–1,150)*
Ne	(0.2–600)	Al	(1,160–2,100)
Ar	(0.2–800)	Ga	(320–1,000)*, (1,480–2,100), (1,400–2,400)*
Kr	(1.5–890)	In	(320–1,000)*, (1,500–2,200), (1,500–2,300)*
Xe	(0.08–1,030)	Tl	(10–200)*, (320–2,000)*, (1,400–2,100)
s^2s^2, p^6s^2		s^2p^2	
Be	(60–110)*, (500–2,000)*	C	(700–920)*, (980–1,250)
Mg	(500–2,000)*, (1,450–1,650)	Si	(1,250–1,550)
Ca	(10–120)*, (1,080–2,100)	Ge	(320–1,000)*
Sr	(40–95)*, (1,640–2,200)	Sn	(320–1,000)*
Ba	(10–200)*, (455–912)*, (1,560–1,770)*, (1,700–2,382)	Pb	(10–120)*, (320–2,500)*, (1,470–1,671)
$ds^2, f^nd^0s^2$		s^2p^3	
La	(320–1,000)*	N	(400–900)
Ce	(10–120)*		
Pr	(320–1,000)*	s^2p^4	
Nd	(320–1,000)*	O	(400–1,000)
Sm	(320–1,000)*, (1,950–2,350)	S	(900–1,830)
		s^2p^5	
Eu	(40–300)*, (320–1,000)*, (1,350–2,200)		
Dy	(320–1,000)*	F	(600–1,500)*
Ho	(320–1,000)*	Cl	(320–1,000)*
Er	(320–1,000)*	Br	(600–1,500)*
Tm	(320–1,000)*, (1,350–2,200)	I	(600–1,500)*
Yb	(320–1,000)*, (1,350–2,00)		
		s^2s, p^6s	
ds^2		Li	(170–210)*, (500–2,500)
U	(320–1,000)*	Na	(50–2,500)
		K	(10–210)*, (600–2,860)
d^5s^2		Rb	(40–120)*, (525–900)*, (1,200–3,000)
Mn	(178–300)*, (1,305–2,040)*	Cs	(70–177)*, (1,200–3,200)
d^6s^2		d^5s	
Fe	(138–258)*	Cr	(30–400)*
d^7s^2		$d^{10}s$	
Co	(138–238)*	Cu	(138–200)*, (350–1,400)*, (1,200–1,600)
		Ag	(350–1,600)*, (1,200–1,600), (300–730)
d^8s^2		Au	(1,200–1,600)
Ni	(138–214)*		
$d^{10}s^2$			
Zn	(70–150)*, (247–1,320), (390–835)*		
Cd	(17–31)*, (50–1,450)		
Hg	(20–120)*, (172–1,188)		

(continued, page 210)

Table 5. (Continued)

Excited neutrals	Ions
He* (2,400–3,100)	Xe^+ (555)
Ar* (2,820–2,950)	Na^+ (150–300)*, (10,600)
Kr* (2,690–3,250)	Li^+ (50–220)*
Xe* (2,700–4,622)	Be^+ (60–110)
Cs* (2,500–5,000)	
Mg* (2.960–3,060)	
Ba* (2,360–3,200)	
Si* (1,647)	

[a] References to the original publications of the above can be found in Kieffer, L.J., Bibliography of Low Energy Electron and Photon Cross Section Data (through Dec. 1974), Nat. Bur. Stand. Special Publication No. 426, and Gallagher, J.W., E.C. Beaty, and J.R. Rumble, Jr., Supplement (1975–1977) to NBS Special Publication 426, Superintendent of Documents, U.S. Government Printing Office, Washington, D.C. 20402, 1976 and 1978 (except where identified).

The following references, listed alphabetically, were not included in the above bibliography. They contain photographic absorption spectra with densitometer traces that give a qualitative account of the absorption probability.

Ag: Krause, M.O. (1980): J. Chem. Phys. *72*, 6474 (photoelectric); Connerade, J.P., Baig, M.A., Mansfield, M.W.D., Radtke, E. (1978): Proc. R. Soc. London A *361*, 379

Ba: Connerade, J.P., Mansfield, M.W.D. (1974): Proc. R. Soc. London. A *341*, 267; Connerade, J.P., Tracy, D.H. (1977): J. Phys. B *10*, L235

Ca: Mansfield, M.W.D. (1976): Proc. R. Soc. London A *348*, 143

Cd: Codling, K., Hamley, J.R., West, J.B. (1978): J. Phys. B. *11*, 1713 (photoelectric).

Co: Bruhn, R., Sonntag, B., Wolff, H.W. (1978): DESY Report SR-78/14 (Aug. 1978)

Cu: Bruhn, R., Sonntag, B., Wolff, H.W. (1978): DESY Report SR-78/14 (Aug. 1978)

Eu: Mansfield, M.W.D., Connerade, J.P. (1976): Proc. R. Soc. London A *352*, 125

Fe: Bruhn, R., Sonntag, B., Wolff, H.W. (1978): DESY Report SR-78/14 (Aug. 1978)

Ga: Connerade, J.P., Baig, M.A. (1981): J. Phys. B *14*, 29

Ge: Connerade, J.P., Martin, M.A.P. (1977): Proc. R. Soc. London A *375*, 103

Hg: Connerade, J.P., Mansfield, M.W.D. (1973): Proc. R. Soc. London A *335*, 87; Mansfield, M.W.D. (1973): Astrophys. J. *180*, 1011

In: Connerade, J.P. (1977): Proc. R. Soc. London A *352*, 561; Connerade, J.P., Baig, M.A. (1981): J. Phys. B *14*, 29

K: Sander, W., Gallagher, T.F., Safinya, K.A., Gounand, F. (1981): Phys. Rev. A *23*, 2732 (photoelectric)

La-Yb: Connerade, J.P.: Private communication

Mn: Connerade, J.P., Mansfield, M.W.D., Martin, M.A.P. (1976): Proc. R. Soc. London A *350*, 405; Bruhn, R., Sonntag, B., Wolff, H.W. (1978): DESY Report SR-78/16 (Sept. 1978); Brown, C.M., Ginter, M.L. (1978): J. Opt. Soc. Am. *68*, 1541

Ni: Bruhn, R., Sonntag, B., Wolff, H.W. (1978): DESY Report SR-78/14 (Aug. 1978)

Pb: Connerade, J.P., Garton, W.R.S., Mansfield, M.W.D., Martin, M.A.P. (1977): Proc. R. Soc. London A *357*, 499; Connerade, J.P. (1981): J. Phys. B. *14*, L141

Rb: Mansfield, M.W.D., Connerade, J.P. (1975): Proc. R. Soc. London A *344*, 303; Mansfield, M.W.D. (1973): Astrophys. J. *183*, 691

Se: Connerade, J.P., Mansfield, M.W.D. (1977): Proc. R. Soc. London A *356*, 135

Sn: Connerade, J.P., Martin, M.A.P. (1977): Proc. R. Soc. London A *357*, 103

Sr: Mansfield, M.W.D., Connerade, J.P. (1975): Proc. R. Soc. London A *342*, 421

Tl: Connerade, J.P., Baig, M.A. (1981): J. Phys. B *14*, 29

Zn: Mansfield, M.W.D., Connerade, J.P. (1978): Proc. R. Soc. London A *359*, 389

Be^+: Mehlman, G., Estava, J.M. (1974): Astrophys. J. *188*, 191

Li^+: Carroll, P.K., Kennedy, E.T. (1977): Phys. Rev. Letters *38*, 1068; McIlrath, T.J., Lucatorto, T.B. (1977): Phys. Rev. Lett. *38*, 1390

Na^+: Lucatorto, T.B., McIlrath, T.J. (1976): Phys. Rev. Letters *37*, 428; Smith, A.V., Goldsmith, J.E.M., Nitz, D.E., Smith, S.J. (1980): Phys. Rev. A *22*, 577

Table 6. Photoionization studies to be undertaken

1. Total photoionization cross sections σ_{tot} for open- and closed-shell atoms.
2. Partial cross sections σ_j and branching ratios R_j (where the branching ratio is defined as the number of ions formed in a given state j relative to the total number of ions produced), and $\sigma_j = R_j \cdot \sigma_{tot}$. Measurements should be made
   (a) within the continuum and
   (b) across resonances.
3. Multiple photoionization cross sections.
4. Angular distribution of photoelectrons from
   (a) the ionization continuum
   (b) autoionizing resonances
   (c) Auger and shake-off transitions
   (d) multiple ionization.
5. Fluorescence efficiency of excited ionic states.
6. Total photoionization cross sections of excited atoms and ions (negative and positive).
7. Polarization state of the photoelectrons.
8. Multiphoton ionization.
9. The relation of atomic photoionization to the photoionization of molecules and solids
10. Production of an atlas of absorption spectra with wavelength identification for atoms, molecules, and solids.

where the spin-orbit interaction produces the greatest degree of spin polarization in the photoelectrons. Recently, HEINZMANN has extended these studies to lead and has shown that a measurement of the spin polarization within an autoionizing resonance along with total photoionization cross-section measurements can yield information regarding the transition probabilities into the various underlying continua.[287] He has also extended spin-polarization measurements into the vacuum uv.[288] It appears that, in general, where spin-orbit interactions are important we can expect to produce spin-polarized electrons through the photoionization process.[289, 290]

In Table 6 a list is given of the various photoionization studies that should be undertaken or continued. Many of these items already have received considerable attention whereas some have barely been examined. This is particularly true for items 4d–9. Item 10, the production of an atlas of absorption spectra, would be very useful in atomic and molecular physics experiments when it is important to know if one is studying a region of pure continuum ionization or if the experimental results are being influenced by the proximity of autoionizing resonances.

The ultimate goal in photoionization is to understand how photons interact with atoms and molecules and even with solids. We would like to reach the point where we can tabulate and catalog precisely how the energy of a photon is apportioned to all the accessible channels in the photoionization process. A thorough understanding of the atomic photoionization process is the first and most important step. To understand the photoionization of molecules and

[287] Heinzmann, U. (1978): J. Phys. B *11*, 399
[288] Heinzmann, U. (1977): J. Phys. E *10*, 1001
[289] Cherepkov, N.A. (1974): Sov. Phys. J.E.T.P. *38*, 463
[290] Lee, C.M. (1974): Phys. Rev. A *10*, 1598

solids we need to know what effects the proximity of other atoms have on the photoionization cross section. This, apparently, is most severe near the ionization threshold where the valence electrons are involved in the ionization process. These electrons have lost their atomic character because they are shared with the neighboring atoms. However, in the region where inner-shell ionization is predominant we have seen that the total absorption cross section of solid Na (Fig. 37) is very similar to that in the vapor phase, although the discrete structure seen with free atoms is usually broadened and shifted in energy in the solid state. This is true for the other alkali metals and for the solid rare gases.[148, 150, 153, 291] SONNTAG[292] has given a brief review of the atomic and molecular effects in the absorption spectra of solids.

From the point of view of basic physics we should investigate photon interaction with a representative number of atoms (for example, closed-shell versus open-shell atoms, low-Z versus high-Z atoms, etc.) at various wavelengths in order to test the different theoretical approximations. The goal in this case is to arrive at a unified theory of the photoionization process; a theory that will take into account threshold phenomena, resonances, multiple excitation and ionization, and predict partial cross sections.

Acknowledgement. It is a pleasure to acknowledge the U.S. Department of Energy, the National Aeronautics and Space Administration, under Grant #NGR 28-004-021, and the Atmospheric Sciences Section of the National Science Foundation for their support of different aspects of our program in vacuum uv radiation physics, which made this review possible. I would also like to express my appreciation to Professor Starace for the many discussions on photoionization, to the various authors for permission to reproduce some of their published and unpublished data, and to Professor Mehlhorn and Dr. Inokuti for their helpful suggestions in the preparation of this review.

General references

[*1*] Starace, A.F. This Encyclopedia, Vol. 31, pp. 1–121, also (1980) Appl. Opt. *19*, 4051
[*2*] Samson, J.A.R.: Techniques of vacuum ultraviolet spectroscopy. Lincoln, N.E.: Pied Publications, reprinted 1980; and Far ultraviolet region. In: Methods of experimental physics, spectroscopy, Part A, Vol. 13, pp. 204–252. New York: Academic Press 1976
[*3*] Weissler, G.L. (1956): Photoionization in gases and photoelectric emission from solids. In: This Encyclopedia, Vol. 21, pp. 304–382
[*4*] Vilessov, F.I. (1963): Photoionization of gases and vapors by vacuum ultraviolet radiation. In: Usp. Fiz. Nauk *81*, 669–738 [English translation: Sov. Phys. Usp. *6*, 888–929 (1964)]
[*5*] Samson, J.A.R. (1966): The measurement of the photoionization cross sections of the atomic gases. In: Advances in atomic and molecular physics, Vol. 2, pp. 177–261. New York: Academic Press
[*6*] Samson, J.A.R. (1976): Photoionization of atoms and molecules. In: Phys. Rep. *28*, 303–354
[*7*] Marr, G.V. (1967): Photoionization processes in gases. New York: Academic Press
[*8*] Amusia, M.Ya., Cherepkov, N.A. (1975): Many-electron correlations in scattering processes. In: Case studies in atomic phys. *5*, 47–179
[*9*] Franklin, J.L., Dillard, J.G., Rosenstock, H.M., Herron, J.T., Draxl, K. (1969): Ionization potentials, appearance potentials, and heats of formation of gaseous positive ions, NSRDS-NBS 26, Washington, DC 20402: Superintendent of Documents

[291] Haensel, R., Keitel, G., Kosuch, N., Nielsen, U., Schreiber, P. (1971): J. Phys. *32*, C4-236
[292] Sonntag, B. (1977): DESY report SR-77/17, October 1977

[10] Siegbahn, K., et al. (1967): ESCA, atomic, molecular, and solid state structure studied by means of electron spectroscopy. Stockholm: Almqvist and Wiksells Boktryekeri AB
[11] Siegbahn, K., et al. (1969): ESCA applied to free molecules. Amsterdam-London: North-Holland
[12] Turner, D.W., Baker, C., Baker, A.D., Brundle, C.R. (1970): Molecular photoelectron spectroscopy. New York: Wiley-Interscience
[13] Baker, A.D., Betteridge, D. (1972): Photoelectron spectroscopy. New York: Pergamon Press
[14] Shirley, D.A. (ed.) (1972): Electron spectroscopy. Amsterdam-London: North-Holland
[15] Eland, J.H.D. (1974): Photoelectron spectroscopy. New York: Wiley
[16] Carlson, T.A. (1972): Photoelectron and auger spectroscopy. New York: Plenum Press
[17] Rudd, E.M. (1973): In: Low-energy electron spectroscopy, Sevier, K.D. (ed.). New York: Wiley
[18] Brundle, C.R., Baker, A.D. (eds.) (1978): Electron spectroscopy: theory, techniques and applications. Vols. I, II, and III. London: Academic Press
[19] Manson, S.T. (1976): Atomic photoelectron spectroscopy, Part I. In: Adv. electronics and electron phys., Vol. 41, pp. 73–111. New York: Academic Press
[20] Manson, S.T. (1977): Atomic photoelectron spectroscopy. II. In: Adv. electronics and electron phys., Vol. 44, pp. 1–32. New York: Academic Press
[21] Ghosh, P.K. (1978): A whiff of photoelectron spectroscopy. New Delhi: Swarm Printing Press
[22] Vacuum ultraviolet radiation physics, proceedings IV International Conf. on VUV Rad. Phys., Hamburg, Koch, E.E., Haensel, R., Kunz, C. (eds.) (1974): Braunschweig: Pergamon/Vieweg
[23] Photoionization and other probes of many-electron interactions, proceedings of NATO Adv. Study Institute, Carry-le Rouet, 1975, Wuilleumier, F.J. (ed.) (1976): New York: Plenum Press
[24] Kieffer, L.J. (1974): Bibliography of low energy electron and photon cross section data, Nat. Bur. Stand. Special Publication No. 426: Superintendent of Documents, U.S. Government Printing Office, Washington, DC 20402, March 1976; and Gallagher, J.W., Beaty, E.C., Rumble, J.R., Jr.: Supplement (1975–1977) to NBS Special Publication 426

Photoelectron Spectroscopy *

By

H. SIEGBAHN and L. KARLSSON

With 139 Figures

I. Introduction

1. Scope and abstract. Photoelectron spectroscopy was essentially developed during the nineteenfifties and sixties. Earlier attempts, following the discovery of the photoelectric law, were too crude and it was then largely impossible to make use of photoelectron spectra for the study of atoms, molecules and solids (see further below). The present treatise will mainly concern the developments during the last decade. During this time period photoelectron spectroscopy has found applications in many different directions. The close relationships between phenomena observed for atoms, molecules and condensed matter have also become clearer.

The basic procedure of photoelectron spectroscopy is illustrated in Fig. 1. The sample (a gas, a liquid or a solid) is irradiated with monochromatic photons. The expelled photoelectrons are then analyzed by means of an elec-

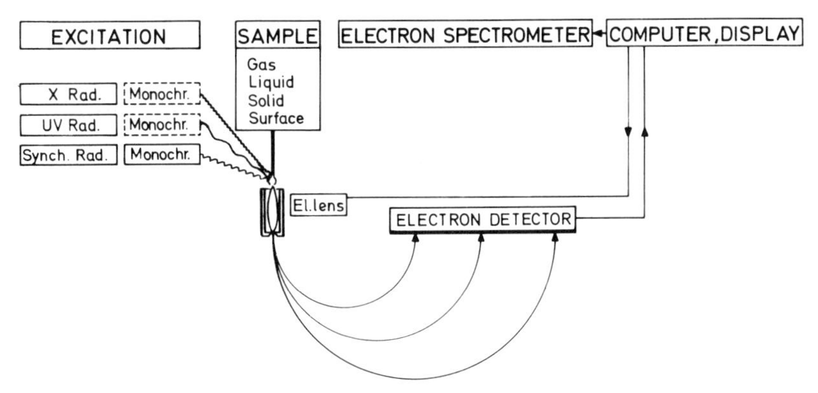

Fig. 1. Basic experimental arrangement for photoelectron spectroscopy for atoms, molecules and condensed matter

* The manuscript was completed in October 1981

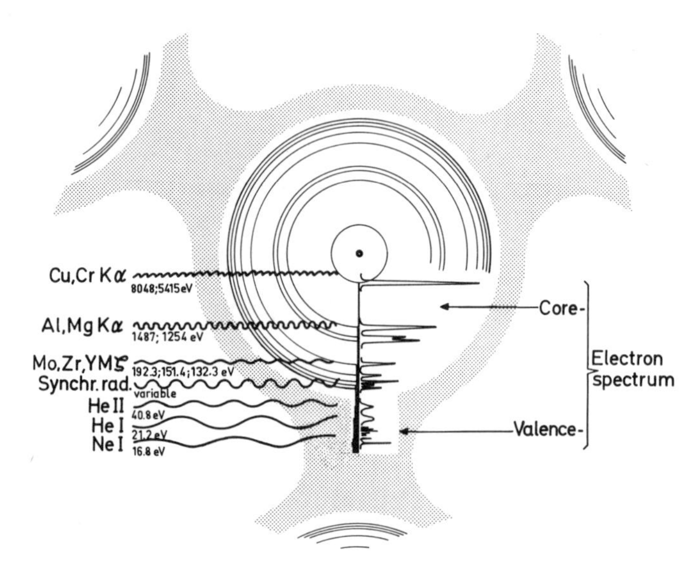

Fig. 2. Schematic illustration of the regions of binding energy accessible with different photon sources

tron spectrometer often combined with a preretardation electron lens. The energies and intensities of the photoelectrons are recorded and also in many cases their angular distributions (i.e. intensity vs. impact angles of the photons and ejection angles of the outgoing photoelectrons).

We have divided the material to be presented below into four main parts (aside from this introductory part). The first of these (Part II) deals with the experimental procedures in general use and the various features observed in photoelectron spectra. The four basic units: the photon source, the sample arrangement, the electron energy analyzer and the electron detector are dealt with consecutively and the different possibilities presented. Figure 2 is a schematic illustration of the binding energy regions which are accessible with the most common discrete line photon sources and the general appearance of the recorded photoelectron spectrum. The increased use now being made of synchrotron radiation for excitation of spectra is exemplified in the Parts III–V by several case studies. The final section of Part II deals with the basic theoretical concepts in current use for interpretation of photoelectron spectra.

Atomic photoelectron spectra is the subject of Part III. Studies of free atomic species have been made mostly for the noble gases, but an increasing number of investigations is currently made on e.g. atomic metallic vapours as a result of improved sample preparation techniques. The atomic systems furnish ideal test cases for the basic processes in photoelectron spectroscopy, which are illustrated in Fig. 3. These may be subdivided into primary processes (a), which produce either main photoelectron lines or satellite lines and the secondary decay transitions (b), which result from the photoexcitation of the system. The decay transitions lead to the emission of electrons or photons. The electron emission decay transitions appear as natural ingredients in the photoelectron

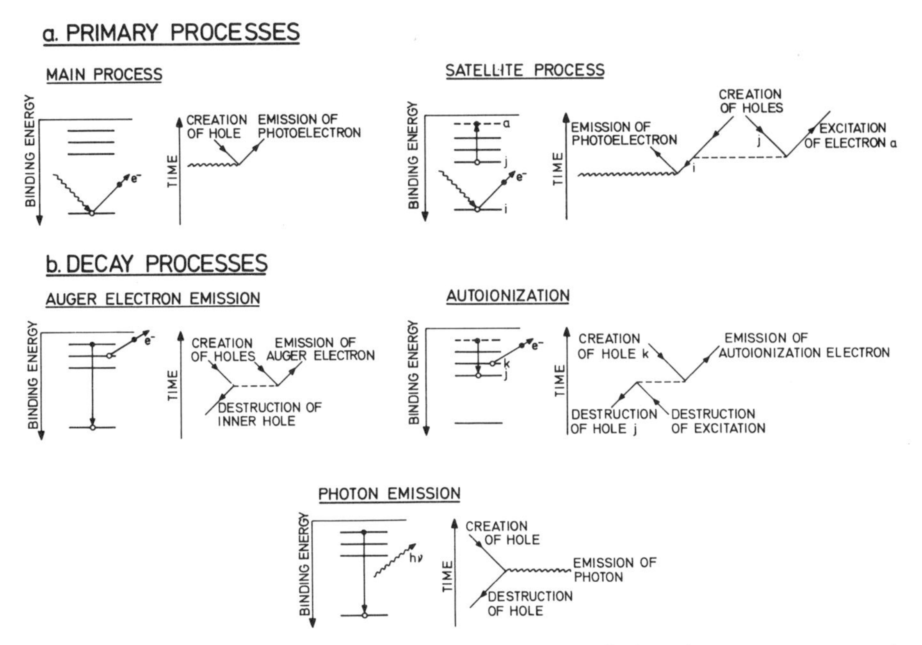

Fig. 3a, b. Schematic representations of the basic processes of photoelectron spectroscopy in terms of energy level and many-body diagrams (**a**) primary processes (**b**) secondary decay processes

spectra, in terms of Auger and autoionization lines. We will not, however, be explicitly concerned with these decay spectra since they are treated by other authors in this volume. For atoms, accurate theoretical calculations can be made to be compared with the experimental results, leading to a deeper understanding of the role of relaxation and correlation in deciding the appearance of the photoelectron spectra.

Part IV treats core level studies in molecules and condensed matter. In the introductory section the various models for description of chemical shifts of core levels are discussed. For large classes of systems, the ground state properties decide the magnitude of the shifts. In other cases, such as molecules adsorbed on surfaces, the relaxation (and possibly also correlation) contributions to the shifts may be at least as important. Close formal relationships exist between core shifts and data obtained via other experiments, e.g. NMR and heat of formation measurements. The relationship with the latter type of data can be effectively used for the discussion of measured core shifts between the vapour and condensed phases. The following section of Part IV is centered on other sources of information found in core electron spectra; the occurrence of multiline features and the factors influencing the shapes and widths of the lines. Additional lines may be either an effect of intrinsic processes, e.g. multielectron excitations or extrinsic, i.e. result from processes where the outgoing photoelectron looses energy due to inelastic scattering. The inherent mechanisms responsible for the shapes and widths of core lines are due to the decay processes, vibrational excitations, multiplet splittings and Fermi surface excitations, the relative importance of which vary from one type of material to another. Part IV is concluded by a section where a number of case studies have been collected to exemplify the applications of core photoelectron spectroscopy to molecules adsorbed on surfaces.

The final Part V deals with valence level photoelectron spectroscopy of molecules and condensed matter. For free molecules, studies have been performed primarily with uv line sources (He, Ne). Molecular valence electron spectra contain in general a wealth of information. In this energy region a large number of molecular orbital levels generally occur, each giving rise to a photoelectron band. The electron bands corresponding to each molecular orbital also commonly exhibit rich vibrational structure. These two aspects are discussed in the first two sections of Part V. Next, the angular distribution of photoelectrons from free molecules is considered. Owing to the multicenter molecular potential the motion of the outgoing photoelectron is a vastly more complicated problem than for free atoms. Therefore, angular distribution measurements from free molecules are generally less well understood although considerable progress has been made in terms of new calculational schemes, such as the multiple scattering $X\alpha$ method. Section 17 contains a selection of topics in relation to molecular valence photoelectron spectra, starting with a short review of MO methods for interpretation of spectra. Advances have been made in the investigation of species, which are not stable under normal conditions. Examples are given of studies of gas-phase reactions, ionically bonded molecules and negative ions.

We conclude Part V by a review of valence level studies of solids and surfaces. These studies have been divided into two separate sections. The first section pertains to investigations of single crystal specimens, where use is made of momentum conservation laws to investigate band structure. Angular distribution measurements and synchrotron radiation give complementary degrees of freedom in mapping the dispersion surfaces. Further, the geometrical structure of adsorbates on single crystal surfaces can be elucidated by means of the study of level shifts and angular distributions. The second and final section is concerned with higher photon energy studies of solid valence bands primarily using monochromatized Al$K\alpha$. In this excitation regime, the observed valence spectra correspond closely to the initial density of states. This is exemplified for different types of solids.

2. Historical background. The time period covered by the present article concerns essentially the last ten or fifteen years. It is therefore appropriate to give some historical notes related to the various steps which led to the present status of the field. A comprehensive review of the early history of electron spectroscopy has been given by JENKIN et al.[1a]. CARLSON[1b] has collected a number of "Benchmark Papers" in the field with comments to several of the

[1a] Jenkin, J.G., Leckey, R.C.G., Liesegang, J. (1977): J. Electron Spectrosc. Relat. Phenom. *12*, 1

[1b] Carlson, T.A. (ed.) (1978): X-ray photoelectron spectroscopy, Benchmark Papers in Phys. Chem. and Chem. Phys., vol. 2: Dowden, Hutchinson and Ross

phases in the developments. K. SIEGBAHN[2] has given a thorough historical account in his Nobel lecture with special emphasis on the work in Uppsala. The following very brief summary may hopefully serve the more limited purpose to provide some background in order to elucidate the situation at the beginning of the seventies.

The photoelectric effect was discovered by HERTZ in 1887[3] when he observed that a spark could be produced when the negative electrode was exposed to ultraviolet radiation. Subsequent experimental work by HALLWACHS[4], ELSTER and GEITEL[5] and LENARD[6] finally laid the ground in 1905 for Einstein's formulation of the photoelectric effect in terms of photons[7].

During the decades after EINSTEIN's paper, experiments on the photoelectric effect developed essentially along two independent lines. The first line was concerned with photoelectric yield phenomena near threshold. In the other line of research investigations were made on x-ray excited photoelectron spectra. These studies were initiated and carried out by ROBINSON[8], DE BROGLIE[9] and VAN DER AKKER and WATSON[10]. The investigations were, however, hampered by severe experimental difficulties concerning resolution and electron detection, among other things. As a consequence of these experimental shortcomings, the physical and chemical applications of electron spectroscopy where high resolution and well defined line structures were required, could not be realized.

During the 1950s work was performed by SIEGBAHN and coworkers following experience from nuclear spectroscopy, in particular internal conversion in radioactive decay, to improve on the resolution in the energy analysis of photoelectrons. A spectrometer of a magnetic double-focusing type[11,12] was developed[13-17], which was capable of a resolution of a few parts in 10^4 with an accuracy of $3:10^5$ over a wide energy range covering both radioactive decay electrons of some hundred keV and electrons in the keV region. This instrument was especially designed for the purpose of studying low energy photoelectrons expelled by x-radiation at very high resolution. It was then found

---

[2] Siegbahn, K. (1982): Les Prix Nobel, Stockholm: Norstedt & Söner; Rev. Mod. Phys. *54*, 709; Science *217*, 111

[3] Hertz, H. (1887): Ann. Phys. *31*, 983

[4] Hallwachs, W. (1888): Ann. Phys. *33*, 301

[5] Elster, J., Geitel, H. (1889): Ann. Phys. *38*, 40 and 497

[6] Lenard, P. (1902): Ann. Phys. *8*, 149

[7] Einstein, A. (1905): Ann. Phys. *17*, 132

[8a] Robinson, H., Rawlinson, W.F. (1914): Philos. Mag. *28*, 277

[8b] Robinson, H. (1923): Proc. Soc. A *104*, 455

[8c] Robinson, H. (1925): Philos. Mag. *50*, 241

[9] Broglie, M. de (1921): Cptes. Rdue. *172*, 274, 527, 806

[10] Van der Akker, J.A., Watson, E.C. (1931): Phys. Rev. *37*, 1631

[11] Svartholm, N., Siegbahn, K. (1946): Ark. Mat. Astron. Fys. *33* A, 21

[12] Siegbahn, K., Svartholm, N. (1946): Nature *157*, 872

[13] Siegbahn, K. (1952): in: Conf. of Swedish Nat. Comm. for Phys., Ark. Fys. 7, 86 (1954)

[14] Siegbahn, K. (1952): Physica *18*, 1043

[15] Siegbahn, K. (1953): in: Conf. of Swedish Nat. Comm. for Phys., Ark. Fys. *8*, 19 (1954)

[16] Siegbahn, K. (1955): in: Beta and gamma-ray spectroscopy, Siegbahn, K. (ed.), Amsterdam: North-Holland, p. 52

[17] Siegbahn, K., Edvarson, K. (1956): Nucl. Phys. *1*, 137

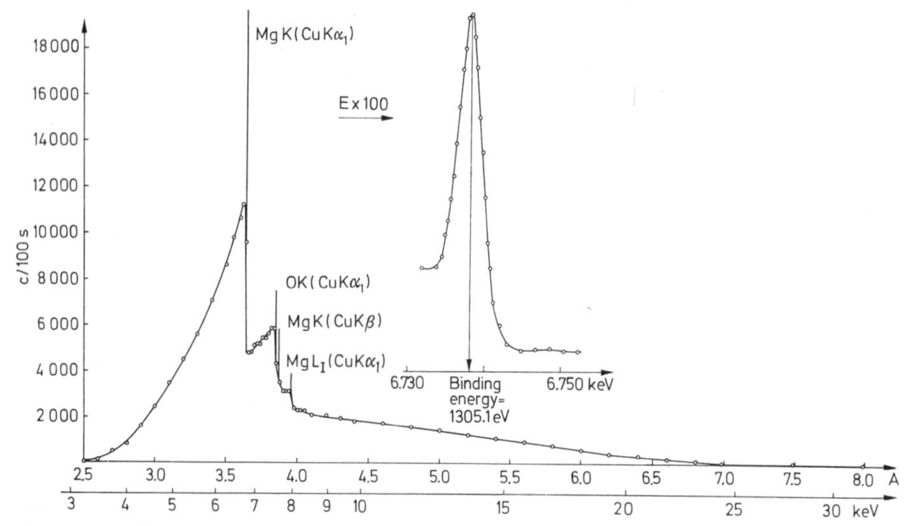

Fig. 4. An early photoelectron spectrum of MgO excited with Cu $K\alpha$ radiation, showing low-energy tails for each electron line (from the introductory chapter of ref. [29])

that under certain sample conditions essentially symmetric photoelectron lines with very small widths could actually be resolved from the high energy edges of the electron loss continuum (see Fig. 4)[18-20]. The use of soft x-ray sources, such as Al ($h\nu = 1486.6$ eV) and Mg ($h\nu = 1253.6$ eV) was advantageous from the point of view of improved line to continuous background ratios and furthermore, of smaller inherent linewidths than harder characteristic x-rays (Cu$K\alpha$, Mo$K\alpha$ etc.), with narrower photoelectron lines as a consequence.

This development initiated comprehensive studies of electron energy levels of a large number of the elements of the periodic system[18-30]. In many cases these measurements led to substantial revisions of previously accepted values of electron binding energies obtained by x-ray spectroscopy[21]. It was also observed in the course of these measurements that the core electron lines from

[18] Sokolowski, E., Nordling, C., Siegbahn, K. (1957): Ark. Fys. *12*, 301

[19] Nordling, C., Sokolowski, E., Siegbahn, K. (1957): Phys. Rev. *105*, 1676

[20] Sokolowski, E., Nordling, C., Siegbahn, K. (1958): Ark. Fys. *13*, 288

[21] Sokolowski, E. (1959): Ark. Fys. *15*, 1

[22] Nordling, C. (1959): Ark. Fys. *15*, 397

[23] Nordling, C., Hagström, S. (1960): Ark. Fys. *16*, 515

[24] Bergvall, P., Hagström, S. (1960): Ark. Fys. *17*, 61

[25] Fahlman, A., Hörnfeldt, O., Nordling, C. (1962): Ark. Fys. *23*, 75

[26] Hagström, S., Nordling, C., Siegbahn, K. (1965): in: Alpha-, beta- and gamma-ray spectroscopy, appendix 2, Siegbahn, K. (ed.), Amsterdam: North-Holland

[27] Krause, M.O. (1965): Phys. Rev. A*140*, 1845

[28] Carlson, T.A. (1967): Phys. Rev. *156*, 142

[29] Siegbahn, K., Nordling, C., Fahlman, A., Nordberg, R., Hamrin, K., Hedman, J., Johansson, G., Bergmark, T., Karlsson, S.-E., Lindgren, I., Lindberg, B. (1967): ESCA-Atomic, molecular and solid state structure studied by means of electron spectroscopy, Uppsala: Almqvist and Wiksells

[30] Fahlman, A., Hamrin, K., Hedman, J., Nordberg, R., Nordling, C., Siegbahn, K. (1966): Nature *210*, 44

an element shifted in energy upon change in chemical state of the atom[31-36]. Thus, a metal surface covered with a very thin layer of its oxide gave rise to two closely spaced photoelectron lines[31,32]. The various atomic components of a molecular species from a surface layer of molecular dimensions could be analyzed by means of its core electron spectrum and recorded small chemical shifts could be used for information on bonds and charge distribution[29,30,37-43]. Similar chemical shift effects were also found for the Auger electron lines in the spectra[29-34].

In parallel with the work on mapping core electron levels by means of x-ray excited electron spectra, valence level electron spectroscopy was developed. High resolution was achieved by means of uv-excitation, chiefly the resonance radiation of helium ($h\nu_I = 21.2$ eV and $h\nu_{II} = 40.8$ eV)[44-66]. Development work

---

[31] Siegbahn, K., Nordling, C., Sokolowski, E. (1957): Proc. Rehovoth Conf. Nucl. Str. 1957, Amsterdam: North-Holland, p. 291

[32] Sokolowski, E., Nordling, C., Siegbahn, K. (1958): Phys. Rev. *110*, 776

[33] Nordling, C., Sokolowski, E., Siegbahn, K. (1958): Ark. Fys. *13*, 483

[34] Fahlman, A., Hamrin, K., Nordberg, R., Nordling, C., Siegbahn, K. (1966): Phys. Lett. *20*, 159

[35] Nordling, C. (1959): Ark. Fys. *15*, 241

[36] Hagström, S., Nordling, C., Siegbahn, K. (1964): Phys. Lett. *9*, 235; Z. Phys. *178*, 439

[37] Siegbahn, K., Nordling, C., Johansson, G., Hedman, J., Hedén, P.-F., Hamrin, K., Gelius, U., Bergmark, T., Werme, L.-O., Manne, R., Baer, Y. (1969): ESCA applied to free molecules, Amsterdam: North-Holland

[38] Siegbahn, K., Allison, D.A., Allison, J.H. (1974): in: CRC handbook of spectroscopy, vol. 1, Robinson, J.W. (ed.), Cleveland, Ohio: CRC Press, p. 257

[39] Bakke, A.A., Chen, H.-W., Jolly, W.L., Lee, T.H. (1980): Table of core electron binding energies for gaseous compounds. J. Electron Spectrosc. Relat. Phenom. *20*, 333

[40] Gelius, U., Hedén, P.F., Hedman, J., Lindberg, B.J., Manne, R., Nordberg, R., Nordling, C., Siegbahn, K. (1970): Phys. Scr. *2*, 70

[41] Lindberg, B.J., Hamrin, K., Johansson, G., Gelius, U., Fahlman, A., Nordling, C., Siegbahn, K. (1970): Phys. Scr. *1*, 286

[42] Gray, R.C., Carver, J.C., Hercules, D.M. (1976): J. Electron Spectrosc. Relat. Phenom. *8*, 343

[43] Lindberg, B.J., Hedman, J. (1975): Chem. Scr. *7*, 155

[44] Turner, D.W., Al-Joboury, M.I. (1962): J. Chem. Phys. *37*, 3007

[45] Turner, D.W., Al-Joboury, M.I. (1963): J. Chem. Soc. 5141

[46] Al-Joboury, M.I., May, D.P., Turner, D.W. (1965): J. Chem. Soc. 616, 4434, 6350

[47] Radwan, T.N., Turner, D.W. (1966): J. Chem. Soc. 85

[48] Al-Joboury, M.I., Turner, D.W. (1967): J. Chem. Soc. 373

[49] Turner, D.W., May, D.P. (1966): J. Chem. Phys. *45*, 471

[50] Brundle, C.R., Turner, D.W. (1967): Chem. Commun. 314

[51] Turner, D.W., Baker, C., Baker, A.D., Brundle, C.R. (1970): Molecular photoelectron spectroscopy, London: Wiley-Interscience

[52] Vilesov, F.I., Kurbatov, B.L., Terenin, A.N. (1961): Sov. Phys. Dokl. *6*, 490 (1962); ibid., *6*, 883

[53] Price, W.C. (1967): Endeavour *26*, 78

[54] Lempka, H.J., Passmore, T.R., Price, W.C. (1968): Proc. R. Soc. A*304*, 53

[55] Frost, D.C., McDowell, C.A., Vroom, D.A. (1965): Phys. Rev. Lett. *15*, 612

[56] Frost, D.C., McDowell, C.A., Vroom, D.A. (1967): Proc. R. Soc. A, *296*, 516

[57] Frost, D.C., McDowell, C.A., Vroom, D.A. (1967): Chem. Phys. Lett. *1*, 93

[58] Frost, D.C., McDowell, C.A., Vroom, D.A. (1967): J. Chem. Phys. *46*, 4255

[59] Berglund, C.N., Spicer, W.E. (1964): Phys. Rev. *136*, A1030; A1044

[60] Spicer, W.E. (1966): J. Appl. Phys. *37*, 947

[61] Blodgett, A.J., Jr., Spicer, W.E. (1966): Phys. Rev. *146*, 390

(Footnotes 62–66 see p. 222)

in this field was performed by TURNER et al.[44-51], VILESOV et al.[52], PRICE et al.[53,54] and FROST et al.[55-58]. Due to the high resolution attainable in this region for free atoms and molecules, the valence electron states and the vibrational excitations accompanying the photoelectric process could be investigated[44-58]. SPICER et al. and others[59-66] studied solid state band structures. For solids, the ultra-high vacuum conditions necessary for adequate sample cleanliness first restricted the available photon energies to below about 11.8 eV, the cut-off energy of the LiF window used for the uv lamps. The differentially pumped He uv-sources were later adopted, which enabled an extension to these photon energies.

The use of x-radiation for excitation of gaseous species was found to be feasible[27-30,37,67-69]. The two-directional focusing and monochromatization by means of spherically bent quartz crystals[70] could be applied to the study of both core- and valence electron states, also for gaseous species[71]. Closer studies showed that complementary information was obtained by exciting valence electron spectra with radiation of different energies. Thus, while valence electron spectra excited by soft x-rays do not have the same resolution as uv-excited spectra, the photoelectron intensities are more readily interpreted at higher photoelectron energies[72-83]. The use of synchrotron radiation, initiated

[62] Blodgett, A.J., Jr., Spicer, W.E. (1967): Phys. Rev. *158*, 514

[63] Yu, A.Y-C., Spicer, W.E. (1968): Phys. Rev. *167*, 674

[64] Spicer, W.E. (1967): Phys. Rev. *154*, 385

[65] Krolikowski, W.F., Spicer, W.E. (1965): Bull. Am. Phys. Soc. *10*, 1186

[66] Eastman, D.E. (1969): J. Appl. Phys. *40*, 1387

[67] Nordberg, R., Hedman, J., Hedén, P.-F., Nordling, C., Siegbahn, K. (1968): Ark. Fys. *37*, 489

[68] Hamrin, K., Johansson, G., Gelius, U., Fahlman, A., Nordling, C., Siegbahn, K. (1968): Chem. Phys. Lett. *1*, 613

[69] Hedman, J., Hedén, P.-F., Nordling, C., Siegbahn, K. (1969): Phys. Lett. *29*A, 178

[70] Siegbahn, K. (1958): Ann. Acad. Regiae Sci. Ups. *2*, 102

[71] Gelius, U., Siegbahn, K. (1972): Faraday Discuss. Chem. Soc. *54*, 257

[72] Baer, Y., Hedén, P.-F., Hedman, J., Klasson, M., Nordling, C., Siegbahn, K. (1970): Phys. Scr. *1*, 55

[73] Gelius, U. (1972): in: Electron spectroscopy, Shirley, D.A. (ed.), Amsterdam: North-Holland, p. 311

[74] Shirley, D.A. (1972): in: Electron spectroscopy, Shirley, D.A. (ed.), Amsterdam: North-Holland, p. 603

[75] Gelius, U., Allan, C.J., Allison, D.A., Siegbahn, H., Siegbahn, K. (1971): Chem. Phys. Lett. *11*, 224

[76] Nefedov, V.I., Sergustin, N.P., Band, I.M., Trzhaskovskaya, M.B. (1973): J. Electron Spectrosc. Relat. Phenom. *2*, 383

[77] Berndtsson, A., Basilier, E., Gelius, U., Hedman, J., Klasson, M., Nilsson, R., Nordling, C., Svensson, S. (1975): Phys. Scr. *12*, 235

[78] Freeouf, J., Erbudak, M., Eastman, D.E. (1973): Solid State Commun. *13*, 771

[79] Sagawa, T., Kato, R., Sato, S., Watanabe, M., Ishii, T., Nagakusa, I., Kono, S., Suzuki, S. (1974): J. Electron Spectrosc. Relat. Phenom. *5*, 551

[80] McLachlan, A.D., Jenkin, J.G., Liesegang, J., Leckey, R.C.G. (1974): J. Electron Spectrosc. Relat. Phenom. *5*, 593

[81] Smith, N.V., Wertheim, G.K., Hüfner, S., Traum, M.M. (1974): Phys. Rev. B*10*, 3197

[82] McFeely, F.R., Kowalczyk, S.P., Ley, L., Cavell, R.G., Pollak, R.A., Shirley, D.A. (1974): Phys. Rev. B*9*, 5268

[83] Nilsson, R., Berndtsson, A., Mårtensson, N., Nyholm, R., Hedman, J. (1976): Phys. Stat. Sol. *75*, 197

during the early seventies has come into wide use for such photon-energy dependent studies. Due to its surface sensitivity and several other properties electron spectroscopy has during the seventies gradually developed into a powerful surface physics and chemistry method.

The theoretical foundations for interpretation of spectra from atoms, molecules and condensed matter have been refined alongside with the experimental development. Thus, the basic understanding of the various phenomena in terms of one-electron concepts evolved mainly during the sixties and early seventies. An increasing emphasis has been put on many-electron contributions. This is related to the improvements in the experimental techniques, which have enabled also the study of many-electron features in the spectra. The theories for interpretations of electron spectra have been intimately related to the development of computational schemes for many-electron wave functions. In this respect, electron spectra have also furnished ideal test cases for various theoretical methods.

The situation in electron spectroscopy at the end of the sixties and the beginning of the seventies was discussed during four main conferences: The first was assembled by the Royal Society in London 1969[84], the next was held at Uppsala 1970 (being a EUCHEM conference no conference report was issued). The third took place at Asilomar USA 1971[85], and the fourth at Brighton 1972 (a Faraday Discussion conference)[86]. These conference volumes together with the above quoted books on electron spectroscopy[29,37,51] written during the sixties summarized the state of the art in electron spectroscopy at the entry of the seventies. Since then, a large number of books dealing with electron spectroscopy with applications to different fields have been published, including conference proceedings[87-91] and ordinary textbooks[92-103]. For the present status of the field we refer to the separate sections below.

---

[84] A discussion on photoelectron spectroscopy (org. by W.C. Price and D.W. Turner) Phil. Trans. R. S. (London) (1970): *268*, 1-175

[85] Shirley, D.A. (ed.) (1972): Electron spectroscopy, Amsterdam: North-Holland

[86] General discussion on the photoelectron spectroscopy of molecules (1972): Faraday Discuss. Chem. Soc. *54*

[87] Dekeyser, W., et al. (eds.) (1973): Electron emission spectroscopy, Dordrecht: Reidel

[88] Caudano, R., Verbist, J. (eds.) (1974): Electron spectroscopy, Amsterdam: Elsevier

[89] Electron spectroscopy of solids and surfaces (1975): Faraday Discuss. Chem. Soc. *60*

[90] Hedman, J., Siegbahn, K. (eds.) (1977): Proc. Int. Symp. Electron Spectroscopy, Uppsala (May 1977). Phys. Scr. *16* (5-6)

[91] Leckey, R.C.G., Jenkin, J.G., Liesegang, J. (eds.) (1978): Proc. Australian Conf. on Electron Spectroscopy (1979): J. Electron Spectrosc. Relat. Phenom. *15*

[92] Baker, A.D., Betteridge, D. (1972): Photoelectron spectroscopy, London: Pergamon Press

[93] Eland, J.H.D. (1974): Photoelectron Spectroscopy, London: Butterworths

[94] Carlson, T.A. (1975): Photoelectron and Auger spectroscopy, New York: Plenum Press

[95] Brundle, C.R., Baker, A.D. (eds.) (1977, 1978, 1979, 1981): Electron spectroscopy, Vols. 1, 2, 3, 4, New York: Academic Press

[96] Briggs, D. (ed.) (1977): Handbook of x-ray and ultraviolet photoelectron spectroscopy, London: Heyden

[97] Rabalais, J.W. (1977): Principles of ultraviolet photoelectron spectroscopy, New York: Wiley-Interscience

[98] Ghosh, P.K. (1978): A whiff of photoelectron spectroscopy, New Delhi: Swan Printing Press

(Footnotes 99-103 see p. 224)

II. Experimental procedures and general features of photoelectron spectra

3. Photon sources

α) Introduction. The particular requirements on a photon source for photoelectron excitation depend on the nature of application. As a general rule a source of high intensity having as small a width in energy as possible is desired. Also, the source should have a high signal to background ratio and be free of neighbouring, disturbing extra radiation, such as satellites. The means of obtaining such a source largely differ, depending on the photon energy.

In order to excite photoelectrons a minimum photon energy of 5 to 10 eV is required for most known materials. In Table 1 we present photon sources,

Table 1. Photon sources for photoelectron spectroscopy[a]

Photon energy (eV)	Source	Typical intensity at sample (photons s^{-1})	Energy width (meV)	Remark (transition, mode of monochromatization)
~3 eV (several lines)	Ar^+ laser	$1 \cdot 10^{19}$	<1	
~6 eV	Ar^+ laser f-doubled	$1 \cdot 10^{15}$	<1	
$2.5 \leqq h\nu \leqq 7.7$	H_2-continuum	$<10^7$/meV[b]		Grating
$4 \leqq h\nu \leqq 14$	Hinteregger lamp	10^8/meV[b]		Grating
$4 \leqq h\nu \leqq 250$	BRV source	10^8/meV[b]		Grating
$40 \leqq h\nu \leqq 350$	Plasma source			Grating
$6.2 \leqq h\nu \leqq 8.4$	Xe-continuum	$<10^7$/meV[b]		Grating
$6.9 \leqq h\nu \leqq 9.9$	Kr-continuum	$<10^7$/meV[b]		Grating
$8 \leqq h\nu \leqq 11.8$	Ar-continuum	10^8/meV[b]		Grating
$12 \leqq h\nu \leqq 21.2$	He-continuum	10^8/meV[b]		Grating
$12.4 \leqq h\nu \leqq 16.8$	Ne-continuum	10^8/meV[b]		Grating
10.1989	HI (Lyman α)	vs	~1[c]	$2p \to 1s$ Grating
11.6237	ArI	s		$3p^5 4s \to 3p^6$ Grating
11.8282	ArI	s	~1[c]	$3p^5 4s \to 3p^6$ Grating
12.0876	HI (Lyman β)	m		$3p \to 1s$ Grating
13.3024	ArII	m		$3s\,3p^6 \to 3p^5$ Grating
13.4798	ArII	m		$3s\,3p^6 \to 3p^5$ Grating
16.6710	NeIα	s		$2p^5 3s \to 2p^6$ Grating
16.8482	NeIα	s	~1[c]	$2p^5 3s \to 2p^6$ Grating
17.1402*	ArII	w		$3p^4 4s \to 3p^5$ Grating

[99] Feuerbacher, B., Fitton, B., Willis, R.F. (eds.) (1978): Photoemission and the electronic properties of surfaces, New York: Wiley-Interscience

[100] Cardona, M., Ley, L. (eds.) (1978, 1979): Photoemission in solids I and II, Topics in Applied Physics, vols. 26 and 27, Berlin, Heidelberg, New York: Springer

[101] Roberts, M.W., McKee, C.S. (1978): Chemistry of the metal-gas interface, Oxford: Clarendon Press

[102] Ballard, R.E. (1978): Photoelectron spectroscopy and molecular orbital theory, Bristol: Adam Hilger

[103] Berkowitz, J. (1979): Photoabsorption, photoionization and photoelectron spectroscopy, New York: Academic Press

Table 1 (continued)

Photon energy (eV)	Source	Typical intensity at sample (photons s^{-1})	Energy width (meV)	Remark (transition, mode of monochromatization)
17.2661*	Ar II	w		$3p^4 4s \to 3p^5$ Grating
18.2882*	Ar II	w		$3p^4 3d \to 3p^5$ Grating
18.4542*	Ar II	m		$3p^4 4s \to 3p^5$ Grating
18.6568*	Ar II	w		$3p^4 4s \to 3p^5$ Grating
18.7326*	Ar II	m		$3p^4 3d \to 3p^5$ Grating
19.4553*	Ar III	w		$3p^3 3d \to 3p^4$ Grating
20.7439*	Ar II	w		$3p^4 4s \to 3p^5$ Grating
21.2182	He Iα	s	$\gtrsim 1^c$	$1s\,2p \to 1s^2$ Grating
21.3671*	Ar II	w		$3p^4 3d \to 3p^5$ Grating
21.6243*	Ar II	w		$3p^4 3d \to 3p^5$ Grating
23.0872	He Iβ	w		$1s\,3p \to 1s^2$ Grating
23.7423	He Iγ	vw		$1s\,4p \to 1s^2$ Grating
24.0460	He Iδ	vw		$1s5p \to 1s^2$ Grating
25.3289	Ne III	w		$2s2p^5 \to 2p^4$ Grating
26.8141	Ne II	w		$2s\,2p^6 \to 2p^5$ Grating
26.9109	Ne II	w	$\gtrsim 1^c$	$2s\,2p^6 \to 2p^5$ Grating
30.4530	Ne II	w		$2p^4 3s \to 2p^5$ Grating
30.5494	Ne II	w		$2p^4 3s \to 2p^5$ Grating
34.2070	Ne II	vw		$2p^4 4d \to 2p^5$ Grating
34.3044	Ne II	vw		$2p^4 4d \to 2p^5$ Grating
37.4838	Ne II	vw		$2p^4 3d \to 2p^5$ Grating
37.9694	Ne II	vw		$2p^4 3d \to 2p^5$ Grating
40.8138	He IIα	w	$\gtrsim 1^c$	$2p \to 1s$ Grating
48.3718	He IIβ	vw		$3p \to 1s$ Grating
51.0170	He IIγ	vw		$4p \to 1s$ Grating
52.2416	He IIδ	vw		$5p \to 1s$ Grating
132.29	Y$M\zeta$ (0.2 kW)	$3 \cdot 10^{11}$	450	$3d^9 4p^6 \to 3d^{10} 4p^5$
151.42	Zr$M\zeta$ (0.2 kW)	$4 \cdot 10^{11}$	770	$3p^5 3d^{10} \to 3p^6 3d^9$
171.38	Nb$M\zeta$ (0.2 kW)	$3 \cdot 10^{11}$	1,210	$3d^9 4p^6 \to 3d^{10} 4p^5$
192.29	Mo$M\zeta$ (0.2 kW)	$2 \cdot 10^{11}$	1,530	$3d^9 4p^6 \to 3d^{10} 4p^5$
260.1	Rh$M\zeta$ (0.2 kW)	$1 \cdot 10^{11}$	4,000	$3d^9 4p^6 \to 3d^{10} 4p^5$
452.2	Ti$L\alpha$		3,000	$2p^5 3d^2 \to 2p^6 3d$
572.9	Cr$L\alpha$		3,000	$2p^5 3d^5 \to 2p^6 3d^4$
676.4	F$K\alpha$		500	$1s2p^6 \to 1s^2 2p^5$
848.63	Ne$K\alpha$		280	$1s2p^6 \to 1s^2 2p^5$
929.70	Cu$L\alpha$		3,800	$2p^5 3d^{10} \to 2p^6 3d^9$
1,041.01	Na$K\alpha$		420	$1s2p^6 \to 1s^2 2p^5$
1,253.64	Mg$K\alpha$ (0.5 kW)	$5 \cdot 10^{12}$	680	$1s2p^6 \to 1s^2 2p^5$
1,486.65	Al$K\alpha$ (15 kW) (monochromatized rot. anode)	$3 \cdot 10^{12}$ $1 \cdot 10^{13}$ $5 \cdot 10^{13}$	165 200 400	
1,486.65	Al$K\alpha$ (1 kW) (monochromatized fixed anode; dispersion compensation)	$2 \cdot 10^{11}$	165	$1s^2 p^6 \to 1s^2 2p^5$ quartz cryst. (100); $n=1$
1,486.65	Al$K\alpha$ (1 kW) (nonmonochromatized)	$1 \cdot 10^{13}$	830	
2,984.40	Ag$L\alpha_1$ (nonmonochr.)			$2p^5 3d^{10} \to 2p^6 3d^9$ quartz cryst. (100); $n=2$

Table 1 (continued)

Photon energy (eV)	Source	Typical intensity at sample (photons s^{-1})	Energy width (meV)	Remark (transition, mode of monochromatization)
4,460.5	$ScK\beta_{1,3}$ (nonmonochr.)			$1s3p^6 \rightarrow 1s^2 3p^5$ quartz cryst. (100); $n=3$
4,511.00	$TiK\alpha_1$ (nonmonochr.)			$1s2p^6 \rightarrow 1s^2 2p^5$ quartz cryst. (100); $n=3$
5,414.90	$CrK\alpha_1$ (2 kW) (nonmon.)	$1 \cdot 10^{13}$	2,100	$1s2p^6 \rightarrow 1s^2 2p^5$
5,898.93	$MnK\alpha_1$ (nonmon.)			$1s2p^6 \rightarrow 1s^2 2p^5$ quartz cryst. (100); $n=4$
5,946.89	$CrK\beta_{1,3}$ (nonmon.)			$1s3p^6 \rightarrow 1s^2 3p^5$ quartz cryst. (100); $n=4$
8,048.04	$CuK\alpha_1$ (5 kW) (nonmon.)	$2 \cdot 10^{13}$	2,550	$1s2p^6 \rightarrow 1s^2 2p^5$

[a] From refs.[1-10]. Wavelengths from refs. [2a, 5]. Conversion factors[1]: 12,398.520 eV · Å, 12,398.266 eV · Å*.

[b] At principal maximum.

[c] Dependent on experimental conditions (see text).

* Produced by means of a duoplasmatron vuv line source.

vs = Very strong ($>10^{12}$).

s = Strong ($10^{11} < I < 10^{12}$).

m = Medium ($10^{10} < I < 10^{11}$).

w = Weak ($10^9 < I < 10^{10}$).

vw = Very weak ($10^8 < I < 10^9$).

---

[1] Cohen, E.R., Taylor, B.N. (1973): J. Phys. Chem. Ref. Data *2*, 661

[2a] Samson, J.A.R. (1967): Techniques of vacuum ultraviolet spectroscopy, New York: Wiley

[2b] Samson, J.A.R. (1969): Rev. Sci. Instr. *40*, 1174

[2c] Martinsson, M.I., Bickel, W.S. (1969): Phys. Lett. *30*A, 524

[2d] Andersson, M.T., Weinbold, F. (1974): Phys. Rev. A*9*, 118

[3] Griem, H.R. (1964): Plasma spectroscopy, New York: McGraw-Hill

[4] Moore, C.E. (1949, 1952, 1958): Atomic energy levels, National Bureau of Standards (US) Circular No. 467. (Conversion factor used: $1.239854 \cdot 10^4$ eV · Å)

[5] Bearden, J.A., Burr, A.F. (1967): Rev. Mod. Phys. *39*, 125

[6] Newburgh, R.G., Haroux, L., Hinteregger, H.E. (1962): Appl. Opt. *1*, 733

[7a] Balloffet, G., Romand, J., Vodar, B. (1961): Cptes. Rdue. Acad. Sci. *252*, 4139

[7b] McCorkle, R.A., Vollmer, H.J. (1977): Rev. Sci. Instr. *48*, 1055

[7c] Hutchinson, M.H.R. (1980): Appl. Opt. *19*, 3883

[7d] Reintjes, J. (1980): Appl. Opt. *19*, 3889

[7e] Nagel, B.J., Peckerar, M.C., Greig, J.R., Pechacek, R.E., Whitlock, R.R. (1978): SPIE Proc. *135*, 46

[8a] Gelius, U., Fellner-Feldegg, H., Wannberg, B., Nilsson, A.G., Basilier, E., Siegbahn, K. (1974): UUIP-855 (Uppsala University Institute of Physics Report)

[8b] Fellner-Feldegg, H., Gelius, U., Wannberg, B., Nilsson, A.G., Basilier, E., Siegbahn, K. (1974): J. Electron Spectrosc. Relat. Phenom. *5*, 643

[8c] Gelius, U., Basilier, E., Svensson, S., Bergmark, T., Siegbahn, K. (1974): ibid. *2*, 405

[9] Wuilleumier, F., Krause, M.O. (1974): Phys. Rev. A*10*, 242

[10] Berndtsson, A., Nyholm, R., Nilsson, R., Hedman, J., Nordling, C. (1978): J. Electron Spectrosc. Relat. Phenom. *13*, 131

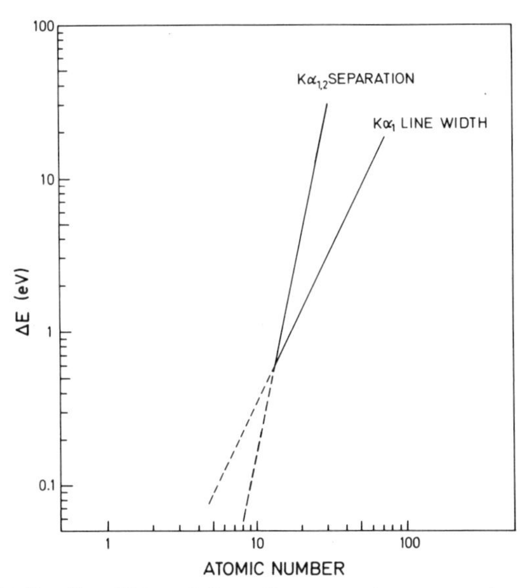

Fig. 5. $K\alpha_1$ linewidth and $K\alpha_{1,2}$ line separation vs. atomic number

which can be used for photoelectron excitation. The indicated intensities can be regarded as guidelines, since only a few of them have actually been measured more quantitatively.

β) X-ray sources. Monochromatization. At x-ray photon energies, sources that combine good intensity with small energy width are not easily accessible. The discrete line sources that would be suitable from one point of view have other drawbacks, which limit their usefulness. Considering first the attainable resolution, the inherent linewidth and the spin-orbit separation for the K-series increase with atomic number in the fashion shown in Fig. 5. If monochromatization is not performed and a total linewidth of below 1 eV is desired one is restricted in the $K\alpha$-series to elements with atomic number below $Z \sim 15$. Among these elements Mg ($h\nu = 1253.6$ eV) and Al ($h\nu = 1486.6$ eV) are those from which the radiation can be excited with the highest intensity by means of ordinary methods. For other elements either difficulties in handling, line broadenings or low yields complicate their utilization as x-ray photon sources. In a few cases the Na $K\alpha$ line has been used for special applications[11,22]. The handling of metallic sodium requires particular precautions. Use of an appropriate alloy, however, or evaporation of the metal in situ may yield a narrow line[11,22]. If the requirements on resolution are relaxed to above 1 eV it is possible to use a much larger number of sources. Since magnesium and aluminum are easy to handle and give narrow lines below 1 eV in the x-ray region, these are still the sources that are used in the overwhelming number of experimental arrangements.

[11] Siegbahn, K., Nordling, C., Fahlman, A., Nordberg, R., Hamrin, K., Hedman, J., Johansson, G., Bergmark, T., Karlsson, S.-E., Lindgren, L., Lindberg, B. (1967): ESCA – Atomic, molecular and solid state structure studied by means of electron spectroscopy, Nova Acta Regiae Soc. Sci. Ups. Ser. IV, vol. 20

The detailed design of the x-ray tube for generation of Mg$K\alpha$ and Al$K\alpha$ radiation varies according to the construction of the source compartment as a whole. The basic features are a water-cooled anode target bombarded by electrons accelerated through a high voltage stage. The high voltage may be applied keeping the anode at ground potential, in which case the electrons are generally focused onto the anode by means of an electron gun[11,13]. A number of workers and commercial instrument builders have also adopted designs where the anode is raised to high voltage[14,15]. Figure 6 shows a photoelectron spectrometer for gases and condensed matter, which incorporates an x-ray source of the former type. This source uses a fine-focus two-stage Pierce electron gun and a rotating Al anode. It is designed for operation in an x-ray monochromator at high power levels (see further below). In the electron gun the purpose of the first stage is to heat the cathode of the second stage, which is a spherically shaped tungsten plate. This tandem construction is necessary in order to keep the size and shape of the focal spot stable. The rotating Al anode target allows an increase of power density by two orders of magnitude (up to 250 kW/cm^2) with respect to a stationary anode. Figure 7 shows a photograph of the source.

An x-ray source used for photoelectron spectroscopy must be shielded from the sample compartment by a metal foil window (Al or Be are the most favourable materials). This is to prevent, firstly, the electrons from the x-ray source to enter the energy analyzer. Secondly, if gases are studied, the window prevents the sample to leak into the x-ray source.

The quality of the radiation sources can be increased if monochromatization of the characteristic emission lines is performed. This results in intensity losses. It was found, however, that the Al$K\alpha$-line may be monochromatized while still maintaining a high intensity[11-13,16-19]. This is achieved by Bragg reflection in the (100)-planes of one or several spherically bent α-quartz crystals. The peak reflectivity of Al$K\alpha$-radiation against a quartz crystal is about 45%[8]. The spherical bending of the crystals is an essential feature of the monochromatization. It results in double (two-dimensional) focusing of the x-radiation, thereby avoiding substantial losses in intensity[8,16,17]. Aside from monochromator aberrations, which can be made small, the attainable resolution is set by the x-ray absorption of the reflecting crystal due to the limited

---

[12] Finnström, B. (1982): UUIP-1069 (Uppsala University Institute of Physics Report)

[13] Siegbahn, K., Nordling, C., Johansson, G., Hedman, J., Hedén, P.F., Hamrin, K., Gelius, U., Bergmark, T., Werme, L.O., Manne, R., Baer, Y. (1969): ESCA applied to free molecules, Amsterdam: North-Holland

[14] Henke, B.L. (1961): Adv. X-Ray Anal. *4*, 244

[15] Barrie, A. (1977): in: Handbook of x-ray and ultraviolet photoelectron spectroscopy, London: Heyden, p. 79

[16a] Siegbahn, K. (1958): Ann. Acad. Reg. Sci. Ups. *2*, 102

[16b] Wassberg, G., Siegbahn, K. (1958): Ark. Fys. *14*, 1

[17a] Gelius, U., Siegbahn, K. (1972): Faraday Discuss. Chem. Soc. *54*, 257

[17b] Siegbahn, K., Hammond, D., Fellner-Feldegg, H., Barnett, E.F. (1972): Science *176*, 245

[18] Siegbahn, K. (1974): J. Electron Spectrosc. Relat. Phenom. *5*, 3

[19] Gelius, U., Siegbahn, K. (private communication)

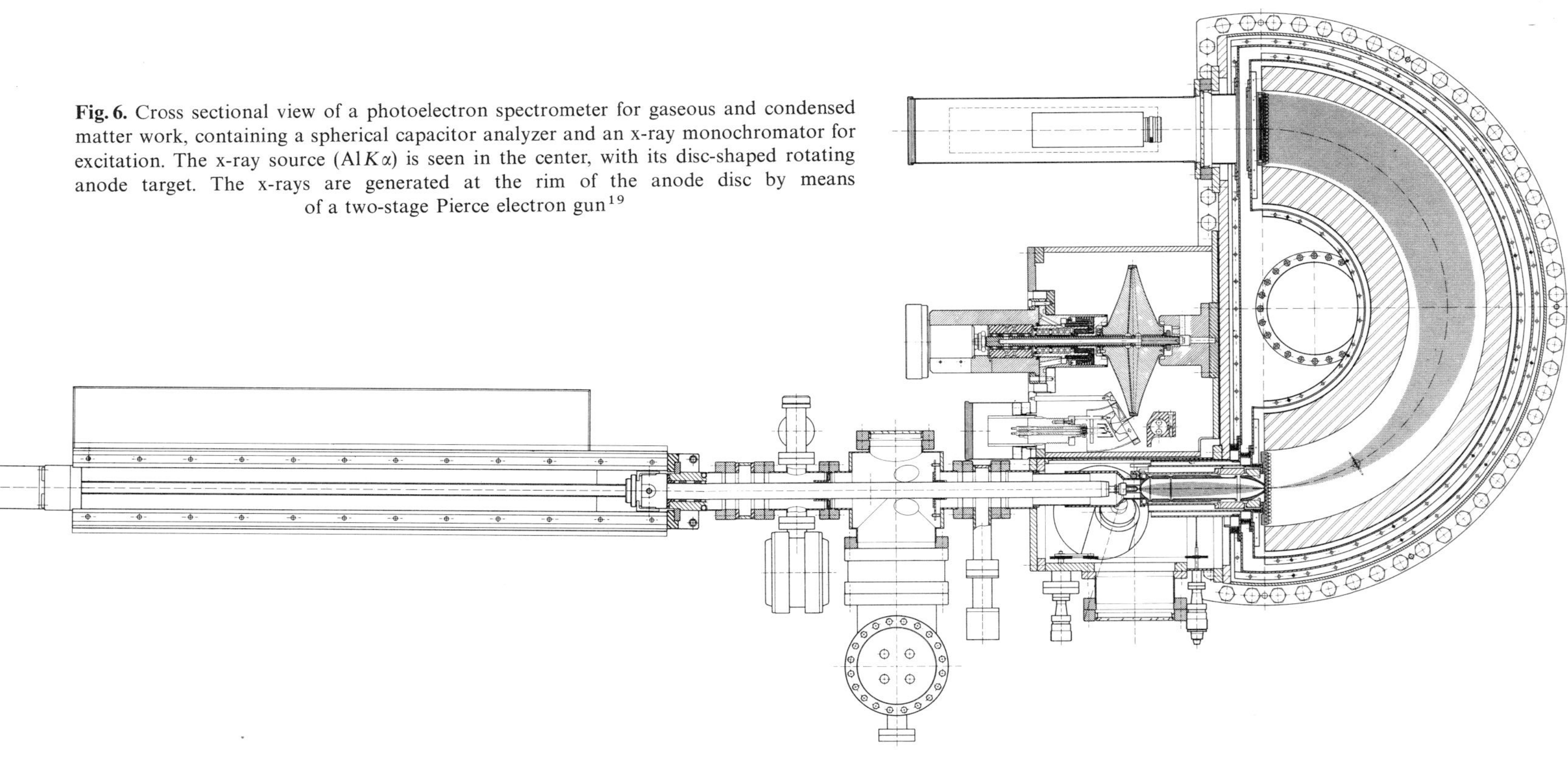

Fig. 6. Cross sectional view of a photoelectron spectrometer for gaseous and condensed matter work, containing a spherical capacitor analyzer and an x-ray monochromator for excitation. The x-ray source (Al $K\alpha$) is seen in the center, with its disc-shaped rotating anode target. The x-rays are generated at the rim of the anode disc by means of a two-stage Pierce electron gun[19]

Fig. 7. Photograph of the x-ray source in Fig. 6[19]. The crystal monochromator is mounted in the compartment to the right (cf. also Fig. 8)

number of crystal planes taking part in the diffraction. It is found to be 0.16 eV (cf. Fig. 8)[8].

Figure 9 shows three alternatives to create monochromatized Al$K\alpha$-radiation based on somewhat different principles. In the left-hand alternative, the x-rays are generated in an ordinary x-ray tube and the monochromatized Al$K\alpha$-profile is focused on the Rowland circle. The desired linewidth is chosen simply by means of a narrow slit. It results, however, in an unnecessary loss of intensity, since a large part of the generated x-rays in the desired energy range is outside the Bragg criterion due to the large electron beam focal spot. By making the size of the focal spot on the anode correspond to the desired energy range, one makes optimum use of the available x-ray intensity (right-hand scheme; the fine-focus method). This leads to a very high power density load on the anode, and to avoid overheating a rotating anode must be used.

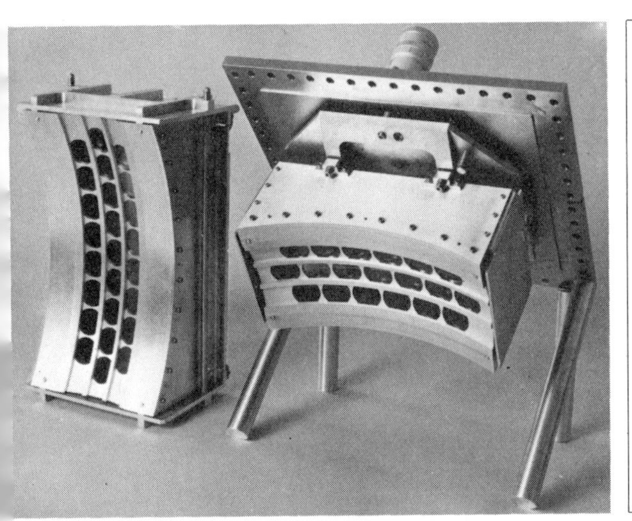

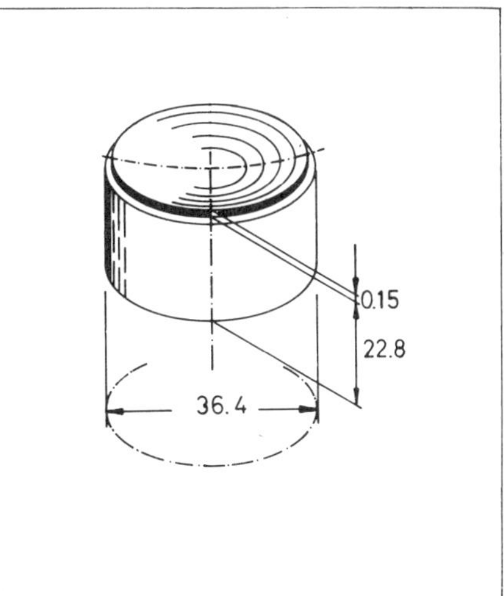

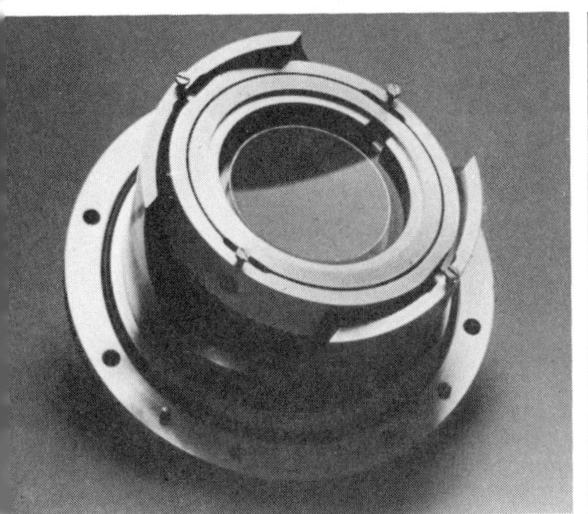

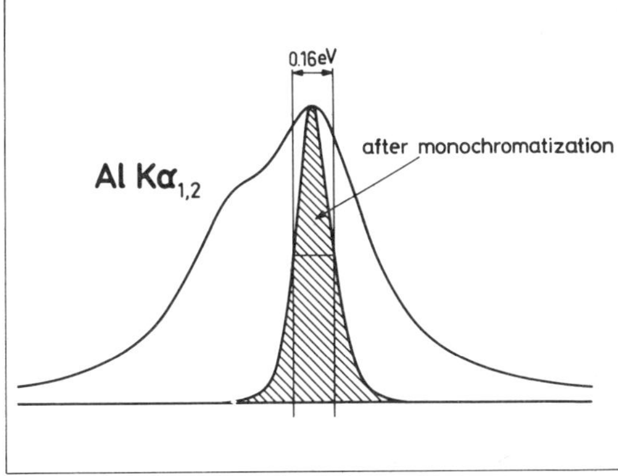

Fig. 8. The photographs show double focusing quartz single crystal x-ray monochromators. The upper left-hand monochromators contain 25 and 19 single crystals of the dimensions (in mm) shown in the upper right figure. The monochromator crystals can be adjusted to optimum positions on the Rowland circle by means of externally mounted micrometer screws. The bent single crystals used may be manufactured by pressing against a carefully ground spherical quartz backing and then welding the edges. The effect of monochromatization on the Al $K\alpha$ line profile is schematically indicated in the bottom figure[8,19], where the hatched area shows the experimentally achieved line shape

The middle alternative is the so-called dispersion compensation scheme[11,17b]. In this scheme one makes use of the full Al$K\alpha$ monochromatized profile, which is focused on and dispersed along the surface of the sample (this scheme is generally applicable only to solid or liquid samples). Thus, the photoelectrons emitted from one level in the sample atoms will have different energies depending on the position on the sample surface. The generated photoelectrons are then retarded and focused to form an electronoptical image by means of a

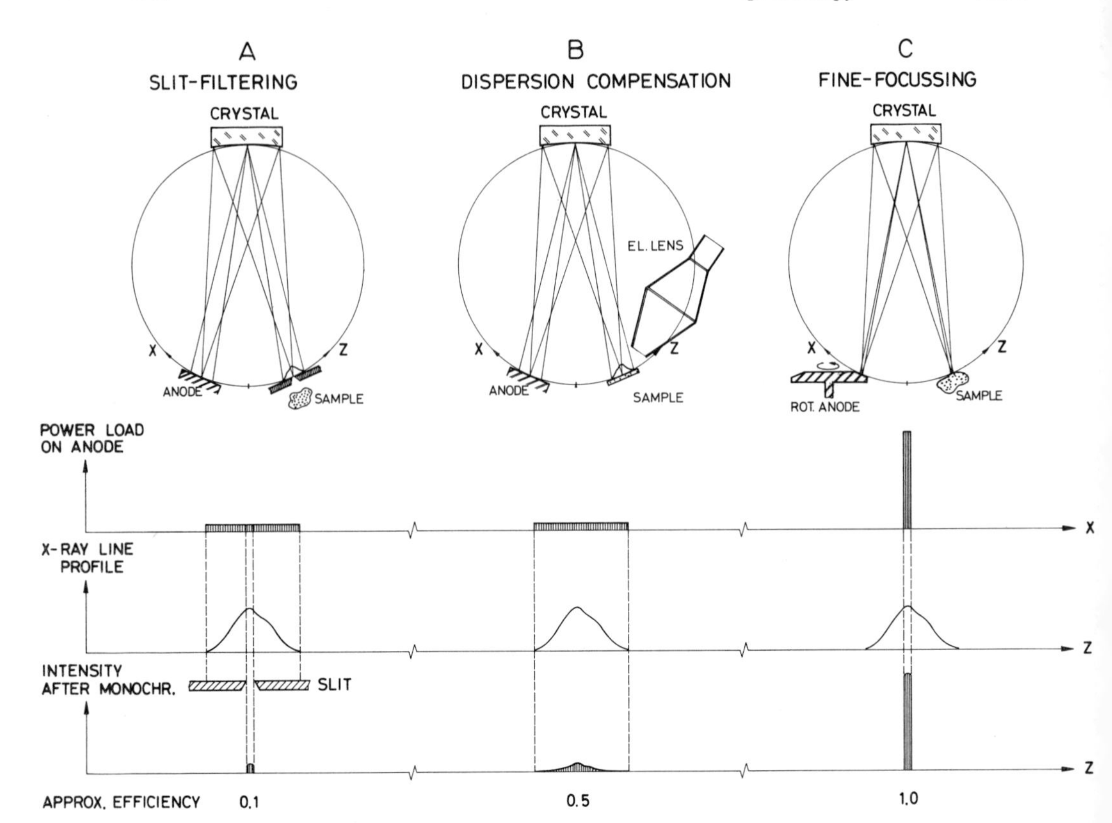

Fig. 9a–c. Three alternative schemes for monochromatization of x-rays and their relative efficiencies [8c]

lens system at the entrance of the electron analyzer. The focusing and the magnification of the lens system is adjusted so that electrons of different energies arrive at different and prescribed points at the entrance slit of the spectrometer. In this way it is possible to transfer and match the crystal dispersion to the electron spectrometer dispersion to achieve a partial cancellation over the slit width. From an intensity point of view the fine-focus method is still to be preferred (cf. Table 1 and Fig. 9). Also, there is no restriction as to the nature of the sample; it may be applied to both solids, gases and liquids.

Monochromatization of the Al$K\alpha$-radiation according to any of the schemes described above does not only have the advantage of giving a more narrow line, but removes also the bremsstrahlung continuum and accompanying satellites from the excitation source. This substantially increases the signal to background ratio in the electron spectra. Figure 10 demonstrates the effect of monochromatization on the x-ray excited core electron spectrum of air [18]. Without monochromatization (top), the three main components nitrogen, oxygen and argon are detected. The increased signal to background ratio in the monochromatized spectrum (bottom) allows also the detection of the carbon

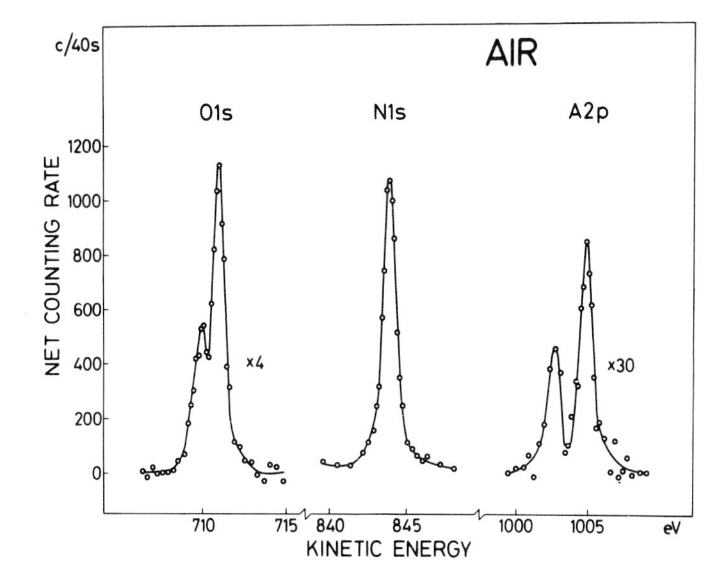

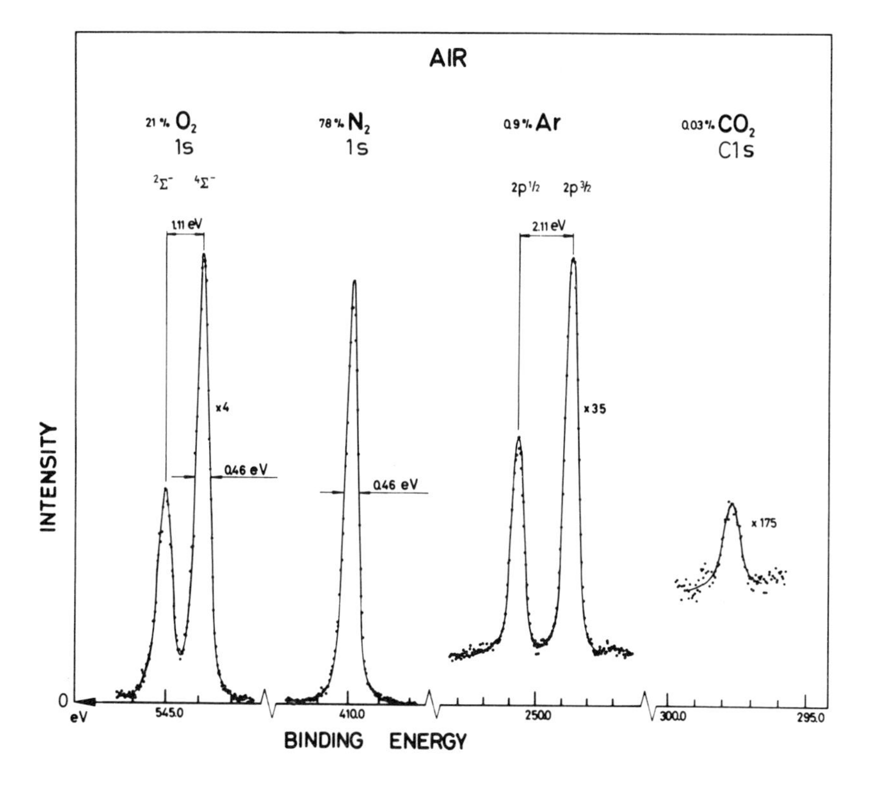

Fig. 10. Comparison between photoelectron spectra of air excited with nonmonochromatized Mg $K\alpha$ radiation (top) and monochromatized Al $K\alpha$ (bottom). The increased signal to background ratio in the latter mode allows detection of the CO_2 content in air. The O1*s* signal is split due to the nonzero total spin of molecular oxygen [18]

dioxide content (0.03 %) (even neon can be traced), in addition to a substantial improvement in resolution.

For the *L*-series, it is found that no line source is available with characteristics comparable to the sources mentioned above. *L*-sources have been used for special applications[20], but the large widths (>2 eV) in general make them less attractive for regular work.

Among the *M* emission spectra $\mathrm{Y}M\zeta(M_V \leftarrow N_{III})$, $\mathrm{Zr}M\zeta$, $\mathrm{Nb}M\zeta$ and $\mathrm{Mo}M\zeta$ have linewidths that are close to and below 1 eV (cf. Table 1). The $M\zeta$-sources are interesting from the point of view of their energy, since they bridge the energy gap in available discrete photon sources between the helium resonance radiations and the $\mathrm{Mg}K\alpha$ and $\mathrm{Al}K\alpha$ soft x-radiation. The $\mathrm{Y}M\zeta$-source has from some points of view the best properties ($h\nu = 132.3$ eV, linewidth $= 0.45$ eV) and has been used the most of the four $M\zeta$-sources[20–28].

γ) UV-sources. Filtering and polarization. A great variety of radiation sources have been designed for the specific purpose of producing uv-radiation to excite photoelectron spectra[11, 29–40]. The emission used to excite the spectra is

---

[20a] Krause, M.O. (1971): Chem. Phys. Lett. *10*, 65
[20b] Krause, M.O., Wuilleumier, F. (1971): Phys. Lett. A*35*, 341
[20c] Wuilleumier, F., Krause, M.O. (1974): Phys. Rev. A*10*, 242
[21] Banna, M.S., Shirley, D.A. (1975): Chem. Phys. Lett. *33*, 441
[22] Banna, M.S., Shirley, D.A. (1976): J. Electron Spectrosc. Relat. Phenom. *8*, 23
[23] Banna, M.S., Shirley, D.A. (1976): J. Electron Spectrosc. Relat. Phenom. *8*, 255
[24] Banna, M.S., Shirley, D.A. (1975): J. Chem. Phys. *63*, 4759
[25] Nilsson, R., Nyholm, R., Berndtsson, A., Hedman, J., Nordling, C. (1976): J. Electron Spectrosc. Relat. Phenom. *9*, 337
[26] Allison, D.A., Cavell, R.G. (1976): J. Chem. Soc. Faraday Trans. II 72, 118
[27] Cavell, R.G., Allison, D.A. (1975): Chem. Phys. Lett. *36*, 514
[28] Allison, D.A., Cavell, R.G. (1978): J. Chem. Phys. *68*, 593
[29a] Samson, J.A.R., Gardner, J.L. (1975): Can. J. Phys. *53*, 1948
[29b] Samson, J.A.R., Gardner, J.L., Haddad, G.N. (1977); J. Electron Spectrosc. Relat. Phenom. *2*, 28
[29c] Åsbrink, L., Lindholm, E., Edqvist, O. (1970): Chem. Phys. Lett. *5*, 609
[29d] Potts, A.W., Price, W.C., Streets, D.G., Williams, T.A. (1972): Discuss. Faraday Soc. *54*, 206
[30] Turner, D.W., Baker, C., Baker, A.D., Brundle, C.R. (1970): Molecular photoelectron spectroscopy, London: Wiley-Interscience
[31] Eland, J.H.D. (1974): Photoelectron spectroscopy, London: Butterworths
[32] Baker, A.D., Betteridge, D. (1972): Photoelectron spectroscopy, London: Pergamon Press
[33] Baker, A.D., Brundle, C.R. (1977): in: Electron spectroscopy, Vol. 1, Brundle, C.R., Baker, A.D. (eds.), London: Academic Press, p. 20
[34] Rabalais, J.W. (1977): Principles of ultraviolet photoelectron spectroscopy, New York: Wiley-Interscience
[35] Poole, R.T., Liesegang, J., Leckey, R.C.G., Jenkin, J.G. (1974): J. Electron Spectrosc. Relat. Phenom. *5*, 773
[36] Evans, S., Orchard, A.F. (1975): J. Electron Spectrosc. Relat. Phenom. *6*, 207
[37] Carlson, T.A. (1975): Photoelectron and Auger spectroscopy, New York: Plenum Press
[38] Kinsinger, J.A., Stebbings, W.L., Valenzi, R.R., Taylor, J.W. (1972): Anal. Chem. *44*, 773
[39] Shevchik, N.J. (1978): J. Electron Spectrosc. Relat. Phenom. *14*, 411
[40] Lee, S.-T., Rosenberg, R.A., Matthias, E., Shirley, D.A. (1977): J. Electron Spectrosc. Relat. Phenom. *10*, 203

in most cases the resonance radiation generated in a low pressure gas discharge.

The most common type of radiation source employs a dc capillary glow discharge. One design is shown in Fig. 11. The essential features are an anode and a cathode separated by an electrically isolating capillary tube which concentrates the discharge into a narrow column. The capillary tube is preferentially made of quartz or boron nitride with an inner diameter of about 1–1.5 mm. The cathode material is often chosen to reduce the amount of sputtered impurities in the discharge. Pure aluminum is a comparatively favourable material, but still better is tantalum.

The important lines for uv-excitation of valence spectra can be produced with the type of source described above. The most frequently employed line (He$I\alpha$) is obtained with high intensity using He gas of 99.995% purity. It is of importance to remove the hydrogen content from the helium gas, since the Hα transition has a much higher oscillator strength than the HeI transition. Even minor amounts of hydrogen could thus produce substantial intensities in the photoelectron spectra. For the same reason the oxygen and nitrogen content should be reduced to a minimum[38]. This can be achieved by passage of the helium gas through a liquid nitrogen cold trap filled with molecular sieve. The He$I\alpha$ line contains almost all of the intensity (98%) of the transitions to ground state He, yielding electron spectra essentially free from interferences due to excitations with other lines.

The characteristics of the radiation emitted from a uv-source, in terms of intensity and energy width, are dependent on the design of the source. Four different designs of the capillary discharge type of source were studied by Samson[2b] for the HeI resonance radiation. Figure 12 shows the line shape of the He radiation obtained at different pressures in this study.

The factors that contribute to the linewidth of the emitted radiation are:

1. The natural linewidth $\Delta E_n \sim 10^{-3}$ meV
2. The pressure (or resonance) width $\Delta E_p \sim 10^{-4}$ meV
3. Stark broadening ΔE_s (small)
4. Doppler broadening given by $\Delta E_D = 1.67 \frac{E}{c} \cdot \left(\frac{2RT}{M}\right)^{1/2}$
5. Self-reversal
6. Self-absorption.

The first two of these amount to effects substantially less than 0.1 meV and can thus not account for either the observed width or the shape of the line in Fig. 12. The natural linewidth is given by the uncertainty relation $\Delta E = \hbar/\tau$ where τ is the lifetime of the state. τ has been measured for both He$I\alpha$ and He$II\alpha$, the values being 0.57 ± 0.03 ns and 0.103 ± 0.007 ns, respectively (Martinson and Bickel, ref. [2c]). This gives $\Delta E_n(\mathrm{He}I\alpha) = 1.2 \cdot 10^{-3}$ meV and $\Delta E_n(\mathrm{He}II\alpha) = 6.4 \cdot 10^{-3}$ meV. The magnitude of the third factor is in general very small and it is thus mainly a combination of the last three contributions that defines the line profile. The photon flux produced in a volume element will have an energy width given essentially by the Doppler broadening.

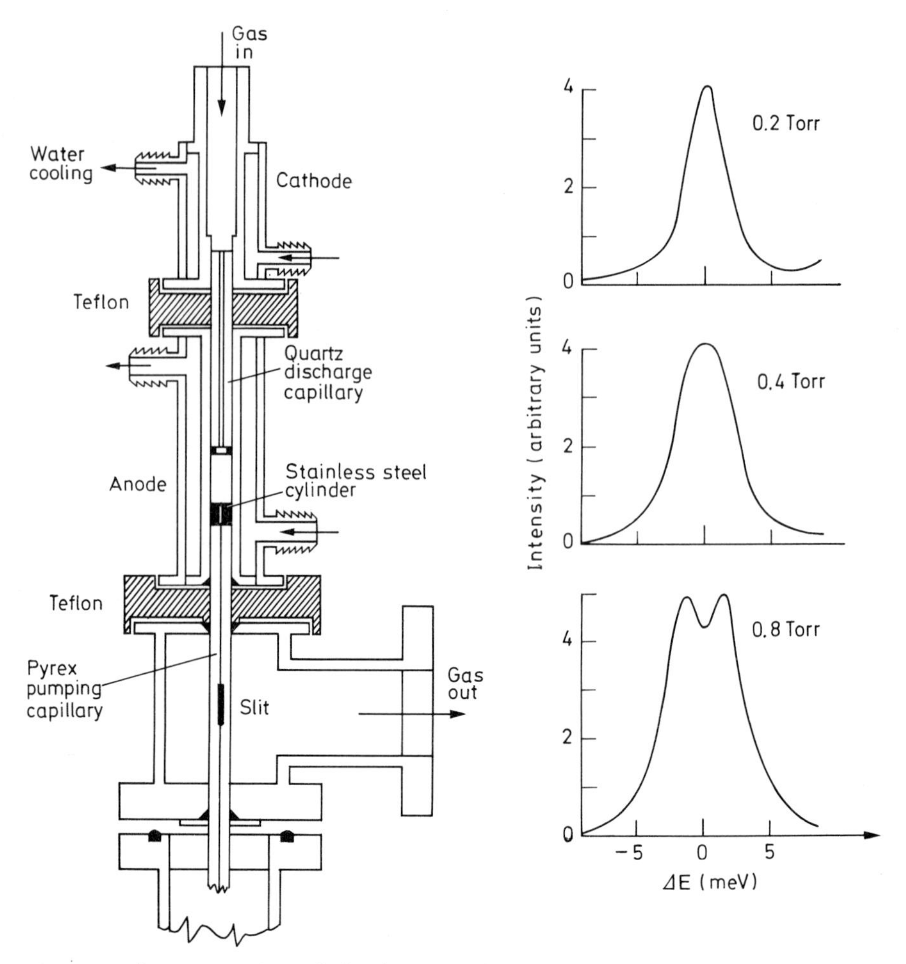

Fig. 11. uv line source. The radiation is produced in a dc glow discharge maintained in the quartz capillary. The design includes a long collimating capillary to comply with demands of uhv conditions [35]

Fig. 12. Typical profiles of the 21.22 eV HeI line obtained with a dc glow discharge at different pressures [2b]. For pressures larger than 0.6 Torr, self-absorption and self-reversal are seen to have a very marked influence on the line shape. The absolute peak height increases slightly with pressure

When this flux passes through the capillary the peak energy is more absorbed than the lower and higher energy wings of the line. The line profile will therefore be flattened out at the central energy. As a consequence the half-width of the line becomes larger ($1\,\text{meV} < \Delta E < 5\,\text{meV}$). This line broadening effect increases with the length of the capillary tube and the pressure. There are thus conflicting demands between the intensity (increasing with pressure) and width of the line. It was found that a suitable length of the capillary was ≈ 3 cm[2b]. Subsequent designs of uv-sources show, however, preference for somewhat longer capillaries, ≈ 5 cm. For high resolution studies, the pressure in the lamp should be as low as possible, since the pressure broadening is directly reproduced on the widths of the electron lines. Lamp pressures lower than 0.5 Torr are desirable (see Fig. 12). The lowering of the pressure must be made with due regard to the difficulty of maintaining the discharge and the demands of a useful intensity.

The capillary discharge has been the standard source for producing radiation to excite outer valence electron spectra (He*I* at 21.22 eV and lower energies from other gases). In order to study the inner valence electrons one must employ radiation with higher energy. The conventional way to extend the accessible region is to use He*II* radiation at 40.8 eV. The usefulness of this radiation for electron spectroscopic studies in ordinary windowless arrangements was first demonstrated by PRICE[41]. Without a filter depressing the high background due to excitation with He*I* radiation an energy interval up to about 30 eV can usually be examined. Studies with He*II* radiation have been performed with the capillary sources using either dc or microwave discharges. The intensity of the He*II* radiation increases when the gas pressure is decreased and the discharge current simultaneously increased. A sufficient amount of He*II* radiation is usually produced at currents between 75 and 100 mA, the normal current for He*I* radiation being of the order of 25 mA. Other sources favourable for production of He*II* radiation are the hollow cathode discharge source[42,43] and the charged particle oscillator[44-47].

Windowed radiation sources have been employed to some extent to reduce unwanted energies in the radiation, arising either due to competing transitions in the discharge gas or from impurity gases. Thin films of aluminum[38], and organic material[48,49] have proved to be useful window materials for electron spectroscopy. Also a gaseous filter has been used[50]. The main drawback with the solid filters is that their handling requires much time, both due to limitations in their lifetime and time-consuming procedures in application to the collimating capillary.

An alternative approach to eliminate unwanted energies in the photon beam is to put a grating monochromator in the beam between the uv-source

---

[41] Price, W.C. (1968): in: Molecular spectroscopy, Hepple, P.W. (ed.), London: Institute of Petroleum, Vol IV, p. 221

[42] Dehmer, J.L., Berkowitz, J. (1974): Phys. Rev. A*5*, 227

[43] Lempka, H.J.: Helectros Developments, Beaconsfield, England

[44] Burger, F., Maier, J.P. (1974, 1979): J. Electron Spectrosc. Relat. Phenom. *5*, 783

[45] Burger, F., Maier, J.P. (1979): J. Electron Spectrosc. Relat. Phenom. *16*, 471

[46] Coatsworth, L.L., Bancroft, G.M., Creber, D.K., Lazier, R.J.D., Jacobs, P.W.M. (1978): J. Electron Spectrosc. Relat. Phenom. *13*, 395

[47] Lancaster, G.M., Taylor, J.A., Ignatiev, A., Rabalais, J.W. (1978): J. Electron Spectrosc. Relat. Phenom. *14*, 143

[48] Potts, A.W., Williams, T.A., Price, W.C. (1972): Discuss. Faraday Soc. *54*, 104

[49] Norton, P.R., Tapping, R.L., Goodale, J.W. (1976): Rev. Sci. Instr. *47*, 777

[50] Ridyard, J.N.A. (1979): Molecular spectroscopy, London: Institute of Petroleum, p. 100

and the ionization region[39,51-55]. This implies that for example Ne*I* and Ar*I* radiation can be used much more conveniently by eliminating the overlapping contributions in spectra arising from excitation with the different spin-orbit split components of the radiation. A combination of a radiation source and a monochromator has been used for studies of intensities of valence electron transitions in order to determine differential cross sections and branching ratios as a function of energy[56-58]. By this means one can obtain a practically continuous covering of the energy interval from ~10 eV up to above 50 eV using appropriate rare gas emission spectra.

Photoelectron spectra have mostly been excited by the unpolarized radiation generated by ordinary gas discharge uv-sources. Studies with polarized radiation from such sources have been enabled in some cases by passing the photon beam through a vuv-polarizer prior to interaction with the target[59-66]. These investigations include both solids[59,65] and gases. The

[51] Samson, J.A.R., Gardner, J.L. (1977): J. Chem. Phys. *67*, 755

[52] Shevchik, N.J. (1976): Rev. Sci. Instr. *47*, 1028

[53] Flodström, S.A., Bachrach, R.Z. (1976): ibid. *47*, 1464

[54a] Siegbahn, K. (1977): in: Molecular spectroscopy, West, A.R. (ed.), London: Institute of Petroleum, p. 227

[54b] UUIP-969 (Uppsala University Institute of Physics Report) (Nov. 1977) (prep. by Siegbahn, K., Gelius, U.)

[55a] Aeppli, G., Conelon, J.J., Eastman, D.E., Johnson, R.W., Pollak, R.A., Stolz, H.J. (1978): J. Electron Spectrosc. Relat. Phenom. *14*, 121

[55b] VG. Scientific, ADES 400 (instrument brochure)

[56] Samson, J.A.R., Gardner, J.L. (1976, 1977): J. Electron Spectrosc. Relat. Phenom. *8*, 35; *12*, 119

[57] Samson, J.A.R., Gardner, J.L., Haddad, G.N. (1977): J. Electron Spectrosc. Relat. Phenom. *12*, 281

[58] Samson, J.A.R., Haddad, G.N., Gardner, J.L. (1977): J. Phys. B*10*, 1749

[59] Becker, H., Dietz, E., Gerhardt, U., Angermüller, H. (1975): Phys. Rev. B*12*, 2084

[60] Karlsson, L., Mattson, Jadrny, R., Siegbahn, K., Thimm, K. (1976): Phys. Lett. *58*A, 381

[61] Hancock, W.H., Samson, J.A.R. (1976): J. Electron Spectrosc. Relat. Phenom. *9*, 211

[62] Siegbahn, K. (1977): in: Proceedings of the Sixth Conference on Molecular Spectroscopy, Durham, March 1976; Molecular spectroscopy, West, A.R. (ed.), London: Heyden, p. 280

[63] Samson, J.A.R., Hancock, W.H. (1977): Phys. Lett. *61*A, 380

[64] Mattsson, L., Karlsson, L., Jadrny, R., Siegbahn, K. (1977): Phys. Scr. *16*, 221

[65] Hansson, G.V., Flodström, S.A. (1978): Phys. Rev. B*17*, 473

[66] Carlson, T.A., Dress, W.B., Ward, F.H., Agron, P.A., Nyberg, G.L. (1978): Rev. Sci. Instr. *49*, 736

[67] Mattsson, L. (1978): UUIP-983 (University of Uppsala Institute of Physics Report)

[68] Cole, T.T., Oppenheimer, F. (1962): Appl. Opt. *1*, 709

[69] Canfield, L.R., Hass, G., Hunter, W.R. (1964): J. Phys. *25*, 124

[70] Hagemann, H.J., Gudat, W., Kunz, C. (1975): Report DESY SR-74/7 (DESY, Hamburg, May 1975)

[71] Leimkühler, K. (1974): Diplomarbeit, PIB 1-235 (Bonn Univ., Feb. 1974)

[72] Schlüter, M. (1972): Z. Phys. *250*, 87

[73] Beaglehole, D. (1965): Proc. Phys. Soc. *85*, 1007

[74] Hass, G., Ramsey, J.B., Hunter, W.R. (1969): Appl. Opt. *8*, 2255

[75] Jacobus, G.F., Madden, R.P., Canfield, L.R. (1963): J. Opt. Soc. Am. *53*, 1084

[76] Rehn, V., Stanford, J.L., Jones, V.O., Choyke, W.J. (1976): Presented at Rome Semiconductor Conference, Aug. 1976

[77] Mattson, L. (1978): UUIP-984 (Uppsala University Institute of Physics Report)

[78] Mattson, L., Wannberg, B., Veenhuizen, H., Reineck, I., Siegbahn, K. (1979): UUIP-1009 (Uppsala University Institute of Physics Report)

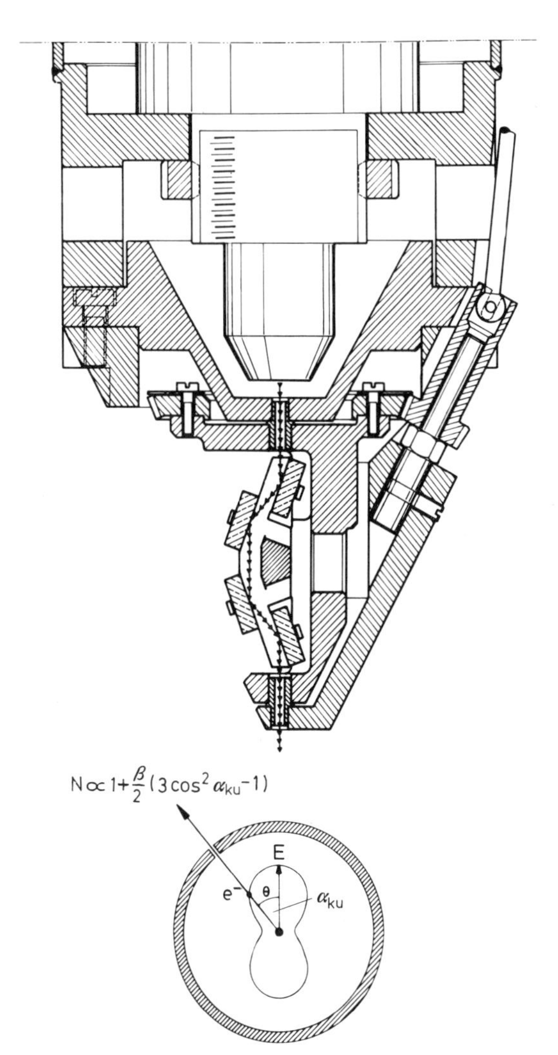

Fig. 13. Four-fold reflection polarizer mounted together with a uv lamp. The polarizer incorporates four polished gold-plated glass mirrors where the radiation is reflected with an angle of incidence of 70°. The path along which the radiation is transmitted through the polarizer is indicated by arrows. The polarizer can be set to any value of α_{ku}, the angle between electric field vector $\boldsymbol{E}$ and outgoing electron. The lower part of the figure indicates a typical angular distribution of the ejected electrons [77]

polarization is achieved at the expense of intensity and it is important to minimize this intensity loss, particularly for gaseous studies. Some polarizers have been introduced which are suitable in such cases. They utilize successive reflections by mirrors to produce a high degree of polarization[68–76] and are rotatable around the direction of the radiation (refs. [67, 77–80]). Figure 13

[67–78] See p. 238

[79] Hoof, H.A. van (1980): Appl. Opt. *19*, 189

[80] Samson, J.A.R. (1978): Nucl. Instr. Methods *152*, 225

shows such a polarizer mounted directly on the output end of a uv gas discharge source. It incorporates four gold-plated mirrors. The reflection of the radiation by these provides an almost completely polarized photon beam. The polarizer is small and forms together with the radiation source a separate unit which can be adapted to any system.

δ) Synchrotron radiation. The term synchrotron radiation refers normally to the radiation emitted by electrons with energies $\gtrsim 100$ MeV which are being accelerated in a circular orbit. The increased availability of synchrotrons and storage rings has led to considerable development during recent years to make use of this radiation in excitation of electron spectra. The intensity of the radiation is strongly related to the energy of the accelerated electrons. The total energy radiated by an electron per revolution in a circular orbit is given by [81-84]:

$$\Delta E(R, E) = \frac{4\pi}{3} \cdot \left(\frac{e^2}{R}\right) \left(\frac{E}{m_0 c^2}\right)^4, \tag{3.1}$$

where R is the radius of the orbit and E the electron energy. The wavelength distribution of the radiation is continuous and it is emitted as a well collimated beam tangentially to the orbit[85]. The radiation emitted in the plane of the orbit is completely linearly polarized and out of plane eliptically polarized. The spectral distribution curves are characterized by a critical energy E_c given by [81-84]:

$$E_c = h\nu_c = \frac{3hc}{2R} \left(\frac{E}{m_0 c^2}\right)^3. \tag{3.2}$$

Half of the total power is radiated above the critical energy and half below. The intensity (in terms of number of photons) decreases rapidly for $h\nu \geqq h\nu_c$ whereas for $h\nu \ll h\nu_c$ the intensity is almost independent of the electron energy E. Figure 14 shows a universal curve for the emitted intensity versus wavelength which is applicable to any orbit radius[83].

Electrons accelerated in synchrotrons and storage rings are grouped in bunches whose length is given by the rf field providing the energy to the system. The electron orbital period must thus be an integral multiple of the rf period. This integer is called the harmonic number of the ring. Typically, the harmonic number is of the order of a few hundred, so that a large number of discrete electron bunches can be used. For example, the storage ring SPEAR operates at an rf frequency of 358 MHz, the 280th harmonic of the 1.28 MHz orbital frequency. The pulse length of the radiation is then 0.2–0.4 ns and up to 280 bunches can be filled[84]. This provides a time structure which is useful e.g. for time-of-flight electron spectroscopy, when only one bunch is used (0.3 ns pulses repeating at 1.28 MHz)[86].

---

[81] Schwinger, J. (1948, 1949): Phys. Rev. *70*, 798; *75*, 1912

[82] Kunz, C. (ed.) (1979): Synchrotron radiation, techniques and applications, Topics in Current Physics, volume 10, Berlin, Heidelberg, New York: Springer

[83] European Synchrotron Radiation Facility, Suppl. 1–3, European Science Foundation, Strasbourg, France (May 1979)

[84] Winick, H., Doniach, S. (eds.) (1980): Synchrotron radiation research, New York: Plenum Press

[85] Jackson, J.D. (1962): Classical electrodynamics, New York: Wiley

[86] White, M.G., Rosenberg, R.A., Gabor, G., Poliakoff, E.D., Thornton, G., Southworth, S.H., Shirley, D.A. (1979): Rev. Sci. Instr. *50*, 1268

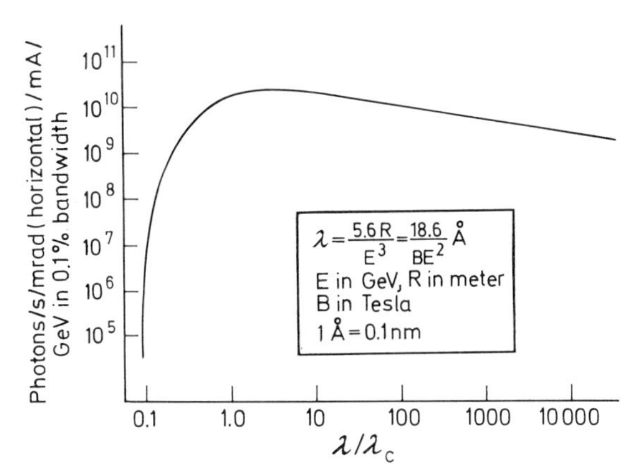

Fig. 14. Universal curve for the synchrotron radiation photon flux[83]

Synchrotron radiation is produced in storage rings and synchrotrons. These have somewhat different characteristics with respect to the production of the radiation. In a storage ring currents of the order of 0.5 A can be stored under stable conditions for several hours, the decay time depending on the average pressure in the ring. The radiation intensity and spectral distribution is thus stable over long time periods. Synchrotrons, on the other hand, are designed to accelerate rapidly (~8–10 ms) groups of electrons to a peak energy. For high-energy physics experiments, these electrons are then made to strike a target. Since the electron energy varies, the spectral distribution will not be constant in time. Moreover, the electron beam current and dimensions may vary from one acceleration cycle to the next. This makes the storage ring much more ideal as a source for SR than synchrotrons.

In order to extract the SR, beam lines must be introduced in tangential positions along the ring. When used as a dedicated SR source the number of beam lines can be made high to accommodate a large variety of different experiments. For use in photoelectron spectroscopy the synchrotron radiation must be monochromatized. From the visible up to about 40 eV photon energy, grating monochromators are used at near normal incidence. Further up in energy grazing incidence grating monochromators (from 20 eV up to 1,000 eV) may be used and above 0.5 keV crystal monochromators can be employed. There is a large variety of arrangements in these different energy regions, leading to different conditions as to resolution and output. For detailed reviews on monochromator design and synchrotron radiation techniques cf. refs. [82, 84, 87–109] and reports on the ESRF project (European Synchrotron Radiation Facility), ref. [83].

[87] Synchrotron radiation instrumentation and new developments, Nuclear Instr. Methods *152* (Conf. Vol., Wuilleumier, F., Farge, Y. (eds.)) 1978

[88a] Synchrotron radiation instrumentation, Nucl. Instr. Methods *172* (Conf. Vol., Ederer, D.L., West, J.B. (eds.)) 1980

[88b] Synchrotron radiation facilities, ibid. *177* (Conf. Vol., Howells, M.R. (ed.)) 1980

[89] Shirley, D.A., Stöhr, J., Wehner, P.S., Williams, R.S., Apai, G. (1977): Phys. Scr. *16*, 398

(Footnotes 90–109 see p. 243)

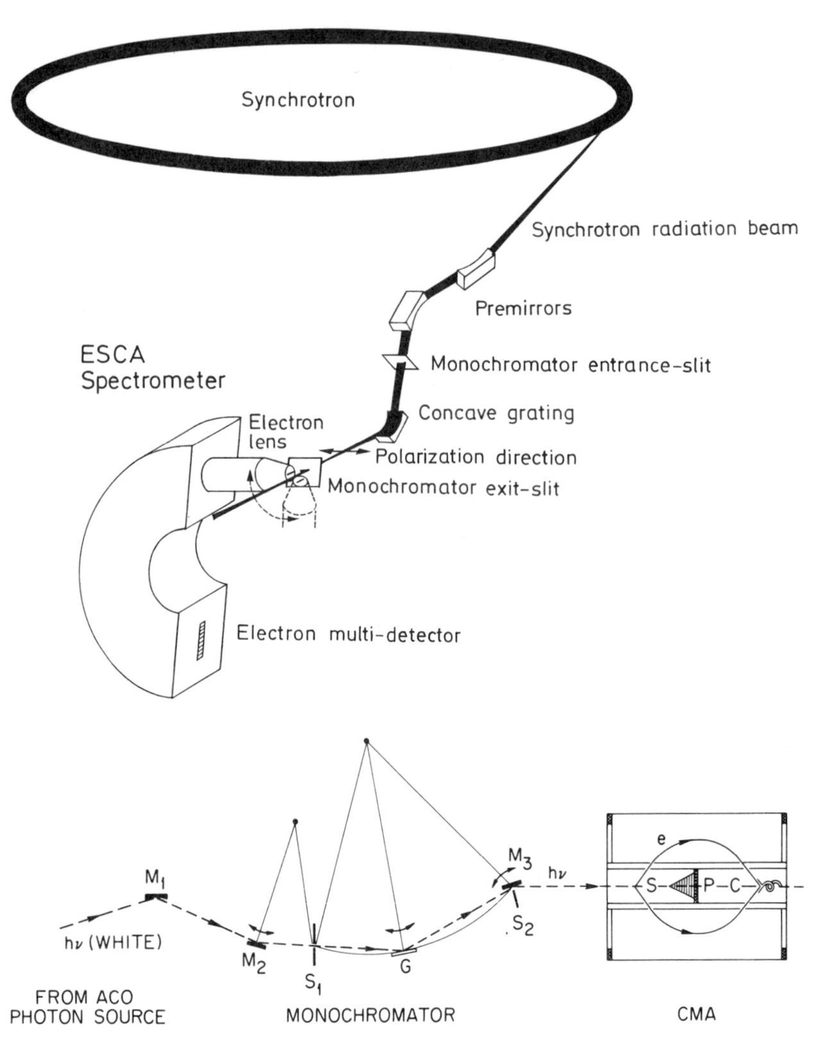

Fig. 15. Scheme (top) for use of synchrotron radiation employing a spherical capacitor analyzer. To make use of the plane polarized radiation for angular studies the analyzer should be rotatable around the photon beam direction[18]. Scheme (bottom) for use of synchrotron radiation in photoelectron spectroscopy at the storage ring ACO[109]. M_1 cylindrical mirror, M_2 spherical mirror, S_1 and S_2 entrance and exit slits, G grating, S electron source, photon (P) and electron (C) detectors

Arrangements attached to synchrotron radiation beam lines for electron spectroscopy are shown schematically in Fig. 15[18, 109]. Clearly, the polarization direction of the radiation is fixed in space, so that in order to make full use of this property the electron analyzer must be rotatable, or the angle between the polarization vector and the photoelectron emission direction be variable inside the electron analyzer.

4. Electron energy analyzers and detectors

α) Introduction. The electron energy analyzers that have been used and are being used in electron spectroscopy represent a variety of constructions, based on different principles[1-5]. The basic requirement put on an analyzer is that it should give a high energy resolution while at the same time transmitting a large electron intensity. Also, the background level in the instrument should be low. The number of electrons in a spectrum which are detected during a given time is a product of several factors:

Footnotes to p. 241

[90] Gudat, W., Kunz, C., in Ref. 82, p. 55

[91a] Haensel, R. (1965): DESY-Report; Haensel, R., Kunz, C. (1967), Z. Angew. Phys. *23*, 276
Sagawa, T., et al. (1974): J. Electron Spectrosc. Relat. Phenom. *5*, 551

[91b] Codling, K., Mitchell, P. (1970): J. Phys. E*3*, 685

[91c] Jaeglé, P., Dhez, P., Wuilleumier, F. (1973): Proceedings of the International Symposium for Synchrotron Radiation Users, Marr, G.V., Munro, I.H. (eds.), Daresbury DNPL/R26, p. 101–105

[91d] Pruett, C.H., Lien, N.C., Steben, J.D. (1971): III. International Conference on Vacuum Ultraviolet Radiation Physics, Tokyo, Conference Digest 31aA2-5

[91e] Thimm, K. (1974): J. Electron Spectrosc. Relat. Phenom. *5*, 755

[91f] Brown, F.C., Bachrach, R.Z., Hagström, S.B.M., Lien, N., Pruett, C.H. (1974): IV. International Conference on Vacuum-Ultraviolet Radiation Physics, Hamburg

[92] Howells, M.R. in Ref. [88a], p. 123

[93] Petersen, H., Baumgärtel, H. (1980): Nucl. Instr. Methods *172*, 191; BESSY Tech. Bericht TB 12/1979

[94] Petersen, H. (1980): BESSY Tech. Bericht 29

[95] Dietrich, H., Kunz, C. (1972): Rev. Sci. Instr. *43*, 434

[96] Eberhardt, W., Kalkoffen, G., Kunz, C. (1978): Nucl. Instr. Methods *152*, 81

[97] Miyake, K.P., Kato, R., Yamashita, H. (1969): Sci. Light, Tokyo *18*, 39

[98] West, J.B., Codling, K., Marr, G.V., (1974): J. Phys. E*7*, 137

[99] Howells, M.R., Norman, D., West, J.B. (1978): J. Phys. E*11*, 199

[100] e.g. Howells, M.R. (1979): Proc. USA-Japan Sem. on Synchrotron Radiation Facilities, Honolulu, Hawaii

[101] Codling, K., Mitchell, P. (1970): J. Phys. E*3*, 685

[102] Jaeglé, P., Dhez, P., Wuilleumier, F. (1974): in: VUV Radiation Physics, Koch, E.E., Haensel, R., Kunz, C. (eds.), p. 788, Braunschweig: Pergamon-Vieweg

[103] Brown, F.C., Bachrach, R.Z., Lien, N. (1978): Nucl. Instr. Methods *152*, 73

[104] Deslattes, R.D. (1980): Nucl. Instr. Methods *172*, 201

[105] Tonner, B.P. (1980): Nucl. Instr. Methods *172*, 133

[106] Pianetta, P., Lindau, J. (1977): J. Electron Spectros. Relat. Phenom. *11*, 13

[107] Bonse, U., Materlik, G., Schröder, W. (1976): J. Appl. Crystallogr. *9*, 223

[108] Jaeglé, P., Dhez, P., Wuilleumier, F. (1977): Rev. Sci. Instr. *48*, 978

[109] Dhez, P., Jaeglé, P., Wuilleumier, F.J., Källne, E., Schmidt, V., Berland, M., Carillon, A. (1978): Nucl. Instr. Methods *152*, 85

Footnotes to Section 4

[1] Siegbahn, K. (1965): in: Alpha-, beta, and gamma-ray spectroscopy, Siegbahn, K. (ed.), Amsterdam: North Holland, vol. 1, p. 79

[2] Steckelmacher, W. (1973): J. Phys. E *6*, 1061

[3a] Wannberg, B., Gelius, U., Siegbahn, K. (1974): J. Phys. E*7*, 149

[3b] Wannberg, B., Nucl. Instr. Methods (to be published)

[4] Eland, J.H.D. (1974): Photoelectron spectroscopy, London: Butterworths

[5a] Roy, R., Carrette, J.D. (1977): Electron spectroscopy for surface analysis, Topics in Current Physics, vol. 4, Ibach, H. (ed.), Berlin, Heidelberg, New York: Springer

[5b] Afanas'ev, V.P., Yavor, S.Ya. (1975): Sov. Phys.-Tech. Phys. *20*, 715

[5c] Matthews, D.L. (1980): in: Methods in Experimental Physics, Marton, L., Marton, C. (eds.), London: Academic Press, vol. 17, p. 439

1. The intensity of the exciting radiation at the sample.
2. The useful area of the spectrometer entrance aperture (S).
3. The useful solid angle of the electron analyzer (Ω).
4. The transmission of possibly existing grids.
5. The number of detectors which can be used simultaneously.

The product of factors 3 and 4 is usually termed the transmission (T). The luminosity of the analyzer is given by $L = T \cdot S$. The requirements on resolution and luminosity are inversely related to one another and each design of an analyzer represents a difficult optimization problem[1–11]. The resolution in the photoelectron spectra is also dependent on other factors, such as the width of the photon source. The resolution of the analyzer should evidently be matched to these other contributions. Thus, when the analyzer does not limit the final resolution in the spectrum, improved electron-optical properties could be used to increase the luminosity.

The different types of analyzers being used can be divided into two categories:

1. Dispersive analyzers, which form an electron-optical image of the photoelectron source, the spatial position of which is dependent on the electron energy.

2. Non-dispersive analyzers, where the electron energies are analyzed in terms of retarding potential grids, time of flight or other means.

In some of the dispersive instruments, several detectors can simultanously be used to collect the data, which greatly increases the speed of information. In the non-dispersive case, only one energy can be registered at one time. This is largely compensated for, however, by the high transmission that can usually be achieved in the latter constructions.

β) Dispersive analyzers. The dispersive electrostatic spectrometers of prism type (the spherical and cylindrical capacitors) have fields, which are special cases of the toroidal field. This general field is characterized by potential surfaces, curved in two orthogonal directions by different radii of curvature (r_0, R_0). It can be shown, that focusing of the electrons moving along these potential surfaces occurs after $\phi_r = \pi/\sqrt{2 - r_0/R_0}$ in the radial direction and after $\phi_z = \pi/\sqrt{r_0/R_0}$ in the direction orthogonal to the radius[12,13]. The radial dispersion at the focus is given by:

$$E\frac{\partial r}{\partial E} = \frac{2r_0}{2 - r_0/R_0}.$$

---

[6] Hafner, H., Arol Simpson, J., Kuyatt, C.E. (1968): Rev. Sci. Instr. *39*, 33
[7] Aksela, S., Karras, M., Pessa, M., Suoninen, E. (1970): Rev. Sci. Instr. *41*, 351
[8] Heddle, D.W.O. (1971): J. Phys. E *4*, 589
[9] Staib, P. (1972): J. Phys. E *5*, 484
[10] Wang, K.L. (1972): J. Phys. E *5*, 1193
[11] Sar-El, H.Z. (1970): Rev. Sci. Instr. *41*, 561
[12] Ewald, H., Liebl, H. (1957): Z. Naturforsch. *12a*, 28
[13] Wollnik, H. (1967): in: Focusing of charged particles, Septier, A. (ed.), London: Academic Press

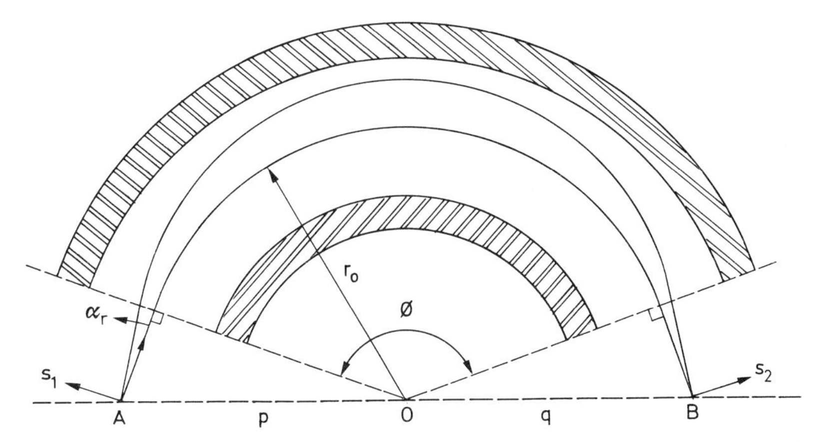

Fig. 16. Geometrical conditions for the spherical sector analyzer. Double focusing occurs with the source and image in the equatorial plane AOB (Barber's rule)

The spherical capacitor, first proposed by PURCELL[14], is defined by $r_0 = R_0$. In this case, space focusing occurs after the angle $\phi = \pi$. The energy of the focused electrons is given by (relativistically)

$$E = k \cdot V + \sqrt{(kV)^2 + E_0^2} - E_0 \tag{4.1}$$

where V is the potential applied between the spherical electrodes and E_0 the rest energy of the electron. k is given by $e \cdot R_1 R_2/(R_2^2 - R_1^2)$, where R_2 and R_1 are the radii of the spherical electrodes. For practical reasons, the actual angle of the field may have to be made less than 180 degrees, since the electron source and the detector have to be placed outside the analyzer field. It turns out that the positions of the source (A), image (B) and spectrometer center (O) even with this reduced field sector lie on a straight line, independent of the distance from the source to the entrance of the spectrometer field (Barber's rule) (see Fig. 16).

WANNBERG et al.[3] have analyzed the resolution of a spherical sector electrostatic spectrometer using:

$$\frac{\Delta E}{E} = \frac{s}{2r_0} + |C_2 \alpha_r^2| + |C_3 \alpha_Z^2| + \left|C_4 \left(\frac{s}{2r_0}\right)^2\right| + \left|C_5 \left(\frac{h}{2r_0}\right)^2\right| + \left|C_6 \alpha_r \frac{s}{2r_0}\right| + \left|C_7 \alpha_Z \frac{n}{2r_0}\right| + \left|\kappa \frac{\alpha_r}{r_0}\right|, \tag{4.2}$$

where s and h are the width and height of the effective electron source, respectively, α_r and α_Z the allowed angular intervals in the radial and axial (orthogonal to the medium plane) directions, r_0 is the spectrometer radius and C_2 to C_7 are aberration coefficients. κ is a relativistic factor depending on β

[14] Purcell, E.M. (1938): Phys. Rev. *54*, 818

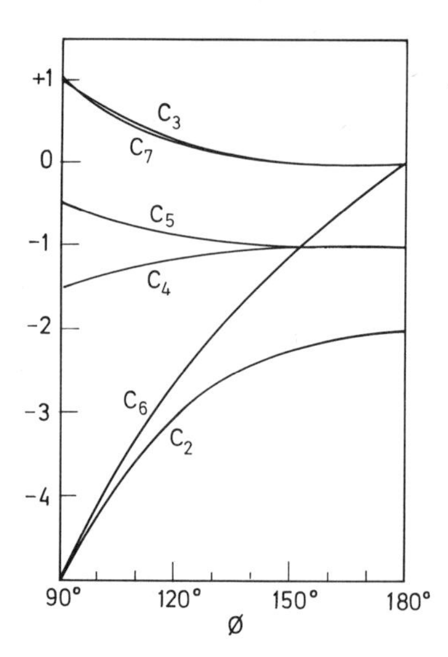

Fig. 17. Variation of aberration coefficients C_2 to C_7 in (4.2) with sector angle ϕ for a spherical capacitor analyzer [3a]

$=v/c$[15]. The calculated variations of the aberration coefficients with respect to sector angle ϕ are shown in Fig. 17. The influence of the fringing fields has been considered in the calculation of these curves. The last term in the expression above describes the relativistic contribution to the resolving power. For the instrument analyzed by these authors, this term amounts to $<2\%$ of C_2 at energies below 5 keV. As evidenced by this figure, the sector angle should be made as close as possible to $\phi=180°$ to minimize aberrations. In order to obtain optimum conditions for the analyzer function, one generally has to correct for the fringing fields by applying extra potentials.

The case of $R_0=\infty$ of one of the toroidal potential surfaces defines the cylindrical capacitor analyzer[16]. It has focusing properties in the radial direction only, according to the above conditions, and it has therefore a lower luminosity than the spherical spectrometer. Its focusing angle is $\phi=\pi/\sqrt{2}=127.3°$. The energy analysis of electrons with energy E is given by:

$$E=1/2\,eV/\ln(R_2/R_1), \tag{4.3}$$

where V is the potential applied between the cylindrical electrodes of radii R_2 (outer) and R_1(inner). The cylindrical analyzer has been extensively used in the

[15] Siegbahn, K., Nordling, C., Fahlman, A., Nordberg, R., Hamrin, K., Hedman, J., Johansson, G., Bergmark, T., Karlsson, S.-E., Lindgren, I., Lindberg, B. (1967): ESCA - Atomic, molecular and solid state structure studied by means of electron spectroscopy, Nova Acta Regiae Soc. Sci. Ups., Ser. IV, Vol. 20

[16] Hughes, A.L., Rojansky, V. (1929): Phys. Rev. *34*, 284

low photon-energy range (He*I*) e.g. [17,18], where the photoelectric cross sections are high, so that the low luminosity is not of critical importance. It is advantageous from the point of view of a simple construction.

Another dispersive instrument which has been widely used is the cylindrical mirror analyzer[19–25]. Figure 18 shows the analyzer schematically. A reflecting potential V is applied between the two cylinders with radii r_0 and r_1. The effective source is defined by a slit at the distance l_1 from the inner cylinder, and this slit is imaged on the detector slit at the distance l_2 inside the cylinder. Both these slits are at the same potential as the inner cylinder. For $(l_1+l_2)=2r_0$, focusing to second order will occur for electrons of energy E such that[21–23]:

$$E=1.3098\,\mathrm{eV}/\ln(r_1/r_0) \tag{4.4}$$

for $\alpha=42.3°$. This means that the angle $\Delta\alpha$ can be made large and the luminosity consequently large. It is found that the resolution at a given luminosity is increased as the opening angle α is made larger than the value for second-order focusing[25]. The practical limit for this angle is about 60°. Although the luminosity in this analyzer may be made high, the shape of the focal surface makes it difficult to implement multichannel detection (the focal surface is approximately a cone with the apex pointing away from the source).

The mirror analyzer has been implemented in a double-pass mode (cf. Fig. 18 bottom)[24,26], which leads to an enhanced resolution and perhaps more importantly, a significant reduction in background. In angular distribution work, the second analyzer may conveniently be used to introduce angle-selecting apertures[26].

Energy analyzers employing magnetic fields were originally introduced in nuclear beta-decay studies e.g. [1]. The iron-free double focusing instruments with $1/\sqrt{r}$-field were developed for low energy electrons and were the first high resolution analyzers in photoelectron spectroscopy[1,27,28]. In present instrument design they have been replaced, however, largely by the spherical capacitor and the cylindrical mirror analyzers. This is due principally to the fact

---

[17] Turner, D.W., Baker, C., Baker, A.D., Brundle, C.R. (1970): Molecular photoelectron spectroscopy, London: Wiley-Interscience

[18] Rabalais, J.W. (1977): Principles of ultraviolet photoelectron spectroscopy, New York: Wiley-Interscience

[19] Blauth, E. (1957): Z. Phys. *147*, 228

[20] Mehlhorn, W. (1960): Z. Phys. *160*, 247

[21] Zashkvara, V.V., Korsunskii, M.I., Kosnenchev, O.S. (1966): Sov. Phys.-Techn. Phys. *11*, 96

[22] Sar-El, H.Z. (1967): Rev. Sci. Instr. *38*, 1210

[23] Aksela, S., Karras, M., Pesoa, M., Suoninen, E. (1970): Rev. Sci. Instr. *41*, 351

[24] Palmberg, P.W. (1974): J. Electron Spectrosc. Relat. Phenom. *5*, 691

[25] Sar-El, H.Z. (1971): Rev. Sci. Instr. *42*, 1601

[26] Wagner, C.D., Riggs, W.B., Davis, L.E., Moulder, J.F., Muilenberg, G.E. (1979): Handbook of x-ray photoelectron spectroscopy, Perkin-Elmer Corp. Phys., Electronics Div.

[27] Nordberg, R., Hedman, J., Hedén, P.F., Nordling, C., Siegbahn, K. (1968): Ark. Fys. *37*, 489

[28a] Siegbahn, K. (1952): Physica *18*, 1043

[28b] Siegbahn, K. (1955): in: Beta- and gamma-ray spectroscopy, Siegbahn, K. (ed.), Amsterdam: North-Holland

[28c] Siegbahn, K., Edvarson, K. (1956): Nucl. Phys. *1*, 137

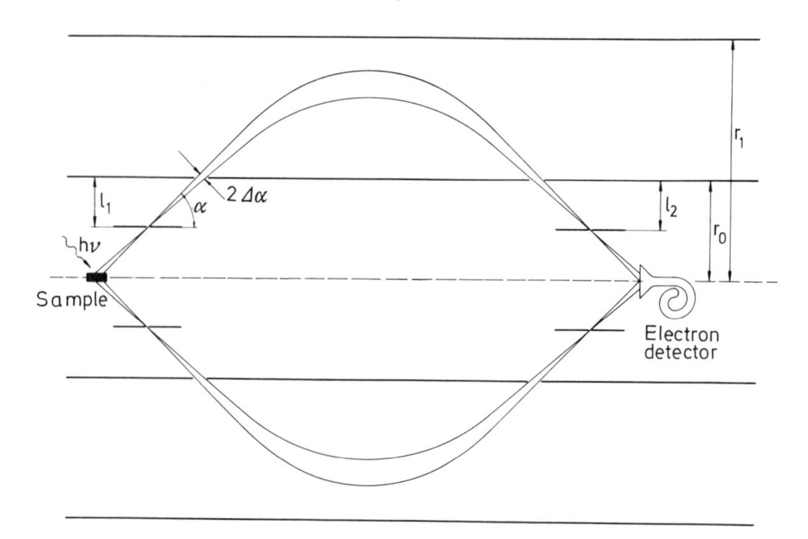

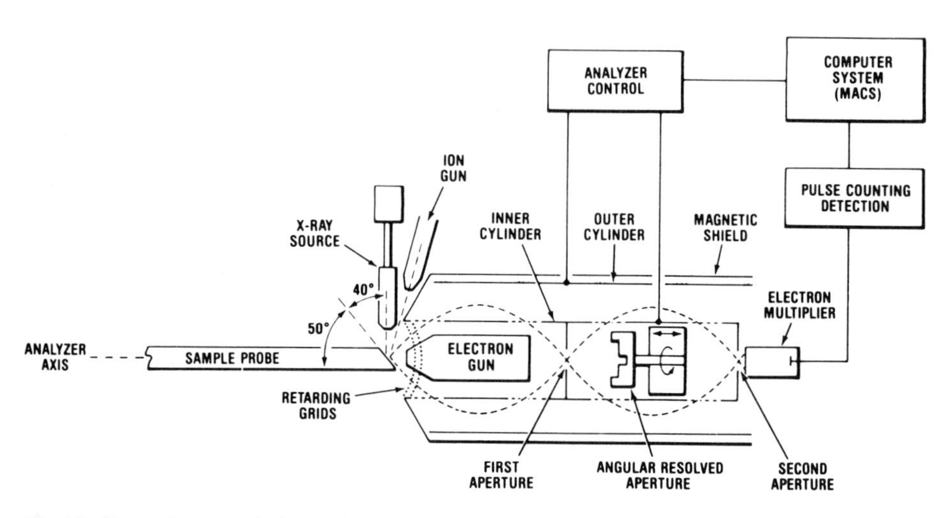

Fig. 18. General geometrical conditions for the cylindrical mirror analyzer (top). Double pass cylindrical mirror analyzer (bottom)[24,26]

that the latter analyzers are much more conveniently shielded from external fields using μ-metal shields, which is not possible with a magnetic analyzer. There are also difficulties in implementing preretardation electron optics in a magnetic analyzer, although this has been done[29,30].

Although the above described analyzers are those which dominate among the dispersive types, several other analyzers have been implemented for photoelectron work. The parallel plate mirror analyzer has been used in particular in

[29] Wannberg, B., Sköllermo, A. (1977): J. Electron Spectrosc. Relat. Phenom. *10*, 45

[30] Asplund, L., Kelfve, P., Blomster, B., Siegbahn, H., Siegbahn, K. (1977): Phys. Scr. *16*, 268

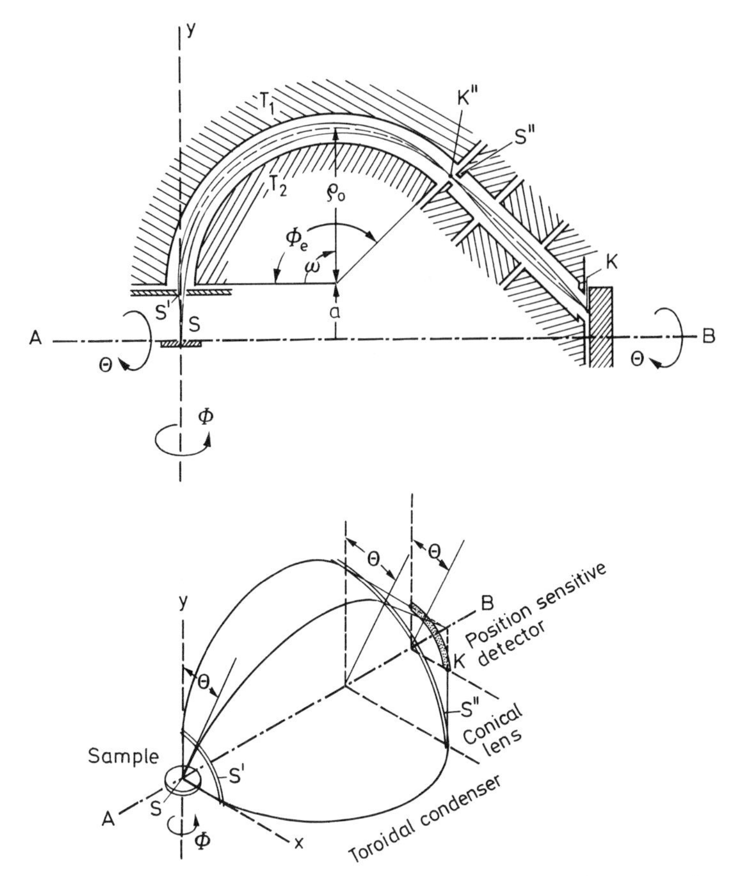

Fig. 19. Toroidal spectrometer designed for photoelectron angular distribution studies from solid surfaces. An extended polar-angle interval is recorded simultaneously at a given azimuthal angle by means of position sensitive detection (multichannel plate-resistive strip combination)[36,37]

connection with angular distribution studies[31-33]. Having a very simple construction, its disadvantage lies in the one-dimensional focusing leading to very low luminosities at high resolution. The Bessel or pill box analyzer[34,35] has good electron-optical properties, but has so far been little used in photoelectron spectroscopy.

Another dispersive analyzer especially designed for angular resolved work has been developed by ENGELHARDT et al.[36,37]. Figure 19 shows the principal

[31] Green, T.S., Proca, G.A. (1970): Rev. Sci. Instr. *41*, 1409

[32] Schmitz, W., Mehlhorn, W. (1972): J. Phys. E *5*, 641

[33] Smith, N.V., Larsen, P.K., Traum, M.M. (1977): Rev. Sci. Instr. *48*, 454

[34] Allen, J.D., Jr., Preston Wolfe, J., Schweitzer, G.K. (1972): Int. J. Mass. Spectrom. Ion Phys. *8*, 81

[35a] Allen, J.D., Jr., Durham, J.D., Schweitzer, G.K., Deeds, W.E. (1976): J. Electron Spectrosc. Relat. Phenom. *8*, 395

[35b] Spinlab Special Instruments Lab. (Model 560)

[36] Engelhardt, H.A., Bäck, W., Menzel, D., Liebl. H. (1981): Rev. Sci. Instr. *52*, 835

[37] Engelhardt, H.A., Zartner, A., Menzel, D. (1981): Rev. Sci. Instr. *52*, 1161

electron paths in the spectrometer, which is of toroidal condenser type. In the first section of the analyzer the electrons in a certain energy window are focused onto a ring-like energy slit. The electrons transmitted through this slit are then focused by a conical lens onto a position sensitive detector in the shape of a ring sector. In this way a polar angle interval between 0° and 100° can be simultaneously measured. The azimuthal angle of emission with respect to the crystal axes is changed by rotating the sample. The angular resolution is of the order of $\delta\theta \sim 2°$ and the energy resolution ~ 150 meV.

γ) Nondispersive analyzers. The most common instrument of the non-dispersive type is the spherical grid retarding potential analyzer. The electron source is placed at the common centre of two spherical grids, between which a retarding potential is applied. If the kinetic energy of the electrons, corresponding to the radial velocity component, exceeds the retarding potential, they will be transmitted through the grids. The actual electron spectrum is evaluated as the derivative of the electron intensity with respect to the retarding potential. The resolution of the instrument is determined partly by the mesh density of the grids, since each mesh hole acts as a small electron lens, and partly by the size of the source, since electrons which are not emitted exactly from the centre of the spheres must have a higher energy to have sufficient radial velocity. The former effect contributes to the base width such that:

$$\Delta E \propto E\, d/D, \tag{4.5}$$

where D is the spacing between the grids and d is the diameter of each mesh hole. The size of the source amounts to an effect on the base width given by:

$$\Delta E = E \cdot (d_s^2/4R_1^2), \tag{4.6}$$

where d_s is the diameter of a circular source, and R_1 is the radius of the inner grid. The signal-to-noise ratio in an instrument of this type is usually low if only a high-pass filter is used for the analysis. These difficulties can be partly overcome, however, if a "post-monochromator" arrangement is used[38-40] (cf. Fig. 20), which focuses the low-energy electrons of interest onto the detector while the unwanted higher energy electrons are collected in the surrounding shield. This type of analyzer is capable of a very high transmission (up to 1 ster.), but the highest obtainable resolving power is significantly lower than for the dispersive instruments.

Instruments of mirror-grid type have been designed in order to introduce a low-pass stage for the energy analysis[41-43]. In these instruments the electrons are first reflected against a low-pass retarding mirror with potential $V - \Delta V$

---

[38] Huchital, D.A., Rigden, J.D. (1972): in: Electron spectroscopy, Shirley, D.A. (ed.), Amsterdam: North-Holland Publ. Co., p. 79

[39] Huchital, D.A., Rigden, J.D. (1972): J. Appl. Phys. *43*, 2291

[40] Staib, P., Dinklage, U. (1977): J. Phys. E: *10*, 914

[41] Lee, J.D. (1972): Rev. Sci. Instr. *43*, 1291

[42] Riggs, W.M., Fedchenko, R.P. (1972): Am. Lab. *4*, 65

[43] Eastman, D.E., Donelon, J.J., Hien, N.C., Himpsel, F.J. (1980): Nucl. Instr. Methods *172*, 327

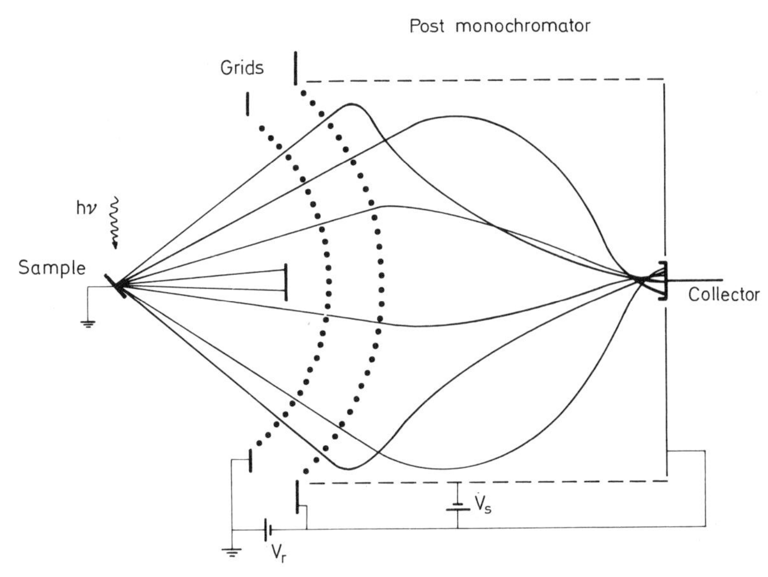

Fig. 20. Retarding grid analyzer including post-monochromator arrangement for improvement of signal-to-background ratio [39]

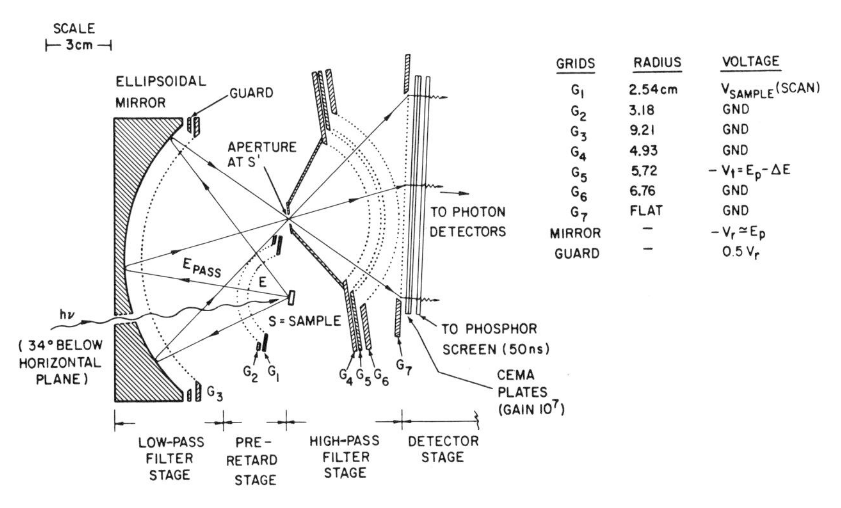

Fig. 21. Ellipsoidal mirror-retarding grid spectrometer for studies of photoelectron angular distributions [43]

and are then allowed to pass through a high-pass retarding grid of potential $V + \Delta V$. Thus, only electrons with an energy between $e(V+\Delta V)$ and $e(V-\Delta V)$ reach the detector. An instrument of this type designed for angular distribution work is shown in Fig. 21 [43]. It is based on the use of a low pass retarding ellipsoidal mirror which reflects the photoelectrons through an aperture into a high pass stage. The transmitted electrons are detected by means of a multichannel plate-phosphor screen combination (cf. below), the output of which is viewed by a TV-camera. The construction of the spectrometer implies a one-to-one correspondence between a certain ejection direction of the photoelectron and a certain position on the phosphor screen. One can thus record the angular distribution of the photoelectrons from a solid sample by scanning the pattern on the screen. The energy resolution with this spectrometer is of the order of $\gtrsim 2\%$ for full angular acceptance. The achievable angular resolution is $\delta\theta \simeq 2°$.

Time-of-flight spectrometers have been discussed by a number of authors[44-46]. Operation of such instruments is dependent on the use of a pulsed photon source. Such sources are available from high power plasma discharges, laser induced plasma sources and electron storage rings. All these sources produce pulses in the nano-second and micro-second range and do not appreciably influence the time resolution of the spectrometer. These instruments display in general high luminosity and the resolution can be made high[44]. One such instrument which has been constructed for use with gaseous samples at the synchrotron radiation source SPEAR is shown in Fig. 22[46]. The resolution of this instrument is decided by the flight path uncertainty due to the finite size of the photon beam cross section. This leads to a resolution of $\sim 5\%$, as evidenced by the spectrum on the right in Fig. 22.

δ) Preretardation in dispersive analyzers. From a resolution expression such as (4.2) it is clear that a gain in absolute resolution, ΔE, is obtained for any dispersive instrument at a given luminosity if the electrons are preretarded before entering the analyzer[3,47-51]. Conversely, the luminosity at a given resolution may be increased by the same means. Instruments of the dispersive type employ preretardation of the electrons for the above reasons. In the retardation process the brightness of the source is decreased, which partly cancels the gain in luminosity of the analyzer. The effects of preretardation in a spherical capacitor can be estimated by considering (4.2) and the Lagrange-Helmholz relations (using the same notation as in (4.2)):

$$\begin{aligned} s \cdot \alpha_r \cdot \sqrt{E} &= \text{const.} \\ h \cdot \alpha_z \cdot \sqrt{E} &= \text{const.} \end{aligned} \qquad (4.7)$$

[44] Siegbahn, K., Fellner-Feldegg, H. (1974): UUIP-858 (Uppsala University Institute of Physics Report)

[45a] Bachrach, R.Z., Brown, F.C., Hagström, S.B.M. (1975): J. Vac. Sci. Technol. *12*, 309

[45b] Tsai, B., Baer, T., Horowitz, M.C. (1974): Rev. Sci. Instr. *45*, 494

[45c] Guyon, P.M., Baer, T., Ferreira, L.F.A., Nenner, I., Tabché-Fonhgilis, A., Botter, R., Grovers, T.R. (1978): J. Phys. B *11*, L141

[46a] Kennerly, R.E. (1977): Rev. Sci. Instr. *48*, 1682

[46b] White, M.G., Rosenberg, R.A., Gabor, G., Poliakoff, E.D., Thornton, G., Southworth, S.H., Shirley, D.A. (1979): ibid, *50*, 1268

[47] Helmer, J.C., Weichert, N.H. (1968): Appl. Phys. Lett. *13*, 266

[48] Siegbahn, K., Hammond, D., Fellner-Feldegg, H., Barnett, E.F. (1972): Science *176*, 245

[49] Weiss, M.J. (1972/73): J. Electron Spectrosc. Relat. Phenom. *1*, 179

[50] Nöller, H.G., Polaschegg, H.D., Schillalies, H. (1974): J. Electron Spectrosc. Relat. Phenom. *5*, 705

[51] Wannberg, B., Sköllermo, A. (1977): J. Electron Spectrosc. Relat. Phenom. *10*, 45

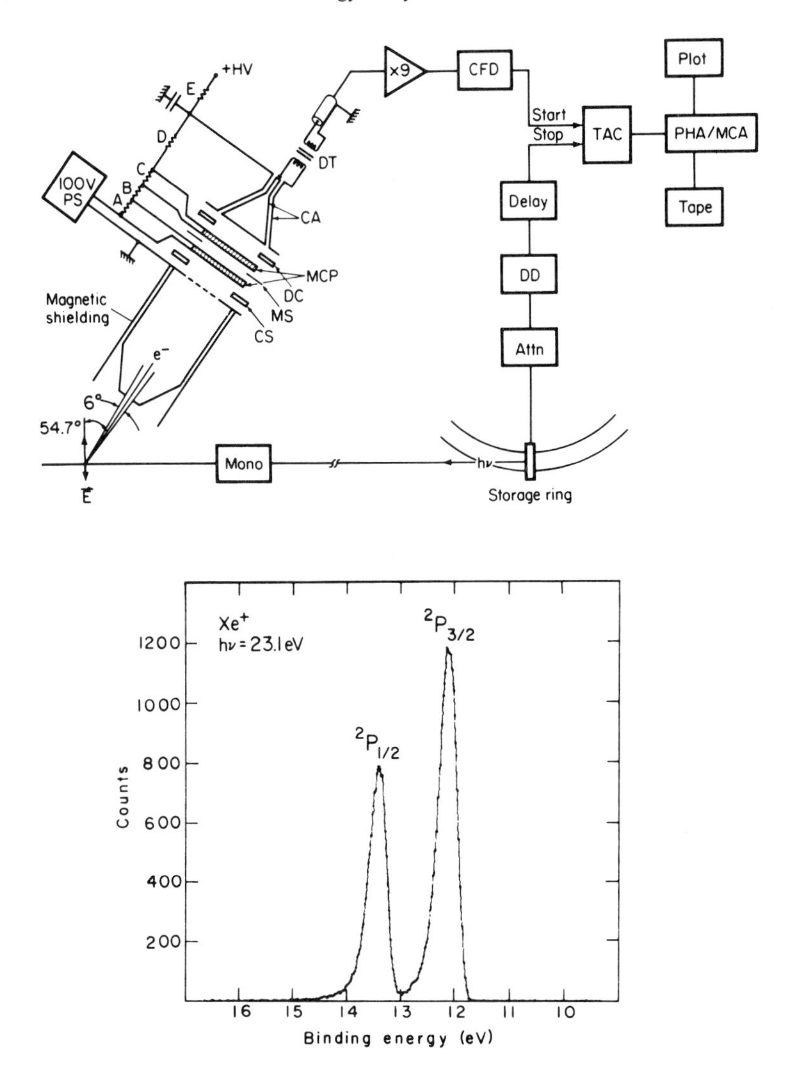

Fig. 22. Time-of-flight spectrometer for photoelectron spectroscopy using the time structure of synchrotron radiation (*top*). Xe 5*p* spin-orbit doublet obtained with the TOF-analyzer[46b] (*bottom*)

From (4.2) it is seen that the gain in luminosity is equal to $(E_1/E_2)^{3/2}$, when the electrons are retarded from energy E_1 to E_2. This is so, since at a given absolute resolution the slit width s can be increased by a factor E_1/E_2 and the radial angle by $(E_1/E_2)^{1/2}$. The slit height h and α_z are generally not limited by their contributions to the resolution expression, but rather by practical considerations, such as the usable detector height. Relations (4.7) imply that when the electrons are retarded, the product of area, $s \cdot h$, and solid angle, $\alpha_r \cdot \alpha_z$, is increased by a factor E_1/E_2. This means that those electrons which are accepted by the analyzer after retardation must have originated in the source with a value of area times solid angle which is E_1/E_2 times smaller than the accep-

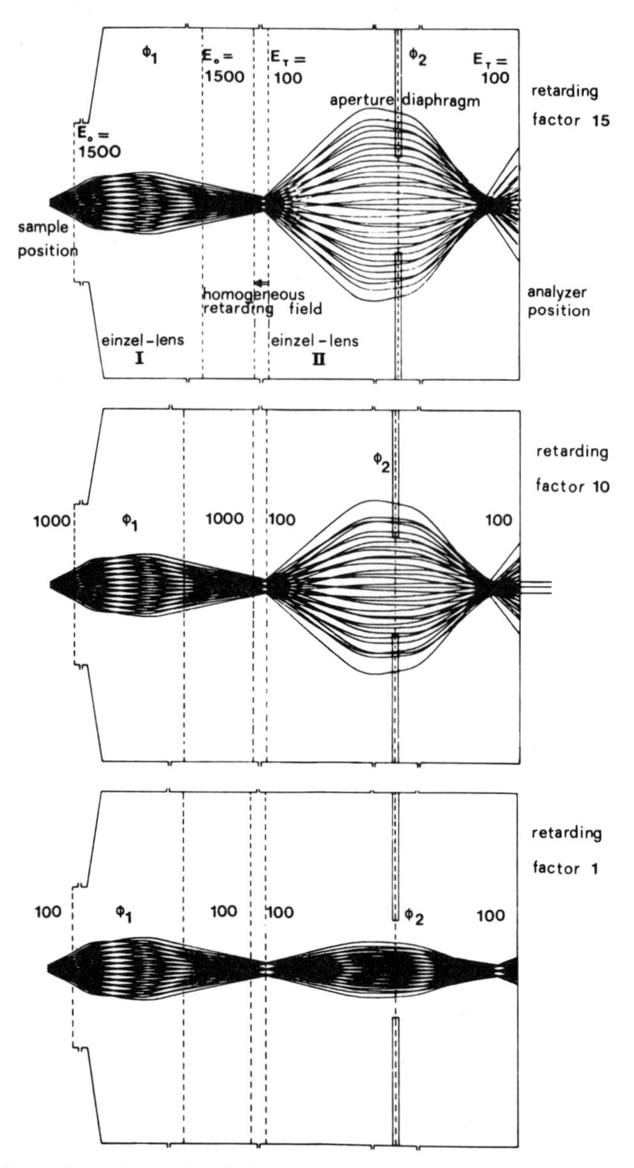

Fig. 23. Electrode configuration and calculated trajectories for various retardation ratios in an electron lens with separated focusing and retardation functions [50]

tance of the analyzer. The practically achievable net gain in intensity for a spherical capacitor will thus be $(E_1/E_2)^{3/2}/(E_1/E_2)=(E_1/E_2)^{1/2}$ at a given absolute resolution when the electrons are retarded from E_1 to E_2.

The use of a lens system for preretardation implies a number of advantages with respect to the simpler method of only applying a retardation field between two closely spaced slits or grids. The lens system creates a virtual source image at the analyzer entrance slit, which means that the excitation arrangement can be removed from the analyzer region. This implies for the spherical capacitor that no restriction is imposed on using the full sector angle (π) of the analyzer, thus minimizing aberrations. Moreover, when studying gases, the distance from the spectrometer entrance slit reduces the gas load inside the analyzer. A number of lens systems have been designed for the above purposes[3, 24, 48–51]. One such system is shown in Fig. 23, with calculated trajectories for different retardation ratios. The retardation stage consists of two closely spaced grids. The consequences of the Lagrange-Helmholz relations (4.7) can here be nicely seen in that the divergence of the beam increases after the retardation.

ε) Electron detection. The electron detectors used in photoelectron spectroscopy today are generally electron multipliers. The channel electron multiplier is a continuous multiplier which consists of a tube of semiconducting glass, the inner surface of which has been pretreated to give high secondary emission. These multipliers usually have the form of a spiral, mainly to reduce noise due to the ions produced. A high voltage applied between the ends of the tube results in a high and saturable gain (10^7–10^8) of the incoming electrons. This gives a narrow pulse height distribution at the output. The pulses are amplified and fed into a discriminator/rate meter. The channel multipliers are smaller than the conventional type of multiplier and also less sensitive to poisoning by air and sample gases.

Multichannel detection is extensively used today to increase the information rate in dispersive analyzers, which have a well-defined focal surface. A position-sensitive electron multiplier is constructed from a bundle of fine capillary tubes made of lead glass, which are reduced in hydrogen atmosphere. This bundle is then sliced into a number of channel plates, whose end surfaces are made conductive by evaporation. The plates are normally circular discs of 25 or 50 mm diameter, but may also be given other shapes of larger dimensions. Their thickness is of the order of 0.5–2.5 mm. The holes in the plates are normally between 10–50 μm wide and have a semiconducting layer covering their entire length through the plate. When this device is used in multichannel counting, two such plates are coupled in series in order to achieve a high gain ($\sim 10^8$). The channels of the respective plates are also biased an angle 5–10° relative to the direction normal to the surface and the pair of plates combined in such a way that the channels are tilted relative to each other. By this means, the noise due to ions is considerably reduced.

Several methods have been used for recording the output from the multichannel plates[48, 52–56]. The electron bunches may be accelerated against

[52a] Basilier, E. (1980): UUIP-1021 (Uppsala University Institute of Physics Report)

[52b] Basilier, E. (1976): Nucl. Instr. Methods *138*, 663

[53] Lawrence, G.M., Stone, E.J. (1975): Rev. Sci. Instr. *46*, 432

[54] Moak, C.D., Datz, S., Garcia Santibanes, F., Carlson, T.A. (1975): J. Electron Spectrosc. Relat. Phenom. *6*, 151

[55a] Kellogg, E., Henry, P., Murray, S., VanSpeybroeck, L. (1976): Rev. Sci. Instr. *47*, 282; Kellogg, E., Murray, S., Briel, U., Bardas, D. (1977): ibid. *48*, 550

[55b] Knapp, G. (1978): ibid. *49*, 982

[55c] Harrison, D., Kubierschky, K. (1979): IEEE Trans. Nucl. Sci. *NS-26*, No. 1, 411

[56] Timothy, J.G., Bybee, R.L. (1975): Rev. Sci. Instr. *46*, 1615

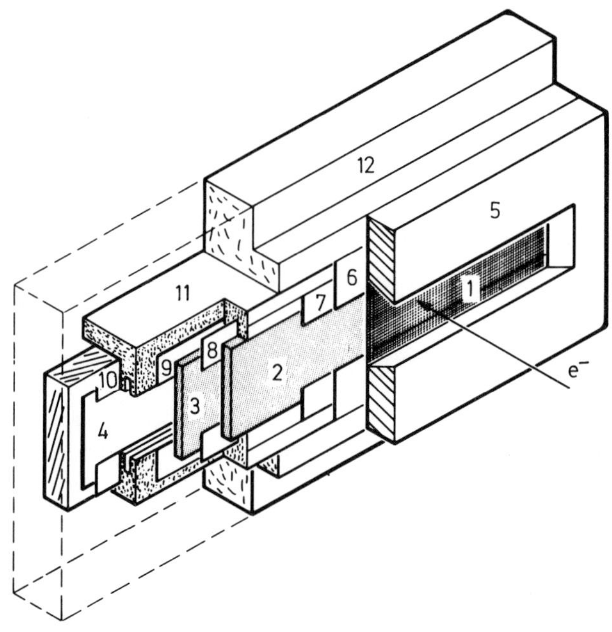

Fig. 24. Cut-away view of rectangular microchannel plate electron detector[57]. *1* Metal grid, *2* 1st microchannel plate, *3* 2nd microchannel plate, *4* phosphor screen, *5* metal frame holding grid, *6* mica frame, *7* 1st metal electrode, *8* 2nd metal electrode, *9* 3rd metal electrode, *10* 4th metal electrode, *11* ceramic spacing frame, *12* ceramic mounting frame

a phosphor screen, which results in conversion into light pulses[48,52]. These pulses can be recorded through a glass window in the vacuum wall. For this purpose a vidicon TV-camera can be used, whose readout is stored in a multichannel analyzer or directly into a computer. Figure 24 shows a multichannel detector assembly based on the above scheme of recording the output pulses[57]. The channel plates are rectangular, 10 cm wide, in order to simultaneously record an extended part of the focal plane in a spherical capacitor.

Another technique to register the multichannel plate output which is frequently used employs a strip of resistive material to collect the electron bunches[53,54]. Two charge pulses appearing at opposite ends of the strip are compared in order to establish the position of a bunch. The relative amounts of charge appearing at the two terminals or a comparison of the pulse shapes can be used to determine position.

5. Sample preparation procedures

α) Vacuum requirements. The analysis of photoelectrons require careful procedures in preparation of the experiments, since local fields and scattering within the sample cause strong perturbations on the motion of the photoelectrons. The vacuum requirements depend on the state of the sample to be

[57] Gelius, U. (private communication)

investigated and the information desired. We shall not discuss pumping equipment used to achieve the appropriate vacuum conditions but refer to other sources for such information[1]. For gaseous samples only moderate vacuum conditions are needed; equipment capable of a vacuum below 10^{-5} Torr is sufficient for most purposes. For liquids, similar vacuum requirements hold, but the handling of the sample necessitates special equipment. Clean solid surfaces and interaction of gases with surfaces, on the other hand, is a field where ultrahigh vacuum (below 10^{-9} Torr) conditions are highly desirably if not imperative. At a pressure of 10^{-6} Torr a monolayer of material can be adsorbed on a surface in one second. The time needed to scan the photoelectron spectrum of the surface ranges in practice up to several hours. This means that in order to obtain reasonably reliable experimental information, the pressure must be kept in the uhv region.

β) Gases. Gases have to be confined to a small volume in the spectrometer. This is to minimize the distance of high-pressure regions through which the photoelectrons have to pass. The usual way of solving this is by means of differential pumping. This means that the gas (which may also be the vapour of a liquid) is leaked into a sample compartment, where the excitation takes place, and in whose side a slit is located, which defines the photoelectron source for the energy analysis. Outside this slit heavy pumping is applied, effectively reducing the pressure before a second slit, leading to the spectrometer. By this means an operating pressure in the sample compartment of e.g. 0.5 Torr is reduced to $\sim 2 \cdot 10^{-4}$ Torr in the differential pumping stage outside the slit and further down to $\sim 1 \cdot 10^{-5}$ Torr inside the analyzer. Several stages of differential pumping can be applied to further decrease the pressure in the analyzer. Alternatively, higher sample pressures can thus be used. Figure 25 shows a standard gas cell arrangement used in an instrument with monochromatized x-ray excitation[2].

The intensities of the photoelectron lines depend on the gas pressure in the sample cell. An increased number of photoelectrons are produced when the pressure is increased due to the larger density of molecules. On the other hand, scattering (elastic and inelastic) also increases with respect to a pressure increase. The optimum pressure is essentially inversely proportional to the scattering cross section times the effective distance that the electrons have to travel through the gas. Since scattering cross sections decrease with increasing electron energy this means that optimum pressures using uv-excitation generally lie at lower pressures than for x-ray excitation (under similar geometrical conditions). Typical operating pressures are ~ 0.1 Torr for the former excitation mode and ~ 0.5 Torr for the latter.

One of the most important factors in deciding the photoelectron linewidths in the uv-excitation mode is the Doppler broadening. It is due to the random

[1a] Power, B.D. (1966): High vacuum pumping equipment, London: Chapman-Hall

[1b] Redhead, P.A., Hobson, J.P., Kornelsen, E.V. (1967): The physical basis of ultrahigh vacuum, London: Chapman-Hall

[1c] Holland, L., et al. (eds.), (1974): Vacuum manual, London: Spon

[2] Gelius, U., Basilier, E., Svensson, S., Bergmark, T., Siegbahn, K. (1973): J. Electron Spectrosc. Relat. Phenom. *2*, 405

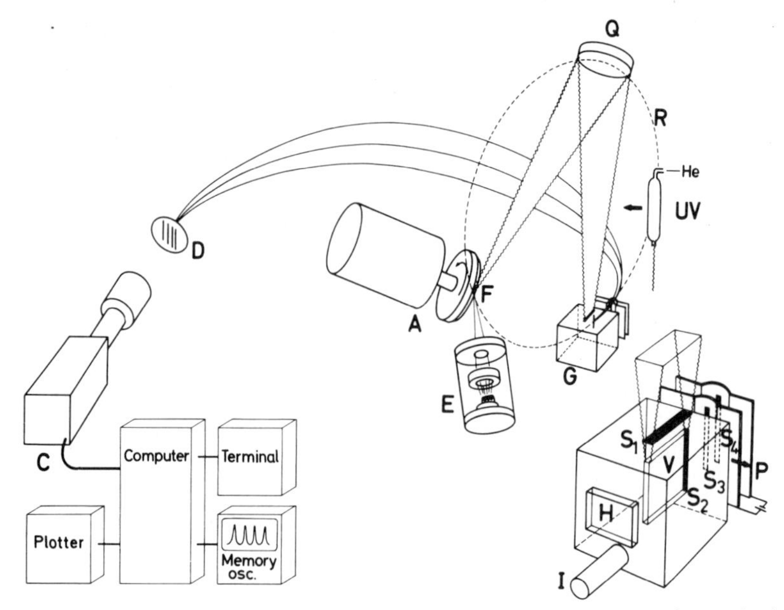

Fig. 25. Electron spectrometer designed for gaseous phase work using monochromatized x-rays. Inset figure shows the gas cell with double differential pumping, reducing the pressure 5 orders of magnitude from the sample cell into the spectrometer[2]

motion of the target molecules. The Doppler broadening can be largely reduced if the target molecules are forced to move uniformly in a molecular beam. A number of studies have been performed with photoelectron spectroscopy using supersonic jet techniques[3,4]. The results so far obtained are encouraging and it is to be expected that these techniques will become increasingly important in gaseous photoelectron work. One disadvantage in the use of molecular beam is the substantial reduction in particle density in the formation of the beam.

The standard methods for heating of materials for vapour studies are resistive, inductive or electron bombardment[5,7–11]. In addition, it has been

[3] Dehmer, P.M., Dehmer, J.L. (1978): J. Chem. Phys. *68*, 3462; *69*, 125

[4a] Nomoto, K., Achiba, Y., Kimura, K. (1980): Inst. Mol. Sci. Japan, Ann. Rev., December 1980, p. 99

[4b] Pollard, J.E., Trevor, D.J., Lee, Y.T., Shirley, D.A. (1981): Rev. Sci. Instr. *52*, 1837

[5] Khodeyev, Y.S., Siegbahn, H., Hamrin, K., Siegbahn, K. (1973): Chem. Phys. Lett. *19*, 16

[6a] Berkowitz, J. (1972): J. Chem. Phys. *56*, 2766

[6b] Berkowitz, J. (1976), in: Electron spectroscopy, techniques and applications, Brundle, C.R., Baker, A.D. (eds.), London: Academic Press

[7] Evans, S. (1972): Faraday Discuss. Chem. Soc. *54*, 143

[8a] Potts, A.W., Williams, T.A., Price, W.C. (1974): Proc. R. Soc. Lond. A *341*, 147

[8b] Williams, T.A., Potts, A.W. (1976): J. Electron Spectrosc. Relat. Phenom. *8*, 331

[9] Allen, J.D., Jr., Boggess, G.W., Goodman, T.D., Wachtel, A.S., Jr., Schweitzer, G.K. (1973): J. Electron Spectrosc. Relat. Phenom. *2*, 289

[10] Bulgin, D., Dyke, J., Goodfellow, F., Jonathan, N., Lee, E., Morris, A. (1977): J. Electron Spectrosc. Relat. Phenom. *12*, 67

[11] Frost, D.C., Lee, S.T., McDowell, C.A., Westwood, N.P.C. (1977): J. Electron Spectrosc. Relat. Phenom. *12*, 95

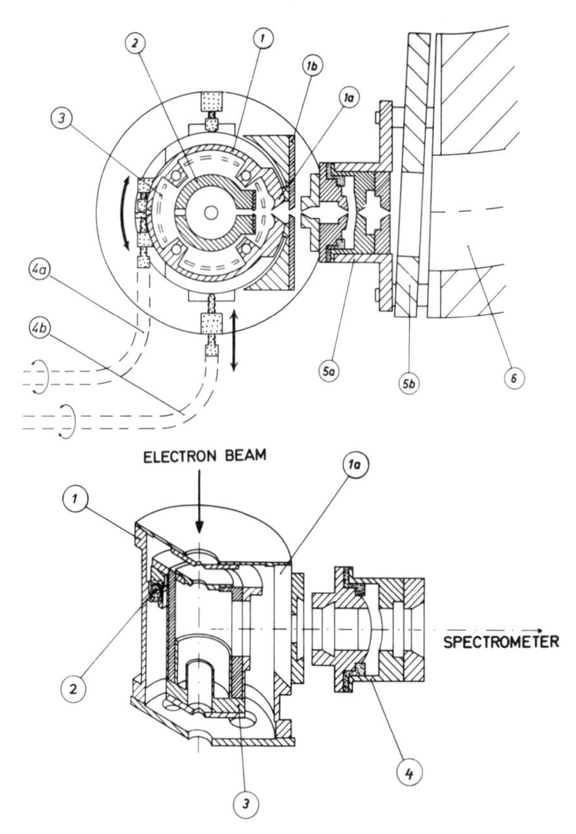

Fig. 26. Oven for high temperature studies of electron spectra from metal vapours using resistive heating[13]. Upper part: (*1*) Cooling jacket, (*1a* and *b*) slits, (*2*) oven, (*3*) heat shield, (*4a+b*) adjustment of the oven, (*5a*) system of entrance slits with retardation, (*5b*) Herzog baffle, (*6*) spectrometer. Lower part: (*1*) Cooling jacket, (*1a*) slit, (*2*) thermal isolation bearing, (*3*) oven, (*4*) baffle system

shown that laser heating is a good alternative for vaporizing solid samples[9–12].

MEHLHORN et al.[13] designed a resistively heated oven for studies of metal vapours (cf. Fig. 26). Excitation of the vapour is performed inside the oven by means of an electron beam traversing the oven through openings in the top and the bottom. When using x-radiation, the opening is closed by a heated metal foil (Al, Be). The operating pressure inside the oven is about 10^{-3} Torr.

A number of workers have used a microwave discharge to produce transient species[11,14–18]. Thus, JONATHAN et al. and SAMSON and PETROSKY[16,18]

[12] Cornford, A.B., Frost, D.C., Herring, F.G., McDowell, C.A. (1971): J. Chem. Phys. *54*, 1872

[13] Mehlhorn, W., Breuckmann, B., Hausamann, D. (1977): Phys. Scr. *16*, 177

[14] Jonathan, N., Morris, A., Ross, K.J., Smith, D.J. (1971): J. Chem. Phys. *54*, 4954

[15] Jonathan, N., Smith, D.J., Ross, K.J. (1970): J. Chem. Phys. *53*, 3758

[16] Jonathan, N., Morris, A., Okuda, M., Smith, D.J., Ross, K.J. (1972): in: Electron spectroscopy, Shirley, D.A. (ed.), Amsterdam: North-Holland

[17] Jonathan, N., Morris, A., Okuda, M., Ross, K.J., Smith, D.J. (1974): Faraday Trans. II *11*, 1810

[18] Samson, J.A.R., Petrosky, V. (1974): Phys. Rev. A *9*, 2449

studied the photoelectron spectra of free atoms. These species were produced in small quantities in the presence of the parent molecules. The former workers have employed phase-sensitive detection for separation of the spectra from the transient and parent species[17]. This technique is useful at low percentages of the transients.

A large number of free molecular radicals have also been investigated using microwave discharges. An alternative method to produce such species which has been frequently employed is pyrolysis[19-21].

γ) Liquids. The handling of liquid samples implies a number of difficulties not encountered for gaseous or solid samples. First, the vapour pressures of many liquids and solvents at room temperature are higher than 0.1 Torr, which means that heavy differential pumping is required to reduce the pressure over a short distance. Second, the liquid sample must continuously be renewed due to pumping off, which is of particular importance if solutions are investigated. In such cases, the concentration of dissolved material may otherwise change during the measurement process. Moreover, it should be possible to emphasize the spectrum of the liquid phase with respect to that from the vapour phase.

The first system designed to meet these requirements was the liquid beam arrangement[22-24]. The liquid to be investigated is in this scheme circulated inside a tubing system held under vacuum. The sample surface is then formed in front of (and parallel to) the spectrometer slit by means of a nozzle hole which creates a thin liquid beam (width $\lesssim 0.5$ mm). This technique was first applied with x-ray excitation[22-24], but has also been used with uv-excitation[25,26].

Another line of development in liquid sample handling is based on the use of wetted metal surfaces[27-34]. A wetted wire method was employed with x-ray

---

[19] Potts, A.W., Glenn, K.G., Price, W.C. (1972): Faraday Discuss. Chem. Soc. *54*, 50

[20] Cornford, A.B., Frost, D.C., Herring, F.G., McDowell, C.A., Faraday Discuss. Chem. Soc. *54*, 56

[21] Dyke, J., Jonathan, N., Lee, E., Morris, A. (1976): J. Chem. Soc. Faraday Trans. II *72*, 1385

[22] Siegbahn, H., Siegbahn, K. (1973): J. Electron Spectrosc. Relat. Phenom. *2*, 319

[23] Siegbahn, H., Asplund, L., Kelfve, P., Hamrin, K., Karlsson, L., Siegbahn, K. (1974): J. Electron Spectrosc. Relat. Phenom. *5*, 1059

[24] Siegbahn, H., Asplund, L., Kelfve, P., Siegbahn, K. (1975): J. Electron Spectrosc. Relat. Phenom. *7*, 411

[25] Ballard, R.E., Barker, S.L., Gunnell, J.J., Hagan, W.P., Pearce, S.J., West, R.H., Saunders, A.R. (1978): J. Electron Spectrosc. Relat. Phenom. *14*, 331

[26] Ballard, R.E., Gunnell, J.J., Hagan, W.P. (1979): ibid. *16*, 435

[27] Fellner-Feldegg, H., Siegbahn, H., Asplund, L., Kelfve, P., Siegbahn, K. (1975): J. Electron Spectrosc. Relat. Phenom. *7*, 421

[28] Lindberg, B., Asplund, L., Fellner-Feldegg, H., Kelfve, P., Siegbahn, H., Siegbahn, K. (1976): Chem. Phys. Lett. *39*, 8

[29] Aulich, H., Delahay, P., Nemec, L. (1973): J. Chem. Phys. *59*, 2354

[30] Aulich, H., Nemec, L., Delahay, P. (1974): J. Chem. Phys. *61*, 4235

[31] Aulich, H., Nemec, L., Chia, L., Delahay, P. (1976): J. Electron Spectrosc. Relat. Phenom. *8*, 27

[32] Nemec, L., Gaehrs, H.J., Chia, L., Delahay, P. (1977): J. Chem. Phys. *66*, 4450

[33] Watanabe, I., Flanagan, J.K., Delahay, P. (1980): ibid. *73*, 2057

[34] Siegbahn, H., Svensson, S., Lundholm, M. (1981): J. Electron Spectrosc. Relat. Phenom. *24*, 205

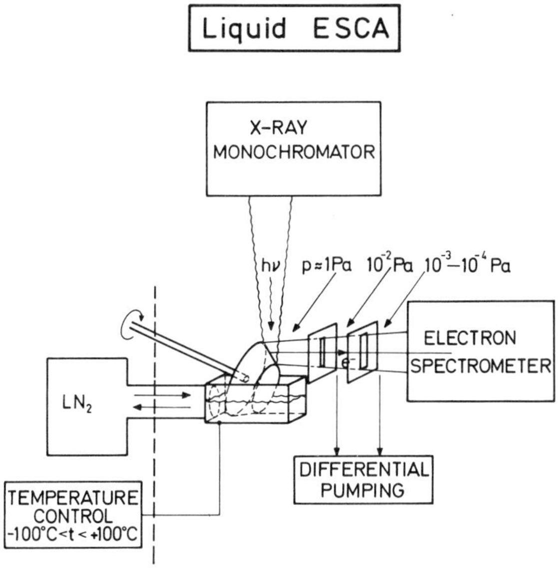

Fig. 27. Liquid ESCA arrangement using the wetted surface principle, including temperature control and two-stage differential pumping. Excitation is performed with monochromatized Al $K\alpha$[34]

excitation[27,28] based on the liquid beam method, the wire replacing the beam thus giving more stable conditions. Workers using uv-excitation have employed a rotating disc submerged into the liquid[29–33]. The photoelectrons excited from the rim of the disc are subsequently analyzed in a retarding field analyzer.

More recent work[34] on wetted surface methods using x-ray excitation has shown the potential of monochromatization in improving the quality of the spectra. Figure 27 shows a schematic figure of the experimental arrangement. The liquid sample is contained in a cup, which is attached to a cold finger. Into this cup a stainless steel truncated cone is submerged, which can be rotated around its axis. During rotation a thin film of the liquid sample will cover the metal sample which is situated in the proper position for photoelectron excitation. By means of liquid nitrogen the liquid sample may be cooled down to temperatures where the vapour pressure becomes low enough so that vapour photoelectron lines are diminished. The temperature may be regulated to within $\pm 1°$ by means of the flow of liquid nitrogen. This facility was introduced for the investigation of various equilibria in solution. The temperature of the sample may alternatively be raised above room temperature by means of resistance heating in order to investigate molten substances. Figure 28 shows a spectrum of liquid methanol recorded with this equipment at 190 K[34].

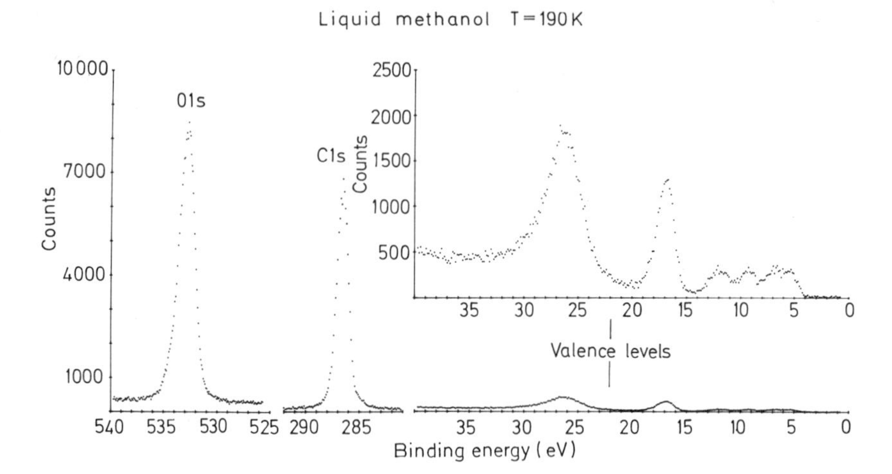

Fig. 28. ESCA spectrum of liquid methanol obtained at 190 K using the arrangement of Fig. 27

Other methods have been suggested for liquid studies which employ solid-phase sample techniques, either by rapid freezing[35,36] or by using very low vapour pressure solvents ($p \sim 10^{-5}$ Torr)[37].

δ) Solids and surfaces. The requirements put on solid sample handling are:

1. It should be possible to study all types of solids, whether a conductor or non-conductor, a single crystal or a polycrystalline material.

2. It should be possible to vary the temperature of the sample from say, liquid nitrogen (or even liquid helium) temperatures up to above 1,000 °C.

3. Preparation of the sample as well as cleaning and perhaps chemical conversion should be possible *in situ.*

4. Arrangements for studies of angular distributions of the photoelectrons are highly desirable.

5. One should have facilities to investigate interactions of the surface with gas phase atomic or molecular species.

To meet these requirements, methods have been developed, which attempt to combine flexibility with careful control of conditions[38]. Instruments have been designed, which to varying extent include the above features or which have concentrated on particular aspects.

The solid samples may be prepared and mounted on the sample probe outside the instrument before they are introduced into measurement position.

---

[35] Burger, K., Tshismarov, F., Ebel, H. (1977): ibid. *10*, 461

[36] Burger, K. (1978): ibid. *14*, 405

[37] Avanzino, S.C., Jolly, W.L. (1978): J. Am. Chem. Soc. *100*, 2228

[38] For a detailed review of sample preparation techniques in electron spectroscopy cf. Siegbahn, H., Karlsson, L. Electron Spectroscopy for Atoms, Molecules and Condensed Matter, Amsterdam: North-Holland (to be published)

Depending on the form of the sample, different mounting arrangements can be used. For powders, pressing onto a soft metal or wire mesh backing is commonly used. If the sample is in the form of a strip or sheet, it can be clamped onto the probe. For metal single crystal specimens, spot welding onto supports may be performed which can then be used for mounting. When a thin sample of an insulating sample (such as a salt) is required, it may be dissolved in a suitable solvent, which is subsequently allowed to evaporate.

Another possible means of preparing solid samples consists in deposition of the material onto the probe tip inside the spectrometer vacuum. For metals, deposition *in situ* is generally performed by means of evaporation. Such evaporated films will be polycrystalline. For refractory materials which cannot be evaporated by resistive heating, electron bombardment can be used. MBE (Molecular Beam Epitaxy) is an advanced evaporation technique for growth of single crystals *in situ*, where the substances to be evaporated are placed in accurately temperature stabilized ovens of refractory materials, such as graphite or boron nitride (Knudsen cells). The substrate in an MBE unit can also be stabilized to a uniform temperature. Evaporation at low rates (typically 1 Å/s), using suitable single crystals as substrates, leads to growth of epitaxial layers. Several Knudsen cells containing different evaporants can be operated simultaneously and compounds of accurately controlled stoichiometry can thus be formed. By operating each Knudsen cell with an individual shutter multilayered compounds, so-called sandwich materials, can be produced.

In situ deposition of gases and vapours from liquids and solids can be performed by freezing out on the probe tip, usually at liquid nitrogen temperature.

A variety of methods exist for pretreatment and cleaning of the samples *in situ*. The most common are short period heating (flashing), argon ion etching and subsequent annealing, mechanical scraping or cleaving of single crystals. Various devices for fracturing of the sample have also been developed[39]. The actual use of these methods is largely dependent on the material investigated.

For surface-gas interactions leak valves are required, which ensure a careful control of the gas exposure. It should be possible to adsorb submonolayer quantities onto the surface.

During the recordings of the spectra, the temperature of the sample may generally be controlled to a preset value within certain limits by employing a combination of liquid-nitrogen cooling and resistance heating. For angular distribution studies, the sample can be mounted on a manipulator, which allows rotations with respect to the directions defined by the photons and the outgoing electrons.

The instrument schematically shown in Fig. 29 may serve as an example of how an instrument can be designed, incorporating many of the above discussed methods for solid sample investigations[40]. The sample handling system consists of five chambers, in which different operations can be performed.

---

[39] Olefjord, I. (1974): J. Electron Spectrosc. Relat. Phenom. 5, 401

[40] Gelius, U., Siegbahn, K. (private communication)

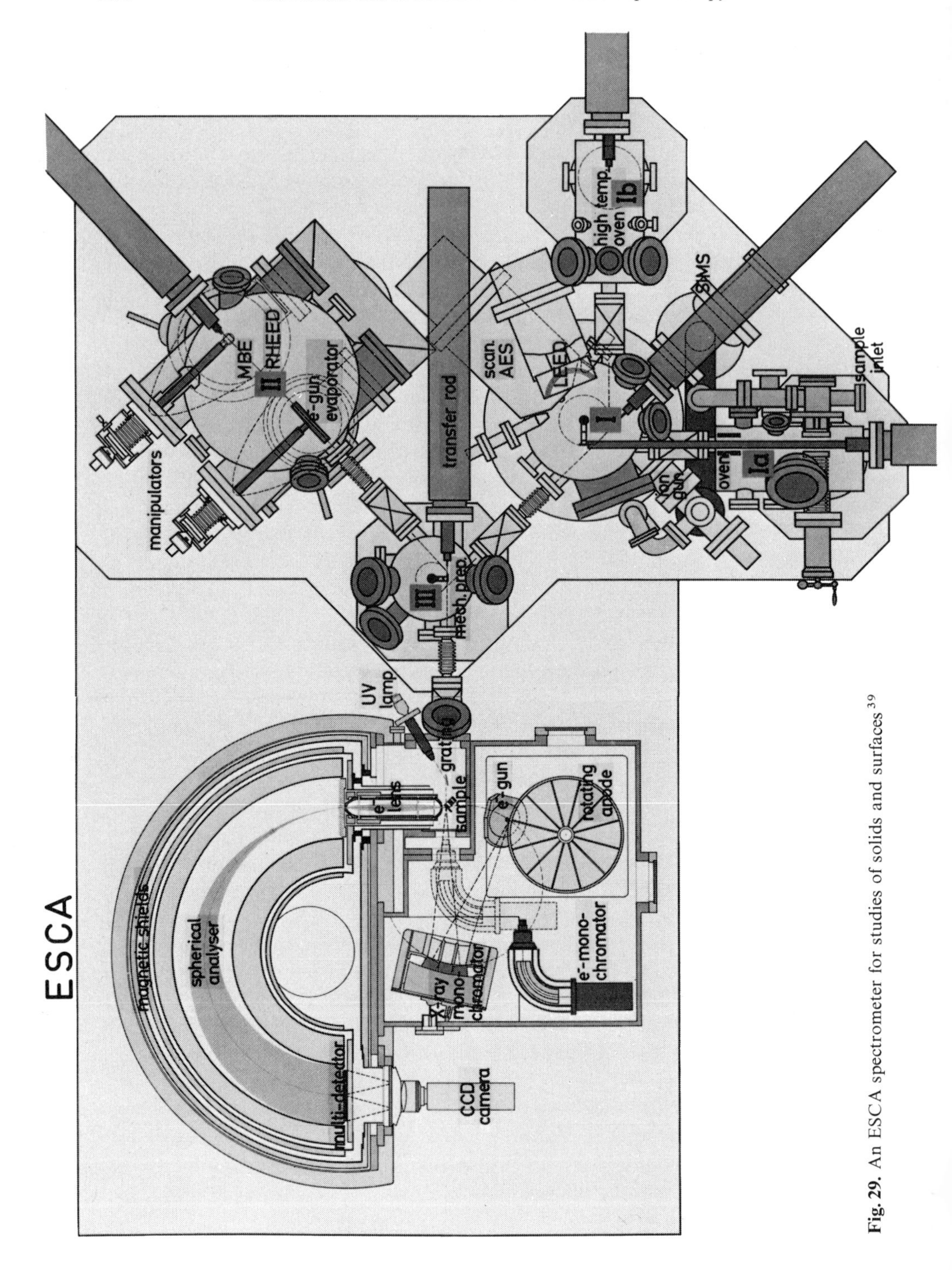

Fig. 29. An ESCA spectrometer for studies of solids and surfaces [39]

ESCA instrument

Monochromator: 25 quartz (100) crystals. uv-source with grating. *e*-gun source and monochrom. *e*-gun source

Quick Introduction chamber Ia

< 2 min for introducing a sample. All metal system. Degassing station $\leqq 500\,°C$. *e*-beam evap. with shutter for calibrants. Catalysis and electrolysis cells can be added

Analysis chamber I with scan AES, SIMS, LEED

Universal manipulator with flip rotation. Sample temp: $-120\,°C$–$900\,°C$. Antivibration manipulator. Gas inlet with leak valve. Chamber isolation valve for adsorption studies

Scanning AES instrument

Cylind. mirror analyser: outer diam = 194 mm, spacial resolution: 0.2 μm with internal *e*-gun. Energy resolution: 0.3%, 0.6%, 1.0%. Auger energy range: 0–3000 eV. Excitation energy/current: $\leqq 10\,keV/10^{-10}$–$5 \cdot 10^{-6}\,A$. Ion gun for in situ depth profiling. Peak multiplexer: 6 elements simultaneously

SIMS instrument

Quadrupole mass analyser: 250 mm rod length. Mass resolution: 10 M max. Mass range: 1–600 amu. Surface sensitivity: $< 10^{-6}$ monolayers. Residual gas sensitivity: $< 10^{-15}$ mbar. Ion gun: differentially pumped. Ion energy: 200–5000 eV. Beam size/density: 0.1–$3\,mm/10^{-10}$–$10^{-3}\,A/cm^2$. Beam deflexion: ± 10 mm with rastering option

LEED analyser

Retarding grid analyser: 4 grid optics. Coaxial *e*-gun: 10–3000 eV, 0.1–4 μA. Spot diam: 0.2–1 mm

High temp. prep. chamber Ib

Heating station: $\leqq 2000\,°C$, *e*-gun bomb. Ion etching, reactive gas cleaning. Watercooled double wall design

Evaporation chamber II with MBE

A. Carrousel manipulator for 7 samples. Station with *e*-beam heating $\leqq 800\,°C$. Shutter, quartz film thickness sensor. Differential pumping of evap. chamber. Resistive evaporation from wire, boat. 5 kW *e*-beam evap. jumping beam. 4 × 2 crucibles
B. MBE manipulator, molybdenum. Azimuthal sample rotation. Regulated temp. $\pm 1\,°C/24\,h$ $\leqq 800\,°C$. Quartz film thickness sensor. Quadrupole mass analyser

RHEED analyser:

Electron energy: $\leqq 10$ keV. Beam size: 0.1 mm. Incident angle: 4°

Mechanical sample prep. chamber III

Single crystal cleaver. Rod fracturer with LN_2 cooling. Micro miller with diamond tip. Scraping device for soft materials

Vacuum systems

Diode ion pumps with internal Ti-sublimation pumps, LN_2 shrouds. uhv $< 1 \cdot 10^{-10}$ mbar. Liquid He cryopumps on ESCA. 35 l filling, $\geqq 4$ weeks duration. Automatic LN_2 distribution system

6. General features of photoelectron spectra

α) Introduction. In photoelectron spectra, there are in principle three sources in the energy distribution of the emitted electrons from which information is normally extracted: line energies, line intensities (which are angular dependent) and line shapes (in particular the widths).

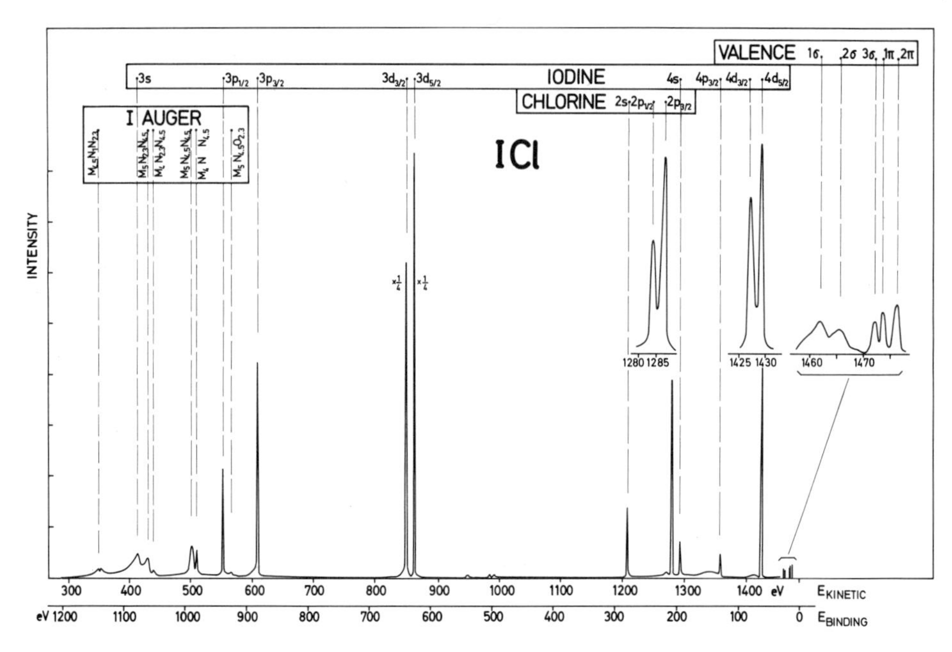

Fig. 30. Electron spectrum of iodine chloride excited with monochromatized Al $K\alpha$-radiation[1]

Figures 30 to 32 illustrate some of the features that we will henceforth discuss. First, in Fig. 30, an x-ray excited scan spectrum of ICl, the relative energy span of the two basic types of levels, i.e. core levels and valence levels (not always mutually exclusive), are clearly visualized. Second, although we will not explicitly discuss such spectra, iodine Auger electron lines are present spanning an energy interval of ~ 300 eV and containing a multitude of transitions due to M_4 and M_5 initial vacancies. Third, apart from the sharp intense line features, additional structure is always present in electron spectra. This aspect is more clearly seen on the expanded energy scale of Fig. 31 (Hg $4f$ spectra excited with Al$K\alpha$). At the low kinetic energy side of each line low intensity lines and sometimes broad continua are present. These features are due to processes inherent to the photoexcitation (intrinsic processes) as well as to inelastic scattering events when the excited electrons traverse the region from the creation site into the spectrometer (extrinsic processes). The intrinsic processes are denoted by "shake-up" and the extrinsic processes by "inelastic scattering" for gaseous Hg. For solid Hg, the distinction between extrinsic and intrinsic processes is less clear-cut, since they both give rise to the same transitions, viz. interband transitions and plasmon excitations. The extrinsic scattering processes will limit the effective sample volume from which "no-energy-loss" electrons can be extracted, i.e. those electrons that contribute to the sharp main lines. This will be further discussed below.

The uv-excited spectrum of Fig. 32 from CO_2 gas illustrates the high

[1] Siegbahn, K. (1974): J. Electron Spectrosc. Relat. Phenom. 5, 3

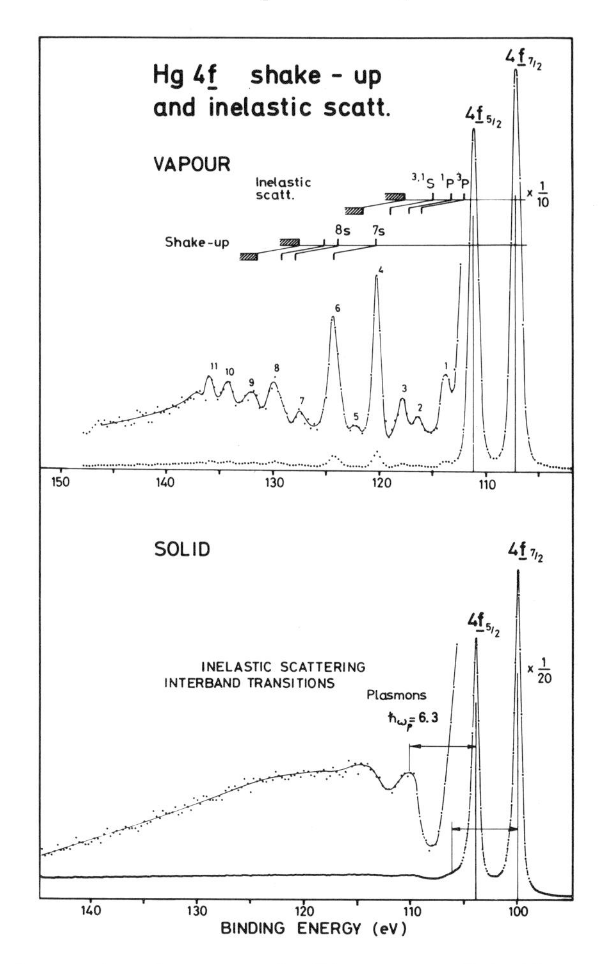

Fig. 31. $4f$ photoelectron spectra of gaseous and solid mercury excited with monochromatized Al $K\alpha$[2]

resolution capability of this excitation mode in the valence level region for free molecules. This is due to the small inherent widths of the uv line sources (~ 1 meV for in this case He *I*) as well as to the low kinetic energies of the ejected photoelectrons. With typical kinetic energies of ~ 10 eV, spectrometer resolution of less than 10 meV is normal in this type of spectra. This means that vibrational and occasionally rotational levels in free molecules can be resolved. The signal to background in uv-excited spectra is generally high due to the high photoelectric cross sections at these photon energies. The background rises steeply close to the photon energy threshold due to scattered low energy electrons.

[2] Svensson, S., Mårtensson, N., Basilier, E., Malmquist, P.-Å., Gelius, U., Siegbahn, K. (1976): J. Electron Spectrosc. Relat. Phenom. 9, 51

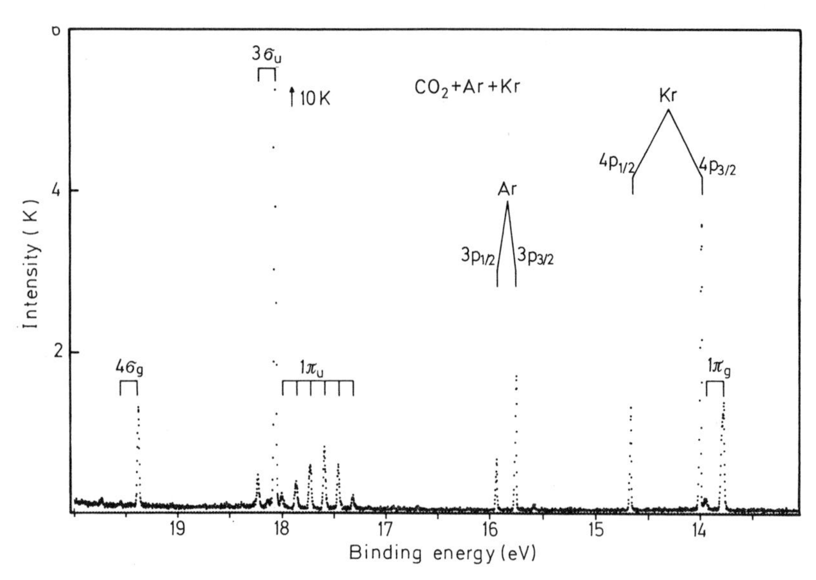

Fig. 32. CO_2 electron spectrum excited with He I radiation. Calibration of the spectrum is performed with a rare gas mixture[3]

β) Calibration of photoelectron spectra. In the photoelectric process, conservation of energy and momentum requires that the kinetic energy of the excited electron is given by:

$$E^{V}_{kin} = h\nu - E^{V}_{B} - E_{r}, \tag{6.1}$$

where $h\nu$ is the photon energy, E^V_B the electron binding energy and E_r the translational energy of the remaining system due to the recoil. We have labelled the kinetic energy and the binding energy by a superscript V to indicate that they refer to the vacuum level potential. Generally E_r may be neglected, due to the small mass of the electron compared to most molecular masses. For very accurate work these corrections may not be entirely negligible, however. (For Ne $2s$ ionization with Al $K\alpha$, $E_r = 0.04$ eV.)

The kinetic energy in (6.1) is that of the electron as it leaves the residual ion. For the different types of sample additional factors involved in the actual measurement of the electron kinetic energy in the spectrometer must be considered. This is illustrated in Fig. 33 for a metallic sample using x-ray excitation. If the sample and spectrometer are in good electrical contact, the Fermi levels will line up and a contact potential will be established. This potential is equal to $(\phi_{sample} - \phi_{sp})$, where ϕ_{sample}, ϕ_{sp} are the sample and spectrometer work functions. Thus, the electron kinetic energy in (6.1) will be changed by this amount as it enters the analyzer:

$$E_{kin} = E^{V}_{kin} + (\phi_{sample} - \phi_{sp}), \tag{6.2}$$

[3] Reineck, I., Maripuu, R., Al-Shamma, S.H., Karlsson, L., Siegbahn, K. (to be published)

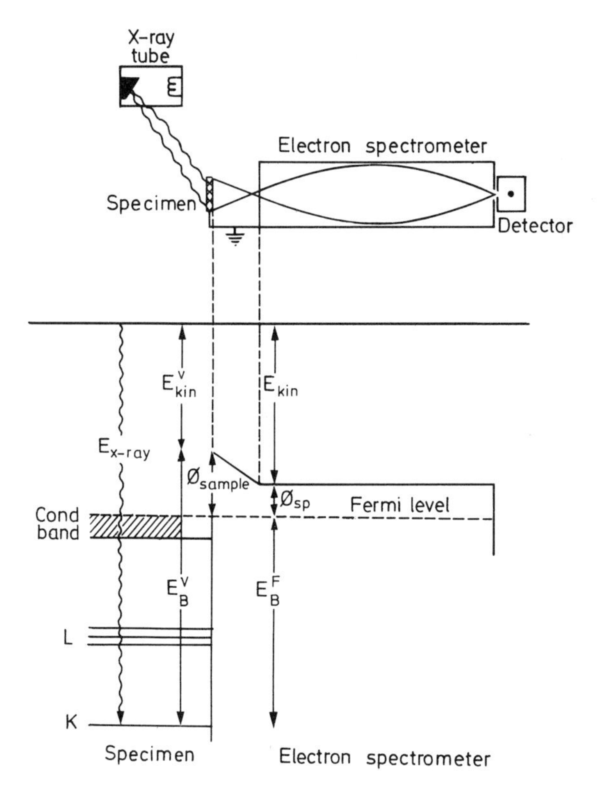

Fig. 33. Schematic representation of the measurement of energies of photoelectrons from a metal sample. The sample is assumed to be in perfect contact with the spectrometer. E_B^V is the binding energy of the K-level with respect to the vacuum level of the sample and E_B^F the binding energy referred to the Fermi level. These two binding energies differ by the work function of the sample, ϕ_{sample}. E_{kin}^V and E_{kin} are the kinetic energies of the electron just outside the sample and inside the spectrometer, respectively

where E_{kin} is the kinetic energy measured in the analyzer. The relation (6.1) must then be restated:

$$E_{kin} = h\nu - E_B^F - \phi_{sp}, \tag{6.3}$$

where the binding energy is now referred to the Fermi level.

For a non-conducting sample, a charging of the sample surface will generally occur as an effect of the photoionization. For such cases a correction must be added to (6.3):

$$E_{kin} = h\nu - E_B^F - \phi_{sp} - C. \tag{6.4}$$

The Fermi level for a pure non-conducting material is defined to lie at the center of the energy gap between the occupied valence band and the unoccupied conduction band. The position within the band gap may be sensitive with respect to doping and band-bending at the surface of the material[4-7]. Using x-ray excitation the Fermi level has been found to lie in the center of the gap in most cases so far examined, however[5-7].

The binding energies of liquid samples are in principle defined analogously to non-conducting solids[8,9] and will thus follow relation (6.4). In view of the way in which liquid samples are generally measured (cf. Sect. 5), with a continuously renewed sample surface, the charging C is expected to be very small (if slits and sample cell walls are kept from condensing liquid).

For gaseous samples, the above discussed Fermi level alignment is inapplicable. A charging of the sample cell walls may occur, however, and this will affect the kinetic energy of the electrons entering the analyzer. The proper form corresponding to (6.1) is then (neglecting E_r):

$$E_{\mathrm{kin}} = h\nu - E_B^V(\mathrm{gas}) - C, \tag{6.5}$$

where C is the charging correction.

Determination of a photoelectron line energy is generally achieved by means of measurement with respect to a calibration substance simultaneously present with the investigated sample. This substance provides photoelectron or Auger electron lines, whose energies have been accurately determined in previous experiments. Depending on the mode of operation of the spectrometer there are two common methods for calibration which we state for a conducting sample without charging:

1. If the spectrometer deflecting field is swept to record the spectrum a spectrometer constant k is obtained from the calibrant kinetic energy and the corresponding field setting, e.g. for a spherical capacitor analyzer (nonrelativistically for simplicity):

$$k = \frac{E_{\mathrm{kin}}(\mathrm{cal})}{V(\mathrm{cal})}, \tag{6.6}$$

where $V(\mathrm{cal})$ is the potential between the spheres needed to focus the calibrant electron line and $E_{\mathrm{kin}}(\mathrm{cal})$ is the kinetic energy of the calibration line inside the spectrometer. The energy of the recorded sample line is thus given by:

$$E_{\mathrm{kin}}(x) = k \cdot V(x) = \frac{E_{\mathrm{kin}}(\mathrm{cal})}{V(\mathrm{cal})} \cdot V(x). \tag{6.7}$$

---

[4] Siegbahn, K., Nordling, C., Fahlman, A., Nordberg, R., Hamrin, K., Hedman, J., Johansson, G., Bergmark, T., Karlsson, S.-E., Lindgren, I., Lindberg, B. (1976): ESCA - Atomic, molecular and solid state structure studied by means of electron spectroscopy, Nova Acta Regiae Soc. Sci. Ups., Ser. IV, Vol. 20

[5] Vesely, C.J., Hengehold, R.L., Langer, D.W. (1972): in: Electron spectroscopy, Shirley, D.A. (ed.), Amsterdam: North-Holland and Phys. Rev. B *5*, 2296

[6] Ley, L., Pollak, R.A., McFeely, F.R., Kowalczyk, S.P., Shirley, D.A. (1974): Phys. Rev. B *9*, 600

[7] Evans, S. (1977): in: Handbook of x-ray and ultraviolet photoelectron spectroscopy, Briggs, D. (ed.), London: Heyden

[8] Gurevich, Yu.Ya., Pleskov, Yu.V., Rotenberg, Z.A. (1980): Photoelectrochemistry, Cons. Bur., New York: Plenum Press

[9] Siegbahn, H., Svensson, S., Lundholm, M. (1981): J. Electron Spectrosc. Relat. Phenom. *24*, 205

2. If the spectrometer operates in the constant deflecting field mode and a retarding voltage is applied and swept to record the spectra, the energy of an unknown line is given by adding the difference of the retarding potential between this line and the calibration line energy.

If charging is present (which in general must be assumed to be the case for all samples except metallic ones) the kinetic energies of all lines are shifted with an amount C. For Method 1 it is necessary to correct the calibration line energy with this amount, otherwise an error will arise in the determination of the spectrometer constant k. Thus, one must use at least two calibration lines in such a case. For Method 2 this is not necessary, since one uses only *differences* in voltage settings. Therefore, and since other potential errors of the same type are avoided in Method 2 this procedure is to be preferred for calibration purposes.

The basic assumption underlying the calibration of a spectrum according to the above is of course that the calibration substance is in perfect electrical contact with the investigated sample. For metals and gases this is no problem, in the former case it is done by evaporation of some reference metal (e.g. gold) onto the surface, in the latter case by mixing with a calibration gas (e.g. a noble gas). One illustration of a gas calibration is shown in Fig. 32: the CO_2 valence electron spectrum calibrated against a mixture of Kr and Ar. For nonconducting solids and liquids the problem is more complicated. The most reliable method[10] has been found to be the use of a hydrocarbon overlayer, where the C1*s* signal can be accurately precalibrated against a metal reference line. Such an overlayer is inevitably present in non-uhv instruments. In uhv instruments a controlled coverage can be performed. Charging of nonconducting solid samples and the procedures used for the calibration of such substances have been discussed by several authors[6,11-24].

Ultimate accuracies for the determination of binding energies can be stated as follows for the different types of sample:

---

[10] Madey, T.E., Wagner, C.D., Joshi, A. (1977): J. Electron Spectrosc. Relat. Phenom. *10*, 359

[11] Johansson, G., Hedman, J., Berndtsson, A., Klasson, M., Nilsson, R. (1973): J. Electron Spectrosc. Relat. Phenom. *2*, 295

[12] Ascarelli, P., Missoni, G. (1974): J. Electron Spectrosc. Relat. Phenom. *5*, 417

[13] Freund, H.J., Gonska, H., Lohneiss, H., Hohlneicher, G. (1977): J. Electron Spectrosc. Relat. Phenom. *12*, 425

[14] Gonska, H., Freund, H.J., Hohlneicher, G. (1977): ibid. *12*, 435

[15] Hnatowitch, D.J., Hudis, J., Perlman, M.L., Ragaini, R.C. (1971): J. Appl. Phys. *42*, 4883

[16a] Dianis, W.P., Lester, J.E. (1973): Anal. Chem. *45*, 1416

[16b] Wagner, C.D. (1980): J. Electron Spectrosc. Relat. Phenom. *18*, 345

[16c] Wamino, Y.U., Ishizuka, T., Yamatera, H. (1981): ibid. *23*, 55

[17] Dickerson, T., Povey, A.F., Sherwood, P.M.A. (1973): J. Electron Spectrosc. Relat. Phenom. *2*, 441

[18] Urch, D.S., Webster, M. (1974): J. Electron Spectrosc. Relat. Phenom. *5*, 791

[19] Ebel, M.E., Ebel, H. (1974): J. Electron Spectrosc. Relat. Phenom. *3*, 169

[20] Clark, D.T. (1973): in: Electron emission spectroscopy, Dekeyser, W., et al. (eds.), Dordrecht: Reidel

[21] Counts, M.E., Jen, J.S.C., Wightman, J.P. (1973): J. Phys. Chem. *77*, 1024

[22] Huchital, D.A., McKeon, R.T. (1972): Appl. Phys. Lett. *20*, 158

[23] Clark, D.T. (1977): Phys. Scr. *16*, 307

[24] Lewis, K.T., Kelly, M.A. (1980): J. Electron Spectrosc. Relat. Phenom. *20*, 105

Table 2. Binding energies and Auger kinetic energies for gases suitable for calibration of electron spectra[a]

Compound	Levels or line	Energy (eV)	Compound	Levels or line	Energy (eV)
Gases			*Gases*		
He	$1s\,^2S_{1/2}$	24.586 (0)	Xe	$5p\,^2P_{1/2}$	13.436 (0)
Ne	$1s\,^2S_{1/2}$	870.21 (5)		$^2P_{3/2}$	12.130 (0)
Ne	$2s\,^2S_{1/2}$	48.47 (0)	N_2	$1s$	409.93 (5)
Ne	$2p\,^2P_{3/2}$	21.564 (0)	N_2	$3\sigma_g\,^2\Sigma_g^+$	15.580 (1)
Ar	$2p\,^2P_{3/2}$	248.629 (10)	H_2O	$1b_1\,^2B_1$	12.615 (1)
	$^2P_{1/2}$	250.777 (10)			(peak maximum)
Ar	$3p\,^2P_{1/2}$	15.937 (0)	CF_4	F $1s$	695.57 (5)
	$^2P_{3/2}$	15.759 (0)	CO_2	O $1s$	541.08 (5)
Kr	$3p\,^2P_{3/2}$	214.55 (15)	CO_2	C $1s$	297.651 (10)
Kr	$3d\,^2D_{5/2}$	93.80 (10)	Ne	$KL_2L_3(^1D_2)$	804.50 (3)
Kr	$4p\,^2P_{1/2}$	14.665 (0)	Ar	$L_3M_{2,3}^2(^1S_0)$	201.10 (2)
	$^2P_{3/2}$	13.999 (0)	Kr	$M_5N_{1,2}N_{2,3}(^1P_1)$	37.87 (3)
			Xe	$N_5O_{2,3}(^1D_2)$	32.28 (4)

[a] From refs.[11,25–27,30,34]. Errors in parentheses refer to last digit.

Table 3. Binding energies referred to the Fermi level for solid suitable for calibration of electron spectra[a]

Compound	Level or line	Energy (eV)	Compound	Level or line	Energy (eV)
Solids[a]			*Solids*[a]		
Cu	$2p_{3/2}$	932.53 (15)	Graphite	C $1s$	284.3 (3)
Cu	$3s$	122.9 (2)	Hydrocarbon contamination	C $1s$	285.0[b] (4)
Ag	$3p_{3/2}$	573.3 (3)			
Ag	$3d_{5/2}$	368.22 (10)			
Pd	$3d_{5/2}$	335.20 (5)	Au	$4f_{7/2}$	84.07 (2)
Pd	Leading edge	0.0 (1)	Pt	$4f_{7/2}$	71.0 (2)

[a] From refs.[11,28,29,31–33]. Errors in parentheses refer to last digit.
[b] Part of the error is due to the variable composition of this layer.

---

[25] Moore, C.E. (1949, 1952, 1958): Atomic energy levels, National Bureau of Standards (US) Circular No. 467. (Conversion factor used: $1.239854 \cdot 10^4$ eV · Å.)

[26] Ågren, H., Nordgren, J., Selander, L., Nordling, C., Siegbahn, K. (1978): J. Electron Spectrosc. Relat. Phenom. *14*, 27

[27] Karlsson, L., Mattsson, L., Jadrny, R. (unpublished results)

[28] Shirley, D.A., Martin, R.L., Kowalczyk, S.P., McFeely, F.R., Ley, L. (1977): Phys. Rev. B *15*, 544

[29] Fuggle, J.C., Mårtensson, N. (1980): J. Electron Spectrosc. Relat. Phenom. *21*, 275

[30] Thomas, T.D., Shaw, R.W., Jr. (1974): J. Electron Spectrosc. Relat. Phenom. *5*, 1081

[31] Nyholm, R., Mårtensson, N. (1980): J. Phys. C *13*, L279

[32] Asami, K. (1976): J. Electron Spectrosc. Relat. Phenom. *9*, 469

[33] Ohtani, S., et al. (1976): Phys. Rev. Lett. *36*, 863

[34] Pettersson, L., Nordgren, J., Selander, L., Nordling, C., Siegbahn, K., Ågren, H.: J. Electron Spectrosc. Relat. Phenom. (to be published)

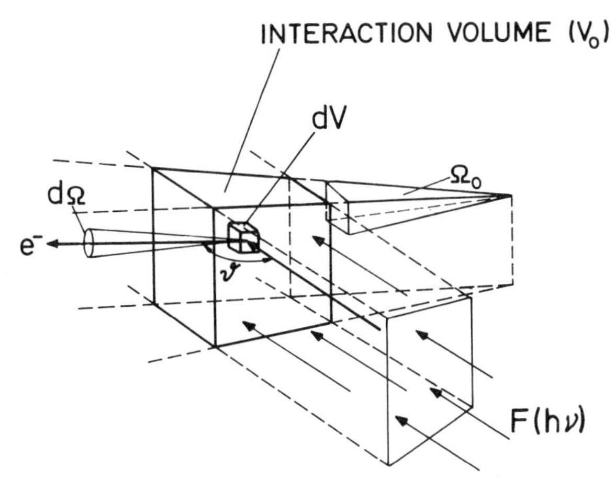

Fig. 34. Geometry of a general photoelectron experiment and relevant quantities for expressing the electron intensities according to (6.8)

Metallic samples: 0.02–0.03 eV (given by uncertainty in determination of peak position ~0.02 eV, uncertainty in the calibration line energy ~0.02 eV (Au4*f* cf. Table 2) and uncertainty due to experimental equipment (can normally be made <0.01 eV)).

Nonconducting condensed samples: 0.2–0.3 eV (given by the same uncertainties as for metallic samples, but with additional uncertainty due to possible lack of proper contact between sample and calibration substance).

Gaseous samples: 0.01–0.02 eV for core lines using x-ray excitation and optical and x-ray data for the calibration line energy. For valence levels excited with uv-radiation linewidths are in the range of 10–20 meV and the 1 meV accuracy is not too difficult to reach.

We give in Tables 2 and 3 energy levels and Auger lines suitable for calibration of photoelectron spectra.

γ) Intensities of photoelectron lines. The number of photoelectrons detected during a given time in an electron spectrum is a function of several factors. Referring to Fig. 34 the intensity (number of electrons/unit time) at a kinetic energy E_{kin} of the electronic ejected into solid angle $d\Omega$ from interaction volume element dV with excitation energy $h\nu$ can be written in the following form:

$$I(E_{\mathrm{kin}}, h\nu)_{\mathrm{F}}\, d\Omega\, dV dE_{\mathrm{kin}}$$
$$= F \cdot \rho(dV) \cdot \frac{d\sigma}{d\Omega}(E_{\mathrm{kin}}, h\nu, \vartheta) \cdot T(E_{\mathrm{kin}}, dV, d\Omega) \cdot D(E_{\mathrm{kin}})\, d\Omega\, dV dE_{\mathrm{kin}}, \quad (6.8)$$

where F is the flux of incident photons (photons/sec · unit area; F assumed constant within the interaction volume), $\rho(dV)$ the number density of the sample in the volume element dV, $\frac{d\sigma}{d\Omega}(E_{\text{kin}}, h\nu, \vartheta)$ the differential cross section for ejection of an electron into solid angle $d\Omega$, $T(E_{\text{kin}}, dV, d\Omega)$ the probability that the electron will be transmitted through the sample from dV into $d\Omega$ and $D(E_{\text{kin}})$ the probability that the electron will be detected in the spectrometer. To get the total detected intensity, integration should be performed over the interaction volume (V_0) and the acceptance solid angle (Ω_0).

Considering the case of a solid sample without surface contamination Eq. (6.8) may be integrated to give[35]

$$I(E_{\text{kin}}, h\nu)\, dE_{\text{kin}} = F \cdot \rho \cdot \frac{d\sigma}{d\Omega} \cdot D(E_{\text{kin}}) \cdot S_0\, \Omega_0\, \lambda(E_{\text{kin}})\, dE_{\text{kin}}, \tag{6.9}$$

where S_0 is the area of the spectrometer slit, $\lambda(E_{\text{kin}})$ the electron inelastic mean free path in the sample and the other quantities were defined above. For a gaseous sample the corresponding equation is:

$$\begin{aligned} I(E_{\text{kin}}, h\nu)\, dE_{\text{kin}} = F \cdot \rho \cdot \frac{d\sigma}{d\Omega} \cdot D(E_{\text{kin}}) \\ \cdot S_0\, \Omega_0 \frac{1}{\sigma_{\text{sc}}(E_{\text{kin}})} \cdot \{1 - \exp[-d \cdot \rho \cdot \sigma_{\text{sc}}(E_{\text{kin}})]\} \\ \times \exp[-d^{\text{sp}} \cdot \rho \cdot \sigma_{\text{sc}}(E_{\text{kin}})]\, dE_{\text{kin}} \end{aligned} \tag{6.10}$$

where ρ is the gas number density, $\sigma_{\text{sc}}(E_{\text{kin}})$ is the molecular inelastic electron scattering cross section at energy $E_{\text{kin}}(\lambda = 1/\sigma_{\text{sc}} \cdot \rho)$, d is the length of the interaction volume in the direction of the spectrometer optical circle and d^{sp} is the distance through the "passive" gas volume between the interaction volume and the spectrometer slit.

The mode of operation of the spectrometer will in itself affect the measured line intensities. Consider two lines (1 and 2) of equal intensities but with different kinetic energies $E_{\text{kin},1}$ and $E_{\text{kin},2}$. For a dispersive analyzer which operates with fixed entrance and exit slits and sweeps the spectrometer field, the energy width accepted is theoretically given by $dE_{\text{kin}}/E_{\text{kin}} = \text{const}$. This means that the measured intensities of lines 1 and 2 are related by:

$$\frac{I_1}{I_2} = \frac{E_{\text{kin},1}}{E_{\text{kin},2}}. \tag{6.11}$$

For an analyzer operating with retardation optics and a fixed analyzer field, on the other hand, the decrease in source brightness results in the opposite behaviour (cf. Sect. 4.4):

$$\frac{I_1}{I_2} = \frac{E_{\text{kin},2}}{E_{\text{kin},1}}. \tag{6.12}$$

---

[35] Fadley, C.S. (1974): J. Electron Spectrosc. Relat. Phenom. 5, 725

It should be emphasized that the above arguments are theoretical and that substantial corrections may be necessary to the above relations[36]. These deviations from theoretical expectations are due to influence of stray fields, the design of the retardation optics etc. In order to obtain reliable true intensity measurements it is therefore adviseable to measure the spectrometer transmission as a function of energy[36].

The magnitude of $\lambda(E_{\text{kin}})$ for solids and its energy dependence is determined by the scattering mechanisms in the material under study. In general, it is expected that the relative importance of these mechanisms will vary from one material to another. It has been found, however, that $\lambda(E)$ for different materials approximately follows the same functional behaviour with energy (the so-called universal curve of inelastic mean free paths)[37–43] (cf. Fig. 35). A minimum in λ generally occurs at an electron energy of ~ 100 eV, corresponding to a λ of ~ 5–10 Å. Below 10 eV λ increases rapidly, and above 100 eV it varies roughly as $(E)^{1/2}$. The mechanisms mainly responsible for inelastic electron scattering in condensed matter are plasmon excitations ($5\,\text{eV} \lesssim \Delta E \lesssim 30\,\text{eV}$), excitation of intra- and interband transitions ($1\,\text{eV} \lesssim \Delta E \lesssim 10\,\text{eV}$ for valence levels) and phonon excitations ($\leqq 1$ eV). At electron energies lower than 10 eV, the rise in mean free path is due to the smaller number of energy-allowed inelastic scattering mechanisms.

For the definitions and measurements of photoelectric cross sections for gaseous species we refer to the articles by Samson and Starace in this volume. For solids, expressions such as (6.9) have been used by a number of workers to compile core electron subshell cross sections[44–50]. Compilations of data have

[36a] Woodruff, P.R., Torop, L., West, J.B. (1977): J. Electron Spectrosc. Relat. Phenom. *12*, 133
[36b] Castle, J.E., West, R.H. (1980): ibid. *19*, 409
[36c] Hall, S.M., Andrade, J.D., Ma, S.M., King, R.N. (1979): ibid. *17*, 181
[36d] Cross, Y.M., Castle, J.E. (1981): ibid. *22*, 53
[36e] Gardner, J., Samson, J.A.R. (1973): ibid. *2*, 267
[37] Powell, C.J. (1974): Surf. Sci. *44*, 29
[38] Lindau, I., Spicer, W.E. (1974): J. Electron Spectrosc. Relat. Phenom. *3*, 409
[39] Seah, M.P., Dench, W.A. (1979): Surface and Interface Anal. *1*, 2
[40] Klasson, M., Berndtsson, A., Hedman, J., Nilsson, R., Nyholm, R., Nordling, C. (1974): J. Electron Spectrosc. Relat. Phenom. *3*, 427
[41] Penn, D.R. (1976): J. Electron Spectrosc. Relat. Phenom. *9*, 29
[42] Lotz, W. (1967): Z. Phys. *206*, 205
[43a] Szajman, J., Jenkin, J.G., Liesegang, J., Leckey, R.C.G. (1978): J. Electron Spectrosc. Relat. Phenom. *14*, 41
[43b] Szajman, J., Leckey, R.C.G. (1981): ibid. *23*, 83
[43c] Szajman, J., Liesegang, J., Jenkin, J.G., Leckey, R.C.G. (1981): ibid. *23*, 97
[44a] Wagner, C.D. (1972): Anal. Chem. *44*, 1050
[44b] Berthou, H., Jorgensen, C.K. (1975): ibid. *47*, 482
[44c] Nefedov, V.I., Sergushin, N.P., Band, I.M., Trzhaskovskaya, M.B. (1973): J. Electron Spectrosc. Relat. Phenom. *2*, 383
[45] Kemeny, P.C., Jenkin, J.G., Liesegang, J., Leckey, R.C.G. (1974): Phys. Rev. B *9*, 5307
[46] Leckey, R.C.G. (1976): Phys. Rev. A *13*, 1043
[47] Brillson, L.J., Ceasar, G.P. (1975): Surf. Sci. *58*, 457
[48] Evans, S., Pritchard, R.G., Thomas, J.M. (1977): J. Phys. C *10*, 3483, (1978): J. Electron Spectrosc. Relat. Phenom. *14*, 341
[49] Wagner, C.D. (1977): Anal. Chem. *49*, 1282
[50] Szajman, J., Jenkin, J.G., Leckey, R.C.G., Liesegang, J. (1980): J. Electron Spectrosc. Relat. Phenom. *19*, 393

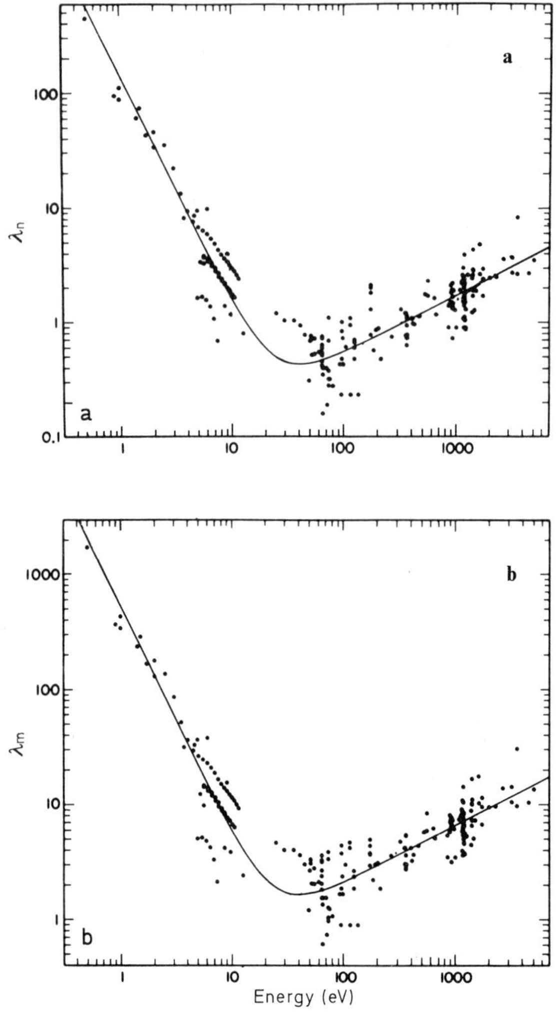

Fig. 35a, b.

been made by EVANS et al.[48] for Mg$K\alpha$ cross sections and SZAJMAN et al.[50] for Al$K\alpha$ cross sections. In these compilations peak areas were used corrected for the inelastic background. Also, corrections were made for spectrometer acceptance, surface contamination and mean free paths. Comparisons with theoretical predictions showed systematic discrepancies[48] of the order of ~20% on the average for cross section ratios, whereas experimental errors were estimated to be no larger than 12%. EVANS et al.[48] thus suggest that the origin of the discrepancies lies rather in the theoretical treatment (Hartree-Slater atomic one-electron cross sections[51]) than in the neglect of experimental circumstances.

[51] Scofield, J.A. (1976): ibid. *8*, 129

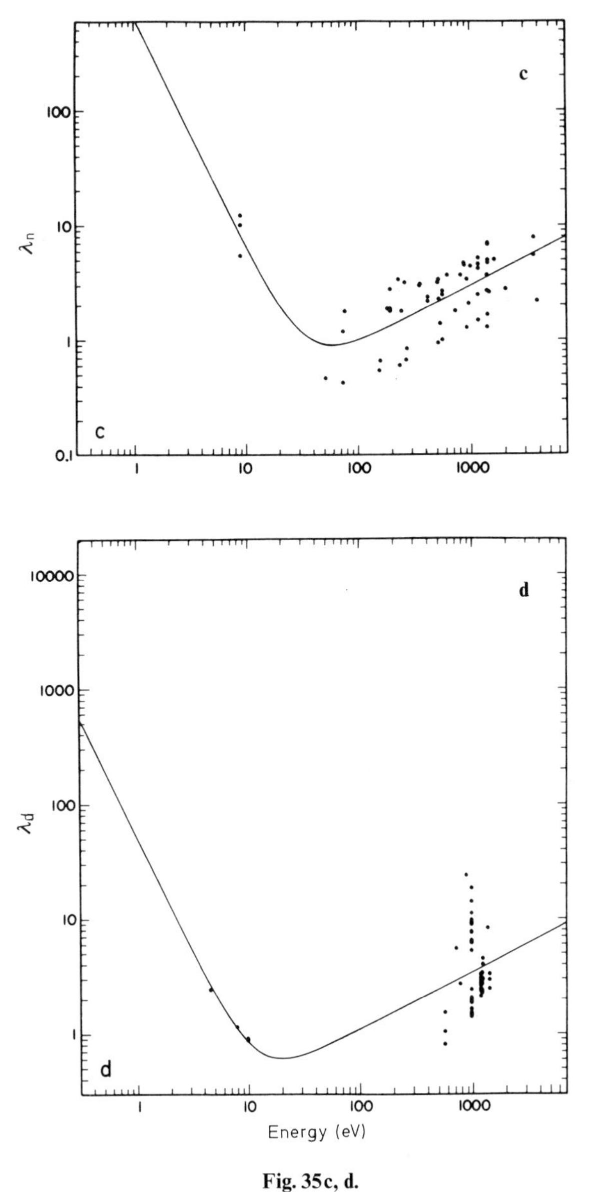

Fig. 35c, d.

Fig. 35a–d. Compilation for elements of mean free path measurements (**a**) in nanometres, (**b**) in monolayers, as a function of energy above the Fermi level. (**c**) Compilation for inorganic compounds of mean free path measurements in nanometres. (**d**) Compilation for organic compounds of mean free path measurements in $mg\,m^{-2}$. The solid curves are fitted to the data by means of the relation:

$$\lambda = \frac{A}{E_{kin}^2} + B \cdot E_{kin}^{1/2},$$

where A and B are constants [39]

This implies that the use of theoretical cross sections in e.g. quantitative analytical estimations by means of core electron line intensities may amount to errors of this order of magnitude, i.e. 20%. In such applications of ESCA, suitable experimental values should then preferably be chosen as references. A similar conclusion is reached by SZAJMAN et al.[50].

The most comprehensive compilations of theoretically calculated atomic partial cross sections are those of SCOFIELD[51], GOLDBERG et al.[52] and BAND et al.[53]. Nonrelativistically, one-particle cross sections are obtained from[54]:

$$\sigma_{nl} = \frac{4\pi\sigma_0 a_0^2}{3} \cdot \frac{N_{nl} \cdot h\nu}{2l+1} [l R_{l-1}^2(E_{\text{kin}}) + (l+1) R_{l+1}^2(E_{\text{kin}})], \tag{6.13}$$

where σ_{nl} is in cm^2, σ_0 is the fine-structure constant, a_0 is the Bohr radius, N_{nl} is the number of electrons in the nl subshell and $R_{l\pm1}(E_{\text{kin}})$ are radial integrals:

$$R_{l\pm1}(E_{\text{kin}}) = \int P_{nl}(r)\, r\, P_{E_{\text{kin}}, l\pm1}(r)\, dr, \tag{6.14}$$

where $P_{nl}(r)/r \equiv R_{nl}(r)$ is the radial part of the initial state orbital and $P_{E_{\text{kin}}, l\pm1}(r)$ that of the continuum orbital. As pointed out above existing experimental data show deviations from one-particle calculations. Further work is needed on both the theoretical and experimental side to reveal the nature of these discrepancies.

δ) Widths of photoelectron lines. The widths and shapes of photoelectron lines depend on several factors. It is useful to divide these into different categories: 1. Experimental factors. 2. The decay width of the final state of the photoionization process (inherent linewidth) and 3. The existence of several final states, which lie close in energy. The last category may consist in several discrete or a continuous band of final states, in which case the photoelectron line is more properly referred to as a band or an envelope.

The most important experimental factors governing the line shapes and widths are:

1. The width and shape of the exciting radiation
2. Thermal or Doppler broadening
3. Inhomogeneous charging of the sample region
4. The imaging properties of the analyzer.

If characteristic, non-monochromatized x-ray emission lines are used for excitation, the first contribution will have the shape of a Lorentzian or a sum of Lorentzians (e.g. Mg$K\alpha_{1,2}$ or Al$K\alpha_{1,2}$). The output of an x-ray monochromator is more complicated and no general analytical shape such as a Gaussian or Lorentzian can be ascribed to the line profile. Actual measurements in-

---

[52] Goldberg, S.M., Fadley, C.S., Kono, S. (1981): J. Electron Spectrosc. Relat. Phenom. *21*, 285
[53] Band, I.M., Kharitonov, Yu.I., Trzhaskovskaya, M.B. (1979): At. Data Nucl. Data Tables *23*, 443
[54] Cooper, J.W. (1962): Phys. Rev. *128*, 681

dicate, however, that the shape of the Al $K\alpha$ line is approximately Gaussian[55]. For a grazing incidence monochromator, used in e.g. synchrotron radiation studies, the output profile has been found to be closely Gaussian[56]. The line shape of the radiation emitted from uv-discharge lamps was discussed in Sect. 3 (cf. also Table 1 for widths of photon sources).

The effects of the analyzer on the line shape and width are dependent on the type of analyzer, photoelectron energy, slit widths and angle-defining apertures. Detailed statements as to the actual imaging properties obtained in a particular analyzer cannot be made without a careful analysis. It is usually assumed, however, that the analyzer contribution is Gaussian with a width proportional to the energy of the transmitted electrons (cf. Sect. 4)[56-60].

The observed photoelectron spectrum may be represented by the convolution integral:

$$I(E)=\int K(E-E')\,S(E')\,dE', \tag{6.15}$$

where $K(E)$ is the instrumental function and $S(E)$ the unbroadened photoelectron spectrum. In $K(E)$ are contained the factors listed above. When the instrumental function is a Gaussian and one photoelectron line of Lorentzian shape is considered, the convolution integral takes the form:

$$V(E)=\int G(E-E')\,L(E')\,dE'. \tag{6.16}$$

This special function (the Voigt function) has been analyzed and put into tabular form[61].

Deconvolution methods, aiming at the inversion of the convolution integral, can be used to artificially enhance the resolution in photoelectron spectra. In such methods, some choice of $K(E)$ must be made. The usual deconvolution methods are Fourier analysis or iterative procedures such as van Citterts deconvolution procedure[62]. In this method the original spectrum is convoluted with the instrumental function, resulting in a further broadening. The first-order deconvoluted solution is then constructed by subtracting from the original spectrum the difference between the broadened data and original data. Repeating this procedure with the first-order solution yields the second-order deconvoluted spectrum etc. Several authors[63-68] have used these meth-

---

[55] Gelius, U., Fellner-Feldegg, H., Wannberg, B., Nilsson, A.G., Basilier, E., Siegbahn, K. (1974): UUIP-855 (Uppsala University Institute of Physics Report)

[56] Fellner-Feldegg, H. (1974): UUIP-857 (Uppsala University Institute of Physics Report)

[57] Siegbahn, K. (1964): in: Alpha-, beta- and gamma-ray spectroscopy, Amsterdam: North-Holland, p. 88

[58] Roy, D., Carrette, J.-D. (1971): J. Appl. Phys. *42*, 3601

[59] Wannberg, B., Sköllermo, A. (1977): J. Electron Spectrosc. Relat. Phenom. *10*, 45

[60] Polaschegg, H.D. (1976): Appl. Phys. *9*, 223

[61] Hummer, O.G. (1965): Mem. R. Astron. Soc. *70*, 1

[62] Cittert, P.H. van (1975): Z. Phys. *69*, 239

[63] Wertheim, G.K. (1975): J. Electron Spectrosc. Relat. Phenom. *6*, 239

[64] Ebel, H., Gurker, N. (1974): J. Electron Spectrosc. Relat. Phenom. *5*, 799

[65] Ebel, H., Gurker, N. (1975): Phys. Lett. *50A*, 449

[66] Carley, A.F., Joyner, R.W. (1978): J. Electron Spectrosc. Relat. Phenom. *13*, 411

[67] Hollinger, G. (1979): Ph.D. thesis (Lyon)

[68a] Beatham, N., Orchard, A.F. (1976): J. Electron Spectrosc. Relat. Phenom. *9*, 129

[68b] Vasquez, R.P., Klein, J.D., Bargon, J.J., Grunthaner, F.J. (1981): ibid. *23*, 63

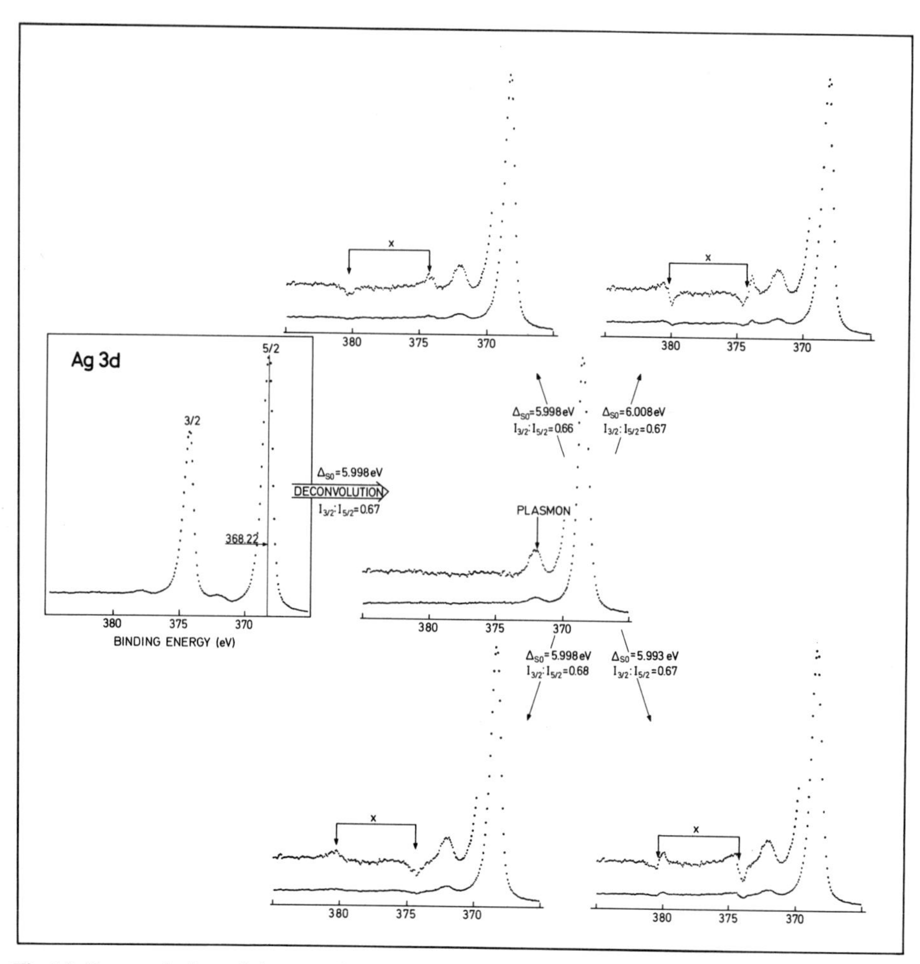

Fig. 36. Deconvolution of the Ag $3d$ spin-orbit doublet using the procedure of ref. [69]. The spin-orbit splitting and intensity ratio are used as input parameters and the doublet separated into subspectra. When these parameters deviate from the optimum values ($\Delta_{so} = 5.998$ eV and $I_{3/2}:I_{5/2} = 0.67$) spurious structures (marked by $\times$) immediately appear in the deconvoluted solutions

ods with various approaches for the instrumental function to deconvolute photoelectron spectra.

A useful deconvolution procedure[69] has been designed especially for separation of a spectrum into subspectra, where the subspectra are assumed to be of equal shape but shifted with respect to each other and differing by a constant factor (e.g. spin-orbit split core photoelectron lines). The procedure results in a subtraction of the low kinetic energy component from the total spectrum. The virtue of this method lies in its extreme sensitivity to the use of

[69a] Nyholm, R., Mårtensson, N. (1980): Chem. Phys. Lett. *74*, 337

[69b] Mårtensson, N., Nyholm, R. (1980): UUIP-1037 (Uppsala University Institute of Physics Report)

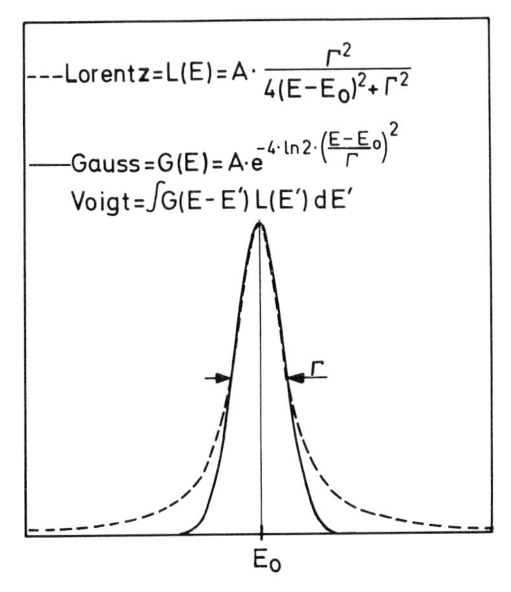

Fig. 37. The Lorentz and Gauss line profiles

incorrect values of spin-orbit splittings and/or spin-orbit intensity ratios. A test case is illustrated in Fig. 36 showing the accuracy attainable for the values of these parameters for the Ag$3d$ spin-orbit doublet.

The inherent or natural linewidth of a photoelectron line corresponding to a single final ionic state is decided by the decay processes of the ionic state. In a first order treatment, the final state broadens by the interaction with the decay channels, electron emission, photon emission, predissociation, into a Lorentzian distribution (cf. Fig. 37):

$$L(E)=A\cdot\frac{\Gamma_i^2}{4(E-E_0)^2+\Gamma_i^2}. \tag{6.17}$$

The natural line profile of the photoelectron line is thus equal to this distribution. The width Γ_i of level i is given by the sum of the widths of all decay processes:

$$\Gamma_i=\sum_x \Gamma_i^x. \tag{6.18}$$

By recording the photoelectron line corresponding to a certain ionic final state, and making appropriate corrections for the experimental factors mentioned above, the inherent level width of that state can be obtained. Values of the level widths may be obtained also from other spectroscopies: Auger electron spectroscopy, x-ray or optical spectroscopy. In these decay spectroscopies, the level widths of both the involved levels will contribute to the linewidth of the observed transition. Thus, for instance for a $K\alpha_1$ x-ray line:

$$\Gamma_{K\alpha_1}=\Gamma_{1s}+\Gamma_{2p_{3/2}}. \tag{6.19}$$

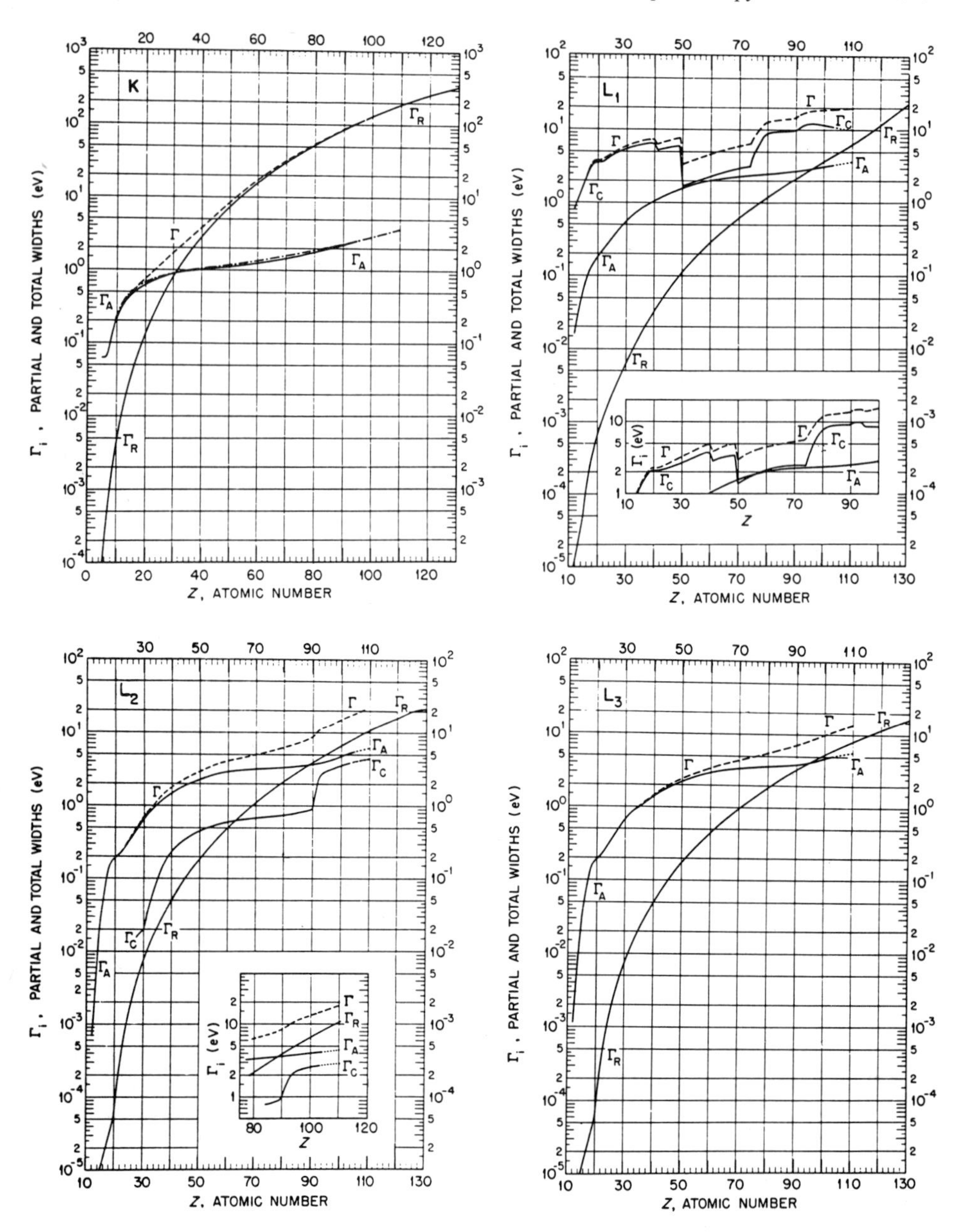

Fig. 38. Total and partial inherent level widths for K- and L-shells of the elements [71]. Γ=total width, Γ_A=Auger width, Γ_C=Coster-Kronig width, Γ_R=radiative width. Inserts in L_1 and L_2 curves are semiempirical widths used in the evaluation (cf. ref. [71])

The widths of the atomic subshells of the elements are most completely covered experimentally for the K- and L-levels. However, a large number of calculations have been made during recent years which cover most elements and levels. Compilations of existing experimental and theoretical results for K-,

L-, M- and N-levels up to $Z=120$ have been made by KESKI-RAHKONEN and KRAUSE[70], KRAUSE[71] and KRAUSE and OLIVER[72]. The K- and L-level data are summarized in Fig. 38. The electron emission rates are usually divided into Auger and Coster-Kronig rates, the latter processes leading to at least one final vacancy with the same main quantum number as the primary vacancy. Coster-Kronig decay is characterized by large rates and rapid changes in the total level widths may occur at the onset or cutoff of such processes. This is the case for the L_1 widths in the region $50 \lesssim Z \lesssim 75$. A sharp increase in the Coster-Kronig rate occurs also for the M_4 shell in the region $60 \lesssim Z \lesssim 80$, which leads to a significant difference in widths between the M_4 and M_5 levels. For K-shells the Auger width completely dominates over the radiative width up to $Z \sim 30$, whereas the reverse is true for heavier elements. For $L_{2,3}$-shells, the dominance of electron emission rates extends up to $Z \sim 90$. The Coster-Kronig rate completely decides the total width for M_1 and $M_{2,3}$ shells above $Z \sim 20$. Electron emission is the dominating decay process also for N-shells.

Since Coster-Kronig transitions usually lead to emitted electrons of low energy only minor changes of the energy levels of an atom can cause such transitions to become energy-allowed or forbidden. The photoelectron line-width associated with the Coster-Kronig primary vacancy may then be appreciably affected. Such effects have been identified when atoms change phase from gas to solid[73] or change chemical state (from a metal to an inorganic compound[74]).

7. Theoretical models for photoelectron spectra

α) Koopmans' theorem and ΔSCF binding energies. The kinetic energies of the electrons measured in photoelectron spectroscopy are related to the binding energies of the system through the photoelectric law, stated in (6.1). The binding energies are given by:

$$E_B = E_{\text{Tot}}^{\text{fi}+} - E_{\text{Tot}}^{\text{in}}, \tag{7.1}$$

where $E_{\text{Tot}}^{\text{fi}+}$ and $E_{\text{Tot}}^{\text{in}}$ are the total energies of the residual ionic state and the initial state, respectively.

Photoionization leads in general to an excited final ionic state. The *main* photoelectron peaks or bands observed in the spectrum are connected to those final electronic states of the ion, which can be described as hole ground states. In a one-electron picture these ionic states are derived from the neutral ground state by removal of a one-particle state from the initial electronic configuration, while keeping the other electrons in their respective states. Satellite peaks, associated to the main photoelectron lines, are thus connected to excitations of the other electrons above each hole ground state.

---

[70] Keski-Rahkonen, O., Krause, M.O. (1974): At. Data Nucl. Data Tables *14*, 139
[71] Krause, M.O. (1979): J. Phys. Chem. Ref. Data *8*, 308
[72] Krause, M.O., Oliver, J.H. (1979): ibid. *8*, 329
[73] Väyrynen, J., Aksela, S., Aksela, H. (1977): Phys. Scr. *16*, 452
[74] Mårtensson, N., Nyholm, R. (1981): Phys. Rev. B *24*, 7121

In the Born-Oppenheimer approximation[1] the excitation energy of any system can be subdivided into electronic and nuclear contributions. The binding energy of the ejected electron can thus be written:

$$E_B = E_B^{ad} + \Delta E_{vib} + \Delta E_{rot}, \tag{7.2}$$

where $\Delta E_{vib} = E_{vib}^{fi+} - E_{vib}^{in}$ and $\Delta E_{rot} = E_{rot}^{fi+} - E_{rot}^{in}$ are the difference in vibrational and rotational excitation energies between final and initial states. E_B^{ad} is termed the adiabatic binding energy and refers to the binding energy in a photoelectric process, where the residual ion is left in its vibrational and rotational ground state. It corresponds thus to a complete nuclear relaxation in the ionic hole state concerned. The binding energy in a hypothetical process in which no such nuclear rearrangement occurs is termed the vertical binding energy, E_B^{vert}.

The basis of most descriptions of the ionic states created in photoionization is the one-electron picture. In its usual form the total wave function of a closed shell system containing N electrons and P nuclei is expressed in terms of a Slater determinant of one-electron orbitals, ψ_i ($\mathscr{A}$ = antisymmetrizer):

$$\Psi^{in} = \frac{1}{\sqrt{N!}} \begin{vmatrix} \psi_1(1) & \dots & \psi_1(N) \\ \vdots & \ddots & \vdots \\ \psi_N(1) & \dots & \psi_N(N) \end{vmatrix} = \mathscr{A} \prod_{j=1}^{N} \psi_j. \tag{7.3}$$

Minimizing the total energy with the above wave function leads to the one-electron Hartree-Fock (HF) equations of the canonical form, e.g.[2-12]:

$$F\psi_i(1) = \varepsilon_i \psi_i(1) \qquad (i = 1, 2, \dots N), \tag{7.4}$$

where F is the one-electron Fock operator and ε_i the ith orbital energy.

A first approximation for the ith ionic state is obtained by removing the ith spin-orbital from the ground state determinant and leaving the other orbitals

---

[1] Born, M., Oppenheimer, R. (1927): Ann. Phys. *84*, 457

[2] Slater, J.C. (1960): Quantum theory of atomic structure, vols. I and II, New York: McGraw-Hill

[3] Slater, J.C. (1963–1975): Quantum theory of molecules and solids, vols. I–IV, New York: McGraw-Hill

[4] Froese Fischer, Ch. (1977): The Hartree-Fock method for atoms, New York: Wiley-Interscience

[5] McWeeny, R., Sutcliffe, B.T. (1969): Methods of molecular quantum mechanics, New York: Academic Press

[6] Cook, D.B. (1974): Ab initio valence calculations in chemistry, London: Butterworths

[7] Hurley, A.C. (1976): Introduction to the electron theory of small molecules, New York: Academic Press

[8] Hurley, A.C. (1976): Electron correlation in small molecules, New York: Academic Press

[9] Salem, L. (1966): Molecular orbital theory of conjugated systems, New York: Benjamin

[10] Murrell, J.N., Harget, A.J. (1972): Semi-empirical self-consistent molecular-orbital theory of molecules, London: Wiley-Interscience

[11] Pople, J.A., Beveridge, D.L. (1970): Approximate molecular orbital theory, New York: McGraw-Hill

[12] Wagnière, G.H. (1976): Lecture Notes in Chemistry, vol. 1, Berthier, G., et al. (eds.), Berlin, Heidelberg, New York: Springer

unchanged (frozen orbitals):

$$\Psi^{\text{fi}+}(i) \approx \mathscr{A} \prod_{j \neq i}^{N} \psi_j. \tag{7.5}$$

For a molecule this ionic state corresponds to vertical ionization, since no change in molecular geometry occurs. Calculating the binding energy as the difference in total energies and using (7.4) leads to:

$$\begin{aligned} E_{\text{B}}^{\text{vert}}(i) &\equiv \langle H(N-1)\rangle - \langle H(N)\rangle = -\langle \psi_i | f | \psi_i \rangle \\ &\quad -\sum_j \left[\langle \psi_i \psi_j | \frac{1}{r_{12}} | \psi_i \psi_j \rangle - \langle \psi_i \psi_j | \frac{1}{r_{12}} | \psi_j \psi_i \rangle \right] \\ &\equiv -\langle \psi_i | F | \psi_i \rangle \equiv -\mathcal{E}_i, \end{aligned} \tag{7.6}$$

where f is the one-electron operator consisting of the kinetic energy and the electron-nuclear attraction potential, and $1/r_{12}$ is the interelectronic repulsion. This result is known as Koopmans' theorem[13]. The simplicity of this concept has rendered it a prime position in the interpretation of photoelectron spectra, e.g. [14–18]. It leads to a one-to-one correspondence between the position of the prominent photoelectron peaks and orbital energies and to simple expressions for the photoelectric cross sections. Due to this correspondence the peaks are most often given their one-electron orbital designations.

The vertical binding energies, given by Koopmans' theorem, are generally higher than those observed experimentally. The discrepancy between these two sets of values increases as the binding energy is increased and amounts to more than 20 eV for Ne 1s ionization. The orbital description of photoionization is not, however, restricted to this level of approximation of the ionic state. The next step of improving the treatment is by allowing the remaining electron orbitals to relax as the photoelectron is removed; a feature which is obviously absent in the 'Koopmans' picture. This implies a reoptimization of the orbitals in the ionic determinant. The vertical binding energies obtained by subtraction of these two total energies are termed ΔSCF binding energies ($E_{\text{B}}(\Delta SCF)$; we drop henceforth the "vert" superscript)[19]. The difference in binding energy with respect to the use of the orbital energy is referred to as the reorganization or relaxation energy, i.e.:

$$E_{\text{Relax}} = -\mathcal{E}_i - E_{\text{B}}(\Delta SCF). \tag{7.7}$$

[13] Koopmans, T. (1934): Physica *1*, 104

[14] Siegbahn, K., et al. (1967): ESCA – Atomic, molecular and solid state structure studied by means of electron spectroscopy, Uppsala: Almqvist and Wiksell

[15] Siegbahn, K., et al. (1969): ESCA applied to free molecules, Amsterdam: North-Holland

[16] Turner, D.W., et al. (1970): Molecular photoelectron spectroscopy, London: Wiley-Interscience

[17] Shirley, D.A. (ed.) (1972): Electron spectroscopy, Amsterdam: North-Holland

[18] Caudano, R., Verbist, J. (eds.) (1974): Electron spectroscopy, Amsterdam: Elsevier

[19] Bagus, P.S. (1965): Phys. Rev. *139*, A619

Table 4. Comparison of vertical binding energies computed by different methods for neon and water (eV)[a]

Hole state	Koopmans' theorem ($-\mathcal{E}_i$)	$E_B(\Delta SCF)$	Transition operator ($-\mathcal{E}_i^T$)	Experiment
Ne1s	891.7	868.6	868.6	870.2
Ne2s	52.5	49.3	49.3	48.5
Ne2p	23.1	19.8	19.9	21.6
$H_2O\,1a_1$	559.5	539.1	539.1	539.9
$H_2O\,2a_1$	37.1	34.5	34.4	32.1
$H_2O\,1b_2$	19.1	17.9	17.8	18.7[b]
$H_2O\,3a_1$	14.9	13.3	13.2	14.8[b]
$H_2O\,1b_1$	13.2	11.2	11.0	12.8[b]

[a] From refs. [20–23].
[b] Corrected for vibrational effects [35].

By including relaxation in the description of the final state, the calculated binding energies agree with the experimental values within a few electron volts. This is seen for the cases of Ne and H_2O in Table 4. Fig. 39 shows the relaxation energies for the *ns* subshells of the elements.

So-called transition methods have been derived for the calculation of ΔSCF binding energies for atoms and molecules. In the Hartree-Fock scheme, GOSCINSKI et al.[23,25] have shown that a good approximation to the ΔSCF binding energy (for orbital *i*) is obtained from:

$$E_B(\Delta SCF) \approx -\mathcal{E}_i^T, \tag{7.8}$$

where $\mathcal{E}_i^T$ is the orbital energy eigenvalue to a Fock operator where half an electron has been removed from orbital *i*. ("The transition operator method".) In Table 4 transition operator results have been included for comparison with the ordinary ΔSCF values. As can be seen, these two sets of values agree very well.

A corresponding transition formalism has been shown by SLATER[26] to be valid for the $X\alpha$-, or Hartree-Slater, SCF method (commonly referred to as

[20] Ågren, H., Nordgren, J., Selander, L., Nordling, C., Siegbahn, K. (1978): J. Electron Spectrosc. Relat. Phenom. *14*, 27
[21] Moore, C.E. (1949, 1952, 1958): Atomic energy levels, National Bureau of Standards (US) Circular No. 467
[22] Mårtensson, N., Malmquist, P.-Å., Svensson, S., Basilier, E., Pireaux, J.J., Gelius, U., Siegbahn, K. (1977): Nov. J. de Chim. *1*, 191
[23a] Goscinski, O., Howat, G., Åberg, T. (1975): J. Phys. (London) B *8*, 11
[23b] Goscinski, O., Hehenberger, M., Roos, B., Siegbahn, P. (1975): Chem. Phys. Lett. *33*, 427
[24a] Huang, K.-N., Aoyagi, M., Chen, M.H., Crasemann, B., Mark, H. (1976): At. Data Nucl. Data Tables *18*, 243
[24b] Desclaux, J.P. (1973): At. Data Nucl. Data Tables *12*, 312
[25] Pickup, B., Goscinski, O. (1973): Mol. Phys. *26*, 1013
[26a] Slater, J.C., Wood, J.H. (1978): Int. J. Quantum Chem. Symp. *4*, 3
[26b] Slater, J.C. (1972): Adv. Quantum Chem. *6*, 1
[26c] Slater, J.C., Johnsson, K.H. (1972): Phys. Rev. B *5*, 844

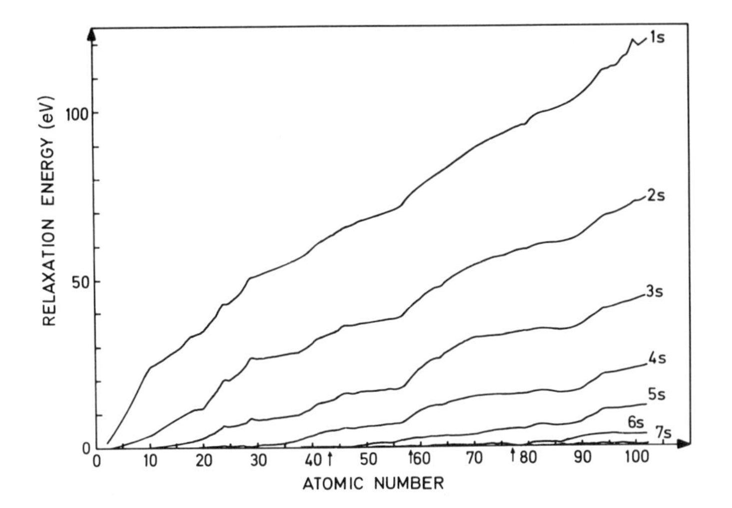

Fig. 39. Relaxation energies for the s-subshells of the elements. The energies are obtained as $-\varepsilon_i(R)-E_B(\Delta SCFR)$, where the orbital energies are taken from DESCLAUX[24b] and the ΔSCE energies from HUANG et al.[24a]. Arrows on the Z-axis indicate elements, for which different configurations were used in the two calculations (interpolated)

"the transition state method"). In this case the $X\alpha$ ΔSCF binding energy is given by:

$$E_B^{X\alpha}(\Delta SCF)=E_{\mathrm{Tot}}^{X\alpha}(n_i=0)-E_{\mathrm{Tot}}^{X\alpha}(n_i=1)\simeq-(\varepsilon_i^{X\alpha})_{n_i=1/2}, \tag{7.9}$$

where $E_{\mathrm{Tot}}^{X\alpha}$ are $X\alpha$ total energies (i.e. calculated using the statistical approximation to the exchange energy) and $(\varepsilon_i^{X\alpha})_{n_i=1/2}$ is the $X\alpha$ orbital energy when half an electron has been removed from orbital i. In this method, problems may arise when calculating core binding energies due to the use of the same value of α for the initial and final states[27]. The procedure has been mostly used for valence states.

β) Correlation in initial and final states. Electron correlation may be treated within the framework of configuration interaction (CI). The basis of this treatment of many-electron effects is the expansion of the total wave function in terms of a set of Slater determinants:

$$\Psi=\sum_{\mu=0}^{\mu_{\max}} C_\mu \Phi_\mu. \tag{7.10}$$

The many-electron functions Φ_μ are formed from sets of one-electron functions, such that the functions with $\mu>0$ constitute excited configurations with respect to a reference configuration Φ_0 which for example may be the determinant constructed from the Hartree-Fock orbitals. The CI-coefficients (C_μ) are determined by minimization of the total energy. In principle, if $\mu_{\max}\to\infty$ and the

[27] Connolly, J.W.D., Siegbahn, H., Gelius, U., Nordling, C. (1973): J. Chem. Phys. *58*, 4265

one-electron basis functions constitute a complete set, the expansion (7.10) approaches the exact wave function of the electron system. For discussion of CI-procedures for atoms and molecules the reader is referred to other sources[2, 3, 5, 8, 28–30].

For photoionization purposes, the effects of correlation are usually divided into two parts; configuration interaction in the initial state (ISCI) and configuration interaction in the final state (FSCI), e.g. [31, 32]. The latter CI may in turn be divided into final ionic state CI (FISCI) and continuum state CI (CSCI). ISCI and FISCI thus refer to discrete bound electron systems (N electrons and $N-1$ electrons, respectively). In CSCI the continuum photoelectron is also included, and it is consequently an N-particle CI.

On the basis of CI-expansions of the total wavefunctions the correlation energy (defined as the difference between the Hartre-Fock total energy and the exact nonrelativistic total energy of the system) can to a good approximation be written as a sum of pair correlation energies, e.g.[8]

$$E_{\text{Correlation}} = \sum_{i<k}^{N} \varepsilon_{ik}, \tag{7.11}$$

where i, k runs over all spin-orbitals of the system.

The pair concept leads to a simple interpretation of correlation energy corrections to binding energies. If it can be assumed that the pair energies of the passive pairs (not containing the ionized electron) are unchanged from initial to final state, the correlation energy correction to the ΔSCF binding energy can be written:

$$\Delta E(i)_{\text{Correlation}} = E^{\text{in}}_{\text{Correlation}} - E^{\text{fi}+}_{\text{Correlation}} = \sum_{k} \varepsilon_{ik}, \tag{7.12}$$

i.e. $\Delta E(i)_{\text{Correlation}}$ is made up of the sum of energies for the pairs which are broken in the ionization. This picture implies that the correction to the binding energy should be expected to be a positive quantity, i.e. the ΔSCF binding energy should always be lower than the true value. From Table 4 it may be seen, however, that this expectation does not generally hold (Ne $2s$ and $H_2O\,2a_1$). The reason is that the correlation energies of the passive pairs change from initial to final state and thus the assumption underlying (7.12) is not valid. Expressed differently, the creation of the extra hole in the final state allows for additional types of configuration interactions in the final state that are not possible in the initial state. For a full consideration of the correlation correction to the binding energy, the initial and final states should therefore be

---

[28] Öksuz, I., Sinanoglu, O. (1969): Phys. Rev. *181*, 42 and 54

[29] Lefebvre, R., Moser, C. (eds.) (1969): Correlation effects in atoms and molecules, London: Wiley-Interscience

[30] Schaefer, H.F. (ed.) (1977): Modern theoretical chemistry, vols. 3 and 4, New York: Plenum

[31] Martin, R.L., Shirley, D.A. (1977): in: Electron spectroscopy, vol. I, Brundle, C.R., Baker, A.D. (eds.), New York: Academic Press, p. 75

[32] Manson, S.T. (1976): in: Adv. Electronics Electron Phys. *41*, 73

treated separately accounting for these additional types of correlations in the ion (pair relaxation[25]). For instance, in the cases of Ne $2s$ and Ar $3s$ ionization, the configuration interactions $nsnp^6 \to ns^2np^4 3d$ are important contributions which make the final ionic state correlation energy larger than that of the initial state[33-35].

γ) Intensities of one-electron and multi-electron transitions. The expression for the differential photoelectric cross section of a randomly oriented atomic or molecular system (a gas or a polycrystalline solid) can be stated in the dipole approximation according to (N electrons)[36-38]

$$\frac{d\sigma}{d\Omega}(i) \propto \frac{1}{h\nu} \cdot \frac{1}{g^{\text{in}}} \sum_{\text{in, fi}} |\boldsymbol{u} \cdot \langle \Psi_N^{\text{fi}}(i, \boldsymbol{k})| \sum_{s=1}^{N} \boldsymbol{p}_s | \Psi_N^{\text{in}} \rangle|^2, \tag{7.13}$$

where g^{in} is the degeneracy of the initial state, $\boldsymbol{u}$ is the photon polarization vector, $\Psi_N^{\text{fi}}(i, \boldsymbol{k})$ is the final state containing a hole in orbital i and a continuum electron with momentum $\hbar \cdot \boldsymbol{k}$, $\sum_{s=1}^{N} \boldsymbol{p}_s$ is the momentum operator and Ψ_N^{in} is the initial state. The expression is assumed averaged over all possible orientations of the system with respect to the photon beam. It is also necessary to sum in (7.13) for all degenerate final states and average over initial degenerate states, such as for instance the magnetic sublevels in the absence of a magnetic field.

It may be shown for any system using only the dipole approximation[38], that the expression for $d\sigma/d\Omega$ may be written in the general form (for linearly polarized radiation):

$$\frac{d\sigma(i)}{d\Omega} = \frac{\sigma(i)}{4\pi} [1 + \beta(E_{\text{kin}}) P_2(\cos \alpha_{ku})], \tag{7.14a}$$

where $P_2(x) = \frac{1}{2}(3x^2 - 1)$ and α_{ku} is the angle between the polarization vector and the electron directions (cf. Fig. 40). $\beta(E_{\text{kin}})$ is the asymmetry parameter and may take values between -1 and $+2$ (to insure non-negativity of the differential cross section). For non-polarized radiation the corresponding dipole approximation expression is:

$$\frac{d\sigma(i)}{d\Omega} = \frac{\sigma(i)}{4\pi} \left[1 - \frac{1}{2} \beta(E_{\text{kin}}) P_2(\cos \theta)\right], \tag{7.14b}$$

where θ is now the angle between the incoming photon beam and the direction of the outgoing electron.

Table 5 gives β-values for the rare gases obtained with discrete UV line sources.

[33] Hedin, L., Johansson, G. (1969): J. Phys. (London) B *2*, 1336
[34] Moser, C.M., Nesbet, R.K., Verhaegen, G. (1971): Chem. Phys. Lett. *12*, 230
[35] Meyer, W. (1971): Int. J. Quant. Chem. *5*, 341
[36] Messiah, A. (1962): Quantum mechanics, Amsterdam: North-Holland, vol. II, p. 722
[37] Breit, G., Bethe, H.A. (1954): Phys. Rev. *93*, 888
[38] Yang, C.N. (1948): Phys. Rev. *74*, 764

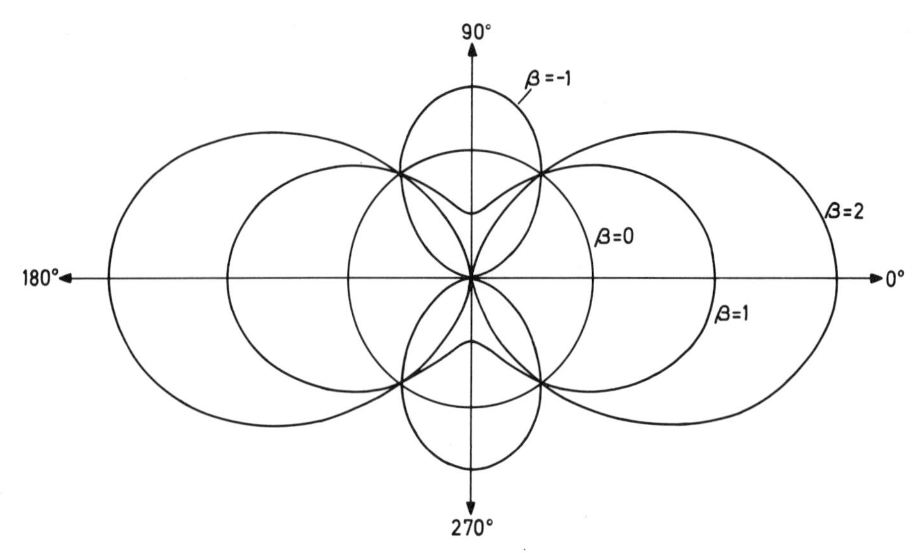

Fig. 40. Angular distributions of photoelectrons excited by plane polarized photons for values of the asymmetry parameter $\beta = -1$, 0, $+1$ and $+2$. The polarization vector is in direction $\alpha_{ku} = 0°$

The initial and final states in the photoionization process can be described at several levels of approximation. If the (closed shell) initial state is represented by a Hartree-Fock determinant, the residual ionic state can be taken as the Koopmans' frozen state (i.e. Eq. (7.5)), by removing the relevant one-electron orbital from the determinant. The total N-electron initial and final states are thus given by:

$$\Psi_N^{\text{in}} = \mathscr{A}\,\psi_i\,\Psi_{N-1}^{\text{KT}}(i), \qquad \Psi_N^{\text{fi}} = \mathscr{A}\,\psi(\boldsymbol{k})\,\Psi_{N-1}^{\text{KT}}(i), \tag{7.15}$$

where $\mathscr{A}$ is the antisymmetrizer, ψ_i the spin-orbital removed from the determinant, $\psi(\boldsymbol{k})$ the continuum orbital describing the outgoing photoelectron and $\Psi_{N-1}^{\text{KT}}(i)$ the frozen wave function. Only the electron making the photoelectric transition changes its wave function in this representation, the wave function of the other electrons remaining unchanged. In the N-particle transition matrix element the contribution from the passive electrons will then integrate out to unity and the differential cross section expression (7.13) reduces to:

$$\frac{d\sigma(i)}{d\Omega} \propto \frac{1}{h\nu}\,|\boldsymbol{u}\cdot\langle\psi_i|\boldsymbol{p}|\psi(\boldsymbol{k})\rangle|^2. \tag{7.16}$$

For open-shell atoms and molecules, additional considerations are required to deduce intensities for each final ionic state. In these cases the ground state does not in general transform according to the totally symmetric representation. It may, for instance, be a state arising from Russel-Saunders coupling of the configuration. Ionization of such a ground state generates in general several ionic state terms of a particular configuration. Rules for predicting the

Table 5. Values of the asymmetry parameter β for the rare gases obtained in studies using line sources for excitation of spectra

Atom	Final state of the ion	Ionizing radiation		Ejected electron		β-values			
						unpolarised radiation		polarised radiation	
		λ (Å)	E (eV)	Kinetic energy (eV)	Binding energy (eV)	NIEHAUS and RUF[41]	DEHMER et al.[40]	KARLSSON et al.[39]	HANCOCK and SAMSON[42]
He	$^2S_{1/2}$	462	26.86	2.28	24.58		1.96 ± 0.14		
He	$^2S_{1/2}$	304	40.81	16.23	24.58		1.97 ± 0.04		
Ne	$^2P_{1/2,3/2}$	462	26.86	5.25	21.61		0.12 ± 0.05		
Ne	$^2P_{1/2,3/2}$	304	40.81	19.20	21.61		0.81 ± 0.05		
Ne	$^2P_{1/2,3/2}$	256	48.37	26.76	21.61		1.00 ± 0.06		
Ar	$^2P_{1/2}$	744	16.67	0.73	15.94	0.25 ± 0.05			
Ar	$^2P_{1/2}$	736	16.85	0.91	15.94	0.26 ± 0.03	0.08 ± 0.05		
Ar	$^2P_{3/2}$	744	16.67	0.91	15.76				
Ar	$^2P_{3/2}$	736	16.85	1.09	15.76	0.31 ± 0.03			
Ar	$^2P_{1/2}$	584	21.22	5.28	15.94	0.95 ± 0.02	0.89 ± 0.04		0.95 ± 0.02
Ar	$^2P_{3/2}$	584	21.22	5.46	15.76	0.95 ± 0.02	0.89 ± 0.04		0.95 ± 0.02
Ar	$^2P_{1/2,3/2}$	462	26.86	11.01	15.85		1.35 ± 0.06		
Ar	$^2P_{1/2,3/2}$	304	40.81	24.96	15.85		1.77 ± 0.04		
Kr	$^2P_{1/2}$	744	16.67	2.00	14.67	0.72 ± 0.04	0.93 ± 0.05		
Kr	$^2P_{1/2}$	736	16.85	2.18	14.67	0.79 ± 0.03			
Kr	$^2P_{3/2}$	744	16.67	2.67	14.00	0.83 ± 0.04	1.05 ± 0.05		
Kr	$^2P_{3/2}$	736	16.85	2.85	14.00	0.91 ± 0.03			
Kr	$^2P_{1/2}$	584	21.22	6.55	14.67	1.37 ± 0.03	1.25 ± 0.05	1.23 ± 0.05	
Kr	$^2P_{3/2}$	584	21.22	7.22	14.00	1.37 ± 0.02	1.29 ± 0.05	1.30 ± 0.05	
Kr	$^2P_{1/2}$	462	26.86	12.19	14.67		1.50 ± 0.06		
Kr	$^2P_{3/2}$	462	26.86	12.86	14.00		1.60 ± 0.05		
Kr	$^2P_{1/2}$	304	40.81	26.14	14.67		1.93 ± 0.06		
Kr	$^2P_{3/2}$	304	40.81	26.81	14.00		1.90 ± 0.07		
Xe	$^2P_{1/2}$	744	16.67	3.23	13.44	1.20 ± 0.08	1.23 ± 0.04		
Xe	$^2P_{1/2}$	736	16.85	3.41	13.44	1.29 ± 0.04			
Xe	$^2P_{3/2}$	744	16.67	4.55	12.12	1.37 ± 0.06	1.39 ± 0.04		
Xe	$^2P_{3/2}$	736	16.85	4.73	12.12	1.45 ± 0.03			
Xe	$^2P_{1/2}$	584	21.22	7.78	13.44	1.64 ± 0.06	1.63 ± 0.05		1.64 ± 0.04
Xe	$^2P_{3/2}$	584	21.22	9.10	12.12	1.71 ± 0.02	1.77 ± 0.05		1.78 ± 0.04
Xe	$^2P_{1/2}$	462	26.86	13.42	13.44		1.86 ± 0.05		
Xe	$^2P_{3/2}$	462	26.86	14.74	12.12		1.98 ± 0.04		
Xe	$^2P_{1/2}$	304	40.81	27.37	13.44		1.58 ± 0.03		
Xe	$^2P_{3/2}$	304	40.81	28.69	12.12		1.28 ± 0.03		

(Footnotes 39–42 see p. 292)

ionic states formed and the intensities have been derived[43-45], which are briefly outlined in Sect. 10β along with a table for a large number of atomic configurations. The formula for the one-particle atomic subshell partial cross section was given in (6.13) (i.e. the sum over all possible final state terms).

The one-electron description above is only a first level of approximation to the ionization process. In particular, such processes are excluded, where the remaining electrons are simultaneously excited as the photoelectron leaves the ion. Processes of this type give rise to satellite lines in the photoelectron spectra and contribute substantially to the photoionization cross sections. For Ne 1s photoionization by Al$K\alpha$, for instance, more than 25% of the events are due to satellite transitions[46]. Such processes are commonly referred to as shake-up (discrete excitation) or shake-off (excitation into continuum states). In order to account for these, the total N-electron wave functions must be brought into consideration. The first step in such an improvement is to allow for reorganization in the final ionic state. The initial and final N-electron states can then be written:

$$\Psi_N^{\text{in}} = \mathscr{A}\,\psi_i\,\Psi_{N-1}^{\text{FR}}(i) \tag{7.17}$$

$$\Psi_N^{\text{fi}}(ij, \boldsymbol{k}) = \mathscr{A}\,\psi(\boldsymbol{k})\,\Psi_{N-1}^{\text{fi}+}(ij). \tag{7.18}$$

Here, $\Psi_{N-1}^{\text{FR}}(i)$ designates a $(N-1)$-electron frozen function (obtained from the N-particle Hamiltonian) where the one particle state ψ_i has been removed. Ψ_N^{in} is not necessarily equal to the Hartree-Fock determinant, but can consist of a configuration expansion. $\Psi_{N-1}^{\text{fi}+}(ij)$ designates the final ionic state, where the ith orbital has been removed and a reoptimization has been performed over the $(N-1)$-particle Hamiltonian. The index j refers to additional excitation above the ith hole state ($j=0$). This function may also be in the form of a configuration expansion.

The differential cross section for photoionization from the ground state to the ij final ionic state is then given by:

$$\frac{d\sigma(ij)}{d\Omega} \propto \frac{1}{h\nu} |\boldsymbol{u}\cdot\langle \mathscr{A}\,\psi_i\,\Psi_{N-1}^{\text{FR}}(i)|\sum_s \boldsymbol{p}_s|\mathscr{A}\,\psi(\boldsymbol{k})\,\Psi_{N-1}^{\text{fi}+}(ij)\rangle|^2. \tag{7.19}$$

To evaluate the many-electron matrix element for high photon electron energies, the sudden approximation can be invoked[36]. It is then assumed, that the photoionization process occurs in such a fashion that the photoelectron is

---

[39a] Karlsson, L., Mattsson, L., Jadrny, R., Siegbahn, K., Thimm, K. (1976): Phys. Lett. A *58*, 381

[39b] Mattson, L., Karlsson, L., Jadrny, R., Siegbahn, K. (1977): Phys. Scr. *16*, 221

[40] Dehmer, J.L., Chupka, W.A., Berkowitz, J., Jivery, W.T. (1975): Phys. Rev. A *12*, 1966

[41] Niehaus, A., Ruf, M.W. (1972): Z. Phys. *252*, 84

[42] Hancock, W.H., Samson, J.A.R. (1976): J. Electron Spectrosc. Relat. Phenom. *9*, 211

[43] Cox, P.A., Orchard, A.F. (1970): Chem. Phys. Lett. *7*, 273

[44] Cox, P.A., Evans, S., Orchard, A.F. (1972): Chem. Phys. Lett. *13*, 386

[45] Cox, P.A. (1975): in: Structure and Bonding, vol. 24, Dunits, J.D., et al. (eds.), Berlin, Heidelberg, New York: Springer

[46] Gelius, U. (1974): J. Electron Spectrosc. Relat. Phenom. *5*, 985

instantaneously removed beyond the effective interaction with the rest of the system. The differential cross section thus reduces to[47-54]:

$$\frac{d\sigma(ij)}{d\Omega} \propto \frac{1}{h\nu} |\boldsymbol{u} \cdot \langle \psi_i | \boldsymbol{p} | \psi(\boldsymbol{k}) \rangle|^2 \cdot |\langle \Psi^{\mathrm{FR}}_{N-1}(i) | \Psi^{\mathrm{fi}+}_{N-1}(ij) \rangle|^2. \tag{7.20}$$

For the cross section to be non-vanishing in this approximation, it is required that the 'frozen state', $\Psi^{\mathrm{FR}}_{N-1}(i)$, and the final ionic state, $\Psi^{\mathrm{fi}+}_{N-1}(ij)$, should have the same symmetry. The selection rule implied by (7.20) is therefore referred to as monopole.

The above description of multielectron transitions at high energies leading to (7.20) is applicable in particular when ψ_i is a core state or Ψ^{in} consists of a single configuration, e.g. [49]. Initial state configuration interaction may lead to the formation of final ionic states which have different symmetries. Also, at lower photoelectron energies when the sudden approximation becomes invalid, CSCI may result in non-vanishing intensity for transitions which are negligible at higher energies. We will discuss some atomic cases below.

Two interesting sum rules can be derived on the basis of the sudden approximation[55-57]

(i) The summed cross sections for all ionization events leading to a hole in state ψ_i is equal to the one-particle cross section, i.e. (cf. (7.16, 7.20)):

$$\sum_j \frac{d\sigma(ij)}{d\Omega} = \frac{d\sigma(i)}{d\Omega} \propto \frac{1}{h\nu} |\boldsymbol{u} \cdot \langle \psi_i | \boldsymbol{p} | \psi(\boldsymbol{k}) \rangle|^2. \tag{7.21}$$

(ii) The first moment of the satellite spectrum with respect to the main line ($j=0$) is equal to the relaxation energy (in the HF-approximation) i.e.:

$$\begin{aligned} E_{\mathrm{Relax}} &= -\mathcal{E}_i - E_{\mathrm{B}}(i0) \\ &= \sum_j \left[\frac{d\sigma(ij)}{d\Omega} \bigg/ \frac{d\sigma(i)}{d\Omega} \right] [E_{\mathrm{B}}(ij) - E_{\mathrm{B}}(i0)]. \end{aligned} \tag{7.22}$$

The existence of shake-up satellites is a manifestation of multi-electron transitions. The conceptual division of the intensities of these transitions into different contributions is not unambigous and depends on the choice of repre-

---

[47] Åberg, T. (1967): Phys. Rev. *156*, 35, (1969): Ann. Acad. Sci. Fenn. A *VI*, 308
[48] Meldner, H.W., Perez, J.D. (1971): Phys. Rev. A *4*, 1388
[49] Martin, R.L., Shirley, D.A. (1977): in: Electron spectroscopy, vol. I, Brundle, C.R., Baker, A.D. (eds.), New York: Academic Press, p. 75
[50] Gadzuk, J.W., Sunjic, M. (1975): Phys. Rev. B *12*, 524
[51] Martin, R.L., Shirley, D.A. (1976): Phys. Rev. A *13*, 1475
[52] Martin, R.L., Shirley, D.A. (1976): J. Chem. Phys. *64*, 3685
[53] Martin, R.L., Mills, B.E., Shirley, D.A. (1976): J. Chem. Phys. *64*, 3690
[54] Williams, R.S. (unpublished result, private communication in Ref. [49])
[55] Fadley, C.S. (1974): J. Electron Spectrosc. Relat. Phenom. *5*, 895
[56] Manne, R., Åberg, T. (1970): Chem. Phys. Lett. *7*, 282
[57] Lundqvist, B.I. (1969): Phys. Kondens. Mater. *9*, 236

sentation of the participating states. It is not always necessary to include CI in the initial and final states in order to account for the occurrence of the satellite transitions. Their presence may be inferred already at the ΔSCF level of approximation, i.e. independent particle initial and final states. This is apparent from the sudden approximation expression for the cross section, Eq. (7.20). For a more quantitative account, however, it is generally a necessity to include CI-contributions. Depending on the way many-electron wave functions are normally calculated, it has become customary in the interpretations of satellite spectra to distinguish between relaxation, initial state CI (ISCI) and final state CI (FSCI) contributions to the intensities. This is in line with the corresponding division of one-electron and many-electron effects on the binding energies.

Above we have outlined the effects of correlation on photoelectron spectra in CI-language. For detailed descriptions of the Greens function and Many-Body Perturbation Theory (MBPT) methods to treat correlation the reader is referred to refs. [58-63] and [64-68], respectively. These methods have been extensively used to calculate for example inner valence shell photoelectron spectra of atoms and molecules which will be further discussed in Sects. 8, 9 and 14. In atomic photoionization, the RPA (Random Phase Approximation) and RPAE (Random Phase Approximation with Exchange) have been extensively employed to calculate the photon energy dependence of the cross sections and angular distributions with correlation contributions included. We refer to AMUSIA[63] and the article by STARACE in this volume.

δ) Vibrational excitations, Franck-Condon principle. The reorganization which occurs as a consequence of ionization involves both electrons and nuclei. In order to discuss the effects on the nuclear motion, the Born-Oppenheimer approximation is assumed and the total molecular wave functions of initial and final states can be written:

$$\begin{aligned} \Psi_{\mathrm{mol}}^{\mathrm{in}}(\boldsymbol{r}_l, \boldsymbol{R}_n) &= \Psi_N^{\mathrm{in}}\, \chi_{\mathrm{vib}}^{v_{\mathrm{in}}}(\boldsymbol{R}_n) \\ \Psi_{\mathrm{mol}}^{\mathrm{fi}}(\boldsymbol{r}_l, \boldsymbol{R}_n) &= \Psi_N^{\mathrm{fi}}(i, \boldsymbol{k})\, \chi_{\mathrm{vib}}^{v_{\mathrm{fi}}}(\boldsymbol{R}_n), \end{aligned} \tag{7.23}$$

---

[58] Cederbaum, L.S., Domcke, W. (1977): Adv. Chem. Phys. *36*, 205

[59] Niessen, W. von, Cederbaum, L.S., Domcke, W. (1978): in: Excited states in quantum chemistry, Nicolaides, C.A., Beck, D.R. (eds.), Dordrecht: Reidel, p. 183

[60] Linderberg, J., Öhrn, Y. (1973): Propagators in quantum chemistry, New York: Academic Press

[61] Wendin, G. (1976): in: Photoionization and other probes of many-electron interactions, Wuilleumier, F.J. (ed.), New York: Plenum Press, p. 61

[62] Csanak, Gy., Taylor, H.S., Yaris, R. (1971): Adv. At. Mol. Phys. *7*, 287

[63] Amusia, M.Ya. (1980): Appl. Optics *19*, 4042

[64] Brandow, B.H. (1967): Rev. Mod. Phys. *39*, 771, (1977): Adv. Quantum Chem. *10*, 187

[65] Kelly, H.P. (1976): in: Photoionization and other probes of many-electron interactions, Wuilleumier, F.J. (ed.), New York: Plenum Press, p. 83

[66] Robb, M.A., Hegarty, D., Prime, S. (1978): in: Excited states in quantum chemistry, Nicolaides, C.A., Beck, D.R. (eds.), Dordrecht: Reidel, p. 297

[67a] Chong, D.P., Herring, F.G., McWilliams, D. (1975): J. Electron Spectrosc. Relat. Phenom. *7*, 445

[67b] Chong, D.P., Langhoff, S.R. (1982): Chem. Phys. *67*, 153

[68] Kelly, H.P. (1975): Phys. Rev. A *11*, 556

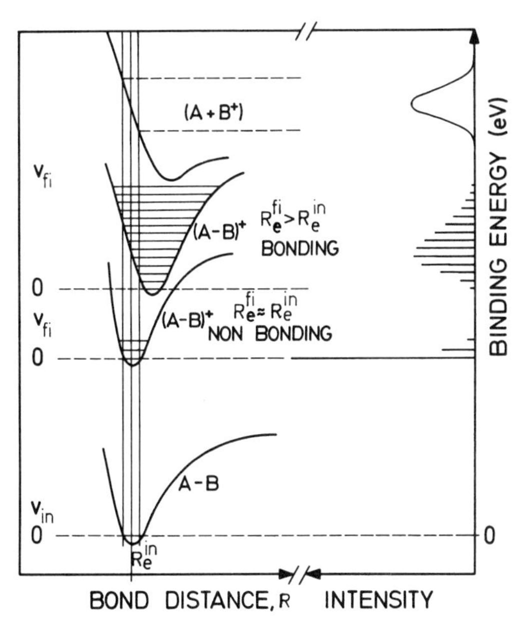

Fig. 41. Schematic illustration of vibrational excitation in photoelectron spectra. The right-hand part of the figure indicates the vibrational progressions expected for the various cases

where $\boldsymbol{r}_l$, $\boldsymbol{R}_n$ represents the set of electron and nuclear coordinates and $\chi_{\text{vib}}^{v_{\text{in}}}$, $\chi_{\text{vib}}^{v_{\text{fi}}}$ are the vibrational eigenfunctions of initial and final states, with quantum numbers v_{in} and v_{fi}, respectively. The electronic wave functions Ψ_N^{in} and Ψ_N^{fi} depend parametrically on the nuclear coordinates and were defined in (7.13).

The cross section of the transition is thus given by, e.g.[69]

$$\left[\frac{d\sigma}{d\Omega}(i)\right]_{\text{mol}}^{v_{\text{in}}\rightarrow v_{\text{fi}}} = \left[\frac{d\sigma}{d\Omega}(i)\right]_{\text{el}} \cdot |\langle\chi_{\text{vib}}^{v_{\text{in}}}|\chi_{\text{vib}}^{v_{\text{fi}}}\rangle|^2. \tag{7.24}$$

$\left[\frac{d\sigma}{d\Omega}(i)\right]_{\text{el}}$ is the cross section for the electronic transition (expressed e.g. through (7.16) taken e.g. at the equilibrium nuclear configuration of the ground state). The quantities $|\langle\chi_{\text{vib}}^{v_{\text{in}}}|\chi_{\text{vib}}^{v_{\text{fi}}}\rangle|^2$ are commonly referred to as the Franck-Condon factors. The *Franck-Condon principle* states that the intensities of the vibrational lines should be directly proportional to the Franck-Condon factors[70] (cf. Fig. 41).

The vertical binding energy can be expressed in terms of the Franck-Condon factors. In the harmonic approximation and under the assumption of equal force constants in the initial and final states, the vertical binding energy is given by[71]:

$$E_{\text{B}}^{\text{vert}} = \sum_{v_{\text{fi}}} |\langle\chi_{\text{vib}}^{v_{\text{in}}=0}|\chi_{\text{vib}}^{v_{\text{fi}}}\rangle|^2\, E_{\text{B}}^{v_{\text{fi}}}. \tag{7.25}$$

[69] Siegbahn, H., Karlsson, L. Electron Spectroscopy for Atoms, Molecules and Condensed Matter, Amsterdam: North-Holland (to be published)

[70] Condon, E.U. (1928): Phys. Rev. *32*, 858

[71] Schwartz, W.H.E. (1975): J. Electron Spectrosc. Relat. Phenom. *6*, 377

Assuming the validity of the Franck-Condon principle, the right-hand side is expressed in terms of the vibrational intensities, $I^{v_{fi}} \propto |\langle \chi_{vib}^{v_{in}=0} | \chi_{vib}^{v_{fi}} \rangle|^2$

$$E_B^{vert} = \sum_{v_{fi}} \frac{I^{v_{fi}}}{I} E_B^{v_{fi}}, \tag{7.26}$$

where $I = \sum_{v_{fi}} I^{v_{fi}}$. This expression is analogous to the relation (7.22) for the spectrum of *electronic* satellite transitions. Within the assumptions mentioned above, the vertical binding energy is thus obtained from the observed vibrational intensities by forming the weighted average according to (7.26).

III. Atomic photoelectron spectra

Below we will discuss some atomic cases, which will serve to illustrate the physical mechanisms responsible for the occurence of various features observed in photoelectron spectra.

We will not concern ourselves specifically with such studies which aim at measurement of cross sections and angular distributions as a function of photon energy. These are covered by SAMSON and STARACE in this volume and also in other sources[1-3].

8. Core electron ionization

α) *The* Ne 1*s spectrum.* The most thoroughly studied atomic photoelectron spectra are those from the noble gases. The main 1*s* line of neon, excited by monochromatized Al$K\alpha$ radiation, is shown in Fig. 42. The spectrometer resolution in this case was high enough to yield a line shape which was nearly the natural Lorentzian. The recorded line has a width of 0.39 eV and if this profile is fitted to a Voigt function, the natural width of the Ne 1*s* level is found to be 0.27(2) eV with a Gaussian spectrometer window of 0.24 eV[4-6]. A many-body perturbation theory (MBPT) calculation gives 0.22 eV for the width of this level[7].

[1] Manson, S.T. (1976): in: Adv. Electron. Electron Phys. *41*, 73

[2a] Amusia, M.Ya. (1980): Appl. Opt. *19*, 4042

[2b] Schmidt, V. (1980): Appl. Opt. *19*, 4080

[3] Codling, K. (1979): in: Synchrotron Radiation, Topics in Current Physics, vol. 10, Kunz, C. (ed.), Berlin, Heidelberg, New York: Springer, p. 231

[4] Gelius, U., Basilier, E., Svensson, S., Bergmark, T., Siegbahn, K. (1973): J. Electron Spectrosc. Relat. Phenom. *2*, 405

[5] Gelius, U., Svensson, S., Siegbahn, H., Basilier, E., Faxälv, Å., Siegbahn, K. (1974): Chem. Phys. Lett. *28*, 1

[6] Svensson, S., Mårtensson, N., Basilier, E., Malmquist, P.Å., Gelius, U., Siegbahn, K. (1976): Phys. Scr. *14*, 141

[7] Kelly, H.P. (1975): Phys. Rev. A *11*, 556

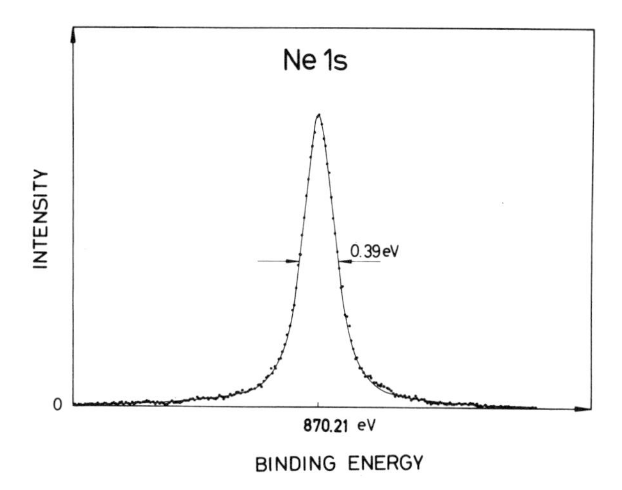

Fig. 42. Ne 1s photoelectron line excited by monochromatized Al $K\alpha$. The solid line is a fitted Voigt profile[5]

Inclusion of interchannel coupling between the different NeKLL Auger channels does not significantly affect the total Auger rate and thus the Ne 1s linewidth[7,8]. Theoretical Ne 1s level widths of 0.29 eV, 0.27 eV and 0.25 eV were found in such calculations[8] using HF-functions (first two values) and HS-functions. It was noted, however, that the partial KLL-Auger rates were substantially influenced by interchannel coupling.

Experimental and theoretical binding energy values for the noble gases are given in Tables 4 and 6–8. For Ne 1s, a number of calculations including correlation corrections have been made[7,9–12]. VERHAEGEN et al.[9] estimated $\Delta E(\text{Ne}\,1s)_{\text{Correlation}}$ from a sum of pair correlation energies obtained for the ground state, i.e. according to (7.12). One obtains for this case:

$$\Delta E(\text{Ne}\,1s)_{\text{Correlation}} = \varepsilon_{1s\alpha 1s\beta} + \varepsilon_{1s\alpha 2s\alpha} + \varepsilon_{1s\alpha 2s\beta} + 3\varepsilon_{1s\alpha 2p\alpha} + 3\varepsilon_{1s\alpha 2p\beta}. \tag{8.1}$$

Using the values $\varepsilon_{1s\alpha 1s\beta} = 1.09$ eV, $\varepsilon_{1s\alpha 2s\alpha} = 0.01$ eV, $\varepsilon_{1s\alpha 2s\beta} = 0.06$ eV, $\varepsilon_{1s\alpha 2p\alpha} = 0.04$ eV and $\varepsilon_{1s\alpha 2p\beta} = 0.05$ eV[13], the correlation correction $\Delta E(\text{Ne}\,1s)_{\text{Correlation}} = 1.43$ eV. Combining this value with a relativistic ΔSCF binding energy of $E_{\text{B}}(\Delta SCFR) = 869.4$ eV[9] [$\Delta E(\text{Ne}\,1s)_{\text{Relativistic}} = 0.8$ eV] leads to a theoretical Ne 1s binding energy of $E_{\text{B}}(\text{Ne}\,1s) = 870.8$ eV. If the correlation energies are calculated separately for the initial and final states, i.e. additional correlations for the final ionic state are included, $\Delta E_{\text{B}}(\text{Ne}\,1s) = 870.0$ eV[10]. KELLY[7] calculated

[8] Howat, G., Åberg, T., Goscinski, O., Soong, S.C., Bhalla, C.P., Ahmed, M. (1977): Phys. Lett. *60A*, 404
[9] Verhaegen, G., Berger, J.J., Desclaux, J.P., Moser, C.M. (1971): Chem. Phys. Lett. *9*, 479
[10] Moser, C.M., Nesbet, R.K., Verhaegen, G. (1971): Chem. Phys. Lett. *12*, 230
[11] Chase, R.L., Kelly, H.P., Köhler, H.S. (1971): Phys. Rev. A *3*, 1550
[12] Martin, R.L., Shirley, D.A. (1976): Phys. Rev. A *13*, 1475
[13] Nesbet, R.K. (1968): Phys. Rev. *175*, 2

Table 6. Non-relativistic (*NR*) and Relativistic (*R*) orbital eigenvalues and ΔSCF binding energies compared to experimental binding energies for Ar (eV)[a]

Shell	$-\varepsilon_i(NR)$	E_B(ΔSCFNR)	$-\varepsilon_i(R)$	E_B(ΔSCFR)	Experiment
$1s$	3,227.4	3,195.2	3,240	3,209	3,205
$2s$	335.3	324.8	336	327	326.3
$2p_{1/2}$	260.4	248.9	261	250	250.6
$2p_{3/2}$	260.4	248.9	259	248	248.5
$3s$	34.8	33.2	34.9	33.3	29.3
$3p_{1/2}$	16.1	14.8	16.1	14.8	15.94
$3p_{3/2}$	16.1	14.8	15.9	14.6	15.76

[a] From refs. [15, 21, 26].

Table 7. Experimental and relativistic ΔSCF binding energies for the *M*- and *N*-shells of krypton (eV)[a]

Shell	E_B(exp)	E_B(ΔSCFR)	$E_B(\text{exp}) - E_B$(ΔSCFR)
$3s$	292.8	296.06	−3.26
$3p_{1/2}$	222.2	225.10	−2.90
$3p_{3/2}$	214.4	216.99	−2.59
$3d_{3/2}$	94.9	93.28	+1.61
$3d_{5/2}$	93.7	91.98	+1.72
$4s$	27.51	30.91	−3.40
$4p_{1/2}$	14.665	13.55	+1.10
$4p_{3/2}$	14.000	12.89	+1.10

[a] From refs. [6] and [21].

Table 8. Experimental and relativistic ΔSCF binding energies for the *M*-, *N*- and *O*-shells of xenon (eV)[a]

Shell	E_B(exp)	E_B(ΔSCFR)	$E_B(\text{exp}) - E_B$(ΔSCFR)
$3s$	1,148.7	1,153.27	−4.6
$3p_{1/2}$	1,002.1	1,007.14	−5.0
$3p_{3/2}$	940.6	944.21	−3.6
$3d_{3/2}$	689.35	689.40	−0.05
$3d_{5/2}$	676.70	676.41	+0.29
$4s$	213.32	222.76	−9.44
$4p_{1/2}$	–	169.16	–
$4p_{3/2}$	145.51*	156.64	−11.13
$4d_{3/2}$	69.48	68.36	+1.12
$4d_{5/2}$	67.50	66.35	+1.15
$5s$	23.40	26.20	−2.80
$5p_{1/2}$	13.436	12.33	+1.10
$5p_{3/2}$	12.130	10.99	+1.14

[a] From refs. [6] and [21].
* Leading peak of the $4p$-spectrum (cf. Figs. 44 and 45).

(Footnotes 21, 26 see p. 302)

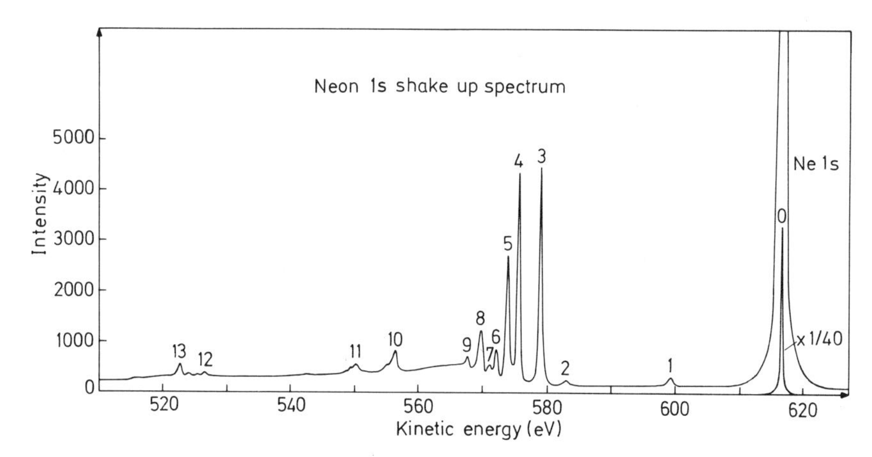

Fig. 43. Neon $1s$ shake-up spectrum excited by monochromatized Al $K\alpha$. Interpretations of the lines are given in Table 9 [14]

$E_B(\text{Ne}\,1s) = 870.5$ eV, using MBPT including also a correction for the Lamb shift.

The full Ne $1s$ satellite spectrum excited by monochromatized Al $K\alpha$ radiation is shown in Fig. 43. The peaks 3–9 in the spectrum correspond to transitions to final states in which, besides the $1s$ hole, a $2p$ electron is excited into unoccupied orbitals. The monopole selection rule, implied by (7.20), leads to ^2S symmetry for the final ionic state. The final ionic states are thus of the type $1s\,2s^2\,2p^5\,np(^2S)$. The description of the coupling in these final states requires the consideration of several possibilities. As a first approximation, two ^2S states may be formed from each configuration by coupling the valence shell to either a singlet or a triplet and then coupling with the core electron to form two doublets. In this scheme, however, the state where the valence electrons couple to a triplet acquires an intensity which is identically zero. Since two multiplet states are indeed observed for each configuration this approximation is unsatisfactory; peaks 3 and 4 correspond to a $2p \rightarrow 3p$-excitation, 5 and 8 to a $2p \rightarrow 4p$-excitation and 6 and 9 to a $2p \rightarrow 5p$-excitation [14]. Ambiguity also arises in this description, because equally acceptable couplings are $1s\,2s^2\,2p^5(^1P)\,np(^2S)$ and $1s\,2s^2\,2p^5(^3P)\,np(^2S)$, which in general will give energies different from those of the other coupling possibility. Thus, in order to remove the above shortcomings in the representation of the states, the next level of approximation is to allow the two doublets to interact. Bagus and Gelius [16] performed MCHF calculations for the satellite transitions and found the energies to be well represented by such a procedure (cf. Table 9, column 4). The intensities are not, however, accounted for in such a scheme [12, 14]. If instead of starting from MCHF functions, initial and final state correlation is introduced by means of general CI methods (i.e. optimization of CI coefficients only) the

[14] Gelius, U. (1974): J. Electron Spectrosc. Relat. Phenom. 5, 985
[15] Siegbahn, K., et al. (1969): ESCA applied to free molecules, Amsterdam: North-Holland
[16] Bagus, P.S., Gelius, U. (results listed in refs. [14, 15])

Table 9. Line energies and intensities in the

Line no.	Final ionic state	Experimental[a] relative energy (eV)	Calculated relative energy (eV)	
			MCHF[a,b]	FISCI[b]
0	$1s\,2s^2p^6(^2S)$	0 (E_B=870.21)	0	0
1	$1s\,2s^2\,2p^6(^2S)$[c]	16.89 (6)	–	–
2	$1s\,2s^2\,2p^5\,3s(^2P)$ lower	33.35 (9)	–	–
3	$1s\,2s^2\,2p^5\,3p(^2S)$ lower	37.35 (2)	35.59	36.4
4	$1s\,2s^2\,2p^5\,3p(^2S)$ upper	40.76 (3)	39.46	39.9
5	$1s\,2s^2\,2p^5\,4p(^2S)$ lower	42.34 (4)	40.50	41.9
6	$1s\,2s^2\,2p^5\,5p(^2S)$ lower	44.08 (5)	42.38	43.0
7	$1s\,2s^2\,2p^5\,6p(^2S)$ lower	45.10 (7)	–	45.2
8	$1s\,2s^2\,2p^5\,4p(^2S)$ upper	46.44 (5)	44.62	46.0
9	$1s\,2s^2\,2p^5\,5p(^2S)$ upper	48.47 (7)	46.60	47.4
10	$1s\,2s\,2p^6\,3s(^2S)$ lower	59.8 (1)	57.95	–
11	$1s\,2s\,2p^6\,3s(^2S)$ upper	65.9 (1)	64.13	–
12		93.14 (7)	–	–
13	$1s\,2s^2\,2p^4\,3p^2(^2S)$	95.9 (1)	–	–
14		97.23 (5)	–	–
	$1s\,2s^2\,2p^5(^3P)$[d]	47.4 (5)	45.16	–
	$1s\,2s^2\,2p^5(^1P)$[d]	51.7 (5)	49.47	–

[a] Ref. [14].
[b] Ref. [12].
[c] Discrete energy loss.
[d] Onset for shake-off.
* Orthogonalized MCHF.

calculated intensities can be further improved as discussed by MARTIN and SHIRLEY[12]. It is found that final ionic state CI leads to calculated intensities which are in good agreement with the experimental data as far as the relative intensities between the satellites are concerned. The total intensity of the satellites with respect to the $1s$-hole peak is, however, underestimated by roughly a factor of two (cf. column 9, Table 9). Introducing CI into the initial ground state of Ne results in intensities, for which this discrepancy is largely accounted (cf. column 10). It should be noted, however, that the relative importance of initial and final state CI for the Ne $1s$ satellites depends on the choice of basis set in the calculations.

The low intensity line 2 cannot, on energetic grounds, be ascribed to a monopole shake-up associated to the $1s$ hole. It is instead interpreted as due to a final ionic state of ^2P-symmetry[14], with the configuration $1s\,2s^2\,2p^5\,3s$. There are a number of possible contributions to the intensity of this transition. Considering the full dipole matrix element [i.e. (7.19)] between the initial $1s^2\,2s^2\,2p^6(^1S)$ state and the final $1s\,2s^2\,2p^5\,3s\,\mathcal{E}s(^1P)$ state, the latter described by relaxed orbitals, basically two types of contributions arise. The first of these corresponds to a dipole transition of a $2p$ electron into the continuum ($\mathcal{E}s$) and a monopole excitation of a $1s$ electron into the $3s$ state. As pointed out by GELIUS[14], however, this term will probably not contribute significantly to the satellite intensity. The second type can be described as a dipole transition of

Ne 1s(Al$K\alpha$) spectrum referred to the main line

Experimental[a] relative intensity (%)	Calculated relative intensity			
	MCHF[b]	MCHF[b*]	FISCI[b]	FISCI + ISCI[b]
100	100	100	100	100
–	–	–	–	–
0.06 (1)	–	–	–	–
3.15 (8)	9.25	3.35	1.26	2.47
3.13 (10)	5.45	1.97	1.68	2.60
2.02 (10)	2.31	0.74	0.85	1.48
0.42 (6)	0.95	0.29	0.24	0.43
0.50 (15)	–	–	0.05	0.09
0.96 (11)	0.89	0.36	0.46	0.70
0.17 (5)	0.31	0.13	0.07	0.11
0.57 (5)	–	–	–	–
0.49 (6)	–	–	–	–
0.08 (2)	–	–	–	–
0.10 (4)	–	–	–	–
0.24 (4)	–	–	–	–
–	–	–	–	–
–	–	–	–	–

the $2p$ electron into the $3s$ state and a monopole excitation of the $1s$ electron into the ($\mathcal{E}s$) continuum (there are additional terms, but these will generally be smaller[17]). The presence of this second part of the dipole matrix element ($\sim\langle 2p|\underline{p}|3s'\rangle\langle 1s|\mathcal{E}s'\rangle$) constitutes a breakdown of the sudden approximation and this term will in general vanish in the sudden limit[17]. This type of breakdown can be considered a special case of CSCI. This latter CI will lead also to other possible contributions of the satellite intensity. In a general CSCI are contained, aside from the above terms, couplings of the type

$$1s\,2s^2\,2p^5\,3p(^2S)\,\mathcal{E}p(^1P) \rightarrow\!\leftarrow 1s\,2s^2\,2p^5\,3s(^2P)\,\mathcal{E}s(^1P)$$

and these may also lead to population of the satellite final ionic state. The magnitudes of these different contributions have not yet been estimated, so a complete analysis is not possible at this stage.

The Ne 1s shake-off continua have been studied theoretically by CHATTARJI et al.[18] who conclude that CSCI may be of considerable importance in deciding their relative intensities.

[17] Martin, R.L., Shirley, D.A. (1977): in: Electron spectroscopy, Brundle, C.R., Baker, A.D. (eds.), New York: Academic Press, p. 76

[18] Chattarji, D., Mehlhorn, W., Schmidt, V. (1978): J. Electron Spectrosc. Relat. Phenom. *13*, 97

A conclusion from the study of the Ne1*s* spectrum is that substantial efforts in terms of the description of the wave functions are needed for a full account of the satellite intensities and that both relaxation, ISCI and FSCI effects are important.

For the alkali metal and magnesium atoms, strong shake-up satellites have been observed in core electron spectra, which however are related to the excitation of the outermost *ns* electrons[24,25].

β) The Xe4*s*, 4*p spectrum.* The normal Lorentzian photoelectron line shape, as exemplified by the Ne1*s* line, is expected from a first-order treatment, involving the interaction between a single discrete level and the continuum. When several discrete and continuum configurations contribute to the description of the final ionic state, however, significant distortions of the line shape may occur[6,14,20-23]. Strong effects of this type have been observed for the 4*s* and 4*p*-levels in the elements $48 \leqq Z \leqq 73$[6,14,22]. Early studies of the 4*p*-levels in these elements showed only one component, where one would expect to find a well resolved spin-orbit split doublet[15,19]. It was not until the quality of the photoelectron spectra could be improved that the origin of these anomalies could be satisfactorily explained[14]. The experimental improvement necessary to this end was the introduction of monochromatized x-rays for excitation of the spectra. Figure 44 shows the photoelectron spectrum of xenon up to 170 eV binding energy excited by monochromatized Al$K\alpha$. As mentioned, all lines appear that are expected from one-particle considerations except in the 4*p*-region of the spectrum. Instead of a Gaussian shaped line, as earlier recorded, the '$4p_{3/2}$' line is revealed in high resolution to have a complex discrete structure. Also, the '$4p_{1/2}$' line is completely smeared out into a broad continuous distribution. The relativistic ΔSCF binding energies of Table 8 for xenon show large discrepancies with respect to experiment for the 4*s* and 4*p* hole states. Since the ΔSCF values are substantially larger than experimental values, this is indicative of strong configuration mixing in the final ionic states (FISCI). Comparison with Auger electron spectra shows that the leading configurations involved are related to excitations from the 4*d* shell. From the $M_{4,5}N_{4,5}N_{4,5}$ Auger electron energies it can be concluded that the $4d^2$-states (super Coster-Kronig threshold) have total energies starting 6.0 eV above the

---

[19] Siegbahn, K., et al. (1967): ESCA – Atomic, molecular and solid state structure studied by means of electron spectroscopy, Uppsala: Almqvist o. Wiksell

[20a] Wendin, G., Ohno, M. (1976): Phys. Scr. *14*, 148

[20b] Wendin, G., Ohno, M. (1976): in: Proc. 2nd Int. Conf. Inner Shell Ionization Phen. Mehlhorn, W. (ed.), Freiburg

[21] Rosén, A., Lindgren, I. (1968): Phys. Rev. *176*, 114

[22] Kowalczyk, S.P., Ley, L., Martin, R.L., McFeely, F.R., Shirley, D.A. (1975): Faraday Discuss. Chem. Soc. *60*, 7

[23] Ohno, M., Wendin, G. (1977): Phys. Scr. *16*, 299

[24] Banna, M.S., Wallbank, B., Frost, D.C., McDowell, C.A., Perera, J.S.H.Q. (1978): J. Chem. Phys. *68*, 5459

[25] Martin, R.L., Davidson, E.R., Banna, M.S., Wallbank, B., Frost, D.C., McDowell, C.A. (1978): J. Chem. Phys. *68*, 5006

[26] Bagus, P.S. (1965): Phys. Rev. *139*, A619

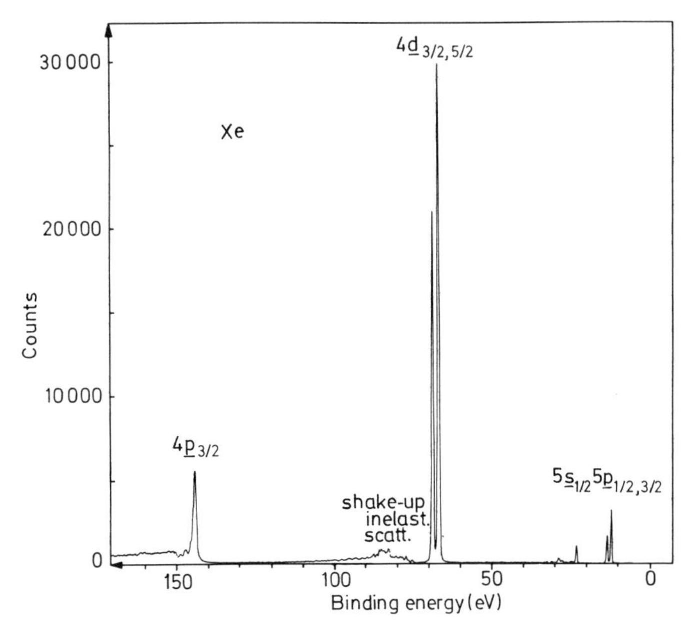

Fig. 44. Al $K\alpha$-excited spectrum of Xe in the binding energy region up to 170 eV [14]

energy of the leading peak of the $4p$-spectrum (the multiplet energies are indicated in Fig. 45). Thus, in the case of a $4\underline{p}$-hole, correlations of the type $4\underline{p} \leftrightarrow 4\underline{d}^2 mf$, where m is discrete or continuous, are expected to be the most important in the description of the spectrum [6, 14, 20]. WENDIN and OHNO have studied this spectrum using the RPAE formalism [20]. They included interactions of the types:

$$\left.\begin{array}{l} 4\underline{p} \leftrightarrow 4\underline{d}^2\, mf \\ 4\underline{p} \leftrightarrow 4\underline{d}\, 5\underline{s}\, mp \\ 4\underline{p} \leftrightarrow 4\underline{d}\, 5\underline{p}\, md \end{array}\right\} \quad (m \text{ discrete or continuous})$$

According to this calculation the leading peak of the '$4p$'-spectrum should correspond not to a $4p_{3/2}$ final ionic state, but rather to a $4\underline{d}^2\, 4f$ state. It is predicted to have 25% of the intensity of the total spectrum. The fine-structure in this peak is suggested to be due to multiplet structure within the $(4\underline{d}^2\, 4f)_{3/2}$ configuration. Experimentally three peaks can be partly resolved (Fig. 45) with relative intensities 1:0.33:0.11 [6, 14]. Furthermore, about 50% of the unperturbed $4p_{3/2}$ intensity is transferred to continuum states of type $4\underline{d}^2\, \varepsilon f$ which leads to the broad distribution on the low energy side of the leading peak.

For the $4s$-hole, interactions with $4\underline{d}\, 4\underline{p}\, mf$ configurations leads to the strong shift of the line with respect to the ΔSCF value. In this case, however, most of the spectral strength (80%) resides in one peak [20]. This line can thus still be characterized as a $4s$-peak, in spite of the strong correlation effects. The remaining part of the $4s$ intensity goes mainly into the super Coster-Kronig $4\underline{p}\, 4\underline{d}$ continuum [20].

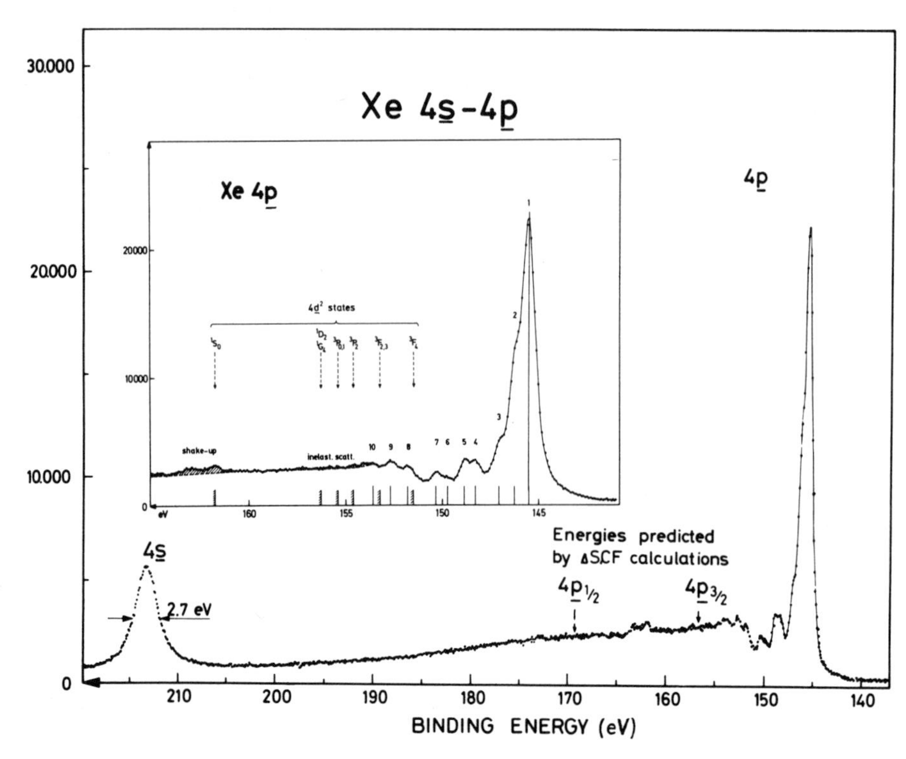

Fig. 45. $4\underline{s}-4\underline{p}$ region of the photoelectron spectrum of Xe[14]. The insert figure shows the spectrum on an expanded scale. The energies of the $4\underline{d}^2$ states have been derived from Auger electron spectra[15]

The very strong distortion and shift of the Xe$4\underline{s}$, $4\underline{p}$ spectrum with respect to the ΔSCF picture is due not only to the energy match between the interacting configurations, but also to the magnitude of the CI matrix elements. Since these elements involve radial wave functions from the same main shell (e.g. $4p$, $4d$ and $4f$ in the leading peak) which have large spatial overlap between each other, the matrix element and thus FISCI will be very strong. Since the properties of wave functions and energy match change smoothly in the periodic system, effects similar to those observed in xenon are expected for the $4s-4p$ spectra in neighbouring elements. Figure 46 shows the spectra obtained for the series of elements $Z=52$–56 excited by Al$K\alpha$ radiation[6,14]. The systematic changes in the spectra can be qualitatively understood on the basis of the position of the $4\underline{d}^2$ super Coster-Kronig threshold. For Te$(Z=52)$ the threshold lies at a lower energy than the onset of the $4\underline{p}$ spectrum. The spectrum will thus be dominated by strongly broadened structures due to interactions of the $4\underline{p}$ unperturbed states with the super Coster-Kronig continuum. With increasing Z the $4\underline{d}^2$ threshold increases more rapidly in energy than do the (unperturbed) $4p$-states, which means that discrete structures will gradually appear at the low binding energy side of the spectrum. The 'normal' spectrum of the spin-orbit split $4p$-doublet will thus be eventually restored at

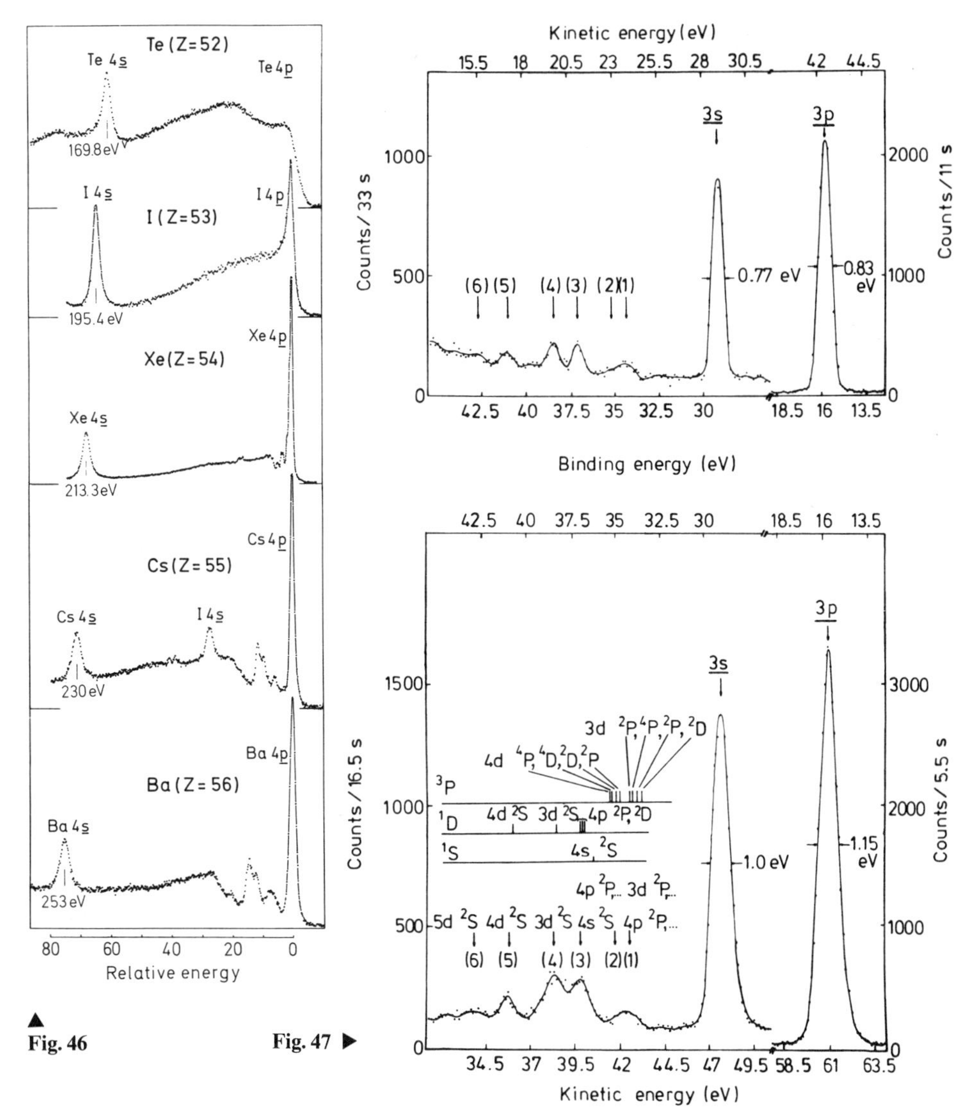

▲ **Fig. 46** **Fig. 47** ▶

Fig. 46. $4\underline{s}-4\underline{p}$ spectra of the elements $52 \leqq Z \leqq 56$[14]. The leading $4\underline{p}$-peak has been normalized to the same energy. The figure shows how the spectra gradually change with Z due to changes in the relative positions of the $4\underline{p}$ hole state and the $4\underline{d}^2$ super Coster-Kronig threshold (cf. text)

Fig. 47. Ar $3s$, $3p$ photoelectron spectra excited by synchrotron radiation[2] at photon energies of $h\nu = 58.3$ eV (top) and $h\nu = 77.2$ eV (bottom). On the left hand side of the figure, the vertical bars give the positions of the excited states of Ar^+ from optical data[53]. The terms ^1S, ^1D and ^3P refer to the core configurations $3s^2 3p^4$. For the assignment of the observed satellites cf. Table 10

(Footnotes 2, 53 see p. 306, 310)

sufficiently high Z. Further studies have shown a distorted $4\underline{p}$-spectrum for the elements between Cd($Z=48$) and Ta($Z=73$)[22].

In krypton, an energy match similar to the xenon case exists between the $3\underline{p}$ and $3\underline{d}^2$ hole states[6, 23]. However, the spatial overlap between the $3p$, $3d$

Table 10. Experimental and theoretical energies and intensities

Final ionic state	Experimental			
	Binding energies[a]	Intensities		
		$h\nu=1{,}487$ eV[b]	$h\nu=58$ eV[a]	$h\nu=43$ eV[a]
$3s^2\,3p^5(^2P^0)$	−13.42	145 (8)	540 (12)	1.0 (4) × 10^4
$3s\,3p^6(^2S)$	0.00	100	100	100
$3s^2\,3p^4(^3P)\,3d(^2D)$	5.16	–	} 14 (2)	2.4 (1.0) × 10^2
$(^3P)\,4p(^2P^0)$	6.06	–		5.0 (2.5) × 10^2
$(^1S)\,4s(^2S)$				
$(^1D)\,4p(^2P^0)$	7.96	3 (2)	17 (1)	
$(^1D)\,3d(^2D)$				} 34 (2)
$(^1D)\,3d(^2S)$	9.29	19 (2)	17 (1)	
$(^3P)\,5p(^2P^0)$				
$(^1D)\,4d(^2S)$	11.91	6 (3)	10 (2)	
$(^1D)\,5d(^2S)$				
$(^1S)\,5s(^2S)$				
$(^1S)\,6d(^2S)$	13.8	–	4 (2)	

[a] Ref. [2] [c] Ref. [5] * High photon energy limit.
[b] Ref. [1] [d] Ref. [6]

electrons and the mf electrons is substantially smaller in this case, leading to much smaller FISCI effects than in the xenon case.

9. Valence electron ionization

α) Inner valence ionization. As previously pointed out, the inner valence region of photoelectron spectra may show substantial effects mainly related to final ionic state CI (cf. Sect. 7). In atomic photoelectron spectra these effects have been investigated for the noble gases. As an example, we shall discuss the Ar$3\underline{s}$ spectrum in some detail. This spectrum was investigated by SPEARS et al.[1] and by ADAM et al.[2]. The latter authors used synchrotron radiation in the 40 eV–80 eV photon energy region to excite the spectra, whereas the former used nonmonochromatized Al$K\alpha$, Mg$K\alpha$ and Zr$M\zeta$ radiations. Spectra recorded at $h\nu=58.3$ eV and $h\nu=77.2$ eV are shown in Fig. 47. A sequence of six satellites are observed in the spectrum on the high binding energy side of the Ar$3s$ peak. It is observed by these authors that the relative intensities of the satellites vary with photon energy (in particular lines 1 and 2). For assignment of the spectra comparison can be made with optical data (inserted in the figure) and theoretical calculations[3–6]. The experimental and theoretical values

[1] Spears, D.P., Fischbeck, H.J., Carlson, T.A. (1974): Phys. Rev. A *9*, 1603
[2] Adam, M.Y., Wuilleumier, F., Krummacher, S., Schmidt, V., Mehlhorn, W. (1978): J. Phys. B *11*, L413
[3] Kjöllerström, B., Möller, N.H., Svensson, H. (1965): Ark. Fys. *29*, 167
[4] Luyken, B.F., Heer, F.J. de, Baas, R.C. (1972): Physica *61*, 200
[5] Martin, R.L., Kowalczyk, S.P., Shirley, D.A. (1977): Lawrence Berkeley Lab. Rep. LBL-5445
[6] Dyall, K.G., Larkins, F.P. (1979): J. Electron Spectrosc. Relat. Phenom. *15*, 165

in the Ar $3s$ spectrum, referred to the main Ar $3\underline{s}(^2S)$ line

Theoretical			
Binding energies		Intensities	
FISCI[c]	FISCI[d]	FISCI[c*]	ISCI + FISCI[d*]
	−14.07		150
0.00	0.00	100	100
	4.91		
	5.87		0.4
8.8	7.41	2.5	2.1
7.3	7.65	1.2	1.6
	7.12		
10.4	9.56	14.4	17.3
	9.60		0.5
13.0	11.89	15.1	13.5
	12.97		5.5
14.8	13.21	0.6	3.5
	13.83		16.5

are contained in Table 10 where the assignment is based on the most comprehensive calculation[6]. On the basis of dipole selectrion rules within *LS*-coupling, the only final ionic states which can be reached have ^2S, $^2P^0$, ^2D, $^2F^0$ or ^2G symmetries. Out of these, the first three manifolds are expected to give the major contributions to the satellite spectrum. Peaks 4, 5 and 6 are straightforwardly assigned from the theoretical data as members of the ^2S manifold with main configurations $3s^2\,3p^4\,3d$, $3s^2\,3p^4\,4d$ and $3s^2\,3p^4\,6d$, respectively. These transitions thus gain their intensity from the $3s\,3p^6$ configuration through final ionic state CI(FISCI). Peaks 1, 2 and 3 display, however, some features which are not expected considering FISCI in the ^2S manifold as the only mechanism. The relative intensities of peaks 1 and 2 are strongly energy-dependent with respect to the $3\underline{s}$ peak, which would not be the case for ^2S states. The Ar $3\underline{s}$ photoelectric cross section exhibits a minimum at $h\nu \approx 43$ eV (Cooper minimum)[7−9]. At this Cooper minimum peaks 1 and 2 are clearly observed by Adam et al., while the main line is very weak. Moreover, there is no state of ^2S-symmetry in the region of these satellite peaks, according to both the optical data and calculations. These data show, however, a substantial number of states in this region of other than ^2S symmetry, viz. ^2P, ^2D and ^2F-states. The population of the ^2P-states may occur through relaxation from ionization of the $3p$ shell. Calculations indicate that this is the

---

[7] Cooper, J.W. (1964): Phys. Rev. Lett. *13*, 762

[8] Houlgate, R.G., West, J.B., Codling, K., Marr, G.V. (1976): J. Electron Spectrosc. Relat. Phenom. *9*, 205

[9] Amusia, M.Y., Ivanov, V.K., Cherepkov, N.A., Chernysheva, L.V. (1972): Phys. Lett. A*40*, 361

probable assignment of satellite peak no. 2, where the relevant final ionic state is $3s^2\,3p^4(^3P)\,4p(^2P^0)$. The relative intensity change of this peak with photon energy is then also explained as due to the variation in the $3s/3p$ cross section ratio. On such a basis, this satellite is expected to be relatively more important at low photon energies which is also experimentally observed. The $3s^2\,3p^4(^1D)\,4p(^2P^0)$ state may also contribute to satellite peak number 3, also through relaxation.

The ^2D final ionic states cannot be populated either by relaxation or FISCI, but can be reached through ISCI. Thus, the ground state $3s^2\,3p^6$ configuration may be mixed with $3s^2\,3p^4\,n\,l\,n'\,l'$ configurations of ^1S symmetry. Calculations indicate that mixing of such configurations with the ground state affects appreciably the variation of the photoionization cross section as a function of photon energy. Ionization of these ISCI states leads to transitions to states of type $3s^2\,3p^4\,m\,l(^2L)$, where L may be higher than $L=1$. The calculations of DYALL and LARKINS[6] indicate that satellite peak number 1, and partly 3, are due to ISCI, where mixing with the $3s^2\,3p^4\,3d^2$ configuration into the ground state leads to the observed intensity. Thus, peak no. 1 is assigned to the $3s^2\,3p^4(^3P)\,3d(^2D)$ final ionic state, whereas the $3s^2\,3p^4(^1D)\,3d(^2D)$ state probably partly contributes to peak no. 3. This latter peak also has contributions from ^2S and ^2P states through FISCI and relaxation (cf. Table 10). Finally, the possibility should be mentioned of CSCI contributing to the satellite intensities at low photoelectron energies.

In the spectra of the heavier noble gases[1, 10–13], the spin-orbit interaction increases in importance, which means that FISCI may mix states, which do not have the same LS-symmetry. This has been noted in particular for Xe $5s$ ionization where e.g. the $5s^2\,5p^4\,6d(^4P)$-state has been predicted to get an appreciable intensity from FISCI with the $5\underline{s}(^2S)$ hole state[13].

Manifestations of the strong inner valence FISCI effects for the noble gases are observed also in the Auger electron and x-ray emission spectra[14–18]. These spectra yield complementary information to photoelectron spectra. In the ArKLM Auger spectrum strong interactions are observed between the $\underline{L}$ $3s^2\,3p^4\,3d$ and $\underline{L}$ $3s\,3p^6$ configurations[14]. In the x-ray emission spectra, the symmetries of the final states reached are decided by the dipole selection rules. For example, P-type final states do not occur in the Ar$L_{2,3}$-emission spectrum, which may serve as an additional aid in the assignment of the Ar photoelectron spectrum[15].

---

[10] Adam, M.Y., Wuilleumier, F., Sandner, N., Schmidt, V., Wendin, G. (1978): J. Phys. *39*, 129

[11] Wendin, G. (1977): Phys. Scr. *16*, 296

[12] Svensson, S., Mårtensson, N., Basilier, E., Malmquist, P.-Å., Gelius, U., Siegbahn, K. (1976): Phys. Scr. *14*, 141

[13] Hansen, J.E., Persson, W., Phys. Rev. A *20*, 364

[14a] Asplund, L., Kelfve, P., Blomster, B., Siegbahn, H., Siegbahn, K. (1977): Phys. Scr. *16*, 268

[14b] Darko, T., Siegbahn, H., Kelfve, P. (1979): UUIP-1004 (Uppsala University Institute of Physics Report), (1981): Chem. Phys. Lett. **81**, 475

[15] Nordgren, J., Ågren, H., Nordling, C., Siegbahn, K. (1979): Phys. Scr. *19*, 5

[16] Werme, L.O., Bergmark, T., Siegbahn, K. (1973): Phys. Scr. *8*, 149

[17] McGuire, E.J. (1975): Phys. Rev. A *11*, 17; A *11*, 1880

[18] Froese-Fischer, C., Ridder, D. (1978): J. Phys. B *11*, 2267

β) Outer valence ionization. In outer valence ionization, the relative importance of mechanisms other than FISCI is generally larger than in inner valence ionization. Aside from the noble gases, studies of the outer energy region have been performed, primarily with uv-radiation, for the group *II*-elements[19–39], the rare earths Sm, Eu, Yb[19a] the transition metals Cr, Mn, Cu, Ag and Au[40,41,55,56], the alkali metals[55,57–59], the halogens[60–62] and Ga[63], Ge and Sn[64]. Below we will discuss the cases of He and Ba[19,42–54].

At photon energies larger than 65.4 eV ionization of He may lead to final states of the residual electron with main quantum number larger than $n=1$. These states lead to satellite lines in the photoelectron spectrum. The relative ($n=2$) satellite intensity has been investigated as a function of photon energy by WUILLEUMIER and co-workers[43] using synchrotron radiation from threshold up to $h\nu=150$ eV. A typical spectrum is shown in Fig. 48 for $h\nu=93$ eV. These

---

[19a] Lee, S.-T., Süzer, S., Matthias, E., Rosenberg, R., Shirley, D.A. (1977): J. Chem. Phys. *66*, 2496

[19b] Rosenberg, R., White, M.G., Thornton, G., Shirley, D.A. (1979): Phys. Rev. Lett. *43*, 1384

[20] Shannon, S.P., Codling, K. (1978): J. Phys. B*11*, 1193

[21] Harrison, H. (1970): J. Chem. Phys. *52*, 901

[22] Niehaus, A., Ruf, M.W. (1972): Z. Phys. *252*, 84

[23] Walker, T.E.H., Berkowitz, J., Dehmer, J.L., Waber, J.T. (1973): Phys. Rev. Lett. *31*, 678

[24] Nilsson, R., Nyholm, R., Berndtsson, A., Hedman, J., Nordling, C. (1976): J. Electron Spectrosc. Relat. Phenom. *9*, 337

[25] Breuckman, B. (1977): Ph. D. Thesis, Univ. Freiburg

[26] Frost, D.C., McDowell, C.A., Vroom, D.A. (1967): Chem. Phys. Lett, *1*, 93

[27] Fuchs, V., Hotop, H. (1969): Chem. Phys. Lett. *4*, 71

[28] Dehmer, J.L., Berkowitz, J. (1974): Phys. Rev. A*10*, 484

[29] Mehlhorn, W., Breuckmann, B., Hausamann, D. (1977): Phys. Scr. *16*, 177

[30] Berkowitz, J., Dehmer, J.L., Kim, H.-K., Desclaux, J.P. (1974): J. Chem. Phys. *61*, 2556

[31] Süzer, S., Shirley, D.A. (1974): J. Chem. Phys. *61*, 2481

[32] Süzer, S., Lee, S.T., Shirley, D.A. (1976): Phys. Rev. A*13*, 1842

[33] Hush, N.S., Süzer, S. (1977): Chem. Phys. Lett. *46*, 411

[34] Hotop, H., Mahr, D. (1975): J. Phys. B*8*, L301

[35] Brehm, B., Höfler, K. (1975): Int. J. Mass Spectrosc. Ion Phys. *17*, 3711

[36] Blake, A.J. (1971): Proc. R. Soc. (London) A*325*, 555

[37] Mitchell, P., Wilson, M. (1969): Chem. Phys. Lett. *3*, 389

[38] Svensson, S., Mårtensson, N., Basilier, E., Malmquist, P.Å., Gelius, U., Siegbahn, K. (1976): J. Electron Spectrosc. Relat. Phenom. *9*, 51

[39] Banna, M.S., Frost, D.C., McDowell, C.A., Wallbank, B. (1978): J. Chem. Phys. *68*, 696

[40] Krause, M.O. (1980): J. Chem. Phys. *72*, 6474

[41] Chandesris, D., Guillot, C., Chauvin, G., Lecante, J., Petroff, Y. (1981): Phys. Rev. Lett. *47*, 1273, 1568

[42] Krause, M.O., Wuilleumier, F. (1972): J. Phys. B*5*, L143

[43a] Wuilleumier, F., Adam, M.Y., Sandner, N., Schmidt, V., Mehlhorn, W. (1980): J. Phys. Lett. *41*, L373

[43b] Adam, M.Y., Wuilleumier, F., Krummacher, S., Sandner, N., Schmidt, V., Mehlhorn, W. (1979): J. Electron Spectrosc. Relat. Phenom. *15*, 211

[43c] Bizau, J.M., Wuilleumier, F., Ederer, D., Koch, P., Dhez, P., Krummacher, S., Schmidt, V. (1981): LURE Research report II-9

[43d] Bizau, J.M., Wuilleumier, F., Ederer, D., Dhez, P., Krummacher, S., Schmidt, V. (1981): ibid. II-10

[43e] Wuilleumier, F., Dhez, P., Krummacher, S., Schmidt, V. (1980): VIth Int. Conf. on VUV Rad. Phys., Ext. Abstr., vol. II-8, Charlottesville, Virg.

(Footnotes 44–64 see p. 310)

data were obtained in the 'magic angle' geometry, i.e. $\alpha_{ku}=54°44'$ in (7.14a). This implies that the angular term vanishes and that thus the partial cross section is directly measured. The ratio between $n=2$ to $n=1$ events ($\equiv R$ in %) as obtained in these measurements is plotted in Fig. 49a. The results of the most comprehensive calculations are also included in this figure. Contributions to the $n=2$ satellite intensity arise both due to ISCI (Initial State CI) and CSCI (Continuum State CI) (FISCI is of course precluded) e.g.[45]. Of the three calculated curves, those of JACOBS and BURKE[44] and CHANG[48] contain both of these contributions whereas that of SALPETER and ZAIDI[47] contains only ISCI. Agreement is seen to be best between the curve of Chang and the experimental data, thus confirming the importance of CSCI in this energy region. At photon energies near the $n=3$ threshold and in the region slightly above all calculations deviate markedly from experiment. This results from the neglect of the $n=3, 4$, He^+ states in the expansion of the final state wave functions.

SAMSON and WOODRUFF[46b] have obtained the $n=2$ satellite cross section by measuring the fluorescence decay of the $He^+(n=2)$ ions through the 40.8 eV-transition. Measurements were made in the photon energy range from the $n=2$ threshold (69.4 eV) up to $h\nu=160$ eV. In the energy region converging to the $n=3$ threshold (73.0 eV) autoionization structures in the $n=2$ cross section were observed. Above the double ionization threshold (78.0 eV) good agreement with Chang's calculation was found[48].

For a final ionic state of main quantum number n of the He ion, different continuum channels are in principle open for the outgoing electron. Thus, for $n=2$ there are the two possibilities:

[44a] Jacobs, V.L. (1971): Phys. Rev. A*3*, 289
[44b] Jacobs, V.L., Burke, P.G. (1972): J. Phys. B*5*, L67
[45] Manson, S.T. (1976): Adv. Electron. Electron Phys. *41*, 73
[46a] Samson. J.A.R. (1969): Phys. Rev. Lett. *22*, 693
[46b] Woodruff, P.R., Samson, J.A.R. (1980): Phys. Rev. Lett. *45*, 110
[47] Salpeter, E.E., Zaidi, M.H. (1962): Phys. Rev. *125*, 248
[48] Chang, T.N. (1980): J. Phys. B*13*, L551
[49] Connerade, J.P., Mansfield, M.W.D., Thimm, K., Tracy, D. (1974): in: VUV radiation physics, Koch, E.E., Haensel, R., Kunz, C. (eds.), New York: Pergamon
[50] Ederer, D.L., Lucatorto, T.B., Salomon, E.B. (1974): ibid. 245
[51] Brehm, B., Bucher, A. (1974): Int. J. Mass. Spectrosc. Ion Phys. *15*, 463
[52] Hansen, J.E. (1975): J. Phys. B*8*, L403
[53] Moore, C.E. (1949, 1952, 1958): Atomic energy levels, National Bureau of Standards (US) Circular No. 467, vols. 1-3
[54] Cooper, J., Zare, R.N. (1968): J. Chem. Phys. *48*, 942
[55] Süzer, S. (1979): J. Chem. Phys. *71*, 2730
[56] Dyke, M., Fayad, N.K., Morris, A., Trickle, R. (1979): J. Phys. B*12*, 2985
[57] Süzer, S., Menzel, W., Breuckmann, B., Theodosiou, C.E., Mehlhorn, W. (1980): J. Phys. B*13*, 2061
[58] Lee, E.P.F., Potts, A.W. (1979): Chem. Phys. Lett. *66*, 553
[59] Potts, A.W., Lee, E.P.F. (1979): ibid. *67*, 93
[60] DeLeuw, D.M., Mooyman, R., DeLange, C.A. (1978): ibid. *54*, 231
[61] Kimura, K., Yamazaki, T., Achiba, Y. (1978): ibid. *58*, 104
[62] Berkowitz, J., Goodman, G.L. (1979): J. Chem. Phys. *71*, 1754
[63] Dyke, J.M., Josland, G.D., Lewis, R.A., Morris, A. (1981): Mol. Phys. *44*, 967
[64] Dyke, J.M., Fayad, N.K., Josland, G.D., Morris, A. (1980): ibid. *41*, 1051

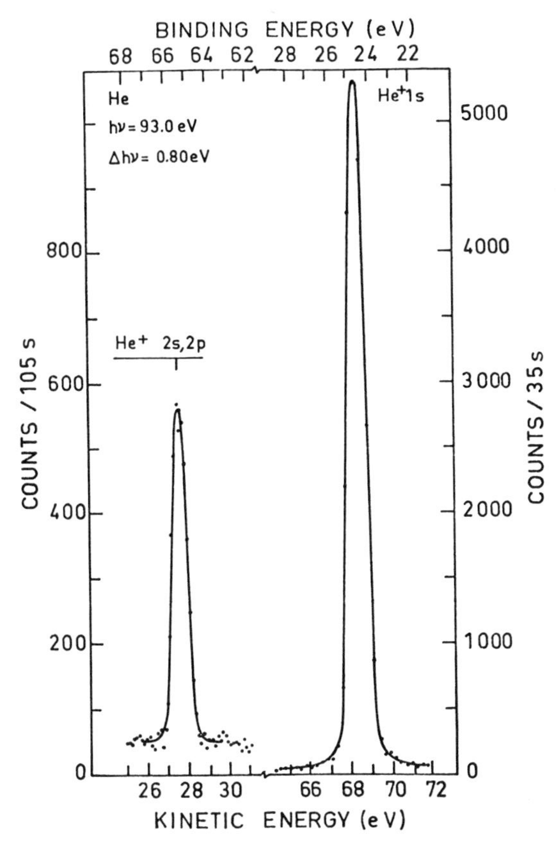

Fig. 48. Spectrum of He excited by synchrotron radiation ($h\nu = 93$ eV)

$$\mathrm{He} + h\nu \rightarrow \mathrm{He}^+(2s) + e^- \tag{9.1}$$

$$\mathrm{He} + h\nu \rightarrow \mathrm{He}^+(2p) + e^-. \tag{9.2}$$

This is analogous to the satellite transition to the $1s\,2s^2\,2p^5\,3s(^1P)$ final ionic state discussed in the Ne$1s$ (cf. discussion on page 300).

It is not possible to distinguish on energetic grounds the peaks corresponding to different final ionic state angular momenta for each n. For $n=2$, the $2s$ and $2p$ final ionic states differ by only 0.7 meV. However, by measuring the angular distributions of the ejected electrons one can gain insight into the nature of the excitation. In an early study, KRAUSE and WUILLEUMIER[42] investigated the $n=2$ satellite in He using five discrete photon sources and obtained the asymmetry parameter β for this peak. If the photoionization leading to the $n=2$ satellite proceeds entirely according to (9.1) then the asymmetry parameter should be the same as that of the main ($n=1$) peak, viz. $\beta_s = 2$. (In a nonrelativistic treatment with no correlation this is always the case for ns subshell ionization, irrespective of photon energy[54]. Inclusion of relativistic and correlation effects leads to deviation from $\beta_{ns} = 2$, however; cf. e.g. the article by STARACE in this volume.) Any deviation of the β-value from the

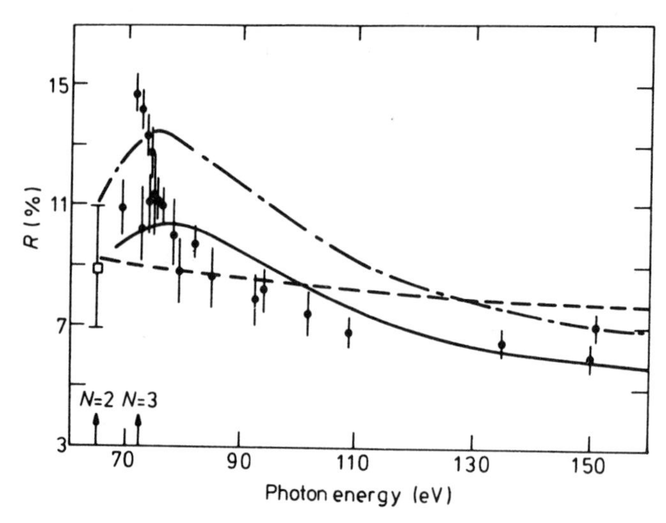

Fig. 49a. The intensity ratio between the $n=2$ satellite and the main line ($n=1$) in photoionization of He vs. photon energy. The filled circles are measurements of WUILLEUMIER and coworkers[43]. The unfilled square is due to SAMSON[46a]. The calculated curves are from (——) CHANG[48], (– – – –) SALPETER and ZAIDI[47] and (–·–·–) JACOBS and BURKE[44]. The four experimental values showing the largest ratio R are due to resonances of doubly excited He atoms[43e]

value 2 implies thus that processes according to (9.2) occur. In the study of ref. [42] such deviations were found for photon energies in the energy range $110\,\text{eV} \leqq h\nu \leqq 190\,\text{eV}$, confirming the existence of $He^+(2p)$ final states. The more recent measurements of BIZAU et al. using synchrotron radiation are summarized in Fig. 49b[43d]. These data show a photon energy dependence of the asymmetry parameter for the $n=2$ satellite which lies between that of $\beta_s=2$ and the calculated β_{2p} dependence of JACOBS and BURKE[44] (Fig. 49 top). If the calculated β_{2p}-curve is used along with the measured intensities of the $n=2$ satellite, the ratio between processes (9.1) to processes (9.2) can be calculated. The results of this procedure is seen in the bottom diagram of Fig. 49b[48d]. Here, good agreement is again found with the calculation of CHANG[48] whereas the curve of JACOBS and BURKE[44] deviates strongly for energies below 100 eV. It can be noted that an appreciable portion of the ions are in fact left in the $2p$ state in this photon energy range.

In several atomic spectra strong autoionization effects have been observed at the photon energy commonly used, He*I*[19]. The Ba spectrum recorded with He*I* radiation is considerably different from the Ne*I* spectrum. The two spectra are compared in Fig. 50. In particular, the region above 10 eV binding energy which contains a large number of lines in the He*I* spectrum, corresponding to final ionic states of type $5p^6nl$, is almost completely lacking of lines in the Ne*I* spectrum. The intensities of these satellite lines are very high in the former spectrum in comparison with the main $6s$-line. In view of the strong photon energy dependence of the spectrum it is likely that the transitions are due to either CSCI or autoionization. Based on the absorption

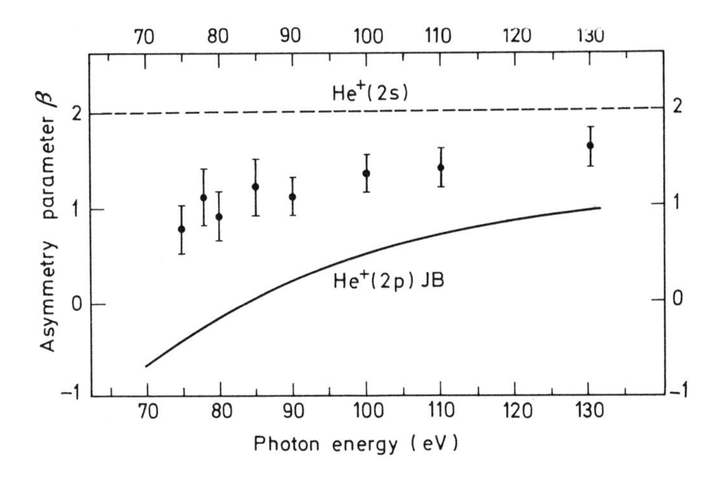

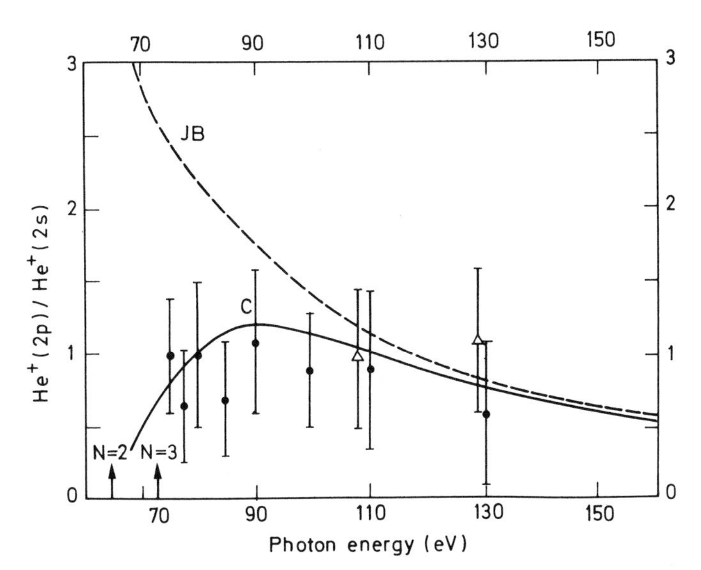

Fig. 49b. Asymmetry parameter of the $n=2$ satellite in He photoionization vs. photon energy[43d] (top). Full-drawn curve shows calculated behaviour of the asymmetry parameter when the He^+-ion is left in the $2p$ state (JACOBS and BURKE, ref. [44]). Ratio between He^+-ions in the $2p$- and $2s$-state vs. photon energy (bottom). The filled circles were obtained from intensity measurements of the $n=2$ satellite coupled with the theoretical β_{2p}-curve of JACOBS and BURKE[44] (see top figure). The calculated curves are due to (----) JACOBS and BURKE[44] and (——) CHANG[48]

spectrum of Ba*I*, LEE et al.[19a] proposed that the latter mechanism is the leading one in the population of these states. The absorption spectrum[49,50] shows strong peaks in the region 19 eV–25 eV converging to the binding energies of the $5p_{3/2}$- and $5p_{1/2}$-states at 22.7 eV and 24.8 eV, respectively. At the Ne*I* photon energy no such absorption maximum is observed.

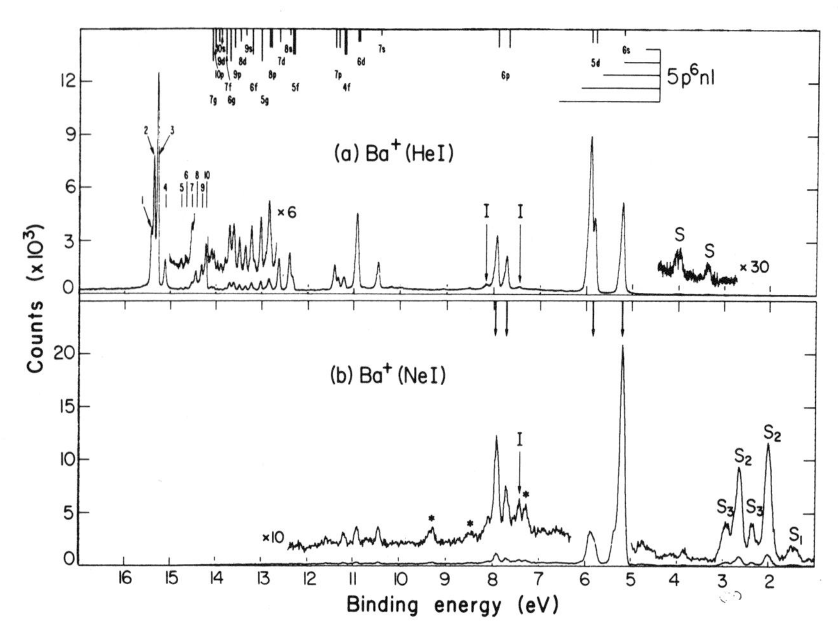

Fig. 50. Ba photoelectron spectra excited by He*I* and Ne*I* radiation. The satellite peaks beyond 10 eV binding energy in the He*I* spectrum are interpreted as due to autoionization effects (cf. text)[19a]

Further insight into the mechanism of population of the Ba satellite states is gained by the use of synchrotron radiation to excite the spectra. This was done by ROSENBERG et al.[19b] and results obtained at two photon energies $h\nu = 19.94$ eV and $h\nu = 21.48$ eV are shown in Fig. 51. These two photon energies correspond to double-electron excitation from the Ba ground state to $\{5p^5\,5d\,6s\}\,nd$ configurations, n being equal to 5 and 6, respectively. The following autoionization transition will predominantly proceed via the Coster-Kronig process (where the nd electron plays a spectator role): $\{5p^5\,5d\,6s\}\,nd \rightarrow \{5p^6 + e^-\}\,nd$. The predominance of this transition explains the resonant enhancement in the two spectra of the $5p^6\,5d$ (upper spectrum) and $5p^6\,6d$ (lower spectrum) peaks. The same type of satellite population mechanism is likely to prevail in the He*I*-excited spectrum.

Autoionization may also occur to the final ionic configuration $5p^5\,6s\,5d$ using the He*I* photon energy. These autoionization electrons were first observed by HOTOP and MAHR[34] and have very low energies (~ 0.1 eV). The $5p^5\,6s\,5d$ Ba^+ states can further undergo Auger transitions to the $5p^6$ ground state of Ba*III*. These Auger electrons correspond to the peaks numbered 1–3 in the spectrum of Fig. 50. The ground state of Ba*III* may be reached also by direct double autoionization. This latter process, along with the two-step autoionization-Auger process, provides a mechanism to explain the fact that double ionization is more probable than single ionization for Ba at the He*I* photon energy, whereas the converse is true at the Ne*I* energy[51, 52].

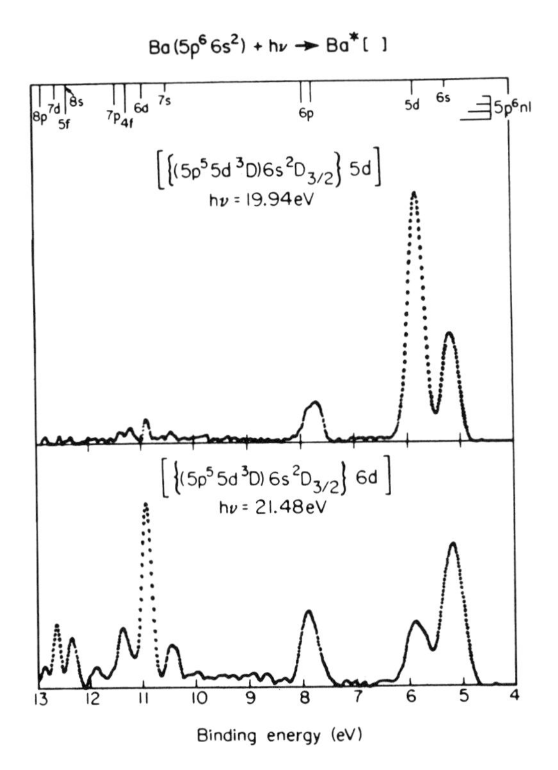

Fig. 51. Photoelectron spectra of atomic Ba taken at two autoionizing resonances (cf. text)[19b]

10. Tables of electron binding energies and relative multiplet intensities for atoms

α) Atomic binding energies. The following Table 11 of binding energies refers to values experimentally obtained for the free atomic state. Values are given for the main hole states, i.e. those final ionic configurations which correspond to the removal of one orbital from the ground state configuration. Satellite states, which result from additional orbital excitation, are thus not considered. It should be noted that several of the values given correspond to final ionic states, which should have zero or a very low intensity according to one-particle considerations in *LS*-coupling (cf. Sect. 10β). Correlation in the initial and final states as well as resonant photoionization (such as in Ba) may, however, lead to population of these states.

Unless otherwise indicated, the values and designations have been taken from MOORE[1] (conversion coefficient: $1{,}239854 \cdot 10^4$ eV · Å). Aside from photoelectron spectroscopy, core electron binding energies have been obtained from photoabsorption, x-ray emission and Auger electron spectroscopy.

[1] Moore, C.E. (1949, 1952, 1958): Atomic energy levels, National Bureau of Standards Circular No. 467, vols. 1–3

Table 11. Electron binding energies for free atoms

Atom	Shell	Term	J	Energy
H	1s			13.599
He	1s	^2S	1/2	24.587
Li	1s	^1S	0	66.31
		^3S	1	64.41 64.41[x1] 64.39[a,b]
	2s	^1S	0	5.392
Be	1s	^2S	1/2	123.6 (1)[b]
	2s	^2S	1/2	9.323
B	1s	$^{1,3}P$	av	200.8 (5)[x]
	2s	^1P	1	17.398
		^3P	2	12.930
			1	12.928
			0	12.927
	2p	^1S	0	8.298
C	1s	^4P	av	296.2 (5)[x]
	2s	^2P	3/2	24.981
			1/2	24.975
		^2S	1/2	23.223
		^2D	5/2	20.550
			3/2	20.550
		^4P	5/2	16.597
			3/2	16.594
			1/2	16.591
	2p	^2P	3/2	11.268
			1/2	11.260
N	2s	^1P	1	35.210
		^3S	1	33.767
		^1D	2	32.410
		^3P	0	28.075
			1	28.074
			2	28.074
		^3D	1	25.971
			2	25.971
			3	25.969
		^5S	2	20.333
	2p	^1S	0	18.586
		^1D	2	16.432
		^3P	2	14.549
			1	14.539
			0	14.533
O	2s	^2P	1/2	39.998
			3/2	39.977
		^2S	1/2	37.883
		^2D	3/2	34.199
			5/2	34.198
		^4P	1/2	28.507
			3/2	28.497
			5/2	28.476
	2p	^2P	1/2	18.636
			3/2	18.635
		^2D	3/2	16.945
			5/2	16.942
		^4S	3/2	13.618
F	2s	^1P	1	47.131
		^3P	0	37.915
			1	37.894
			2	37.856
	2p	^1S	0	22.992
		^1D	2	20.011
		^3P	0	17.484
			1	17.465
			2	17.423
Ne	1s	^2S	1/2	870.21 (5)[z1] 870.31 (2)[y1]
	2s	^2S	1/2	48.476 48.42 (5)[w]
	2p	^2P	1/2	21.662
			3/2	21.565
			1/2, 3/2	21.59[w]
Na	1s	$^{1,3}S$	0, 1	1,078.6 (3)[c] 1,079.1[y9]
	2s	^1S	0	71.27 (10)[d]
		^3S	1	70.86 (10)[d]
	2p	^1P	1	38.462
		^3P	0	38.155
			1	38.069
			2	37.986
	3s	^1S	0	5.139
Mg	1s	^2S	1/2	1,310.9 (3)[e,f] 1,311.5[y9]
	2s	^2S	1/2	96.5 (3)[g]
	2p	^2P	1/2	57.85 (10)[g] 57.82[h]
			3/2	57.57 (10)[g] 57.54[h]
	3s	^2S	1/2	7.646
Al	3s	^1P	1	13.406
		^3P	2	10.645
			1	10.630
			0	10.622
	3p	^1S	0	5.986
Si	3s	^2P	3/2	18.566
			1/2	18.541
		^2S	1/2	17.656
		^2D	5/2	15.010
			3/2	15.008
		^4P	5/2	14.496
			3/2	13.474
			1/2	13.461
	3p	^2P	3/2	8.187
			1/2	8.151
P	3s	^3S	1	24.157
		^1P	1	23.233

Table 11 (continued)

Atom	Shell	Term	J	Energy
		^1D	2	20.122
		^3P	0	20.012
			1	20.011
			2	20.005
		^3D	3	18.584
			2	18.580
			1	18.577
		^5S	2	16.152
	$3p$	^1S	0	13.162
		^1D	2	11.588
		^3P	2	10.545
			1	10.507
			0	10.487
S	$3s$	^2P	1/2	23.508
			3/2	23.453
		^4P	1/2	20.275
			3/2	20.249
			5/2	20.204
	$3p$	^2P	3/2	13.407
			1/2	13.401
		^2D	5/2	12.206
			3/2	12.202
		^4S	3/2	10.360
Cl	$3s$	^1P	1	27.358
		^3P	0	24.714
			1	24.672
			2	24.594
	$3p$	^1S	0	16.477
		^1D	2	14.463
		^3P	0	13.141
			1	13.104
			2	13.018
Ar	$1s$	^2S	1/2	3,205.9 (5)[w]
	$2s$	^2S	1/2	326.3 (1)[w] 326.37 (5)[y1]
	$2p$	^2P	1/2	250.56 (7)[w] 250.55 (5)[y3] 250.777 (10)[y2] 250.770 (20)[kk]
			3/2	248.45 (7)[w] 248.52 (5)[y3] 248.629 (10)[y2] 248.62 (8)[y6] 248.623 (17)[kk]
	$3s$	^2S	1/2	29.240 29.3 (1)[w]
	$3p$	^2P	1/2	15.937
			3/2	15.760
			1/2,3/2	15.81[w]
K	$2p_{1/2}\,4s_{1/2}$		0, 1	303.4 (1)[e] 303.2[y9]
	$2p_{3/2}\,4s_{1/2}$		1, 2	300.9 (1)[e] 300.5[y9]
	$3p_{1/2}\,4s_{1/2}$		1	24.979
			0	24.817
	$3p_{3/2}\,4s_{1/2}$		1	24.580
			2	24.489
	$4s$	^1S	0	4.341
Ca	$2p$	^2P	1/2	360.5 (5)[l] 360.3 (10)[e]
			3/2	357.1 (5)[l] 356.6 (10)[e]
	$3p$	^2P	1/2	34.67 (7)[k] 34.66 (1)[m]
			3/2	34.33 (7)[k] 34.31 (1)[m]
	$4s$	^2S	1/2	6.113 6.11 (2)[aa]
Sc	$3d$	^1S	0	7.995
	$4s$	^1D	2	6.855
		^3D	3	6.562
			2	6.549
			1	6.540
Ti	$3d$	^2D	5/2	9.944
			3/2	9.915
	$4s$	^2D	5/2	7.904
			3/2	7.900
		^2F	7/2	7.428
			5/2	7.394
		^4F	9/2	6.869
			7/2	6.848
			5/2	6.832
			3/2	6.820
V	$4s$	^3D	1	12.220
			2	12.215
			3	12.208
		^1F	3	10.984
		^3F	2	10.543
			3	10.539
			4	10.536
		^1D	2	9.863
		^1H	5	9.640
		^1P	1	9.502
		^3D	3	9.297
			2	9.296
			1	9.284
		^3H	6	9.265
			5	9.254
			4	9.250
		^3P	2	9.262
			1	9.231
			0	9.239
		^3P	0	9.116

Table 11 (continued)

Atom	Shell	Term	J	Energy	Atom	Shell	Term	J	Energy
			1	9.116				7/2	13.047
			2	9.112			^2D	5/2	12.435
		^1G	4	9.110				3/2	12.428
		^3G	5	8.790			^2D	3/2	12.095
			4	8.776				5/2	12.084
			3	8.766			^2S	1/2	11.776
		^5P	3	8.444			^2I	11/2	11.753
			2	8.426				13/2	11.750
			1	8.415			^2G	9/2	11.703
		^3F	4	7.868				7/2	11.685
			3	7.836			^2G	9/2	11.547
			2	7.811				7/2	11.540
		^5F	5	7.132			^4D	1/2	11.526
			4	7.108				3/2	11.522
			3	7.088				5/2	11.516
			2	7.073				7/2	11.510
			1	7.063			^2F	7/2	11.180
Cr	4s	^2G	9/2	13.252				5/2	11.175
			7/2	13.249			^2P	3/2	11.149
		^2D	3/2	12.639				1/2	11.063
			5/2	12.637			^2H	11/2	11.082
		^2S	1/2	12.259				9/2	11.059
		^2F	7/2	11.709			^4G	11/2	10.943
			5/2	11.693				9/2	10.934
		^2G	9/2	11.263				7/2	10.921
			7/2	11.241				5/2	10.909
		^2H	9/2	11.192			^4F	9/2	10.636
			11/2	11.180				7/2	10.630
		^4F	9/2	10.839				5/2	10.623
			7/2	10.837				3/2	10.619
			5/2	10.839			^4P	5/2	10.592
			3/2	10.832				3/2	10.523
		^2F	5/2	10.808				1/2	10.479
			7/2	10.777			^4H	13/2	10.533
		^2D	3/2	10.675				11/2	10.522
			5/2	10.652				9/2	10.512
		^2I	13/2	10.503				7/2	10.504
			11/2	10.503			^4D	7/2	9.248
		^4D	1/2	9.869				5/2	9.220
			3/2	9.870				3/2	9.199
			5/2	9.871				1/2	9.186
			7/2	9.869			^6D	9/2	8.315
		^4P	1/2	9.471				7/2	8.291
			3/2	9.471				5/2	8.271 } 8.29[ii]
			5/2	9.471				3/2	8.257
		^4G	11/2	9.309				1/2	8.248
			9/2	9.309	Mn	3d	^5D	4	14.340
			7/2	9.309				3	14.299
			5/2	9.309				2	14.268 } 14.26[ii]
		^6S	5/2	6.765				1	14.246
				6.76[ii]				0	14.234
	3d	^2F	9/2	13.050		4s	^5F	5	12.831

Table 11 (continued)

Atom	Shell	Term	J	Energy
			4	12.832
			3	12.815
			2	12.808
			1	12.804
		^3D	1	12.372
			2	12.370
			3	12.370
		^3P	0	11.951
			1	11.943
			2	11.932
		^3G	3	11.560
			4	11.556
			5	11.544
		^5D	0	11.503
			1	11.505
			2	11.508
			3	11.508
			4	11.499
		^5P	1	11.148
			2	11.144
			3	11.140
		^5G	2	10.855
			3	10.855
			4	10.854
			5	10.853
			6	10.850
		^5S	2	8.609; 8.61 [ii]
		^7S	3	7.34; 7.43 [ii]
Fe	$3d$	^6S	5/2	19.074
	$4s$	^2P	1/2	19.522
			3/2	19.380
		^4G	5/2	19.414
			7/2	19.404
			9/2	19.383
			11/2	19.336
		^4F	3/2	19.039
			5/2	19.027
			7/2	19.011
			9/2	18.990
		^4H	7/2	18.875
			9/2	18.859
			11/2	18.840
			13/2	18.818
		^4P	1/2	18.963
			3/2	18.887
			5/2	18.766
		^4D	1/2	17.280
			3/2	17.259
			5/2	17.224
			7/2	17.169
		^6D	1/2	16.304
			3/2	16.290
			5/2	16.266
			7/2	16.231
			9/2	16.183
Co	$3d$	^5D	0	–
			1	13.074
			2	13.040
			3	12.988
			4	12.911
	$4s$	^3D	1	11.286
			2	11.351
			3	11.273
		^3P	2	10.951
			1	–
			0	–
		^3P	0	10.892
			1	10.874
			2	10.850
		^5P	1	10.139
			2	10.101
			3	10.069
		^3F	2	9.269
			3	9.193
			4	9.082
		^5F	1	8.511
			2	8.479
			3	8.431
			4	8.365
			5	8.281
Ni	$4s$	^2G	7/2	11.667
			9/2	11.664
		^2P	1/2	11.304
			3/2	11.239
		^4P	5/2	10.739
			3/2	10.708
			1/2	10.714
		^2D	3/2	10.585
			5/2	10.500
		^2F	5/2	9.494
			7/2	9.315
		^4F	3/2	8.957
			5/2	8.889
			7/2	8.792
			9/2	8.676
Cu	$3d$	^1D	2	10.983; 10.98 [ii]
		^3D	1	10.702; 10.69 [ii]
			2	10.559; 10.55 [ii]

Table 11 (continued)

Atom	Shell	Term	J	Energy
			3	10.445 10.44[ii]
	$4s$	^1S	0	7.726 7.72[ii]
Zn	$2p$	^2P	1/2	1052.3 (2)[v] 1051.6 (3)[e]
			3/2	1029.1 (2)[v] 1028.7 (3)[e]
	$3p$	^2P	1/2	98.7 (7)[n]
			3/2	96.1 (5)[n]
	$3d$	^2D	3/2	17.508 17.50 (2)[aa]
			5/2	17.171 17.17 (2)[aa]
	$4s$	^2S	1/2	9.394 9.39 (2)[aa]
Ga	$4s$	^1P	1	14.764
		^3P	2	12.043
			1	11.927
			0	11.872
	$4p$	^1S	0	5.998
Ge	$4s$	^2P	3/2	19.307
			1/2	19.170
		^2S	1/2	18.535
		^2D	5/2	15.967
			7/2	15.946
		^4P	5/2	14.502
			3/2	14.369
			1/4	14.280
		^2P	3/2	8.105
			1/2	7.885
As	$4s$	^3P	2	20.369
			1	20.309
			0	20.289
		^1P	1	20.118
		^3D	3	19.020
			2	18.973
			1	18.959
	$4p$	^1S	0	12.617
		^1D	2	11.067
		^3P	2	10.130
			1	9.947
			0	9.815
Se	$4s$		1/2	22.892
			1/2	22.803
			3/2	22.755
			3/2	22.733
			5/2	22.353
		^2P	1/2	–
			3/2	22.188
	$4p$	^4P	1/2	20.469
			7/2	20.363
			5/2	20.152
		^2P	3/2	12.715
			1/2	12.609
		^2D	5/2	11.464
			3/2	11.385
		^4S	3/2	9.752
Br	$4s$	^3P	0	–
			1	24.097
			2	23.804
	$4p$	^1D	2	13.261
		^3P	0	12.323
			1	12.236
			2	11.847
Kr	$2s$	^2S	1/2	1924.6 (8)[w] 1921.2 (8)[y4]
	$2p$	^2P	1/2	1730.9 (5)[w] 1731.8 (1)[y5] 1727.2 (6)[y4]
			3/2	1678.4 (5)[w] 1679.3 (1)[y5] 1674.8 (6)[y4]
	$3s$	^2S	1/2	292.8 (3)[w]
	$3p$	^2P	1/2	221.8[z2]
			3/2	214.55 (15)[y6]
	$3d$	^2D	3/2	94.9 (2)[w] 95.04[y7]
			5/2	93.7 (2)[w] 93.80 (10)[y6] 93.83[y7]
	$4s$	^2S	1/2	27.515 27.51 (2)[w, z2]
	$4p$	^2P	1/2	14.666
			3/2	14.000
Rb	$3\underline{p}_{1/2}\ 5s_{1/2}$		0, 1	254.3[y9]
	$3\underline{p}_{3/2}\ 5s_{1/2}$		1, 2	245.4[y9]
	$3\underline{d}_{3/2}\ 5s_{1/2}$		1, 2	117.4[o]
	$3\underline{d}_{5/2}\ 5s_{1/2}$		2, 3	116.0[o]
	$4\underline{p}_{1/2}\ 5s_{1/2}$		1	21.965
			0	21.768
	$4\underline{p}_{3/2}\ 5s_{1/2}$		1	20.900
			2	20.710
	$5s$	^1S	0	4.177
Sr	$3p$		1/2	288.3 (7)[e]
			3/2	278.3 (7)[e]
	$3d$		3/2	144.0 (2)[n]
			5/2	142.3 (2)[n]
			3/2, 5/2	142.9 (7)[e, p]
	$4p$		1/2	29.17 (10)[k]
			3/2	28.21 (10)[k]
			1/2, 3/2	28.7 (7)[e]
	$5s$	^2S	1/2	5.694 5.69 (2)[aa]

Table 11 (continued)

Atom	Shell	Term	J	Energy	Atom	Shell	Term	J	Energy
Y	5*s*	^1D	2	6.787				4	7.322
		^3D	3	6.558				3	7.258
			2	6.508				2	7.209
			1	6.483				1	7.175
	4*d*	^1S	0	6.379		5*s*	^1D	2	10.734
Zr	4*d*	^2D	5/2	8.663 (8.596)			^1G	4	10.557
			3/2	8.610 (8.502)			^3P	0	10.352
	5*s*	^2G	7/2	8.596				1	10.329
			9/2	8.580				2	10.265
		^4P	5/2	7.932			^3F	4	9.545
			3/2	7.796				3	9.501
			1/2	7.768				2	9.444
		^2F	7/2	7.639			^1F	3	9.177
			5/2	7.550			^1S	0	9.015
		^2P	3/2	7.594			^1I	6	8.791
			1/2	7.546			^1G	4	8.716
		^2D	3/2	7.395			^3F	4	8.577
			5/2	7.363				3	8.580
		^4F	9/2	7.001				2	8.470
			7/2	6.931			^3D	1	8.509
			5/2	6.876				2	8.554
			3/2	6.837				3	8.501
Nb	4*d*	^1F	3	10.821			^1D	2	8.403
		^3F	2	10.034			^3G	5	8.236
			3	10.026				4	8.197
			4	10.027				3	8.153
		^1D	2	9.900			^3H	6	8.146
		^3P	2	9.550				5	8.099
			1	9.491				4	8.062
			0	9.405			^3P	2	7.783
		^1H	5	9.495				1	7.650
		^1P	1	9.417				0	7.572
		^3D	3	9.324			^5D	4	7.034
			2	9.282				3	6.982
			1	9.216				2	6.937
		^3H	4	9.049				1	6.902
			5	9.027				0	6.883
			6	9.043	Mo	4*d*	^4F	9/2	12.581
		^1G	4	8.893			^2F	7/2	12.430
Nb	4*d*	^3G	5	8.873				5/2	12.422
			4	8.860			^2S	1/2	12.291
			3	8.811			^2D	5/2	12.235
		^3P	0	8.703				3/2	12.220
			1	8.696			^2D	5/2	12.048
			2	8.700				3/2	11.965
		^5P	3	8.289			^2G	9/2	11.818
			2	8.226				7/2	11.740
			1	8.203			^2F	7/2	11.655
		^3F	4	7.914				5/2	11.599
			3	7.862			^2I	13/2	11.451
			2	7.813				11/6	11.413
		^5F	5	7.397			^2P	3/2	11.367

Table 11 (continued)

Atom	Shell	Term	J	Energy	Atom	Shell	Term	J	Energy
			1/2	11.082			^2D	5/2	9.882
		^4D	7/2	11.259				3/2	9.830
			5/2	11.219			^4D	7/2	9.201
			3/2	11.262				5/2	9.250
			1/2	11.216				3/2	9.229
		^2H	11/2	11.265				1/2	9.182
			9/2	11.197			^4P	5/2	9.045
		^2G	9/2	11.222				3/2	9.044
			7/2	11.209				1/2	9.020
		^4G	11/2	10.867			^4G	11/2	9.015
			9/2	10.845				9/2	9.012
			7/2	10.821				7/2	9.000
			5/2	10.782				5/2	8.984
		^4F	9/2	10.699			^6S	5/2	7.099
			7/2	10.694	Tc	5*s*	^5G	6	10.254
			5/2	10.680				5	10.257
			3/2	10.681				4	10.251
		^4P	5/2	10.698				3	10.243
			3/2	10.525				2	10.234
			1/2	10.398			^5S	2	8.842
		^4H	13/2	10.461			^7S	3	7.278
			11/2	10.415	Ru	4*d*	^2D	5/2	12.773
			9/2	10.384				3/2	12.666
			7/2	10.328			^2S	1/2	12.353
		^4D	7/2	10.241			^2D	3/2	12.290
			5/2	10.213				5/2	12.199
			3/2	10.157			^2G	7/2	12.075
			1/2	10.121				9/2	12.007
		^3D	9/2	8.768			^2G	7/2	11.893
			7/2	8.699				9/2	11.831
			5/2	8.639			^2I	11/2	11.858
			3/2	8.591				13/2	11.822
			1/2	8.560			^2F	5/2	11.742
	5*s*	^2G	9/2	12.345				7/2	11.586
			7/2	12.328			^2P	1/2	11.841
		^2D	5/2	11.176				3/2	11.549
			3/2	11.150			^4D	7/2	11.499
		^2S	1/2	10.689				5/2	11.460
		^2H	11/2	10.525				3/2	11.444
			9/2	10.513				1/2	11.453
		^2F	7/2	10.498			^2H	9/2	11.419
			5/2	10.440				11/2	11.411
		^2G	9/2	10.332			^4G	5/2	11.122
			7/2	10.290				7/2	11.140
		^2F	7/2	10.138				9/2	11.098
			5/2	10.098				11/2	10.964
		^4F	9/2	10.054			^4P	1/2	11.146
			7/2	10.052				3/2	10.973
			5/2	10.042				5/2	10.703
			3/2	10.017			^4F	3/2	10.899
		^2I	13/2	9.982				5/2	10.855
			11/2	9.949				7/2	10.831

Table 11 (continued)

Atom	Shell	Term	J	Energy
			9/2	10.781
		^4H	7/2	10.648
			9/2	10.604
			11/2	10.603
			13/2	10.584
		^4D	1/2	10.050
			3/2	10.000
			5/2	9.910
			7/2	9.769
		^6D	1/2	8.805
			3/2	8.767
			5/2	8.711
			7/2	8.624
			9/2	8.501
	5*s*	^2D	5/2	11.684
			3/2	11.680
		^2F	7/2	10.129
			5/2	10.039
		^2H	9/2	9.365
			11/2	9.184
		^2D	3/2	9.476
			5/2	9.172
		^2P	1/2	9.201
			3/2	8.972
		^2G	7/2	8.890
			9/2	8.713
		^4P	1/2	8.528
			3/2	8.417
			5/2	8.390
		^4F	3/2	7.751
			5/2	7.675
			7/2	7.555
			9/2	7.366
Rh	4*d*	^1D	2	12.779
		^1H	5	12.553
		^3D	1	12.260
			2	12.252
			3	12.043
		^3H	4	12.292
			5	12.026
			6	11.901
		^3P	0	–
			1	12.015
			2	11.972
		^1G	4	11.987
		^3P	0	11.805
			1	11.769
			2	11.709
		^3G	3	11.660
			4	11.506
			5	11.398
		^5P	1	11.068
			2	10.951
			3	10.911
		^3F	2	11.039
			3	10.866
			4	10.611
		^5F	1	10.091
			2	10.023
			3	9.918
			4	9.762
			5	9.557
	5*s*	^1G	4	9.305
		^3P	2	8.907
			1	8.767
			0	8.798
		^1D	2	8.476
		^3F	2	7.908
			3	7.761
			4	7.464
Pa	4*d*	^2D	3/2	8.775
			5/2	8.336
Ag	4*d*	^1D	2	13.285 13.28[ii] 13.286 (4)[jj]
		^3D	1	12.999 12.80[ii] 13.003 (4)[jj]
			2	12.628 12.62[ii] 12.631 (4)[jj]
		^3D	3	12.432 12.43[ii] 12.434 (3)[jj]
	5*s*	^1S	0	7.576 7.57[ii] 7.57 (1)[jj]
Cd	3*p*	^2P	1/2	659.2 (6)[e]
			3/2	625.0 (4)[e]
	3*d*	^2D	3/2	418.9 (3)[e] 418.8 (2)[v]
			5/2	412.2 (3)[e] 412.0 (2)[v]
	4*d*	^2D	3/2	18.279 18.27 (2)[aa] 18.28 (2)[hh]
			5/2	17.581 17.58 (2)[aa] 17.57 (2)[hh]
	5*s*	^2S	1/2	8.994 8.99 (2)[aa] 8.98 (2)[hh]
In	5*s*	^1P	1	13.602
		^3P	2	11.468

Table 11 (continued)

Atom	Shell	Term	J	Energy
			1	11.161
			0	11.028
		^1S	0	5.786
Sn	$5s$	^2P	3/2	17.476
			1/2	17.319
		^2S	1/2	17.288
	$5s$	^2D	5/2	14.716
			3/2	14.640
	$5s$	^4P	5/2	13.634
			3/2	13.341
			1/2	13.105
	$5p$	^2P	3/2	7.871
			1/2	7.344
Sb	$5s$	^1P	1	22.677
		^3S	1	22.487
		^1D	2	18.361
		^3P	0	–
			1	18.206
		^3D	3	17.059
			2	16.887
			1	16.861
	$5p$	^1S	0	11.606
		^1D	2	10.288
		^3P	2	9.343
			1	9.021
			0	8.642
Te	$5s$	^4P	1/2	–
			3/2	18.295
			5/2	17.837
	$5p$	^2P	3/2	11.989
			1/2	11.557
		^2D	5/2	10.550
			3/2	10.277
		^4S	3/2	9.010
I	$5s$	^1P	1	22.355
		^3P	0	21.044
			1	20.900
			2	20.613
	$5p$	^1S	0	14.502
		^1D	2	12.159
		^3P	0	11.257
			1	11.336
			2	10.457
Xe	$2s$	^2S	1/2	5453.2 (4)[w]
	$2p$	^2P	1/2	5107.2 (4)[w]
			3/2	4787.3 (4)[w]
	$3s$	^2S	1/2	1148.7 (5)[w]
	$3p$	^2P	1/2	1002.1 (3)[w]
			3/2	940.6 (2)[w]
	$3d$	^2D	3/2	689.35 (8)[z2]
				689.0 (2)[w]
			5/2	676.70 (8)[z2]
				676.4 (1)[w]
	$4s$	^2S	1/2	213.32 (3)[z2]
				213.2 (2)[w]
	$4p$	^2P	3/2	145.51 (2)[z2]
				145.5 (2)[w]
	$4d$	^2D	3/2	69.48 (1)[z2]
				69.5 (1)[w]
				69.52 [y7]
			5/2	67.50 (1)[z2]
				67.5 (1)[w]
				67.55 [y7]
	$5s$	^2S	1/2	23.397
				23.40 (1)[z2]
				23.3 (1)[w]
	$5p$	^2P	1/2	13.437
			3/2	12.130
Cs	$3\underline{d}_{3/2}\ 6s_{1/2}$		1, 2	745.8[y9]
	$3\underline{d}_{5/2}\ 6s_{1/2}$		2, 3	731.8[y9]
	$5\underline{p}_{1/2}\ 6s_{1/2}$		1	19.128
			0	19.066
	$5\underline{p}_{3/2}\ 6s_{1/2}$		1	17.273
			2	17.209
	$6s$	^1S	0	3.894
Ba	$4d$	^2D	3/2	101.0 (1)[e]
				101.0 (2)[q]
			5/2	98.5 (1)[e]
				98.3 (2)[q]
	$5p$	^2D	1/2	24.71 (10)[e]
			3/2	22.76 (10)[e]
	$6s$	^2S	1/2	5.211
				5.21 (3)[aa]
La	$5d$	^1S	0	6.532
	$6s$	^3D	3	6.019
			2	5.937
			1	5.851
		^1D	2	5.789
Sm	$4f$	^6H	5/2, 7/2	8.8 (1)[bb]
		^6F	5/2, 7/2	9.6 (1)[bb]
		^6P	5/2	11.9 (1)[bb]
	$6s$	^8F	1/2	5.64 (2)[bb]
				5.63 [cc]
		^6F	1/2	5.80 (2)[bb]
				5.82 [cc]
Eu	$4f$	^7F	6	10.55 (2)[bb]
			5	10.46 (2)[bb]
			4	10.37 (2)[bb]
			3	10.25 (2)[bb]
			2	10.15 (2)[bb]
			1	10.07 (2)[bb]
			0	9.99 (3)[bb]

Table 11 (continued)

Atom	Shell	Term	J	Energy
	$6s$	^7S	3	5.88 (2)[bb] 5.877[dd]
		^9S	4	5.67 (2)[bb] 5.670[dd]
Yb	$4f$	^2F	7/2	8.91 (2)[bb] 8.910[ee]
			5/2	10.17 (2)[bb] 10.168[ee]
	$6s$	^2S	1/2	6.25 (2)[bb] 6.254[ee]
Hf	$6s$	^2G	7/2	9.201
			9/2	9.161
		^2P	3/2	9.216
			1/2	8.896
		^2D	5/2	9.159
			3/2	8.786
		^2F	7/2	8.875
			5/2	8.502
		^4P	5/2	8.677
			3/2	8.607
			1/2	8.487
		^4F	9/2	8.042
			7/2	7.792
			5/2	7.613
			3/2	7.457
	$5d$	^2D	5/2	7.383
			3/2	7.005
Re	$5d$	^5D	1	9.715
			2	9.656
			3	9.728
			4	9.722
	$6s$	^3G	5	11.099
			4	11.130
			3	11.016
			3	10.947
			4	10.839
			5	10.776
			1	10.747
			2	10.672
		^5P	1	10.818
			2	10.771
			3	10.558
		^5G	6	10.608
			5	10.478
			4	10.414
			3	10.250
			2	10.213
		^5S	2	10.012
		^7S	3	7.877
Os	$6s$		5/2	11.890
			7/2	11.832
			5/2	11.768
			9/2	11.412
			7/2	11.213
			5/2	11.164
			11/2	10.928
			5/2	10.913
			3/2	10.895
			7/2	10.873
			9/2	10.670
			5/2	10.398
			7/2	10.372
			3/2	10.364
			5/2	10.180
			7/2	10.156
			5/2	9.713
		^6D	1/2	9.558
			3/2	9.428
			5/2	9.222
			7/2	9.180
			9/2	8.735
Ir	$6s$	^3F	4	9.0
Pt	$5d$	^2P	3/2	12.961
			1/2	12.344
		^2D	5/2	13.046
			3/2	11.924
		^2G	9/2	12.592
			7/2	12.564
		^2F	5/2	11.873
			7/2	11.208
		^4P	1/2	11.657
			3/2	11.589
			5/2	11.050
		^4F	3/2	10.922
			5/2	10.617
			7/2	10.124
			9/2	9.558
	$6s$	^2D	3/2	10.008
			5/2	8.964
Au	$5d$	^1D	2	12.898 12.89[ii]
		^3D	1	12.668 12.66[ii]
			2	11.414 11.41[ii]
			3	11.090 11.08[ii]
	$6s$	^1S	0	9.226 9.22[ii]
Ta	$5d$	^1G	4	9.647
		^1D	2	9.567
		^3F	4	9.094
			3	8.732
			2	8.280

Table 11 (continued)

Atom	Shell	Term	J	Energy
		^3P	2	8.587
			1	8.546
			0	8.397
	$6s$	^1F	3	10.969
		^1P	1	10.787
		^3P	2	–
			1	–
			0	10.784
		^1D	2	10.774
		^3F	4	10.747
			3	10.814
			2	10.728
		^3P	2	10.179
			1	10.040
			0	–
		^3D	3	10.186
			2	10.014
			1	9.699
		^3H	6	9.851
			5	–
			4	–
		^3F	4	10.178
			3	9.693
			2	9.087
		^3G	5	9.476
			4	9.461
			3	9.344
		^5P	3	9.427
			2	9.358
			1	9.214
		^5F	5	8.653
			4	8.433
			3	8.213
			2	8.013
			1	7.886
W	$6s$	^2H	11/2	11.324
			9/2	11.228
		^4D	7/2	10.842
		^2P	3/2	10.775
		^4H	9/2	10.561
		^4H	13/2	10.395
		^2P	1/2	10.390
		^4H	7/2	10.216
		^4G	11/2	10.147
			9/2	10.037
			7/2	10.042
			5/2	9.998
		^4D	7/2	9.863
			5/2	9.840
			3/2	9.799
			1/2	9.618
		^4F	9/2	9.827

Atom	Shell	Term	J	Energy
			7/2	9.648
			5/2	9.386
			3/2	9.065
		^4P	5/2	9.650
			3/2	9.298
			1/2	9.080
		^6D	9/2	8.747
			7/2	8.569
			5/2	8.378
			3/2	8.173
			1/2	7.985
Hg	$4s$	^2S	1/2	809.4[x2]
	$4p$	^2P	1/2	686.4[x2]
			3/2	583.8[x2]
	$4d$	^2D	3/2	385.4[x2]
			5/2	366.0[x2]
	$5s$	^2S	3/2	134.0[x2]
	$4f$	^2F	5/2	111.1[x2] 111.0 (3)[u]
			7/2	107.1[x2] 107.2 (3)[u]
	$5p$	^2P	1/2	90.3[x2]
			3/2	71.7[x2]
	$5d$	^2D	3/2	16.705 16.7[x2] 16.70 (2)[aa]
			5/2	14.841 14.9[x2] 14.84 (2)[aa]
	$6s$	^2S	1/2	10.438 10.43 (2)[aa]
Tl	$5d$	1°	2	13.686
	$6s$	^1P	1	9.381
		^3P	2	7.653
			1	6.496
			0	6.131
	$6p$	^1S	0	6.108
Pb	$4\underline{f}_{5/2}$	$6p^2\,(^3P_0)$	5/2	148.5 (1)[s]
	$4\underline{f}_{7/2}$	$6p^2\,(^3P_0)$	7/2	144.1[r]
				143.6 (1)[s]
	$5\underline{d}_{3/2}$	$6p^2\,(^3P_0)$	3/2	28.25 (5)[t]
	$5\underline{d}_{5/2}$	$6p^2\,(^3P_0)$	5/2	25.28 (5)[t]
	$6s$	^2S	1/2	20.348
		^2P	3/2	20.394
			1/2	18.358 18.35 (1)[ff]
		^2D	5/2	18.448
			3/2	17.718
		^4P	5/2	16.580 16.57 (1)[ff]
			3/2	15.615 15.61 (1)[ff]

Table 11 (continued)

Atom	Shell	Term	J	Energy
			1/2	14.597 14.59 (1)[ff]
	$6p$	^2P	3/2	9.163 9.16 (1)[ff]
			1/2	7.417 7.42 (1)[ff]
Bi	$4f_{7/2}\,6p^3\,(^4S_{3/2})$		2–5	164.9 [r]
	$6s$	^3D	3	19.199
			2	19.059
			1	18.998
		^5S	2	16.730 16.73 (3)[gg]
	$6p$	^1S	0	12.766
		^1D	2	11.497 11.49 (3)[gg]

Atom	Shell	Term	J	Energy
		^3P	2	9.401 9.40 (3)[gg]
			1	8.941 8.94 (3)[gg]
			0	7.289 7.29 (3)[gg]
Po	$6p$	^4S	3/2	8.429
Rn	$6p$	^2P	3/2	10.749
Ra	$7s$	^2S	1/2	5.279
Ac	$7s$	^1D	2	8.027
		^3D	3	7.821
			2	7.553
			1	7.488
	$6d$	^1S	0	6.9 (6)

References for Table 11

[a] H. Prömpeler, Diplom-Thesis, University of Freiburg (1976)
[b] D.L. Ederer, T. Lacatorto, R.P. Madden, Phys. Rev. Lett. *25*, 1537 (1970)
[c] H. Hillig, B. Cleff, W. Mehlhorn and W. Schmitz, Z. Physik *268*, 225 (1974)
[d] E. Breuckmann, B. Breuckmann, W. Mehlhorn and W. Schmitz, J. Phys. B *10*, 3135 (1977)
[e] B. Breuckmann, Ph.D. Thesis, University of Freiburg (1977)
[f] B. Breuckmann, V. Schmidt, Z. Physik *268*, 235 (1974)
[g] B. Breuckmann, V. Schmidt and W. Schmitz, J. Phys. B *9*, 3037 (1976)
[h] G.H. Newson, Astrophys. J. *166*, 243 (1971)
[i] G. Mehlmann, J.M. Esteva, Astrophys. J. *188*, 191 (1974)
[k] W. Schmitz, B. Breuckmann and W. Mehlhorn, J. Phys. B *9*, L493 (1976)
[l] M.W.D. Mansfield, Proc. Roy. Soc. London *A 348*, 143 (1976)
[m] M.W.D. Mansfield, G.H. Newson, Proc. Roy. Soc. London *A 357*, 77 (1977)
[n] D. Hausaman. Dipl.-Thesis. University of Freiburg (1977)
[o] M.W.D. Mansfield and J.P. Connerade, Proc. Roy. Soc. London *A 344*, 303 (1975)
[p] M.W.D. Mansfield and J.P. Connerade, Proc. Roy. Soc. London *A 342*, 421 (1975)
[q] D.L. Ederer, T.B. Lucatorto, E.B. Saloman, R.P. Madden and J. Sugar, J. Phys. B *8*, L21 (1975)
[r] Y.S. Khodeyev, H. Siegbahn, K. Hamrin and K. Siegbahn, Chem. Phys. Lett. *19*, 16 (1973)
[s] J.P. Connerade and M.W.D. Mansfield, Proc. Roy. Soc. London *A 348*, 235 (1976)
[t] A. Reiter, Dipl.-Thesis, University of Freiburg (1977)
[u] J.P. Connerade and M.W.D. Mansfield, Proc. Roy. Soc. London *A 335*, 87 (1973)
[v] M.S. Banna, D.C. Frost, C.A. McDowell and B. Wallbank, J. Chem. Phys. *68*, 696 (1978)
[w] K. Siegbahn, C. Nordling, G. Johansson, J. Hedman, P.F. Heden, K. Hamrin, U. Gelius, T. Bergmark, L.O. Werme, R. Manne and Y. Baer, ESCA Applied to Free Molecules, North-Holland Publ. Co., Amsterdam (1969)
[x] P. Bisgaard, R. Bruch, P. Dahl, B. Fastrup and M. Rødbro, Phys. Scr. *17*, 49 (1978)
[x1] S. Bashkin and J.O. Stoner, Atomic Energy Levels and Grotrian Diagrams, Vol. I, North-Holland Publ. Co., Amsterdam (1975)
[x2] S. Svensson, N. Mårtensson, E. Basilier, P.Å. Malmqvist, U. Gelius and K. Siegbahn, J. Electron Spectrosc. *9*, 51 (1976)
[y1] T.D. Thomas and R.W. Shaw Jr., J. Electron Spectrosc. *5*, 1081 (1974)
[y2] L. Pettersson, J. Nordgren, L. Selander, C. Nordling, K. Siegbahn and H. Ågren, J. Electron Spectrosc. (to be published)

[y3] M. Nakamura, M. Sasanuma, S. Sato, M. Watanabe, H. Yamashita, Y. Iguchi, A. Ejiri, S. Nakai, S. Yamaguchi, T. Sagawa, Y. NaKai and T. Oshio, Phys. Rev. Lett. *21*, 1303 (1968)

[y4] M.O. Krause, Phys. Rev. *140*, A1845 (1965)

[y5] F. Wuilleumier, Thesis, University of Paris (1969)

[y6] G. Johansson, J. Hedman, A. Berndtsson, M. Klasson and R. Nilsson, J. Electron Spectrosc. *2*, 295 (1973)

[y7] K. Codling and R.P. Madden, Phys. Rev. Lett. *12*, 106 (1964)

[y8] K. Codling and R.P. Madden, Applied Optics *4*, 1431 (1965)

[y9] M.S. Banna, B. Wallbank, D.C. Frost, C.A. McDowell and J.S.H.Q. Perera, J. Chem. Phys. *68*, 5459 (1978)

[z1] H. Ågren, J. Nordgren, L. Selander, C. Nordling and K. Siegbahn, J. Electron Spectrosc. Rel. Phen. *14*, 27 (1978)

[z2] S. Svensson, N. Mårtensson, E. Basilier, P.Å. Malmquist, U. Gelius and K. Siegbahn, Phys. Scr. *14*, 141 (1976)

[aa] S. Süzer, S.-T. Lee and D.A. Shirley, Phys. Rev. *A 13*, 1842 (1976)

[bb] S.-T. Lee, S. Süzer, E. Matthias, R.A. Rosenberg and D.A. Shirley, J. Chem. Phys. *66*, 2496 (1977)

[cc] J. Blaise, C. Monillon, M.-G. Schweighofer and J. Verges, Spectrochim. Acta *24 B*, 405 (1962); W. Albertson, Astrophys. J. *84*, 26 (1936)

[dd] A.C. Parr, J. Chem. Phys. *54*, 3161 (1971); H.M. Russel, W. Albertson and D.N. Davies, Phys. *60*, 641 (1941)

[ee] V. Kaufman and J. Sugar, J. Opt. Soc. Am. *63*, 1168 (1973); W.F. Meggers, J. Res. NBS, 396 (1967); A.C. Parr and F.A. Elder, J. Chem. Phys. *49*, 2665 (1968)

[ff] S. Süzer, M.S. Banna and D.A. Shirley, J. Chem. Phys. *63*, 3473 (1975)

[gg] S. Süzer, S.-T. Lee and D.A. Shirley, J. Chem. Phys. *65*, 412 (1976)

[hh] S. Süzer and D.A. Shirley, J. Chem. Phys. *61*, 2481 (1974)

[ii] J.M. Dyke, N.K. Fayad, A. Morris and I.R. Trickle, J. Phys. B *12*, 2985 (1979)

[jj] M.O. Krause, J. Chem. Phys. *72*, 6474 (1980)

[kk] G.C. King, M. Tronc, F.H. Read and R.C. Bradford, J. Phys. B *10*, 2479 (1977)

β) Relative multiplet intensities for ionization of open shell atoms. The following Table 12 contains multiplet intensities calculated in the one-particle framework by COX[2] and BEATHAM et al.[3]. Full Russel-Saunders (LS) coupling is assumed throughout, except for p^n- and f^3-, f^{13}-configurations where calculations have also been performed in jj- and intermediate coupling. In LS-coupling, the relative multiplet intensity for ionic state $S'L'J'$ when ionizing the nl subshell of the SLJ initial state is given by [2]:

$$I(n\,l\,\alpha\,S'L'J';\,SLJ)=I(n\,l\,\alpha\,S'L';\,SL)\cdot(2J'+1)\cdot(2S+1)(2L+1)\sum_j(2j+1)\begin{pmatrix}1/2 & l & j\\ S & L & J\\ S' & L' & J'\end{pmatrix}^2. \quad (10.1)$$

The summation is over the two values $j=l-1/2$, $j=l+1/2$ and the last bracket is a $9j$-coefficient. α takes the values 1, 2, 3, ... to distinguish between repeated states with the same values of $S'L'J'$ which may occur in the ionized configuration. $I(nl\alpha S'L';\,SL)$, which is the relative intensity of the state $|nlS'L'\rangle$ without spin-orbit coupling, is obtained from fractional parentage coefficients

[2a] Cox. P.A. (1975): in: Structure and Bonding, vol. 24, Dunitz, J.D., et al. (eds.), Berlin, Heidelberg, New York: Springer

[2b] Cox, P.A.: Private communication. We are greatly indebted to Dr. Cox for performing the calculations on which this table is based

[3] Beatham, N., Cox, P.A., Orchard, A.F., Grant, I.P. (1979): Chem. Phys. Lett. *63*, 69

e.g.[4]:

$$I(nl\alpha S'L'; SL) = N_l |\langle \alpha S'L'l | \} SL \rangle|^2. \tag{10.2}$$

The intensities are given in units of the average one-electron cross section for the shell being ionized, i.e. the total summed intensity for a configuration is equal to the number of electrons. Cases marked by * contain repeated SL-

Table 12. Relative multiplet intensities for ionization of open-shell configurations

a) p^n-configurations, LS-coupling

Initial state		Final (LS) state and intensity		Multiplet components (J) and intensities	
p^2	$^3P_0$	^2P	2.000	1/2	1.333
				3/2	0.667
p^3	$^4S_{3/2}$	^3P	3.000	0	0.333
				1	1.000
				2	1.667
p^4	$^3P_2$	^4S	1.333	3/2	1.333
		^2D	1.667	3/2	0.167
				5/2	1.500
		^2P	1.000	1/2	0.167
				3/2	0.833
p^5	$^2P_{3/2}$	^2P	3.000	0	0.167
				1	0.750
				2	2.083
		^1D	1.667	2	1.667
		^1S	0.333	0	0.333

b) p^n-configurations, jj-coupling

	Initial (jj) state		Final (jj) state	Multiplet intensities	
p^2	$(1/2)^2$	0	$(1/2)^1$	1/2	2.000
p^3	$(1/2)^2\ (3/2)$	3/2	$(1/2)^2$	0	1.000
			$(1/2)^1\ (3/2)^1$	1	0.750
				2	1.250
p^4	$(1/2)^2\ (3/2)^2$	2	$(1/2)^2\ (3/2)^1$	3/2	2.000
			$(1/2)^1\ (3/2)^2$	3/2	0.800
				5/2	1.200
p^5	$(1/2)^2\ (3/2)^3$	3/2	$(1/2)^2\ (3/2)^2$	0	0.500
				2	2.500
			$(1/2)^1\ (3/2)^3$	1	0.750
				2	1.250
p^6	$(1/2)^2\ (3/2)^4$	0	$(1/2)^2\ (3/2)^3$	3/2	4.000
			$(1/2)^1\ (3/2)^4$	1/2	2.000

[4] Griffith, J.S. (1962): The irreducible tensor method for molecular symmetry groups, London: Prentice-Hall

Table 12 (continued)

c) d^n-configurations, LS-coupling

Initial state		Final (LS) state and intensity		Multiplet components (J) and intensities	
d^2	$^3F_2$	^2D	2.000	3/2	1.600
				5/2	0.400
d^3	$^4F_{3/2}$	^3F	2.400	2	1.646
				3	0.720
				4	0.034
		^3P	0.600	0	0.280
				1	0.280
				2	0.040
d^4	$^5D_0$	^4F	2.800	3/2	1.120
				5/2	1.680
		^4P	1.200	3/2	1.080
				5/2	0.120
d^5	$^6S_{5/2}$	^5D	5.000	0	0.200
				1	0.600
				2	1.000
				3	1.400
				4	1.800
d^6	$^5D_4$	^6S	1.200	5/2	1.200
		^4G	1.800	9/2	0.268
				11/2	1.500
		^4F	1.400	5/2	0.058
				7/2	0.357
				9/2	0.982
		^4D	1.000	3/2	0.058
				5/2	0.331
				7/2	0.612
		^4P	0.600	3/2	0.240
				5/2	0.360
d^7	$^4F_{9/2}$	^5D	2.500	2	0.107
				3	0.625
				4	1.768
		^3H	1.571	5	0.122
				6	1.444
		^3G	1.286	4	0.177
				5	1.100
		^3F	1.000*	3	0.174
				4	0.818
		^3D	0.214	3	0.179
		^3P	0.429*	2	0.429
d^8	$^3F_4$	^4F	3.200	5/2	0.209
				7/2	0.794
				9/2	2.183

Table 12 (continued)

Initial state		Final (*LS*) state and intensity		Multiplet components (*J*) and intensities	
		^4P	0.800	3/2	0.080
				5/2	0.720
		^2H	1.571	9/2	0.071
				11/2	1.500
		^2G	1.286	7/2	0.107
				9/2	1.179
		^2F	0.200	7/2	0.179
		^2D	0.714	3/2	0.071
				5/2	0.643
		^2P	0.229	3/2	0.229
d^9	$^2D_{5/2}$	^3F	4.200	2	0.333
				3	1.167
				4	2.700
		^3P	1.800	0	0.133
				1	0.500
				2	1.167
		^1G	1.800	4	1.800
		^1D	1.000	2	1.000
		^1S	0.200	0	0.200
d^{10}	^1S	^2D	10.000	3/2	4.000
				5/2	6.00

* Summed intensity for repeated (SL) states.

d) $d^n s^1$ configurations, *LS*-coupling

Initial state		Final (*LS*) state and intensity		Multiplet components (*J*) and intensities	
$d^4 s^1$	$(^5D)^a\ ^6D_{1/2}$	$(^4F)^a\ ^5F$	2.800	1	0.448
				2	1.680
				3	0.672
		$(^4P)\ ^5P$	1.200	1	0.072
				2	1.053
				3	0.075
$d^5 s^1$	$(^6S)\ ^7S_3$	$(^5D)\ ^6D$	5.000	1/2	0.333
				3/2	0.667
				5/2	1.000
				7/2	1.333
				9/2	1.667
$d^6 s^1$	$(^5D)\ ^6D_{9/2}$	$(^6S)\ ^5S$	0.033	2	0.033
		^7S	1.167	3	1.167
		$(^4G)\ ^5G$	1.800	5	0.306
				6	1.444
		$(^4F)\ ^5F$	1.400	3	0.083
				4	0.393
				5	0.917

Table 12 (continued)

Initial state		Final (*LS*) state and intensity		Multiplet components (*J*) and intensities	
		$(^4D)\ ^5D$	1.000	2	0.082
				3	0.357
				4	0.561
		$(^4P)\ ^5P$	0.600	2	0.267
				3	0.333
d^7s^1	$(^4F)\ ^5F_5$	$(^5D)\ ^4D$	0.100	7/2	0.079
		^6D	2.400	5/2	0.086
				7/2	0.648
				9/2	1.667
		$(^3H)\ ^4H$	1.571	11/2	0.160
				13/2	1.400
		$(^3G)\ ^4G$	1.286	9/2	0.225
				11/2	1.040
		$(^3F)\ ^4F$	1.000	7/2	0.218
				9/2	0.764
		$(^3D)\ ^4D$	0.214	7/2	0.168
		$(^3P)\ ^4P$	0.429	5/2	0.429
d^8s^1	$(^3F)\ ^4F_{9/2}$	$(^4F)\ ^3F$	0.200	4	0.164
		^5F	3.000	3	0.196
				4	0.766
				5	2.017
		$(^4P)\ ^3P$	0.050	2	0.050
		$(^2H)\ ^3H$	1.571	5	0.122
				6	1.444
		$(^2G)\ ^3G$	1.286	4	0.177
				5	1.100
		$(^2F)\ ^3F$	0.200	4	0.164
		$(^2D)\ ^3D$	0.714	2	0.119
				3	0.595
		$(^2P)\ ^3P$	0.228	2	0.228
d^9s^1	$(^2D)\ ^3D_3$	$(^3F)\ ^2F$	0.467	5/2	0.067
				7/2	0.400
		^4F	3.733	5/2	0.316
				7/2	0.991
				9/2	2.381
		$(^3P)\ ^2P$	0.200	3/2	0.156
		^4P	1.600	1/2	0.089
				3/2	0.551
				5/2	0.960
		$(^1G)\ ^2G$	1.800	7/2	0.133
				9/2	1.667
		$(^1D)\ ^2D$	1.000	3/2	0.200
				5/2	0.800
		$(^1S)\ ^2S$	0.200	1/2	0.200
$d^{10}s^1$	$(^1S)\ ^2S_{1/2}$	$(^2D)\ ^1D$	2.500	2	2.500
		^3D	7.500	1	1.500
				2	2.500
				3	3.500

[a] d^n parentage

Table 12 (continued)

e) $d^n s^2$-configurations, *LS*-coupling

Initial state		Final (*LS*) state and intensity		Multiplet components (*J*) and intensities	
$d^1 s^{2\dagger}$	$^2D_{3/2}$	^1D	0.500	2	0.500
		^3D	1.500	1	0.750
				2	0.750
$d^2 s^{2\dagger}$	$^3F_2$	^2F	0.667	5/2	0.667
		^4F	1.333	3/2	0.800
				5/2	0.533
$d^3 s^{2\dagger}$	$^4F_{3/2}$	^3F	0.750	2	0.750
		^5F	1.250	1	0.750
				2	0.500
$d^4 s^{2\dagger}$	$^5D_0$	^4D	0.800	1/2	0.800
		^6D	1.200	1/2	1.200
$d^5 s^{2\dagger}$	$^6S_{5/2}$	^5S	0.833	2	0.833
		^7S	1.167	3	1.167
$d^6 s^{2\dagger}$	$^5D_4$	^4D	0.800	7/2	0.800
		^6D	1.200	7/2	0.089
				9/2	1.111
$d^7 s^{2\dagger}$	$^4F_{9/2}$	^3F	0.750	4	0.750
		^5F	1.250	4	0.150
				5	1.100
$d^8 s^{2\dagger}$	$^3F_4$	^2F	0.667	7/2	0.667
		^4F	1.333	7/2	0.222
				9/2	1.111
$d^9 s^2$	$^2D_{5/2}$	^1D	0.500	2	0.500
		^3D	1.500	2	0.333
				3	1.167

† Ionization of *s* electron

f) f^n-configurations, *LS*-coupling

Initial state		Final (*LS*) state and intensity		Multiplet components (*J*) and intensities	
f^2	$^3H_4$	^2F	2.000	5/2	1.714
				7/2	0.286
f^3	$^4I_{9/2}$	^3H	2.333	4	1.890
				5	0.424
		^3F	0.667	2	0.563
f^4	$^5I_4$	^4I	2.545	9/2	1.903
				11/2	0.599

Table 12 (continued)

Initial state		Final (*LS*) state and intensity		Multiplet components (*J*) and intensities	
		^4G	0.955	5/2	0.658
				7/2	0.263
		^4F	0.500	3/2	0.371
				5/2	0.114
f^5	$^6H_{5/2}$	^5I	2.758	4	1.755
				5	0.919
		^5G	1.266	2	0.513
				3	0.575
				4	0.165
		^5F	0.500	1	0.168
				2	0.234
		^5D	0.476	0	0.149
				1	0.224
f^6	$^7F_0$	^6H	3.143	5/2	0.898
				7/2	2.245
		^6F	2.000	5/2	1.428
				7/2	0.571
		^6P	0.857	5/2	0.816
f^7	$^8S_{7/2}$	^7F	7.000	0	0.143
				1	0.429
				2	0.714
				3	1.000
				4	1.286
				5	1.571
				6	1.857
f^8	$^7F_6$	^8S	1.143	7/2	1.143
		^6I	1.857	15/2	0.303
				17/2	1.500
		^6H	1.571	11/2	0.101
				13/2	0.424
				15/2	1.030
		^6G	1.286	9/2	0.134
				11/2	0.434
				13/2	0.694
		^6F	1.000	7/2	0.138
				9/2	0.383
				11/2	0.460
		^6D	0.714	5/2	0.107
				7/2	0.307
				9/2	0.301
		^6P	0.429	5/2	0.230
				7/2	0.199
f^9	$^6H_{15/2}$	^7F	2.333	5	0.583
				6	1.674
		^5L	1.545	9	0.136
				10	1.400
		^5K	1.364	8	0.211
				9	1.131

Table 12 (continued)

Initial state		Final (*LS*) state and intensity		Multiplet components (*J*) and intensities	
		^5I	1.182*	7	0.240
				8	0.913
		^5H	1.000	6	0.233
				7	0.739
		^5G	0.817*	5	0.195
				6	0.605
		^5F	0.310*	5	0.232
		^5D	0.454*	4	0.454
f^{10}	$^5I_8$	^6H	2.800	11/2	0.138
				13/2	0.646
				15/2	2.004
		^6F	0.800	11/2	0.727
		^4M	1.462	21/2	1.375
		^4L	1.307	17/2	0.134
				19/2	1.167
		^4K	1.189*	15/2	0.161
				17/2	1.018
		^4I	1.000*	13/2	0.154
				15/2	0.837
		^4H	0.379*	13/2	0.318
		^4G	0.692*	11/2	0.611
		^4F	0.405*	9/2	0.405
f^{11}	$^4I_{15/2}$	^5I	3.182	6	0.169
				7	0.736
				8	2.263
		^5G	1.193	5	0.241
				6	0.913
		^5F	0.625	5	0.550
		^3M	1.462	10	1.400
		^3L	1.307	9	0.206
		^3K	1.154*	7	0.117
				8	1.033
		^3I	0.364*	7	0.320
		^3H	0.846*	6	0.747
		^3G	0.454*	5	0.415
		^3F	0.382*	4	0.382
f^{12}	$^3H_6$	^4I	3.677	11/2	0.212
				13/2	0.914
				15/2	2.536
		^4G	1.688	7/2	0.106
				9/2	0.419
				11/2	1.157
		^4F	0.667	7/2	0.151
				9/2	0.486
		^2L	1.545	17/2	1.500
		^2K	1.364	15/2	1.288
		^2I	0.263	13/2	0.242
		^2H	1.000	11/2	0.909
		^2G	0.396*	9/2	0.359
		^2F	0.470*	7/2	0.436
		^2D	0.296*	5/2	0.296

Table 12 (continued)

Initial state		Final (*LS*) state and intensity		Multiplet components (*J*) and intensities	
f^{13}	$^2F_{7/2}$	^3H	4.714	4	0.321
				5	1.375
				6	3.018
		^3F	3.000	2	0.357
				3	0.875
				4	1.768
		^3P	1.286	0	0.107
				1	0.375
				2	0.804
		^1I	1.857	6	1.857
		^1G	1.286	4	1.286
		^1D	0.714	2	0.714
		^1S	0.143	0	0.143

g) f^3-configuration, intermediate- and *jj*-coupling

Initial state		Final (*LS*) state		Multiplet components (*J*) and intensities Intermediate coupling		
				$Nd^{3+}(4f^3)$	$U^{3+}(5f^3)$	*jj*-limit
f^3	$^4I_{9/2}$	^3H	4	2.0110	2.1374	2.3571
			5	0.3086	0.1872	0
			6	0.0099	0.0050	0
		^3F	2	0.5866	0.6116	0.6429
			3	0.0747	0.451	0
			4	0.0026	0.0102	0

h) f^{13}-configuration, intermediate- and *jj*-coupling

Initial state		Final (*LS*) state		Multiplet components (*J*) and intensities Intermediate coupling		
				$Yb^{3+}(4f^{13})$	$No^{3+}(5f^{13})$	*jj*-limit
f^{13}	$^2F_{7/2}$	^3H	4	2.2081	2.2462	2.2500
			5	1.3750	1.3750	1.3750
			6	3.1071	3.1688	3.2500
		^3F	2	0.7782	1.1398	1.2500
			3	0.8750	0.8750	0.8750
			4	0.9503	1.0909	1.1250
		^3P	0	0.1635	0.2102	0.2500
			1	0.3750	0.3750	0.3750
			2	0.3578	0.1194	0.000
		^1I	6	1.7679	1.7060	1.6250
		^1G	4	0.2166	0.0379	0.000
		^1D	2	0.7390	0.6158	0.6250
		^1S	0	0.0865	0.0398	0.000

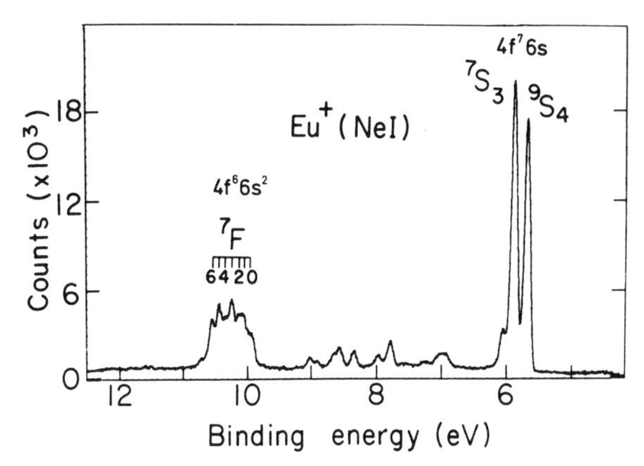

Fig. 52. Photoelectron spectrum of atomic Eu excited by NeI radiation[5]. Designations refer to final ionic states reached from the $4f^7\ 6s^2\ (^8S)$ ground state. Each photoelectron line is accompanied by a low energy satellite due to the NeI photon energy doublet

states. Without full diagonalization of the electrostatic matrix (which would give results varying from element to element) it is only possible to give summed intensities for these cases. Intensities are only given which are larger than 0.05 units ($\geqq 0.1$ for f^n-configurations).

As an example, we take the spectrum of atomic Eu shown in Fig. 52. In general, ionization of any of the closed shells in an open-shell species will only lead to final ionic states, in the one particle picture, where the coupling in the open shell remains unaltered (cf. e.g. the $d^n s^2$-cases below). Thus, removal of a $6s$-electron from Eu will result in two final ionic states of $^7S_3$ and $^9S_4$ symmetry, respectively. Inspection of the f^n-table shows that ionization of the Eu $4f$-shell is expected to lead to final ionic states of ^7F symmetry with J-values $0 \leqq J \leqq 6$. These states are observed as the band at binding energies slightly above 10 eV in the spectrum of Fig. 52. Qualitatively, one-particle considerations thus predict the correct multiplet structure, although the calculated relative J-intensities do not appear to agree with experiment. Discrepancies with respect to (10.1) may be due to deviations from pure LS-coupling[2]. The additional peaks observed in the Eu spectrum are satellite transitions, probably due to ISCI[5].

IV. Core level studies in molecules and condensed matter

11. Chemical shifts of core electron lines

α) Introduction. When an element changes its chemical state, the valence electron charge distribution is rearranged so as to provide a minimum total

[5] Lee, S.-T., Süzer, S., Matthias, E., Rosenberg, R., Shirley, D.A. (1977): J. Chem. Phys. *66*, 2496

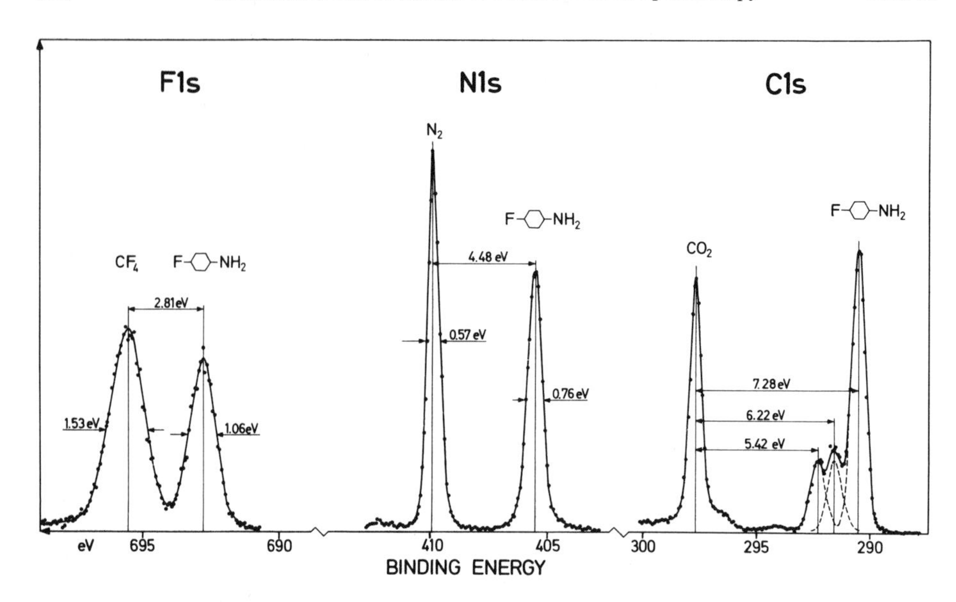

Fig. 53. Core electron spectrum of paraaminofluorobenzene, mixed with tetrafluoromethane (F 1*s*), nitrogen (N 1*s*) and carbon dioxide (C 1*s*)[1]

energy. The shape of the core electron orbitals, on the other hand, are essentially unchanged by chemical transformations. This has some important consequences for the core photoelectron spectra. First, the changes in core binding energies which occur (chemical shifts) can be treated and systematized by means of simple models, which relate observations to chemically relevant information. Second, the core photoelectron cross sections will be largely unchanged, which means that the core lines can be used for analytical applications.

The chemical shifts are small on a *relative* scale ($\Delta E_B/E_B \leqq 5\%$) but easily measurable with high accuracy on an absolute scale. As an example, in Fig. 53 the core photoelectron spectrum of paraaminofluorobenzene, three C1*s* peaks are resolved, corresponding to the different positions in the benzene ring. As will be discussed below, the positions of these peaks provide information on the charge distribution in the molecule. It can be noted from this example that core electron lines from one element in different chemical situations may have different widths. This feature, which is due to vibrational excitations in the core ionization, can be used to provide data on core hole potential energy surfaces.

The example shown in Fig. 54 demonstrates the quantitative and qualitative analytical aspect inherent in core electron spectra. The polar ligands in polymeric substances give rise to substantial shifts of the C1*s* levels, as in these cases of viton polymers. Measurement of the intensities of the peaks allows

---

[1] Siegbahn, K. (1977): in: Molecular spectroscopy, West, A.R. (ed.), London: Heyden, p. 227

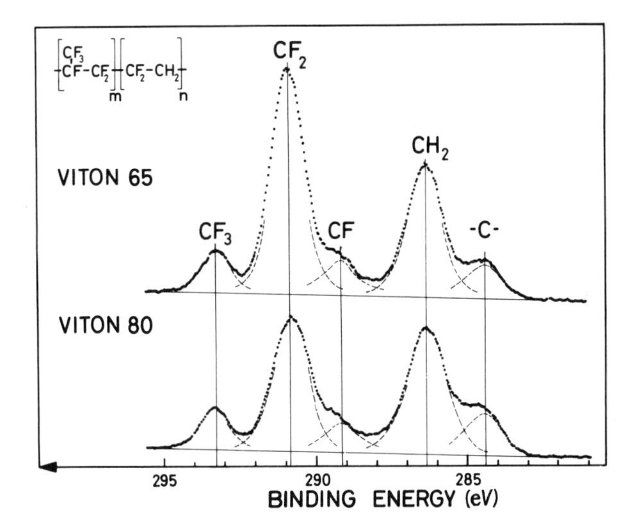

Fig. 54. C 1*s* spectra of viton 65 and viton 80 polymers[2]

determination of the ratios $m:n$. Polymers have become a major area of application for core electron spectroscopy e.g.[3-6].

In the following we shall consider first the binding energy aspect, i.e. chemical shifts, and the various models underlying the current interpretations of the shifts. Next, the formation and observation of multiline features in core electron spectra will be discussed, both due to many-electron transitions, vibrational excitations and multiplet splittings. These aspects affect intensities and widths of core electron lines and thus their use for quantitative chemical analysis. Finally, a number of studies will be presented illustrating the use of core level photoelectron spectroscopy in surface applications.

β) Basic model considerations. Models of chemical shifts have been discussed at several levels of approximation. The starting point of most models is to assume the ΔSCF-level of approximation:

$$\Delta E_{\mathrm{B}}(i) = -\Delta \mathcal{E}_i - \Delta E^{(i)}_{\mathrm{Relax}}. \tag{11.1}$$

In doing so, relativistic effects, the Breit interaction, radiative contributions and also correlation corrections have been neglected. Of these contributions the neglect of the first three are justified to within ~ 0.1 eV. The correlation

[2] Siegbahn, K. (1974): J. Electron Spectrosc. Relat. Phenom. 5, 3

[3] Clark, D.T., Dilks, A., Shuttleworth, D., Thomas, H.R. (1978): J. Electron Spectrosc. Relat. Phenom. *14*, 247

[4] Clark, D.T., Feast, W.J., Musgrave, W.K.R., Ritchie, I. (1975): J. Polymer Sci. *13*, 857

[5] Clark, D.T., Dilks, A. (1978): J. Polymer Sci. *16*, 911

[6] Clark, D.T., Shuttleworth, D. (1978): J. Polymer Sci. *16*, 1093

Table 13. Shifts in C1*s* energies for the fluoromethanes

Molecule	Experiment[a]	$-\Delta\mathcal{E}$[b]	$\Delta E_B(\Delta SCF)$[c]	$\Delta E_B(\Delta SCF+CI)$[d]
CH_4	0	0	0	0
CH_3F	2.8	2.89	2.88	2.72
CH_2F_2	5.55	5.93	6.02	5.69
CHF_3	8.1	8.81	9.23	8.68
CF_4	11.1	12.11	12.36	11.43

[a] Refs. [7, 8].
[b] Ref. [7].
[c] Ref. [9].
[d] Ref. [10].

correction to the shift may not be entirely negligible, as discussed in ref.[10] for the C1*s* shifts in the fluoromethanes, cf. Table 13. Comparison between experiment and theory suggests, however, that the ΔSCF-level of approximation is adequate for the overwhelming majority of cases.

The development of the description of shifts was initially centered on those methods, which treat only the shift in ground state orbital energy. The further development has led to models, which incorporate the relaxation term. *Ab initio* ΔSCF methods for the calculation of core binding energy shifts have been used for atoms and a fairly large number of small molecules e.g.[9, 11–23]. For larger molecular systems, however, such calculations are in general prohibitive in terms of computer time. The virtues of *ab initio* ΔSCF-calculations in this context lie mainly in the detailed consideration of special cases for tests of model assumptions and a better understanding of the physical mechanisms involved. For larger molecules and more extended classes of compounds, one is generally forced to rely on further approximations to (11.1), often involving semiempirical calculations. Below we will briefly review the basis of these more approximate treatments.

---

[7] Gelius, U., Hedén, P.F., Hedman, J., Lindberg, B.J., Manne, R., Nordberg, R., Nordling, C., Siegbahn, K. (1970): ibid. *2*, 70
[8] Davis, D.W., Shirley, D.A. (1974): J. Electron Spectrosc. Relat. Phenom. *3*, 137
[9] Clementi, E., Routh, A. (1972): Int. J. Quantum Chem. *6*, 525
[10] Levy, B., Millie, Ph, Ridard, J., Vink, J. (1974): J. Electron Spectrosc. Relat. Phenom. *4*, 13
[11] Clark, D.T., Scanlan, I.W., Müller, J. (1974): Theor. Chim. Acta *35*, 341
[12] Clark, D.T., Cromarty, B.J. (1977): Chem. Phys. Lett. *49*, 137
[13] Pireaux, J.J., Svensson, S., Basilier, E., Malmqvist, P.Å., Gelius, U., Caudano, R., Siegbahn, K. (1976): Phys. Rev. A*14*, 2133
[14] For references prior to 1973 cf. Ref. [33], p. 100
[15] Guest, M.F., Hillier, I.H., Saunders, V.R., Wood, M.H. (1973): Proc. R. Soc. A*333*, 201
[16] Ortenburger, I.B., Bagus, P. (1975): Phys. Rev. A*11*, 1501
[17] Ågren, H., Svensson, S., Wahlgren, U. (1975): Chem. Phys. Lett. *35*, 336
[18] Eade, R.H.A., Robb, M.A., Theodorakopoulos, G., Csizmadia, I. (1977): Chem. Phys. Lett. *52*, 526
[19] Goscinski, O., Hehenberger, M., Roos, B., Siegbahn, P. (1975): Chem. Phys. Lett. *33*, 427
[20] Meyer, W. (1971): Int. J. Quantum Chem. *5*, 34
[21] Broughton, J.Q., Bagus, P.S. (1980): J. Electron Spectrosc. Relat. Phenom. *20*, 127
[22] Butscher, W., Buenker, R.J., Peyerimhoff, S.D. (1977): Chem. Phys. Lett. *52*, 449
[23] Ewig, C.S., Mathews, R.D., Banna, M.S. (1982): J. Am. Chem. Soc. *103*, 5002

Table 14. Quantities relevant to calculation of orbital energy shifts in second row elements when a $2s$ electron is removed ($\Delta\mathcal{E}_{1s}^{2s}$). J_{1s2s}^0, K_{1s2s}^0 are direct and exchange integrals between $1s$ and $2s$ orbitals in the ground state and ΔJ_{1sj}^{2s} and ΔK_{1sj}^{2s} shifts in the integrals when a $2s$ electron is removed. $\Delta\langle h\rangle_{1s}^{2s}$ is the shift in kinetic energy plus nuclear attraction when a $2s$ electron is removed[24]

Atom	$\Delta\mathcal{E}_{1s}^{2s}$	J_{1s2s}^0	K_{1s2s}^0	$\Delta\langle h\rangle_{1s}^{2s}$	ΔJ_{1s1s}^{2s}	ΔJ_{1s2s}^{2s}	ΔK_{1s2s}^{2s}	ΔJ_{1s2p}^{2s}	ΔK_{1s2p}^{2s}	$\Delta[\sum_j(J_{1sj} - 1/2K_{1sj})]$	$\Delta\sum_j(J_{1sj})$
B	−13.1	17.6	1.1	0.0	0.1	1.1	0.1	2.9	0.2	−13.3	−13.6
C	−15.2	22.2	1.5	0.0	0.1	1.0	0.2	2.8	0.2	−15.3	−15.7
N	−17.0	26.7	1.9	0.1	0.1	1.0	0.2	2.6	0.2	−17.2	−17.6
O	−18.8	31.8	2.2	0.2	0.1	0.8	0.2	2.6	0.2	−19.1	−19.5
F	−20.6	35.3	2.5	0.2	0.1	1.0	0.2	2.6	0.2	−20.9	−21.4

γ) Orbital energy shifts. The orbital energy shift has several contributing terms and it can be expressed as (assuming doubly occupied spatial orbitals):

$$\Delta\mathcal{E}_i = \Delta\left[\langle i|h|i\rangle + \sum_j (2J_{ij} - K_{ij}) - \sum_{B \neq A} \langle i|\frac{Z_B}{r_B}|i\rangle\right], \tag{11.2}$$

where $\langle i|h|i\rangle$ is the kinetic and one-center nuclear attraction energy for core orbital i situated on atom A $\left(h = -\frac{1}{2}\nabla^2 - \frac{Z_A}{r_A}\right)$. J_{ij} and K_{ij} are the direct and exchange integrals between the core orbital i and the other molecular orbitals. The last electron-nuclear attraction sum extends over all nuclear centers B in the molecule except A. Successive approximations in this expression[7,24] show that the orbital energy shift is well described by considering only the shift in the classical electrostatic potential of the valence electrons and the other nuclei, i.e.:

$$\Delta\mathcal{E}_i \simeq \Delta\left[\sum_{\substack{j=\text{val.}\\ \text{orb.}}} 2J_{ij} - \sum_{B \neq A} \frac{Z_B^*}{R_{AB}}\right] = \Delta\langle V\rangle_i^{\text{val}}, \tag{11.3}$$

where the effective nuclear charge $Z_B^* = Z_B -$ no. of core electrons on B. We illustrate this by Table 14 showing quantities relevant to (11.2) for second row element atoms obtained from optimized HFS calculations.

Further approximations to (11.3) are made by the assumption that the electrostatic potential of the core orbital may be taken equal to that evaluated at the nucleus of the atom concerned[25–28] i.e.:

$$\Delta\mathcal{E}_i \simeq \Delta\langle V\rangle_{\text{nucl.}}^{\text{val.}}. \tag{11.4}$$

[24] Siegbahn, H., Karlsson, L. Electron Spectroscopy for Atoms, Molecules and Condensed Matter, Amsterdam: North-Holland (to be published)

[25] Basch, H. (1970): Chem. Phys. Lett. *5*, 337

[26] Schwartz, M.E. (1970): Chem. Phys. Lett. *6*, 631

[27] Gelius, U. (1974): Phys. Scr. *9*, 133

[28] Gelius, U., Roos, B. Siegbahn, P. (1970): Chem. Phys. Lett. *4*, 471

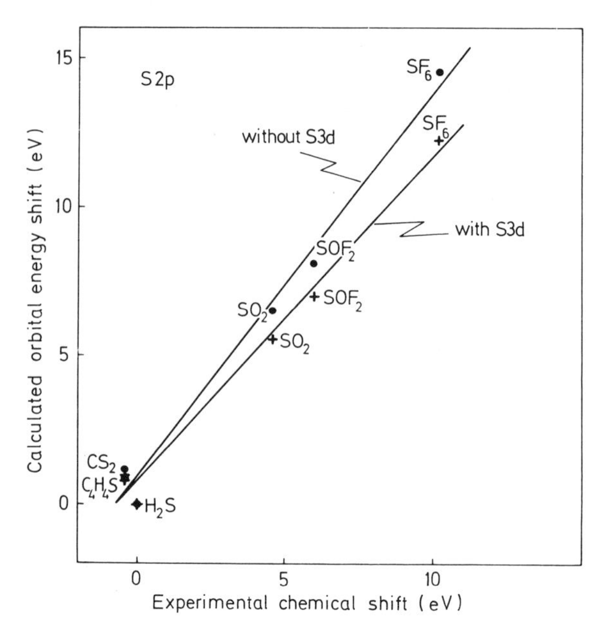

Fig. 55. Calculated sulphur $2p$ ground state orbital energy shifts vs. measured shifts[28]. (+) Basis set including S $3d$ functions (●) not including S $3d$ functions

In this way one disposes entirely of the dependence of core orbital shape on the orbital energy shift.

GELIUS et al.[28], using a mediumsized basis set, investigated the correlation between experimental binding energy shifts and orbital and potential energy shifts obtained in *ab initio* calculations on a series of sulphur-containing molecules. Figures 55 and 56 summarize the results. First, it is found (Fig. 55) that the $2p$ orbital energy shifts agree quite well with the experimental data. This indicates the near constancy of relaxation energies for this series of molecules. This observation has in general been found to hold for large numbers of compounds of different elements[7,25,27–29] (cf. also Table 13 for the fluoromethane series). The condition for this is that the compounds are similar in nature of bonding, so that they respond in essentially the same way on core electron ionization. This question will be considered further below.

The validity of the approximation (11.4) depends clearly on the degree of diffuseness of the core orbital in question. This point is illustrated in Fig. 56 which shows the shell effect in the orbital energy shifts for the sulphur compounds. The $1s$ and $2p$ orbital energy shifts are plotted versus the potential shifts at the nucleus. At higher charge transfers from the sulphur to the ligands, such as in SF_6, significant deviations occur between the values of these three quantities. The important property of the core orbital in this context is not its absolute binding energy, but rather its shape and localization. The radial expectation value may be taken as a measure of the effective localization of the

[29] Basch, H., Snyder, L.C. (1969): Chem. Phys. Lett. *3*, 333

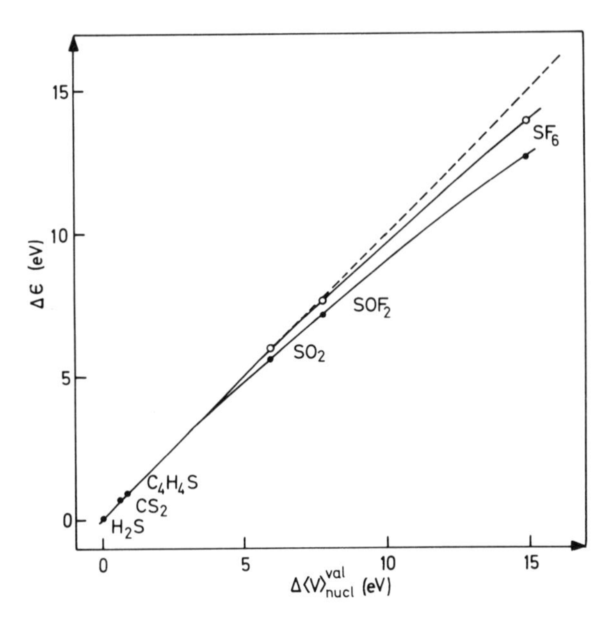

Fig. 56. Calculated ground state orbital energy shifts vs. shift in electrostatic potential at the nucleus (after[28]) (○) S 1*s* shifts (●) S 2*p* shifts. Dashed line has unity slope

orbital within the atom. Observed differential shifts for the 1*s*, 2*s* and 2*p* core levels seem to follow the trends of the radial expectation values[27] but relaxation effects are also likely to be of importance in this context.

δ) Ground state potential models (GPM). The most simplified treatment of the models (11.3) and (11.4) consists in the assumption that molecular valence charge distributions can be subdivided into spherical shells centered on each atomic site. Using simple electrostatics these expressions then reduce to the so-called ground state potential model (GPM), which has the form (cf. Fig. 57):

$$-\Delta\mathcal{E}_i = k_A q_A + \sum_{A \neq B} \frac{q_B}{R_{AB}} + l_A. \tag{11.5}$$

Although simple in structure and neglecting relaxation, this ground state potential model (GPM) has been found to describe the observed shifts in binding energies to a high accuracy for the compounds of a large number of elements[7, 30-40]. Figure 58 shows an example of a correlation between ob-

[30] Siegbahn, K., et al. (1969): ESCA applied to free molecules, Amsterdam: North-Holland

[31] Lindberg, B.J., Hamrin, K., Johansson, G., Gelius, U., Fahlman, A., Nordling, C., Siegbahn, K. (1970): Phys. Scr. *1*, 286

[32] Allison, D.A., Johansson, G., Allan, C.J., Gelius, U., Siegbahn, H., Allison, J., Siegbahn, K. (1972/73): J. Electron Spectrosc. Relat. Phenom. *1*, 269

[33] Shirley, D.A. (1973): Adv. Chem. Phys. *23*, 85

[34] Clark, D.T. (1973): in: Electron emission spectroscopy, Dekeyser et al. (eds.), Dordrecht: Reidel

(Footnotes 35–40 see p. 344)

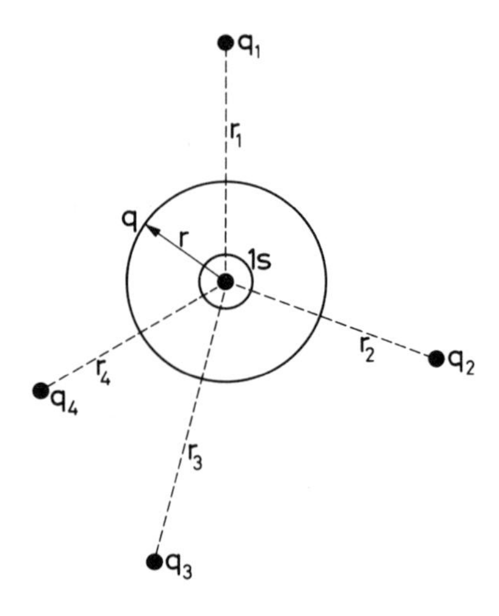

Fig. 57. Schematic illustration of the potential model of chemical shifts in the point charge approximation [cf. (11.5)]

served binding energy shifts and calculated shifts for C1*s* according to (11.5) by means of charges obtained from ground state CNDO calculations[7]. The deviations from the straight line are for most cases not larger than ~1 eV. In this correlation the one-center electron repulsion integral, k_A, and the constant to fix the energy scale, l_A, were taken as adjustable parameters.

Equation (11.5) is valid both for the orbital energy shift (11.3) and the potential at nucleus shift (11.4). In the former case the factor k refers to the average core-valence electron repulsion and in the latter case it corresponds to the average of the inverse radial expectation value of the valence electrons on the atom concerned. As indicated above, these will be slightly different, the "orbital energy"-k value being slightly smaller than the "potential at nucleus"-k value.

The form (11.5) of the potential shift implies that no distinction is made between different valence orbital characters, i.e. *s*, *p*, *d*, ...-type basis orbitals are all regarded as equivalent in the calculation of the shift. Refinements to (11.5) have been proposed which include such distinctions[41,42]. The introduction of higher multipole terms in the calculation of the potential has been

[35] Davis, D.W., Banna, M.S., Shirley, D.A. (1974): J. Chem. Phys. *60*, 237

[36] Wyatt, J.F., Hillier, I.H., Saunders, V.R., Connor, J.A., Barber, M. (1971): J. Chem. Phys. *54*, 5311

[37] Jolly, W.L., Perry, W.B. (1974): Inorg. Chem. *13*, 2686

[38] Jolly, W.L., Perry, W.B. (1974): J. Am. Chem. Soc. *95*, 5442

[39] Carver, J.C., Gray, R.C., Hercules, D.M. (1974): J. Am. Chem. Soc. *96*, 6851

[40] Perry, W.B., Jolly, W.L. (1974): Inorg. Chem. *13*, 1211

[41] Ellison, F.O., Larcom, L.L. (1971): Chem. Phys. Lett. *10*, 580

[42] Ellison, F.O., Larcom, L.L. (1972): Chem. Phys. Lett. *13*, 399

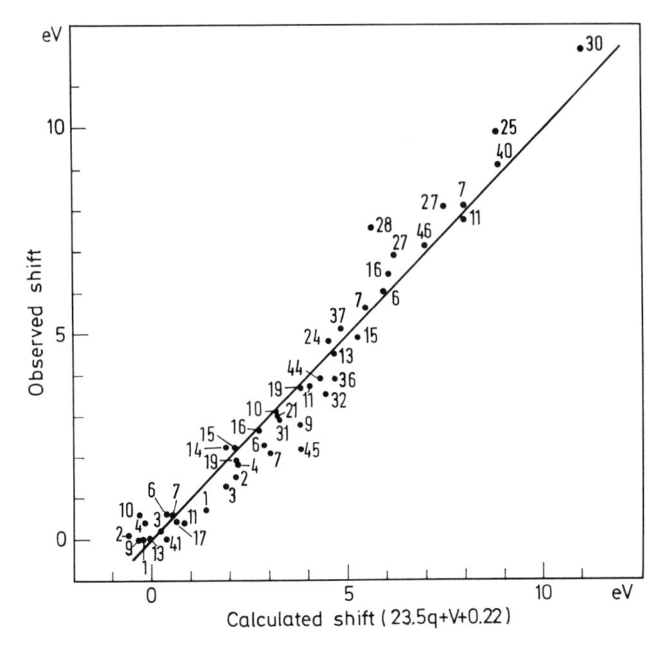

Fig. 58. Observed vs. calculated binding energy shifts for carbon compounds. The calculated shifts were obtained according to the ground state potential model with adjusted parameters k_C and l_C[7]

discussed[8, 43–45] as well as the possibility to introduce a fitting parameter in front of the Madelung term $(\sum_{A \neq B} q_B/R_{AB})$[45, 46]. These refinements should naturally be made with due consideration of the inherent errors in the computational scheme for the molecular wave functions.

ε) *Core shift models including relaxation.* Although the ground state models have been found to describe observed binding energy shifts to a good approximation for large classes of systems, the inclusion of relaxation contributions does lead to a better overall agreement with experiment[8, 11–13, 35, 38, 47–49]. In several cases it has been concluded that relaxation effects play a dominant role in deciding the directions of binding energy shifts. Core shifts connected with changes of phase provide examples in this category, such as metal to free metal atom shifts.

---

[43] Maksic, Z.B., Rupnik, K. (1977): Croat. Chem. Acta *50*, 307

[44] Maksic, Z.B., Kovacevic, K., Metiu, H. (1974): ibid. *46*, 1

[45] Blomster, B., Siegbahn, H., Sanhueza, J.E., Goscinski, O. (1981): Tech. Rep. 630, Quant. Chem. Group, Uppsala

[46] Schwartz, M.E., Switalski, J.D., Stronski, R.E. (1972): in: Electron spectroscopy, Shirley, D.A. (ed.), Amsterdam: North-Holland

[47] Howat, G., Goscinski, O. (1975): Chem. Phys. Lett. *30*, 87

[48] Siegbahn, H., Medeiros, R., Goscinski, O. (1976): J. Electron. Spectrosc. Relat. Phenom. *8*, 149

[49] Lindberg, B., Svensson, S., Malmqvist, P.Å., Basilier, E., Gelius, U., Siegbahn, K. (1975): UUIP-910 (Uppsala University Institute of Physics Report)

One case of particular consideration in the discussion of relaxation contributions to the core binding energies of molecules and solids is when there are several equivalent atomic centra. In such a case the question arises, whether or not the core hole should be considered localized to one specific atomic center. One would normally require on the basis of molecular symmetry, that all electronic states (initial and final) should have equal amplitudes on equivalent atomic centra. At the Hartree-Fock ΔSCF level this may be done by removing the molecular core electron orbital and reoptimizing the orbitals retaining their ground state symmetry. It is found, however, that the binding energies obtained by this approach are much too high. For instance, in the case of $1s$ ionization in O_2, symmetry-required ($D_{\infty h}$) calculations give for the multiplet components $E_B(^2\Sigma_g)=556.6$ eV and $E_B(^4\Sigma_g)=554.4$ eV [50a]. These values should be compared to the experimental ones, which are $E_B(^2\Sigma_g)=544.2$ eV and $E_B(^4\Sigma_g)=543.1$ eV [30]. In order to remedy for this apparent error, Bagus and Schaefer [50a] in the case of $1s$ ionization in O_2, broke the $D_{\infty h}$ symmetry requirement on the final state molecular orbitals and performed the calculation in $C_{\infty v}$ symmetry. The binding energy values obtained in this calculation for the two multiplet components were $E_B(^2\Sigma_g)=542.6$ eV and $E_B(^4\Sigma_g)=542.0$ eV. The calculated values are thus brought into normal agreement (at the ΔSCF level) with experiment. It was concluded that this sort of final state, where the core hole is localized on either of the atomic centers corresponds to physical reality. A symmetric total wave function ($D_{\infty h}$) may be formed by combining the two equivalent $C_{\infty v}$ solutions (i.e. where the core hole is either on center 1 or center 2). An alternative approach is to view the problem of core hole localization as a correlation problem, i.e. when two core electrons are at one center the valence electrons should have a larger tendency of being found at the other. Addition of ($\sigma_u \rightarrow \sigma_g, \pi_u \rightarrow \pi_g^*$) configurations to the $D_{\infty h}$ HF hole state leads to polarization of the π-electrons towards the core hole in the same way as with a symmetry-broken ΔSCF-calculation [50b], The relation between these two approaches have been discussed by several authors [50-54].

The transition operator formalism (cf. Sect. 7) provides a convenient scheme for extending the ground state models of the previous subsection to include relaxation. In this formalism, the transition orbital energies closely approximate ΔSCF-values of binding energies (cf. (7.8)). Using the same arguments as for the ground state orbital energy shifts, the transition orbital energies may be approximated by a potential model:

$$-\varepsilon_A^T = k_A^T q_A^T + \sum_{B \neq A} \frac{q_B^T}{R_{AB}} + l_A^T = k_A^T q_A^T + V_A^T + l_A^T. \tag{11.6}$$

[50a] Bagus, P.S., Schaefer, H.F. (1972): J. Chem. Phys. *56*, 224
[50b] Ågren, H., Bagus, P.S., Roos, B. (to be published)
[51] Snyder, L.C. (1971): J. Chem. Phys. *55*, 95
[52] Cederbaum, L.S., Domcke, W. (1977): J. Chem. Phys. *66*, 5084
[53] Lozes, R.L., Goscinski, O., Wahlgren, U.I. (1977): Technical Note 495, Dept. Quant. Chem., Uppsala University
[54] Siegbahn, H. (1975): UUIP-891 (Uppsala University Institute of Physics Report)

The atomic charges in this transition potential model (TPM) are then calculated using a quasi-atom, which has a core electron population between the initial and final states. This implies increasing the effective nuclear charge by half an electronic charge (i.e. effective nuclear charge $=Z_A^*+0.5$, where $Z_A^* = Z_A -$ no. of core electrons in the ground state). Under the assumption that (11.6) describes the full binding energy including relaxation effects, the relaxation energy is obtained as the difference between (11.6) and the ground state orbital energy:

$$E_{\text{Relax}} = -\varepsilon_A^G - (-\varepsilon_A^T) = (k_A^G q_A^G - k_A^T q_A^T) - (V_A^T - V_A^G) + (l_A^G - l_A^T). \tag{11.7}$$

This can be written in terms of the valence electron populations on atom A, P_A^G and $P_A^T (P_A^{G,T} = Z_A^* - q_A^{G,T})$:

$$\begin{aligned} E_{\text{Relax}} &= k_A^T P_A^T - k_A^G P_A^G - (V_A^T - V_A^G) \\ &\quad + [(l_A^G - l_A^T) + (k_A^G - k_A^T) Z_A^* - k_A^T \cdot 0.5] \end{aligned} \tag{11.8}$$

which reduces to:

$$E_{\text{Relax}} = (k_A^T - k_A^G) P_A^G + k_A^T \cdot \Delta P - (V_A^T - V_A^G) + l'_A \tag{11.9}$$

if one introduces the quantity:

$$\Delta P = P_A^T - P_A^G \quad \text{and} \tag{11.10}$$

and

$$l'_A = [(l_A^G - l_A^T) + (k_A^G - k_A^T) Z_A^* - k_A^T \cdot 0.5].$$

This division of the relaxation energy into different terms has proved useful in discussing the physics in the core ionization for molecules[13, 48, 49]. The first term represents the energy gained by contraction of the valence charge density on the atom where the ionization takes place. The second and third terms represent the relaxation energy due to the reorganization of charge in the molecule as a whole. Specifically, the second term is the energy gained by the increase of charge, ΔP, on the ionized atom. This energy is proportional to the one-center electron repulsion integral, k_A^T. Since this charge is taken from the rest of the molecule, the third term reflects from where the charge ΔP is effectively taken and transferred to the ionized atom. One may then write:

$$E_{\text{Relax}} = E_{\text{Contr.}} + E_{\text{Flow}}, \tag{11.11}$$

where the "contraction part" is equal to

$$E_{\text{Contr.}} = (k_A^T - k_A^G) \cdot P_A^G + l'_A \tag{11.12}$$

and the "flow part":

$$E_{\text{Flow}} = k_A^T \cdot \Delta P - (V_A^T - V_A^G). \tag{11.13}$$

The importance of the relaxation energy in deciding binding energy shifts has been analyzed on the basis of this model for a number of different

Table 15. Ground-state populations and potentials (eV), their respective differences with transition operator and relaxation energies (eV) for boron 1s ionization[a]

A			P_A^G	ΔP_A	V_A^G	ΔV_A	E_{Relax}
1	BF_3		2.23	0.46	−8.52	5.04	8.5
2	BCl_3		2.54	0.48	−3.86	4.06	10.6
3	$B(CH_2CH_3)_3$		2.84	0.43	−2.10	3.03	11.3
4	$ClB(CH_2CH_3)_2$		2.74	0.42	−2.75	2.98	10.9
5	$B(OCH_3)_3$		2.38	0.49	−8.12	3.94	10.5
6	$B(OCH_2CH_3)_3$		2.42	0.49	−8.11	3.80	10.8
7	B_2H_6		3.12	0.29	2.03	3.02	9.1
8	$B_2H_5N(CH_3)_2$		2.97	0.42	−0.89	3.65	10.7
9	1,5-$C_2B_3H_5$		2.98	0.35	0.29	2.75	10.3
10	1,6-$C_2B_4H_6$		3.02	0.39	0.29	2.90	11.0
11 (a)	2,4-$C_2B_5H_7$	(1, 7)	2.89	0.41	−0.45	3.00	11.0
(b)		(3)	3.07	0.35	1.98	2.33	10.9
(c)		(5, 6)	3.07	0.36	1.00	2.44	11.0
12 (a)	1,2-$C_2B_{10}H_{12}$	(3, 6)	2.96	0.39	1.36	2.47	11.3
(b)		(4, 5, 7, 11)	2.97	0.39	0.60	2.44	11.3
(c)		(8, 10)	2.99	0.37	−0.06	3.09	10.3
(d)		(9, 12)	2.98	0.39	−0.53	2.42	11.4
13 (a)	1,7-$C_2B_{10}H_{12}$	(2, 3)	2.96	0.39	1.42	2.50	11.3
(b)		(4, 6, 8, 11)	2.97	0.39	0.62	2.48	11.3
(c)		(5, 12)	2.96	0.39	0.15	2.52	11.2
(d)		(9, 10)	2.99	0.39	−0.07	2.41	11.4
14 (a)	B_5H_9	(1)	3.15	0.36	0.67	2.76	10.8
(b)		(2, 3, 4, 5)	2.99	0.38	0.38	2.96	10.7

[a] Ref. [48].

elements[13,47–49]. In Table 15 are contained the relaxation energies as well as the relevant quantities to calculate them according to (11.9) for a series of boron compounds. Figure 59 shows the correlation between theory and experiment using both the ground state potential model (GPM) (open circles) (i.e. (11.5)) and the transition potential model (TPM) (filled circles). It is seen that the TPM relaxation-corrected results are in better overall agreement with experiment than the GPM ones. It appears also that for particular compounds the improvement with respect to GPM is especially large, namely compounds nos. 1 and 7, BF_3 and B_2H_6. These constitute two examples of different relaxation behaviour, which can be traced back to the electronic structure. The relaxation energies calculated from (11.9) for these compounds are found to be low with respect to the other compounds, 8.5 eV and 9.1 eV respectively vs. ~11.0 eV on the average for the other compounds. For BF_3 it is seen that the charge compensation, ΔP, is close to complete (0.46) and that the main part of the difference in relaxation energy with respect to the other molecules comes from the large value of ΔV. A comparison with the simplest of the other molecules of this type, BCl_3, explains this figure as being due to the magnitude of the B-F bond length. The difference in ΔV of 1 eV obtained between these two molecules is accounted for by the difference in bond length (1.30 Å for BF_3 vs. 1.73 Å for BCl_3). For B_2H_6, however, the low charge compensation of 0.29

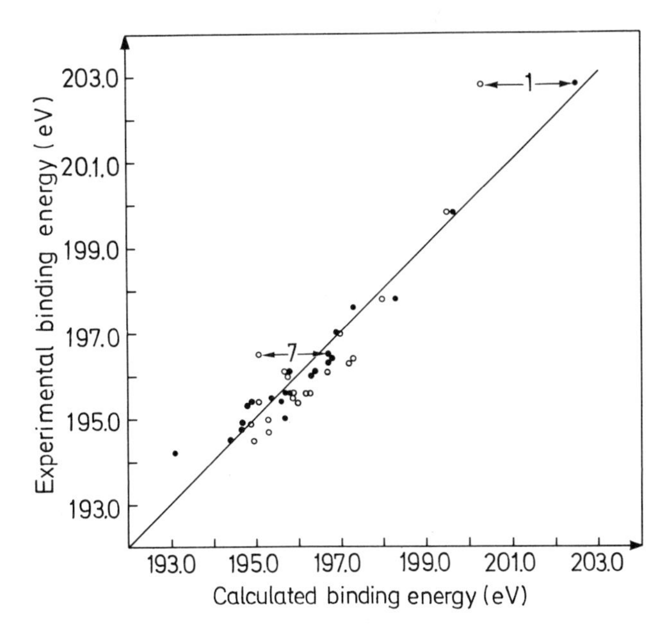

Fig. 59. Experimental vs. calculated core binding energies for boron compounds. Calculations were performed according to the transition potential model (TPM), (11.6)[48]. Compounds no. 1 and no. 7 refer to BF_3 and B_2H_6, respectively (cf. Table 15)

electrons is apparently the dominating factor in deciding the relaxation energy. This is a reflection of the typical chemistry of boron compounds, whose hydrides are characterized in their chemical behaviour by their electron-deficiency. In this context of ionization it results in a low "conductivity" of the B-H-B bonds. It is interesting to note that if one replaces one of the bridging hydrogens in B_2H_6 with a dimethylamino group (compound no. 8), the possibilities of charge flow increase in the system by virtue of the B-N-B π-bonds. The ΔP-value as well as the relaxation energy thus increase substantially.

Relaxation corrected binding energy shifts may be obtained by means of another scheme proposed by Davis and Shirley[8]. Their model is based on the result of Libermann[55] and Hedin and Johansson[56]. According to this scheme accurate predictions of ΔSCF binding energies are given by:

$$E_B(\Delta SCF) = 1/2(\varepsilon + \varepsilon^*), \tag{11.14}$$

where ε is the ground state orbital energy and ε^* the orbital energy of the fully ionized state. This result is similar to that underlying the transition operator method. On the level of potential models they are equivalent as far as accuracy is concerned. The potential model resulting from (11.14) consists in replacing the shifts in orbital energies, ε and ε^*, by their electrostatic potential

[55] Liberman, D. (1964): Bull. Am. Phys. Soc. *9*, 731
[56] Hedin, L., Johansson, G. (1969): J. Phys. B (London) *2*, 1336

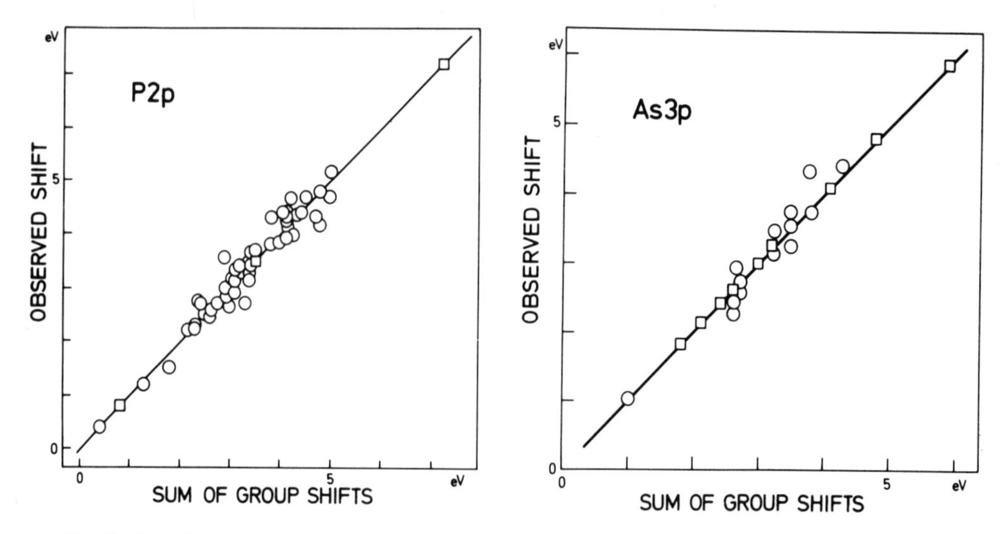

Fig. 60. Correlations between experimental and calculated binding energy shifts according to the group shift model for phosphorous and arsenic compounds. Calculated shifts were obtained from $\Delta E_B(\text{calc}) = \sum_{\text{groups}} \Delta E_{gr}$, where the group shifts were empirically obtained through fitting to experimental data [57]

counterparts (cf. (11.3, 4)):

$$\Delta E_B = 1/2(\Delta V + \Delta V^*). \tag{11.15}$$

In the equivalent cores and semiempirical regime the ΔV^* term is obtained simply by replacing the core charge of Z by $Z+1$:

$$\Delta E_B = 1/2[\Delta V(Z) + \Delta V(Z+1)]. \tag{11.16}$$

A model for obtaining relaxation-corrected binding energy shifts, which becomes quite useful for large systems is the so-called group shift model. The general idea behind this model is that one should be able to assign a characteristic shift, ΔE_{gr} to a certain group that attaches to a specific element, such that the addition of all these group shifts should give the total shift on the atom concerned. This assumes that the bond geometry and the transition charge distribution of the group are always the same when it is bonded to the element, irrespective of what other groups are bonded to it. This approximation may be more or less good, depending on the nature of the molecular groups and the element studied. The model has been found, however, to reproduce shifts to a high accuracy. The example shown in Fig. 60 of a group shift analysis of phosphorous and arsenic compounds demonstrates this fact quite convincingly [57]. The deviations from perfect agreement are small and quite acceptable, considering the nature of the underlying assumptions. In

---

[57] Lindberg, B.J., Hedman, J. (1975): Chem. Scr. 7, 155

Table 16. Group shifts for groups bound to carbon [7]

Group	N^c	ΔE_{gr} (eV)	χ_{gr}	Group	N^c	ΔE_{gr} (eV)	χ_{gr}
$-H$	37	0.01	2.6	$-CCl_2F$	1	0.46	3.0
$-CH_3$	13	−0.32	2.0	$-CClF_2$	1	0.48	3.1
$-CH_2NH_2$	1	−0.52	1.9	$-CF_3$	2	0.59	3.1
$-CH_2C$	1	−0.06	2.3	$-NH_2$	2	0.25	2.9
$-CH_2OH$	1	−0.42	2.0	$-N(CH_2)_2$	1	0.79	3.2
$-CH_2OCH_2C$	2	−0.23	2.1	$-OH$	7	1.51	3.5
$-CH_2OC(O)X^a$	3	0.08	2.7	$=O$	14	3.03	4.1
$-CH_2Cl$	1	−0.32	2.0	$-OCH_2C$	6	1.37	3.5
$-CH\overset{..}{-}C^b$	2	−0.18	2.2	$-OCH_3$	2	1.46	3.5
$-CH(OH)C$	1	−0.06	2.3	$-OCH(CH_3)O$	1	1.31	3.5
$-CH(CH_3)OC$	1	−0.52	1.9	$-OC(O)C$	2	1.91	3.7
$-CH(OC)_2$	1	−0.52	1.9	$-OC(O)X^a$	2	2.14	3.8
$\overset{..}{-}C(OH)\overset{..}{-}C^b$	1	−0.25	2.1	$-OCH(OCH_3)_2$	1	1.68	3.6
$-C(O)C$	3	−0.17	2.2	$=S$	1	0.85	3.3
$-C(O)OH$	3	−0.15	2.3	$-Br$	1	0.89	3.3
$-C(O)OCH_2X^a$	2	−0.35	2.1	$-Cl$	7	1.56	3.6
$\overset{..}{-}CF\overset{..}{-}C^b$	1	0.31	2.9	$-F$	8	2.79	4.0

[a] X is a more electronegative group than carbon if it is two atoms removed from the carbon atom considered, otherwise it is arbitrary.
[b] Aromatic compounds.
[c] The number of compounds by which a group is represented in the calculation.

general, the group shifts are treated as unknowns and least-squares fitted to the experimental data. Table 16 contains group shifts and electronegativities for groups bound to carbon [7].

ζ) Relations to other experimental data. The thermodynamic model. Core shifts have been related to a large variety of other measured quantities such as NMR [58–64], IR and Mössbauer shifts [66–69]. The relation to shifts in the Auger electron lines provides insight into the mechanism of relaxation in the photo-

---

[58] Basch, H. (1970): Chem. Phys. Lett. *5*, 337
[59] Block, R.E. (1971): J. Magn. Res. *5*, 155
[60] Clark, D.T., Kilcast, D. (1971): J. Chem. Soc. (A) 3286
[61] Swartz, W.E., Hercules, D.M. (1971): Anal. Chem. *43*, 1066
[62] Mateescu, G.D., Riemenschneider, J.L. (1972): in: Electron spectroscopy, Shirley, D.A. (ed.), Amsterdam: North-Holland
[63] Gelius, U., Johansson, G., Siegbahn, H., Allan, C.J., Allison, D.A., Allison, J., Siegbahn, K. (1972/73): J. Electron Spectrosc. Relat. Phenom. *1*, 285
[64] Lindberg, B.J. (1974): J. Electron Spectrosc. Relat. Phenom. *5*, 149
[65] Clark, D.T. (1973): in: Electron emission spectroscopy, Dekeyser et al. (eds.), Dordrecht: Reidel
[66] Yatsimirskii, K.B., Nemoshkalenko, V.V., Nazarenko, Yu.P., Aleshin, V.C., Zhilinskaya, V.V., Tomashevsky, N.A. (1977): J. Electron Spectrosc. Relat. Phenom. *10*, 239
[67] Adams, I., Thomas, J.M., Bancroft, G.M., Butler, K.D., Barber, M. (1972): J. Chem. Soc. D751
[68] Barber, M., Seift, P., Cunningham, D., Fraser, M.Y. (1970): Chem. Commun. 1938
[69] Holsboer, F., Beck, W., Bortunic, H.D. (1973): Chem. Phys. Lett. *18*, 217

ionization process[30,45,70–85]. It was noted by JOLLY and HENDRICKSON[86] that binding energy shifts could be related to measurable heats of reaction under certain assumptions. As an example, consider the ionization of CO and CO_2:

$$\begin{aligned} CO + h\nu &\rightarrow (C^*O)^+ + e^-[E_{kin}(CO)] \\ CO_2 + h\nu &\rightarrow (C^*O_2)^+ + e^-[E_{kin}(CO_2)], \end{aligned} \tag{11.17}$$

where the asterisk denotes C1s-hole states. Subtracting these two equations yields:

$$CO + (C^*O_2)^+ \rightarrow (C^*O)^+ + CO_2 + \Delta E_B, \tag{11.18}$$

where ΔE_B is the C1s binding energy shift between CO_2 and CO. Thus, in this formulation, the binding energy shift is equal to the heat of reaction in forming $(C^*O)^+ + CO_2$ from $CO + (C^*O_2)^+$. Obviously, such a reaction cannot occur, since the core hole decays within a time very short compared to ordinary reaction times. However, the reaction (11.18) may be substituted by a process of approximately the same heat of reaction using the concept of equivalent cores. In this case the $(C^*O)^+$ and $(C^*O_2)^+$ states could, according to this principle, be subjected to the core exchange reactions:

$$(C^*O)^+ + N^{5+} \rightarrow (NO)^+ + C^{*5+} + \delta_{CO} \tag{11.19}$$

$$(C^*O_2)^+ + N^{5+} \rightarrow (NO_2)^+ + C^{*5+} + \delta_{CO_2}, \tag{11.20}$$

where N^{5+} is a nitrogen atom stripped of all its valence electrons. With these substitutions (11.18) becomes:

---

[70] Asplund, L., Kelfve, P., Siegbahn, H., Goscinski, O., Fellner-Feldegg, H., Hamrin, K., Blomster, B., Siegbahn, K. (1976): Chem. Phys. Lett. *40*, 353

[71] Keski-Rahkonen, O., Krause, M.O. (1976): J. Electron Spectrosc. Relat. Phenom. *9*, 371

[72] Siegbahn, H., Goscinski, O. (1976): Phys. Scr. *13*, 225

[73] Asplund, L., Kelfve, P., Blomster, B., Siegbahn, H., Siegbahn, K., Lozes, R.L., Wahlgren U.I. (1977): Phys. Scr. *16*, 273

[74] Kelfve, P., Blomster, B., Siegbahn, H., Siegbahn, K., Sanhueza, E., Goscinski, O. (1980): Phys. Scr. *21*, 75

[75] Kowalczyk, S.P., Pollak, R.A., McFeely, F.R., Shirley, D.A. (1973): Phys. Rev. B*8*, 3583

[76] Ley, L., Kowalczyk, S.P., McFeely, F.R., Pollak, R.A., Shirley, D.A. (1973): Phys. Rev. B*8*, 2392

[77] Kowalczyk, S.P., Pollak, R.A., McFeely, F.R., Ley, L., Shirley, D.A. (1973): Phys. Rev. B*8*, 2387

[78] Kowalczyk, S.P., Ley, L., McFeely, F.R., Pollak, R.A., Shirley, D.A. (1974): Phys. Rev. B*9*, 381

[79] Wagner, C.D., Biloen, P. (1973): Surf. Sci. *35*, 82

[80] Matthew, J.A.D. (1973): Surf. Sci. *40*, 451, (1979): ibid. *89*, 596

[81] Utriainen, J., Linkohao, M., Åberg, T. (1973): in: Proc. Int. Symp. X-ray Sp. and El. Str. of Mat. vol. 1, p. 382

[82] Hoogewijs, R., Fiermans, L., Vennik, J. (1977): Surf. Sci. *69*, 273

[83] Wagner, C.D. (1975): Far. Disc. Chem. Soc. *60*, 291

[84] Wagner, C.D. (1977): in: Handbook of x-ray and ultraviolet photoelectron spectroscopy, Briggs, D. (ed.), London: Heyden, chap. 7

[85] Thomas, T.D. (1980): J. Electron Spectrosc. Relat. Phenom. *20*, 117

[86] Jolly, W.L., Hendrickson, D.N. (1970): J. Am. Chem. Soc. *92*, 1863

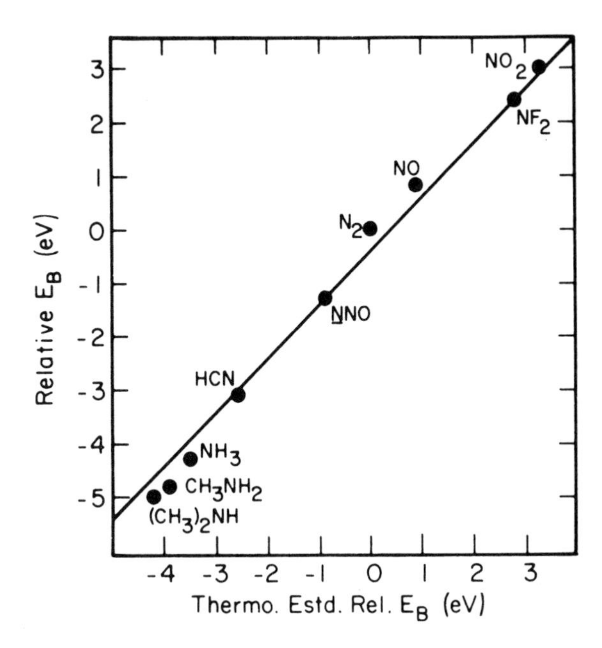

Fig. 61. Observed N 1*s* shifts vs. shifts calculated from thermodynamical data[86]

$$CO + (NO_2)^+ \rightarrow (NO)^+ + CO_2 + \Delta E_B + (\delta_{CO} - \delta_{CO_2}). \quad (11.21)$$

The heat of reaction of this process may be obtained using experimental thermodynamical data, which thus leads to an estimate of the shift. Clearly, in order to make these replacements in (11.18), it is not necessary that the heats of reaction δ_{CO} and δ_{CO_2} in (11.19) and (11.20) are zero. The only requirement is that they should be equal, since they will then cancel in (11.21). In fact, estimates made of δ-values for second row elements indicate that they are of the order of 5–15 eV[87–91].

A correlation between experiment and theory according to the thermodynamical model is shown in Fig. 61. The agreement is seen to be excellent and similar observations have been made for other elements[92]. One virtue of this model is that shifts may be predicted, including relaxation contributions, directly from other experimental data. The estimates of shifts are thus independent of model assumptions, such as the partitioning of the molecular charge distribution in the charge potential model (11.5). A definite drawback exists, however, in that thermodynamical data do not exist for other than a

---

[87] Shirley, D.A. (1972): Chem. Phys. Lett. *15*, 325

[88] Jolly, W.L., Gin, C. (1977): Int. J. Mass Spect. Ion Phys. *25*, 27

[89] Adams, D.B. (1976): J. Chem. Soc. Faraday Trans. II, *72*, 383

[90] Adams, D.B. (1976): J. Electron Spectrosc. Relat. Phenom. *9*, 251

[91] Jolly, W.L., Gin, C., Adams, D.B. (1977): Chem. Phys. Lett. *46*, 220

[92] Jolly, W.L. (1972): in: Electron spectroscopy, Shirley, D.A. (ed.), Amsterdam: North-Holland, p. 629

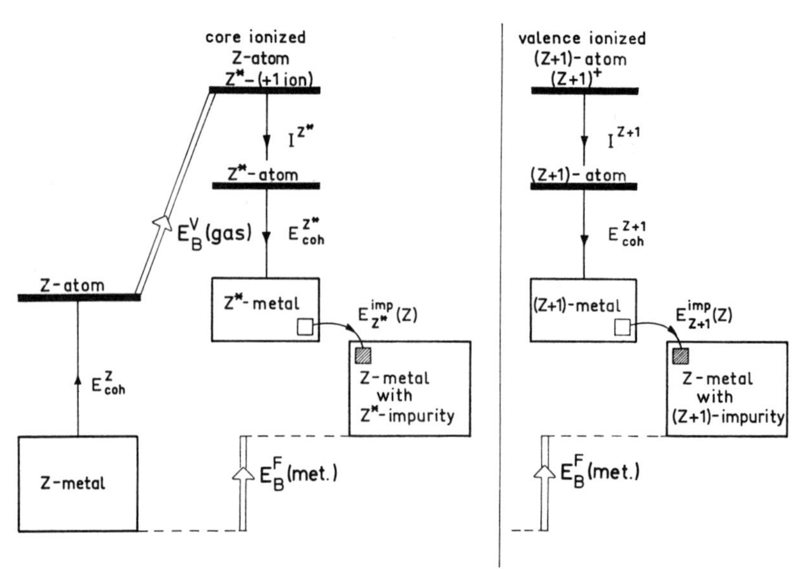

Fig. 62. Born-Haber cycle for calculation of the relation between the atomic gaseous phase binding energy E_B^V (referred to the vacuum level) and the solid metallic phase binding energy E_B^F (referred to the Fermi level). The left hand cycle yields:

$$E_B^F(\text{metal}) = E_B^V(\text{gaseous atom}) + E_{\text{coh}}^Z - E_{\text{coh}}^{Z^*} - I^{Z^*} + E_{Z^*}^{\text{imp}}(Z)$$

where E_{coh}^Z, $E_{\text{coh}}^{Z^*}$, $E_{Z^*}^{\text{imp}}(Z)$, I^{Z^*} are, respectively, the cohesive energy of the Z-metal, the (hypothetical) cohesive energy of the Z^* (Z-element with a core hole and one extra valence electron) metal, the solution energy of the Z^*-metal into the Z-metal and the first ionization potential of the Z^*-atom. In the $(Z+1)$ - or equivalent core approximation (Z^* element $\rightarrow Z+1$ element) the left-hand Born-Haber cycle is modified according to the right hand part. This leads to (11.22) for the relation between E_B^F (metal) and E_B^V (atom)[94]

rather limited set of molecules. Of course, the model may be turned around, such that if shifts have been measured they may be used to predict thermodynamic quantities. This has been done by JOLLY and GIN[88], who estimated heats of formation of more than 150 gaseous cations containing second and third row elements.

Special use of the above thermodynamic model of chemical shifts has been made in conjunction with metal atom - metal shifts[93-96]. Applying the equivalent cores arguments to a Born-Haber cycle (involving the Z- and $Z+1$-metals and the corresponding atoms) (cf. Fig. 62) yields an expression for the relation between the vacuum-level referenced atomic binding energy and the Fermi-level referenced metallic binding energy, which has the form (for the Z-element):

$$E_B^F(\text{metal}) = E_B^V(\text{atom}) + E_{\text{coh}}^Z - E_{\text{coh}}^{Z+1} - I^{Z+1} + E_{Z+1}^{\text{imp}}(Z), \qquad (11.22)$$

[93] Jen, J.S., Thomas, T.D. (1976): Phys. Rev. B*13*, 5284
[94] Johansson, B., Mårtensson, N. (1980): Phys. Rev. B*21*, 4427
[95] Broughton, J.Q., Perry, D.L. (1979): J. Electron Spectrosc. Relat. Phenom. *16*, 45
[96] Aksela, S., Kumpula, R., Aksela, H., Väyrynen, J. (1982): Phys. Scr. *25*, 45

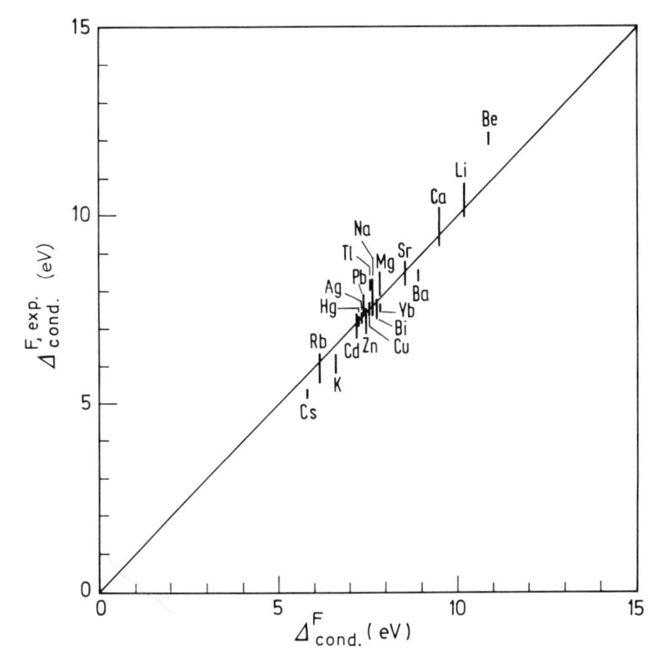

Fig. 63. Correlation between experimental and calculated values of $E_B^V(\text{gas}) - E_B^F(\text{metal})$ ($\equiv \Delta_{\text{cond}}^F$) for elemental results according to the thermodynamic model, (11.22)[94]

where E_{coh}^Z and E_{coh}^{Z+1} are the cohesive energies of the Z and $Z+1$ metals, respectively, I^{Z+1} the first ionization potential of the $Z+1$ atom, and $E_{Z+1}^{\text{imp}}(Z)$ the solution energy of the $Z+1$ metal into the Z metal. The last term is usually small[94]. The validity of this relation was tested[94] for the elemental metals, and the agreement between measured values of $(E_B^V - E_B^F)$ and those calculated from (11.22) is seen to be excellent (Fig. 63). The method has further applications for Auger shifts[96,97], binding energy shifts for alloys[97,98] and shifts between surface and bulk metal atoms[94,99,100]. In the last case one obtains the following expression[94]:

$$E_B^F(\text{surf.}) - E_B^F(\text{bulk}) = 0.2(E_{\text{coh}}^{Z+1} - E_{\text{coh}}^Z - E_{Z+1}^{\text{imp}}(Z)) \tag{11.23}$$

using the same notation as above. Such shifts have been positively identified for Au[99], W[100], Ir[101], Si[102] and GaAs and GaSb[103]. The W data are shown in Fig. 64; the $W4f_{7/2}$ line excited with synchrotron radiation ($h\nu = 70$ eV).

---

[97] Mårtensson, N., Nyholm, R., Calén, H., Hedman, J., Johansson, B. (1981): Phys. Rev. B*24*, 1725

[98] Steiner, P., Hüfner, S., Mårtensson, N., Johansson, B. (1981): Solid State Commun. *37*, 73

[99] Citrin, P.H., Wertheim, G.K., Baer, Y. (1978): Phys. Rev. Lett. *41*, 1425

[100] Duc, T.M., Guillot, C., Lassailly, Y., Lecomte, J., Jugnet, Y., Vedrine, J.C. (1979): Phys. Rev. Lett. *43*, 789

[101] Van der Veen, J.F., Himpsel, F.J., Eastman, D.E. (1980): Phys. Rev. Lett. *44*, 189

[102] Brennan, S., Stöhr, J., Jaeger, R., Rowe, J.E. (1980): ibid. *45*, 1414

[103] Eastman, D.E., Chang, T.-C., Heimann, P., Himpsel, F.J. (1980): ibid. *45*, 656

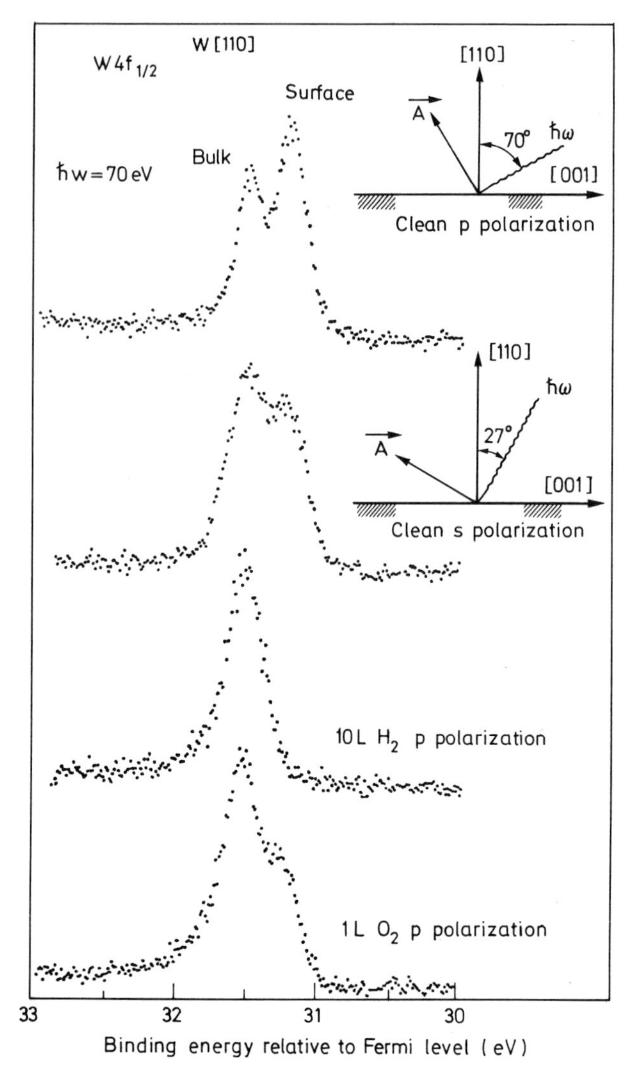

Fig. 64. W $4f_{7/2}$ spectra of a W(110) surface using synchrotron radiation showing surface-bulk atom splitting. The two uppermost spectra were recorded with a clean surface and the two lower spectra with adsorbed layers of H_2 and O_2, respectively[100]. The latter spectra show the sensitivity of the surface peak to adsorbed species

It has been pointed out that the surface-bulk shift may affect binding energy values measured at high accuracies[104]. Thus, if the two components are not resolved a shift in the observed peak position from the true bulk value may occur depending on the degree of surface character of the spectrum. This shift may be larger than the ultimate accuracy with which the binding energies are measured.

[104] Chadwick, D., Karolewski, M.A. (1981): J. Electron Spectrosc. Relat. Phenom. *24*, 181

12. Multiple excitations in core ionization of molecules and solids

α) Molecular core electron spectra. The investigation of multielectron satellite transitions in molecular photoelectron spectra is generally a more complicated problem than in atomic spectra. The same basic considerations as for atoms apply, e.g. the sudden approximation expression (7.20) for the satellite intensities. The spectra are more difficult to delineate, however, since the number of transitions are generally larger. Moreover, in molecular calculations for interpretations of these spectra, the quality of the wave functions requires considerably more elaborate methods than for atoms. It is found that quite extensive CI's are needed for the final states to obtain agreement with experiment. Detailed interpretations have been done only for a fairly limited number of shake-up spectra in small molecules e.g. [1-12].

From the experimental point of view, the recording of the shake-up spectra is a matter of signal-to-background ratio, since the excitation probability into any transition is generally small compared to the main line. The advent of monochromatized radiation has, however, substantially improved the quality of the spectra [1,2].

The shake-up spectra of N_2 and CO(C1*s* hole state), shown in Fig. 65 are expected to show strong similarities. This is due to the fact that the $(C^*O)^+$ ion (C1*s* core hole) is the equivalent core species of the $(N^*N)^+$ ion. Thus, these two systems should have similar excitation energies. The intensities observed in the two shake-up spectra will be different, however, as observed in Fig. 65. This is expected, since the excitation probability is dependent also on the form of the initial wave functions, which are quite different for CO and N_2. The results of extensive CI-calculations by CLARK [11] and by GUEST et al. [7] on CO agree on the interpretation that the first and second peaks are strongly dominated by transitions from 1π to $2\pi^*$. The interpretations of the other peaks are different for the two calculations, however. CLARK suggests the third and fourth shake-up peaks to be due mainly to single electron shake-up from the 5σ orbital, while GUEST et al. find double electron processes dominant for the third peak. In general, the classification in terms of orbital transitions in this context is difficult to make and each final state will be more or less mixed. The configuration mixing will depend on the one-electron basis used in the calculations.

As systems get larger interpretations in terms of large scale CI's are difficult

---

1 Siegbahn, K. (1974): J. Electron Spectrosc. Relat. Phenom. *5*, 3
2 Gelius, U. (1974): J. Electron Spectrosc. Relat. Phenom. *5*, 985
3 Hillier, I.H., Kendrick, J. (1975): J. Chem. Soc. Faraday II *71*, 1369, 1654
4 Butscher, W., Buenker, R.J., Peyerimhoff, S.D. (1977): Chem. Phys. Lett. *52*, 449
5 Cederbaum, L.S., Domcke, W., Schirmer, J. (1980): Phys. Rev. *22*, 206
6 Rodwell, W.R., Guest, M.F., Darko, T., Hillier, I.H., Kendrick, J. (1977): Chem. Phys. *22*, 467
7 Guest, M.F., Rodwell, W.R., Darko, T., Hillier, I.H., Kendrick, J. (1977): J. Chem. Phys. *66*, 5447
8 Arneberg, R., Müller, J., Manne, R. (1982): Chem. Phys. *64*, 249
9 Svensson, S., Ågren, H., Wahlgren, U.I. (1976): Chem. Phys. Lett. *38*, 1
10 Wahlgren, U.I. (1977): Mol. Phys. *33*, 1190
11 Clark, D.T. (private communication)
12 Martin, R.L., Mills, B.E., Shirley, D.A. (1976): J. Chem. Phys. *64*, 3690

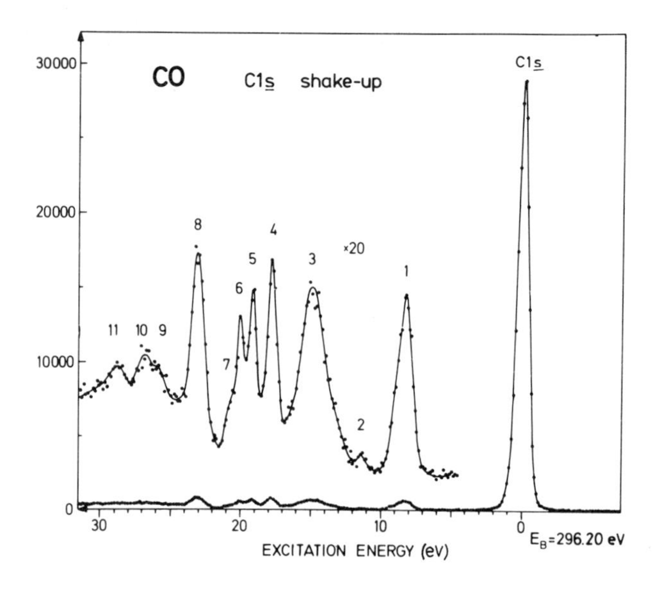

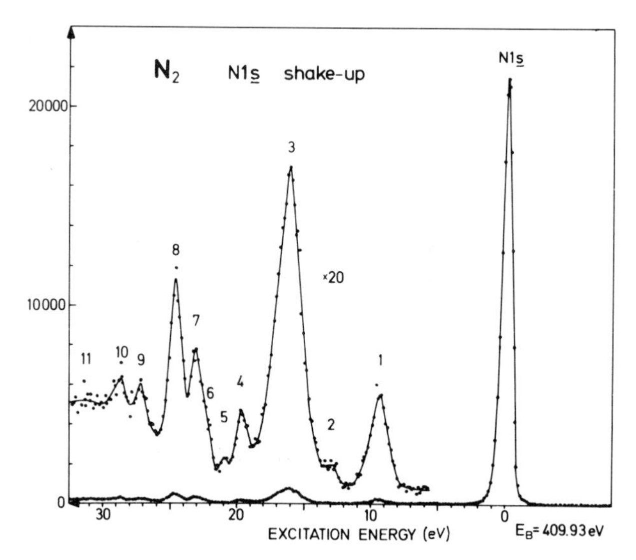

Fig. 65. Core electron spectra of CO(C 1*s*) and N_2[1,2] showing multiline shake-up features. Note the similarity between the excitation energies of the two spectra

to implement. Simpler schemes, which allow overall assignments to be made, have been used by a number of authors[13–28]. These involve semiempirical calculations and also empirical correlations with e.g. uv-absorption data.

[13] Basch, H. (1975): Chem. Phys. *10*, 157

[14] Basch, H. (1976): Chem. Phys. Lett. *37*, 447

[15] Carroll, T.X., Thomas, T.D. (1974): J. Electron Spectrosc. Relat. Phenom. *4*, 270

(Footnotes 16–28 see p. 359)

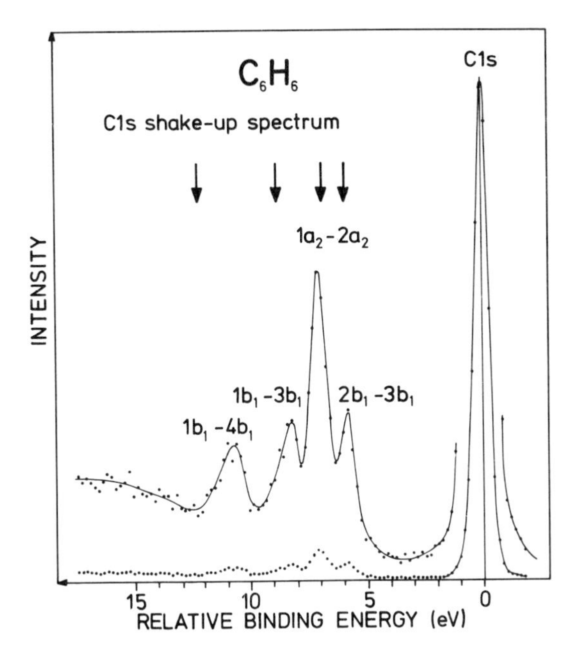

Fig. 66. C 1*s* shake-up spectrum of C_6H_6[28]. Arrows indicate calculated positions of shake-up transitions (cf. Table 17)

Some of the features of the shake-up spectra are of a qualitative nature and lend themselves to simple interpretations. This concerns, in particular, the $\pi \to \pi^*$ transitions, which are generally strong for unsaturated systems. Examples of this are provided by the CO and N_2 spectra, where the $\pi \to \pi^*$ transition occurs at a comparatively low excitation energy, below 10 eV.

In the C 1*s* shake-up spectrum of benzene, C_6H_6 (Fig. 66)[28] the $\pi \to \pi^*$ type transitions occur at even lower energies. They are split into several com-

[16] Mills, B.E., Shirley, D.A. (1976): Chem. Phys. Lett. *39*, 236

[17] Aarons, L.J., Guest, M.F., Hillier, I.H. (1972): J. Chem. Soc. Trans. Far. Soc. *68*, 1866

[18] Clark, D.T., Adams, D.B. (1975): J. Electron Spectrosc. Relat. Phenom. *7*, 401

[19] Clark, D.T., Adams, D.B. (1975): Theor. Chim. Acta *39*, 321

[20] Clark, D.T., Dilks, A., Peeling, J., Thomas, H.R. (1975): Faraday Discuss. Chem. Soc. *60*, 183

[21] Clark, D.T., Adams, D.B., Dilks, A., Peeling, J., Thomas, H.R. (1976): J. Electron Spectrosc. Relat. Phenom. *8*, 51

[22] Clark, D.T., Dilks, A. (1977): J. Polym. Sci. *15*, 15

[23] Carlson, T.A., Dress, W.B., Grimm, F.A., Haggerty, J.S. (1977): J. Electron Spectrosc. Relat. Phenom. *10*, 147

[24] Chambers, S.A., Thomas, T.D. (1977): J. Chem. Phys. *67*, 2596

[25] Allan, C.J., Gelius, U., Allison, D.A., Johansson, G., Siegbahn, H., Siegbahn, K. (1972/73): J. Electron Spectrosc. Relat. Phenom. *1*, 131

[26] Spears, D.P., Fischbeck, H.J., Carlson, T.A. (1975): J. Electron Spectrosc. Relat. Phenom. *6*, 411

[27] Pireaux, J.J., Svensson, S., Basilier, E., Malmqvist, P.Å., Gelius, U., Caudano, R., Siegbahn, K. (1976): Phys. Rev. A*14*, 2133 and private communication

[28] Lunell, S., Svensson, S., Malmqvist, P.Å., Gelius, U., Basilier, E., Siegbahn, K. (1978): Chem. Phys. Lett. *54*, 420

Table 17. Comparison of a theoretical and experimental results for *K*-shell hole states and satellites in benzene[a]

State	Assignment	Without CI		With CI		Experiment	
		Energy (eV)	Intensity[b]	Energy (eV)	Intensity	Energy (eV)	Intensity
1	$2b_1 \rightarrow 3b_1$	6.18	14	5.92	7	5.9	2
2	$1a_2 \rightarrow 2a_2$	6.76	0.2	6.87	4	7.0	6
3	$1b_1 \rightarrow 3b_1$	8.68	7	8.76	11	8.3	3
4	$2b_1 \rightarrow 4b_1$	9.18	0.7	9.21	0.03	–	–
5	$1b_1 \rightarrow 4b_1$	12.09	0.1	12.13	0.4	10.7	~1

[a] From ref. [28].
[b] Relative to the main peak.

ponents, which are resolved when monochromatized Al$K\alpha$ is employed for excitation. The orbital notations refer to the symmetry of the core ionized species (C_{2v}). Thus, the first two shake-up peaks are assigned to transitions from the outermost (e_{1g}) orbital of the neutral molecule and the third and fourth to transitions from the a_{2u} orbital[28], LUNELL et al.[28] have made semi-empirical calculations on this spectrum and the values from their analysis are compared to experiment in Table 17. The method used was PPP (Pariser-Parr-Pople)[29] including also a CI among the shake-up states. Calculations were performed for both C_6H_6 and $C_5H_5NH^+$, the latter being the equivalent core species of the core hole ion. The results using this simple approach are seen to compare quite favourably with experiment, particularly concerning the energies. The calculated intensities do not predict the correct relative ordering, but they still single out the peaks of major intensity in the spectrum. The ability of this simple computational method to describe the features of the C_6H_6 spectrum is not unexpected, in view of the fact that it has been developed specifically to treat optical excitations in aromatic systems.

The general difference between the shake-up spectra from saturated and unsaturated systems (with respect to the $\pi \rightarrow \pi^*$ transitions) seems to be valid in a large number of cases. Thus the shake-up spectrum can be used as a fingerprint in identifying the existence of unsaturated character in a system. CLARK et al.[18–22] have discussed and applied this aspect of shake-up spectra in connection with core electron spectra of polymeric substances.

β) Many-electron effects in solid core photoelectron spectra. It has been observed that under certain conditions very strong satellite peaks may occur in core electron spectra of solids. Such effects have been bound for transition metal compounds and adsorbed molecular species on metal surfaces. Examples of such spectra are shown in Fig. 67 for N_2 and CO adsorption[30]. The

[29] Pople, J.A., Beveridge, D.L. (1970): Approximate molecular orbital theory, New York: McGraw-Hill
[30] Fuggle, J.C., Umbach, E., Menzel, D., Wandelt, K., Brundle, C.R. (1978): Solid State Commun. *27*, 65

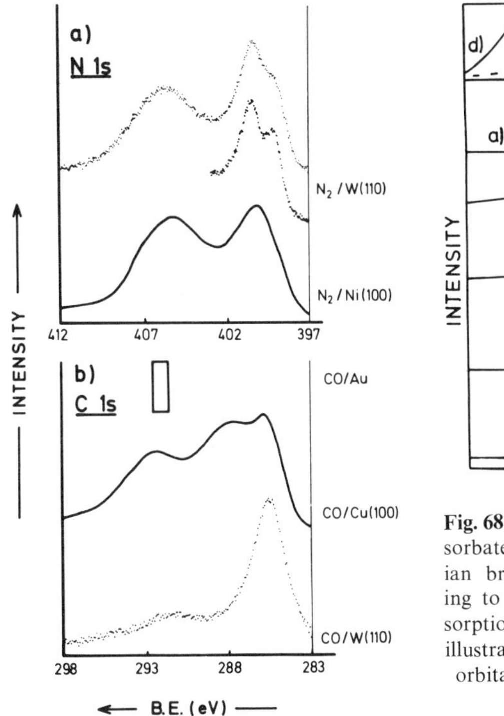

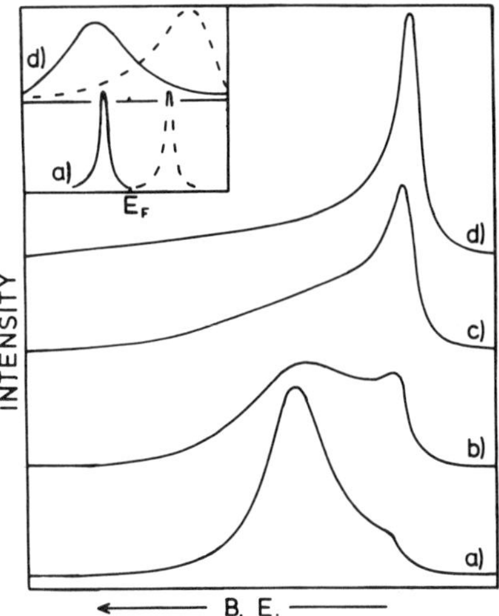

Fig. 68 a–d. Results of model calculations of adsorbate core line shapes (folded with Lorentzian broadening)[30, 32]. The strength of bonding to the surface increases from (**a**) ($\simeq$physisorption) to (**d**) strong chemisorption). The inset illustrates the electron density of the screening orbital in initial (----) and final (——) states

Fig. 67. N 1*s* and C 1*s* spectra for N_2 and CO adsorption on transition metals[30]

interpretation of these spectra in terms of different adsorption sites on the surface (cf. further below) is ruled out, since the relative intensities of the peaks are invariant with coverage. The theoretical framework to describe the shapes of these spectra have been discussed by LANG and WILLIAMS[31] and SCHÖNHAMMER and GUNNARSSON[32]. The basic ingredient in these models is that the unoccupied valence levels of the adsorbate are pulled below the Fermi level of the metal when the adsorbate is core ionized. As a result, the screening of the core hole on the adsorbate may proceed via a charge transfer from the metal to the adsorbate.

One can distinguish three different situations, depending on the bond strength between metal and adsorbate, each corresponding to a different shape of the core spectrum. In a physisorbed state, when the bond strength is small, one expects little orbital overlap between the adsorbate and metal. In the final ionic state, no charge transfer has thus occurred between metal and adsorbate and a single peak will be observed. This is illustrated in part (a) of Fig. 68,

[31] Lang, N.D., Williams, A.R. (1977): Phys. Rev. B*16*, 2408

[32a] Schönhammer, K., Gunnarsson, O. (1978): Solid State Commun. *23*, 691 and *26*, 399 and Z. Phys. B*30*, 297

[32b] Gunnarsson, O., Schönhammer, K. (1979): Surf. Sci. *80*, 471

[32c] Gunnarsson, O., Schönhammer, K. (1978): Phys. Rev. Lett. *41*, 1608

which shows results of model calculations[32] where the bond strength has been used as a variable parameter. As the bond strength increases, i.e. the adsorption can be increasingly characterized as chemisorption, the spectral weight is more shifted towards a lower binding energy. This corresponds to a transfer of charge between metal and adsorbate when the latter is ionized and thus to a higher relaxation energy. The appearance of the spectrum in the transition from the physisorption case to strong chemisorption is illustrated in going from (a) to (d) in Fig. 68. In the extreme chemisorption case, a single peak is observed with a continuum on the high binding energy side, leading to an asymmetry of this peak. In the case of intermediate coupling strength [(b) and (c)], a double peak structure is obtained. The strong satellites in the spectra of Fig. 67 are hence interpreted in terms of this intermediate situation (N_2/W(110), N_2/Ni(100))[30]. For CO on Cu(100) the interaction of the adsorbate level with the metal *sp*- and *d*-bands must be considered and calculations show that a three-peak structure may be theoretically obtained, in agreement with observations[32c].

For transition metal and lanthanide complexes, strong satellites are commonly observed in the metal core spectra[33-50] and their occurrence may be rationalized on the same grounds as above[51,52]. In this case, the charge transfer occurs between a ligand orbital and a transition metal nd or nf orbital. Typical spectra are shown in Fig. 69, the $2p$ spectra of the copper oxides. These illustrate one interesting aspect, which is in line with the above description in terms of interaction between metal- and ligand-centered orbitals, viz. the observation that for Cu_2O there are no satellites, whereas they are strong in CuO. This is readily understood from the fact that in Cu_2O, the

---

[33] Yin, L., Adler, I., Tsang, T., Mantienzo, L.J., Grim, S.O. (1974): Chem. Phys. Lett. *24*, 81

[34] Hamer, A.D., Tisley, D.G., Walton, R.A. (1974): J. Inorg. Nucl. Chem. *36*, 1771

[35] Pignataro, S., Foffani, A., Distefano, G. (1973): Chem. Phys. Lett. *20*, 350

[36] Tricker, M.J. (1974): J. Inorg. Nucl. Chem. *36*, 1543

[37] Johansson, L.Y., Larsson, R., Blomquist, J., Cederström, C., Grapengiesser, S., Helgeson, U., Moberg, L.C., Sundbom, M. (1974): Chem. Phys. Lett. *24*, 508

[38] Barber, M., Connor, J.A., Hillier, I.H. (1971): Chem. Phys. Lett. *9*, 570

[39] Hüfner, S., Wertheim, G.K. (1973): Phys. Rev. B *7*, 5086

[40a] Rosencwaig, A., Wertheim, G.K., Guggenheim, H.J. (1971): Phys. Rev. Lett. *27*, 479

[40b] Rosencwaig, A., Wertheim, G.K. (1973): J. Electron Spectrosc. Relat. Phenom. *1*, 493

[41] Frost, D.C., McDowell, C.A., Woolsey, I.S. (1972): Chem. Phys. Lett. *17*, 320

[42] Kim, K.S. (1974): Chem. Phys. Lett. *26*, 234

[43] Kim, K.S., David, R.E. (1972/73): J. Electron Spectrosc. Relat. Phenom. *1*, 251

[44] Tolman, C.A., Riggs, W.M., Linn, W.J., King, C.M., Wendt, R.C. (1973): Inorg. Chem. *12*, 2770

[45] Matienzo, L., Yin, L., Grim, S.O., Schwartz, W. (1973): Inorg. Chem. *12*, 2764

[46] Kim, K.S. (1974): J. Electron Spectrosc. Relat. Phenom. *3*, 217

[47] Frost, D.C., Ishitani, A., McDowell, C.A. (1972): Mol. Phys. *24*, 861

[48] Carlson, T.A., Carver, J.C., Saethre, L.J., Santibanez, F.G., Vernon, G.A. (1974): J. Electron Spectrosc. Relat. Phenom. *5*, 247

[49] Jørgensen, C.K., Berthou, H. (1972): Chem. Phys. Lett. *13*, 186

[50] Suzuki, S., Ishii, T., Sagawa, T. (1974): J. Phys. Soc. Japan *37*, 1334

[51] Larsson, S. (1977): Phys. Scr. *16*, 378

[52] Domcke, W., Cederbaum, L.S., Schirmer, J., von Niessen, W. (1979): Phys. Rev. Lett. *42*, 1238

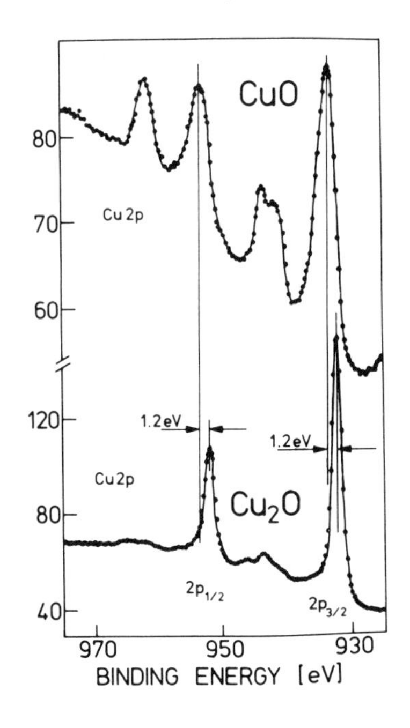

Fig. 69. Cu $2p$ spectra of CuO and Cu_2O showing strong satellites only for the former compound (cf. text)[40b]

Fig. 70. Al $2\underline{s}$, $2\underline{p}$ spectrum and associated bulk plasmon loss structure from a partly oxidized aluminum metal surface[1]

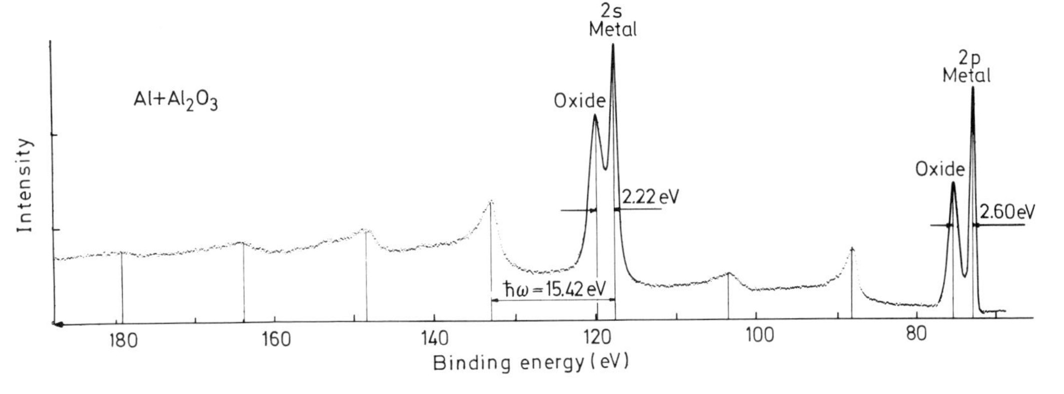

copper ion has a closed $3d^{10}$ configuration which is not the case for divalent Cu. Hence, no shake-up transitions of the type discussed above can occur in Cu_2O. This feature may be used as a probe for investigating the metal atom valency in metal complexes.

In solids, core electron excitation may give rise to additional processes not encountered in molecular spectra. Multielectron transitions which produce lines of substantial intensity in metal core electron spectra, are due to excitation of plasmons. Fig. 70 shows a spectrum of a partly oxidized aluminum sample. Following the $2s$- and $2p$-electron lines a sequence of volume plasmon peaks occur, whose separations are given by the fundamental plasmon frequency in aluminum metal ($h\omega_p = 15.42$ eV). The excitations of the plasmons may have two formally different origins. Either it is excited in the process of core hole creation (intrinsic plasmon creation) or in an inelastic scattering event accompanying the passage of the photoelectron out of the material. These two types of contributions cannot be energetically separated in the photoelectron spectrum, but the *intensities* of the peaks may yield information as to the relative weight of extrinsic and intrinsic processes. MAHAN[53] has suggested a model based on random spatial emission of extrinsic plasmons. According to this model, the average number of extrinsic bulk plasmons produced by a photoelectron originating at a depth Z is $Q(Z) \approx Z/l$, where l is the mean free path for plasmon emission. The probability of creating n bulk plasmons in the photoelectron emission is then given by:

$$P(n, Z) = \exp(-Z/l)(Z/l)^n/n!.$$

If this expression is integrated over Z to the total photon penetration depth, which is much larger than l, one finds that the total probability of producing n extrinsic bulk plasmons is independent of n. This means that the area under each plasmon satellite peak should be constant, independent of n. Including other inelastic processes, characterized by an attenuation length L, causes a slow decrease of the plasmon satellite intensities with n according to:

$$I(n) = a \cdot I(n-1), \tag{12.1}$$

where $a = 1/(1 + l/L) = \lambda/l$ (λ = the total inelastic mean free path). The intrinsic part of the spectrum can be shown theoretically to vary with n according to[53,54]:

$$I(n) = I(0)\, \mathrm{e}^{-\beta} \beta^n/n!, \tag{12.2}$$

where β is a constant parameter. (12.1) and (12.2) taken together then yield for the plasmon intensities:

$$I(n) = I(0)\left[\mathrm{e}^{-\beta} \cdot \beta^n/n! + a\frac{I(n-1)}{I(0)}\right]. \tag{12.3}$$

This model was used by PARDEE et al.[55] in order to distinguish the extrinsic from the intrinsic contributions for Na, Mg and Al. In Mg and Al the recorded spectra could be fitted with only (12.1), i.e. intrinsic contributions were found negligible for these cases. For Na, on the other hand, a measurable intrinsic contribution could be detected using (12.3). A more recent analysis by STEINER et al.[56] gave larger intrinsic fractions for these metals than those predicted by PARDEE et al.[55]. For the first plasmon loss peak in these metals ($n = 1$) the intrinsic fraction was found to be 44%, 29% and 14% in Na, Mg and

---

[53] Mahan, G.D. (1973): Phys. Status Solidi B*55*, 703

[54] Chang, J.J., Langreth, D.C. (1972): Phys. Rev. B*5*, 3512

[55] Pardee, W.J., Mahan, G.D., Eastman, D.E., Pollak, R.A., Ley, L., McFeely, F.R., Kowalczyk, S.P., Shirley, D.A. (1975): Phys. Rev. B*10*, 3614

[56] Steiner, P., Höchst, H., Hüfner, S. (1977): Phys. Lett. *61* A, 410

Al respectively. Similar results have been obtained by PENN[57] (Na, Mg, Al) and ATTEKUM and TROOSTER[58] (Al).

Under high spectrometer resolution metal core photoelectron lines show a marked asymmetry, which has been attributed to collective excitations[59-69]. The mechanism leading to this line shape is the creation of low-energy electron-hole pairs in the conduction band. This phenomenon was first discussed in connection with x-ray absorption edges in metals[70]. DONIACH and SUNJIC[61] have proposed a theory, which leads to a formula for the shape of metal core electron lines. Thus, the primary peak at energy E_0 is replaced by a distribution of the form:

$$A(E) = 1/(E - E_0)^{1-\alpha}, \tag{12.4}$$

where

$$\alpha = \sum_{l=0}^{\infty} 2(2l+1)(\delta_l/\pi)^2 \tag{12.5}$$

is termed the singularity index. The δ_l are the Friedel phase shifts, which satisfy the sum rule[71]:

$$\sum_{l=0}^{\infty} 2(2l+1)\,\delta_l/\pi = 1. \tag{12.6}$$

If the spectral function (12.4) is convoluted with the Lorentzian lifetime of the core hole state one obtains for the line shape:

$$I(E) \propto \frac{\Gamma(1-\alpha)\cos[\pi\alpha/2 + (1-\alpha)\arctan(E-E_0)/\gamma]}{[(E-E_0)+\gamma^2]^{(1-\alpha)/2}}, \tag{12.7}$$

where Γ denotes the gamma function and γ is the Lorentzian *half* width at half maximum. This expression has been found to represent the line shapes to a very good approximation in a variety of metals[1,65-69,72].

[57a] Penn, D.R. (1977): J. Vac. Sci. Technol. *14*, 300
[57b] Penn, D.R. (1977): Phys. Rev. Lett. *38*, 1429
[58] van Attekum, P.M.Th.M., Trooster, J.M. (1978): Phys. Rev. B*18*, 3872
[59] Langreth, D.C. (1969): Phys. Rev. *182*, 973
[60] Lundquist, B.I. (1969): Phys. Kondens. Mater. *9*, 236
[61] Doniach, S., Sunjic, M. (1970): J. Phys. C*3*, 285
[62] Minnhagen, P. (1977): J. Phys. F*7*, 2441
[63] Almbladh, C.O., von Barth, U. (1976): Phys. Rev. B*13*, 3307
[64] Mahan, G.D. (1975): Phys. Rev. B*11*, 4814
[65] Citrin, P.H. (1973): Phys. Rev. B*8*, 5545
[66] Hüfner, S., Wertheim, G.K., Buchanan, D.N.E., West, K.W. (1974): Phys. Lett. *46* A, 420
[67] Hüfner, S., Wertheim, G.K., Wernick, J.H. (1975): Solid State Commun. *17*, 417
[68] Citrin, P.H., Wertheim, G.K., Baer, Y. (1975): Phys. Rev. Lett. *35*, 885; (1977): Phys. Rev. B*16*, 425
[69] Baer, Y., Citrin, P.H., Wertheim, G.K. (1976): Phys. Rev. Lett. *37*, 49
[70a] Mahan, G.D. (1967): Phys. Rev. *163*, 612
[70b] Anderson, P.W. (1967): Phys. Rev. Lett. *18*, 1049
[70c] Nozieres, P., De Dominicis, C.T. (1969): Phys. Rev. *178*, 1097
[71] Friedel, J. (1954): Adv. Phys. *3*, 446
[72] Wertheim, G.K., Citrin, P.H. (1978): in: Photoemission in solids I, Topics in Applied Physics, vol. 26, Cardona, M., Ley, L. (eds.), Berlin, Heidelberg, New York: Springer, p. 196

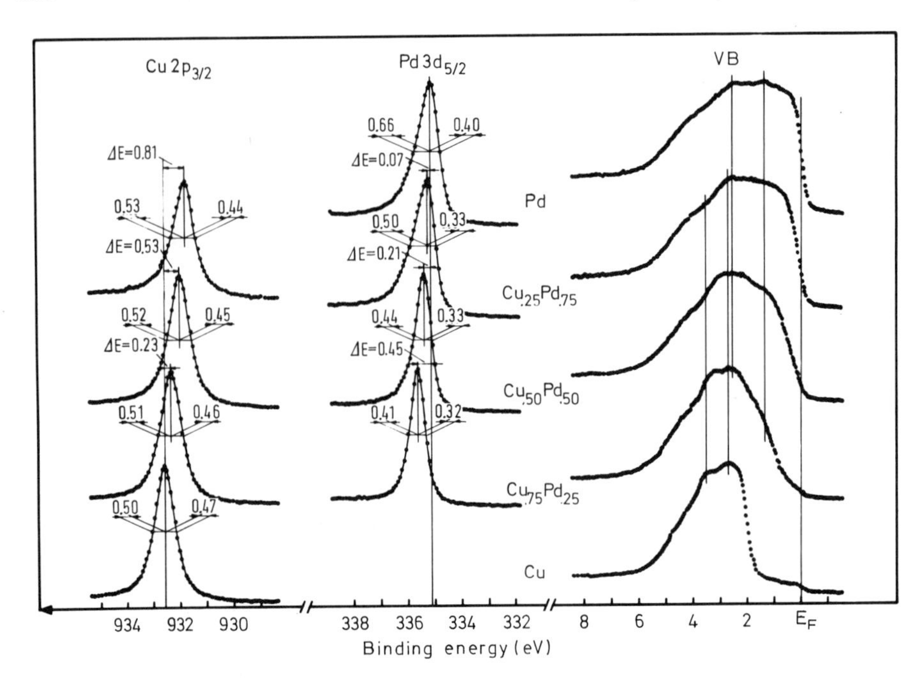

Fig. 71. Cu $2p$, Pd $3d$ and conduction band spectra of Cu and Pd metals and CuPd alloys, showing asymmetry in the Pd $3d$ lines due to excitation of soft electron-hole pairs in the conduction band (cf. text)[73]

One can intuitively expect, and it has been found[66] that the peak asymmetry is largest for those metals, which have a large density of states extending to the Fermi edge in the band, e.g. Rh, Pd, Ir and Pt, whereas e.g. Au with a low density at the edge has symmetric core lines. In Fig. 71 is shown a study of the effect of alloying on the peak shapes for the Cu-Pd system. Aside from the shifts of the lines, the Pd $3d$ peak shapes are seen to be significantly affected in going from the pure metal to the alloy. Since Cu metal has a substantially lower density of states at the Fermi edge than has Pd metal, the Cu lines are almost symmetric. The Pd lines, which are seen to be strongly asymmetric in the pure metal, are observed to become nearly symmetric as well as significantly more narrow in the alloy as a result of the decreasing Pd local density of states at the Fermi edge[73].

γ) Vibrational excitations in core electron spectra. In the early stages of gas-phase photoelectron spectroscopy of core levels the lines appeared as symmetric structureless peaks. A chemical dependence of the widths was observed[74-76], but could not be given any definite explanation. In particular, it

---

[73] Mårtensson, N., Nyholm, R., Calén, H., Hedman, J. (1981): Phys. Rev. B *24*, 1725

[74] Siegbahn, K., et al. (1967): ESCA – Atomic, molecular and solid state structure studied by means of electron spectroscopy, Uppsala: Almqvist and Wiksell

[75] Gelius, U., Allan, C.J., Allison, D.A., Siegbahn, H., Siegbahn, K. (1971): Chem. Phys. Lett. *11*, 224

[76] Thomas, T.D. (1970): J. Am. Chem. Soc. *92*, 4184

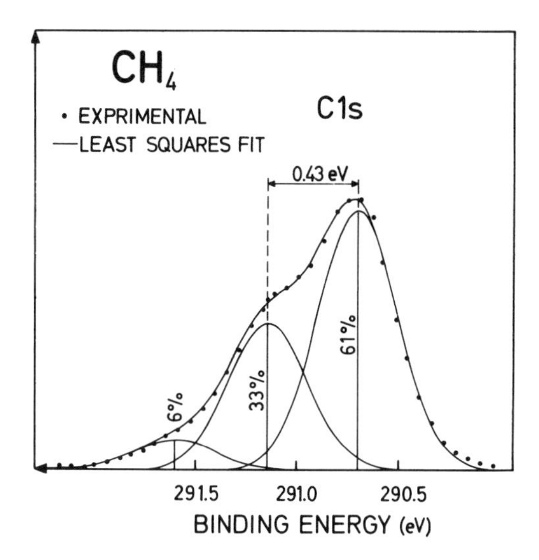

Fig. 72. C 1*s* spectrum of CH_4 obtained by monochromatized Al $K\alpha$ excitation[78]

was noted that the F 1*s* lines were generally broad[76]. With the advent of high resolution core electron spectra by means of x-ray monochromatization Gelius et al. observed that the peaks were in general neither symmetric nor lacking in structure[1,2,77,78]. The previously observed broadenings could thus be ascribed to vibrational excitations.

The case of C 1*s* in methane is shown in Fig. 72. The peak is resolved into three vibrational components of relative intensities 61%, 33% and 6%, respectively. The fundamental frequency was found to be 0.43 eV, which is an increase of 0.07 eV with respect to the ground state, indicating a narrowing of the potential curve of the ion. The shortening of the C-H bond length concomitant with the core ionization can be calculated from the spectrum and is found to be $\Delta R_{\text{C-H}} = -0.06$ Å in the harmonic approximation. The spectrum has been theoretically studied by Meyer[79] and Cederbaum and Domcke[80] who both obtained excellent agreement with experiment by means of extensive CI and Greens' function methods, respectively.

A large number of cases have been studied, where core line asymmetry and broadening may be attributed to vibrational excitation[1,2,77,78,81–88].

---

[77] Gelius, U. (1973): Acta Universitatis Upsaliensis 242 (Ph.D. Thesis)

[78] Gelius, U., Svensson, S., Siegbahn, H., Basilier, E., Faxälv, Å., Siegbahn, K. (1974): Chem. Phys. Lett. *28*, 1

[79] Meyer, W. (1973): J. Chem. Phys. *58*, 1017

[80a] Domcke, W., Cederbaum, L.S. (1975): Chem. Phys. Lett. *31*, 582

[80b] Cederbaum, L.S., Domcke, W. (1977): Adv. Chem. Phys. *36*, 205

[81] Saethre, L.J., Svensson, S., Mårtensson, N., Gelius, U., Malmquist, P.Å., Basilier, E., Siegbahn, K. (1977): Chem. Phys. *20*, 431

[82] Saethre, L.J., Mårtensson, N., Svensson, S., Malmquist, P.Å., Gelius, U., Siegbahn, K. (1980): J. Am. Chem. Soc. *102*, 1783

[83] Ågren, H., Selander, L., Nordgren, J., Nordling, C., Siegbahn, K. (1975): Z. Phys. A*272*, 131

(Footnotes 84–88 see p. 368)

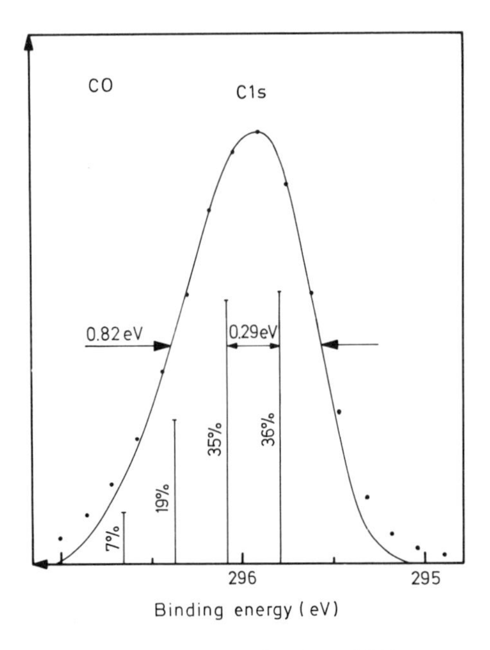

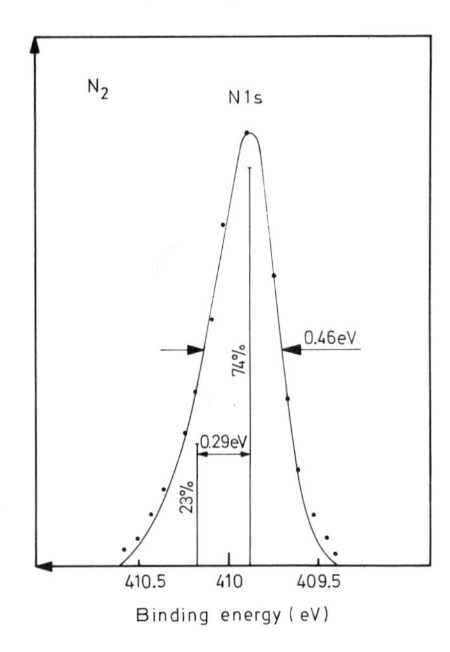

Fig. 73. C 1*s* spectrum of CO and N 1*s* spectrum of N_2 obtained by monochromatized Al $K\alpha$ excitation[78]. Theoretical spectra are indicated by bars and have been obtained from optical data using the equivalent core species (NO^+) for the core hole potential curves. The solid lines are least squares fitted to experiment, adjusting the widths of the vibrational components

Among these are carbon monoxide, CO, and molecular nitrogen, N_2[78] (Fig. 73). The C1*s* peak is clearly asymmetric and broader than would be expected from lifetime considerations only.

The core hole potential curves of the CO and N_2 cases as well as the paramagnetic molecules NO and O_2 have been studied extensively theoretically with *ab initio* theoretical methods[80, 83, 86, 87, 89–92]. CO and N_2 were investigated by CLARK and MÜLLER[17] in MO-LCAO-SCF calculations with various choices of basis sets. These authors conclude that the vibrations are excited in the C*O and N_2 cases due to a shortening of the bond length upon ionization. For the O1*s* ionization in CO the theoretical results of these authors suggest, however, that the vibrations occur due to an increase in bond

[84] Ågren, H., Müller, J., Nordgren, J. (1980): J. Chem. Phys. *72*, 4078

[85] Ågren, H., Müller, J. (1980): J. Electron Spectrosc. Relat. Phenom. *19*, 285

[86] Ågren, H., Selander, L., Nordgren, J., Nordling, C., Siegbahn, K., Müller, J. (1979): Chem. Phys. *37*, 161

[87] Müller, J., Ågren, H. (1980): J. Electron Spectrosc. Relat. Phenom. *18*, 235

[88] Gelius, U., Siegbahn, K. (unpublished)

[89] Clark, D.T., Müller, J. (1976): Theor. Chim. Acta *41*, 193

[90] Goscinski, O., Palma, A. (1977): Chem. Phys. Lett. *47*, 322

[91] Goscinski, O. (1976): in: Quantum science, Calais, J.L., et al. (eds.), New York: Plenum, p. 427

[92] Palma, A., Poulain, E., Goscinski, O., Wahlgren, U. (1978): Technical Note 538, Dept. Quant. Chem., Uppsala University

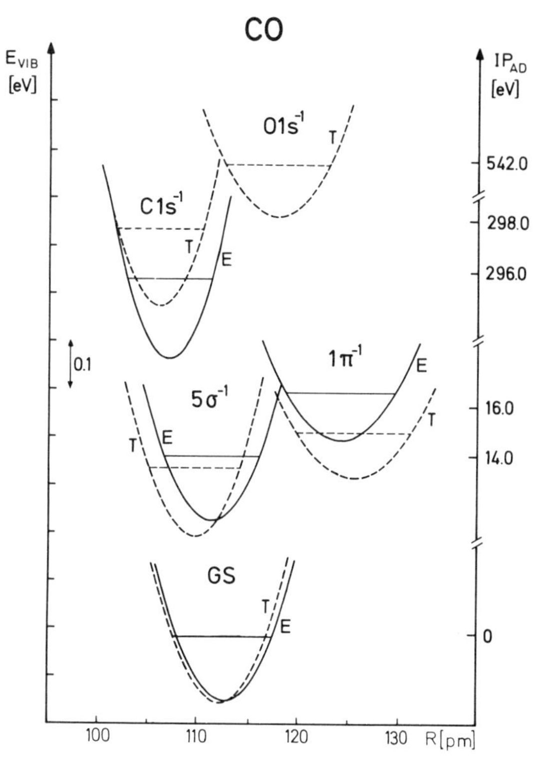

Fig. 74. Experimental and theoretically computed harmonic energy curves for CO[86]. The left vertical axis for each molecule is scaled for the vibrational levels (difference between two marks =0.1 eV) and the right axis for the adiabatic ionization potentials. E=Experimental curves. T =Theoretical (Hartree-Fock) curves. The horizontal lines for each curve marks the zero vibrational level

length. This theoretical result is consistent with the observed oxygen x-ray emission spectrum[93]. Experimental and theoretical potential energy curves for the CO case are shown in Fig. 74[86].

The differing behaviour of the two core hole states in CO was further studied by GOSCINSKI et al.[90–92]. In order to decide whether a core ionization leads to a shortening or lengthening of the bond, they consider the gradient $(dE_B/dR)_{R=R_0}$ where R_0 is the ground state bond length. This quantity is positive when the bond is shortened and negative in the opposite case. In the usual way it may be expressed in terms of orbital and relaxation energy contributions (the correlation and relativistic effects assumed negligible):

$$\left(\frac{dE_B}{dR}\right)_{R_0} = -\left(\frac{d\varepsilon}{dR}\right)_{R_0} - \left(\frac{dE_{\text{Relax}}}{dR}\right)_{R_0}. \tag{12.8}$$

Table 18[90,91] shows these quantities as obtained from ab initio MO-LCAO-SCF calculations[87]. It is seen that for C*O and N_2 the orbital energy contri-

[93] Nordgren, J. (private communication)

Table 18. Derivatives of orbital and relaxation energies with respect to internuclear distance for core ionization in CO and N_2 (in eV/a.u.) evaluated at optimized internuclear distance of the neutral ground state[a]

Molecule	$-\left(\frac{d\varepsilon_i}{dR}\right)_{R_0}$	$-\left(\frac{dE_{\text{Relax}}}{dR}\right)_{R_0}$	$\left(\frac{dE_B}{dR}\right)_{R_0} = -\frac{d\varepsilon}{dR} - \frac{dE_{\text{Relax}}}{dR}$
C*O	5.8193	−0.6394	5.1799
CO*	0.1436	−3.1337	−2.9901
NN*	4.6782	−2.1744	2.3352

[a] Ref. [87].

bution decides the direction of the bond length change. In the CO* case, however, the relaxation energy slope is large enough to outweigh the orbital energy slope, thus making the binding energy slope negative. The relaxation energy plays thus an important role in the bond length change of CO*. GOSCINSKI and PALMA also show by perturbation theory arguments, that the relaxation contribution, dE_{Relax}/dR, has the same sign as the relaxation energy itself, i.e. positive[90]. This means that relaxation will always tend to increase the bond lengths.

The use of the potential curve gradients for calculation of geometry changes and Franck-Condon factors has been the subject of several studies[80, 84, 85, 87]. It is found that gradient approaches generally yield better results for the Franck-Condon factors than for the geometry changes. This is due to the fact, as discussed by CEDERBAUM and DOMCKE[80], that the overall shape of the spectrum depends primarily on the behaviour of the final state energy surface within the FC-region (cf. also Sect. 15δ).

Polyatomic small molecules, such as H_2O and NH_3, have been investigated by the same procedures as for CO and N_2 with respect to core hole vibrational excitation[84, 85]. In these cases, core ionization may result in excitations of both stretching and bending vibrations. Calculations will naturally be considerably more complicated due to the multidimensional potential surfaces. Core ionization for NH_3 and H_2O is predicted to lead to considerable increase in the HXH (X=N, O) bond angle according to the calculations of refs. [84, 85].

Extreme cases of vibrational excitation have been found for the trithiapentalene molecule, whose S $2p$ spectrum is shown in Fig. 75, and similar molecules[2, 81, 82]. The broad structure interpreted as the signal from the $S_{1,6}$ positions is vibrationally broadened. Semiempirical calculations for the potential curves for the two sulphur positions has been made by SAETHRE et al.[81]. These are shown in Fig. 76. They constitute only a first approximation in describing the hole state formation in this molecule, since firstly semiempirical calculations of potential curves may not be particularly reliable and secondly they only consider one vibrational coordinate. Nevertheless the calculations explain satisfactorily the unusual shape of the core electron spectrum. From these curves it appears that ionization of the middle sulphur does not lead to any appreciable geometry change, as expected, but it is rather free to move between the adjacent sulphurs. Ionization of any of these outer sulphurs will,

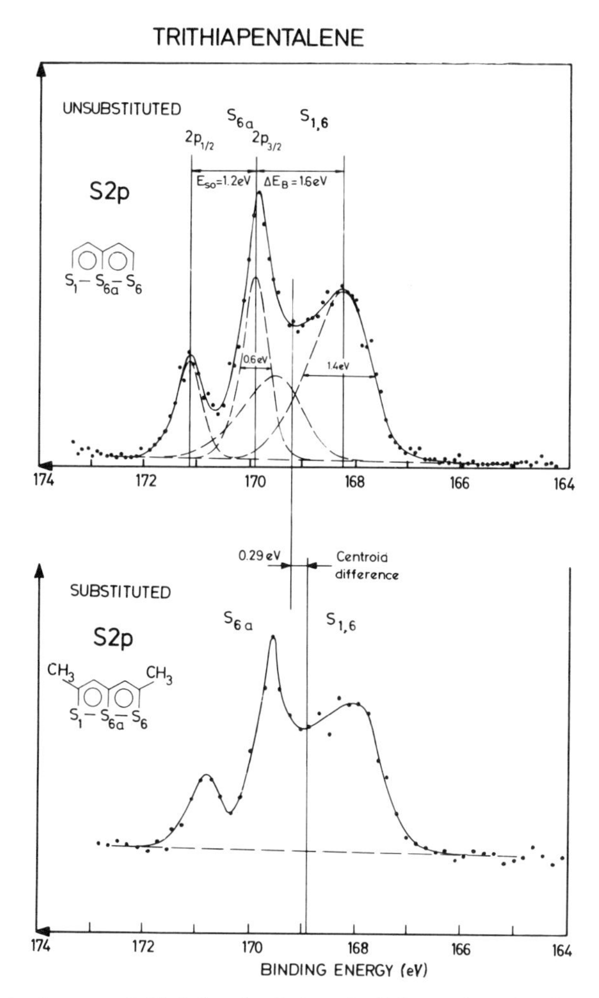

Fig. 75. S 2p spectra from unsubstituted and substituted trithiapentalene showing strong vibrational broadening of the $S_{1,6}$ lines[81]

however, result in a substantial change of equilibrium distance to the middle sulphur. This latter ionization will give rise to considerable vibrational structure, which is indeed observed experimentally. SAETHRE et al.[82] have investigated molecules of similar structure, containing Se and Te and obtained results in terms of line broadenings, which could be rationalized on the same basis.

Independent experimental evidence for vibrational excitation in core ionization is given by consideration of soft x-ray emission spectra[83–87, 94]. Thus, whereas in photoionization the spectrum is given by the FC factors in one transition (cf. Fig. 77) the observed x-ray emission spectrum is decided both by the FC factors in the initial core hole creation and the subsequent x-ray

[94] Werme, L.O., Nordgren, J., Ågren, H., Nordling, C., Siegbahn, K. (1975): Z. Phys. A*272*, 131

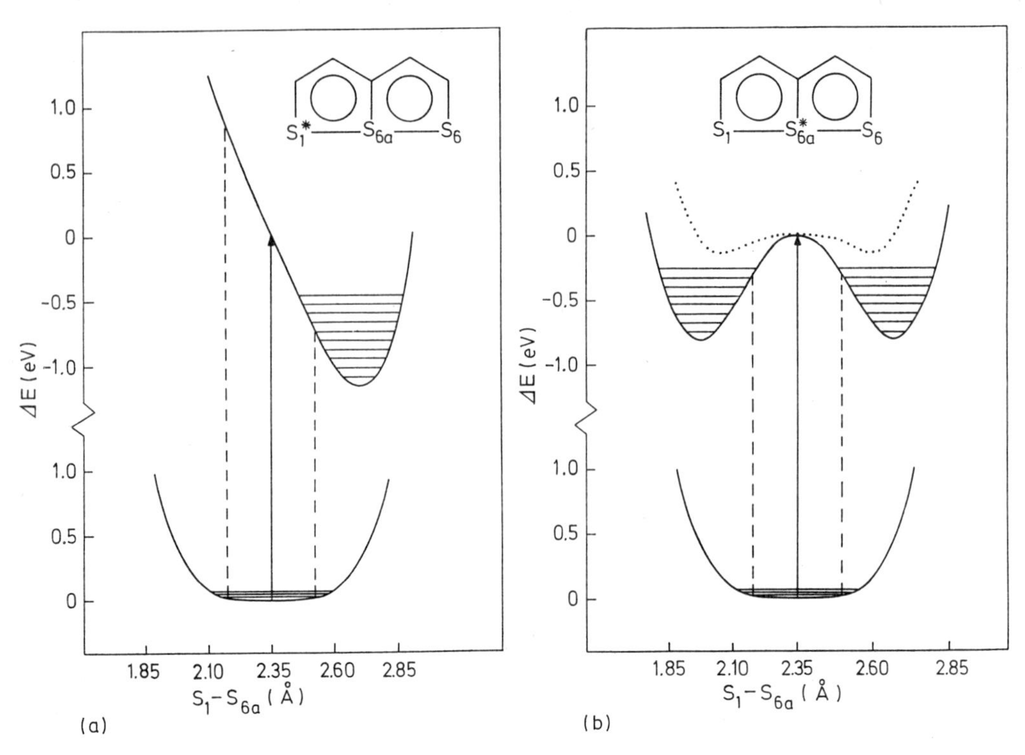

Fig. 76a, b. Potential curves obtained from semi-empirical calculations for sulphur core ionization in trithiapentalene. Only motion along the sulphur bonds is considered [81] (cf. text)

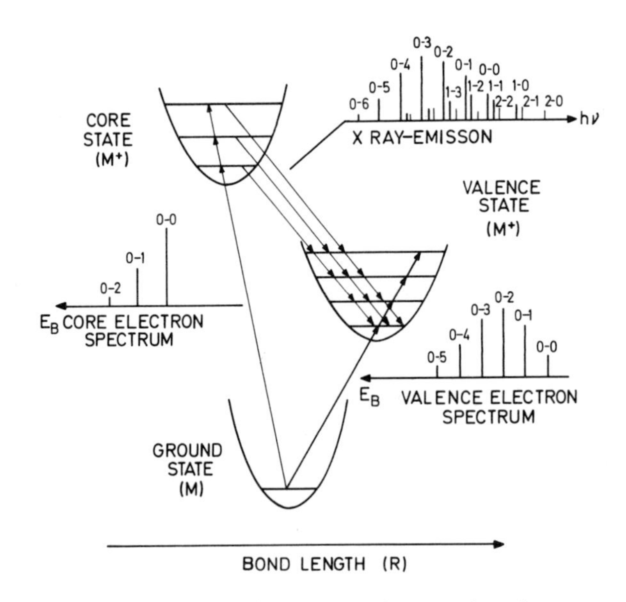

Fig. 77. Schematic illustration of the relation between the core photoelectron spectrum, the x-ray emission spectrum and the valence photoelectron spectrum for a hypothetical molecule

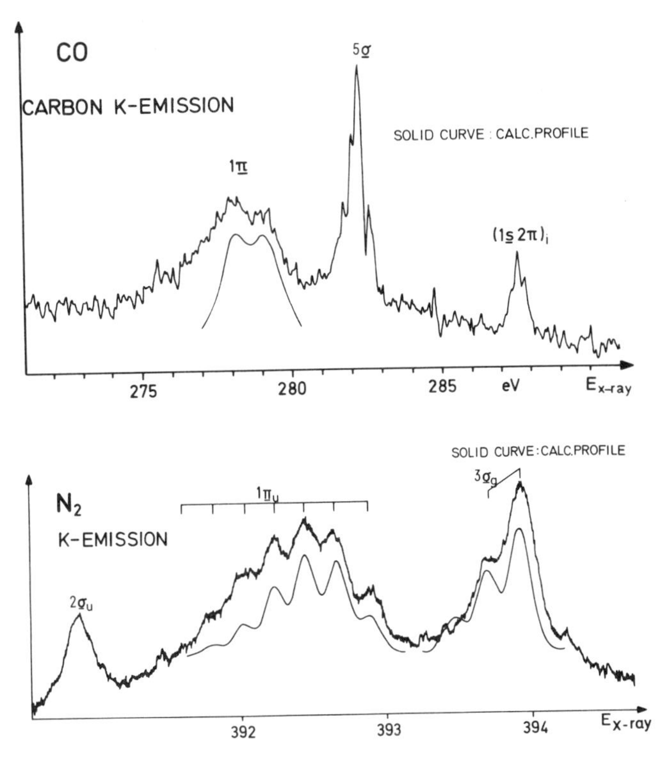

Fig. 78. Soft x-ray emission spectra of CO and N_2 [83]. Calculated profiles (solid lines) have been obtained from optical data using equivalent core arguments for the core hole states. Note difference in energy scales

emission. The vibrational excitation in the initial core hole state consequently decides the difference between the x-ray emission and valence photoelectron spectra. The $1\underline{s} \rightarrow 1\underline{\pi}_u$ transition in the x-ray emission spectrum of N_2, shown in Fig. 78 (bottom), shows a clearly resolved vibrational progression. In the corresponding π-transition in CO (Fig. 78, top), no clear vibrational peaks can be observed. This qualitative difference between these π-bands can be explained by the core hole vibrational excitations. Thus, since the N1s vibrational excitation is fairly weak, the resulting x-ray spectrum shows a vibrational progression which closely resembles that of direct valence ionization. As discussed above, for C1s ionization in CO, the vibrational excitation is large. This leads to a large difference with respect to the 1π valence photoelectron spectrum and the corresponding π-transition in N_2.

High resolution soft x-ray emission spectra have been shown to yield accurate quantitative information on the geometry of core hole states [83, 84, 86, 94]. The $C1\underline{s} \rightarrow 5\underline{\sigma}$ band (Fig. 79) contains well resolved vibrational structure, which can be employed to determine the potential curve of the C1s hole state. The ground state and $5\underline{\sigma}$ hole state potential curves are well known from optical data and the intermediate C1s hole state curve can thus

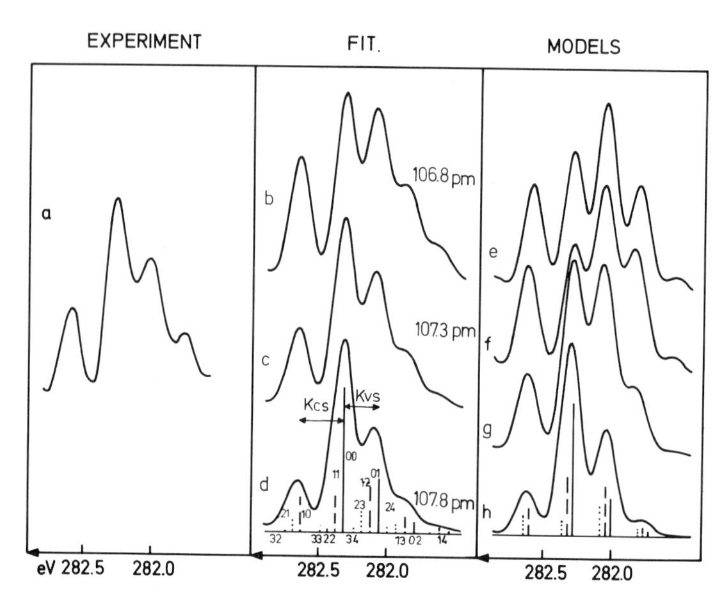

Fig. 79. X-ray emission spectra of CO (C 1s→5σ). (*a*) experimental spectrum (*b*)–(*d*) fitted spectra, obtained by using optical data for neutral (ground state) and valence hole states and a parametrized (anharmonic) curve for the core hole states. (*e*) theoretical spectrum; equivalent core for hole state and valence hole state. (*f*) theoretical spectrum; Koopmans' approximation for core hole state and optical data for the two other states (*g*) Hartree-Fock (relaxed) potential curve for core hole state and optical data for the two other states. (*h*) Hartree-Fock potential curves for all states. The theoretical spectra (*e*)–(*h*) were all calculated in the harmonic approximation[86]

be parametrized and fitted to the observed spectrum. Figure 79 shows a sequence of simulated spectra, which differ in their assumed C*O bond length. As can be seen, the sensitivity of the spectra to bond length changes allows a determination accurate to within 0.005 Å. The value found in this case was $R_{C^*O} = 1.073 \pm 0.005$ Å. Similar analyses have been performed for N_2 and CO_2[86,95] (cf. Fig. 80).

For solid core photoelectron spectra, the excitation of phonons may lead to additional broadening of the lines. In molecular crystals, similar considerations as in the preceding discussion may be used, viz. the excitation can be described in terms of local molecular vibrational modes. One example is the N1*s* spectrum of NH_4NO_3 where the N1*s* line from the ammonium group was found considerably broader than the N1*s* line from the nitrate group due to excitation of N-H stretching vibrations in the former case[2].

Phonon broadening in polar materials was investigated by CITRIN et al.[96]. For such materials one finds in general considerably broader core lines than in

[95] Nordgren, J., Selander, L., Pettersson, L., Nordling, C., Siegbahn, K., Ågren, H. (to be published)

[96] Citrin, P.H., Eisenberger, P., Hamann, D.R. (1974): Phys. Rev. Lett. *33*, 965

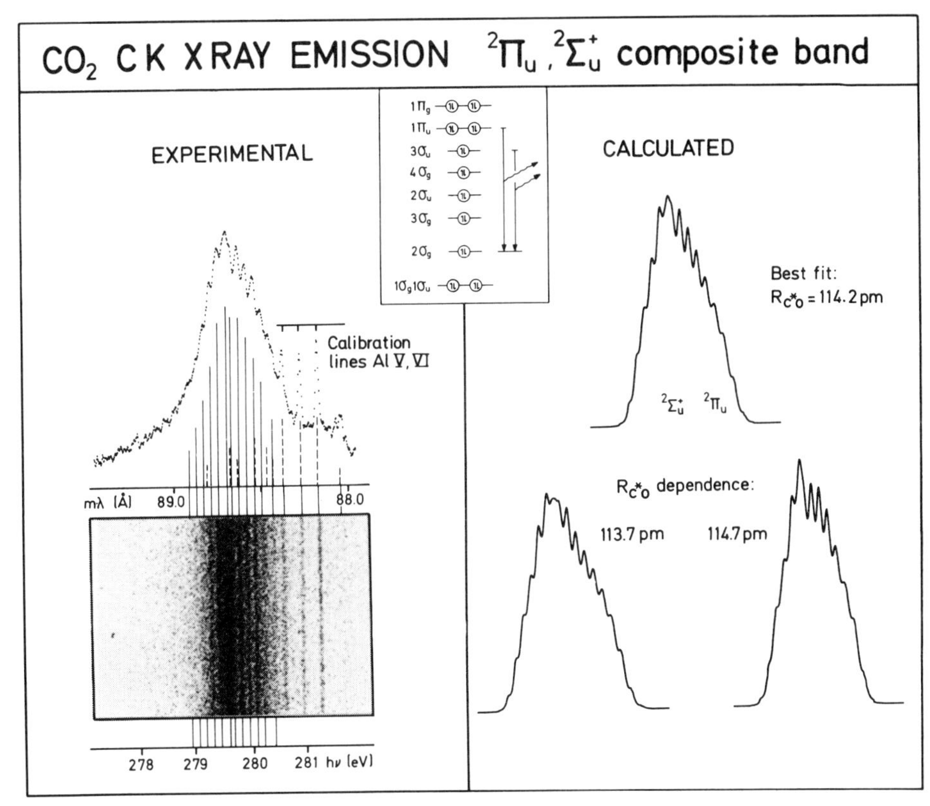

Fig. 80. Experimental and calculated C K x-ray emission ($^2\Pi_u$, $^2\Sigma_u^+$) bands of CO_2. The best fit to experiment is obtained with a Voigt profile for the individual vibrational transitions having a Lorentzian lifetime width of 0.07 eV[95]

e.g. metals, even when account has been taken of inhomogeneous charge-up of the sample surface. A simple model[96] leads to the following expression for the width of the Gaussian phonon excitation spectrum at temperature T:

$$\Gamma_{\text{ph}} = 2.35[\hbar\,\omega_{\text{LO}}\,\Delta E_{\text{nucl}} \cdot \coth(\hbar\,\omega_{\text{LO}}/kT)]^{1/2}, \tag{12.9}$$

where ω_{LO} is the longitudinal optical frequency and ΔE_{nucl} is the nuclear relaxation energy, i.e. corresponding to the energy $E_B^{\text{vert}} - E_B^{\text{ad}}$ in the molecular case (cf. (7.26)). A condition for the validity of this expression is that $\Delta E_{\text{nucl}} \gg \hbar\,\omega_{\text{LO}}$, which means that a large number of phonons are excited in the core ionization. This is expected to be a common situation for polar materials where the screening of the core hole by electronic reorganization is generally small. The nuclear relaxation energy for polar compounds can be expressed using the static (ϵ_0) and dynamic (ϵ_∞) dielectric constants (in au):

$$\Delta E_{\text{nucl}} = \left(\frac{6}{\pi V}\right)^{1/3} \left(\frac{1}{\epsilon_\infty} - \frac{1}{\epsilon_0}\right), \tag{12.10}$$

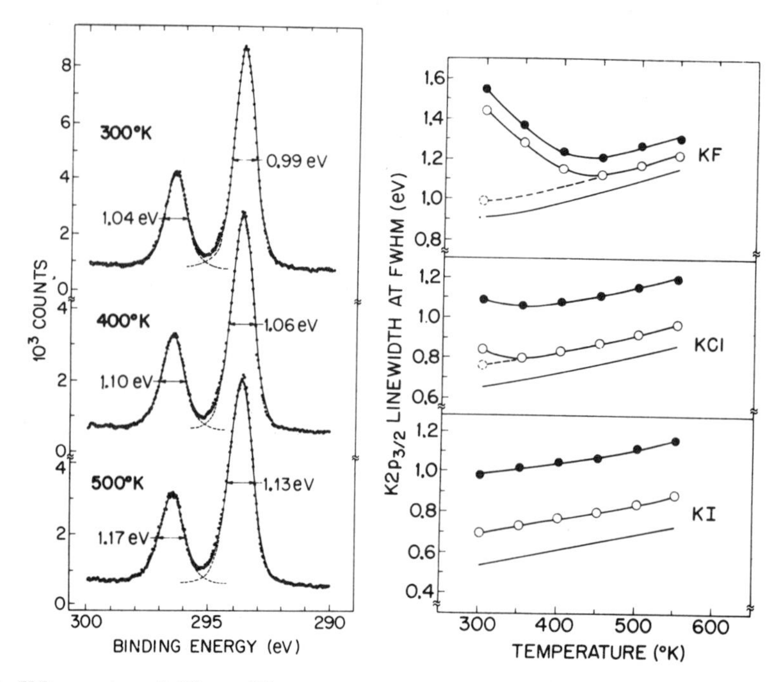

Fig. 81. K $2p$ spectra of KI at different temperatures (left) K $2p_{3/2}$ linewidth as a function of temperature for potassium halides. Filled circles are measured widths, open circles are the widths after removing lifetime and instrumental contributions, solid line is theoretical (Eqs. 12.9, 12.10)[96]

where V is the volume of the primitive cell. Fig. 81 shows the results by Citrin et al. for the K $2p_{3/2}$ linewidths in potassium halides as a function of temperature. The solid lines are estimated phonon widths using relations (12.9, 10). It is seen that appreciable line broadening are expected for these materials. The experimental results are found to be in good agreement with the temperature dependence derived from (12.9).

δ) Core line multiplet structure in molecules and solids. Splitting and broadening of core electron lines may occur whenever a system possesses unpaired electron angular momenta. This effect is largely atomic in character, and is often discussed in these terms. The observation of its chemical dependence leads to information on the shape of the system wave function.

Multiplet splittings were first observed in the paramagnetic molecules O_2 and NO by HEDMAN et al.[97]. Figure 82 shows the O $1s$ and N $1s$ spectra of these molecules along with the N $1s$ line of N_2. For O_2, the two O $1s$ components are due to the coupling in the final ionic state between the unpaired $1s$-electron and the triplet π valence shell. This results in a statistical intensity ratio of 1:2 between the two components. An estimate of the energy splittings in these cases may be obtained from van Vleck's theorem, which states that the multiplet splitting in the configuration $s\,a^k$, where a^k couples to spin S, is equal

[97] Hedman, J., Hedén, P.F., Nordling, C., Siegbahn, K. (1969): Phys. Lett. A *29*, 178

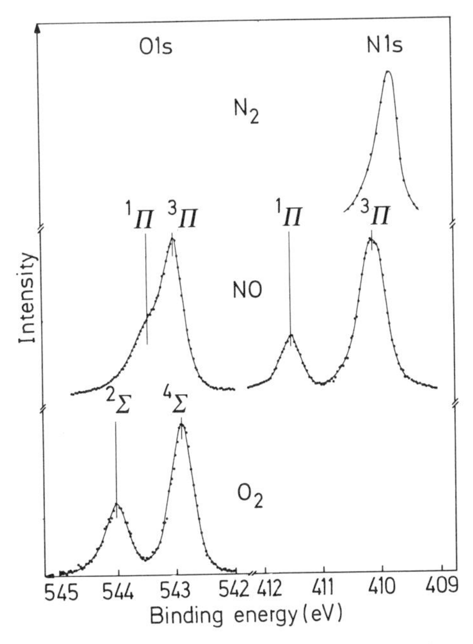

Fig. 82. O 1*s* and N 1*s* spectra of N_2, NO and O_2 showing multiplet splitting in the latter two molecules due to unpaired spins

to[98]:

$$\Delta E=(2S+1)\,K_{as}, \tag{12.11}$$

where K_{as} is the exchange integral between the *s* subshell and the open a^k-shell. Specifically, for O_2 this leads to:

$$\Delta E=3K_{1s\pi g}. \tag{12.12}$$

The assumption underlying (12.11) is that the same molecular orbitals are used for both final ionic multiplet states. For NO (ground state: $^2\Pi$) one obtains:

$$\Delta E=2K_{1s\pi} \tag{12.13}$$

between the two final ionic states of $^1\Pi$ and $^3\Pi$ symmetry. SCHWARZ[99] calculated the splittings in NO using (12.13) with wave functions obtained from ground state HF calculations and obtained 1.26 eV for the N 1*s* splitting and 0.77 eV for the O 1*s* splitting (experimental values: 1.42 eV and 0.55 eV, respectively[100]).

More exact treatments of the splitting and multiplet intensity ratios require that total wave functions are used which are separately optimized for the

[98] Van Vleck, J.H. (1934): Phys. Rev. *45*, 405
[99] Schwartz, M.E. (1970): Chem. Phys. Lett. *5*, 50
[100] Davis, D.W., Martin, R.L., Banna, M.S., Shirley, D.A. (1973): J. Chem. Phys. *59*, 4235

ground and final multiplet states. Such calculations for O_2 and NO[101–106] indicate that correlation in initial and final states are important to decide both splittings and intensity ratios.

Strong manifestations of multiplet splitting occur in the transition and rare earth elements and their salts[107, 108–125]. FADLEY et al.[107] first observed the effect in Mn compounds for the Mn $3s$ and Mn $3p$ levels. The usual treatment of these spectra is based on atomic considerations, i.e. it is assumed that the local wave function around the metal atom in these compounds is well characterized in terms of some atomic configuration. In MnF_2, the Mn atom is divalent and the spectrum is hence interpreted on the basis of the ground state configuration of the Mn^{2+} ion, which is $3d^5(^6S)$. For ns-ionization van Vleck's theorem (12.11) in the atomic case can be stated:

$$\Delta E(ns) = \frac{2S+1}{2l+1} G^l(ns; n'l), \tag{12.14}$$

where $n'l$ are the open subshell quantum numbers and G^l the radial atomic exchange integral. The intensity ratio between the $S+1/2$ and $S-1/2$ final

---

[101] Bagus, P.S., Schaefer, H.F. (1972): J. Chem. Phys. *56*, 224

[102] Bagus, P.S., Schaefer III, H.F. (1972): J. Chem. Phys. *55*, 1474

[103] Basch, H. (1976): Paper presented at 9th Jerusalem Symp. on Metal-Ligand Interactions in Organic and Biochemistry

[104] Darko, T., Hillier, I.H., Kendrick, J. (1977): Chem. Phys. Lett. *45*, 188

[105] Darko, T., Hillier, I.H., Kendrick, J. (1977): J. Chem. Phys. *67*, 1792

[106] Bagus, P.S., Schrenk, M., Davis, D.W., Shirley, D.A. (1974): Phys. Rev. A *9*, 1090

[107] Fadley, C.S., Shirley, D.A., Freeman, A.G., Bagus, P.S., Mallov, J.V. (1969): Phys. Rev. Lett. *23*, 1397

[108] Davis, D.W., Shirley, D.A. (1972): J. Chem. Phys. *56*, 669

[109] Clark, D.T., Adams, D.B. (1971): Chem. Phys. Lett. *10*, 121

[110] Zeller, M.V., Hayes, R.G. (1971): Chem. Phys. Lett. *10*, 610

[111] Wertheim, G.K., Rosencwaig, A. (1971): Chem. Phys. Lett. *54*, 3235

[112] Frost, D.C., McDowell, C.A., Woolsey, I.S. (1972): Chem. Phys. Lett. *17*, 320

[113] Carver, J.C., Schweitzer, G.K., Carlson, T.A. (1972): J. Chem. Phys. *57*, 973

[114] McFeely, F.R., Kowalczyk, S.P., Ley, L., Shirley, D.A. (1974): Phys. Lett. *49* A, 301

[115] Kowalczyk, S.P., Ley, L., Pollak, R.A., McFeely, F.R., Shirley, D.A. (1973): Phys. Rev. B *7*, 4009

[116] Bagus, P.S., Freeman, A.J., Sasaki, F. (1973): Phys. Rev. Lett. *30*, 850

[117] Kowalczyk, S.P., Edelstein, N., McFeely, F.R., Ley, L., Shirley, D.A. (1974): Chem. Phys. Lett. *29*, 491

[118a] Kowalczyk, S.P., Ley, L., McFeely, F.R., Shirley, D.A. (1975): Phys. Rev. B *11*, 1721

[118b] Kowalczyk, S.P. (1976): PhD Thesis, Lawrence Berkeley Laboratory, Univ. Calif. LBL-4319

[119] Wertheim, G.K., Cohen, R.L., Rosencwaig, A., Guggenheim, H.J. (1972): in: Electron spectroscopy, Shirley, D.A. (ed.), Amsterdam: North-Holland, p. 813

[120] Carlson, T.A. (1975): Far. Disc. Chem. Soc. *60*, 30

[121] Gupta, R.P., Sen, S.K. (1974): Phys. Rev. B *10*, 71

[122] Fadley, C.S. (1973): in: Electron emission spectroscopy, Dekeyser, W., et al. (eds.), Dordrecht: Reidel, p. 151

[123] Bagus, P.S., Wahlgren, U.I. (1974): Proc. Int. Conf. X-ray Processes in Matter, Otaniemi, Finland, 295

[124] Yamaguchi, T., Sugano, S. (1977): J. Phys. Soc. Japan *42*, 1949

[125] Gupta, R.P., Sen, S.K. (1973): in: Electron emission spectroscopy, Dekeyser, W., Fiermans, L., Vanderkelen, G., Vennik, J. (eds.), Dordrecht: Reidel, p. 225

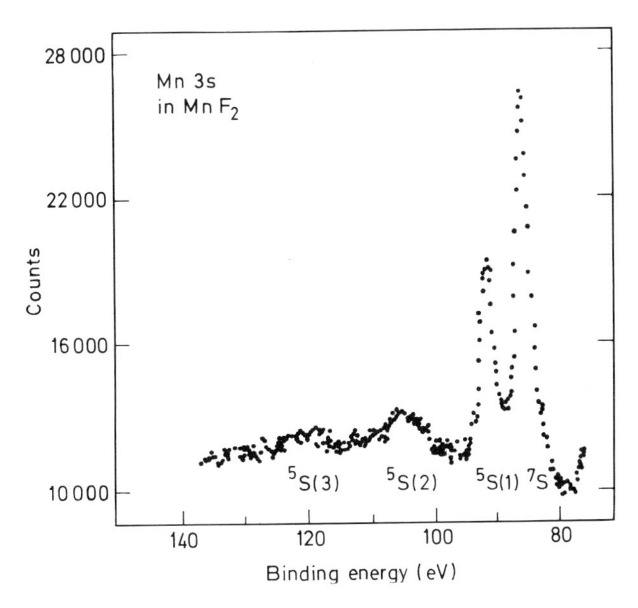

Fig. 83. The Mn 3*s* spectrum of MnF_2[115]. The two satellite ^5S states at ~15 eV and ~30 eV below the main ^5S line are consequences of final ionic state correlation

ionic terms is given by:

$$\frac{I(L, S+1/2)}{I(L, S-1/2)}=\frac{2S+2}{2S}. \tag{12.15}$$

For Mn^{2+} one thus expects two lines of intensity ratio 7:5. Van Vleck's theorem yields for 3*s* ionization of this ion a multiplet splitting of 13.0 eV. The experimentally observed 3*s*-spectrum of MnF_2 is shown in Fig. 83[115]. Two peaks are obtained with an intensity ratio substantially larger than the expected ratio 7:5. Also, the splitting between the peaks is found to be $\Delta E = 6.5$ eV, i.e. a factor of two smaller than the prediction of van Vleck's theorem. The explanation for this discrepancy between theory and experiment is largely due to the neglect of correlation. This may be explained in simple terms. Since the ^7S state corresponds to a parallel coupling of the spin of the 3*s*-electron with that of the 3*d*-shell, a substantial amount of correlation between the 3*s*- and 3*d*-electrons is already present at the one-particle level through the Pauli principle. This situation does not pertain to the ^5S state, which is consequently lowered more in energy than the ^7S state when correlation is included. This description is confirmed by free-ion calculations including correlation, which gave a value for the splitting of $\Delta E = 8.2$ eV[116]. The FISCI in this spectrum is mainly due to the excitations $|3s^1\,3p^6\,3d^5(^5S)\rangle \leftrightarrow |3s^2\,3p^4\,3d^6(^5S)\rangle$[116]. This FISCI also leads to the occurrence of the satellites in the spectrum at ~15 eV and ~30 eV below the ^5S line. When the intensities of these lines are added to the main ^5S line the multiplet intensity ratio ^7S:^5S becomes in far better agreement with the statistical ratio[115,116].

13. Core electron spectra of molecules on metal surfaces. ESCA diffraction. As indicated in the introduction to Sect. 11, the properties of core electron spectra imply that they can be used for qualitative and quantitative analysis in terms of shifts and intensities. Naturally, the observation and measurement of the multiple excitation features discussed in the preceeding section leads to further information on the electronic structure of the material. The space of an article of this kind is too limited for a comprehensive review of all the implications of core electron spectroscopy in applied research e.g.[1]. Here, we shall contend ourselves to mention some studies from only one field which has rapidly expanded during recent years; viz. the study of adsorbed species on metal surfaces. These and the examples already given in previous sections may hopefully serve to illustrate the potential of applied core electron spectroscopy.

The surface sensitivity of photoelectron spectroscopy implied by the small magnitudes of the electron scattering mean free paths makes it a versatile tool for studies of solid surfaces. One particular advantage of photoelectron studies is the nondestructive feature. Equilibria in adsorption systems involving weakly bound molecules are often sensitive to the action of highly ionizing radiation, such as electrons or ions. Therefore, at many instants core photoelectron spectra may yield information which is more characteristic of the unperturbed adsorption situation than that obtained from other techniques. For a fuller understanding of the phenomena involved in adsorption and in the processes on surfaces in general, photoelectron spectrometers are often combined with other techniques such as SIMS (Secondary Ion Mass Spectroscopy), LEED (Low Energy Electron Diffraction), FFMS (Flash Filament Mass Spectroscopy) etc.

One of the most intensely studied species in adsorption investigations is that of carbon monoxide. CO adsorbed on tungsten may exist in a variety of bound states. The distribution of CO molecules among these states is a function of the temperature at which the adsorption occurs. At low temperatures ($\sim$100 K) on the W(100) surface the adsorbed layer consists mainly of a α-CO state and a so-called virgin CO-state. The structure of the virgin-CO state is not completely clear, although several studies[2-4,22] are consistent with the structure indicated in Fig. 84 which is a simplified picture of the states involved on the W(100) surface. The α-CO state is characterized by a carbonyl-like bond to the tungsten surface atoms with the molecules attached in an upright position, the oxygen pointing outward. Upon heating, the "virgin-layer" of CO is transferred into a number of new states denoted β-CO. Evidence from several investigations[2-4,22] show that these latter states are dissociated on the W-surface.

Umbach et al.[3] studied the adsorption of CO on the W(110) surface. These

---

[1] Siegbahn, H., Karlsson, L. Electron Spectroscopy for Atoms, Molecules and Condensed Matter, Amsterdam: North-Holland (to be published)

[2] Yates Jr., J.T., Erickson, N.E., Worley, S.D., Madey, T.E. (1975): in: The physical basis for heterogeneous catalysis, Dragulis, E., Jaffee, R.I. (eds.), New York: Plenum Press, p. 75

[3] Umbach, E., Fuggle, J.C., Menzel, D. (1977): J. Electron Spectrosc. Relat. Phenom. *10*, 15

[4] Menzel, D. (1978): in: Photoemission and the electronic properties of surfaces, Feuerbacher, B., Fitton, B., Willis, R.F. (eds.), New York: Wiley

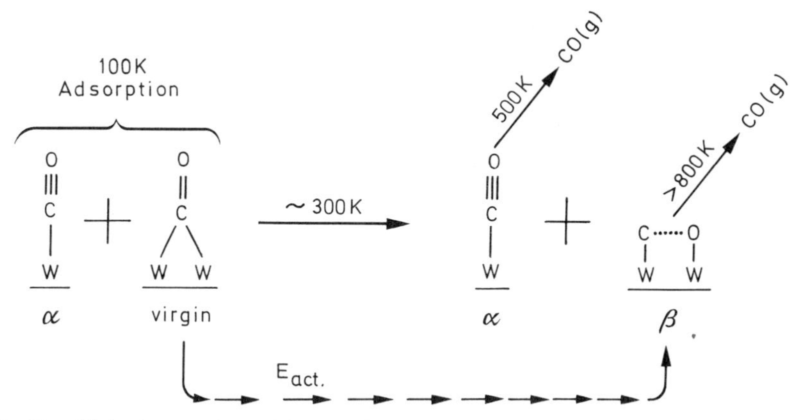

Fig. 84. Simplified picture of possible bonding structures for chemisorbed CO on a tungsten surface (W(100))[2]. Several β-CO states may co-exist on the surface with possible large variations in the strength of the C...O bond

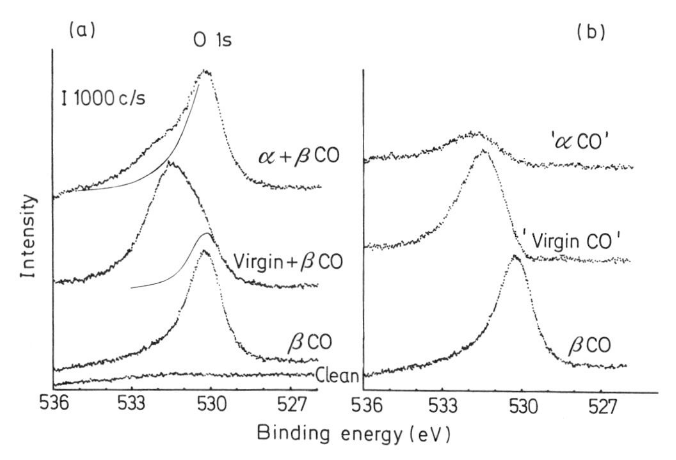

Fig. 85a, b. O 1*s* spectra from adsorbed CO layers on a W(110) surface[3]. (**a**) Direct spectra. Virgin +β CO: Surface covered with CO at room temperature. βCO: same layer after heating to 600 K. α+βCO: same layer during reexposure to $\sim 10^{-6}$ Torr CO at room temperature. (**b**) Isolation of contributions of individual CO states obtained by taking difference of the direct specta in (**a**). "Virgin CO": (virgin+βCO)−0.5βCO "αCO": (α+βCO)−βCO

authors made their adsorption experiments starting at room temperature and then heating to 600 K after which the system was reexposed to CO at room temperature. The O 1*s* spectra obtained are shown in Fig. 85. The middle spectrum of the a) figure is that obtained directly after the first exposure at room temperature. Here, the dominating state is ascribed to virgin-CO. The same designations for the states are used for the different crystal faces, although the bonding characteristics may not be quite the same. After heating to 600 K the bottom spectrum was obtained, completely dominated by β-CO. The re-exposure at room temperature introduces states whose O 1*s* binding energies

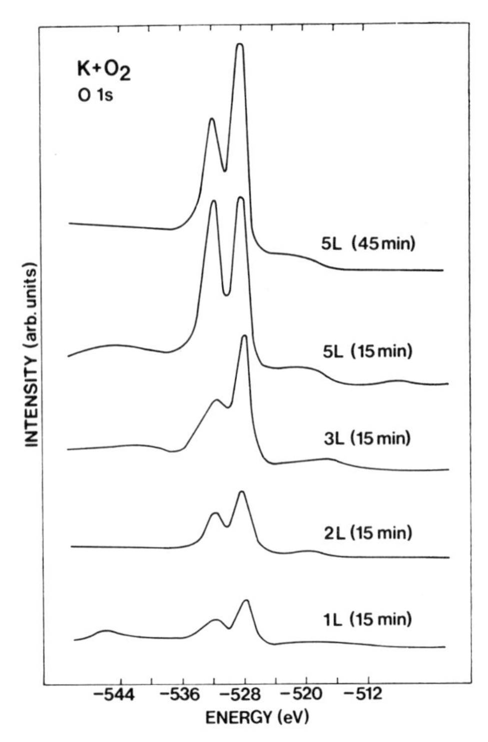

Fig. 86. O 1*s* spectra for different amounts of oxygen exposure on a potassium metal surface. The low binding energy peak is due to potassium oxide and the other peak to physisorbed oxygen ($1L = 10^{-6}$ Torr · *s*)[5]

lie close to those for virgin-CO, assigned as α-CO states. These α-CO states differ in O 1*s* binding energy by ~1 eV from the α-CO states on the W(100) surface[2]. The binding energy shifts of surface-adsorbed molecules both with respect to different states on the surface and with respect to the gaseous phase have a large contribution associated with the relaxation of the metal conduction electrons ($\Delta E_{\text{Relax}} \sim 1$–5 eV for the shift between the gaseous phase and adsorbed state for e.g. CO on a typical metal surface[4]). Direct evidence for this relaxation contribution is provided by the formation of satellite structure in the adsorbate ionization as previously discussed in Sect. 12β. As mentioned there, CO on the W(110) surface should be classified as a strong chemisorption case with charge transfer relaxation from the metal substrate (cf. Figs. 67, 68).

PETERSSON and KARLSSON[5] investigated the oxidation of potassium films using uv- and Al$K\alpha$-excited photoelectron spectra. O 1*s* spectra from a O_2-exposed clean surface recorded at 77 K are shown in Fig. 86. The observed doublet structure clearly indicates the existence of oxygen in two distinctly different chemical environments. The low binding energy peak is interpreted by

[5] Petersson, L.-G., Karlsson, S.-E. (1977): Phys. Scr. *16*, 425

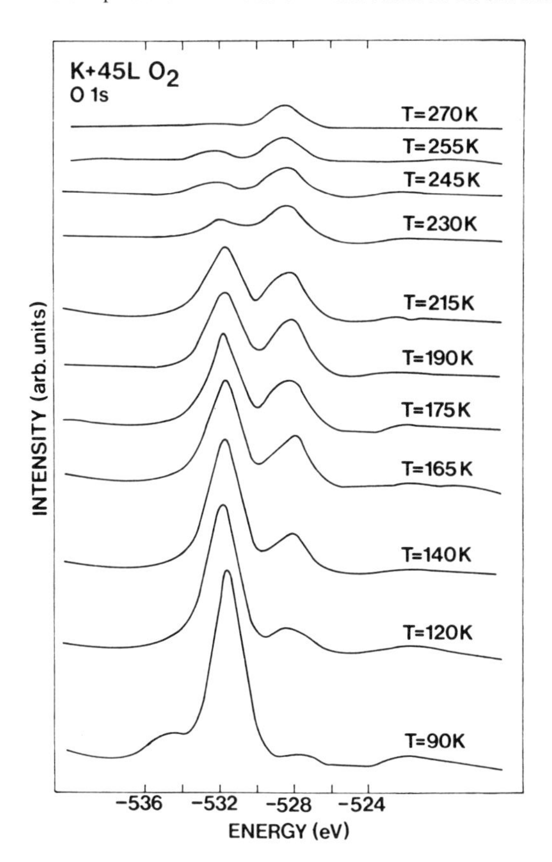

Fig. 87. O 1s spectra after an exposure of $45 L\, O_2$ on potassium recorded with increasing temperature. Intensity is transferred from the molecular-associated peak to the oxide-associated peak[5]

these authors as due to oxygen in oxide form while the other peak originates from oxygen physisorbed on the potassium oxide, probably in molecular form. At small exposures the oxide peak dominates, while the other peak grows in relative intensity with increasing exposure. This is due to the fact that molecular oxygen can only physisorb as a molecule on the potassium oxide, which is formed initially. The above conclusion was substantiated by recording the O1*s* spectra with increasing temperature (cf. Fig. 87). The spectra were obtained at an oxygen exposure sufficiently high, that the physisorbed oxygen peak dominates at 90 K. As the temperature increases, potassium diffuses out to the surface and forms potassium oxide with the available surface oxygen. Thus, the relative intensity of the surface bonded oxygen peak will decrease with respect to the bulk oxide peak. Due to a low heat of adsorption for the molecular oxygen physisorbed on the oxide surface oxygen molecules may also leave the surface and be pumped off. This leads to an overall decrease of the intensity in the spectrum.

A phenomenon, which has a high potential for investigation of surface structure is that of ESCA diffraction. As the name implies it refers to the

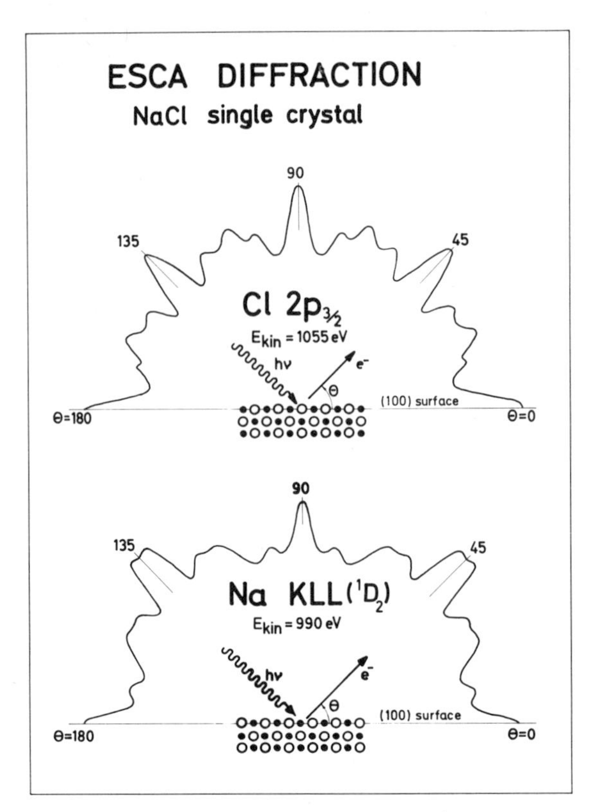

Fig. 88. Angular distributions of $C1\,2p_{3/2}$(Mg $K\alpha$) and Na $KLL(^1D_2)$ photoelectrons from a NaCl single crystal[6]

elastic diffraction of the photoelectrons as they leave the surface region of the sample. Such effects were first discovered for single crystals of NaCl[6] where strong intensity modulations were found as a function of angle between the photoelectron path and the crystal normal. These results are shown in Fig. 88 for the Cl $2p$ line excited with Mg $K\alpha$ radiation and the Na $KLL(^1D_2)$ Auger line. Prominent intensity maxima occur along the direction of the crystal planes, in particular for those of low Miller index.

The conceptual framework within which these phenomena are to be treated is that of dynamic diffraction theory. In its most general form this implies the complete solution of the Schrödinger equation of the outgoing photoelectron. The problem is thus related to that of LEED for which a variety of methods of solution have been developed e.g. [7] and references therein. The boundary conditions for the photoelectron experiment is different from that of LEED, however, in that there is no incoming electron wavefield, but the photoelectron waves originate at the crystal sites.

[6a] Siegbahn, K., Gelius, U., Siegbahn, H., Olson, E. (1970): Phys. Lett. *32* A, 221

[6b] Siegbahn, K., Gelius, U., Siegbahn, H., Olson, E. (1970): Phys. Scr. *1*, 272

[7] Pendry, J.B. (1974): Low energy electron diffraction, New York: Academic Press

There are similarities between photoelectron emission and other observations of electron emission, such as Kikuchi bands in electron diffraction or electron channeling in radioactive decay from single crystals. In view of this, SIEGBAHN et al.[6] considered treatments of the type used for these latter phenomena[8] in order to explain the above observed angular distribution features in NaCl. In spite of the fact that such models were developed for much higher electron energies (10–10,000 keV) this was found to represent both the observed shape and widths of the angular profiles in the NaCl experiments. Subsequent measurements[9–13] by other workers on other single crystals have shown agreement with the qualitative conclusions of SIEGBAHN et al.[6].

One particular case of interest, where ESCA diffraction studies could provide a tool for obtaining geometrical information is that of adsorption of atoms and molecules on surfaces. The virtue of using core photoelectrons for such systems lies in the localized nature of the core electrons, with the consequence that the primary electron wave field originates at one particular atomic site. This is opposed to ordinary electron scattering experiments, where all atomic centers act as secondary emitters. This implies that in order to obtain information on adsorbed geometry from photoelectron intensities, it is not necessary to create periodical order of the adsorbate which is required in LEED studies.

KONO et al.[14] and PETERSSON et al.[15] studied the adsorption systems O/Cu(001) and CO/Ni(001) using core line angular distributions. The measurements allowed full variation of the azimuthal and polar angles of the photoelectron direction with respect to the sample surface. For comparison with the experimental data, the angular distributions were calculated by means of a single scattering model. This model implies[16] that the adsorbate acts as an emitter of primary waves which are subsequently scattered against the neighbouring atoms in the surface (cf. Fig. 89). In constructing the total wavefield as a sum of the primary and secondary contributions one neglects the multiple scattering of the electrons.

Results for the O1*s* ionization of the system O/Cu(001) obtained by these workers are shown in Fig. 90[15]. The original data were subject to a fourfold averaging procedure and a subtraction of the minimum intensity. The resulting "flower" patterns emphasized hence the anisotropic effects. It can be noted for these data (Fig. 90), that the anisotropy is a rather strong function of the polar angle of photoelectron emission. Thus, the anisotropy decreases from $\sim 24\%$ at $\theta = 7°$ to $\sim 6\%$ for $\theta = 45°$. This confirms the dominance of forward scattering events at these energies. Comparison of the O1*s* experimental data and model

[8] De Wames, R.E., Hall, W.F. (1968): Acta Crystallogr. A*24*, 206
[9] Fadley, C.S., Bergström, S.Å.L. (1971): Phys. Lett. *35* A, 375
[10] Baird, R.J., Fadley, C.S., Wagner, L.F. (1977): Phys. Rev. B*15*, 666
[11] Zehner, D.M., Noonan, J.R., Jenkins, L.H. (1977): Phys. Lett. *62* A, 267
[12] Briggs, D., Marbrow, R.A., Lambert, R.M. (1978): Solid State Commun. *25*, 40
[13] Erickson, N.E. (1977): Phys. Scr. *16*, 462
[14] Kono, S., Goldberg, S.M., Hall, N.F.T., Fadley, C.S. (1978): Phys. Rev. Lett. *41*, 1831
[15] Petersson, L.-G., Kono, S., Hall, N.F.T., Fadley, C.S., Pendry, J.B. (1979): Phys. Rev. Lett. *42*, 1545
[16] McDonnell, L., Woodruff, D.P., Holland, B.W. (1975): Surf. Sci. *51*, 249

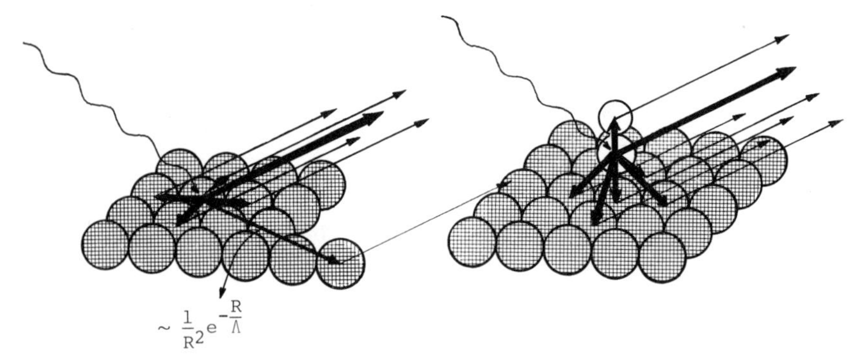

Fig. 89. Elastic scattering of photo- and Auger-electrons leading to ESCA diffraction from a local source in or on a single crystal surface

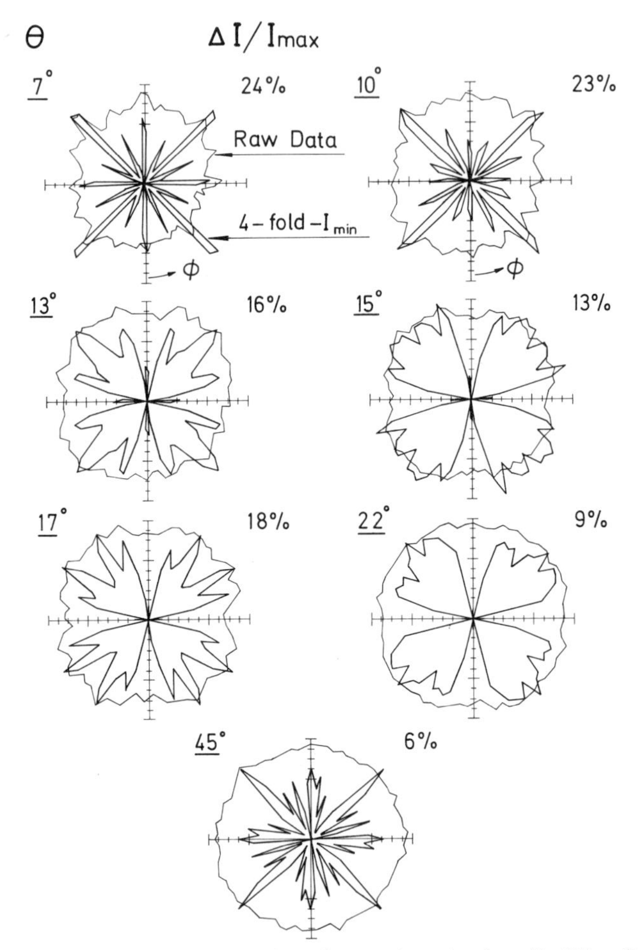

Fig. 90. Azimuthal distributions of O 1s photoelectron intensity for $c(2\times2)$/on Cu(001) at seven grazing angles between 7° and 45°. θ is measured with respect to the surface. Both the raw data (dashed curves) and data subjected to a fourfold averaging and minimum subtraction (solid curves) are shown. The percentage anisotropy as determined from $\Delta I/I_{max}$ is indicated for each angle[15]

calculations according to the single scattering scheme showed that the oxygen atoms are adsorbed in "fourfold hole" positions which are coplanar with the surface Cu atoms. The above study was made using Al$K\alpha$ radiation. Using synchrotron radiation[17–21] with photon energies closer to threshold other surface structures can be emphasized. A recent such study[21] of the O/Cu(001) system indicated that the coplanar geometry is a minority configuration and that another configuration exists where the oxygen sits 0.8 Å above the surface.

V. Valence level studies in molecules and condensed matter

14. General features of molecular valence electron spectra

α) Introduction. Valence photoelectron spectra of molecules are usually more complicated than those of atoms, due to the larger number of electrons. The general features observed are similar, however, regarding electronic transitions. As previously discussed there are a number of main bands each of which usually can be associated with ionization from a given molecular orbital. This correspondence between photoelectron spectra and the occupied molecular orbitals of independent particle models is expressed via Koopmans' theorem, Eq. (7.6). The straightforward basic relationship thus existing between photoelectron spectra and MO theory has been applied to numerous molecules for assignments of spectra and tests of theoretical models. As in the atomic cases discussed above, the relative influence of relaxation and correlation on molecular valence spectra is generally different for the outer and inner valence part.

The geometrical and electronic structure of molecules is discussed in terms of symmetry by employing basic group theory. The molecular orbitals as well as the various electronic states considered in photoelectron spectra are thus commonly represented by irreducible representations of the point group to which the molecule or molecular ion belongs. Figure 91 shows a typical outer valence spectrum obtained from the benzene molecule by using He*I* excitation

---

[17] Woodruff, D.P., Norman, D., Holland, B.W., Smith, N.V., Farrell, H.H., Traum, M.M. (1978): Phys. Rev. Lett. *41*, 1130

[18] Kevan, S.D., Rosenblatt, D.H., Denley, D.R., Lee, B.-C., Shirley, D.A. (1979): Phys. Rev. B*20*, 4133; (1978): Phys. Rev. Lett. *41*, 1565

[19] McGovern, I.T., Eberhardt, W., Plummer, E.W. (1979): Sol. St. Comm. *32*, 963

[20] Guillot, C., Jugnet, Y., Lasailly, Y., Lecante, J., Spanjaard, D., Duc, T.M. (private communication)

[21] Tobin, J.G., Klebanoff, L.E., Rosenblatt, D.H., Davis, R.F., Umbach, E., Baca, A.G., Shirley, D.A., Huang, Y., Kang, W.M., Tong, S.Y. (to be published)

[22] Vorburger, T.V., Sandstrom, D.R., Waclawski, B.J. (1976): Surf. Sci. *60*, 211

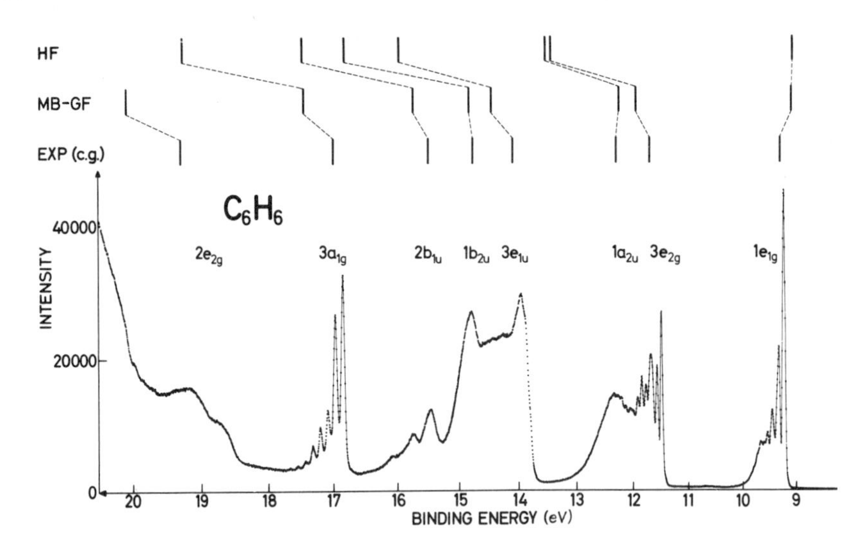

Fig. 91. The He*I*-excited electron spectrum of benzene (C_6H_6)[1]. The main bands are associated with transitions to primary hole states, the vacant orbital indicated above the corresponding band. The electron configuration is (core) $2a_{1g}^2\ 2e_{1u}^4\ 2e_{2g}^4\ 3a_{1g}^2\ 2b_{1u}^2\ 1b_{2u}^2\ 3e_{1u}^4\ 1a_{2u}^2\ 3e_{2g}^4\ 1e_{1g}^4$. Calculated vertical electron binding energies were obtained with the Hartree-Fock (HF; $-\varepsilon_i$) and many-body Green's function (MB-GF) methods[2]

at 21.22 eV ($\lambda = 584$ Å)[1]. The main spectral bands can be associated with ionization from the twelve outermost occupied molecular orbitals (including degeneracy) as indicated in the figure by means of irreducible representations. The overwhelming majority of all valence photoelectron spectra of molecules have been obtained by He*I* excitation, due to the favourable properties of this radiation (cf. Sect. 3).

The complete set of valence orbitals can usually only be reached by the complementary use of higher energy excitation, for example He*II* at 40.8 eV ($\lambda = 304$ Å) or monochromatized Al*Kα* at 1486.7 eV ($\lambda = 8.340$ Å). Figure 92 shows the complete spectrum of the benzene molecule obtained by the latter radiation[3]. X-ray excitation gives lower resolution than for example He*II* but provides on the other hand additional means for interpretation of spectra due to the very different cross sections generally obtained for different types of orbitals, as is clearly seen in the figure (cf. also Sect. 14γ). Without the filtering of other energies, He*II* excitation is in general useful only up to at most 30 eV, which is in many cases insufficient. Another powerful alternative radiation source for studies of molecular valence spectra is synchrotron radiation. It has been used to increasing extent, in most cases to study variations in the intensity as function of photon energy, angular distributions and satellite structure in the inner valence region (cf. Sects. 14γ, δ and 16).

---

[1] Karlsson, L., Mattsson, L., Jadrny, R., Bergmark, T., Siegbahn, K. (1976): Phys. Scr. *14*, 230
[2] Von Niessen, W., Cederbaum, L.S., Kraemer, W.P. (1976): J. Chem. Phys. *65*, 1378
[3] Gelius, U. (1974): J. Electron Spectrosc. Relat. Phenom. *5*, 985

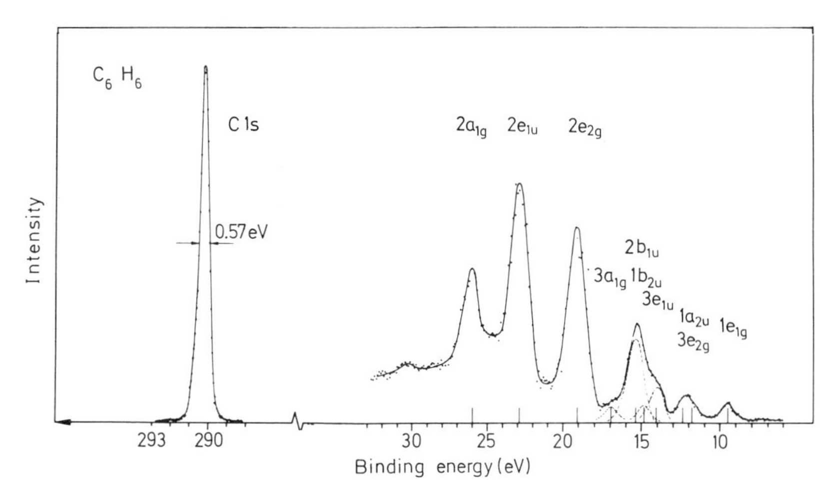

Fig. 92. The electron spectrum of benzene excited by means of monochromatized Al $K\alpha$ radiation[3]

The satellite structure can be associated with relaxation and correlation effects. The outer valence spectrum of C_6H_6, for example, shows such a satellite band at 18.7 eV which has been associated with correlation in the final ionic state (FISCI satellite)[4-6]. A similar feature is observed at 30 eV in the Al$K\alpha$ excited spectrum. Although these structures usually do not appear with the same intensity as the primary hole states, they can often be clearly observed and studied in detail. Photoelectron spectroscopy is thus an excellent technique for studies of both one-electron and many-electron aspects of molecules.

In addition to many-electron effects there exist two other cases when there may be more bands in the photoelectron spectrum than the number of molecular orbitals. In molecules with high symmetry, orbitally degenerate final states may be split by spin-orbit and Jahn-Teller interactions to give as many bands as there would be orbitals if they were all non-degenerate (cf. Sect. 14ε). Ionization from open-shell systems can produce more than one final state for each molecular orbital (cf. also Sect. 14δ).

β) Electron binding energies of main bands. As was indicated above, independent particle models are usually appropriate to describe the prominent features of spectra. As for atoms, the deviations from the measured binding energies occurring due to the approximations implied by Koopmans' theorem are typically of the order of 1–2 eV in the outer valence region and tend to increase with increasing binding energy. To this must be added errors due to approximations in the evaluation of the HF equations. In ab initio SCF MO-

[4] Cederbaum, L.S., Domcke, W., Schirmer, J., von Niessen, W., Diercksen, G.H.F., Kraemer, W.P. (1978): J. Chem. Phys. *69*, 1591

[5] Lindholm, E., Åsbrink, L. (1980): J. Electron Spectrosc. Relat. Phenom. *18*, 121

[6] Snyder, L.C., Basch, H. (1972): Molecular wave functions and properties, New York: Wiley-Interscience

Table 19. Molecular orbitals and vertical binding energies (eV) of benzene

Orbital	Exp. (max. int.) [1]	CNDO/S2 $-\mathcal{E}_i$ [7]	INDO/S $-\mathcal{E}_i$ [8]	SPINDO $-\mathcal{E}_i$ [9]	HF $-\mathcal{E}_i$ [10]	HF $-\mathcal{E}_i$ [11]	MB-GF [2]
$1e_{1g}(\pi)$	9.24	10.04	9.07	9.34	9.08	9.23	9.10
$3e_{2g}(\sigma)$	11.49	12.34	11.86	11.97	13.46	13.39	11.95
$1a_{2u}(\pi)$	12.3	13.64	13.42	12.38	13.55	13.73	12.26
$3e_{1u}(\sigma)$	13.95	15.20	15.16	13.92	15.98	16.09	14.46
$1b_{2u}(\sigma)$	14.80	16.15	15.67	15.06	16.83	16.90	14.83
$2b_{1u}(\sigma)$	15.46	17.48	17.55	15.15	17.49	17.42	15.75
$3a_{1g}(\sigma)$	16.84	19.65	22.58	16.87	19.27	19.43	17.28
$2e_{2g}(\sigma)$	19.1		25.13		22.39	22.46	19.97

LCAO methods the errors can be made small so that the HF limit is approached. The accuracy of various methods is indicated by Table 19.

The difference between the energies obtained by Koopmans' theorem and the experimental energies can be corrected empirically. One expression which has been used for molecules consisting of first and second row elements is[12]

$$E_{\mathrm{B}}(i) = -0.92\,\mathcal{E}_i. \tag{14.1}$$

Corrections which to some extent account for individual variations among the orbitals have also been introduced, for example in a study of benzene, pyridine and other azines[13].

As discussed in Sect. 7, the effects responsible for the deviations from Koopmans' theorem are electronic relaxation and correlation. We show in Fig. 93 the result of ΔSCF and many-electron calculations for N_2 along with the energies obtained from Koopmans' theorem[14–18].

Calculations which include relaxation and a large part of the correlation energy (like CI, Green's function and perturbation methods) give in general good agreement with observed spectra. As an example, Table 20 summarizes results

[7] Lipari, N.O., Duke, C.B. (1975): J. Chem. Phys. *63*, 1748
[8] Krogh-Jespersen, K., Ratner, M. (1976): J. Chem. Phys. *65*, 1305
[9] Fridh, C., Åsbrink, L., Lindholm, E. (1972): Chem. Phys. Lett. *15*, 282
[10] Ermler, W.C., Kern, C.W. (1973): J. Chem. Phys. *58*, 3458
[11] Fuss, I., McCarthy, I.E., Minchinton, A., Weigold, E., Larkins, F.P. (1981): Flinders Univ. Report, FIAS-R-77, Adelaide, Australia
[12] Brundle, C.R., Robin, M.B., Basch, H., Pinsky, M., Bond, A. (1970): J. Am. Chem. Soc. *92*, 3863
[13] Almlöf, J., Roos, B., Wahlgren, U., Johansen, H. (1973): J. Electron Spectrosc. Relat. Phenom. *2*, 51
[14] Cederbaum, L.S., Domcke, W. (1977): Adv. Chem. Phys. *36*, 205
[15] Turner, D.W., Baker, C., Baker, A.D., Brundle, C.R. (1970): Molecular photoelectron spectroscopy, London: Wiley-Interscience
[16] Hall, D., Maier, J.P., Rosmus, P. (1977): Chem. Phys. *24*, 373
[17] Rosmus, P., Solouki, B., Bock, H. (1977): Chem. Phys. *22*, 453
[18] Bock, H., Solouki, B., Bert, B., Rosmus, P. (1977): J. Am. Chem. Soc. *99*, 1663

Table 20. Calculated and experimental vertical binding energies for CH_4, H_2O and N_2

CH_4

Method	Orbital	
	$2a_1$	$1t_2$
Koopmans' theorem[19]	25.66	14.77
ΔSCF[19]	24.31	13.64
QD-MBPT[19]	24.00	14.40
ΔCEPA[20]	23.38	14.29
EXP[a, 20]	22.86	14.37

H_2O

Method	Orbital			
	$1a_1$	$1b_2$	$2a_1$	$1b_1$
Koopmans' theorem[19]	36.83	19.44	15.75	13.89
ΔSCF[19]	–	17.62	13.22	11.14
Transition operator[21]	34.4	17.8	13.2	11.0
QD-MBPT[19]	35.19	19.24	14.74	12.73
RSPT[22]		18.97	14.73	12.42
ΔCEPA[23]	32.35	18.85	14.68	12.48
MB-GF[24]	–	18.96	14.91	12.69
EXP[a, 23]	32.2	18.72	14.83	12.78

N_2

Method	Orbital		
	$2\sigma_u$	$1\pi_u$	$3\sigma_g$
Koopmans' theorem[25]	20.92	17.10	17.36
ΔSCF[25]	19.93	15.67	16.01
CI[b]	19.68	17.11	16.04
RSPT[22]	18.39	16.67	14.90
Propagator[26]	17.55	17.23	14.91
Third order propagator[27]	19.20	18.29	15.94
EOM[28]	18.63	18.03[c]	15.69
EOM[29]	18.76	16.89	15.57
EOM[30]	19.01	17.20	15.94
MB-GF[31]	18.59	16.83	15.50
EXP[d, 32]	18.78	16.98	15.60

Abbreviations

QD-MBPT stands for quasi-degenerate many-body perturbation theory, RSPT for Rayleigh-Schrödinger perturbation theory, EOM for equations of motion, CEPA for coupled electron pair approximation and MB-GF for many-body Greens function.

[a] Experimental energies corrected by MEYER[23] to fit the theoretically obtained quantities.

[b] W. ERMLER and A.C. WAHL, cited from ref. [30].

[c] Cited from ref. [30].

[d] Maximum intensity of the electron band.

(Footnotes 19–32 see p. 392)

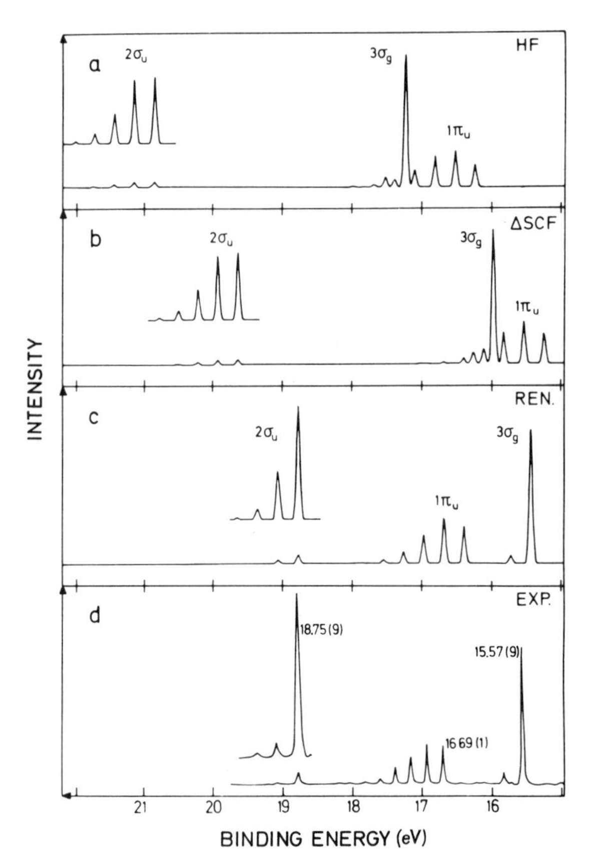

Fig. 93a–d. Calculated (**a**, **b**, **c**) and experimental (HeI-excited[5]) (**d**) valence electron spectra of N_2. The spectra of **a**, **b** and **c** were obtained in the frozen one-particle picture (HF), by the ΔSCF method and by including many-electron effects, respectively. The latter result was obtained by the Green's function method of CEDERBAUM et al.[14] where the electron binding energies were obtained from the Dyson equation $\boldsymbol{G}^{-1} = \omega \mathbf{1} - \mathcal{E} - \boldsymbol{\Sigma}(\omega)$. The last term is the self-energy which was calculated by using a series expansion. Terms of high order were included by a renormalization procedure (REN). The vibrational structure was calculated by the gradient method (cf. Sect. 15) and by drawing the lines as Gaussians of width 30 meV. From ref. [14]

---

[19] Prime, S., Robb, M.A. (1977): Chem. Phys. Lett. *47*, 527
[20] Meyer, W. (1973): J. Chem. Phys. *58*, 1017
[21] Goscinski, O., Hehenberger, M., Roos, B., Siegbahn, P. (1975): Chem. Phys. Lett. *33*, 427
[22] Chong, D.P., Herring, F.G., McWilliams, K. (1974): J. Chem. Phys. *61*, 3567
[23] Meyer, W. (1971): Int. J. Quant. Chem. *55*, 341
[24] Cederbaum, L.S., Holneicher, G., von Niessen, W. (1973): Mol. Phys. *26*, 1405
[25] Cade, P.E., Sales, K.D., Wahl, A.C. (1966): J. Chem. Phys. *44*, 1973
[26] Purvis, G.D., Öhrn, Y. (1974): J. Chem. Phys. *60*, 4063
[27] Purvis, G., cited from Ref. [30]
[28] Chen, T.T., Smith, W.D., Simons, J. (1974): Chem. Phys. Lett. *26*, 296
[29] Bacskay, G.B., Hush, N.S. (1976): Chem. Phys. *16*, 219
[30] Herman, M.F., Yeager, D.L., Reed, K.F., McKoy, V. (1977): Chem. Phys. Lett. *46*, 1
[31] Cederbaum, L.S., Holneicher, G., von Niessen, W. (1973): Chem. Phys. Lett. *18*, 503
[32] Karlsson, L., Mattsson, L., Jadrny, R. (unpublished)

from various approaches for CH_4, H_2O and N_2. Two additional examples are given in Figs. 91 and 107 which include Green's function results.

Koopmans' theorem provides in most cases the proper binding energy ordering. However, predictions which give incorrect ordering (often referred to as a break-down of Koopmans' theorem) are common. One example is provided by Fig. 93, others are given in refs.[33-36]. This type of break-down may occur for example as a result of the creation of a localized valence hole (as for the metallocenes[37,38] and pyridine[39]) or by other correlation contributions (as for N_2 of Fig. 93). A thumb rule pertaining to the break-down of Koopmans' theorem has been given[33].

Relativistic effects can become significant when heavy atoms are present. These effects are manifested both as binding energy shifts and state splittings. Inner valence levels can become appreciably shifted. In the case of TeH_2, for example, this shift is 1 eV[40]. Splittings can in certain cases be important in the whole valence energy region and even dominate the appearance of the spectrum. In such cases the spectrum must be interpreted in terms of the molecular spin-orbital structure. One example of this is given in Fig. 94 for UF_6[41]. Relativistic splittings are further discussed in Sect. 14ε.

γ) Intensities of main bands. Ionization of a subshell with degeneracy $g(i)$ leads to a differential cross section in the independent particle framework according to (cf. (7.16)):

$$\frac{d\sigma(i)}{d\Omega} \propto \frac{2g(i)}{h\nu} |\boldsymbol{u} \cdot \langle \psi_i | \boldsymbol{p} | \psi(\boldsymbol{k}) \rangle|^2, \tag{14.2}$$

where ψ_i and $\psi(\boldsymbol{k})$ represent the ionized and the continuum orbitals, respectively. In a first approximation one may consider the photoionization matrix element as a constant, that is, the spectral intensities reflect the orbital degeneracy. This, however, is not a reliable assumption and effects due to variations of the matrix element between different orbitals, many-electron effects and variations in the spectrometer transmission must generally also be taken into consideration[42-45].

---

[33] Chong, D.P., Herring, F.G., McWilliams, D. (1975): J. Electron Spectrosc. Relat. Phenom. *7*, 445

[34] Berkowitz, J., Dehmer, J.L., Appelman, E.H. (1973): Chem. Phys. Lett. *19*, 334

[35] Cederbaum, L.S., Domcke, W., von Niessen, W. (1975): Chem. Phys. *10*, 459

[36] Von Niessen, W., Diercksen, G.H.F., Cederbaum, L.S. (1977): J. Chem. Phys. *67*, 4124

[37] Coutiére, M.M., Demuynck, J., Veillard, A. (1972): Theor. Chim. Acta *27*, 281

[38] Rohmer, M.M., Veillard, A. (1973): J.C.S. Chem. Commun. 250

[39] Von Niessen, W., Diercksen, G.H.F., Cederbaum, L.S. (1975): Chem. Phys. *10*, 345

[40] Rosén, A., Ellis, D.E. (1975): J. Chem. Phys. *62*, 3039

[41] Rosén, A. (1978): Chem. Phys. Lett. *55*, 311

[42] Cox, P.A. (1975): in: Structure and Bonding, vol. 24, Dunitz, J.D., et al. (eds.), Berlin, Heidelberg, New York: Springer

[43] Cox, P.A., Orchard, A.F. (1970): Chem. Phys. Lett. *7*, 273

[44] Cox, P.A., Evans, S., Orchard, A.F. (1972): Chem. Phys. Lett. *13*, 386

[45] Schirmer, J., Cederbaum, L.S., Domcke, W. (1977): Vth Conf. on VUV Radiation Phys., Montpellier

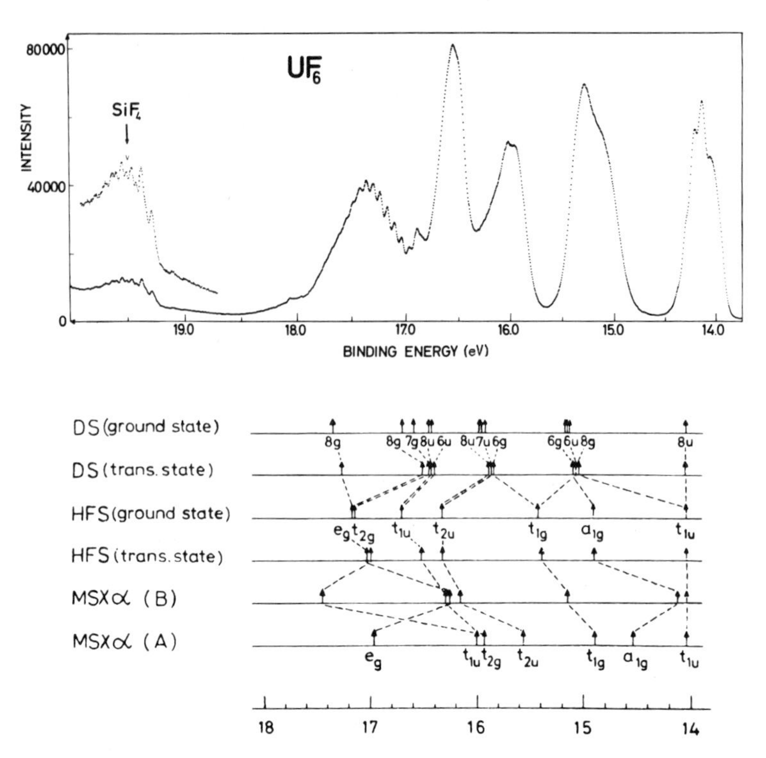

Fig. 94. The He*I*-excited spectrum of UF_6[1] (the innermost band centred at 19.5 eV is due to an impurity of SiF_4). Theoretical electron binding energies were obtained from MS $X\alpha$, Hartree-Fock-Slater (HFS) and Dirac-Slater (DS) calculations. From ref.[41]

In general, the photoionization cross section increases to a maximum near threshold and falls successively with increasing photon energy (e.g.[46,47]) but the detailed behaviour can vary substantially between different types of orbitals. Such variations have been studied extensively experimentally and typical behaviour patterns have been found which are useful for discussing molecular valence spectra[48-50]. In the uv excitation region comparison between He*I* and He*II* excited spectra can give valuable information in this respect. For example, the intensity of s, p-levels having $n=3$ is decreased with respect to s, p-levels having $n=2$ when going from He*I* to He*II* excitation. This fact has been

[46] Rabalais, J.W. (1977): Principles of photoelectron spectroscopy, New York: Wiley, p. 140

[47] Berkowitz, J. (1979): Photoabsorption, photoionization and photoelectron spectroscopy, New York: Academic Press

[48] Siegbahn, K., Nordling, C., Johansson, G., Hedman, J., Hedén, P.F., Hamrin, K., Gelius, U., Bergmark, T., Werme, L.O., Manne, R., Baer, Y. (1969): ESCA applied to free molecules, Amsterdam: North-Holland

[49] Price, W.C., Potts, A.W., Streets, D.G. (1972): in: Electron Spectroscopy, Shirley, D.A. (ed.), Amsterdam: North-Holland, p. 187

[50] Price, W.C. (1974): Adv. Atom. Mol. Phys. *10*, 131

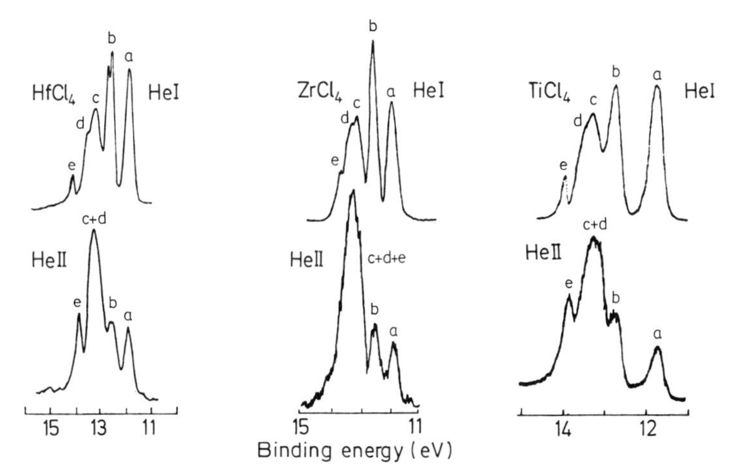

Fig. 95. He*I*- and He*II*-excited electron spectra of the group IV chlorides of Ti, Zr and Hf. The electron configuration of these molecules is (core) $2a_1^2\ 2t_2^6\ 1e^4\ 3t_2^6\ 1t_1^6$. The bands labelled *a*, *b*, *c* and *d*, *e* have been associated with ionization from the $1t_1$, $3t_2$, $2t_2+e$, and $2a_1$ orbitals, respectively. The three latter bands show a strong He*II* intensity indicating that the corresponding orbitals contain large metal *d* contributions. From ref.[55]

used in assignments of spectra like chloroethenes[51,52], phosphoryl chloride[53] and benzoyl chloride[54]. Figure 95 shows another example from a study of MCl_4 molecules[55]. Here the metal *d*-orbital contributions can be selected through much higher cross sections than the Cl *p* orbitals at He*II* excitation. Several similar studies have been performed[56–60]. Empirical rules have been derived for intensities of photoelectron bands from studies of closely related pyridines[61] and iron tetracarbonyl-olefin complexes[62].

X-ray excited spectra exhibit in general much higher intensity for ionization from *s*-orbitals than *p*-orbitals. This behaviour applies to the atomic basis set of molecular orbitals in a way which is most useful in assignments of spectra. In the plane wave approximation of the continuum electron wave function[63,64] an intensity model has been obtained which accounts quantitatively for the

[51] Robin, M.B., Kuebler, N.A., Brundle, C.R. (1972): in: Electron Spectroscopy, Shirley, D.A. (ed.), Amsterdam: North-Holland, p. 351

[52] Katrib, A., Debies, T.P., Colton, R.J., Lee, T.H., Rabalais, J.W. (1973): Chem. Phys. Lett. *22*, 196

[53] Gan, T.H., Peel, J.B., Willett, G.D. (1977): Chem. Phys. Lett. *48*, 483

[54] Gan, T.H., Livett, M.K., Peel, J.B. (1979): J. Electron Spectrosc. Relat. Phenom. *15*, 71

[55] Egdell, R.G., Orchard, A.F. (1978): J.C.S. Faraday Trans. II *74*, 485

[56] Egdell, R.G., Orchard, A.F., Lloyd, D.R., Richardson, N.V. (1977): J. Electron Spectrosc. Relat. Phenom. *12*, 415

[57] Egdell, R.G., Orchard, A.F. (1978): J. Electron Spectrosc. Relat. Phenom. *14*, 277

[58] Egdell, R.G., Fragala, I., Orchard, A.F. (1978): J. Electron Spectrose. Relat. Phenom. *14*, 467

[59] Egdell, R.G., Orchard, A.F. (1978): J.C.S. Faraday Trans. II *74*, 1179

[60] Thornthon, G., Edelstein, N., Rösch, N., Egdell, R.G., Woodwack, D.R. (1979): J. Chem. Phys. *70*, 5218

[61] Daamen, H., Oskam, A. (1978): Inorg. Chem. Acta *27*, 209

[62] Van Dam, H., Oskam, A. (1979): J. Electron Spectrosc. Relat. Phenom. *16*, 307

[63] Gelius, U. (1972): in: Electron spectroscopy, Shirley, D.A. (ed.), Amsterdam: North-Holland, p. 311

[64] Huang, J.-T.J., Ellison, F.O. (1974): J. Electron Spectrosc. Relat. Phenom. *4*, 233

relative intensities of x-ray excited spectra. The original formulation by Gelius was[65]

$$\left(\frac{d\sigma}{d\Omega}\right)_i = \frac{1}{2} n_i \sum_{A,\lambda} \left(C_{A\lambda}^2 \frac{d\sigma_{A\lambda}}{d\Omega}\right), \tag{14.3}$$

where A, λ refers to the atomic subshell of the LCAO description and i represents the molecular orbital. $C_{A\lambda}^2$ is the population and $d\sigma_{A\lambda}/d\Omega$ the appropriate atomic cross section. By using calculated atomic populations and empirically determined atomic cross sections sensitive tests could be obtained of the electron binding energy ordering for several molecules.

The model has been further developed and influences of the angular asymmetry of the photoelectrons have also been considered[66,67]. Extensive investigations have been made by Nefedov et al.[68] using atomic subshell cross sections from both Dirac-Slater and RPAE calculations. Generally good agreement has been obtained with experimental data and the model has been found useful also for the rather low energies of $YM\zeta$ and $RhM\zeta$ and even $HeII$. Hilton et al.[69,70] found that the model is applicable down to threshold if the interference terms appearing in the cross section are included. Their treatment incorporated an adjustment of the wave vector of the photoelectron to account for the molecular potential. This was done also by Beerlage and Feil[71] who by this means improved results obtained by plane wave calculations.

Measurements of partial photoionization cross sections concern so far rather few molecules. Determinations have been made as a function of photon energy by employing discrete lines from a uv-source and by synchrotron radiation. An accurate technique of the former kind by Samson (cf. article in this volume) using a monochromator to select various energies was employed in investigations of CO[72], O_2 and N_2[73,74]. By means of another method using reference photoelectron lines with predetermined cross sections, rare gases, several small molecules and aliphatic compounds could be investigated[75-77].

---

[65] Gelius, U., Allan, C.J., Johansson, G., Siegbahn, H., Allison, D.A., Siegbahn, K. (1971): Phys. Scr. *3*, 237

[66] Berndtsson, A., Basilier, E., Gelius, U., Hedman, J., Klasson, M., Nilsson, R., Nordling, C., Siegbahn, K. (1975): Phys. Scr. *12*, 235

[67] Nefedov, V.I., Sergushin, N.P., Band, I.M., Trzhaskovskaya, M.B. (1973): J. Electron Spectrosc. Relat. Phenom. *2*, 283

[68] Yarzhemsky, V.G., Nefedov, V.I., Amusia, M.Ya., Cherepkov, N.A., Chernysheva, L.V. (1981): J. Electron Spectrosc. Relat. Phenom. *23*, 175

[69] Hilton, P.R., Nordholm, S., Hush, N.S. (1980): J. Electron Spectrosc. Relat. Phenom. *18*, 101

[70] Hilton, P.R., Nordholm, S., Hush, N.S. (1976): Chem. Phys. *15*, 345

[71] Beerlage, M.J.M., Feil, D. (1977): J. Electron Spectrosc. Relat. Phenom. *12*, 161

[72] Samson, J.A.R., Gardner, J.L. (1976): J. Electron Spectrosc. Relat. Phenom. *8*, 35; (1977): *12*, 119

[73] Samson, J.A.R., Gardner, J.L., Haddad, G.N. (1977): J. Electron Spectrosc. Relat. Phenom. *12*, 281

[74] Samson, J.A.R., Haddad, G.N., Gardner, J.L. (1977): J. Phys. B*10*, 1749

[75] Kimura, K., Achiba, Y., Morishita, M., Yamazaki, T. (1979): J. Electron Spectrosc. Relat. Phenom. *15*, 269

[76] Kemeny, P.C., Leckey, R.C.G., Jenkin, J.G., Liesegang, J. (1974): J. Electron Spectrosc. Relat. Phenom. 5, 881

[77] Kemeny, P.C., Poole, R.T., Jenkin, J.G., Liesegang, J., Leckey, R.C.G. (1974): Phys. Rev. A*10*, 190

Using synchrotron radiation, studies have been performed on small molecules like N_2[78,79], CO[78], O_2[80], SF_6[81] and CO_2[82] and $Cr(CO)_6$[82a]. Large variations in the cross sections, associated with the presence of shape resonances have been observed in several cases for energies up to about 10 eV above threshold. The shape resonances, which are quasi-stationary states in the continuum, occur due to a potential barrier formed by an attractive molecular potential and a centrifugal repulsion of the type $l(l-1)/r^2$. Figure 96 shows a drawing of the effect on the wave function of such a barrier in three dimensions[83] and Fig. 97 shows the effect on the photoionization cross section for CO. The shape resonance occurs in the $l=3$ partial wave (f-wave) and is observed for all σ primary hole states, approximately 10 eV above corresponding thresholds at 14 eV, 19.7 eV and 38.3 eV, respectively. It is lacking in the π primary hole state due to symmetry restrictions. $MSX\alpha$ calculations have been performed showing generally good agreement with experiment such as for CO and N_2[84], O_2, NO, CO_2, BF_3 and SF_6[85].

The $MSX\alpha$ calculations and also the bound state method by Rescigno, Langhoff et al.[86–90] have the most successful ones in predicting molecular photoionization cross sections. Simpler methods like the plane wave and orthogonal plane wave e.g.[46,91] methods as well as one- and two-centre Coulomb expansions[92–94] often tend to give less satisfactory results.

The position of the shape resonance has been found to be sensitive to the internuclear distance, that is, the energy varies with the vibrational quantum number. Photoelectron studies near shape resonances may therefore lead to vibrational intensity distributions which deviate significantly from normal Franck-Condon distributions[85,95].

---

[78] Plummer, E.W., Gustafsson, T., Gudat, W., Eastman, D.E. (1977): Phys. Rev. A *15*, 2339

[79] Woodruff, P.R., Marr, G.V. (1977): Proc. R. Soc. A *358*, 293

[80] Codling, K., West, J.B., Parr, A.C., Dehmer, J.L., Cole, B.E., Ederer, D.L., Stockbauer, R.L. (1980): in: VIth Int. Conf. on VUV Radiat. Phys., Ext. Abstr., vol. II, Charlottesville, Virginia

[81] Gustafsson, T. (1978): Phys. Rev. A *18*, 1481

[82] Gustafsson, T., Plummer, E.W., Eastman, D.E., Gudat, W. (1978): Phys. Rev. A *17*, 175

[82a] Loubriel, G., Plummer, E.W. (1979): Chem. Phys. Lett. *64*, 234

[83] Loomba, D., Wallace, S., Dill, D., Dehmer, J.L. (1981): J. Chem. Phys. *75*, 4546

[84] Davenport, J.W. (1976): Phys. Rev. Lett. *36*, 945

[85a] Wallace, S., Swanson, J.R., Dill, D., Dehmer, J.L. (1980): in: VIth Int. Conf. on VUV Radiat. Phys., Ext. Abstr., vol. II, Charlottesville, Virginia

[85b] Stephens, J.A., Dill, D., Dehmer, J.L. (1981): J. Phys. B *14*, 3911

[86] Langhoff, P.W., Gerwer, A., Asaro, C., McKoy, B.V. (1979): Int. J. Quantum Chem. Symp. No. 13, 645

[87] Gerwer, A., Asaro, C., McKoy, B.V., Langhoff, P.W. (1980): J. Chem. Phys. *72*, 713

[88] Orel, A.E., Rescigno, T.N., McKoy, B.V., Langhoff, P.W. (1980): J. Chem. Phys. *72*, 1265

[89] Rescigno, T.N., Langhoff, P.W. (1977): Chem. Phys. Lett. *51*, 65

[90] Padial, N., Csanak, G., McKoy, B.V., Langhoff, P.W. (1978): J. Chem. Phys. *69*, 2992

[91] Lohr, L.L. (1972): in: Electron spectroscopy, Shirley, D.A. (ed.), Amsterdam: North-Holland, p. 245

[92] Iwata, S., Nagakura, S. (1974): Mol. Phys. *27*, 425

[93] Schneider, B., Berry, R.S. (1969): Phys. Rev. *182*, 141; *186*, 265

[94] Hirota, F. (1976): J. Electron Spectrosc. Relat. Phenom. *9*, 149

[95] Unwin, R., Khan, I., Richardson, N.V., Bradshaw, A.M., Cederbaum, L.S., Domcke, W. (1981): Chem. Phys. Lett. *77*, 242

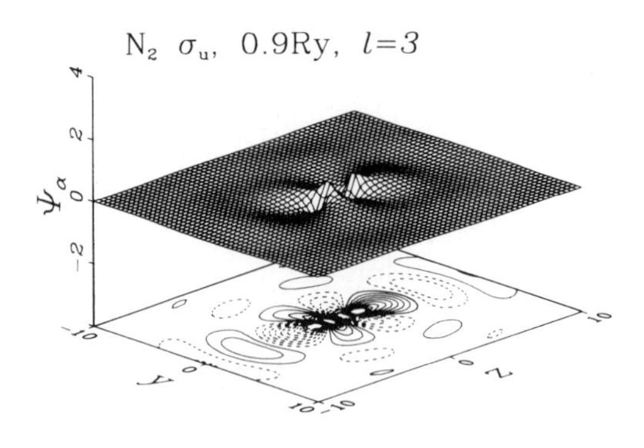

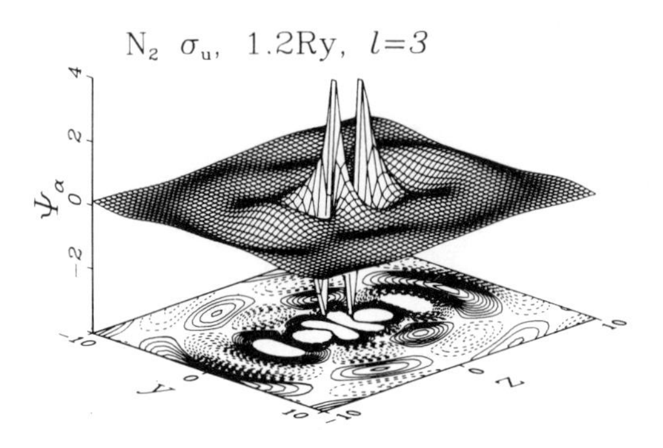

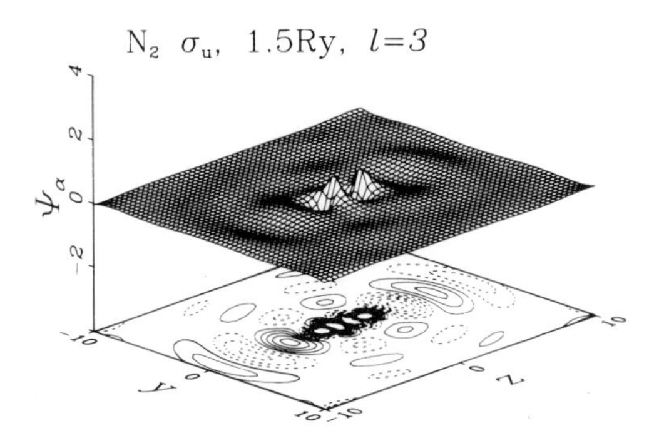

Fig. 96. $N_2 \sigma_u$ continuum f-wave ($l=3$) eigenchannel wave functions for three different electron kinetic energies. The molecule is situated in the yz plane, directed along the z-axis and centered at $y=z=0$. Contours mark steps of 0.03 from 0.02 to 0.29; solid = positive and dashed = negative. From ref. [83]

Considerable deviations from Franck-Condon intensity distributions may also occur due to autoionization processes. This can happen if the energy of the incident photons coincides with the energy of an excited state (to within the natural widths). The intensity of a photoelectron line can then be written (for a diatomic molecule assuming that the vibrational constant is the same

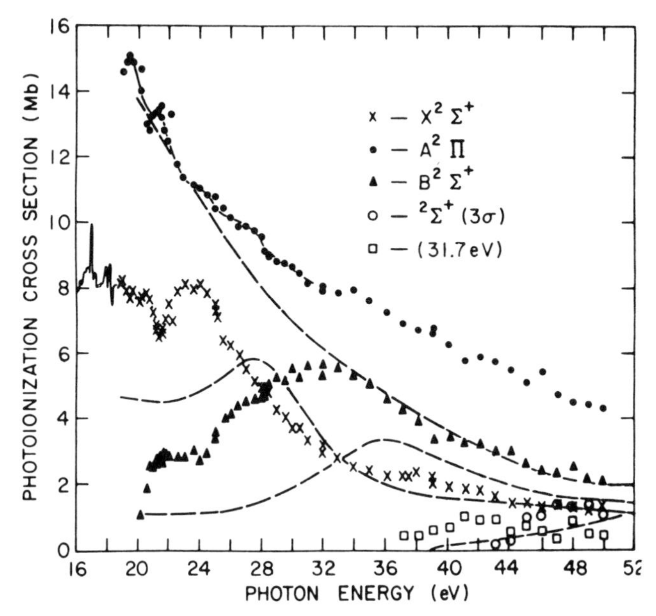

Fig. 97. Partial photoionization cross sections for transitions to the four lowest primary hole states as well as one many-electron excited state of CO^+ as a function of photon energy. The dashed lines are calculated results by DAVENPORT[84]. From ref. [78]

in the autoionizing and ionic states)

$$I \propto C_1 F_0^{V_{fi}} + C_2 F_0^{V_a} F_{V_a}^{V_{fi}}, \tag{14.4}$$

where the first term represents the intensity in the absence of autoionization. An example of application is given in ref. [96].

Autoionization effects are best studied experimentally by varying the photon energy. Initially, several studies were performed by He*I* and Ne*I* excitation[50, 97–100]. In later studies synchrotron radiation has been used extensively[78, 79, 101].

Photoionization cross sections have been determined as function of energy for several molecules by using electron impact ionization. In these studies the incident electron simulates a photon in its interaction with the molecule and is detected in coincidence with the "photoelectron" (e.g.[102, 103]).

---

[96] Kinsinger, J.A., Taylor, J.W. (1973): J. Mass Spectrom. Ion Phys. *11*, 461

[97] Collin, J.E., Delwiche, J., Natalis, P. (1972): in: Electron Spectroscopy, Shirley, D.A. (ed.), Amsterdam: North-Holland, p. 401

[98] Natalis, P., Delwiche, J., Collin, J.E. (1972): Discuss. Faraday Soc. *54*, 98

[99] Natalis, P., Delwiche, J., Collin, J.E., Caprace, G., Praet, M.Th. (1977): Phys. Scr. *16*, 242

[100] Gardner, J.L., Samson, J.A.R. (1973): J. Electron Spectrosc. Relat. Phenom. *2*, 153

[101] West, J.B., Codling, K., Parr, A.C., Ederer, D.L., Cole, B.E., Stockbauer, R. (1980): Daresbury Report DL/SCI/P254E

[102] Brion, C.J., Hammett, A., Wright, G.R., Van der Wiel, M.J. (1977): J. Electron Spectrosc. Relat. Phenom. *12*, 323

[103] Hammett, A., Stoll, W., Branton, G., Brion, C.E., Van der Wiel, M.J. (1976): J. Phys. B*9*, 945

δ) Satellite structure. In a one-electron description of the ionization process only the hole ground states are expected. These states correspond to the removal of a single electron from each of the occupied orbitals of the initial state. Although such one-electron transitions are strongly predominant, many-electron transitions involving excitations from occupied to unoccupied orbitals in addition to the ionization can give rise to a complex satellite structure in the spectra, particularly in the inner valence region ($E_B > 20$ eV).

As discussed in Sect. 7, contributions to the satellite structure are usually divided into relaxation and correlation (ISCI and FSCI) parts. The examples given for atoms (Sects. 8–10) display trends in the relative importance of these two mechanisms which apply also to molecules. Thus, for core electron ionization, the relaxation contributions are usually more important than the correlation contributions (we emphasize again that the detailed division into relaxation and correlation depends on the choice of representation of the initial and final states; we adopt the qualitative terminology currently in use, cf. refs. cited below). In the valence electron region, correlation contributions are often dominant in deciding the satellite structures. For example, the well-known intense $\tilde{C}\,^2\Sigma^+$ state in CO^+ and N_2^+ [104–109] (at approximately 25 eV) could in principle be interpreted as an $\tilde{X}\,^2\Sigma^+$ relaxation satellite where the 5σ ionization is accompanied by a $1\pi \rightarrow 2\pi$ excitation, but is more properly described as a correlation satellite which essentially borrows its intensity from the $\tilde{B}\,^2\Sigma^+\,4\sigma$ single hole state (the electron configuration of CO is [inner shell] $3\sigma^2\,4\sigma^2\,5\sigma^2\,1\pi^4$) [110–114]. With a configuration interaction description the $\tilde{C}$ state interacts with the single hole $\tilde{B}$ state (as do all many-electron states of $^2\Sigma^+$ symmetry) and thereby acquires some one-electron character. Hence the $\tilde{C}$ state can be reached by allowed one-electron transitions and the intensity reflects the percentage of one-electron character as obtained from the CI wave function expansion.

Some of the intensity of the $\tilde{B}$ state of CO^+ is thus distributed among other states which cannot be reached if correlation effects are neglected [in this case FISCI (cf. Sect. 7)]. In fact, most satellite features in valence electron spectra have been associated with such correlation states, expressed in CI or related terms. ISCI satellites which owe their intensity to correlation in the initial state

[104] Potts, A.W., Williams, T.A. (1974): J. Electron Spectrosc. Relat. Phenom. *3*, 3
[105] Åsbrink, L., Fridh, C. (1974): Phys. Scr. *9*, 338
[106] Åsbrink, L., Fridh, C., Lindholm, E., Codling, K. (1974): Phys. Scr. *10*, 183
[107] Gelius, U., Basilier, E., Svensson, S., Bergmark, T., Siegbahn, K. (1974): J. Electron Spectrosc. Relat. Phenom. *2*, 405
[108] Krummacher, S., Schmidt, V., Wuilleumier, F. (1980): J. Phys. B *13*, 3993
[109] Krummacher, S., Schmidt, V., Bizau, J.Z., Ederer, D., Wuilleumier, F. (1981): Europhysics Conf. Abstr. 5 A, 1029
[110] Domcke, W., Cederbaum, L.S. (1976): J. Chem. Phys. *64*, 612
[111]. Hillier, I.H., Kendrick, J. (1976): J. Electron Spectrosc. Relat. Phenom. *8*, 239
[112] Bagus, P.S., Viinikka, E.-K. (1977): Phys. Rev. A *15*, 1486
[113] Schirmer, J., Cederbaum, L.S., Domcke, W., von Niessen, W. (1977): Chem. Phys. *26*, 149
[114a] Honjou, N., Saoajima, T., Sasaki, F. (1981): Chem. Phys. *57*, 475
[114b] Langhoff, P.W., Langhoff, S.R., Rescigno, T.N., Schirmer, J., Cederbaum, L.S., Domcke, W., von Niessen, W. (1981): Chem. Phys. *58*, 71

may nevertheless be important[115,116]. The intensities of these satellites reflect the admixture of doubly excited electron configurations in the ground state wave function.

The many-electron satellite structure can in certain cases be so intense that it is not meaningful or even possible to select a specific band which corresponds to an expected single hole state. This situation has been referred to as a break-down of the orbital picture. One example of this is the spectrum of CS where a break-down can be observed already in the outer valence region[117,118]. This molecule possesses the electron configuration [core] $5\sigma^2 6\sigma^2 2\pi^4 7\sigma^2$. The three outermost single hole states are observed in the He*I*-excited spectrum. However, the 6σ ionization is accompanied by another band of similar intensity, which can be associated essentially with a $2\pi^3 7\sigma 3\pi$ configuration. Examples of inner valence spectra are CO_2, N_2O[119], CS_2[120] and COS[121] where the intensity of two single hole states is completely distributed among a manifold of states over a wide energy region.

Many-electron transitions are in general favoured by the presence of low-lying unoccupied molecular orbitals. This is manifested for example by differences between spectra of saturated and unsaturated molecules[4]. Intense satellite bands are observed chiefly in spectra of the latter type whereas in the former the transitions to the primary hole states are strongly dominating. There is an interesting interpretation of the interactions leading to strong CI satellites for unsaturated systems. As one specific example we discuss the strong satellite centred at 27.6 eV in the spectrum of ethylene (Fig. 98). The high binding energy of the peak suggests that it involves excitations with respect to a hole state having either a $1a_g$ or $1b_{3u}$ vacancy. The most likely are $1b_{1u}(\pi) \rightarrow 1b_{2g}(\pi^*)$ transitions yielding $^3B_{3u}$ and $^1B_{3u}$ final states. Coupling of these to the single hole states yields

$$\begin{aligned} &1a_g[1b_{1u}1b_{2g}^*(^{1,3}B_{3u})]\,^2B_{3u} \\ &1b_{3u}[1b_{1u}1b_{2g}^*(^{1,3}B_{3u})]\,^2A_g. \end{aligned} \tag{14.5}$$

Since the ^2A_g state will be close to the single hole $(1a_g)\,^2A_g$ state the final state configuration mixing most likely occurs in this manifold as was also confirmed by MARTIN and DAVIDSON[122]. The a_g and b_{3u} orbitals can be considered as the bonding and antibonding σ-orbitals described as $\sigma \sim s_A + s_B$ and $\sigma^* \sim s_A - s_B$ and the π-orbitals as $\pi \sim p_A + p_B$ and $\pi^* \sim p_A - p_B$. The configuration interaction

---

[115] Åsbrink, L., Chong, D.P., Fridh, C., Lindholm, E. (to be published)
[116] Harcourt, R.D. (1979): Chem. Phys. Lett. *61*, 25
[117] Schirmer, J., Domcke, W., Cederbaum, L.S., von Niessen, W. (1978): J. Phys. B*11*, 1901
[118] Okuda, M., Jonathan, N. (1974): J. Electron Spectrosc. Relat. Phenom. *3*, 19
[119] Domcke, W., Cederbaum, L.S., Schirmer, J., Brion, C.E., Tan, K.H. (1979): Chem. Phys. *40*, 171
[120] Schirmer, J., Domcke, W., Cederbaum, L.S., von Niessen, W., Åsbrink, L. (1979): Chem. Phys. Lett. *61*, 30
[121] Cook, J.P.D., White, M.G., Brion, C.E., Cederbaum, L.S., Domcke, W., von Niessen, W. (1981): J. Electron Spectrosc. Relat. Phenom. *22*, 261
[122] Martin, R.L., Davidson, I.A. (1977): Chem. Phys. Lett. *51*, 237

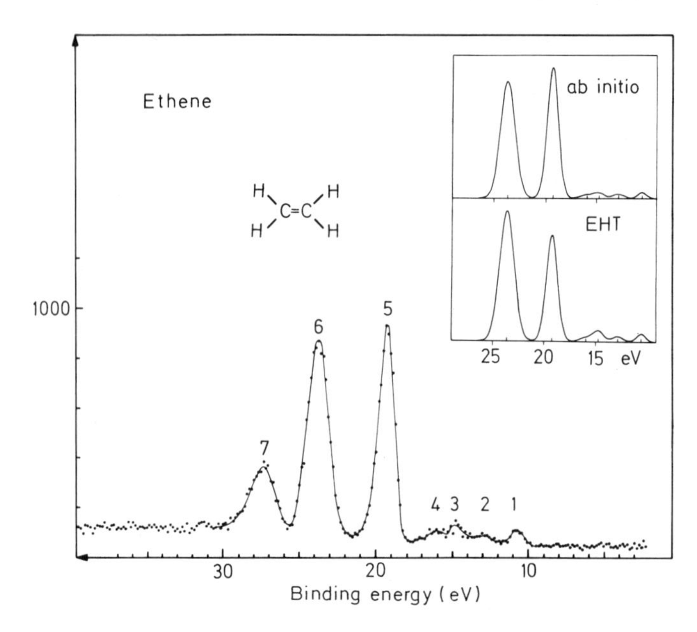

Fig. 98. Valence electron spectrum of ethylene obtained by monochromatized Al$K\alpha$ radiation. The inserted figure shows spectra calculated by the intensity model (cf. previous subsection) employing ab initio and extended Hückel orbital populations. The electron configuration is (core) $1a_g^2\ 1b_{3u}^2\ 1b_{2u}^2\ 2a_g^2\ 1b_{1g}^2\ 1b_{1u}^2(\pi)$. The peaks labelled 6–1 are associated with one-electron transitions from these valence orbitals in the given order. Peak 7 is associated with a many-electron transition. From ref.[66]

thus amounts to a mixing of the hole state determinant $|\ldots\sigma^*\pi|$ (a_g hole) with the doubly excited determinant $|\ldots\sigma\pi^*|$ (b_{3u} hole, $\pi\to\pi^*$ exc.). We may consider this mixing with a coefficient α by means of the determinant identity:

$$\Psi_{N-1}^{\text{fi}+} \sim |\ldots\sigma^*\pi| - \alpha|\ldots\sigma\pi^*| = [|\ldots s_A(\pi-\alpha\pi^*)| - |\ldots s_B(\pi+\alpha\pi^*)|]. \quad (14.6)$$

With the simplified molecular orbital description above for the π and π^* orbitals this equation becomes

$$\Psi_{N-1}^{\text{fi}+} \sim [|\ldots s_A[(1+\alpha)\,p_B+(1-\alpha)\,p_A]| - |\ldots s_B[(1+\alpha)\,p_A+(1-\alpha)\,p_B]|]. \quad (14.7)$$

This expression leads to a qualitative understanding of the nature of the CI state. It is seen that increased CI mixing amounts to a correlation of the motions of the hole and the π-electrons; the motion of the s electrons from center A to center B is accompanied by a motion of the p electrons in the opposite directions. The CI state thus represents two fluctuating dipoles which are oscillating out of phase with one another. One would expect to encounter manifestations of this effect whenever the ($\sigma^*\to\sigma\ \pi\to\pi^*$) excited configuration lies close in energy to the primary hole state. This is likely to occur for other unsaturated systems similar to ethylene, such as substituted derivatives.

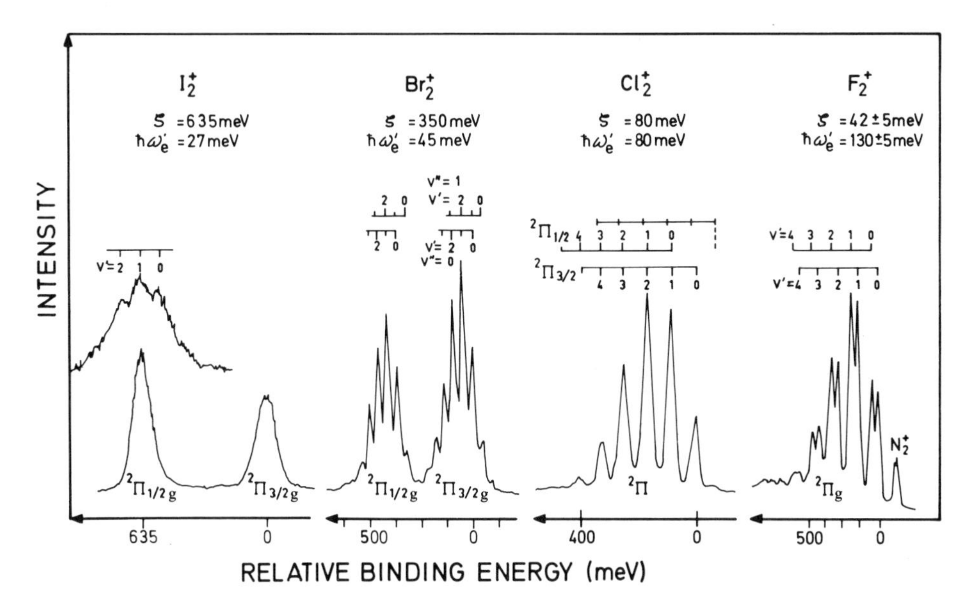

Fig. 99. He*I*-excited spectra of the diatomic halogen molecules showing transitions to the $\tilde{X}^2\Pi_g$ ionic state. The spin-orbit splitting and vibrational energies are given in meV. The vibrational quantum number v'' given for Br_2^+ represents excitations in the initial (neutral ground) state of Br_2. Transitions involving $v''>0$ are referred to as hot bands. Similar hot band excitations can be inferred for I_2^+ from the weak features on the low binding energy side of the $v'_2=0$ level. The scaling factor for the expanded $^2\Pi_{1/2}$ line of I_2^+ is 3.65. From ref.[123]

ε) Spin-orbit and Jahn-Teller interactions. Both these interactions tend to split electronic states which possess an orbital degeneracy. Such splittings can be important ingredients in photoelectron spectra.

The spin-orbit splitting increases with the atomic number Z. The magnitude is indicated in Fig. 99 which shows spectra of the diatomic halogen molecules. The spin-orbit split states are usually represented by the symbol $^{2S+1}\Gamma$, where Γ is the irreducible representation for the electronic state. For linear or approximately linear molecules the component of the total angular momentum along the internuclear axis, Ω, is also included to give $^{2S+1}\Gamma_\Omega$, as used in Fig. 99. If the spin-orbit interaction is very large, as in the case of UF_6^+ of Fig. 94, this classification may become inadequate. One should rather consider the spin-orbital degeneracy by employing, as far as symmetry is concerned, the irreducible representations of the extended point groups (cf. ref.[124]).

When Russell-Saunders coupling applies, the energy of the spin-orbit split components of linear or nearly linear molecules following Hund's case a) can be written

$$E_i = E_0 + \zeta \cdot \Lambda\Sigma, \tag{14.8}$$

[123] Cornford, A.B., Frost, D.C., McDowell, C.A., Ragle, J.A., Stenhouse, I.A. (1971): J. Chem. Phys. *54*, 2651

[124] Herzberg, G. (1966): Electronic spectra of polyatomic molecules, Princeton: Van Nostrand

where Λ and Σ are the axial components of the electronic and spin angular momenta, respectively, and $\Omega\hbar = |\Lambda + \Sigma|\hbar$. ζ is the molecular spin-orbit coupling constant for the given term. The splitting can be calculated by means of the atomic orbital coefficients C_A of the LCAO MO expansion and the corresponding atomic subshell spin-orbit coupling parameters ζ_A as

$$\zeta = \sum_A C_A^2 \zeta_A^2. \tag{14.9}$$

Coupling parameters have been given for several atoms in various states (cf. refs.[46, 125–127]). Additional rules for spin-orbit splittings expressed in these terms are given in ref.[128]. An improved treatment which includes consideration of the initial state[129] has been applied in several cases[127, 130, 131].

The intensities of the two spin-orbit split components due to ionization from a π^4 configuration are expected to be equal, as is also commonly observed (cf. Fig. 99). They are approximately equal also for ionization from the π^2 open shell of O_2. For the heavier dichalcogens, however, the $^2\Pi_{3/2}/^2\Pi_{1/2}$ ratio decreases and is only 0.1 for Te_2^+. This behaviour reflects an increasing $\omega\omega$-coupling[132] (the analogue to *jj*-coupling in atoms) and is expected to occur for molecules where a π^2 configuration is studied.

The above interpretations apply also to non-linear molecules belonging to axial point groups. Ionization from the $2e$ halogen lone-pair orbital of CH_3Br and CH_3I, for example, shows the expected spin-orbit splitting (cf. Fig. 100). The reduced intensity of the second peak of CH_3Br and CH_3Cl as well as the reduced energy spacing in the CH_3Cl spectrum have been associated with vibronic interactions where the inner peak corresponding to the $^2E_{1/2}(v'=0)$ state is mixed with vibrationally excited levels of the $^2E_{3/2}$ state[133] (cf. Sect. 15). In molecules with cubic or higher symmetry e.g. CX_4 (X = halogen atom) the spin-orbit splitting of ^2E states is zero. The ^2T states, on the other hand, show a splitting pattern which may resemble that of atomic ^2P states[134, 135]. Additional splittings may be observed due to Jahn-Teller instability.

Other aspects of spin-orbit interaction show up in ionically bonded mol-

---

[125] Wittel, K., Manne, R. (1974): Theor. Chim. Acta, *33*, 347

[126] Eland, J.H.D., (1974): Photoelectron spectroscopy, London: Butterworths

[127] Grimm, F.A. (1973): J. Electron Spectrosc. Relat. Phenom. *2*, 475

[128] Bieri, G., Heilbronner, E., Jones, T.B., Kloster-Jensen, E., Maier, J.P. (1977): Phys. Scr. *16*, 202

[129] Leach, S. (1968): Acta Phys. Pol. *34*, 705

[130] Anderson, C.P., Mamantov, G., Bull, W.E., Grimm, F.A., Carver, J.C., Carlson, T.A. (1971): Chem. Phys. Lett. *12*, 137

[131] Ishiguro, E., Kobori, M. (1967): J. Phys. Soc. Japan, *22*, 263

[132] Berkowitz, J. (1975): J. Chem. Phys. *62*, 4074; Streets, D.G., Berkowitz, J. (1976): J. Electron Spectrosc. Relat. Phenom. *9*, 269

[133] Karlsson, L., Jadrny, R., Mattsson, L., Chau, F.T., Siegbahn, K. (1977): Phys. Scr. *16*, 225

[134] Green, J.C., Green, M.H.L., Joachim, P.J., Orchard, A.F., Turner, D.W. (1970): Phil. Trans. R. Soc. Lond. A*268*, 111

[135] Dixon, R.N., Murrell, J.N., Narayan, B. (1971): Mol. Phys. *20*, 611

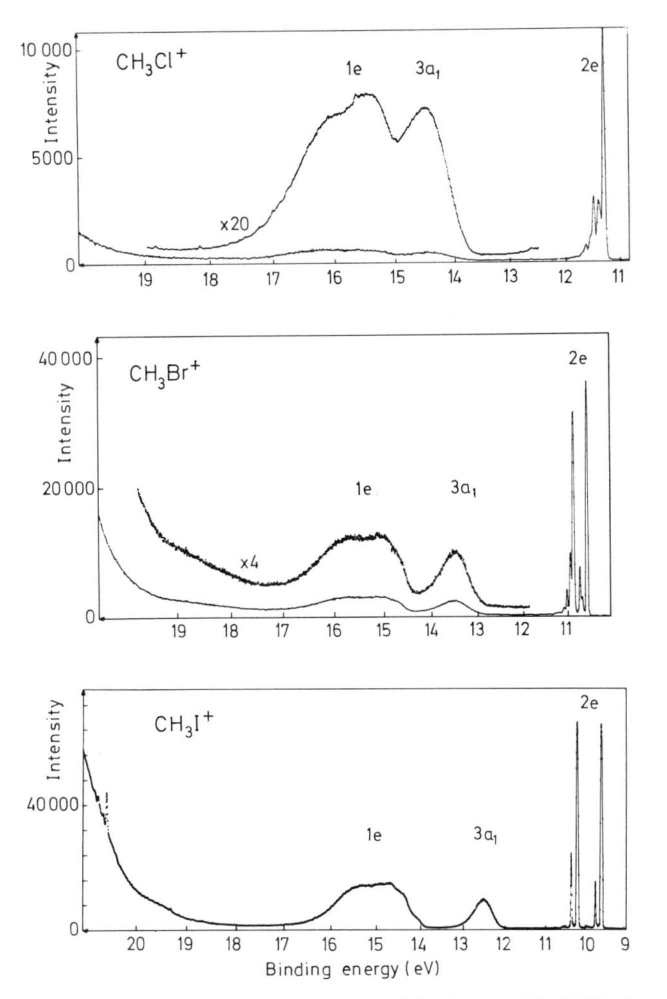

Fig. 100. He *I*-excited electron spectra of the methyl halides (except CH_3F). The features observed at 20.5 eV in the spectrum of CH_3I are due to excitation with $HL\alpha$ (hydrogen was an impurity of the He gas). From ref. [133]

ecules[136–138] (cf. Sect. 17) and low symmetry molecules. The latter case includes molecules with weak fields of low symmetry which break configurational degeneracies. For example, the splitting of the *p*-orbital of Br (cf. Fig. 100) should be described by including both spin-orbit interaction and molecular orbital effects like conjugation in molecules of lower symmetry than CH_3Br[139].

[136] Potts, A.W., Price, W.C. (1974): Proc. R. Soc. Lond. A*341*, 147

[137] Potts, A.W., Price, W.C. (1977): Phys. Scr. *16*, 191

[138] Berkowitz, J. (1977): in: Electron spectroscopy, Brundle, C.R., Baker, A.D. (eds.), vol. I, New York: Academic Press, p. 355

[139] Broglie, F., Heilbronner, E. (1971): Helv. Chim. Acta *54*, 1423

Calculations of spin-orbit splittings have been performed in several cases by approaches which include higher order terms through diagonalization of the spin-orbit interaction matrix. Usually a one-electron spin-orbit interaction operator of the form (for electron i)

$$H_{\text{S-O}} = \sum_A \xi(r_{Ai})\, \boldsymbol{l}_{Ai} \cdot \boldsymbol{s}_i \tag{14.10}$$

has been used, where the summation goes over all nuclei A. Good results have been obtained by introducing this operator into the extended Hückel formalism[140–145]. To some extent also Dirac-Slater calculations have been performed (cf. e.g. Fig. 94).

According to the Jahn-Teller theorem[146–149], the most symmetrical conformation of a non-linear molecule is unstable towards at least one asymmetrical normal coordinate if the molecule is in an orbitally degenerate state. This is the so-called Jahn-Teller instability. The stable state is reached by lowering the symmetry so that the degeneracy is removed, cf. Fig. 101 (the only exception is the Kramers' degeneracy[124]). The potential function is split in a first order expansion of the Hamiltonian

$$H = H_0 + \sum_k \frac{\partial H}{\partial Q_k} Q_k + \dots \tag{14.11}$$

Table 21 gives the possible pathways for such distortions. The vibrational modes which can give rise to such distortion are selected by the basic group theoretical condition

$$\Gamma_i \times \Gamma_{Q_k} \times \Gamma_i \supset \Gamma_{\text{TS}}, \tag{14.12}$$

where Γ_i and Γ_{Q_k} are the irreducible representations of the molecular orbital and the normal coordinate, respectively. The product must contain the totally symmetric representation Γ_{TS}.

The actual distortion will in each given case be determined by the electronic localization properties and of the bonding strength. Ionization from non-bonding orbitals does not in general give rise to any observable splitting, cf. the $\tilde{X}\,^2E$ state of the methyl halides of Fig. 100. Ionization from bonding orbitals can on the other hand lead to substantial splittings of up to about 1 eV as observed in the $\tilde{B}\,^2E$ state of the methyl halides. Other examples of

---

[140] Manne, R., Wittel, K., Mohanty, B.S. (1975): Mol. Phys. *29*, 485
[141] Wittel, K., Mohanty, B.S., Manne, R. (1974): J. Electron Spectrosc. Relat. Phenom. *5*, 1124
[142] Wittel, K., Manne, R. (1975): J. Chem. Phys. *63*, 1322
[143] Peel, J.B., Willett, G.D. (1976): J. Electron Spectrosc. Relat. Phenom. *9*, 175
[144] Berkosky, J.L., Ellison, F.O., Lee, T.H., Rabalais, J.W. (1973): J. Chem. Phys. *59*, 5342
[145] Lee, T.H., Rabalais, J.W. (1974): J. Chem. Phys. *60*, 1172
[146] Jahn, H.A., Teller, E. (1937): Proc. R. Soc. Lond. A*161*, 220
[147] Liehr, A.D. (1963): Adv. Inorg. Chem. *5*, 385; Adv. Chem. Phys. *5*, 241; J. Phys. Chem. *67*, 389
[148] Sturge, M.D. (1967): Solid State Phys. *20*, 92
[149] Englman, R. (1972): The Jahn-Teller effect in molecules and crystals, London: Wiley-Interscience

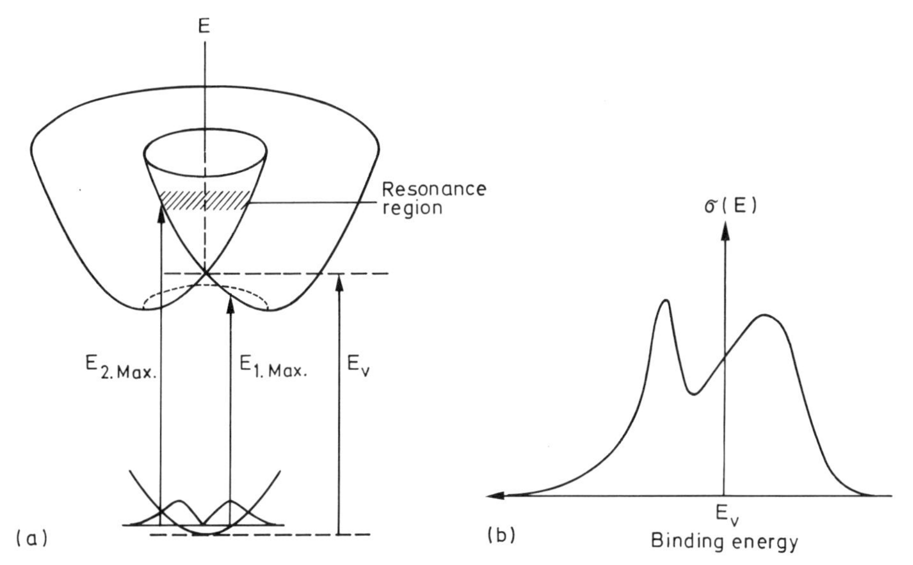

Fig. 101 a, b. Configuration coordinate diagram for linear Jahn-Teller splitting of an electronic E state and the expected shape of the corresponding photoelectron band. $E_{1\,max}$ and $E_{2\,max}$ give the energies of the two intensity maxima. From ref. [150]

large splittings are found in the spectra of AH_3 (A=N, P, As, Sb)[152-154], BH_4 (B=C, Si, Ge, Sn)[155-157], C_2H_6[158,159] and C_3H_6[160,161] molecules.

The splitting between the two band maxima of a ^2E state in a C_{3v} type molecule can be approximately calculated by

$$\Delta E_{JT} = \hbar\omega\sqrt{6D}, \tag{14.13}$$

where D is the Jahn-Teller coupling parameter and $\hbar\omega$ is the vibrational energy. The spectrum gives thus in principle information about D, which otherwise can be obtained from the vibrational fine structure (cf. Sect. 15).

A number of states and coupling cases can arise which are more complicated than that discussed above, such as for methane[157,162]. We refer to detailed discussions in other sources[147-149].

---

[150] Schwarz, W.H.E. (1975): J. Electron Spectrosc. Relat. Phenom. *6*, 377
[151] Jotham, R.W., Kettle, S.F.A. (1971): Inorg. Chim. Acta *5*, 184
[152] Potts, A.W., Price, W.C. (1972): Proc. R. Soc. A*326*, 181
[153] Rabalais, J.W., Karlsson, L., Werme, L.O., Bergmark, T., Siegbahn, K. (1973): J. Chem. Phys. *58*, 3370
[154] Haller, E., Cederbaum, L.S., Domcke, W., Köppel, H. (1980): Chem. Phys. Lett. *72*, 427
[155] Potts, A.W., Price, W.C. (1972): Proc. R. Soc. A*326*, 165
[156] Rabalais, J.W., Katrib, A. (1974): Mol. Phys. *27*, 923
[157] Meyer, W. (1973): J. Chem. Phys. *58*, 1017
[158] Rabalais, J.W., Bergmark, T., Werme, L.O., Karlsson, L., Siegbahn, K. (1971): Phys. Scr. *3*, 13
[159] Richartz, A., Buenker, R.J., Bruna, P.J., Peyerimhoff, S.D. (1977): Mol. Phys. *33*, 1345
[160] Haselbach, E. (1970): Chem. Phys. Lett. *7*, 428
[161] Buenker, R.J., Peyerimhoff, S.D. (1974): Chem. Rev. *74*, 127
[162] Dixon, R.N. (1971): Mol. Phys. *20*, 81

Table 21. Jahn-Teller distortions for important point groups[151]

Parent point group	Jahn-Teller active vibrations	Electronic states split	Ground-state symmetries consistent with the operation of a Jahn-Teller effect
O_h	E_g	$E_g, E_u, T_{1g}, T_{1u}, T_{2g}, T_{2u}, G_{3/2g}, G_{3/2u}$	D_{4h}, D_{2h} (rhombus)
	T_{2g}	$T_{1g}, T_{1u}, T_{2g}, T_{2u}, G_{3/2g}, G_{3/2u}$	D_{3d}, D_{2h} (rectangle), C_{2h}, C_i
T_d	E	$E, T_1, T_2, G_{3/2}$	
	T_2	$T_1, T_2, G_{3/2}$	C_{3v}, C_{2v}, C_s, C_1
T_h	E_g	$E_g, E_u, T_g, T_u, G_{3/2g}, G_{3/2u}$	D_{2h}
	T_g	$T_g, T_u, G_{3/2g}, G_{3/2u}$	C_{2h}, S_6, C_i
O	E	$E, T_1, T_2, G_{3/2}$	D_4, D_2
	T_2	$T_1, T_2, G_{3/2}$	D_3, D_2, C_2, C_1
T	E	$E, T, G_{3/2}$	D_2
	T	$T, G_{3/2}$	C_3, C_2, C_1
D_{6h}	E_{2g}	$E_{1g}, E_{2g}, E_{1u}, E_u$	D_{2h}, C_{2h}
D_{4h}	B_{1g}	E_g, E_u	D_{2h} (rhombus)
	B_{2g}	E_g, E_u	D_{2h} (rectangle)
D_{3h}	E'	E', E''	C_{2v}, C_s
C_{6h}	E_{2g}	$E_{1g}, E_{1u}, E_{2g}, E_{2u}$	C_{2h}
C_{4h}	$2B_g$	E_g, E_u	C_{2h}
C_{3h}	E'	E', E''	C_s
C_{6v}	E_2	E_1, E_2	C_{2v}, C_2
C_{4v}	B_1	E	C_{2v}
	B_2	E	C_{2v}
C_{3v}	E	E	C_s, C_1
D_{3d}	E_g	E_g, E_u	C_{2h}, C_i
D_{2d}	B_1	E	D_2
	B_2	E	C_{2v}
D_6	E_2	E_1, E_2	D_2, C_2
D_4	B_1	E	D_2
	B_2	E	D_2
D_3	E	E	C_2, C_1
C_6	E_2	E_1, E_2	C_2
C_4	$2B$	E	C_2
C_3	E	E	C_1
S_6	E_g	E_g, E_u	C_i
S_4	$2B$	E	C_2
I_h	G_g	$G_g, G_u, H_g, H_u, I_{5/2g}, I_{5/2u}$	$T_h, D_{3d}, C_{2h}, S_6, C_i$
	$2H_g$	$T_{1g}, T_{1u}, T_{2g}, T_{2u}, G_g, G_u, H_g, H_u$ $G_{3/2g}, G_{3/2u}, I_{5/2g}, I_{5/2u}$	$D_{5d}, D_{3d}, D_{2h}, C_{2h}, C_i$
I	G	$G, H, I_{5/2}$	T, D_3, C_3, C_2, C_1
	$2H$	$T_1, T_2, G, H, G_{3/2}, I_{5/2}$	D_5, D_3, D_2, C_2, C_1
$D_{\infty h}$	none[1]		
D_{5h}	E_1'	E_2', E_2''	C_{2v}, C_s
	E_2'	E_1', E_1''	C_{2v}, C_s
C_{5h}	E_1'	E_2', E_2''	C_s
	E_2'	E_1', E_1''	C_s
$C_{\infty v}$	none[1]		
C_{5v}	E_1	E_2	C_s, C_1
	E_2	E_1	C_s, C_1
D_{6d}	B_1	E_3	D_6
	B_2	E_3	C_{6v}
	E_2	E_1, E_5	D_2, C_{2v}, C_2
	E_4	E_2, E_4	D_{2d}, S_4

Table 21 (continued)

Parent point group	Jahn-Teller active vibrations	Electronic states split	Ground-state symmetries consistent with the operation of a Jahn-Teller effect
D_{5d}	E_{1g}	E_{2g}, E_{2u}	C_{2h}, C_i
	E_{2g}	E_{1g}, E_{1u}	C_{2h}, C_i
D_{4d}	B_1	E_2	D_4
	B_2	E_2	C_{6v}
	E	E_1, E_3	D_2, C_{2v}, C_2
D_5	E_1	E_2	C_2, C_1
	E_2	E_1	C_2, C_1
C_5	E_1	E_2	C_1
	E_2	E_1	C_1
S_{12}	$2B$	E_3	C_6
	E_2	E_1, E_5	C_2
	E_4	E_2, E_4	S_4
S_{10}	E_{1g}	E_{2g}, E_{2u}	C_i
	E_{2g}	E_{1g}, E_{1u}	C_i
S_8	$2B$	E_2	C_4
	E_2	E_1, E_3	C_2

The pseudo-Jahn-Teller effect is a further mechanism for state splittings, pertaining to vibronic interaction between two close-lying non-degenerate states. This mechanism has been discussed in connection with studies of alkyl substituted allenes[163].

In principle, there is a competition between spin-orbit and Jahn-Teller interaction in lowering the symmetry. The combined influence of both interactions may thus be observed in spectra[150,164,165] although there is often one predominant term. For example in the case of CH_3Br^+ and CH_3I^+ (cf. Fig. 100) the $\tilde{X}\,^2E$ state is split mainly by spin-orbit interaction whilst the $\tilde{B}\,^2E$ state is split primarily by Jahn-Teller interaction. When both interactions are strong the observed band splittings should be approximately given by

$$\Delta E = (\Delta E_{\mathrm{JT}}^2 + \Delta E_{\mathrm{s-o}}^2)^{1/2}. \qquad (14.14)$$

Application of this formula has been discussed for SbH_3^+[150]. The competing Jahn-Teller and spin-orbit interactions can also be studied via the vibrational structures (cf. Sect. 15).

15. Vibrational and rotational excitations of photoelectron spectra

α) Vibrational selection rules. Vibrational structure is frequently observed in photoelectron spectra, but normally not all vibrational modes are excited. In

[163] Broglie, F., Crandall, J.K., Heilbronner, E., Kloster-Jensen, E., Sojka, S.A. (1973): J. Electron Spectrosc. Relat. Phenom. *2*, 455

[164] Schenk, H.J., Schwartz, W.H.E. (1972): Theor. Chim. Acta *24*, 225

[165] Chau, F.T., Karlsson, L. (1977): Phys. Scr. *16*, 258

fact, in transitions to non-degenerate electronic states only the symmetric modes become appreciably excited. This is explained by (7.24), which selects the possible excitations by the vibrational overlap integral. In terms of basic group theory this integral is non-zero only if the irreducible representations of the vibrational wavefunctions are identical, that is, the direct product obeys

$$\Gamma(\chi_{\text{vib}}^{v_{\text{fi}}}) \times \Gamma(\chi_{\text{vib}}^{v_{\text{in}}}) = \Gamma_{\text{TS}} \quad (15.1)$$

where Γ_{TS} is totally symmetric. In the standard experiment at room temperature the target molecules are generally in the vibrationless totally symmetric ground state. Hence, only totally symmetric modes are observed with high intensity.

The possible excitations also include antisymmetric modes in units of two quanta and various other combinations. These will in general be weak (perhaps a few percent of the 0–0 transition) due to an unfavorable overlap integral for antisymmetric modes which do not shift the equilibrium position of the potential function[1].

In certain cases the separation of the electronic and vibrational wave functions which leads to (7.24) is not possible, but they must be treated as vibronically coupled. The selection rule provided by (15.1) is then relaxed and the vibronically active modes can become appreciably excited.

Two different cases of vibronic interaction have been discussed[1]. In the first case (type a)) the symmetry of the total molecular final ionic state wave function $\Psi_{\text{mol}}^{\text{fi}+}(i, Q_k)$ is considered (i represents here the final ionic electronic configuration and Q_k the nuclear vibrational state). This symmetry is given by $\Gamma_i \times \Gamma_{Q_k}$. In photoionization such a state (which may even be an electronically excited satellite state) can borrow intensity from strongly allowed states with the same total symmetry $\Psi_{\text{mol}}^{\text{fi}+}(j, Q_l)$ i.e.:

$$\Gamma_i \times \Gamma_{Q_k} = \Gamma_j \quad (15.2)$$

(since $\Gamma_{Q_l} = \Gamma_{\text{TS}}$). This may be considered as a more general form of FISCI.

This type of coupling has been discussed in several cases, like for CO_2^+[2,3], NO_2^+[4], SO_3^+[5], HCN^+[6], butatriene[7a] and BF_3^+[7b]. Table 22 gives a list of vibronic symmetries pertaining to linear triatomic molecules.

Type b) coupling is that considered in the Jahn-Teller theorem (cf. Sect. 14). For transitions to degenerate electronic states Jahn-Teller active modes may become singly excited in non-linear molecules and so-called Renner active

---

[1] Herzberg, G. (1966): Electronic spectra of polyatomic molecules, Princeton: Van Nostrand
[2] Eland, J.H.D., Berkowitz, J. (1977): J. Chem. Phys. *67*, 2782
[3] Domcke, W. (1979): Phys. Scr. *19*, 11
[4] Weiss, M.J. (1976): Chem. Phys. Lett. *39*, 250
[5] Alderdice, D.S., Dixon, R.N. (1972): J.C.S. Faraday Trans. II, *72*, 372
[6] Cederbaum, L.S., Domcke, W., Köppel, H., von Niessen, W. (1977): Chem. Phys. *26*, 169
[7a] Köppel, H., Domcke, W., Cederbaum, L.S., von Niessen, W. (1978): J. Chem. Phys. *69*, 4252
[7b] Haller, E., Köppel, H., Cederbaum, L.S. (1982): Chem. Phys. Lett. *85*, 12

Table 22. Symmetry species of vibronic levels in linear molecules

Symmetry of electronic state Γ_{el}	Vibrational excitation v_1	v_2	v_3	Symmetry of vibrational state Γ_{vib}	Symmetry of vibronic state $\Gamma_{el} \times \Gamma_{vib}$
$^2\Sigma_g$	0	1	0	Π_u	$^2\Pi_u$
$^2\Sigma_g$	1	1	0	$\Sigma_g \times \Pi_u = \Pi_u$	$^2\Pi_u$
$^2\Sigma_g$	0	0	1	Σ_u	$^2\Sigma_u$
$^2\Sigma_g$	0	1	1	$\Pi_u \times \Sigma_u = \Pi_g$	$^2\Pi_g$

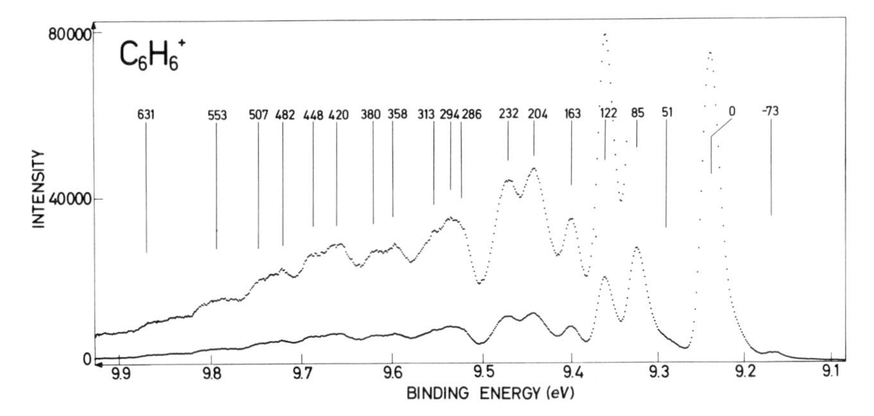

Fig. 102. High resolution He*I*-excited electron spectrum of benzene showing transitions to the $\tilde{X}\,^2E_{1g}$ state of the ion (this is the outermost band in the spectrum of Fig. 91). The numbers at the line positions give the vibrational energies in meV. The weak feature at −73 meV is probably a hot band which corresponds to transitions from a vibrationally excited molecule. From ref.[10]

modes in linear molecules[1]. The latter are the bending modes and such excitations have been discussed for example for CO_2^+ [8] and HCN^+ [9].

For non-linear molecules the $J-T$ active modes are selected according to (14.12) (cf. also Table 21). One example is provided by Fig. 102 which shows ionization from the $1e_{1g}(\pi)$ orbital of benzene. The final state $\tilde{X}\,^2E_{1g}$ is $J-T$ instable and excitations of the e_{2g} modes can be expected. In fact, three of the existing four modes have been identified in addition to the two totally symmetric ones[10–14]. Table 23 summarizes the results for this state and also for the $\tilde{A}\,^2E_{2g}$ state where similar excitations are present.

A vibrational level can by the vibronic interaction be substantially split into components of different angular momentum. These components are char-

[8] Frey, R., Gotchev, B., Kalman, O.F., Peatman, W.B., Pollak, H., Schlag, E.W. (1977): Chem. Phys. *21*, 89

[9] Fridh, C., Åsbrink, L. (1974): J. Electron Spectrosc. Relat. Phenom. *7*, 119

[10] Karlsson, L., Mattsson, L., Jadrny, R., Bergmark, T., Siegbahn, K. (1976): Phys. Scr. *14*, 230

[11] Potts, A.W., Price, W.C., Streets, D.G., Williams, T.A. (1972): Discuss. Faraday Soc. *54*, 206

[12] Åsbrink, L., Lindholm, E., Edqvist, O. (1970): Chem. Phys. Lett. *5*, 609

[13] Duke, C.B., Lipari, N.O. (1975): Chem. Phys. Lett. *36*, 51

[14] Lord, R.C., Marston, A.L., Miller, F.A. (1957): Spectrochim. Acta *9*, 113

Table 23. Vibrational energies (in meV) of the two outermost electronic states $^2E_{1g}(\pi)$ and $^2E_{2g}(\sigma)$ and of the neutral ground state of benzene

Vacated orbital	Adiabatic binding energy eV	Vibrational modes			
		ν_{18} (e_{2g})	ν_2 (a_{1g})	ν_{17} (e_{2g})	ν_{16} (e_{2g})
$1e_{1g}$	9.240	85	122	163	204
$3e_{2g}$	11.488	80	123		196
Neutral molecule		75	123	146	197

acterized by vibrational and ordinary angular momentum quantum numbers[1]. For C_{3v} type molecules the components are labelled by $j=l+\frac{1}{2}\Lambda$, where l is the vibrational angular momentum quantum number, and Λ the electronic angular momentum quantum number. For linear molecules, the relevant quantum number is $K=l+\Lambda$. Selection rules with respect to j and K have been discussed[15,16].

β) Hot bands. Transitions from vibrationally excited initial states give rise to the so-called hot bands. These are normally weak because the population of excited molecules, $N_{v_{\text{in}}}$ is low. $N_{v_{\text{in}}}$ can be estimated from Boltzmann statistics

$$\frac{N_{v_{\text{in}}}}{N_0}=\frac{n_{v_{\text{in}}}}{n_0}\,e^{-\frac{E_{\text{vib}}}{kT}}, \tag{15.3}$$

where n refers to the state degeneracy. Substantial contributions can arise if there exist very low energy vibrations, for example if heavy atoms are present. This can be seen from Fig. 99 where the hot band structure of Br_2^+ and I_2^+ is prominent. Also there is probably a hot band seen in Fig. 102 for benzene at -73 meV. A low energy bending mode can also be present, like in C_3O_2, where thermal excitation in the ground state leads to sequence bands in the spectrum[17].

Studies of hot band structure are usually best performed by temperature variation techniques. This has been done both by using low temperatures[18,19] and high temperatures[20,21].

γ) Qualitative correspondence between vibrational structure and orbital shapes. The vibrational energies may be qualitatively correlated with the orbital bonding properties as follows:

---

[15] Karlsson, L., Jadrny, R., Mattsson, L., Chau, F.T., Siegbahn, K. (1977): Phys. Scr. *16*, 225

[16] Dixon, R.N., Duxbury, G., Horani, M., Rostas, I. (1971): Mol. Phys. *22*, 977

[17] Rabalais, J.W., Bergmark, T., Werme, L.O., Karlsson, L., Hussain, M., Siegbahn, K. (1972): in: Electron spectroscopy, Shirley, D.A. (ed.), Amsterdam: North-Holland, p. 425

[18] Lloyd, D.R., Roberts, P.J. (1975): J. Electron Spectrosc. Relat. Phenom. 7, 325

[19] Dehmer, P.M., Dehmer, J.L. (1978): J. Chem. Phys. *68*, 3462; (1979): *69*, 125; (1979): *70*, 4574 and (1979): in: Electronic and atomic collisions, ICPEAC XI, Oda, N., Takayanagi, K. (eds.), Amsterdam: North-Holland

[20] Dyke, J., Jonathan, N., Morris, A., Sears, T. (1976): J.C.S. Faraday Trans. II *72*, 597

[21] Dechant, P., Schweig, A., Thon, N. (1977): J. Electron Spectrosc. Relat. Phenom. *12*, 443

Table 24. Vibrational energies (0–1 spacings) in meV of the neutral molecules and various ionic states of N_2, PN and P_2. The ratio between the final state and the initial state vibrational energies (ω'/ω'') is indicated in parentheses

Electronic state	Molecule		
	N_2	PN	P_2
$\tilde{X}\ {}^1\Sigma_g^+$	292[a]	166[a]	97[a]
$(3\sigma_g^1)\tilde{X}\ {}^2\Sigma_g^+$	270 (0.92)[b]	153 (0.92)[c]	91 (0.94)[d]
$(1\pi_u^3)\tilde{A}\ {}^2\Pi_u$	234 (0.80)[b]	138 (0.83)[c]	83 (0.86)[e]
$(2\sigma_u^1)\tilde{B}\ {}^2\Sigma_u^+$	296 (1.01)[b]	174 (1.05)[c]	102 (1.05)[e]

[a] Ref. [24].
[b] Ref. [25].
[c] Ref. [26].
[d] Ref. [27].
[e] Ref. [28].

1. Ionization from a non-bonding orbital which has little influence on the molecular geometry leads to essentially the same energies as in the initial state.

2. Ionization from a bonding or antibonding orbital leads to lower or higher, respectively, vibrational energies in the final state.

3. The effects of angle-determining orbitals must be considered separately from case to case. Typical examples of such orbitals are the outermost a_1 orbital of H_2X (X=O, S, Se, Te) and H_3X (X=N, P, As, Sb).

Turner et al.[22] combined the Franck-Condon factor and vibrational energy considerations into an approximate experimental correlation

$$E_B^{\text{vert}}(i) - E_B^{\text{ad}}(i) = 1.2\left(\frac{\omega''}{\omega'} - 1\right), \tag{15.4}$$

where ω'' and ω' refer to the initial and final state vibrational energies, respectively. $E_B^{\text{vert}}(i)$ and $E_B^{\text{ad}}(i)$ are the vertical and adiabatic electron binding energies, respectively, for ionization from the i:th orbital (cf. Sect. 7). Dyke et al.[23] studied diatomic molecules and obtained good correlations between ω and R (the internuclear distance) similarly to results in absorption spectroscopy[24].

Examples of the energy changes for non-bonding and bonding orbitals are given in Table 24 for group V diatomic molecules. The σ_g orbital is essentially non-bonding and the π_u is bonding. The σ_u orbital is anti-bonding and an increase in energy could be expected. For N_2, however, due to a many-electron effect (essentially a mixing between the $\tilde{B}$ state and the $\tilde{C}$ state at ~25 eV) the

[22] Turner, D.W., Baker, C., Baker, A.D., Brundle, C.R. (1970): Molecular photoelectron spectroscopy, London: Wiley-Interscience
[23] Dyke, J.M., Golob, L., Jonathan, N., Morris, A. (1975): J.C.S. Faraday Trans. II, *71*, 1026
[24] Herzberg, G. (1950): Spectra of diatomic molecules, Princeton: Van Nostrand
[25] Maripuu, R., Wu, Nianzu, Ji, Ming-Rong, Reineck, I., Karlsson, L. (unpublished data)
[26] Bulgin, D.K., Dyke, J.M., Morris, A. (1977): J.C.S. Faraday Trans. II *73*, 983
[27] Narasimham, N.A. (1975): Can. J. Phys. *35*, 1242
[28] Bulgin, D.K., Dyke, J.M., Morris, A. (1976): J.C.S. Faraday Trans. II *72*, 2225

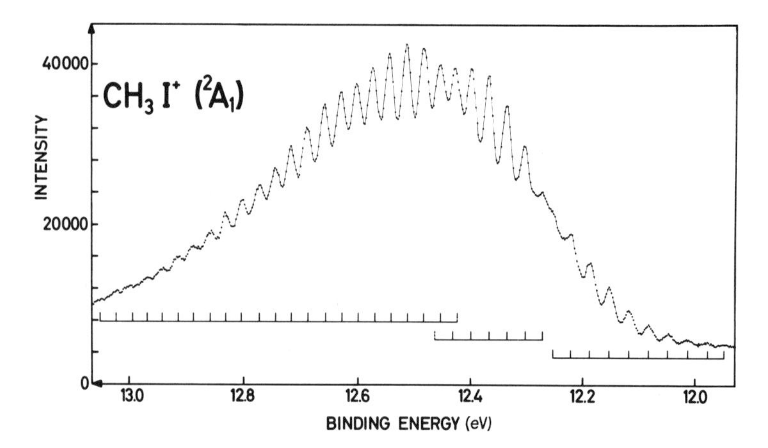

Fig. 103. High resolution (He*I*-excitation) recording of the second band of the electron spectrum of CH_3I corresponding to ionization from the $3a_1$ orbital (cf. also Fig. 100). Three vibrational progressions are indicated. The first one is $n\nu_3$ ($n=0...9$) where ν_3 is the C–I stretching mode with the energy $\hbar\omega_3=34$ meV. From ref. [15]

bond strength is unchanged upon ionization[29]. This many-electron effect was discussed in connection with Fig. 93.

Examples where an increase of the vibrational spacing occurs as expected are found in O_2^+, NO^+ and the diatomic halogen and interhalogen molecules (e.g.[22], and ref.[123] of Sect. 14).

The largest changes of vibrational energies are observed for ionization from localized strongly bonding orbitals in single bonds. An example is given in Fig. 103 which shows the $\tilde{A}A_1$ state of CH_3I^+. The spacing of the ν_3 (C-I stretch) mode is reduced from 65 meV to 34 meV due to ionization of one of the C-I bonding electrons.

For larger molecules the vibrational assignments and correlation with electronic structure are often less clearcut. Exceptions exist, e.g. large molecules of high symmetry such as substituted acetylenes and polyacetylenes[30, 31]. For five- and six-membered (and larger) ring type molecules the vibrational energies are often found to be similar to those of the neutral ground state (cf. e.g. Table 22). Photoelectron spectra of such molecules have been used to assign IR and Raman spectra[32].

The vibrational spacings of progressions are often found to change successively with increasing vibrational quantum number, due to anharmonicity of the potential function. This applies e.g. to the N_2 spectrum (Fig. 93). Anharmonicity introduces higher order terms in the vibrational energies according to

---

[29] Cederbaum, L.S., Domcke, W. (1977): Adv. Chem. Phys. *36*, 205
[30] Heilbronner, E., Hornung, V., Kloster-Jensen, E. (1970): Helv. Chim. Acta, *53*, 331
[31] Heilbronner, E., Hornung, V., Maier, J.P., Kloster-Jensen, E. (1974): J. Am. Chem. Soc. *96*, 4252
[32] Derrick, P.J., Åsbrink, L., Edquist, O., Lindholm, E. (1971): Spectrochim. Acta A*27*, 2525

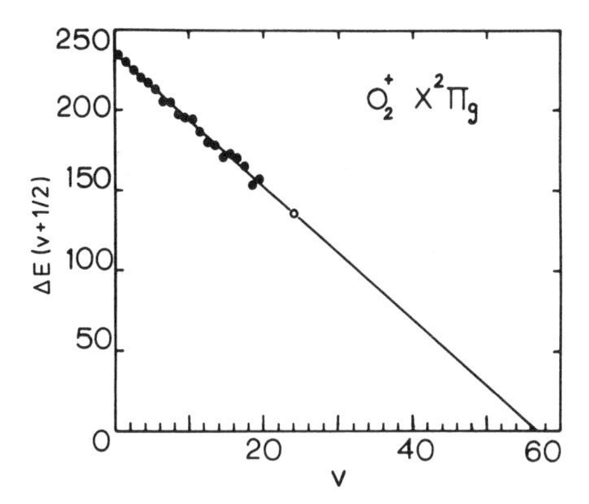

Fig. 104. Vibrational spacings $\Delta E(v+\frac{1}{2})$ in meV for the $\tilde{X}^2\Pi_g$ state of O_2^+ plotted as a function of $v+\frac{1}{2}$. The open circle is an average value around $v=24$. From the area under the extrapolated line the dissociation energy could be determined as $D_e=6.67\pm0.18$ eV. From ref. [33]

(for diatomic molecules)[24]:

$$E_{\text{vib}}=\hbar\omega_e(v+\tfrac{1}{2})-\hbar\omega_e x_e(v+\tfrac{1}{2})^2+\hbar\omega_e y_e(v+\tfrac{1}{2})^3\ldots \tag{15.5}$$

The magnitude of x_e and y_e can in principle be determined from the spectra. From such data the dissociation energy can be obtained by extrapolation to the limit. An example is given in Fig. 104 for O_2^+ [33] where a large number of vibrational levels could be populated via an autoionizing state. Similar enhancements of vibrational state populations can be obtained in threshold photoelectron spectroscopies[34, 35]. Other studies of dissociation energies include HF^+, DF^+, F_2^+ [36] and O_2^+, SO^+, S_2^+ [23].

Predissociative states reached in the photoelectron transition, are sometimes revealed in electron spectra by degraded vibrational structure. One example is provided by a study of HBr^+ [37].

A characteristic vibrational structure can be observed for molecules which possess an inversion barrier in the final state. This is indicated by Fig. 105. Most significant is the halving of the vibrational spacing which occurs above the potential maximum, but also level splittings can be observed. The best known examples of photoelectron spectra showing these features are the angle-determining a_1 orbital bands of the group *V* (XH_3; X=N, P, As, Sb) and group *VI* (YH_2; Y=O, S, Se, Te) hydride molecules. Figure 106 shows the $2a_1$ band of H_2S. Approximately at the maximum intensity the spacings change from about 120 meV to about 70 meV. Also level splittings are observed which can be associated with transitions to levels with different *K*-values (cf. above). The positions of the various levels can be approximately calculated by employ-

---

[33] Samson, J.A.R., Gardner, J.L. (1977): J. Chem. Phys. *67*, 755
[34] Peatman, W.B., Gotchev, B., Gürtler, P., Saile, V. (1978): DESY-Report SR-78/01
[35] Ajello, J.M., Chutjian, A., Winchell, R. (1980): J. Electron Spectrosc. Relat. Phenom. *19*, 197
[36] Guyon, P.-M., Spohr, R., Chupka, W.A., Berkowitz, J. (1976): J. Chem. Phys. *65*, 1650
[37] Delwiche, J., Natalis, P., Momigny, J., Collin, J.E. (1972/73): J. Electron Spectrosc. Relat. Phenom. *1*, 219

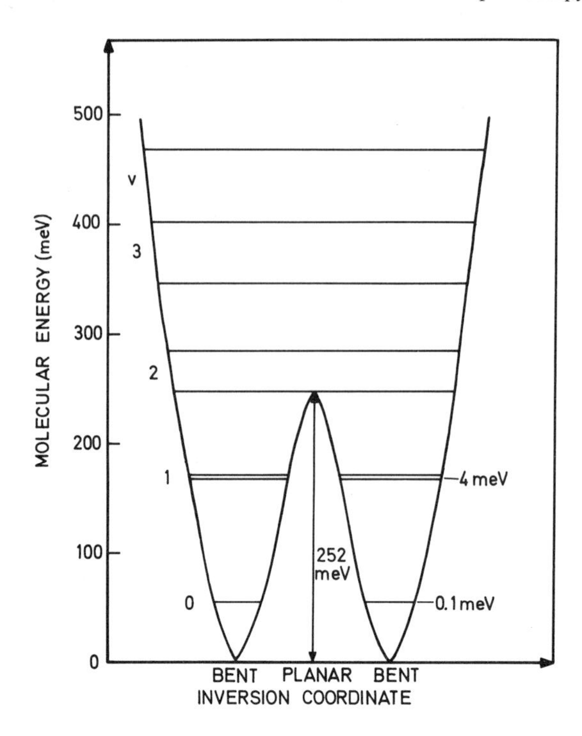

Fig. 105. Schematic potential energy curve for a state which possesses a weak potential maximum for the planar (or linear) molecular geometry. The energy spacings correspond to the ground state of NH_3

ing a harmonic potential perturbed by a Gaussian according to[38-41]

$$V=\tfrac{1}{2}q^2+\alpha e^{-\beta q^2}. \tag{15.6}$$

An accurate treatment also implies consideration of the vibronic interaction (Renner effect) which occurs for the linear or nearly linear molecular states[16,41-43].

For H_2S^+ the spacings between adjacent levels with the same K show a minimum approximately at the band maximum. Such a behaviour of the spacings was predicted in a treatment by DIXON[38] using a potential of the form (15.6) showing that a minimum separation occurs at the maximum of the potential barrier. For H_2Se^+ and H_2Te^+ the minimum is shifted towards higher energies,

[38] Dixon, R.N. (1964): Trans. Far. Soc. *60*, 1363 and (1965): Mol. Phys. *9*, 357
[39] Coon, J.B., Naugle, N.W., McKenzie, R.D. (1966): J. Mol. Spectrosc. *20*, 107
[40] Durmaz, S., Murrell, J.N., Taylor, J.M., Suffolk, R. (1970): Mol. Phys. *19*, 533
[41] Durmaz, S., King, G.H., Suffolk, R.J. (1972): Chem. Phys. Lett. *13*, 304
[42] Barrow, T., Dixon, R.N., Duxbury, G. (1974): Mol. Phys. *27*, 1217
[43] Aarts, J.F.M. (1978): Mol. Phys. *35*, 1785

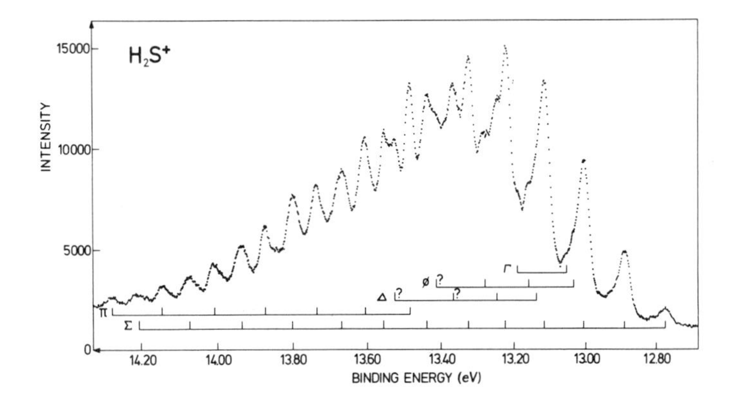

Fig. 106. The second electronic state of H_2S^+ reached by He *I* ionization from the $2a_1$ angle-determining orbital. Vibronic symmetries of the spectral bands are assigned in accordance with data from optical emission spectroscopy. Tentative assignments are indicated by question marks. From ref. [92]

corresponding to a higher barrier. For H_2O^+ [44,45a] no minimum is observed but merely a steep decrease in the spacings towards the lowest levels. This suggests that the potential function is strongly anharmonic but lacks the pronounced maximum of H_2S^+.

Improved agreement with photoelectron spectra has been obtained in calculations involving other potential functions than that of Eq. (15.6). Such examples are studies of H_2O [45a] (using a harmonic potential perturbed by a Lorentzian), NH_3 [45b] (using a Manning potential) and PH_3 [45c] (using the potential of Eq. (15.6) extended by a quartic term).

Substituent effects on the vibrational structure have been studied in several cases, particularly for deuteration which is used to facilitate vibrational assignments. These include for example methyl cyanide[46], acetaldehyde[47], hydrogen cyanide[9], formic acid[48], methanol[49], ketene[50], ethylene[51] and the methyl radical[52]. Isotopic exchange of heavier elements than hydrogen has been studied for H_2O^{16} and H_2O^{18} (ref. [44]).

---

[44] Karlsson, L., Mattsson, L., Jadrny, R., Albridge, R.G., Pinchas, S., Bergmark, T., Siegbahn, K. (1975): J. Chem. Phys. *62*, 4745

[45a] Dixon, R.N., Duxbury, G., Rabalais, J.W., Åsbrink, L. (1976): Mol. Phys. *31*, 423

[45b] Ågren, H., Reineck, I., Veenhuizen, H., Maripuu, R., Arneberg, R., Karlsson, L. (1982): Mol. Phys. *45*, 477

[45c] Maripuu, R., Reineck, I., Ågren, H., Wu Nian Zu, Ji Ming Rong, Veenhuizen, H., Al-Shamma, S.H., Karlsson, L., Siegbahn, K. (1982): UUIP-1053 (Uppsala University Institute of Physics Report)

[46] Lake, R.F., Thompson, H.W. (1971): Spectrochim. Acta *27* A, 783

[47] Cvitas, T., Güsten, H., Klasinc, L. (1976): J. Chem. Phys. *64*, 2549

[48] Watanabe, I., Yokohama, Y., Ikeda, S. (1973): Chem. Phys. Lett. *19*, 406

[49] Macneil, K.A.G., Dixon, R.N. (1977): J. Electron Spectrosc. Relat. Phenom. *11*, 315

[50] Hall, D., Maier, J.P., Rosmus, P. (1977): Chem. Phys. *24*, 373

[51] Cvitas, T., Güsten, H., Klasinc, L. (1979): J. Chem. Phys. *70*, 57

[52] Dyke, J., Jonathan, N., Lee, E., Morris, A. (1976): J.C.S. Faraday Trans. II *72*, 1385

The Teller-Redlich product rule[53]

$$\prod_{i=1}^{N} \frac{\omega_i^*}{\omega_i} = \rho \tag{15.7}$$

has been applied in some cases to discuss vibrational excitations and molecular geometry (e.g.[44, 52]). In this relationship ω_i^* and ω_i are the vibrational energies of the isotopic and normal molecules, respectively, for a given symmetry species. The factor ρ includes a dependence on the atomic masses and molecular geometry. A value of ρ can be calculated explicitly without knowledge of the vibrational energies. Equation (15.7) can therefore serve as a means to confirm vibrational assignments and to investigate the molecular geometry. The derivation of the product rule relies upon the assumption that the force constants are the same in the isotopic molecules, which in general is expected to be a good approximation.

δ) Calculation of vibrational structure. By means of theoretical methods more complete interpretations of vibrational spectra may be obtained along with quantitative information concerning the molecular geometry. Pertinent corrections of the observed vibrational structure with respect to the instrumental function have been discussed[54–56]. It is also important to identify possible effects of autoionization, shape resonances and variations in the asymmetry parameter β.

The simplest theoretical method employs a semi-classical harmonic oscillator description. The maximum intensity is assumed to be reached for a vertical transition at the equilibrium geometry of the initial state. The energy at this point is used as the classical harmonic oscillator turning point for the final state. For a diatomic molecule one obtains

$$\tfrac{1}{2}\mu\,\omega^2(\Delta R)^2 = \hbar\,\omega(v_{\max} + \tfrac{1}{2}), \tag{15.8}$$

where μ is the reduced mass and $v_{\max}$ is the vibrational quantum number at maximum intensity. This approach can be easily extended to larger molecules[57]. The geometry changes ΔR are usually predicted with deviations of less than 10% from the correct value[58].

Most applications have relied upon calculation of the Franck-Condon factors (7.24). Naturally the most accurate studies have been made for diatomic molecules where anharmonicity has been accounted for by using both Morse

[53] Herzberg, G. (1945): Infrared and Raman spectra of polyatomic molecules, New York: Van Nostrand, p. 169

[54] Natalis, P., Delwiche, J., Caprace, G., Collin, J.E., Hubin, M.-J., Praet, M.Th. (1977): J. Electron Spectrosc. Relat. Phenom. *10*, 93

[55] Natalis, P. (1976): J. Phys. Chem. *80*, 2829

[56] Natalis, P., Delwiche, J., Collin, J.E., Caprace, G., Praet, M.Th. (1977): J. Electron Spectrosc. Relat. Phenom. *11*, 417

[57] Eland, J.H.D. (1974): Photoelectron spectroscopy, London: Butterworths; Eland, J.H.D., Frey, R., Kuestler, A., Schulte, H., Brehm, B. (1976): Int. J. Mass Spectrom. Ion Phys. *22*, 155

[58] Price, W.C. (1981): Int. Rev. Phys. Chem. *1*, 1; (1974): Adv. Atom. Mol. Phys. *10*, 131

and Rydberg-Klein-Rees potential functions[59-62]. Autoionization contributions have also been calculated and used to determine bond distances accurately.

For polyatomic molecules several approximations are usually employed, for example

1) The crude adiabatic approximation which implies that the wave functions are written

$$\Psi_{\mathrm{mol}}(\boldsymbol{r}_l, \boldsymbol{R}_n) = \Psi_{N,\mathrm{el}}(\boldsymbol{r}_l; \boldsymbol{R}_0)\, \chi_{\mathrm{vib}}(\boldsymbol{R}_n) \tag{15.9}$$

where $\boldsymbol{r}_l$, $\boldsymbol{R}_n$ are electronic and nuclear coordinates and $\Psi_{N,\mathrm{el}}(\boldsymbol{r}_l; \boldsymbol{R}_0)$ the electronic wave function at the equilibrium geometry ($\boldsymbol{R}_0$).
2) Harmonic oscillator wave functions are used.
3) Thermal excitations in the initial state are neglected.

The errors introduced by the various approximations have been discussed by SCHWARZ[63]. The effect of non-constant electronic transition cross section in (7.24) can be accounted for by expanding in terms of the nuclear coordinates, $\boldsymbol{R}$[64].

Solutions of the vibrational overlap integral are obtained by expressing the normal coordinates $\boldsymbol{Q}$ of one state in terms of the other. A linear approximation is usually employed according to

$$\boldsymbol{Q}' = \boldsymbol{A}\boldsymbol{Q}'' + \boldsymbol{D}, \tag{15.10}$$

where $\boldsymbol{Q}'$, $\boldsymbol{Q}''$ and $\boldsymbol{D}$ are column vectors of the order $3N$-6 [($3N$-5) for linear molecules]. $\boldsymbol{Q}'$ and $\boldsymbol{Q}''$ refer to the final and initial states, respectively, and $\boldsymbol{D}$ represents the shift of equilibrium position. $\boldsymbol{A}$ is in principle a non-orthogonal matrix. Techniques for solving this general problem have been developed[65,66].

Sometimes $\boldsymbol{A}$ may be appropriately chosen as the unit matrix, which simplifies the treatment. If in addition the vibrational spacings are assumed to be equal in the initial and final states the intensities will be derived from Poisson distributions.

In order to correlate the change in the normal coordinate with structural changes a transformation is performed to the symmetry coordinates S_i by

$$\boldsymbol{S} = \boldsymbol{L} \cdot \boldsymbol{Q}, \tag{15.11}$$

where the $\boldsymbol{L}$ matrix involves nuclear masses and force constants. In the approach by COON et al.[67] the elements of $\boldsymbol{S}$ and $\boldsymbol{Q}$ are expressed in terms of the

[59] Chong, D.P., Herring, F.G., McWilliams, D. (1975): J. Electron Spectrosc. Relat. Phenom. 7, 429

[60] Chong, D.P., Takahata, Y. (1977): J. Electron Spectrosc. Relat. Phenom. *10*, 137

[61] Caprace, G., Delwiche, J., Natalis, P., Collin, J.E. (1976): Chem. Phys. *13*, 43

[62] Natalis, P., Delwiche, J., Collin, J.E., Caprace, G., Praet, M.Th. (1977): Phys. Scr. *16*, 242

[63] Schwarz, W.H.E. (1975): J. Electron Spectrosc. Relat. Phenom. *6*, 377

[64] Frank-Kemenetskii, M.D., Lukashin, A.V. (1976): Sov. Phys. - Usp. *18*, 391

[65] Sharp, T.E., Rosenstock, H.M. (1964): J. Chem. Phys. *41*, 3453

[66] Malmqvist, P.-Å. UUIP-1058 (Uppsala University Institute of Physics Report, 1982)

[67] Coon, J.B., de Wames, R.E., Lloyd, C.M. (1962): J. Mol. Spectrosc. *8*, 285

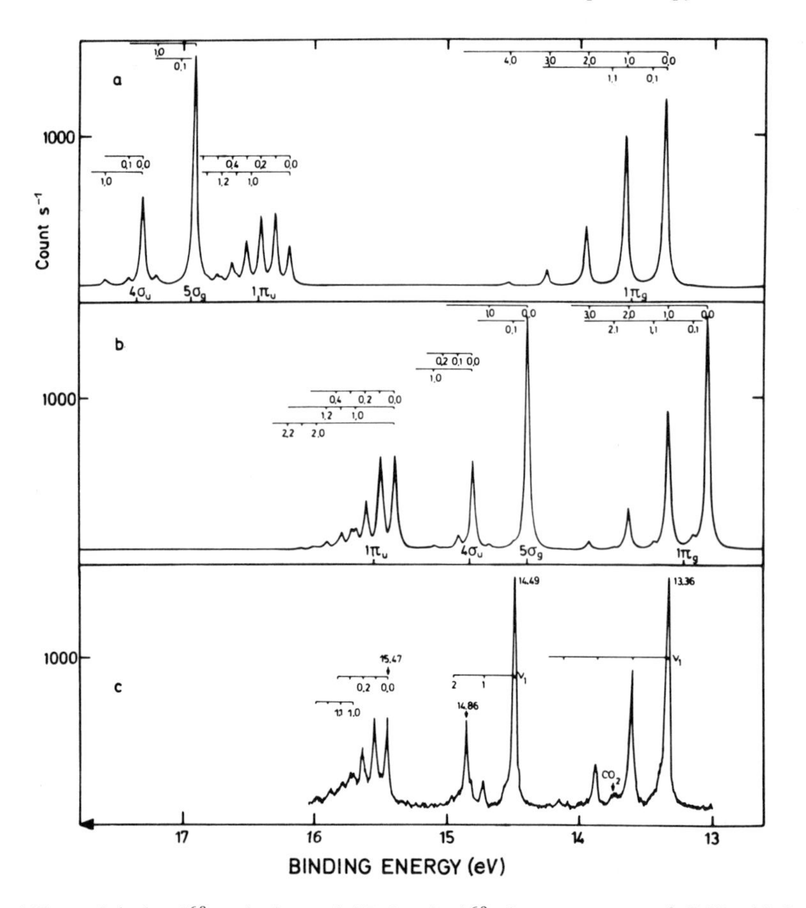

Fig. 107a–c. Calculated[68] and observed (He*I*-excited)[69] electron spectra of C_2N_2. (**a**) Spectrum calculated at the Koopmans' level (HF). (**b**) Spectrum calculated by the Green's function technique including many-electron effects. (**c**) Experimental spectrum including measured electron binding energies. The pair of numbers used to specify each vibrational component in parts a and b correspond to the vibrational quantum numbers of the ν_1 and ν_2 vibrational modes. The same notations are used in part c (the spacings of higher energy are due to the ν_1 mode and of lower energy to the ν_2 mode)

changes and the $\boldsymbol{L}$ matrix is obtained from the observed energies of the ion (denoted $\boldsymbol{L}'$). For example, in the case of C_2N_2 (cf. Fig. 107) one obtains

$$\begin{aligned} \Delta R_1(\text{N-C}) &= L'_{11}\,\Delta Q_1 + L'_{12}\,\Delta Q_2 \\ \Delta R_2(\text{C-C}) &= L'_{21}\,\Delta Q_1 + L'_{22}\,\Delta Q_2. \end{aligned} \tag{15.12}$$

Table 25 gives the numerical results. For the $1\pi_g$ ionization only the ν_1 mode (C-C stretch) is appreciably excited. ΔR_1 and ΔR_2 are still similar in magni-

[68] Cederbaum, L.S., Domcke, W., von Niessen, W. (1975): Chem. Phys. *10*, 459

[69] Turner, D.W., Baker, C., Baker, A.D., Brundle, C.R. (1970): Molecular photoelectron spectroscopy, London: Wiley-Interscience

Table 25. Possible changes of bond lengths from the ground state of C_2N_2 to the $\tilde{X}\ ^2\Sigma_g$ and $\tilde{A}\ ^2\Sigma_g^+$ states of $C_2N_2^+$ calculated by (15.12). The less likely solution is given within the parentheses (from ref. [70])

State of the ion		$\Delta R_2 = \Delta R$(C-C) (Å)	$\Delta R_1 = \Delta R$(N-C) (Å)
$\tilde{X}\ ^2\Sigma_g$	1.	-0.042 ± 0.010	0.043 ± 0.008
	(2.	0.042 ± 0.010	-0.043 ± 0.008)
$\tilde{A}\ ^2\Sigma_g^+$	1.	-0.014 ± 0.005	0.018 ± 0.010
	(2.	0.014 ± 0.005	-0.018 ± 0.010)

tude, due to a large magnitude of the off-diagonal matrix element L_{21}. Two possible solutions are generally obtained and the final choice between them must be based on other evidence (bonding character of the orbital). Similar techniques have been used in calculations of several molecules [70-74].

An alternative to the above is provided by the so-called gradient method [20]. It consists basically in calculating theoretically the potential functions and obtaining vibrational intensities from the slopes of these. If harmonic potentials are assumed one can in a first approximation derive the values of the components of $\boldsymbol{D}$ by calculating the gradient for a given normal coordinate at the equilibrium geometry of the ground state as

$$\left(\frac{\partial E}{\partial Q_k}\right)_0 = \omega_k D_k \tag{15.13}$$

expressed in dimensionless units. This expression may be rewritten in terms of so-called first order vibrational coupling constants, which are related in a simple manner to the Franck-Condon factors. Comparison between the traditional and gradient methods have been made [75], as mentioned in the context of core electron ionization (cf. Sect. 12γ).

Both CNDO/S and ab initio HF methods have been found to produce good results for potential energy calculations [29,76-78]. Many-body calculations give often improved results and may sometimes be necessary (for example for the $\tilde{B}\,\Sigma_u^+$ state of N_2^+ of Fig. 93). Applications include the use of the Green's function method by CEDERBAUM et al. [29] and the RSPT method by CHONG et al. [59,60].

Going beyond the Born-Oppenheimer approximation implies the consideration of vibronic wave-functions obtained through coupling of electronic and

[70] Hollas, J.M., Sutherley, I.A. (1971): Mol. Phys. *21*, 183; (1972): *22*, 213; (1972): *24*, 1123
[71] Chau, F.T., McDowell, C.A. (1975): J. Electron Spectrosc. Relat. Phenom. *6*, 357
[72] Heilbronner, E., Muszkat, K.A., Schäublin, J. (1971): Helv. Chim. Acta *54*, 58
[73] Griebel, R., Hohlneicher, G., Dörr, F. (1974): J. Electron Spectrosc. Relat. Phenom. *4*, 185
[74] Smith, W.L., Warsop, P.A. (1968): Trans. Faraday Soc. *64*, 1165
[75] Ågren, H., Müller, J. (1979): UUIP-996 (Uppsala University Institute of Physics Report)
[76] Duke, C.B., Lipari, N.O. (1975): Chem. Phys. Lett. *36*, 51
[77] Duke, C.B., Yip, K.L., Ceasar, G.P., Potts, A.W., Streets, D.G. (1977): J. Chem. Phys. *66*, 256
[78] Cederbaum, L.S., Domcke, W. (1974): Chem. Phys. Lett. *25*, 357

vibrational functions. For the Jahn-Teller interaction the vibrational wave functions have conventionally been approximated by linear expansions of two- or three-dimensional harmonic oscillator functions specified by v and l (cf. Sect. 15α)[79]. An interaction determinant is obtained which yields both the vibronic energies and expansion coefficients by diagonalization. The latter give in turn the Franck-Condon factors. Values of the Jahn-Teller interaction parameter D (not to be confused with the $\boldsymbol{D}$ vector of (15.10)) have been determined for several molecules[15, 79–84].

A detailed interpretation of the vibrational structure has been facilitated for the $\tilde{X}\,^2E$ state of the methyl halides (cf. Fig. 108) by observation of the spin-orbit splitting. For CH_3I^+ and CH_3Br^+ where the spin-orbit coupling parameter $\zeta \gg \hbar\omega$ or $> \hbar\omega$ it suffices to consider the vibrationless and first excited levels of the $^2E_{1/2}$ and $^2E_{3/2}$ states. The energy shifts are given approximately by[15]

$$\Delta E_{\mp} = \frac{2D(\hbar\omega)^2}{\zeta \mp \hbar\omega} \tag{15.14}$$

and the fractional intensities by

$$I_{\mp} \sim \frac{2D(\hbar\omega)^2}{(\zeta \mp \hbar\omega)^2}. \tag{15.15}$$

Thus, the forbidden transitions $^1A_1(v=0) \rightarrow {}^2E_{1/2,\,3/2}(v=1)$ (where v represents a degenerate vibrational mode) can borrow intensity from the allowed $^1A_1(v=0) \rightarrow {}^2E_{1/2,\,3/2}(v=0)$ transitions through mixing of the $v=0$ and $v=1$ levels in the final state. This explains quantitatively the excitation of the degenerate modes and in particular the substantial loss of intensity of the $^2E_{1/2}(v=0)$ transition of CH_3Br^+ which is transferred to the $^2E_{3/2}(v_4=1)$ transition. In CH_3Cl^+ where $\zeta \approx \hbar\omega$ the full interaction matrix must be used. The vibronic effects are here stronger and the spin-orbit splitting substantially quenched. This has been associated with a so-called Ham effect[85, 86]. Extensive treatments of the Renner effect (vibronic interaction in linear molecules) have been made in similar terms (e.g.[88]). Studies include for example CO_2^+[87], H_2O^+[45] and H_2S^+[16].

ε) Rotational excitations. The resolution commonly achieved in photoelectron spectroscopy is not in general sufficient for separation of the rotational structure. Only in the case of H_2 which possesses the largest rotational spac-

---

[79] Chau, F.T., Karlsson, L. (1977): Phys. Scr. *16*, 248, 258
[80] Alderdice, D.S., Dixon, R.N. (1976): J.C.S. Faraday Trans. II *72*, 372
[81] Rabalais, J.W., Karlsson, L., Werme, L.O., Bergmark, T., Siegbahn, K. (1973): J. Chem. Phys. *58*, 3370
[82] Köppel, H., Cederbaum, L.S., Domcke, W., von Niessen, W. (1978): Mol. Phys. *35*, 1283
[83] Rabalais, J.W., Katrib, A. (1974): Mol. Phys. *27*, 923
[84] Richartz, A., Buenker, R.J., Bruna, P.J., Peyerimhoff, S.D. (1977): Mol. Phys. *33*, 1345
[85] Ham, F.S. (1965): Phys. Rev. A*138*, 1727
[86] Sturge, M.D. (1967): Solid State Phys. *20*, 92
[87] Frey, R., Gotchev, B., Kalman, O.F., Peatman, W.B., Pollak, H., Schlag, E.W. (1977): Chem. Phys. *21*, 89
[88] Duxbury, G., Horani, M., Rostas, J. (1972): Proc. R. Soc. Lond. A*331*, 109

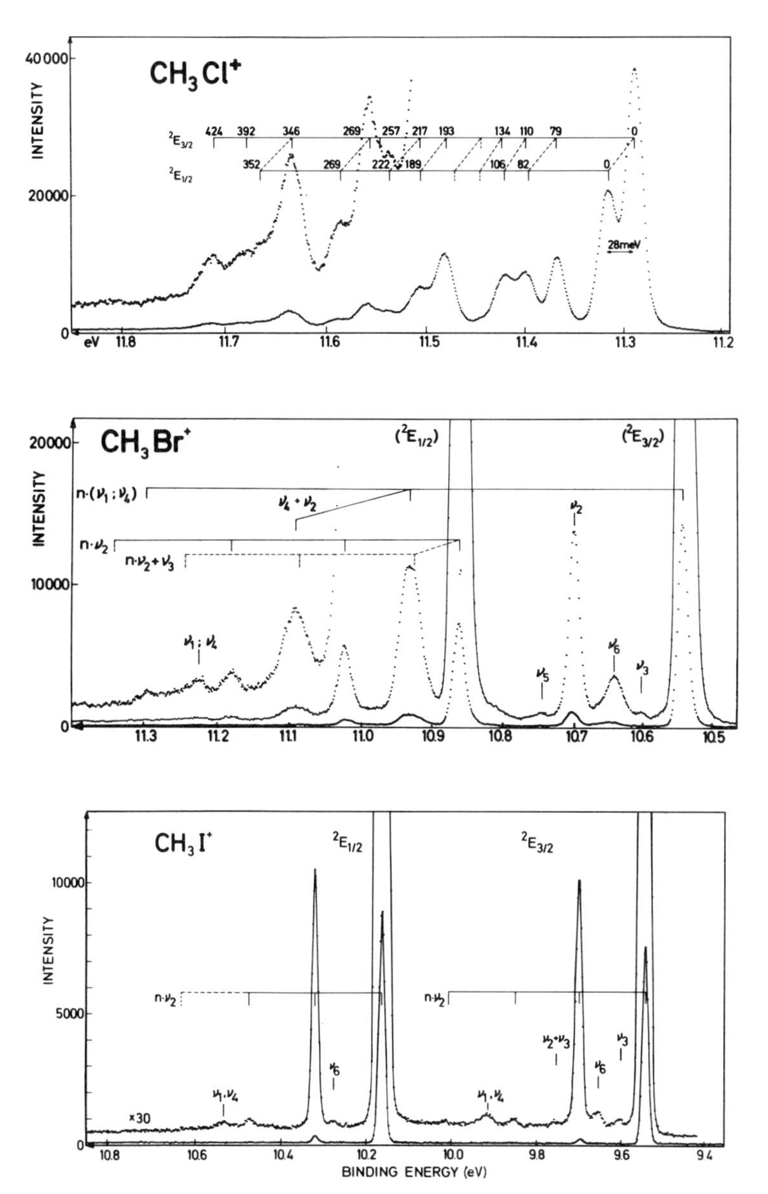

Fig. 108. High resolution He*I*-excited electron spectra of the methyl halides (except CH_3F) showing transitions to the $\tilde{X}\,^2E$ state. The spin-orbit split components are labeled using the total electronic angular momentum quantum number Ω ($=\frac{1}{2}$ or $\frac{3}{2}$). The $\nu_1 \nu_2$ and ν_3 vibrational modes are non-degenerate (a_1) and $\nu_4 \nu_5 \nu_6$ are degenerate (e). From ref. [15] (cf. also Fig. 100)

ings has it been possible to separate the main rotational components by employing Ne*I* excitation[89] (cf. Fig. 112).

The intensity of the various rotational transitions and thus the rotational

[89] Åsbrink, L. (1970): Chem. Phys. Lett. 7, 549

envelope of the vibrational photoelectron peaks is determined mainly by the thermal population of the initial state rotational levels. This contribution to the photoelectron linewidth can be substantial which has also been demonstrated by using a supersonic beam target[19]. Analyses of such broadening in terms of the rotational and spin-orbit coupling constants have been made for HF^+, DF^+[90] and O_2^+[91]. Asymmetries and particularly large broadenings of vibrational peaks associated with the rotational degrees of freedom have been observed in several cases like e.g. H_2O^+[44,45] and H_2S^+[92].

Rotational selection rules for photoelectron transitions have been derived[16] and applied in several of the above studies.

16. Angular distributions of molecular valence photoelectrons

α) General aspects. The angular distributions of photoelectrons from molecules depend on the ionization process, the electronic structure of the target molecule and the photoelectron energy. The basic theory and applications to atoms are discussed by Samson and Starace in this volume.

The angular dependence has been measured both by unpolarized[1-5] and polarized[6,7] radiation from uv line sources (cf. Sect. 7) and by synchrotron radiation[8-10]. Comparisons of various properties of polarized and unpolarized sources have been made[11].

The dependence of the photoelectron intensity on β can be eliminated by a specific choice of angle of analysis (cf. (7.14)), the so-called magic angle. Figure 109 shows an example of a recording of a part of the benzene spectrum at this angle obtained with polarized He*I* radiation ($\alpha_{ku}=54°\,55'$)[6]. For comparison the corresponding spectra obtained at 0° and 90° (between the direction of the electric vector of the radiation and the photoelectrons) are shown in Fig. 110. Such recordings at the magic angle are particularly useful for determination of branching ratios since no correction is needed for the angular asymmetry. Variations of the magic angle due to relativistic and multipole effects have been discussed[12,13].

---

[90] Walker, T.E.H., Dehmer, P.M., Berkowitz, J. (1973): J. Chem. Phys. *59*, 4292

[91] Samson, J.A.R., Gardner, J.L. (1975): Can. J. Phys. *53*, 1948

[92] Karlsson, L., Mattsson, L., Jadrny, R., Bergmark, T., Siegbahn, K. (1976): Phys. Scr. *13*, 229

[1] Carlson, T.A., Jonas, A.E. (1971): J. Chem. Phys. *55*, 4913

[2] Niehaus, A., Ruf, M.W. (1971): Chem. Phys. Lett. *11*, 55

[3] Carlson, T.A., McGuire, G.E. (1972/73): J. Electron Spectrosc. Relat. Phenom. *1*, 209

[4] Sell, J.A., Kupperman, A. (1978): Chem. Phys. *33*, 379

[5] Kreile, J., Schweig, A. (1980): Chem. Phys. Lett. *69*, 71

[6] Mattsson, L., Karlsson, L., Jadrny, R., Siegbahn, K. (1977): Phys. Scr. *16*, 221

[7] Hancock, W.H., Samson, J.A.R. (1976): J. Electron Spectrosc. Relat. Phenom. *9*, 211

[8] White, M.G., Southworth, S.H., Kobrin, P., Shirley, D.A. (1980): J. Electron Spectrosc. Relat. Phenom. *19*, 115

[9] McKoy, D.G., Morton, J.M., Marr, G.V. (1978): J. Phys. B *11*, L547

[10] Carlson, T.A., Krause, M.O., Grimm, F.A., Allen, Jr., J.D., Mehaffy, D., Keller, P.R., Taylor, J.W. (1981): Phys. Rev. A *23*, 3316; Grimm, F.A., Allen, Jr., J.D., Carlson, T.A., Krause, M.O., Mehaffy, D., Keller, P.R., Taylor, J.W. (1981): J. Chem. Phys. *75*, 92

[11] Mattsson, L. (1978): UUIP-983 (Uppsala University Institute of Physics Report)

[12] Cooper, J.W., Manson, S.T. (1969): Phys. Rev. *177*, 159

[13] Ron, A., Pratt, R.H., Tseng, H.K. (1977): Chem. Phys. Lett. *47*, 377

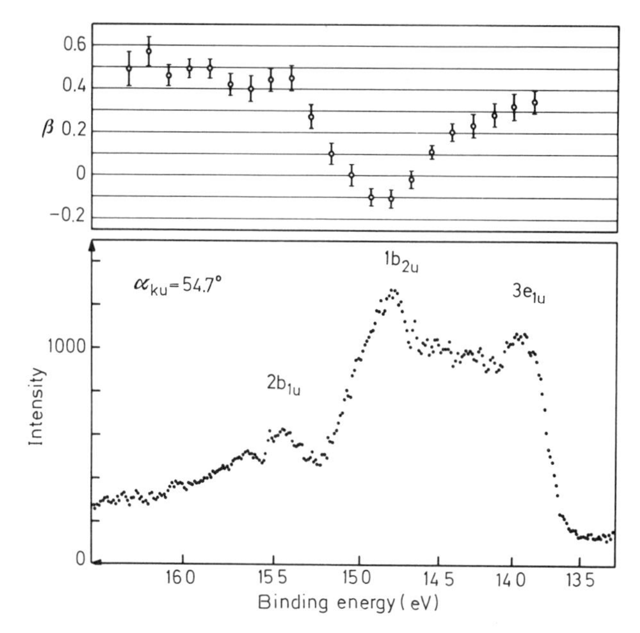

Fig. 109. Part of the valence electron spectrum of benzene (cf. Fig. 91) showing ionization from the $3e_{1u}$, $1b_{2u}$ and $2b_{1u}$ orbitals by means of linearly polarized HeI radiation. This spectrum was recorded at the magic angle 54.7° between the electric field vector and the emitted electron where the dependence of the differential cross section on β vanishes. The β-values are given above the corresponding energies. From ref. [6]

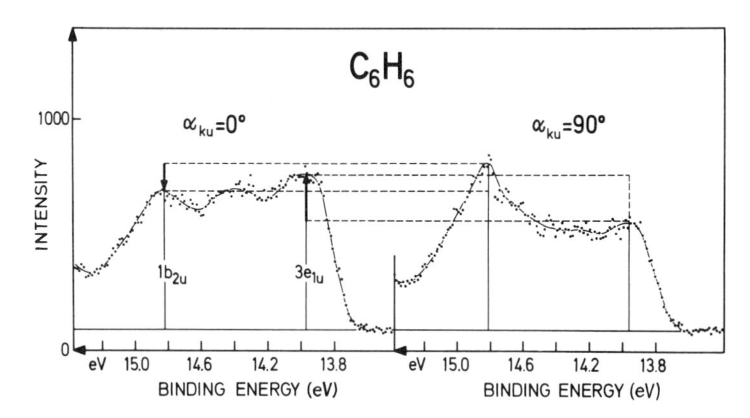

Fig. 110. Ionization from the $3e_{1u}$ and $1b_{2u}$ orbitals of benzene at the angles 0° and 90° between the electric field vector of the ionizing linearly polarized HeI radiation and the emitted electrons. From ref. [6]

Theoretically, the β-value is for atoms expressed in terms of the orbital angular momentum quantum number l. Since this is not in general a good quantum number for molecules the resulting expression in this case takes the more complex form (weighted average)

$$\beta = \frac{\sum \beta_l \sigma_l}{\sum \sigma_l}, \tag{16.1}$$

where σ_l represents the photoionization cross section for a given channel of the molecule. An explicit expression for β in terms of l has been given for diatomic molecules[14].

In connection with investigations of diatomic molecules it has been shown by FANO and DILL[15] that the photoionization can also be described by the angular momentum transfer theory (AMTT). In particular it is possible to select transitions which are parity favoured (F) or unfavoured (U). The photoionization process can be considered in terms of these classes (F, F/U, U) which should in general possess rather different β-values[16-18]. Deviations from the predictions of the AMTT have been observed in some cases[8,9].

β) Photoelectron energy dependence of the asymmetry parameter. Studies of the energy dependence of β have been performed by using both discrete sources at various energies (He*I*, He*II*, Ne*I*, Ar*I*) and the continuous synchrotron radiation. The latter have in general been more detailed. Such studies include for example N_2 and CO[19-21], O_2[20], CH_4[22], H_2[23a,b], NO[23c], CO_2[10], COS and CS_2[23d].

The dependence of β on the photoelectron energy can be substantial. Figure 111 shows one example from the $1t_2$ ionization of CH_4[22]. The corresponding values of Ne[24] are seen to be very similar, particularly at high energies, which suggests that the tetrahedral molecular field essentially approximates the isoelectronic spherically symmetric atomic field. A similar analogy has been drawn for the spin-orbit interaction (cf. Sect. 14ε). A behaviour which parallels that of CH_4 has been found for the ethylene π-orbital ionization which exhibits a β-value similar to that of the carbon atom[25].

Multiple scattering calculations have so far been the most successful in

---

[14] Buckingham, A.D., Orr, B.J., Sichel, J.M. (1970): Phil. Trans. R. Soc. A*268*, 147
[15] Fano, U., Dill, D. (1972): Phys. Rev. A*6*, 185
[16] Dill, D., Fano, U. (1972): Phys. Rev. Lett. *29*, 1203
[17] Chang, E.S., Fano, U. (1972): Phys. Rev. A*6*, 173
[18] Chang, E.S. (1978): J. Phys. B*11*, 293
[19] Marr, G.V., Morton, J.M., Holmes, R.M., McKoy, D.G. (1979): J. Phys. B*12*, 43
[20] Holmes, R.M., Marr, G.V. (1980): J. Phys. B *13*, 945
[21] Cole, B.E., Ederer, D.L., Stockbauer, R., Codling, K., Parr, A.C., West, J.B., Poliakoff, E.D., Dehmer, J.L. (1980): J. Chem. Phys. *72*, 6308
[22] Marr, G.V., Holmes, R.M. (1980): J. Phys. B *13*, 939
[23a] Marr, G.V., Holmes, R.M., Codling, K. (1980): J. Phys. B *13*, 283
[23b] Southworth, S., Brewer, W.D., Truesdale, C.M., Kobrin, P.H., Lindle, D.W., Shirley, D.A. (1982): J. Electron Spectrosc. Relat. Phenom. *26*, 43
[23c] Southworth, S., Truesdale, C.M., Kobrin, P.H., Lindle, D.W., Brewer, W.D., Shirley, D.A. (1982): J. Chem. Phys. *76*, 143
[23d] Carlson, T.A., Krause, M.O., Grimm, F.A., Allen, Jr., J.D., Mehaffy, D., Keller, P.R., Taylor, J.W. (1981): ibid. *75*, 3288
[24] Codling, K., Houlgate, R.G., West, J.B., Woodruff, P.R. (1976): J. Phys. B*9*, L83
[25] Kibel, M.H., Livett, M.K., Nyberg, G.L. (1979): J. Electron Spectrosc. Relat. Phenom. *15*,

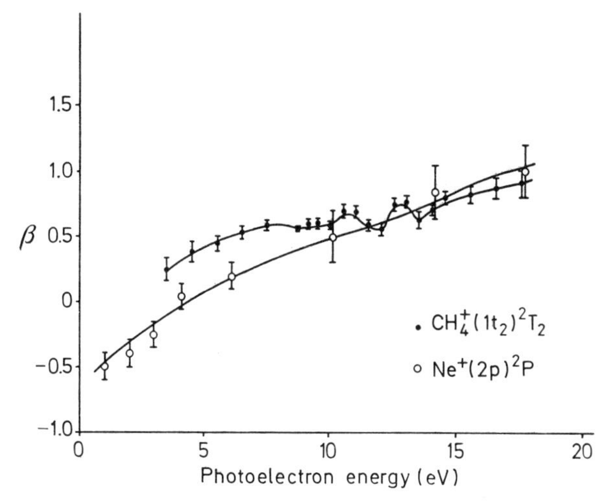

Fig. 111. Energy dependence of the asymmetry parameter β for photoelectron transitions from the neutral ground state to the $\tilde{X}\,^2T_2$ ionic state of methane (full circles). Also included are the corresponding data for Ne $2p$ ionization (unfilled)[24]. In both cases synchrotron radiation was employed for ionization. From ref.[22]

predicting β-values and in particular variations with energy. Good agreement between such calculations and experiment were, for example, obtained for the σ_g and π_u ionization of N_2 and the corresponding orbitals of CO[26]. For the σ_u ionization the agreement is less satisfactory. This can possibly be associated with the strong many-electron effects (which are not considered in the multiple scattering treatment). Further calculations of this kind have particularly emphasized the influence on the β-values of N_2 and CO by the $l=3$ shape resonance at ~ 10 eV (cf. also Sect. 14)[27]. Other multiple scattering calculations have been performed on CO_2, COS and CS_2[26], H_2O and H_2S[28a] and NO[28b].

γ) Vibrational state dependence of the asymmetry parameter. A change of the β-value with the vibrational state for a given electronic transition is frequently observed. One explanation of such changes is the variation of the photoelectron energy in experiments with fixed excitation energy. However, the observed β-variations are often much larger than one could normally expect from only the variation of the photoelectron energy.

A substantial dependence of β on the vibrational wave function is not likely within the Born-Oppenheimer approximation[29–31] although a depen-

---

[26] Grimm, F.A., Carlson, T.A., Dress, W.B., Agron, P., Thomson, J.O., Davenport, J.W. (1980): J. Chem. Phys. *72*, 3041

[27] Wallace, S., Dill, D., Dehmer, J.L. (1979): J. Phys. B *12*, L417

[28a] Roche, M., Salahub, D.R., Messmer, R.P. (1980): J. Electron Spectrosc. Relat. Phenom. *19*, 273

[28b] Wallace, S., Dill, D., Dehmer, J.L. (1981): J. Chem. Phys. *75*, 1971

[29] Ritchie, B. (1974): J. Chem. Phys. *60*, 898

[30] Kalman, O.F. (1976): J. Electron Spectrosc. Relat. Phenom. *8*, 335

[31] Kalman, O.F. (1977): Mol. Phys. *34*, 397

dence of the electronic transition moment on the internuclear distances can lead to a non-negligible influence[32]. Also, it seems that type b) vibronic interaction (cf. Sect. 15α) does not in general give rise to any particular vibrational state dependence of β[33–36]. The same conclusion seems valid for spin-orbit split components of a given state[35, 36].

Large variations of the β-value have been discussed in connection with some different situations

1) The ionization occurs in the vicinity of a shape resonance
2) Autoionizing states contribute to the intensity
3) Type a) vibronic interaction.

The common feature of these is that the ionization involves additional electronic states, which may alter the β-value. The two former depend on the population of a quasibound state in the ionization continuum and are thus very sensitive to the photon energy. Detailed studies have been performed on small molecules both of shape resonances (e.g.[37, 38]) and autoionization effects (e.g.[39–45]).

Type a) vibronic interaction implies the mixing of different electronic states via vibrational excitations of suitable symmetry. If the electronic states possess different β-values this interaction can lead to an observable vibrational state dependence. Such an effect has been discussed for CO_2 and CS_2[46], where considerable variations of β occur.

Vibrational state dependences have been observed also for many larger molecules, like benzene[6] and halogen substituted ring type molecules[35]. It is likely that also in these cases the origin is to be sought among the mechanisms listed above.

δ) Rotational state dependence of the asymmetry parameter. The theory for angular distributions from diatomic molecules of BUCKINGHAM et al.[14] includes rotational transitions explicitly. By assuming that one value of l in (16.1) is

---

[32] Itikawa, Y. (1979): Chem. Phys. Lett. *62*, 261

[33] Carlson, T.A., McGuire, G.E., Jonas, A.E., Cheng, K.L., Anderson, C.P., Lu, C.C., Pullen, B.P. (1972): in: Electron Spectroscopy, Shirley, D.A. (ed.), Amsterdam: North-Holland, p. 207

[34] Leng, F.J., Nyberg, G.L. (1977): J. Electron Spectrosc. Relat. Phenom. *11*, 293

[35] Sell, J.A., Kupperman, A. (1978): Chem. Phys. *33*, 367

[36] Carlson, T.A., White, R.M. (1972): Discuss. Faraday Soc. *54*, 285

[37] Dehmer, J.L., Dill, D., Wallace, S. (1980): Phys. Rev. Lett. *43*, 1005

[38] Unwin, R., Khan, I., Richardson, N.V., Bradshaw, A.M., Cederbaum, L.S., Domcke, W. (1981): Chem. Phys. Lett. *77*, 242

[39] Grimm, F.A. (1980): J. Electron Spectrosc. Relat. Phenom. *20*, 245

[40] Mintz, D.M., Kupperman, A. (1978): J. Chem. Phys. *69*, 3953

[41] Sell, J.A., Kupperman, A., Mintz, D.M. (1979): J. Electron Spectrosc. Relat. Phenom. *16*, 127

[42] Samson, J.A.R., Gardner, J.L. (1976): J. Electron Spectrosc. Relat. Phenom. *8*, 35; (1977): *12*, 119

[43] Plummer, E.W., Gustafsson, T., Gudat, W., Eastman, D.E. (1977): Phys. Rev. A*15*, 2339

[44] Sell, J.A., Kupperman, A., Mintz, D.M. (1979): J. Electron Spectrosc. Relat. Phenom. *16*, 127

[45] Carlson, T.A., Krause, M.O., Mehaffy, D., Taylor, J.W., Grimm, F.A., Allen, J.D., Jr. (1980): J. Chem. Phys. *73*, 6056

[46] Domcke, W. (1979): Phys. Scr. *19*, 11

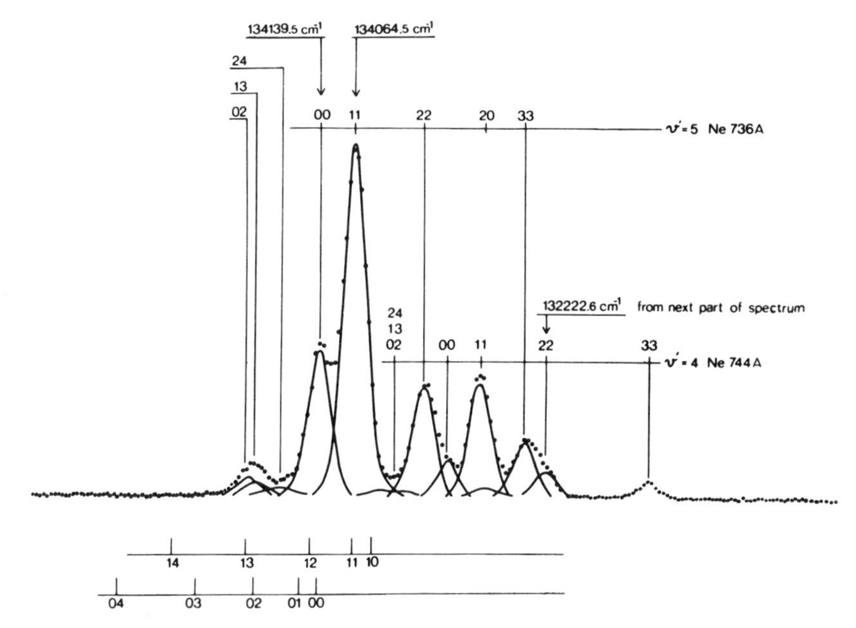

Fig. 112. Part of the Ne*I*-excited spectrum of H_2 showing transitions to the $v'=4$ and $v'=5$ vibrational states of H_2^+. A rotational analysis is given in terms of the Q-branch ($\Delta N=0$), S-branch ($\Delta N=+2$) and O-branch ($\Delta N=-2$). From ref. [47]

dominating and that other values are negligible this theory leads to the atomic expression for β with a correction for molecular rotation. Coupling of angular momenta were considered in terms of Hund's coupling cases a) and b). In addition, expressions were derived for unresolved rotational states, which corresponds to the situation normally encountered in photoelectron spectra.

The only detailed study so far in electron spectroscopy of a rotational state dependence of β concerns H_2. An analysis of the Ne*I* excited spectrum (cf. Fig. 112) has been made in terms of the rotational excitations with $\Delta N=0, \pm 2$, the former components being more intense by an order of magnitude, as predicted by selection rules[48]. Widely different values of β were found for these components ($\Delta N=0$ and $\Delta N=\pm 2$) for transitions to $v'=1^2$ (cf. Table 26). β_Q and β_S correspond to transitions with $\Delta N=0$ (Q-branch) and $\Delta N=+2$ (S-branch), respectively. β_{Q0} is obtained for the transitions $N''=3 \rightarrow N'=3$ and $N''=2 \rightarrow N'=2$ and $\bar{\beta}$ is the average asymmetry parameter for the total number of contributions to the $v''=0 \rightarrow v'=1$ transition. The latter value has also been measured by MARR et al.[23] as a function of energy by using synchrotron radiation.

Theoretical analyses of these results have been given by DILL[49] and CHANG[50] based on the angular momentum transfer theory. Dill considered

---

[47] Åsbrink, L. (1970): Chem. Phys. Lett. *7*, 549

[48] Dixon, R.N., Duxbury, G., Horani, M., Rostas, J. (1971): J. Mol. Phys. *22*, 977

[49] Dill, D. (1972): Phys. Rev. A*6*, 160

[50] Chang, E.S. (1978): J. Phys. B*11*, L69

Table 26. Values of the asymmetry parameter β obtained for Q and S rotational branches of the electron spectrum of H_2. The energy of the photoelectron is of the order of 1.0–1.5 eV

	Experiment	Theory	
	NIEHAUS and RUF[2]	DILL[48]	CHANG[50]
β_Q	1.95 ± 0.03	1.89	1.95 ± 0.01
β_{Q_0}	1.95 ± 0.03	–	1.93 ± 0.01
β_S	0.85 ± 0.14	0.2	0.45 ± 0.87
β	1.93 ± 0.03	1.71	1.87 ± 0.01

dipole interaction and treated the outgoing electrons as p-waves ($l=1$) which accounts for at least the predominant features of the ionization. The selection rule obtained is strictly $\Delta N = 0, \pm 2$. A good agreement with experiment was obtained. The deviations are most pronounced for β_s which could possibly be due to a neglect of $p-f$ coupling in the theoretical treatment. However, the experimental data do not seem to justify any definite conclusions in this respect due to the experimental uncertainty in the value of β_s[50].

ε) Asymmetry parameters for interpretation of valence spectra. In ordinary photoelectron spectra the intensity is plotted versus binding energy. By plotting also the β-values versus energy in the same spectrum one obtains additional independent information on the electronic transitions since these in general are characterized by different values of β. Studies of the asymmetry parameter can thus be helpful for interpretation of spectra, particularly from larger molecules.

Orbitals which give rise to overlapping bands in spectra can be resolved by the different β-values associated with each electronic transition. Indications are also obtained about the degree of overlap and the Franck-Condon contours. An example from benzene is given in Fig. 110. It is important that variations of β within a given electronic transition are identified for reliable conclusions.

For series of related compounds one can make the reasonable assumption that a certain β-value is associated with a given orbital throughout the series. This gives evidence for correlations between spectra of the different molecules. Such similarities were observed by CARLSON et al.[33,36] and later applications include for example benzene and monosubstituted derivatives[35,51–53] phosphabenzene and arsabenzene[54], furan, thiophene and pyrrol[55], pyridine, pentafluoropyridine and 2,6-lutidine[56] and 1,4-cyclohexadiene[57]. Other studies of

---

[51] Sell, J.A., Mintz, D.M., Kupperman, A. (1979): Chem. Phys. Lett. *58*, 601

[52] Kobayashi, T. (1978): Phys. Lett. A*69*, 105

[53] Kobayashi, T., Nagakura, S. (1975): J. Electron Spectrosc. Relat. Phenom. *7*, 187

[54] Ashe, A.J., Burger, F., El-Sheik, M.Y., Heilbronner, E., Maier, J.P., Muller, J.-F. (1976): Helv. Chim. Acta *59*, 1944

[55] Sell, J.A., Kupperman, A. (1979): Chem. Phys. Lett. *61*, 355

[56] Utsunomiya, C., Kobayashi, T., Nagakura, S. (1978): Bull. Chem. Soc. Jap. *51*, 3482

[57] Kibel, M.H., Livett, M.K., Nyberg, G.L. (1978): J. Electron Spectrosc. Relat. Phenom. *14*, 155

systematic behaviour have concerned changes of β-values upon exchange of atoms within a given group of the periodic table[3,36], differences between different types of π-orbitals in multi-π-electron systems[25] and between π and σ orbitals[57] as well as changes of β-values when composite molecules are formed[52]. A factor of uncertainty in correlations of β-values between different molecules is the usually unknown variation of β with photoelectron energy.

17. Molecular structure and bonding

α) MO methods for characterization of spectra and correlation with chemical reaction parameters. The interpretation of a photoelectron spectrum implies that band energies and shapes, vibrational structures, spin-orbit and Jahn-Teller splittings etc. are related to results from theoretical models at various levels of approximation. The interaction between theory and experiment thus leads to an improved understanding of the molecular electronic structure and the photoionization process. PRICE et al.[1,2] have discussed the interpretation of valence photoelectron spectra based on simple molecular orbital models for hydrides and linear and bent triatomic molecules. Figure 113 shows spectra of the isoelectronic hydrides formed by hypothetically separating out hydrogen atoms from Ne. In HF the electron configuration becomes (core, inner valence) $2\sigma^2 1\pi^4$. The former orbital is H-F bonding and the latter is a F lone-pair type as confirmed experimentally by band shapes, vibrational structure and spin-orbit splittings. In H_2O one of the degenerate lone-pair orbitals of HF is employed in the bonding of the additional H atom and hence becomes angle-

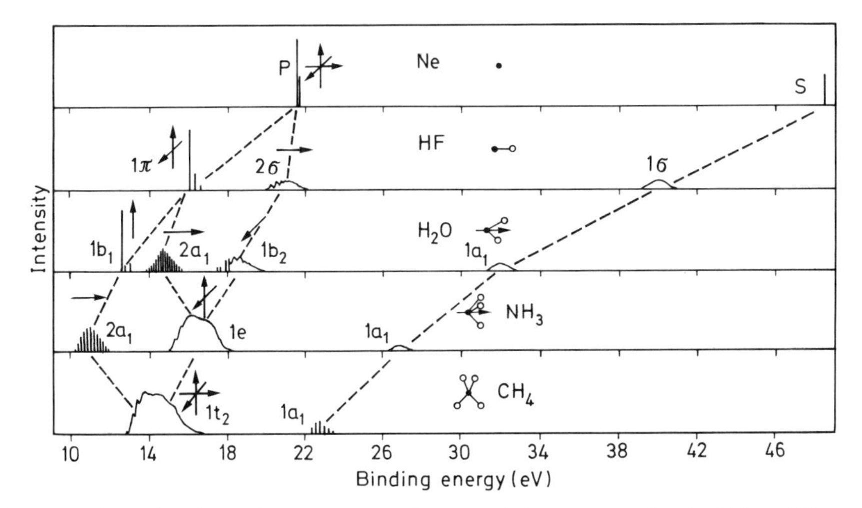

Fig. 113. Schematic valence electron spectra of neon and the corresponding hydrides which can be formed by hypothetically removing protons from the neon core. Only one-electron transitions are considered. From ref. [1]

[1] Price, W.C. (1974): Adv. At. Mol. Phys. *10*, 131
[2] Potts, A.W., Price, W.C. (1972): Proc. R. Soc. A*326*, 181

determining. This property is revealed by a broad rounded band with excitations of the bending vibrational mode. The same is observed for the outermost $2a_1$ lone-pair angle-determining orbital of NH_3 whereas ionization from the $1e$ orbital of NH_3 leads to a very broad band due to strong Jahn-Teller interaction (cf. also Sect. 15). In the CH_4 spectrum only one band is observed in conformity with the (core, inner valence) $1t_2^6$ electron configuration. It has a complex shape due to Jahn-Teller splittings.

Similar arguments can be applied to most small molecules, other examples are provided by studies of N_2 and CO[1,2]. One can also readily understand the results of exchange of hydrogen atoms against halogen atoms or methyl groups[3-9] (for methane cf. e.g. Fig. 100). The halogen atoms (except F) usually contribute with orbitals of lone-pair character which give rise to characteristic sharp bands. This property has been used to facilitate interpretations of spectra also from somewhat larger molecules (e.g. ring type).

By changing the set of hydrogen atoms against fluorine atoms in planar molecules the σ molecular orbitals tend to acquire strongly increased binding energy (several eV) whereas the π-orbitals remain essentially unshifted. This is known as the perfluoro effect[10] which has been widely employed in assignments of spectra (e.g.[11-14]). Limitations concerning the predictive value of this effect have been observed in many cases[15-17].

A fruitful approach for spectra of larger molecules is to consider them as consisting of separate, interacting molecular subunits. The spectra can thus be interpreted in terms of the spectra of such subunits, by simple addition. As an example Fig. 114 shows the experimental energy diagram for ethylene glycol[18]. This molecule and its spectrum can be considered to be formed from two methyl alcohol subunits, each of which is in turn built up from CH_4 and H_2O. In particular, the orbitals of CH_3OH are essentially symmetrically split at the

---

[3] Potts, A.W., Lempka, H.J., Streets, D.G., Price, W.C. (1970): Phil. Trans. R. Soc. Lond. A, *268*, 59

[4] Dixon, R.N., Murrell, J.N., Narayan, B. (1971): Mol. Phys. *20*, 611

[5] Manne, R., Wittel, K., Mohanty, B.S. (1975): Mol. Phys. *29*, 485

[6] Jonas, A.E., Schweitzer, G.K., Grimm, F.A., Carlson, T.A. (1972/73): J. Electron Spectrosc. Relat. Phenom. *1*, 29

[7] Evans, S., Green, J.C., Joachim, P.J., Orchard, A.F., Turner, D.W., Maier, J.P. (1972): J.C.S. Faraday Trans. II *68*, 905

[8] Jadrny, R., Karlsson, L., Mattsson, L., Siegbahn, K. (1977): Phys. Scr. *16*, 235

[8] Chau, F.T., McDowell, C.A. (1975): J. Electron Spectrosc. Relat. Phenom. *6*, 365

[9] Chau, F.T., McDowell, C.A. (1975): J. Electron Spectrosc. Relat. Phenom. *6*, 357

[10] Robin, M.B., Kuebler, N.A., Brundle, C.R. (1972): in: Electron Spectroscopy, Shirley, D.A. (ed.), Amsterdam: North-Holland, p. 351

[11] Wittel, K., Bock, H. (1974): Chem. Ber. *107*, 317

[12] Bieri, G., Heilbronner, E., Jones, T.B., Kloster-Jensen, E., Maier, J.P. (1977): Phys. Scr. *16*, 202

[13] Delwiche, J.P., Praet, M.-Th., Caprace, G., Hubin-Franskin, M.-J., Natalis, P., Collin, J.E. (1977): J. Electron Spectrosc. Relat. Phenom. *12*, 395

[14] Basch, H., Bieri, G., Heilbronner, E., Jones, T.B. (1978): Helv. Chim. Acta *61*, 46

[15] Duke, C.B., Yip, K.L., Ceasar, G.P., Potts, A.W., Streets, D.G. (1977): J. Chem. Phys. *66*, 256

[16] Dougherty, D., McGlynn, S.P. (1977): J. Am. Chem. Soc. *99*, 3234

[17] Åsbrink, L., Bieri, G., Fridh, C., Lindholm, E., Chong, D.P. (1979): Chem. Phys. *43*, 189

[18] Karlsson, L., Åsbrink, L., Fridh, C., Lindholm, E., Svensson, A. (1980): Phys. Scr. *21*, 170

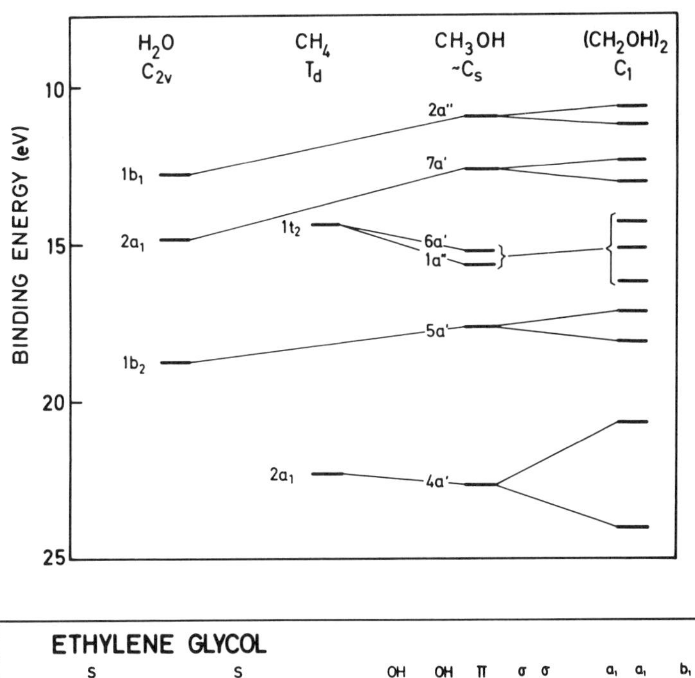

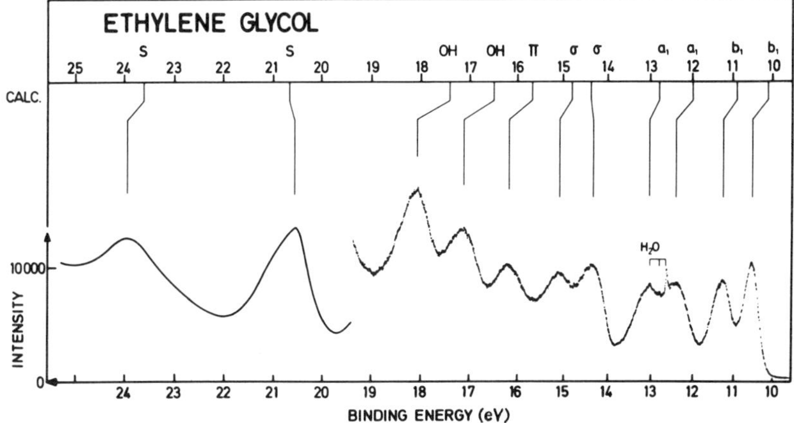

Fig. 114. Correlation diagram for the molecular orbitals of water, methane, methylalcohol and ethylene glycol. Experimental electron binding energies are used. The lower part shows the spectrum of ethylene glycol along with a HAM/3 calculation of the electron binding energies. From ref. [18]

formation of $(CH_2OH)_2$ as would be expected for the interaction of two homologous subunits. A theoretical justification of the approach has been given by RABALAIS[19].

The composite molecule approach relies upon a certain degree of electronic localization. This property can be used also for studies of substituent effects. In such investigations additive shifts (where the electron binding energies change

[19] Rabalais, J.W. (1977): Principles of photoelectron spectroscopy, New York: Wiley-Interscience

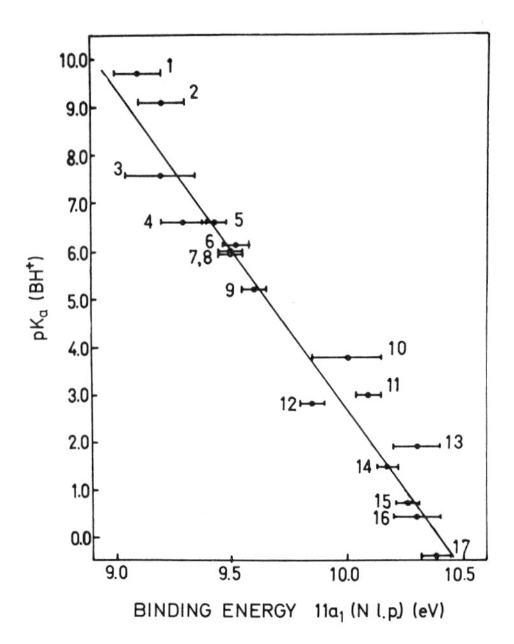

Fig. 115. Correlation between pK_a and the electron binding energy of the nitrogen lone-pair orbital ($11a_1$) for substituted pyridines. The molecules included are 4-Me_2N(1), 4-H_2N(2), 2, 4, 6-Me_3(3), 3, 4-Me_2(4), 4-MeO(5), 3, 5-Me_2(6), 4-Me(7), 2-Me(8), H(9), 4-Cl(10), 3-F(11), 3-Cl(12), 4-CN(13), 3-CN(14), 3, 5-Cl_2(15), 2-Cl(16), 2-F(17). From ref. [27]

linearly with the number of substituents) have been found, for example for π and lone-pair orbitals of carbonyls[20] and pyridines[21, 22]. Apart from the fact that such correlations give information about intramolecular interactions they can be useful for assignment purposes. This particularly applies to larger molecules. For these, assignments can be difficult and it is in general advisable to compare spectra of many closely related molecules. Examples of application have been given by HEILBRONNER et al. (e.g.[23]) including acetylene and polyacetylenes and C_5H_5X (X=N, P, As, Sb and Bi) molecules.

Linear correlations between orbital binding energies and substituent constants have been obtained. Such studies include both atomic[9, 19] and group electronegativities[19, 24], Hammett σ and similar parameters[23, 25, 26], *pKa*

[20] Meeks, J.L., Maria, H.J., Brint, P., McGlynn, S.P. (1975): Chem. Rev. 75, 603

[21] King, G.H., Murrell, J.N., Suffolk, R.J. (1972): J.C.S. Dalton *1*, 564

[22] Karlsson, L., Bergmark, T., Jadrny, R., Siegbahn, K., Gronowitz, S., Mattsson, L. (1974): Chem. Scripta *6*, 214

[23] Heilbronner, E. (1977): in: Molecular Spectroscopy, West, A.R. (ed.), Institute of Petroleum, London: Heyden, p. 422

[24] Baker, A.D., Betteridge, D. (1972): Photoelectron spectroscopy: Chemical and analytical aspects, Oxford: Pergamon Press

[25] Cocksey, B.J., Eland, J.H.D., Danby, C.J. (1971): J. Chem. Soc. B 790

[26] Thompson, M., Hewitt, P.A., Wooliscroft, D.S. (1977): in: Handbook of x-ray and ultraviolet photoelectron spectroscopy, Briggs, D. (ed.), London: Heyden

(which is a measure of acidity cf. also Fig. 115 (e.g.[27]) and even hallucinogenic activity[28]. Linear correlations also exist with the heat of formation[8, 26].

The information obtained about the electronic structure in the above methods is mainly qualitative. It concerns chemically significant concepts such as orbital localization and interaction and substituent effects. The behaviour of molecular orbital binding energies can be quantitatively rationalized in terms of simple models, e.g. semiempirical calculations. Many studies have been based upon the ZDO formalism expressing energy shifts in terms of Coulomb and resonance integrals[29-32]. This has led to an orbital energy sum rule[33-36]. Splittings of non-bonded orbitals can be described in terms of through-space and through-bond interactions[37]. The former refers to direct orbital overlap and the latter to delocalization via other intervening σ-orbitals of suitable symmetry and energy. Hyperconjugation, connected to σ/π interaction by delocalization of alkyl orbitals into the π orbital system of an unsaturated group, has also been studied[38, 39].

The Hückel method[23, 40, 41] has been found to be particularly useful for large conjugated and aromatic molecules, where the π orbitals almost exclusively occupy the outermost region of spectra and commonly give rise to sharp well-defined bands. In such cases the predictive value of this model is high. Applications include also the inner valence region[32, 42, 43].

Standard semiempirical methods like CNDO[44], INDO[45], MINDO[46] and extended Hückel in different forms[47, 48] provide simple and inexpensive means

---

[27] Ramsey, B.G., Walker, F.A. (1974): J. Am. Chem. Soc. *96*, 3314

[28] Domelsmith, L.N., Munchausen, L.L., Houk, K.N. (1977): J. Am. Chem. Soc. *99*, 4311

[29] Heilbronner, E., Hornung, V., Kloster-Jensen, E. (1970): Helv. Chim. Acta *53*, 331

[30] Hainck, H.J., Heilbronner, E., Hornung, V., Kloster-Jensen, E. (1970): Helv. Chim. Acta *53*, 1073

[31] Heilbronner, E., Maier, J.P. (1977): in: Electron spectroscopy, Brundle, C.R., Baker, A.D. (eds.), New York: Academic Press

[32] Bieri, G., Dill, J.D., Heilbronner, E., Schmelzer, A. (1977): Helv. Chim. Acta *60*, 2234

[33] Potts, A.W., Williams, T.A., Price, W.C. (1972): Discuss. Faraday Soc. *54*, 98

[34a] Katsumata, S., Kimura, K. (1975): J. Electron Spectrosc. Relat. Phenom. *6*, 309

[34b] Kimura, K., Katsumata, S. (1978): Monograph Series Res. Inst. of Applied Electricity No. 25

[35] Nagy-Felsobuki, E., Peel, J.B. (1979): J. Electron Spectrosc. Relat. Phenom. *15*, 61

[36] De Leeuw, D.M., De Lange, C.A. (1981): Chem. Phys. *61*, 109

[37] Hoffman, R. (1971): Acc. Chem. Res. *4*, 1

[38] Bock, H., Ramsey, B.G. (1973): Angew. Chem. Int. Ed. Engl. *12*, 734

[39] Schweig, A., Weidner, U., Manuel, G. (1973): J. Organomet. Chem. *54*, 145

[40] Clark, P.A., Broglie, F., Heilbronner, E. (1972): Helv. Chim. Acta, *55*, 1415

[41] Broglie, F., Heilbronner, E. (1972): Theor. Chim. Acta, *26*, 289

[42] Streets, D.G., Potts, A.W. (1974): J.C.S. Faraday Trans. II *70*, 1505

[43] Heilbronner, E. (1977): Helv. Chim. Acta *60*, 2248

[44] Dewar, M.S. (1969): The molecular orbital theory of organic chemistry, New York: McGraw-Hill

[45] Pople, J.A., Beveridge, D.L. (1970): Approximate molecular orbital theory, New York: McGraw-Hill

[46] Baird, N.C., Dewar, M.S. (1969): J. Chem. Phys. *50*, 1262

[47] Hoffman, R. (1963): J. Chem. Phys. *39*, 1397

[48] Larsen, J.S. (1973): J. Electron Spectrosc. Relat. Phenom. *2*, 33

to acquire information about the complete set of orbitals even in fairly large molecules. They have therefore been much used although the deviations from experimental binding energies may become appreciable. Much better results are obtained by spectroscopically parameterized versions like CNDO/S[49,50], INDO/S[51], LNDO/S[52] and SPINDO[53], where electron correlation has been considered in some cases (e.g.[54-55]). This is also done in the more recent HAM/3 method by LINDHOLM et al.[56,57] which has proved its usefulness for a number of molecular properties including electron binding energies and electron affinities. Non-empirical methods (in addition to those discussed in Sect. 14) have been used to some extent, like the multiple scattering MS-$X\alpha$[58,59] and pseudopotential approaches[60-64].

β) Biological molecules. Photoelectron spectra have been recorded and interpreted for a fairly large number of biologically active molecules. Figure 116 shows spectra of some purines[65]. Rather well separated bands are observed in the outer energy region containing π and lone-pair orbitals, as observed also in spectra of other similar molecules. Interpretations suggested by CNDO/S and HAM/3 calculations appear to be the most reliable[65-67].

Experimental difficulties with these compounds are due to dissociation in the gas phase which occurs particularly at the elevated temperatures often necessary to achieve sufficient vapour pressure (e.g.[68]). Various aspects of the molecular stability have been discussed with regard to photoelectron spectra[69]. Studies in the solid phase have furthermore been performed[70].

Biochemical processes are often discussed in terms of charge transfer in acceptor-donor complexes. Electron accepting and donating abilities are given by the lowest electron binding energy and the electron affinity, respectively.

---

[49] Del Bene, J., Jaffé, H.H. (1968): J. Chem. Phys. *48*, 1807; *49*, 1221
[50] Lipari, N.O., Duke, C.B. (1975): J. Chem. Phys. *63*, 1748
[51] Krogh-Jespersen, K., Ratner, M. (1976): ibid. *65*, 1305
[52] Lauer, G., Schulte, K.-W., Schweig, A. (1978): J. Am. Chem. Soc. *100*, 4925
[53] Lindholm, E., Fridh, C., Åsbrink, L. (1972): Far. Disc. *54*, 127
[54] Hase, H.L., Lauer, G., Schulte, K.-W., Schweig, A. (1978): Theor. Chim. Acta *48*, 47
[55] Schultz, R., Schweig, A. (1979): Tetrahedron Lett. 59
[56] Åsbrink, L., Fridh, C., Lindholm, E. (1977): Chem. Phys. Lett. *52*, 63, 69, 72
[57] Åsbrink, L., Fridh, C., Lindholm, E., de Bruijn, S., Chong, D.P. (1980): Phys. Scr. *22*, 475
[58] Connolly, J.W.D., Johnson, K.H. (1971): Chem. Phys. Lett. *10*, 616
[59] Connolly, J.W.D., Siegbahn, H., Gelius, U., Nordling, C. (1973): J. Chem. Phys. *58*, 4265
[60] Murrell, J.N., Scollary, C.E. (1976): J.C.S. Dalton 818 and (1977): 1034
[61] Nagy-Felsobuki, E., Peel, J.B. (1979): J. Electron Spectrosc. Relat. Phenom. *15*, 71
[62] Switalski, J.D., Schwartz, M.E. (1975): J. Chem. Phys. *62*, 1521
[63] Osman, R., Ewig, C.S., van Wazer, J.R. (1976): Chem. Phys. Lett. *39*, 27
[64] Kahn, L.R., Baybutt, P., Truhlar, D.G. (1976): J. Chem. Phys. *65*, 3826
[65] Lin, J., Yu, C., Peng, S., Akiyama, I., Li, K., Li Kao Lee, LeBreton, P.R. (1980): J. Am. Chem. Soc. *102*, 4627
[66] Åsbrink, L., Fridh, C., Lindholm, E. (1977): Tetrahedron Lett. 4627
[67] O'Donell, T.J., LeBreton, P.R., Shipman, L.L. (1978): J. Phys. Chem. *82*, 323
[68] Dougherty, D., Younathan, E.S., Voll, R., Abdulnur, S., McGlynn, S.P. (1978): J. Electron Spectrosc. Relat. Phenom. *13*, 379
[69] McGlynn, S.P., Dougherty, D., Mathers, T., Abdulnur, S. (1977): in: Excited states in organic chemistry and biochemistry, Pullman, B., Goldblum, N. (eds.), Dordrecht: Reidel, p. 247
[70] Sato, N., Kimura, K., Inokuchi, H., Yagi, T. (1980): Chem. Phys. Lett. *73*, 35

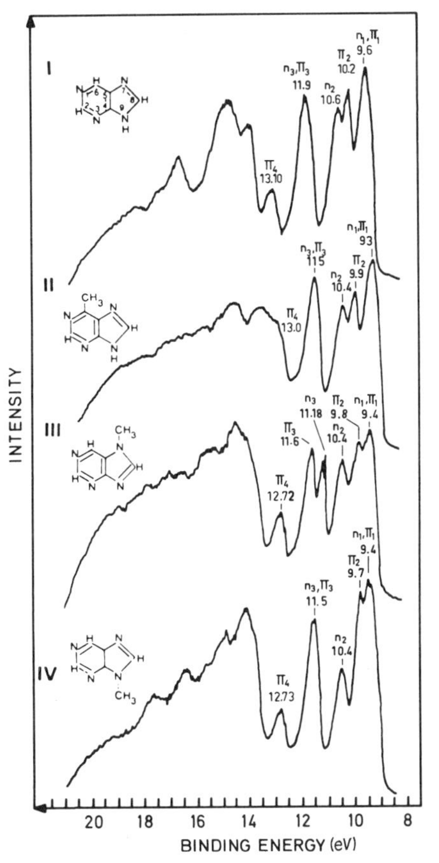

Fig. 116. He*I*-excited electron spectra of purine, 6-methylpurine, 7-methylpurine and 9-methylpurine. The assignments include the seven highest occupied molecular orbitals. From ref. [65]

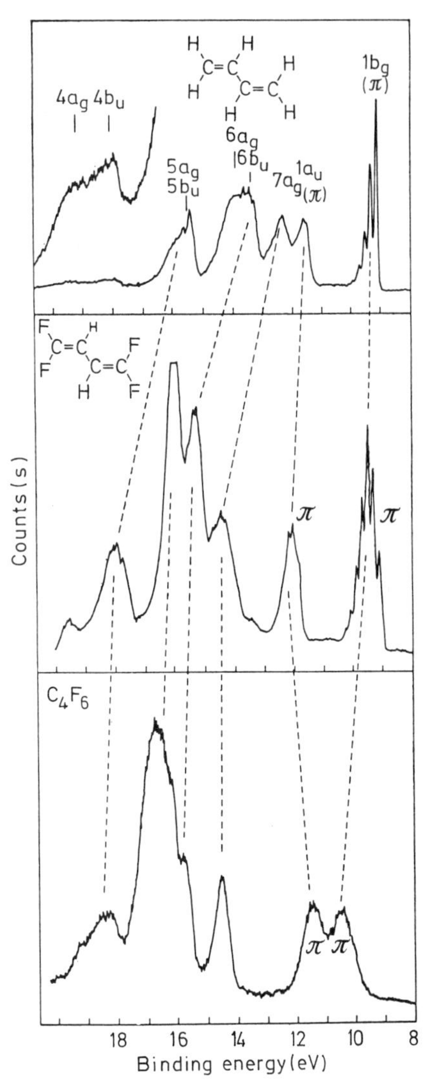

Fig. 117. He*I*-excited spectra of butadiene C_4H_6 and two of its fluorinated derivatives $C_4H_2F_4$ and C_4F_6. The reduced π-orbital splitting in perfluorobutadiene indicates that this molecule has been twisted about the central C–C bond. From ref. [72]. The assignments of butadiene are obtained from ref. [72a]

Diagrams of such electron binding energy data measured by photoelectron spectroscopy have been presented[69, 71]. The results can be rationalized in terms of simple Hückel theory and includes the Pullman's index which gives a measure of the capacity of a molecule to take part in biochemical processes[71].

[71] Pullman, B., Pullman, A. (1963): Quantum Biochemistry, London: Wiley-Interscience

[72] Brundle, C.R., Robin, M.B. (1970): J. Am. Chem. Soc. *92*, 5550

[72a] Bieri, G., Åsbrink, L. (1980): J. Electron Spectrosc. Relat. Phenom. *20*, 149

γ) Molecular conformations, equilibrium systems and chemical reactions. Most conformational studies with photoelectron spectroscopy are concerned with the determination of angles of twist. In the simplest case the splitting of a pair of interacting levels is studied, as indicated in Fig. 117 for butadiene and two of its fluorinated derivatives. The splitting is associated with the through space π-orbital overlap and suggests that the hexafluorobutadiene is twisted about the central C-C bond (to a dihedral angle of 42° [72]) compared to the other molecules which are known to be planar.

Level splittings are usually assumed linearly correlated with the angle of twist according to

$$\Delta E = 2B\cos\theta, \tag{17.1}$$

where B is the resonance integral. Studies have been performed on π and lone-pair orbital splittings [73–75]. More complex relationships than that of (17.1) have been recorded in some cases [76,77], possibly due to $\sigma-\pi$ mixing and deviations from predictions of Koopmans' theorem.

Detailed information about the electronic structure with respect to the molecular conformation is obtained from a more extensive use of theoretical methods (e.g. [78]). In these studies the determination of the conformation is performed by selecting out of a set of theoretical calculations for conceivable conformations the one which shows best general agreement with experimental binding energies. For example hydrazine [79], sulphoxides [80] and ethylene glycol [18] (cf. Fig. 114) have been investigated in this manner.

The presence of different conformers may be revealed in photoelectron spectra by studying the electron binding energies, the intensities giving the relative amounts. Examples are provided by pyridazines [81], ethers and sulphides [82] and diphosphines and diarsines [83,84]. A more definite way of distinguishing between different conformers or determination of an equilibrium constant is to study the temperature dependence of the photoelectron spectrum, as shown in Fig. 118 for thionanisole [85]. Such spectra establish the presence of different conformations and can in the case of thioanisole be interpreted in terms of two stable states with different energy. The quantitative conclusion may not be straightforward, however, particularly when the conform-

---

[73] Maier, J.P., Turner, D.W. (1972): Discuss. Faraday Soc. *54*, 149
[74] Rademacher, P. (1975): Chem. Ber. *108*, 1548
Rademacher, P., Koopman, H. (1975): Chem. Ber. *108*, 1557
[75] Maier, J.P. (1974): Helv. Chim. Acta *57*, 994
[76] Cowling, S.A., Johnstone, R.A.W. (1973): J. Electron Spectrosc. Relat. Phenom. *2*, 161
[77] Baker, A.D., Brisk, M., Gellender, M. (1974): J. Electron Spectrosc. Relat. Phenom. *3*, 227
[78] Schmidt, H., Schweig, A., Vermeer, H. (1977): J. Mol. Struct. *37*, 93
[79] Osafune, K., Katsumata, S., Kimura, K. (1973): Chem. Phys. Lett. *19*, 369
[80] Bock, H., Solouki, B. (1974): Chem. Ber. *107*, 2229; (1972): Angew. Chem. Int. Ed. *11*, 436
[81] Nelsen, S.F., Buschek, J.M., Hintz, P.J. (1973): J. Am. Chem. Soc. *95*, 2014
[82] Dewar, P.S., Ernstbrunner, E., Gilmore, J.R., Godfrey, M., Mellor, J.M. (1974): Tetrahedron *30*, 2455
[83] Elbel, S., Dieck, H.T., Becker, G., Ensslin, W. (1976): Inorg. Chem. *15*, 1235
[84] Schweig, A., Thon, N., Vermeer, H. (1979): J. Electron Spectrosc. Relat. Phenom. *15*, 65 and (1979): J. Am. Chem. Soc. *101*, 80
[85] Schweig, A., Thon, N. (1976): Chem. Phys. Lett. *38*, 482

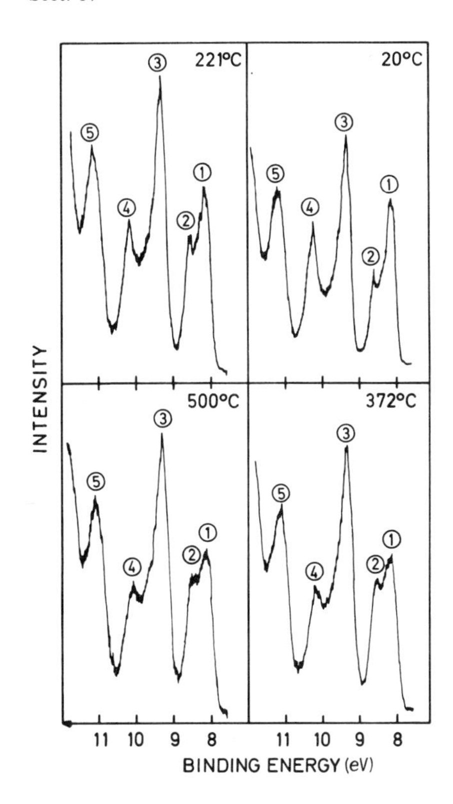

Fig. 118. He*I*-excited electron spectra (outer part) of thioanisole recorded at some different temperatures in the range between 20 °C and 500 °C. From ref. [85]

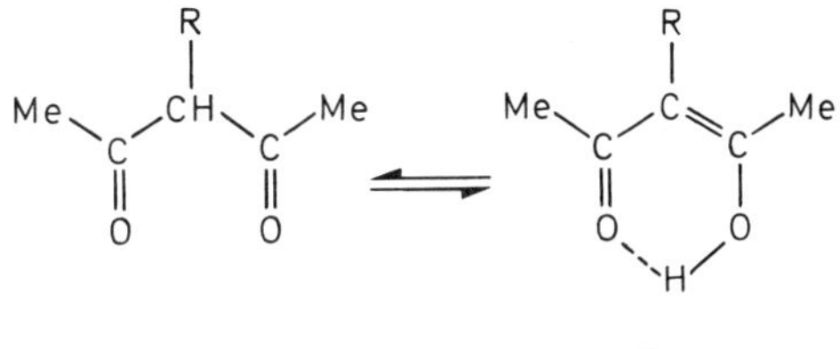

Keto Enol

Fig. 119. The keto and enol forms of acetylacetone. From ref. [87]

ers differ little in relative stability and are separated by rather low activation energy barriers, as emphasized by HONEGGER and HEILBRONNER[86].

Another application of the variable temperature technique concerns the keto-enol tautomerism of acetylacetones[87,88] (cf. Fig. 119) where the equilibrium constant

$$K=\frac{[\text{enol}]}{[\text{keto}]} \tag{17.2}$$

has been determined. Other examples include detection of diazocyclo-hexadienone formed in a gas phase reaction[89] and conformations of halogen substituted ethanes[90].

Photoelectron spectroscopy can also be used in analysis and optimization of gas-phase reactions. This has been shown, particularly by BOCK et al.[91,92] and SCHWEIG et al.[93]. The studies are carried out by connecting the reaction

[86] Honegger, E., Heilbronner, E. (1981): Chem. Phys. Lett. *81*, 615
[87] Schweig, A., Vermeer, H., Weidner, U. (1974): ibid. *26*, 229
[88] Mines, G.W., Thompson, H. (1975): Proc. R. Soc. A*342*, 327
[89] Schultz, R., Schweig, A. (1979): Angew. Chem. Int. Ed. Engl. *18*, 692
[90] Carnovale, F., Gan, T.H., Peel, J.B. (1979): J. Electron Spectrosc. Relat. Phenom. *16*, 87
[91] Solouki, B., Rosmus, P., Bock, H. (1976): J. Am. Chem. Soc. *98*, 6054
[92] Bock, H., Solouki, B., Bert, G., Rosmus, P. (1977): J. Am. Chem. Soc. *99*, 1663
[93] Schultz, R., Schweig, A. (1979): Tetrahedron Lett. 59

chamber on-line to the spectrometer and measuring simultaneously the spectra of all reaction products. The observation of by-products like H_2, N_2, H_2O confirm the reaction process and the spectrum of the desired species is obtained by a spectral stripping procedure. In this manner even rather unstable species have been identified, such as dibromamine by reacting NH_3 and Br_2[94] and dichloramine from aqueous NH_3 and NaOCl and dilute HNO_3[95].

The potential of photoelectron spectroscopy for determination of reaction rates have been demonstrated by BETTERIDGE et al.[96]. Comparisons were made with other spectroscopies employed in studies of degradation of matter, like manometry and gas-liquid chromatography. Electron spectroscopy was found to be superior to manometry in many respects but often less sensitive than GLC. The combination of electron spectroscopy with GLC has been tried by on-line detection of the effluent from the GLC in the electron spectrometer[97].

δ) Short-lived species. Studies of short-lived (transient) species require specific considerations both concerning experimental and theoretical aspects. The former is connected primarily to the production of a high concentration of target molecules in the ionization volume and the latter to the calculation of electron binding energies of open-shell systems.

The relative and absolute concentrations of the transient in a flow system depend mainly on the efficiency of formation and the distance to the ionization volume. This distance should be made as small as possible but must be chosen with due consideration of possible influences from the sample generating system on the spectra. Most systems have employed microwave discharges, chemical reactions and pyrolysis for the production of target species, but also photolysis has been employed[98,99]. Another possibility is the use of dissociative attachment of low energy electrons. This can be a much more efficient process for dissociating molecules than for example photolysis (e.g.[100]). A review of various techniques employed in photoelectron spectroscopy is given in ref.[101]. A more general account of techniques for generation and monitoring of non-stable species in the gaseous phase is given in ref.[102].

An efficient and sensitive system for studies of very short-lived species was introduced by JONATHAN et al.[101] who used an inductively heated oven reach-

---

[94] Nagy-Felsobuki, E., Peel, J.B. (1979): J. Electron Spectrosc. Relat. Phenom. *15*, 61

[95] Colbourne, D., Frost, D.C., McDowell, C.A., Westwood, N.P.C. (1978): J. Chem. Phys. *69*, 1078

[96] Betteridge, D., Joyner, D.J., Greening, F., Shoko, N.R., Cudby, M.E.A., Willis, H.A., Attwood, T.E., Henriksen, L. (1977): Phys. Scr. *16*, 339

[97] Betteridge, D., Hasanuddin, S.K., Rees, D.I. (1976): Anal. Chem. *48*, 1078

[98] Imre, D., Koenig, T. (1980): Chem. Phys. Lett. *73*, 62

[99] Ruf, M.W., Sahr, D., Hotop, H. (1980): in: VIth Int. Conf. on VUV Rad. Phys., Ext. Abstracts Vol. II, Charlottesville

[100] Verhaart, G.J., van Sprang, H.A., Brongersma, H.H. (1980): Chem. Phys. *51*, 389

[101] Dyke, J., Jonathan, N., Morris, A. (1979): J. Electron Spectrosc. Relat. Phenom. *15*, 45 and (1979): in: Electron spectroscopy, vol. 3, Brundle, C.R., Baker, A.D. (eds.), New York: Academic Press, p. 189

[102] Setser, D.W. (ed.) (1979): Reactive intermediates in the gas phase, generation and monitoring, New York: Academic Press

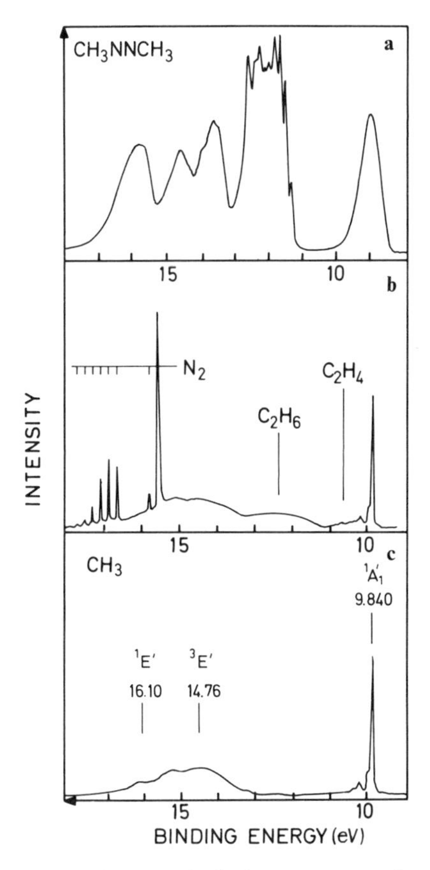

Fig. 120a–c. Schematic drawing of He*I*-excited electron spectra from (**a**) azomethane at room temperature, (**b**) decomposed azomethane at 1,700 K with contributions of e.g. nitrogen, ethane and ethylene, (**c**) the methyl radical after deconvolution of (**b**). From ref. [105]

ing temperatures of up to 3,000 K for generation of the transient species. By employing a pulsed mode of operation this oven could be mounted only a few centimeters from the ionization volume. By this technique reactive radicals such as SiO[103], GeO[104], CH_3[105], *t*-butyl[106a] HO_2[106b] could be studied. Figure 120 shows the spectrum of CH_3 obtained by pyrolysis of azomethane.

Microwave discharges can be very efficient in dissociating molecules. Since the technique is not selective it is best suited for studies of small molecules, where the number of different fragments is by nature small. Examples of

---

[103] Colburn, E.A., Dyke, J.M., Lee, E.P.F., Morris, A., Trickle, I.R. (1978): Mol. Phys. *35*, 873

[104] Colburn, E.A., Dyke, J.M., Fackerell, A., Morris, A., Trickle, I.R. (1978): J.C.S. Faraday Trans. II *74*, 2278

[105] Dyke, J., Jonathan, N., Lee, E., Morris, A. (1976): J.C.S. Faraday Trans. II *72*, 1385

[106a] Dyke, J., Jonathan, N., Lee, E., Morris, A., Winter, M. (1977): Phys. Scr. *16*, 197

[106b] Dyke, J.M., Jonathan, N.B.H., Morris, A., Winter, M.J. (1981): Mol. Phys. *44*, 1059

application include studies of SO[107], NS[108], CH_2 and O_3[109] and N_2H_2[110]. Phase-sensitive detection in combination with frequency-modulated discharge are means to increase the relative concentration of transients[107].

Gas-phase reactions can in certain cases be fast enough to provide a sufficient vapour pressure of a desired radical for studies by photoelectron spectroscopy. Mostly reactions involving hydrogen atoms have been used, for example in studies of ClO[111], BrO[112] and OH[113].

The correspondence between occupied molecular orbitals and main bands, which is the leading feature in spectra of closed-shell molecules, is often lost in spectra of radicals which are generally open-shell systems. Koopmans' theorem does not apply for calculation of binding energies for such molecules but one must employ calculational schemes which lead to the proper initial and final state symmetries, such as OSHF (Open Shell Hartree-Fock). Such procedures have been used in most applications although in simplified forms by using the unrelaxed ground-state wave function to calculate the final state total energies. For assignments of spectra it may be sufficient to use these more qualitative methods which lead to a consideration of ground state Coulomb and exchange integrals. Energy splittings in terms of these integrals have been predicted for several configurations of linear molecules[114].

ε) Ionically bonded molecules. Studies of involatile ionically bonded materials can be performed in the gaseous phase by direct vapourization at high temperatures. Conventional non-inductive resistive heating of the sample has been employed in most cases. This is combined with a molecular beam arrangement where the gas is generated in one end of the system and collected on a cool-trap in the other. Ionization is arranged between these points (e.g.[115-117]). The efficiency of such systems can be sufficient to allow studies with He*II* excitation.

The alkali halides have been the standard molecules for testing experimental facilities and verifying theoretical models[115-119]. Figure 121 shows a diagram of valence electron binding energies of these molecules. The interpretational models include both the conventional molecular orbital and the classical ionic descriptions. The latter relies upon the transfer of the outer *s* electron of the alkali atom (M) to the halogen atom (X) to give an ionically bonded

---

[107] Jonathan, N., Ross, K.J., Smith, D.J. (1971): Chem. Phys. Lett. *9*, 217

[108] Dyke, J.M., Morris, A., Trickle, I.R. (1977): J.C.S. Faraday Trans. II *73*, 147

[109] Dyke, J.M., Golob, L., Jonathan, N., Morris, A., Okuda, M. (1974): J.C.S. Faraday Trans. II *70*, 1828

[110] Frost, D.C., Lee, S.T., McDowell, C.A., Westwood, N.P.C. (1975): Chem. Phys. Lett. *30*, 26

[111] Bulgin, D.K., Dyke, J.M., Jonathan, N., Morris, A. (1976): Mol. Phys. *32*, 1487

[112] Dunlavey, S.J., Dyke, J.M., Morris, A. (1978): Chem. Phys. Lett. *53*, 382

[113] Katsumata, S., Lloyd, D.R. (1977): Chem. Phys. Lett. *45*, 519

[114] Scott, P.R., Raftery, J., Richards, W.G. (1973): J. Phys. B*6*, 881

[115] Potts, A.W., Williams, T.A., Price, W.C. (1974): Proc. R. Soc. Lond. A*341*, 147

[116] Evans, S., Orchard, A.F. (1975): J. Electron Spectrosc. Relat. Phenom. *6*, 207

[117] Berkowitz, J. (1977): in: Electron spectroscopy, Brundle C.R., Baker, A.D. (eds.), New York: Academic Press, p. 355

[118] Goodman, T.D., Allen, J.D., Jr., Cusachs, L.C., Schweitzer, G.K. (1974); J. Electron Spectrosc. Relat. Phenom. *3*, 289

[119] Potts, A.W., Price, W.C. (1977): Phys. Scr. *16*, 191

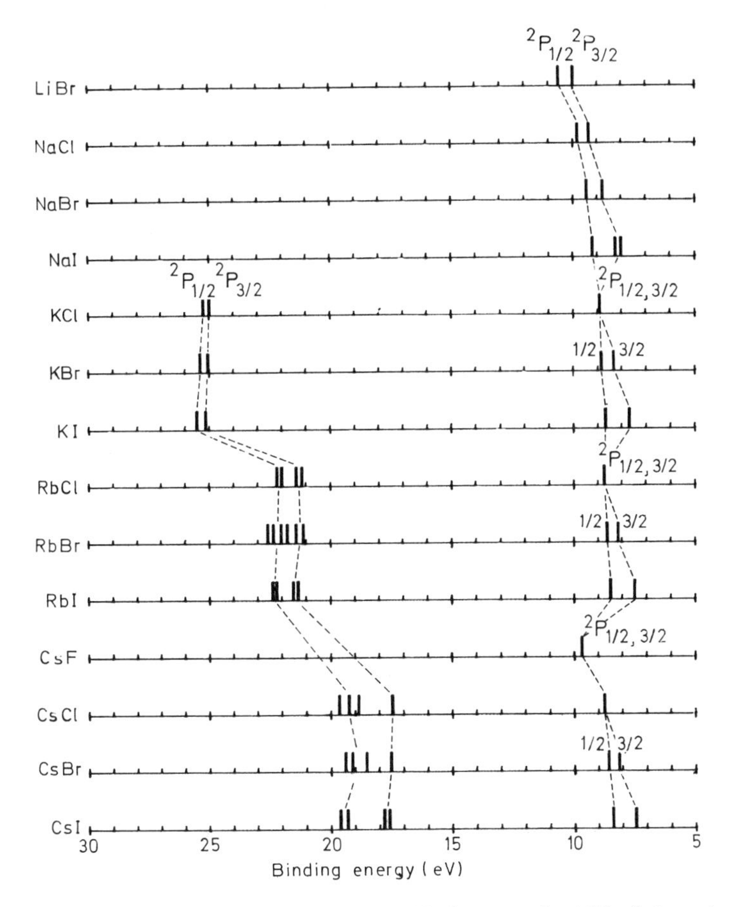

Fig. 121. Correlation diagram showing the electron binding energies (eV) of the valence region of the alkali halide molecules

molecule $M^+X^-(np^6)$. Ionization from the outermost $np^6(X^-)$ orbital leads to the cation $M^+X^0(np^5)$. This state becomes weakly bonded due to polarization of the X^0 atom by the M^+ ion. By inserting experimental values of the adiabatic and vertical binding energies the dissociation energy and also bond distance of the cationic state can be calculated from the ionic bonding model. Table 27 gives results obtained in this way for the lowest state ($^2P_{3/2}$) of the alkali halides. Various other molecules have been treated[120]. Studies of ionic bonding in somewhat larger systems include oxyanion containing compounds[121,122].

[120a] Potts, A.W., Lyus, M.L. (1978): J. Electron Spectrosc. Relat. Phenom. *13*, 327

[120b] Potts, A.W., Law, D., Lee, E.P.F. (1981): J.C.S. Faraday Trans. II *77*, 797

[121] Vick, D.O., Woodley, D.G., Bloor, J.E., Allen, J.D., Jr., Mui, T.C., Schweitzer, G.K. (1978): J. Electron Spectrosc. Relat. Phenom. *13*, 247

[122] Lassiter, T.W., Allen, J.D., Jr., Schweitzer, G.K. (1980): J. Electron Spectrosc. Relat. Phenom. *20*, 47

Table 27. The onset energy of the first photoelectron band (E_B^{on}) and the dissociation energy of the corresponding initial neutral ground state and final ionic state ($^2P_{3/2}$) for some of the alkali halide molecules. Included are also the equilibrium internuclear distances for the neutral and ionic species (from ref. [115])

Molecule	E_B^{on} (eV)	D_0(MX) (eV)	$D_0(M^+X^0)$ (eV)	r_e (Å)	r_e^+ (Å)
NaCl	8.93	4.228	0.33	2.36	2.78
NaBr	8.31	3.760	0.49	2.50	2.86
NaI	7.60	3.153	0.60	2.71	3.05
KCl	8.44	4.393	0.19	2.67	3.29
KBr	7.85	3.942	0.33	2.82	3.31
KI	7.21	3.330	0.37	3.05	3.51
RbCl	8.26	4.367	0.20	2.79	3.45
RbBr	7.75	3.920	0.26	2.94	3.55
RbI	7.12	3.326	0.32	3.18	3.64
CsCl	8.32	4.605	0.06	2.91	3.61
CsBr	7.74	4.185	0.24	3.07	3.58
CsI	7.10	3.573	0.27	3.31	3.94

The splittings of the cationic states are conventionally discussed in terms of the intramolecular field strength and spin-orbit interactions. Intermediate coupling generally applies like for example in the alkali halides[123] and thallium halides (cf. Fig. 122). The final states arising from the $\ldots d^9 s^2$ cationic configuration of the latter ($^2\Sigma_{1/2}$, $^2\Pi_{1/2}$, $^2\Pi_{3/2}$, $^2\Delta_{3/2}$, $^2\Delta_{5/2}$) are all observed in the He*II* excited spectrum. The parameterization of the splittings (cf. ref. [120]) must include consideration of $\Sigma-\Pi$ separation in the absence of spin-orbit interaction, spin-orbit splitting of orbitally degenerate electronic states and interaction between states with the same Ω.

Further aspects have been given by BANCROFT et al.[124–126] who studied the outer, atomic, core levels of molecules, like XeF_2, XeF_4, $(CH_3)_2Cd$, $(C_2H_5)_2Cd$, CH_3Ga, CH_3In and $(C_2H_5)_4Pb$, $(C_6H_5)_4Sn$. The level splittings were treated in terms of ligand-field theory.

ζ) *Negative ions.* The basic process used in photoelectron spectroscopy of negative ions is photodetachment schematically represented by

$$M^- + h\nu \rightarrow M + e^-. \qquad (17.3)$$

If the detachment involves the orbital i, conservation of energy in the photoelectron measurement requires that

$$E_B^-(i) = h\nu - E_{kin}, \qquad (17.4)$$

[123] Berkowitz, J., Dehmer, J.L., Walker, T.E.H. (1973): J. Chem. Phys. *59*, 3645

[124] Bancroft, G.M., Malmquist, P.-Å., Svensson, S., Basilier, E., Gelius, U., Siegbahn, K. (1978): Inorg. Chem. *17*, 1595

[125] Bancroft, G.M., Coatsworth, L.L., Creber, D.K., Tse, J.S. (1977): Phys. Scr. *16*, 217

[126] Bancroft, G.M., Adams, I., Creber, D.K., Eastman, D.E., Gudat, W. (1976): Chem. Phys. Lett. *38*, 83

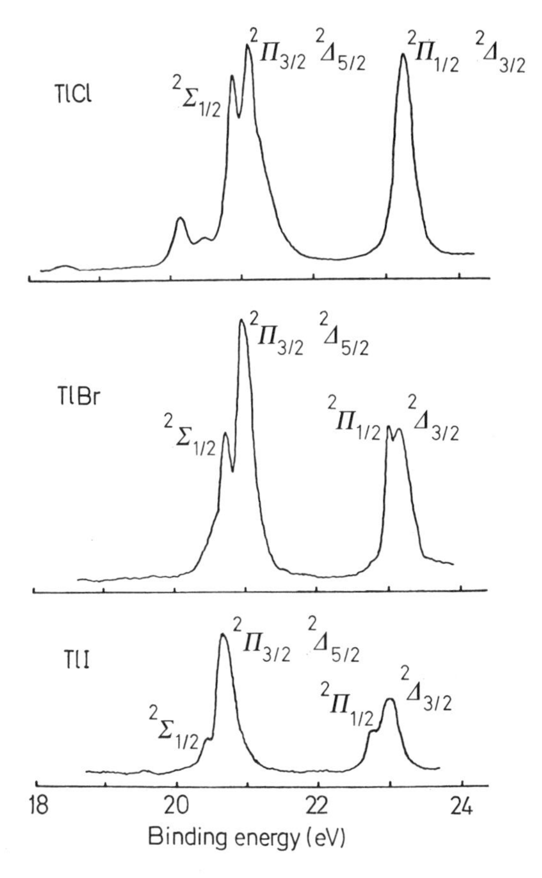

Fig. 122. He*II*-excited electron spectra showing the Tl *d*-orbital ionization in thallium chloride, bromide and iodide. From ref. [119]

where $E_B^-(i)$ is the electron binding energy referred to the anion. An important quantity which may be measured in this way is the electron affinity which is defined as the difference in total energy between the lowest rovibronic levels of the neutral molecule and the anion. This is equal to the lowest adiabatic binding energy.

The main solution to the experimental problem of obtaining a high photoelectron intensity involves two steps. First, the target anions (produced in a discharge) are accelerated and focused into a well collimated beam in the collision volume, according to the technique introduced by BRANSCOMB et al.[127]. Second, a high intensity laser beam is employed for ionization (as originally done by BREHM et al. in studies of He^- [128]). Photoelectron spectra have been recorded in such arrangements using fixed-frequency continuum-

[127] Branscomb, L.M., Fite, W.L. (1954): Phys. Rev. *93*, 651 A
[128] Brehm, B., Gusinow, M.A., Hall, J.L. (1967): Phys. Rev. Lett. *19*, 737

Table 28. Electron affinities of small molecules determined by photodetachment

	E_A (eV)	Ref.		E_A (eV)	Ref.
Diatomic			Triatomic		
CH	1.238 ± 0.008	a	NH_2	0.779 ± 0.037	c
CN	3.82 ± 0.02	b	PH_2	1.271 ± 0.010	e
NH	0.38 ± 0.03	c	SiH_2	1.124 ± 0.020	f
OH	1.829 ± 0.014	c	NO_2	2.36 ± 0.10	g
OD	1.823 ± 0.002	d	O_3	2.1028 ± 0.0025	h
PH	1.028 ± 0.010	e	SO_2	1.097 ± 0.036	c
PO	1.092 ± 0.010	e			
S_2	1.663 ± 0.040	c	Four-atomic		
SiH	1.385 ± 0.005	f	CH_3	0.08 ± 0.03	i
			Five-atomic		
			CH_3O	1.570 ± 0.022	j
			CD_3O	1.552 ± 0.022	j
			CH_3S	1.882 ± 0.024	j

[a] Kasdan, A., Herbst, E., Lineberger, W.C. (1975): Chem. Phys. Lett. *31*, 78
[b] Berkowitz, J., Chupka, W.A., Walker, T.A. (1969): J. Chem. Phys. *50*, 1497
[c] Celotta, R.J., Bennett, R.A., Hall, J.L. (1974): J. Chem. Phys. *60*, 1740
[d] Hotop, H., Patterson, T.A., Lineberger, W.C. (1974): J. Chem. Phys. *60*, 1806
[e] Zittel, P.F., Lineberger, W.C. (1976): J. Chem. Phys. *65*, 1236
[f] Kasdan, A., Herbst, E., Lineberger, W.C. (1975): J. Chem. Phys. *62*, 541
[g] Herbst, E., Patterson, T.A., Lineberger, W.C. (1974): J. Chem. Phys. *61*, 1300
[h] Novick, S.E., Engelking, P.C., Jones, P.L., Futrell, J.H., Lineberger, W.C. (1979): J. Chem. Phys. *70*, 2652
[i] Ellison, G.B., Engelking, P.C., Lineberger, W.C. (1978): J. Am. Chem. Soc. *100*, 2556
[j] Engelking, P.C., Ellison, G.B., Lineberger, W.C. (1978): J. Chem. Phys. *69*, 1826

wave (cw) lasers[129,130]. Studies of the photodetachment cross section as function of energy have been performed near threshold by the use of tunable lasers[131]. High resolution recordings are facilitated by reduction of Doppler broadening due to both the beam target and the low photoelectron energy as well as the good energy definition of the laser radiation. Most photoelectron spectroscopic measurements have been performed at a resolution of about 50 meV, which is good enough for resolution of vibrational structure. Later progress has enabled improvement by one order of magnitude, whereby the rotational structure has become accessible for studies in favourable cases[130]. Even higher resolution has been obtained in studies of the cross sectional behaviour, particularly in detailed investigations of autodetachment resonances (window resonances). Since the laser radiation is polarized angular distribution studies can conveniently be made (for an example cf. ref.[132]).

A large number of the atoms in the periodic system possess positive electron affinity and can have a few long-lived configurationally excited anion

[129] Siegel, M.W., Celotta, R.J., Hall, J.L., Levine, J., Bennett, R.A. (1972): Phys. Rev. A*6*, 607
[130] Breyer, F., Frey, P., Hotop, H. (1981): Z. Phys. A*300*, 7
[131] Lineberger, W.C., Woodward, B.W. (1970): Phys. Rev. Lett. *25*, 424
[132] Engelking, P.C., Lineberger, W.C. (1979): Phys. Rev. A*19*, 149

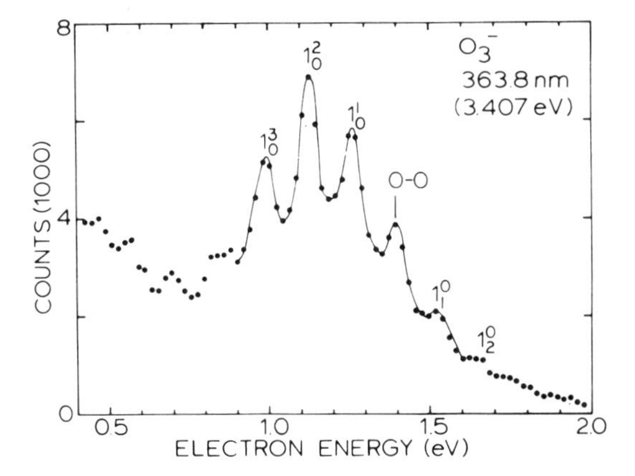

Fig. 123. Photoelectron spectrum of O_3^- obtained by means of a fixed-frequency ArII laser operated at 3.407 eV (3,638 Å). Excitations are observed of the ν_1 symmetric stretching vibrational mode. The vibrational lines on the high binding energy side of the O–O transition correspond to excitations in the final state of O_3 and those on the low-binding energy side are due to excitations in the initial state of O_3^- (hot bands). From ref.[136]

states. Many of these energies and particularly the electron affinities have been determined with high accuracy (cf. e.g. SAMSON in this volume and refs. [133], [134]). Metastable configurationally excited states exist among the group *II* and *VIIIA* atoms with negative electron affinity. Lifetime measurements have been performed for such states, the best known example is He^- [135].

Many molecules, particularly radicals, have positive electron affinities. Table 28 summarizes such data from photodetachment studies. Figure 123 shows a spectrum obtained from O_3^- [136]. The vibrational progression has been associated with excitations of the ν_1 (symmetric stretching) mode in both the initial state of the anion and in the neutral final state. Excitations of the ν_2 (bending) mode were present in the spectrum obtained by tunable laser excitation.

Calculations of electron affinities have been made with several approaches. Good results have been obtained with many-electron methods, for example on C_2 and P_2[137], SO_2 and O_3[138] by application of the Green's function method by CEDERBAUM et al. Also the HAM/3 method, which incorporates many-electron effects through parametrization, has been successful[139].

---

[133] Hotop, H., Lineberger, W.C. (1975): J. Phys. Chem. Ref. Data *4*, 539

[134] Feigerle, C.S., Corderman, R.R., Bobashev, S.V., Lineberger, W.C. (1981): J. Chem. Phys. *74*, 1580

[135] Blau, L.M., Novick, R., Weinflash, D. (1970): Phys. Rev. Lett. *24*, 1268

[136] Novick, S.E., Engelking, P.C., Jones, P.L., Futrell, J.H., Lineberger, W.C. (1979): J. Chem. Phys. *70*, 2652

[137] Cederbaum, L.S., Domcke, W., von Niessen, W. (1977): J. Phys. B*10*, 2963

[138] von Niessen, W., Cederbaum, L.S., Domcke, W. (1977): in: Vth Int. Conf. on VUV Radiat. Phys., Ext. Abstr., vol. I, Montpellier, p. 156

[139] Åsbrink, L., Fridh, C., Lindholm, E. (1977): Chem. Phys. Lett. *52*, 72

18. Valence electron studies of single crystals and adsorbates

α) Selection rules in photoionization from solids. The basic treatment of photoionization in solids may be taken the same as that for atoms and molecules within the framework of the dipole approximation. The translational symmetry of the solid introduces, however, additional selection rules not present for molecules, which make the subsequent development of theory different. The electronic one-electron states in an extended periodic lattice are Bloch functions and are labelled by the value of the crystal wave vector, $\boldsymbol{k}$. In the photoionization process the bound electron with wave vector $\boldsymbol{k}^{\mathrm{in}}$ is transferred into a final state Bloch wave with wave vector $\boldsymbol{k}^{\mathrm{fi}}$. The translational symmetry requires that for the transition matrix element to be nonzero, the following condition must be fulfilled (e.g.[1]):

$$\boldsymbol{k}^{\mathrm{fi}} = \boldsymbol{k}^{\mathrm{in}} + \boldsymbol{G} + \boldsymbol{k}^{h\nu}, \tag{18.1}$$

where $\boldsymbol{G}$ is a reciprocal lattice vector and $\boldsymbol{k}^{h\nu}$ the momentum of the incident photon. Here, the effects of additional excitation of the lattice, such as creation or destruction on phonons, have not been considered. Transitions that fulfil (18.1) are termed direct transitions. For photon energies below 100 eV, $\boldsymbol{k}^{h\nu} \ll \boldsymbol{G}$ and (18.1) reads:

$$\boldsymbol{k}^{\mathrm{fi}} = \boldsymbol{k}^{\mathrm{in}} + \boldsymbol{G} \tag{18.2}$$

which means that the transitions must be vertical in the reduced zone scheme. Furthermore, it must be required that:

$$E^{\mathrm{fi}} - E^{\mathrm{in}} = h\nu \tag{18.3}$$

as for any photoelectric transition, where E^{fi} and E^{in} are now one-electron energies referred to the vacuum level, i.e. E^{fi} is the electron kinetic energy just outside the solid and E^{in} is the negative of the electron binding energy. Equations (18.2) and (18.3) imply that the allowed transitions define a surface in the reduced Brillouin zone given by the condition (cf. Fig. 124):

$$E^{\mathrm{fi}}(\boldsymbol{k}) - E^{\mathrm{in}}(\boldsymbol{k}) = h\nu, \tag{18.4}$$

where $\boldsymbol{k}$ is the reduced zone wave vector. The above relations consider the photoabsorption process associated with photoelectron ejection in an infinite lattice. The presence of the surface will lead to modifications of the final state wave function. As the photoelectron crosses the surface into the vacuum its momentum is decreased. Since the retarding force is directed orthogonal to the surface, however, it is only this momentum component which changes its magnitude. One has thus:

$$\boldsymbol{k}^{\mathrm{fi}}_{\parallel} = \boldsymbol{K}_{\parallel} \tag{18.5}$$

$$\boldsymbol{k}^{\mathrm{fi}}_{\perp} \neq \boldsymbol{K}_{\perp}, \tag{18.6}$$

[1] Ziman, J.M. (1971): Principles of the theory of solids, Cambridge: University Press

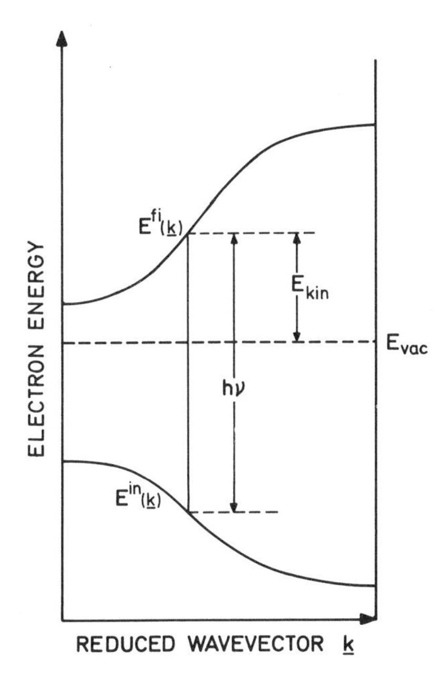

Fig. 124. Direct transition in the reduced zone scheme

where $\boldsymbol{K}_{\parallel}, \boldsymbol{K}_{\perp}$ are the components parallel and orthogonal to the surface of the wave vector outside the solid and $\boldsymbol{k}_{\parallel}^{\mathrm{fi}}$, $\boldsymbol{k}_{\perp}^{\mathrm{fi}}$ those of the final state wave vector inside the solid. The fact that the *orthognal* wave vector component is not conserved across the surface implies that the electron wave vectors associated with the photoabsorption process inside the solid cannot be directly determined from the photoelectron experiment. From (18.5) it is seen, however, that the component *parallel* to the surface is conserved and this leads to a possibility, although limited by (18.6), of probing band energies versus $\boldsymbol{k}$ for solids. This will be discussed further below in connection with angular studies of solid photoelectron spectra.

The usually adopted scheme for interpretation of valence photoelectron spectra from solids is the so-called three-step model[2–5]. According to this model, the photoionization process is divided into a sequence of

1. Optical excitation of the electron from a bound to a free state within the crystal

2. The transport of the photoelectron from the site of creation to the surface and

3. The transmission of the photoelectron through the surface into the vacuum.

The three-step model contains a number of shortcomings and may become insufficient for detailed interpretations of spectra. In particular, it can be noted

[2] Fan, H.Y. (1945): Phys. Rev. *68*, 43

[3] Mayer, H., Thomas, H. (1957): Z. Phys. *147*, 419

[4] Puff, H. (1961): Phys. Status Solidi *1*, 636, 704

[5] Berglund, C.N., Spicer, W.E. (1964): Phys. Rev. *136*, A1030

that the step model actually violates the Heisenberg uncertainty principle in that the transition is assumed to occur between states of definite momenta at a given point in space. More rigourous approaches have been discussed by a number of authors[6-15]. The aim of these treatments is to discuss the photoionization process in the solid directly in terms of a one-step transition from a bound state of the crystal to a continuum state with a vacuum photoelectron. The proper form of the continuum wave function is then the time reversed LEED state[14]. This type of function includes intrinsically the surface transmission characteristics of the final state. The photoelectron intensities in the different directions outside the crystal are evaluated directly in terms of matrix elements between the initial states and these time-reversed LEED states.

β) Two-dimensional band structures. There is one class of compounds, where the above mentioned nonconservation of $\boldsymbol{k}$ can be essentially disregarded, and for which dispersion relations can hence be derived from photoelectron spectroscopy in a direct way[16-24]. These are the so-called layer compounds. They are characterized by a structure where the atoms are bound in twodimensional sheets, which in turn are weakly bound to each other. The valence electrons are confined to individual layers and the band structure thus exhibits small dispersions in a direction orthogonal to the layers. When measurements are made on these compounds, they are cut along the atomic layers and the photoelectrons generally observed for several polar angles at certain selected

[6] Mahan, G.D. (1970): Phys. Rev. B*2*, 4334; (1970): Phys. Rev. Lett. *24*, 1068
[7a] Wehrum, R.P., Hermeking, H. (1974): J. Phys. C*7*, L107
[7b] Hermeking, H. (1972): Z. Phys. *253*, 3791 and (1973): J. Phys. C*6*, 2898
[8] Keldysh, L.V. (1965): Sov. Phys. JETP *20*, 1018
[9a] Caroli, C., Lederer-Rozenblatt, D., Roulet, B., Saint-James, D. (1973): Phys. Rev. B*8*, 4552
[9b] Caroli, C., Roulet, B., Saint-James, D. (1978): in: Handbook of surfaces and interfaces, Dobrzynski, L. (ed.), New York: Garland, vol. 1, 99
[10] Chang, J.J., Langreth, D.C. (1972): Phys. Rev. B*5*, 3512; (1973): B*8*, 4638
[11] Schaich, W.L., Ashcroft, N.W. (1971): Phys. Rev. B*3*, 2452
[12] Cf. also articles by Ashcroft, N.W., Kliewer, K.L., Pendry, J.B., Gadzuk, J.W. (1978): in: Photoemission and the electronic properties of surfaces, Feuerbacher, B., Fitton, B., Willis, R.F. (eds.), London: Wiley-Interscience, p. 21, 45, 87 and 111, respectively
[13] Schaich. W.L. (1978): in: Photoemission in Solids I, Cardona, M., Ley, L. (eds.), Topics in Applied Physics, vol. 26, Berlin, Heidelberg, New York: Springer
[14] Feibelman, P.J., Eastman, D.E. (1974): Phys. Rev. B*10*, 4932
[15] Pendry, J.B. (1976): Surf. Sci. *57*, 679
[16a] Smith, N.V., Traum, M.M., Dickloo, F.J. (1974): Sol. State Commun. *15*, 211
[16b] Smith, N.V., Larsen, P.K. (1978): in: Photoemission and the electronic properties of surfaces, Feuerbacher, B., Fitton, B., Willis, R.F. (eds.), London: Wiley-Interscience, p. 409
[16c] Larsen, P.K., Schlüter, M., Smith, N.V. (1977): Sol. State Commun. *21*, 775
[17] Grandke, T., Ley, L. (1977): Vth International Conf. on VUV Rad. Phys., Montpellier, (France) Ext. Abstr. II-74
[18] Schlüter, M., Comassel, J., Kohn, S., Voitchovsky, J.P., Shen, Y.R., Cohen, M.L. (1978): Phys. Rev. B*13*, 3534
[19] Williams, R.H., Thomas, J.M., Barber, M., Alford, N. (1972): Chem. Phys. Lett. *17*, 142
[20] Smith, N.V., Traum, M.M. (1975): Phys. Rev. B*11*, 2087
[21] Williams, P.M., Latman, D., Wood, J. (1975): J. Electron Spectrosc. Relat. Phenom. 7, 281
[22] Lloyd, D.R., Quinn, C.M., Richardson, N.V., Williams, P.M. (1976): Commun. Phys. *1*, 11
[23] Petroff, Y. (1978): J. Phys. *39*, Coll. C*4*, 149
[24] Larsen, P.K., Chiang, S., Smith, N.V. (1977): Phys. Rev. B*15*, 3200

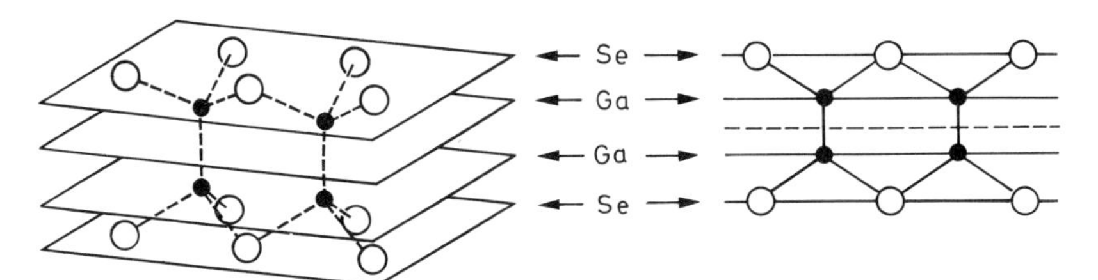

Fig. 125a. Crystal structure of GaSe

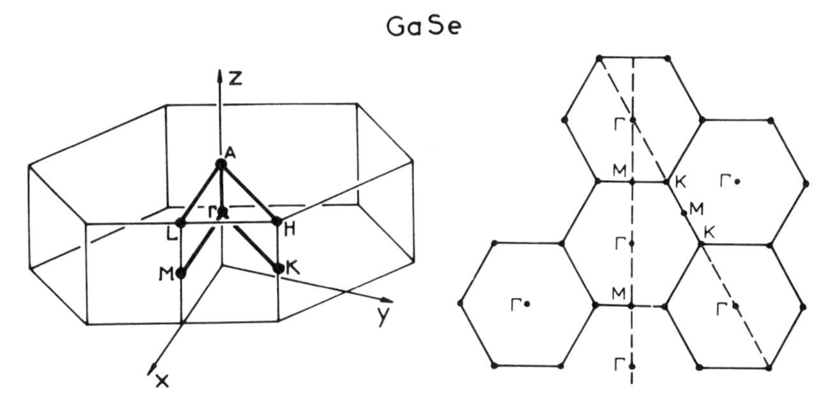

Fig. 125b. Brillouin zone of GaSe

azimuthal angles. In this way the dispersion E^{in} vs. $\boldsymbol{k}_{\parallel}$ can be investigated along selected directions in reciprocal space. The case of GaSe has been studied by PETROFF[23]. The crystal structure is shown in Fig. 125a. The interactions between adjacent layers are of Van der Waals type, and the electrons are strongly localized to the atomic layers. The Brillouin Zone (BZ), shown in Fig. 125b, is hexagonal and one expects that the energy dispersion along the line ΓA (perpendicular to the layers) will be small. The photoelectron spectra were recorded using synchrotron radiation at $h\nu = 27$ eV and were obtained in azimuthal angles, such that $\boldsymbol{k}_{\parallel}$ was confined to lie along $\Gamma M \Gamma$- or $\Gamma K M K$-directions (cf. Fig. 125b). The spectra obtained (Fig. 126) show a considerable movement of the lines with respect to polar angle of detection. The four bands which are observed correspond to different types of bonds between the Ga and Se atoms. The band A (the uppermost valence band) is due to a bonding combination of Ga and Se p_z-orbitals. Band B has mainly p_x, p_y-character in the Ga-Se bond and C and D are mostly due to Ga s-orbitals. In these latter bands is also contained some p_z-character, such that C corresponds to an antibonding and D to a bonding combination in the Ga-Ga bond. Figure 127 shows the band structure derived from the photoelectron spectra. These have been constructed by measuring the kinetic energy of each peak at a given angle. $|\boldsymbol{k}_{\parallel}^{\text{fi}}| = k_{\parallel}^{\text{fi}}$ is then obtained from:

$$k_{\parallel}^{\text{fi}} = K_{\parallel} = \frac{\sqrt{2E_{\text{kin}}}}{\hbar} \sin\theta \tag{18.7}$$

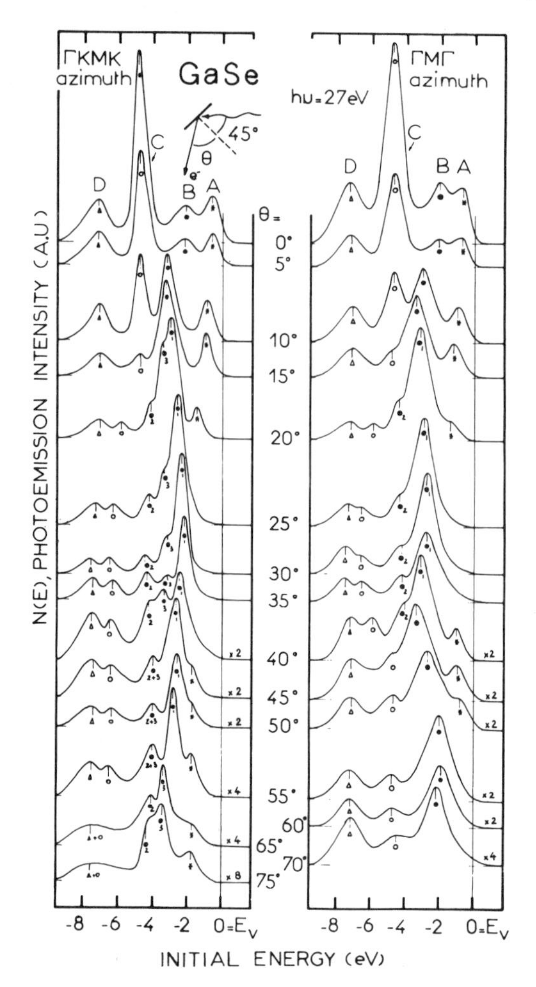

Fig. 126. Valence electron spectra of GaSe obtained with $h\nu = 27$ eV at different polar angles in the ΓKMK and $\Gamma M\Gamma$ azimuths[23]

according to (18.5). The initial state energy (with respect to the top of the valence band) is then plotted vs. $\boldsymbol{k}_{\parallel}^{\text{fi}}$ as in Fig. 127a and b. Along the $\Gamma M\Gamma$ azimuthal, one expects mirror symmetry about the M point and this is indeed verified by the measurements. In Fig. 127c the bands have been plotted in the reduced zone scheme (i.e. vs. $\boldsymbol{k}_{\parallel}^{\text{in}} = \boldsymbol{k}_{\parallel}^{\text{fi}} - \boldsymbol{G}$). The bands as derived from these measurements compare very favourably with the results of band structure calculations on GaSe (Fig. 128), although the theoretical bands are somewhat compressed in energy with respect to experiment. Also, there is some discrepancy concerning the band-width of the B-band.

γ) Three-dimensional band structures. The layer compounds discussed above constitute special cases of two-dimensional band structure and for the general three-dimensional case the problem of non-conservation of $\boldsymbol{k}_{\perp}$ must be taken

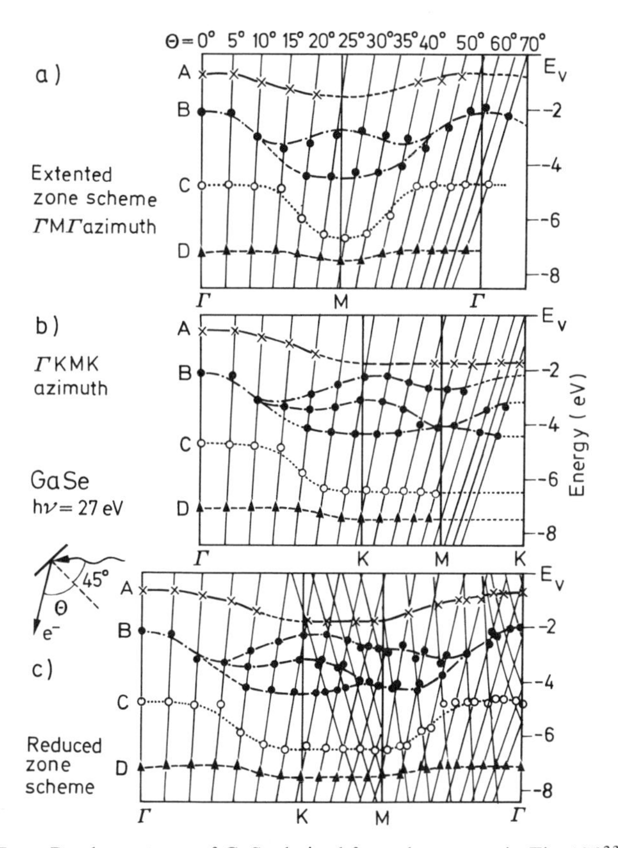

Fig. 127a–c. Band structures of GaSe derived from the spectra in Fig. 126[23]

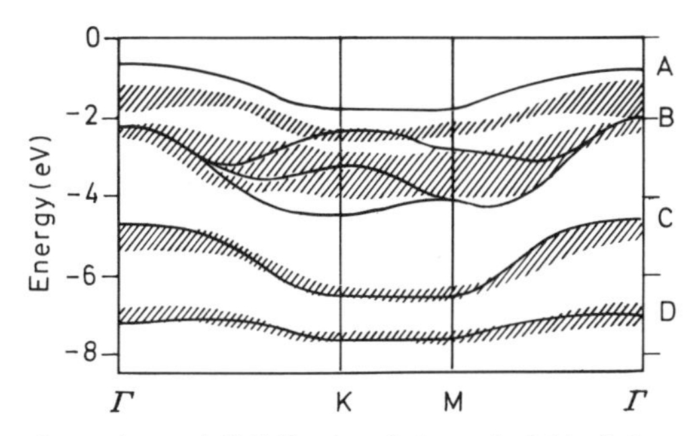

Fig. 128. Comparison of experimental (full lines) and theoretical (shaded areas) band structures of GaSe[23]

into consideration. Various ways have been devised of circumventing this problem (e.g. [25–33]). We will illustrate one of these possibilities due to DIETZ

[25] Stöhr, J., Wehner, P.S., Williams, R.S., Apai, G., Shirley, D.A. (1978): Phys. Rev. B *17*, 587

[26] Thiry, P., Chandesris, D., Lecante, J., Guillot, C., Pinchaux, R., Petroff, Y. (1979): Phys. Rev. Lett. *43*, 82

(Footnotes 27–33 see p. 454)

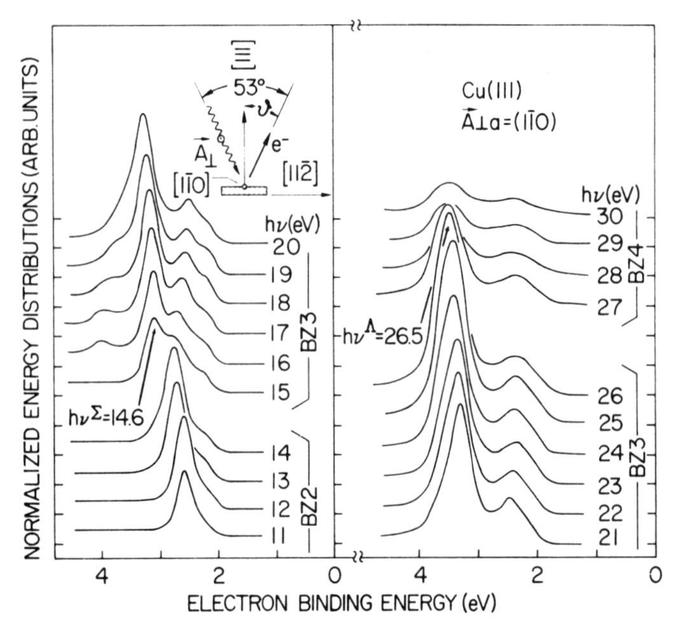

Fig. 129. Valence spectra of a Cu(111) surface obtained using different photon energies in a detection angle of $\theta = -30°$ [28] with respect to the surface normal

and EASTMAN [28], which yields band structures along principal axes. These authors investigated the angular resolved spectra from a Cu(111) surface using synchrotron radiation in the photon energy range $10\,\text{eV} \leqq h\nu \leqq 30\,\text{eV}$. Spectra were measured at polar angles of $-60° \leqq \theta \leqq 60°$ with respect to the surface normal. Figure 129 shows a series of spectra obtained at a fixed angle of detection ($\theta = -30°$). The most prominent feature observed in the variation of the spectra with photon energy is the dramatic change of intensity of the peaks within a very small photon energy range. Such photon energies are, for $\theta = -30°$, $h\nu = 14.6$ eV and $h\nu = 26.5$ eV. These features occur because the final state transitions cross Brillouin zone boundaries. This can be realized by inspection of the momentum space plot of the Cu band structure in the extended zone scheme (cf. Fig. 130). In this plot portions of the first through fourth BZ are outlined. The boundaries between BZ2 and BZ3 and between BZ3 and BZ4 are the Σ and Λ axes respectively (heavy lines). The nearly free-electron final state band structure is given in terms of constant energy contours (thin lines). The dashed line in the figure corresponds to those values of $\boldsymbol{k}^{\text{fi}}$, which can contribute to the spectra at $\theta = -30°$, irrespective of the final

[27] Chiang, T.-C., Knapp, G.A., Aono, M., Eastman, D.E. (1980): Phys. Rev. B *21*, 3513
[28] Dietz, E., Eastman, D.E. (1978): Phys. Rev. Lett. *41*, 1674
[29] Himpsel, F.J., Eastman, D.E. (1978): Phys. Rev. B *18*, 5236
[30] Kane, E.O. (1964): Phys. Rev. Lett. *12*, 97
[31] Turtle, R.R., Calcott, T.A. (1975): Phys. Rev. Lett. *34*, 86
[32] Nilsson, P.O., Dahlbäck, N. (1979): Solid State Commun. *29*, 303
[33] Heimann, P., Miosga, H., Neddermeyer, H. (1979): ibid. *29*, 463

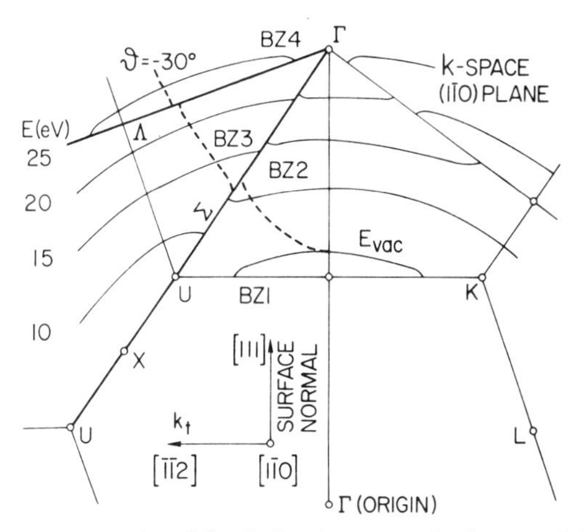

Fig. 130. Momentum space plot of the Cu band structure in the extended zone scheme[28]

state energy. This line has been constructed using the calculated nearly free-electron bands and the conservation of parallel momentum. Thus, given a final state energy $[E^{fi}(\boldsymbol{k}^{fi})=E_{kin}]$ and a certain direction of electron detection, $\boldsymbol{k}_{\parallel}^{fi}$ is given by (18.7). This defines a point on the final state energy contour $E^{fi}(\boldsymbol{k}^{fi})=E_{kin}$. The final state energy of a transition using a certain photon energy, leading to an electron in the direction $\theta=-30°$, is thus given by the intersection of this line with the line defined by $E^{fi}(\boldsymbol{k}^{fi})-E^{in}(\boldsymbol{k}^{fi}-\boldsymbol{G})=h\nu$. Consequently, using a theoretical band structure one can construct the expected spectra in a certain direction of detection. The occurrence of the dramatic intensity changes across zone boundaries offers a possibility of going the other way around, i.e. derive the energy dispersion along the principal axes from the experimental spectra. Starting at a low photon energy transitions will occur into the BZ2. As the photon energy is increased, the possible transitions into $\theta=-30°$ will move upward along the dashed line in Fig. 130 and cross the boundary into BZ3. At this point the strong variations in intensity are observed. Since the final state energy is known from the electron spectrum one can calculate $\boldsymbol{k}_{\parallel}^{fi}$ from E_{kin} and θ. Projecting this $\boldsymbol{k}_{\parallel}^{fi}$ onto the zone boundary one can now find the $\boldsymbol{k}^{fi}$-vector on the Σ-axis which corresponds to the final-state energy $E^{fi}(\boldsymbol{k}^{fi})=E_{kin}$. The initial state energy is given by $E^{in}=E^{fi}-h\nu$. Changing the angle θ one may thus scan the band energies along the axes Σ and Λ. The bands obtained in this way are shown in Fig. 131, plotted in the reduced zone scheme. For comparison a calculated band structure[34] is shown as the solid lines. There is seen to be substantial agreement between the theory and experiment in this case.

δ) Angular studies of adsorbates. Valence photoelectron spectra offer possibilities of obtaining information on the adsorption geometry of molecules on surfaces[35–44]. For this purpose use may be made of polarized radiation and

[34] Janak, J.F., Williams, A.R., Moruzzi, V.L. (1975): Phys. Rev. B *11*, 1522

[35] Smith, R.J., Anderson, J., Lapeyre, G.J. (1976): Phys. Rev. Lett. *37*, 1081

(Footnotes 36–44 see p. 456)

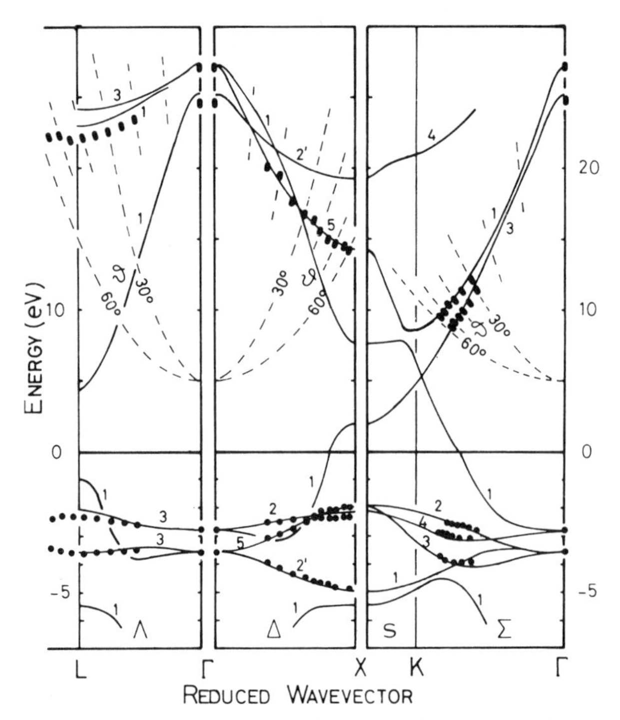

Fig. 131. Band structures of Cu along principal axes obtained from angle-resolved photoelectron spectra [28]

the dipole selection rules in photoionization [35,43]. These rules are based on the symmetry of the initial and final state wave functions and the direction of the photon beam polarization vector with respect to the molecular axis. Considering a σ-state in the CO molecule the rules have some consequences which can be easily tested experimentally. According to the dipole selection rules, photoionization of a σ-orbital leads to both σ- or π-symmetry final state photoelectron waves. The extent to which these respective channels are excited depends on the orientation of the electric field vector with respect to the molecular axis. If this vector is orthogonal to the molecular axis (i.e. the light propagates

[36] Gadzuk, J.W. (1974): Phys. Rev. B *10*, 5030

[37] Shirley, D.A., Stöhr, J., Wehner, P.S., Williams, R.S., Apai, G. (1977): Phys. Scr. *16*, 398

[38a] Liebsch, A. (1976): Phys. Rev. B *13*, 544

[38b] Liebsch, A. (1978): in: Photoemission and the electronic properties of surfaces, Feuerbacher, B., Fitton, B., Willis, R.F. (eds.), London: Wiley-Interscience, p. 167

[39] Apai, G., Wehner, P.S., Williams, R.S., Stöhr, J., Shirley, D.A. (1976): Phys. Rev. Lett. *37*, 1497

[40] Davenport, J.W. (1976): Phys. Rev. Lett. *36*, 945

[41] Lecante, J., Guillot, C., Lassailly, Y., Chandersris, D., Jugnet, Y. (1978–1979): Rapport d'Activité, L.U.R.E. (Jan.–June 30) p. 123

[42] Williams, P.M., Butcher, P., Word, J., Jacobi, K. (1976): Phys. Rev. B *14*, 3215

[43] Allyn, C.L., Gustafsson, T., Plummer, E.W. (1977): Chem. Phys. Lett. *47*, 127

[44] Horn, K., Scheffler, M., Bradshaw, A.M. (1978): Phys. Rev. Lett. *41*, 822

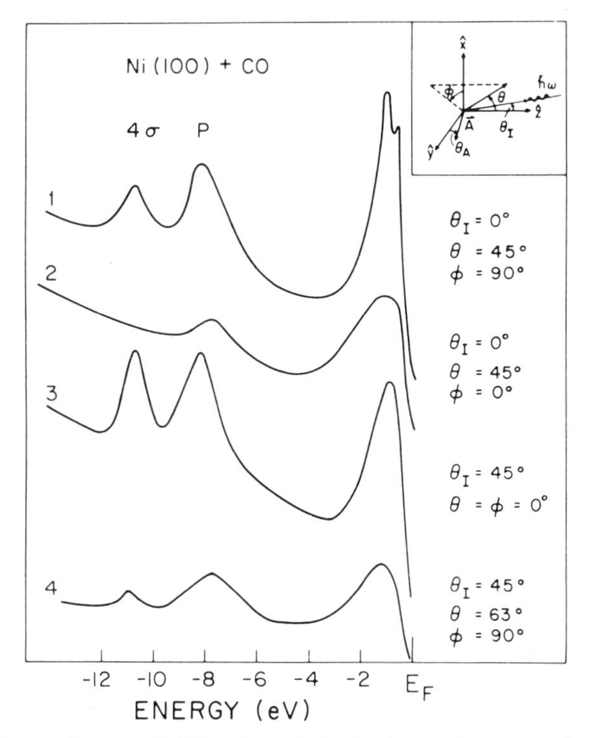

Fig. 132. Spectra from CO on a Ni(100) surface obtained using synchrotron radiation at $h\nu = 32$ eV with different angular parameters[43]. $x-y$ plane = surface plane, θ, ϕ = electron polar, azimuthal angles, A = photon polarization vector

parallel to the axis), then the $\mathcal{E}\sigma$-channel cannot be excited, due to parity reasons. The $\mathcal{E}\pi$-channel can be excited, however, so that the total photoelectron intensity from the initial state σ-orbital will be non-vanishing. The differential cross section is going to show strong variation with angle in this geometry and in particular, the intensity will be zero in the nodal plane of the final state π-electron wave. This nodal plane is then perpendicular to the plane defined by the electric field vector and the molecular axis. This circumstance offers thus a possible means of establishing the orientation of the CO molecule (or any other molecule of $C_{\infty v}$ symmetry) on a substrate. If polarized light is normal to the surface and the electron spectrometer slit placed in the plane defined as above, than a vanishing σ-intensity gives a very strong indication that the molecule is standing straight up.

The above consequence of the dipole selection rules was studied for CO on a Ni(100) surface by SMITH et al.[35] and ALLYN et al.[43]. Spectra from the latter study taken at $h\nu = 32$ eV using synchrotron radiation are shown in Fig. 132. The various angular parameters are defined in the inset figure. The topmost spectrum was recorded with the photon beam along the surface normal (z-axis), electric vector along the y-axis and electron detection in the yz-plane. Here the 4σ-orbital shows a prominent intensity, as expected. However, when electron detection is performed in a plane normal to the yz-plane

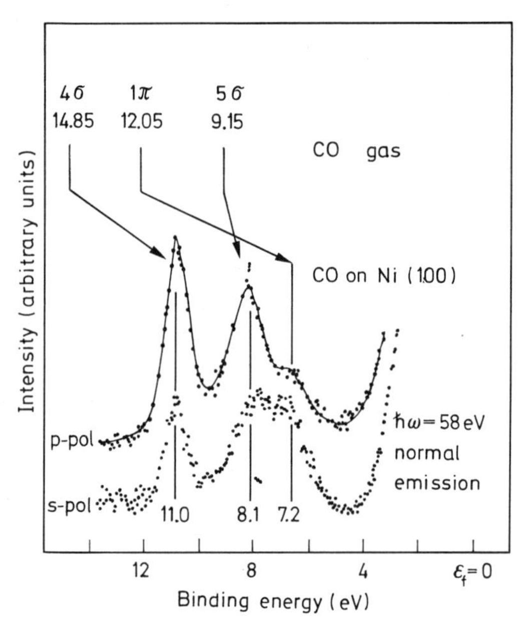

Fig. 133. Spectra from CO on Ni(100) obtained with *s*- and *p*-polarized synchrotron radiation at $h\nu = 58$ eV[41]. The gas and adsorbate spectra have both been referred to the Fermi level by subtracting the Ni work function from the gas phase energies

(the xz-plane) the σ-intensity vanishes (second spectrum from top) thus providing strong evidence for the upright geometry, as discussed above. In the two lowermost spectra the geometries have been chosen to enhance the σ-orbitals (spectrum 3) and the π-orbital (spectrum 4).

This possibility of enhancing or suppressing the intensities of the orbital symmetries provides a means of separating the 5σ- and 1π-orbital contributions to peak P. The data in Fig. 132 yield a $4\sigma - 1\pi$ separation of 3.0 ± 0.1 eV, which is closely the same as that in the free molecule ($\Delta E = 2.9$ eV). The 5σ-level is found at a binding energy 0.5 ± 0.2 eV *higher* than the 1π-level, substantially different from the free-state value, which is 2.7 eV *lower* than the 1π-value.

This conclusion of a level crossing between the 1π- and 5σ-levels is supported by later studies made at higher resolution. Fig. 133 shows spectra taken by LECANTE et al.[41] with synchrotron radiation for two different polarization directions. In both spectra, the 5σ- and 1π-lines can be clearly resolved, the latter lying at the lower binding energy by 0.9 eV.

19. Density of states structure from photoelectron spectra

α) *Solid valence photoelectron spectra in the high photon energy limit.* When higher photon energies (such as e.g. Al$K\alpha$) are used to excite valence photoelectrons from a solid the observed spectrum is generally found to closely represent the initial density of states (IDOS) structure of the material (e.g. [1–3]). There are a number of factors which cooperatively lead to this result. First, as the photon energy increases and hence the photoelectron energy and momentum, it becomes increasingly probable that the direct transition requirement can be fulfilled for a given initial state. This means that the sampling of the Brillouin zone will tend to be more and more complete as the photon energy is increased. In addition, the importance of nondirect transitions increases when the photon energy is higher. Such transitions occur both due to the finite sampling depth of the photoelectrons (final state $\boldsymbol{k}$-smearing) and to the excitation of phonons[4–7]. The net result of the above effects will be that the photoelectron intensity at an energy corresponding to the initial state energy E^{in} of the band can be factorized according to:

$$I(E^{\text{in}}, h\nu) \propto \text{IDOS}(E^{\text{in}}) \times \text{FDOS}(E^{\text{fi}}) \times \sigma(E^{\text{in}}, h\nu), \tag{19.1}$$

where IDOS and FDOS are initial and final densities of states, respectively and $\sigma(E^{\text{in}}, h\nu)$ is the average photoelectric cross section at initial energy E^{in} and photon energy $h\nu$. Equation (19.1) leads to the results experimentally observed if it is recognized that FDOS varies very little over the energy interval of the valence electron spectrum at sufficiently high photon energies (at $h\nu \geqq 1000$ eV this function varies by at most 1% over the extension of any normal metal conduction band).

β) Metal valence bands. The convergence of valence band photoelectron spectra towards the IDOS according to (19.1) has led to a large number of studies, where x-ray excited spectra have provided critical tests of initial state band structure calculations. Also, such spectra may serve as experimental IDOS for cases, where reliable calculations do not exist. x-ray excited valence band spectra from metals belong mostly to the first group of 'test' spectra. Here, good calculations are very often available and detailed comparisons can be made between experiment and theory. Spectra from the fcc metals Au, Ag and Cu are shown in Fig. 134[8]. According to the IDOS interpretation, one should expect structure at initial state energies corresponding to critical points in the band structure, i.e. points in the BZ where the IDOS is high. In Fig. 134

---

[1] Cardona, M., Ley, L. (eds.) (1979): Photoemission in solids I, Topics in Applied Physics, vol. 26, Berlin, Heidelberg, New York: Springer; Ley, L., Cardona, M. (eds.) (1979): Photoemission in solids II, Topics in Applied Physics, vol. 27, Berlin, Heidelberg, New York: Springer

[2] Nemoshkalenko, V.V., Aleshin, V.G. (1979): Electron spectroscopy of crystals, New York: Plenum Press

[3] Siegbahn, H., Karlsson, L. Electron Spectroscopy for Atoms, Molecules and Condensed Matter, Amsterdam: North-Holland (to be published)

[4] Stöhr, J., McFeely, F.R., Apai, G., Wehner, P.S., Shirley, D.A. (1976): Phys. Rev. B*14*, 443

[5] Feibelman, P.J., Eastman, D.E. (1974): Phys. Rev. B*10*, 4932

[6] Shevchik, W.J. (1977): J. Phys. C*10*, L555

[7] Williams, R.S., Wehner, P.S., Stöhr, J., Shirley, D.A. (1977): Phys. Rev. Lett. *39*, 302

[8a] Hüfner, S., Wertheim, G.K., Buchanan, D.N.E. (1974): Solid State Commun. *14*, 1173

[8b] Mueller, F.M. (1967): Phys. Rev. *153*, 659

[8c] Christensen, N.E. (1972): Phys. Stat. Sol. *54*, 551; (1973): ibid. *55*, 117

[8d] Christensen, N.E., Seraphin, B.O. (1971): Phys. Rev. B*4*, 332

[8e] Mueller, F.M., Freeman, A.J., Dimmock, J.O., Furdyna, A.M. (1970): Phys. Rev. B*1*, 4617

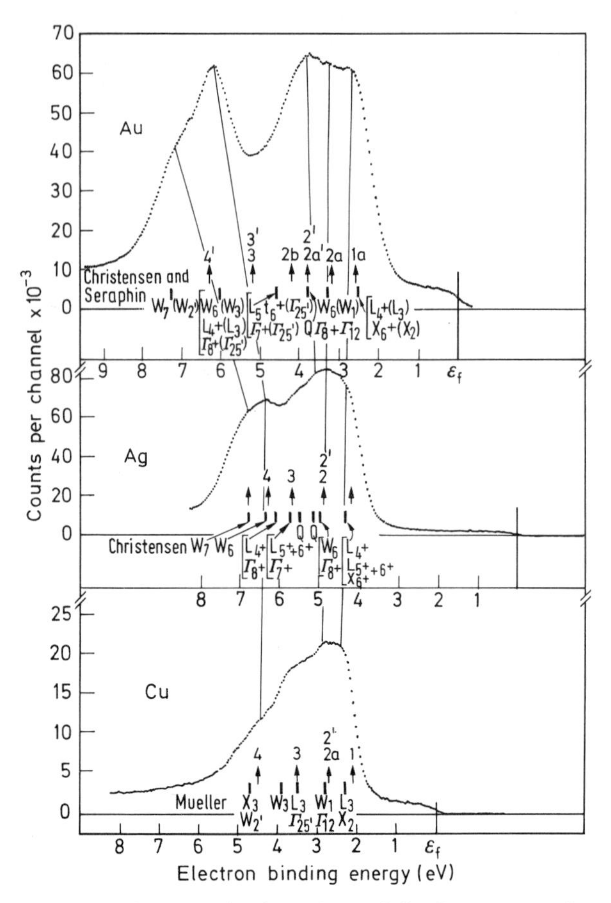

Fig. 134. Al $K\alpha$ excited valence bands of Au, Ag, and Cu. Bar spectra refer to energies of critical points obtained from theoretical band structures. Arrows indicate structures obtained in uv-excited spectra [8]

critical point energies from band structure calculations have been indicated. The valence bands of the noble metals can be considered to be formed from the atomic $(n-1)\,d$ and ns levels, which are nearly degenerate. Bands formed from the $(n-1)\,d$ orbitals show comparatively small energy dispersion due to the contracted character of these orbitals. The ns orbitals on the other hand, extend considerably beyond the average radius of the $(n-1)\,d$ orbitals and have appreciable interatomic overlap. The ns-like bands thus formed are consequently wide. These general features are clearly seen in the spectra of Fig. 134, where the large intensity peaked structures are due to the d-bands and the flat bands extending to the Fermi levels are due to the ns-like bands. The relative intensity between the s-band and the d-band is given both by the relative IDOS and the difference in subshell cross section between the s- and d-atomic symmetries. At the Al $K\alpha$ photon energy the latter dominates strongly over the former ($\sigma_{nd}:\sigma_{(n+1)s} > 15$[9]). A comparatively good correspondence can

[9] Scofield, J.H. (1976): J. Electron Spectrosc. Relat. Phenom. *8*, 129

be noted for these cases between the energy positions of critical points and the observed structure in the spectra.

Other examples of transition metal bands are shown in Fig. 135 for Re and Os[10,11]. Parts (*b*) and (*d*) are the experimentally obtained spectra and the (*a*) and (*c*) figures are IDOS curves for Os and Re, respectively, obtained by two methods, LMTO (Linear Combination of Muffin Tin Orbitals)[11a] and RAPW (Relativistic Augmented Plane Wave)[11b]. In accordance with the results of other *d*-band metals the experimental spectra are very well represented by the calculated curves. From the IDOS curves there is seen to be a substantial degree of rigidity in the band-structure. Thus, the calculated Os IDOS may be very accurately reproduced from the calculated Re IDOS by adding the density of unoccupied states of the Re calculation corresponding to one extra electron. The IDOS curves of Fig. 135 have been obtained by broadening of the calculated IDOS histograms with the spectrometer function.

The rare earth elements are characterized in their atomic ground states by configurations $4f^{n+1}5d^0 6s^2$. In the metallic state the atomic configurations are more properly described as $4f^n(dsp)^3$, i.e. a 4*f*-electron is promoted to form a trivalent (*dsp*) configuration. This promotion energy is regained by the increase in cohesive energy. The 4*f*- and (*dsp*)-electrons are located in two distinct energy regions. This is clearly observed in the photoelectron spectra of these metals. Figure 136 shows the spectra of Sm, Eu and Gd (Sm and Gd excited with monochromalized Al$K\alpha$)[12].

The 4*f*-bands in the spectra of Fig. 136 have much larger intensities than the *spd*-conduction bands. This is in line with the expected atomic cross section ratios ($\sigma_{4f}:\sigma_{5d}:\sigma_{6s}=1.43:0.06:0.03$) for atomic Gd($4f^7 6s^2 5d^1$)[9]. The peaks in the 4*f*-band are due to the multiplet components resulting from ionization of the various initial state $4f^n$-configurations[13] (cf. Table 12). The structure and intensity of the Sm conduction band indicates peaks which are possibly due to initially divalent Sm, i.e. a $4f^6$ configuration[14].

γ) Alloy valence bands. Valence photoelectron spectra of alloys yield information regarding the interaction between the constituent components and their respective band-structures. Several models have been proposed to describe these interactions[15–20]. In the simplest of these, the rigid-band model[15],

---

[10a] Nilsson, R., Berndtsson, A., Mårtensson, N., Nyholm, R., Hedman, J. (1976): Phys. Stat. Sol. *75*, 197

[10b] Berndtsson, A., Nyholm, R., Mårtensson, N., Nilsson, R., Hedman, J. (1979): Phys. Stat. Sol. *93*, K103

[11a] Jepsen, O., Andersen, O.K., Mackintosh, A.R. (1975): Phys. Rev. B*12*, 3084

[11b] Mattheiss, L.F. (1966): Phys. Rev. *151*, 450

[12] Campagna, M., Wertheim, G.K., Baer, Y. (1979): in: Photoemission in solids II, Ley, L., Cardona, M. (eds.), Topics in Applied Physics, vol. 27, Berlin, Heidelberg, New York: Springer, p. 217

[13] Cox, P.A. (1975): Struct. Bonding *24*, 59

[14] Wertheim, G.K., Campagna, M. (1977): Chem. Phys. Lett. *47*, 182

[15] Mott, N.F., Jones, J. (1958): Theory of the properties of metals and alloys, New York: Dover

[16] Friedel, J. (1958): Nuovo Cimento 7, 287

[17] Anderson, P.W. (1961): Phys. Rev. *124*, 41 (Footnotes 18–20 see p. 463)

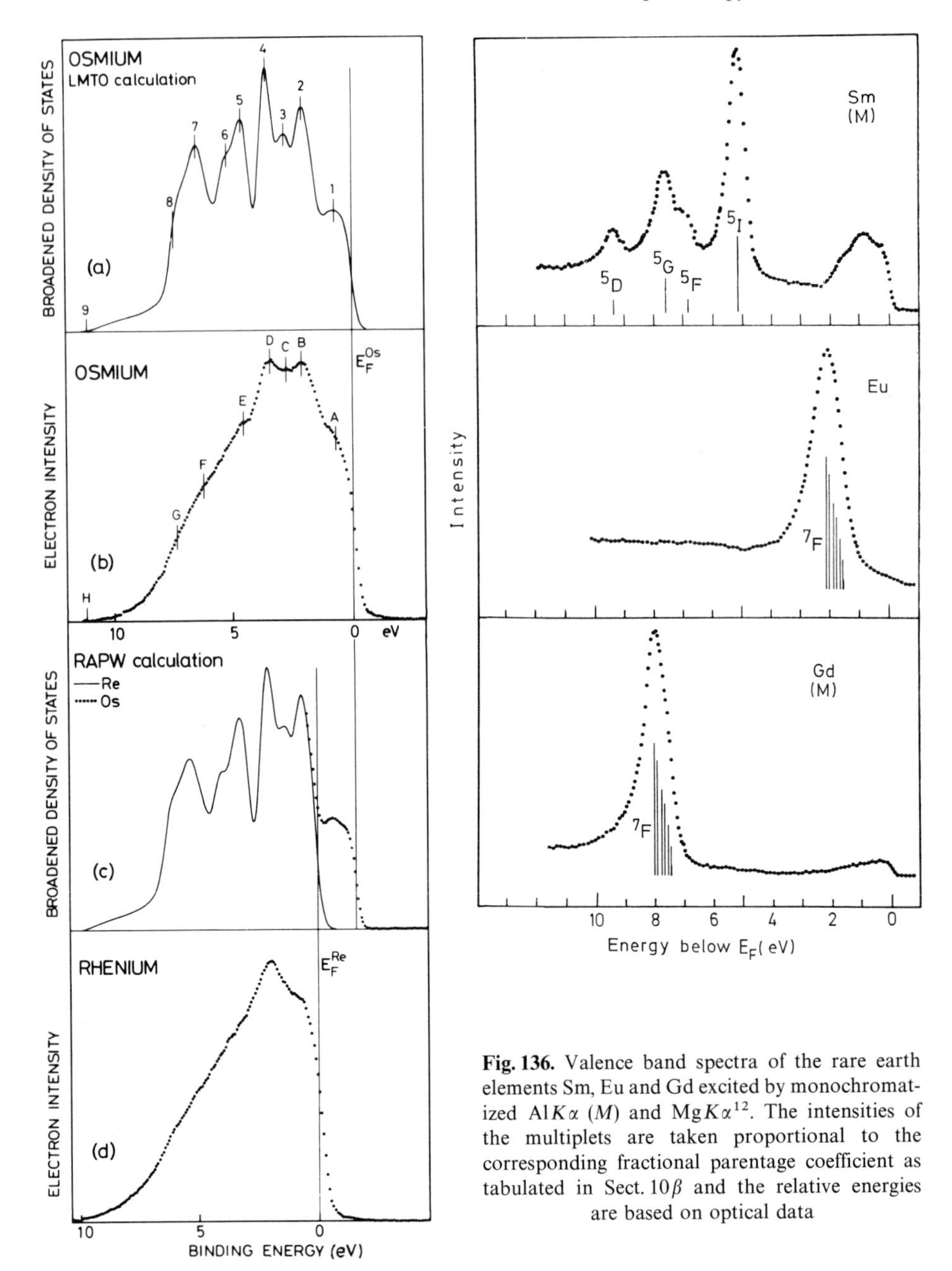

Fig. 136. Valence band spectra of the rare earth elements Sm, Eu and Gd excited by monochromatized Al$K\alpha$ (M) and Mg$K\alpha$[12]. The intensities of the multiplets are taken proportional to the corresponding fractional parentage coefficient as tabulated in Sect. 10β and the relative energies are based on optical data

Fig. 135a–d. Re and Os valence bands[10]. (**a**) LMTO IDOS broadened with the spectrometer function[11]. (**b**) Os valence band. A background proportional to the integrated intensity from the Fermi level to the considered energy has been subtracted. (**c**) RAPW IDOS broadened with the spectrometer function[11]. The IDOS of Os is obtained from the Re IDOS by a rigid band approximation (dotted line). (**d**) Re valence band (same background subtraction as in (**b**))

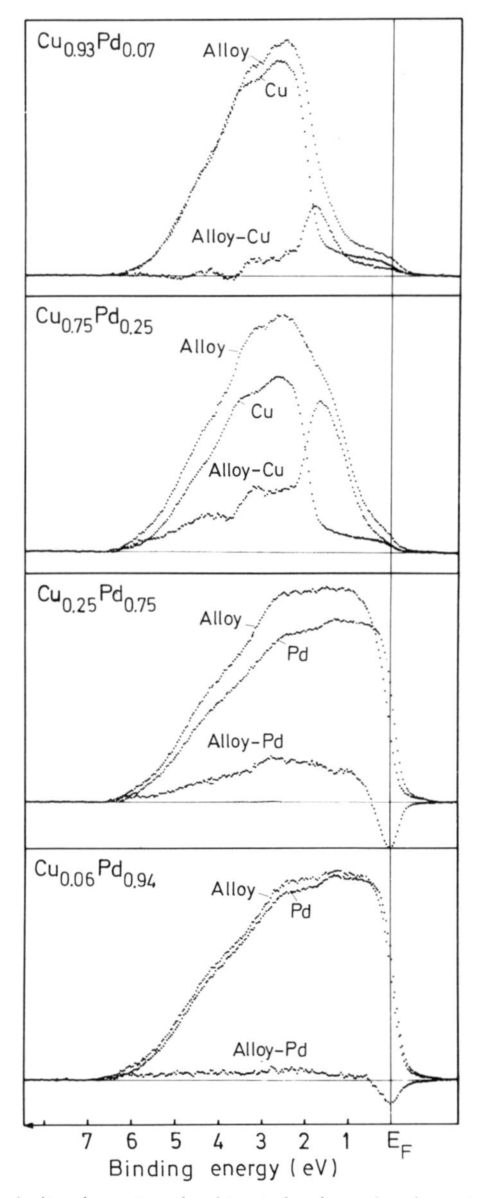

Fig. 137. Alloy, pure majority element and subtracted valence band spectra in the CuPd system (cf. text)[24]

it is assumed that in a transition metal alloy a common *d*-band exists. This *d*-band is constructed from the band structure of one of the components by simply shifting the Fermi level up to the value corresponding to the mean

[18] Soven, P. (1967): Phys. Rev. *156*, 809; (1969): *178*, 1136

[19] Velicky, B., Kirkpatrick, S., Ehrenreich, H. (1968): Phys. Rev. *175*, 747

[20] Stocks, G.M., Williams, R.W., Faulkner, J.S. (1971): Phys. Rev. B*4*, 4390

concentration of valence electrons for the given alloy. Although this model has been successful in several respects, it fails to describe a number of properties exhibited by real systems, in particular for dilute alloys. A more proper model to treat dilute alloys was proposed by FRIEDEL[16] and ANDERSON[17]. This model considers the impurity atom in the alloy to be screened by the host conduction electrons and to approximately retain its individual d-orbital identity. Confirmation of the Friedel-Anderson model has been obtained from photoelectron spectra of several systems, such as AgPd and CuPd[21–25].

Spectra taken for four different concentrations in the CuPd system are shown in Fig. 137[24]. In each figure is shown both the spectrum of the alloy, the spectrum of the pure majority component and the difference spectrum (alloy-pure metal). In the Cu-rich alloys (top spectra of Fig. 137) the subtracted spectrum clearly shows a peak situated at ~ 2 eV from the Fermi level. This peak, which is ~ 1 eV wide, is interpreted as a Pd$4d$ virtual bound state (VBS) within the Friedel-Anderson model. As the Pd concentration increases, the peak broadens and moves up towards the Fermi level. The Pd-rich alloys (lower spectra of Fig. 137) exhibit a high density of states at the Fermi level. In the difference spectrum (with respect to pure Pd) the copper impurity band extends over almost the whole region of the Pd d-band. The shape of the Cu subband will, however, be quite sensitive to the subtraction procedure for these concentrations.

δ) Formation of valence bands in extended systems. A special topic in the study of IDOS structures concerns the comparison of the solid valence spectra with those obtained in the free state. The extent to which the energy level system changes in going from the free to the condensed form is clearly dependent on the cohesive interactions involved. Thus, the study of valence level spectra from the two phases may yield information on these interactions[26–33].

[21] Norris, C., Myers, H.P. (1971): J. Phys. F*1*, 62

[22] Hüfner, S., Wertheim, G.K., Wernick, J. (1975): Solid State Commun. *17*, 1585

[23] Hüfner, S., Wertheim, G.K., Wernick, J.H. (1973): Phys. Rev. B*8*, 4511

[24] Mårtensson, N., Nyholm, R., Calen, H., Hedman, J., Johansson, B. (1981): Phys. Rev. B*24*, 1725

[25a] Nicholson, J.A., Riley, J.D., Leckey, R.C.G., Liesegang, J., Jenkin, J.G. (1977): J. Phys. F7, 351

[25b] Nicholson, J.A., Riley, J.D., Leckey, R.C.G., Jenkin, J.G., Liesegang, J., Azoulay, J. (1978): Phys. Rev. B*18*, 2561

[26] Svensson, S., Mårtensson, N., Basilier, E., Malmqvist, P.Å., Gelius, U., Siegbahn, K. (1976): J. Electron Spectrosc. Relat. Phenom. *9*, 51

[27] Süzer, S., Lee, S.T., Shirley, D.A. (1976): Phys. Rev. A*13*, 1042

[28] Ley, L., Kowalczyk, S.P., McFeely, F.R., Pollak, R.A., Shirley, D.A. (1973): Phys. Rev. B*8*, 2392

[29] Mason, M.G., Baetzold, R.G. (1976): J. Chem. Phys. *64*, 271

[30] Mason, M.G., Gerenser, L.J., Lee, S.-T. (1977): Phys. Rev. Lett. *39*, 288

[31] Pireaux, J.J., Svensson, S., Basilier, E., Malmqvist, P.Å., Gelius, U., Caudano, R., Siegbahn, K. (1976): Phys. Rev. A*14*, 2133

[32] Siegbahn, H., Asplund, L., Kelfve, P., Hamrin, K., Karlsson, L., Siegbahn, K. (1974): J. Electron Spectrosc. Relat. Phenom. *5*, 1059

[33] Mårtensson, N., Malmqvist, P.Å., Svensson, S., Basilier, E., Pireaux, J.J., Gelius, U., Siegbahn, K. (1977): Nov. J. Chim. *1*, 191

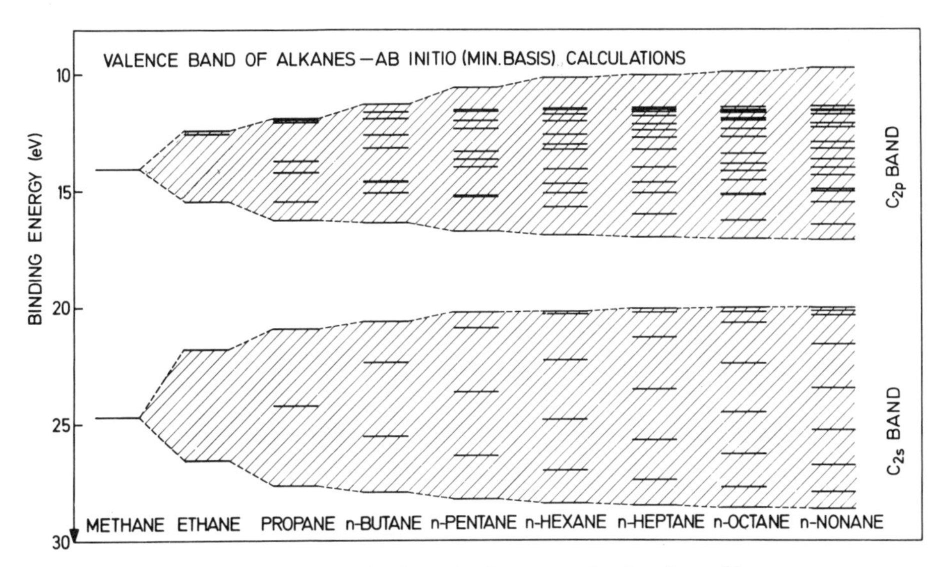

Fig. 138. Theoretical valence level structure for the alkanes[31]

For a hydrocarbon of formula C_nH_{2n+2}, the valence electron region contains $6n+2$ electrons distributed among $3n+1$ energy levels. These are divided into two groups, each of which contains $2n+1$ and n molecular orbitals, respectively. The first, low binding energy group, is mainly composed of orbitals of $C2p$ and $H1s$ character, which form the C-H bonds. The molecules in the higher binding energy group are mainly composed of C-C bonding and antibonding combinations of $C2s$ atomic orbitals. The number of levels in the $C2s$ region is equal to the number of carbon atoms in the alkane chain. The $1a_1$ level in methane splits into two levels in ethane, three levels in propane etc. Since the levels are spread out over a limited energy range of a few electron volts the spacing between each level decreases with an increasing number of carbon atoms. For large n this leads to the formation of a band structure. This is illustrated in Fig. 138 from which it is seen that the increase in bandwidths is small for $n \geqq 6$. PIREAUX et al.[31] studied the successive variation in the valence level photoelectron spectra with increasing chain size. The spectra from selected members of this series are shown in Fig. 139. From these spectra one can observe the limiting number of n when the spectrum becomes indistinguishable from that of the infinite one-dimensional solid polyethylene. This indicates how large the system must be to approximate the band structure of an infinite solid. Already for n-tridecane ($n=13$) no fine substructure could be resolved, as for the case of n-nonane ($n=9$). Also, it was observed that this valence band of gaseous $n-C_{13}H_{28}$ is very similar to the solid phase spectrum of $n-C_{36}H_{74}$ (hexatricontane). This latter spectrum is essentially the same as that of polyethylene. It was thus concluded that 13 atoms in the chain are sufficient to approximate an infinite solid.

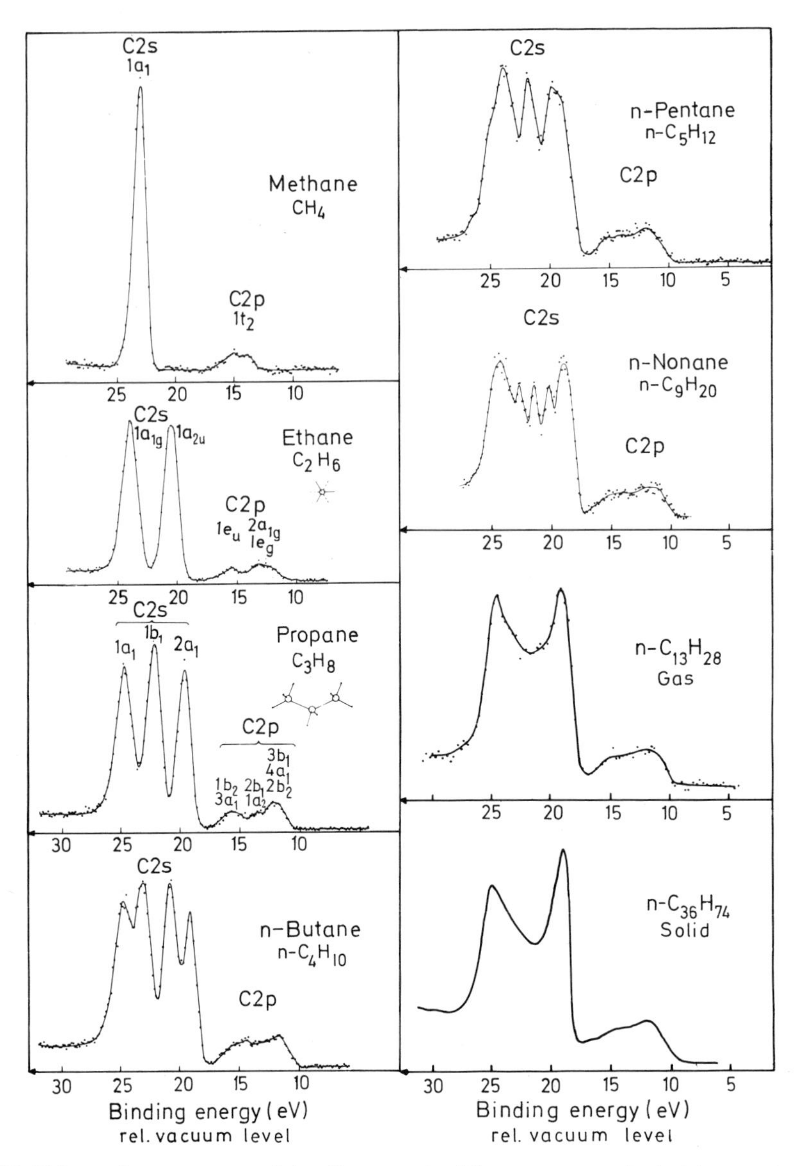

Fig. 139. Valence electron spectra of the alkanes and solid $n-C_{36}H_{74}$ excited with monochromatized Al $K\alpha$[31]

Appendix

Graph of electron binding energies for the elements. The energies were taken from: K. Siegbahn et al., ESCA – Atomic, Molecular and Solid State Structure Studied by Means of Electron Spectroscopy, Almqvist-Wiksell (1967).

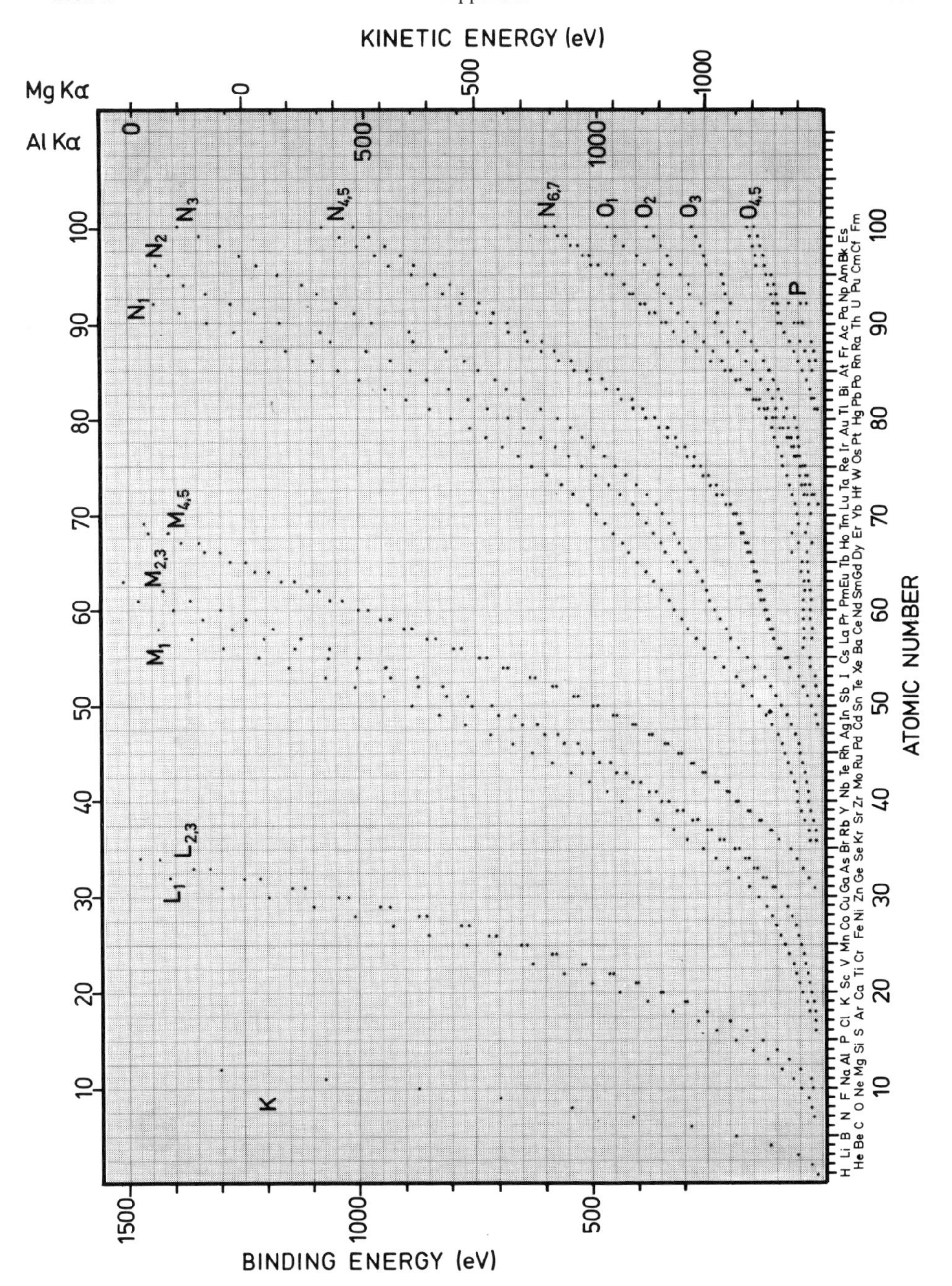

Acknowledgements. We gratefully acknowledge the good cooperation with Professor WERNER MEHLHORN, the editor of this volume. The many detailed and informative discussions we have had with Professor KAI SIEGBAHN have contributed substantially to the contents of our work. Many colleagues from our institute have kindly provided previously unpublished material. Finally, we wish to thank Mrs. GUNNEL INGELÖG for her patient and able work in the preparation of our manuscript.

Theory of the Auger Effect*

By

T. ÅBERG and G. HOWAT

With 14 Figures

Introduction

The ionization of atomic inner shells in matter is not only accompanied by the emission of characteristic x-ray and ultraviolet radiation but also by the emission of electrons with energies which are also characteristic of the emitting atoms. This phenomenon which is closely related to autoionization is known as the Auger effect. Since the nonradiative emission of Auger electrons has been proven to be important for testing the quantum theory of structure and collisions of atomic systems and valuable as a spectroscopic tool for applied physics, a number of monographs[1] and review articles[2,3] has been devoted to various aspects of the Auger effect. Hence it may be legitimate to ask why yet another treatise of the subject. Our answer is that the methods of calculating nonradiative transition energies and amplitudes and the methods of electron spectroscopy seem to have reached such a precision and maturity that it begins to be meaningful to ask for the fundamentals beyond the present perturbation theory of the Auger effect, based on the famous work of WENTZEL.[4] Hence our work may be looked upon as an attempt to treat the Auger effect in the broader context of the theory of resonance scattering, which, in turn, would

* The Manuscript was submitted in August 1979

[1] Burhop, E.H.S. (1952): The Auger effect and other radiationless transitions, Cambridge: University Press; Burhop, E.H.S., Asaad, W.N., (1972) in: Adv. At. Mol. Phys, Vol. 8, Bates, D.R., Esterman, I. (eds.), p. 164. New York: Academic Press; Chattarji, D. (1976): The theory of Auger transitions, New York: Academic Press

[2] Several articles in the following books and proceedings: Crasemann, B. (ed.) (1975): Atomic inner-shell processes, Vols. I and II, New York: Academic Press; Wuilleumier, F.J. (ed.) (1976): Photoionization and other probes of many-electron interactions, New York: Plenum Press; Fink, F.W., Manson, S.T., Palms, J.M., Venugopala Rao, P. (eds.) (1973): Proc. Int. Conf. Inn. Shell Ioniz. Phenomena Future Appl. Atlanta, Vols. I, II and III, Us. At. Energy Comm. Rep. No CONF-720404, Oak Ridge; Mehlhorn, W., Brenn, R. (eds.) (1976): Proc. Sec. Int. Conf. Inn. Shell Ioniz. Phenomena, Invited Papers, Freiburg. See also Bambynek, W., Crasemann, B., Fink, R.W., Freund, H.U., Mark, H., Swift, C.D., Price, R.E., Venugopala Rao, P. (1972): Rev. Mod. Phys. *44*, 716

[3] Mehlhorn, W. (1978): Electron spectroscopy of Auger and autoionizing states: experiment and theory. Lecture Notes, University of Aarhus (unpublished)

[4] Wentzel, G. (1927): Z. Phys. *43*, 524

ultimately lead to a valid framework for any nonradiative process, and at arbitrary impact energies. The second goal has been to display all the known important facts which are needed for consistent and accurate calculations of nonradiative transition energies and amplitudes.

We begin Chap. I by developing the time-independent theory of multichannel resonance scattering with two requirements in mind. Firstly, the electron and photon field should be treated on equal footing since the atoms can decay both radiatively and nonradiatively.[5–7] Secondly, the theory should resemble, following Fano[8] and Fano and Prats[9], the traditional configuration-interaction method as closely as possible since this is the approach which has been used most extensively to treat effects of electron correlation on nonradiative transition energies and probabilities. As a result, the concept of the radiative or nonradiative partial width is introduced quite rigorously in connection with the separability of the ionization and decay processes. The relationship between autoionization and the Auger effect is established, and it is shown how relaxation and final-state channel interaction including photon-electron coupling effects can be taken into account. The additivity of the radiative and nonradiative widths is obtained with photon-electron coupling corrections included in both widths.

In Chap. II we develop the nonrelativistic theory of nonradiative transitions from a many-electron point of view but by using central field one-electron spin orbitals as basis functions. Relaxation and final-state channel mixing corrections are introduced explicitly.[10] Special attention is paid to the phase factors of the various nonradiative amplitudes in the *LS*, *jj*, mixed, and intermediate coupling with somewhat controversial and disturbing results due to the many-electron version[11] of the *LS*- to *jj*-coupling transformation. The perturbation approach of Wentzel[4] and the many-body-perturbation-theory approach of Kelly[12] are examined from the point of view of our theory.

In Chap. III the relativistic two-electron amplitude of nonradiative transitions, given by Möller's formula,[13] is rederived in a manner which shows the relationship between the relativistic theory and the nonrelativistic theory in the independent-electron limit but with photon-electron coupling effects included. A formula of the relativistic amplitude is given in terms of radial integrals in the *jj*-coupling limit. The intermediate-coupling generalizations[14,15] are examined by use of an example.

[5] Åberg, T. (1977) in: Extended abstracts of V Int. Conf. Vac. Ultrav. Rad. Phys., At. Molec. Phys., Vol. I, Castex, M.C., Pouey, M., Pouey, N. (eds.), p. 73

[6] Armstrong, L., Jr., Theodosiou, C.E., Wall, M.J. (1978): Phys. Rev. A *18*, 2538

[7] Åberg, T. (1980): Physica Scripta *21*, 495; (1981) in: Inner-shell and x-ray physics of atoms and solids, Fabian, D.J., Kleinpoppen, H., Watson, L.M. (eds.), New York: Plenum Press, p. 251

[8] Fano, U. (1961): Phys. Rev. *124*, 1866

[9] Fano, U., Prats, F. (1963): Proc. Nat. Acad. Sci. India A *33*, 553

[10] Howat, G., Åberg, T., Goscinski, O. (1978): J. Phys. B *11*, 1575

[11] Condon, E.U., Shortley, G.H. (1964): The theory of atomic spectra, Chapt. XII, Cambridge: University Press

[12] Kelly, H.P. (1975): Phys. Rev. A *11*, 556

[13] Massey, H.S.W., Burhop, E.H.S. (1936): Proc. Roy. Soc. Lond. A *153*, 661

[14] Briançon, Ch., Desclaux, J.P. (1976): Phys. Rev. A *13*, 2157

[15] Asaad, W.N., Petrini, D. (1976): Proc. Roy. Soc. Lond. A *350*, 381

The angular distribution of Auger electrons is reviewed in Chap. IV for the first time with the exception of MEHLHORN's lecture notes.[3] This field has developed rapidly since MEHLHORN's[16] prediction of nonisotropic effects in the emission of Auger electrons. We treat only the Auger electron emission following photon and high-energy electron impact and assume complete separability of the ionization and decay processes. Although not necessary, the density matrix formalism is used since it provides the natural framework for the study of the radiative and nonradiative emission pattern in arbitrary collisions. Our approach as it is developed allows for channel mixing in the final state.

The calculation of nonradiative transition energies and amplitudes forms the subject of the last two chapters. The semiempirical methods for the calculation of transition energies are examined critically. In order to give an idea of the accuracy and consistency of ab initio calculations, we expose all the known factors, ranging from level shifts to correlation effects, which may influence the transition energies. The computation of the radial wave functions of the continuum electron has not received much attention in previous reviews. We have tried to remedy this situation by also discussing the choice of the potential and numerical methods. Our discussion of various factors which influence the nonradiative transition amplitudes remains rather exploratory. We have found it especially difficult to systematize the complex behavior of the radial Slater integrals which ultimately determine the dynamics of the Auger process.

Finally, we would like to list a number of limitations of our work. Firstly, only the case of one isolated resonance is treated although there are nonradiative processes which require a generalization of the theory to many closely spaced resonance states of the same symmetry. Secondly, the Auger electron emission from ion-atom collisions and the cascade of x-rays and Auger electrons are not treated explicitly. Thirdly, post-collision effects, i.e., the interaction between the Auger electron and a slowly escaping electron, is not discussed although it would be possible to develop the theory of these effects on the basis of the results of Chap. 1.[7] It is also clear that we deal solely with nonradiative transitions in free atoms and that multiple Auger electron emission and the radiative Auger effect are not discussed. We would also like to remark that our bibliography should be considered as illustrative rather than complete partly because our task has been to complement rather than repeat what has been said in previous reviews.[17]

I. A nonrelativistic scattering approach to the decay of inner-shell vacancy states

Fundamentally, any excitation process which gives rise to emission of particles with definite energies is an example of a resonance scattering process.

[16] Mehlhorn, W. (1968): Phys. Lett. A *26*, 166

[17] After the manuscript was submitted in 1979 there has been some new development, closely associated to the central themes of this work. This material has been included by bringing a few footnotes and references up to the date of publication in 1981

Inner-shell ionization followed by the emission of characteristic x rays and Auger electrons is no exception. In many cases this process can be succesfully interpreted as a two-step process, where the decay is treated separately from the ionization. Nevertheless, in order to understand the background and limitations of the two-step model it is necessary to adopt a broader view inherent in the scattering theory. This provides also the framework for generalizations of the perturbation theory[1] of the Auger effect to more complicated processes like the radiative Auger effect. Furthermore, it shows how electron correlation effects like final-state channel interaction can be taken into account in a systematic fashion.

In this chapter both characteristic electron and photon emission are considered from a unified point of view using the time-independent multichannel scattering theory. In this respect it differs from a related work of SHORE[2], who utilizes the time-dependent scattering and perturbation theory to describe absorption and autoionization-line profiles. In Sects. 1, 2, and 3 we develop the time-independent theory of multichannel scattering with and without resonances. The derivations are based on the configuration-interaction approach of FANO[3], and FANO and PRATS[4]. In this respect our approach can be looked upon as generalization of the traditional configuration-interaction approach but which involves the mixing of both discrete and continuum configurations. In Sect. 4 the theory is first applied to the autoionization process and then to the Auger effect. This shows the similarities and differences between these two processes. Section 5 is devoted to the application of the multichannel scattering theory to photon emission, and Sect. 6 indicates how this mode intermingles with the Auger electron emission in the decay of inner-shell vacancies.

1. Boundary conditions. Consider elastic or inelastic collisions between a beam of electrons and an ensemble of atoms under the usual scattering conditions (see, for instance, MOTT and MASSEY [1], GOLDBERGER and WATSON [2], RODBERG and THALER [3], TAYLOR [4], or JOACHAIN [5]). The accessible quantum states of the atoms are characterized by a set of quantum numbers denoted by μ. Some of these states may be in the continuum which would make the boundary conditions formally complicated. Hence it is assumed that the states $|\mu\rangle$ are discrete, i.e., $\langle\nu|\mu\rangle=\delta_{\nu\mu}$. In the time-independent scattering theory[5] the information concerning the result of an individual collision is given by the asymptotic behavior of the stationary scattering wave function Ψ_ν^+ which fulfils the outgoing wave boundary condition[6]

[1] The calculation of Auger electron probabilities is usually based on WENTZEL's ansatz (Wentzel, G. (1927): Z. Phys, *43*, 524), which is an example of Fermi's Golden Rule (see Sect. 10.γ)

[2] Shore, B.W. (1967): Rev. Mod. Phys. *39*, 439; (1968): Phys. Rev. *171*, 43; Shore, B.W. (1968): Lecture notes, University of Colorado, unpublished. See also the recent work by Armstrong, L., Jr., Theodosiou, C.E., Wall, M.J. (1978): Phys. Rev. A*18*, 2538

[3] Fano, U. (1961): Phys. Rev. *124*, 1866

[4] Fano, U., Prats, F. (1963): Proc. Nat. Acad. Sci. India A*33*, 553

[5] e.g. Ref. [*4*], Chap. 10

[6] Note that the antisymmetrization is not introduced explicitly in the boundary conditions, although it is assumed that the stationary scattering wave functions and the target wave functions are antisymmetric in the scattering region. (For a discussion see Ref. [*4*], Chap. 22)

$$\lim_{\text{any } r\to\infty} \Psi_\nu^+ = \sqrt{\frac{k}{(2\pi)^3}} \cdot \left[e^{i\boldsymbol{k}\cdot\boldsymbol{r}} \chi_{m_s}(\zeta)\,|\nu\rangle + \sum_{\mu} \sum_{m_s'=-1/2}^{+1/2} \sqrt{\frac{k}{k_\mu}} f_{\nu m_s}^{\mu m_s'}(\boldsymbol{k}, \hat{\boldsymbol{r}})\, \chi_{m_s'}(\zeta)\,|\mu\rangle \frac{1}{r} \cdot e^{ik_\mu r} \right], \tag{1.1}$$

where $\boldsymbol{k}$ is the wave vector of the electron in the incoming beam (Fig. 1). It is assumed that the atom is in the state $|\nu\rangle$ before the collision and that the incoming electron has the spin projection m_s with respect to the atomic axis of quantization. This situation is described by the spin eigenfunction $\chi_{m_s}(\zeta)$ which depends on the spin coordinate ζ. The scattering which leaves the atom in the state $|\mu\rangle$ occurs in the direction $\hat{\boldsymbol{r}}$ with the $\boldsymbol{k}$-dependent probability amplitude $f_{m_s}^{m_s'}(\boldsymbol{k}, \hat{\boldsymbol{r}})$. In the scattering process the spin projection m_s of the scattered electron has eventually changed into m_s' due to the presence of exchange and spin-orbit interactions. Energy conservation requires that

$$E = \tfrac{1}{2}k^2 + E_\nu = \tfrac{1}{2}k_\mu^2 + E_\mu \qquad \mu = 1 \ldots \nu \ldots M, \tag{1.2}$$

where E_μ is the energy of the state $|\mu\rangle$. The normalization constant in (1.1) has been chosen so that

$$\langle \Psi_\mu | \Psi_\nu \rangle = \delta_{\mu\nu}\, \delta(E - E')\, \delta(\hat{\boldsymbol{k}} - \hat{\boldsymbol{k}}'). \tag{1.3}$$

It follows that $|f_{\nu m_s}^{\mu m_s'}(\boldsymbol{k}, \hat{\boldsymbol{r}})|^2$ represents the differential cross section of the transition $\nu m_s \to \mu m_s'$ from which the actual cross section, corresponding to the experimental situation, can be obtained by averaging procedures.

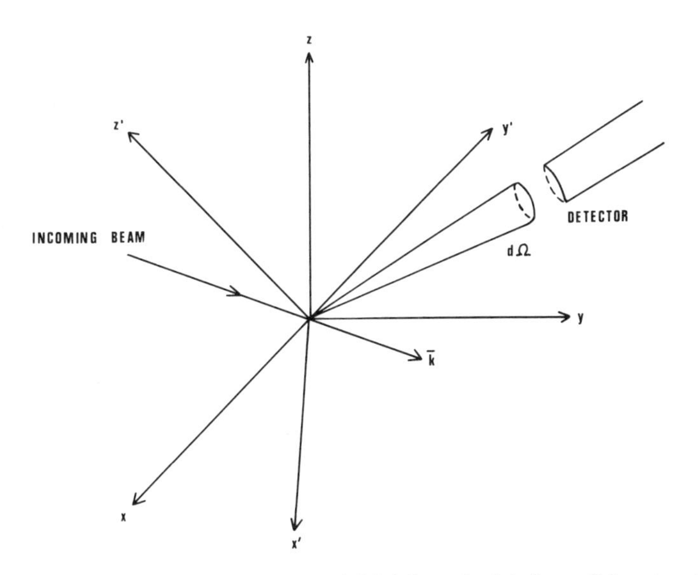

Fig. 1. Scattering geometry. Coordinate system $x'y'z'$ defines the lab frame (laboratory-fixed coordinate system) which may not coincide with the body frame (target-fixed coordinate system xyz) which defines the axes of quantization

The asymptotic form (1.1) can be written in terms of an $\boldsymbol{S}$ matrix by using partial wave expansions of $\exp(\mathrm{i}\boldsymbol{k}\cdot\boldsymbol{r})$ and $f_{\nu m_s}^{\mu m_s'}(\boldsymbol{k},\hat{\boldsymbol{r}})$ in the following form:

$$\mathrm{e}^{\mathrm{i}\boldsymbol{k}\cdot\boldsymbol{r}}=4\pi\sum_{l=0}^{\infty}\sum_{m=-l}^{+l}\mathrm{i}^l j_l(kr)\,Y_{lm}^*(\hat{\boldsymbol{k}})\,Y_{lm}(\hat{\boldsymbol{r}}),\tag{1.4}$$

where the spherical Bessel function $j_l(kr)$ is given asymptotically by

$$\lim_{r\to\infty} j_l(kr)=\frac{\sin\left(kr-l\frac{\pi}{2}\right)}{kr}\tag{1.5}$$

and

$$f_{\nu m_s}^{\mu m_s'}(\boldsymbol{k},\hat{\boldsymbol{r}})=\frac{1}{k}\sum_{lm_l}\sum_{l'm_l'} f_{\nu m_s l m_l}^{\mu m_s' l' m_l'}(k)\,Y_{lm_l}^*(\hat{\boldsymbol{k}})\,Y_{l'm_l'}(\hat{\boldsymbol{r}}).\tag{1.6}$$

The result is

$$\begin{aligned}\lim\Psi_\nu^+ =&\sum_{lm_l}\sum_{\kappa'} a_{lm_l}^+(\hat{\boldsymbol{k}})\,Y_{l'm_l'}(\hat{\boldsymbol{r}})\,\chi_{m_s'}(\zeta)\,|\mu\rangle\sqrt{\frac{2}{\pi k_\mu}}\\ &\cdot\left(S_{\kappa'\kappa}\,\mathrm{e}^{+\mathrm{i}\theta_{\kappa'}}-\delta_{\kappa'\kappa}\,\mathrm{e}^{-\mathrm{i}\theta_{\kappa'}}\right)/2\mathrm{i}r,\end{aligned}\tag{1.7}$$

where we have introduced the abbreviations $\kappa'=(\mu m_s' l' m_l')$ and $\kappa=(\nu m_s l m_l)$. In (1.7)

$$a_{lm_l}^+(\hat{\boldsymbol{k}})=\mathrm{i}^l\,Y_{lm_l}^*(\hat{\boldsymbol{k}})\tag{1.8}$$

and

$$\theta_{\kappa'}=k_\mu r-l'\tfrac{\pi}{2}.\tag{1.9}$$

The coefficients $f_{\kappa'\kappa}(k)$ in the expansion (1.6) are related to the elements $S_{\kappa'\kappa}$ of the scattering matrix $\boldsymbol{S}$ through the relation

$$f_{\kappa'\kappa}(k)=2\pi\mathrm{i}\,(\delta_{\kappa'\kappa}-S_{\kappa'\kappa}).\tag{1.10}$$

This shows that the information concerning the scattering is contained in the scattering matrix $\boldsymbol{S}$ which is given here in the lsm_lm_s representation.

Since we consider a Hamiltonian which is rotationally invariant, that is, which fulfils the commutations relations

$$[J^2,H]=[J_z,H]=0,\tag{1.11}$$

it is adventageous to introduce the total angular momentum representation JM using the unitary transformation

$$\begin{aligned}S_{\nu J_\nu lj}^{\mu J_\mu l'j'}(J)=&\sum_{\substack{m_l'm_s'\\ m_lm_s}}\sum_{\substack{m'M_\mu\\ mM_\nu}}(s'l'm_s'm_l'|j'm')(j'J_\mu m'M_\mu|JM)\\ &\cdot S_{\nu J_\nu M_\nu m_s l m_l}^{\mu J_\mu M_\mu m_s' l' m_l'}(slm_sm_l|jm)(jJ_\nu mM_\nu|JM),\end{aligned}\tag{1.12}$$

where $(j_1j_2m_1m_2|j_3m_3)$ are the Clebsch-Gordan coefficients and where the explicit notation $\mu=\{\mu J_\mu M_\mu\}$ and $\nu=\{\nu J_\nu M_\nu\}$ has been used. This enables us to write (1.7) in the form

$$\lim \Psi_\nu^+ = \sum_{\substack{\beta l \\ JM}} \sqrt{\frac{2}{\pi k_\beta}}\, A_{l\nu}^+(\hat{\boldsymbol{k}})\,|\beta J M\rangle\,(S_{\beta\alpha}\,\mathrm{e}^{+\mathrm{i}\theta_\beta} - \delta_{\beta\alpha}\,\mathrm{e}^{-\mathrm{i}\theta_\beta})/2\mathrm{i}r, \tag{1.13}$$

where the notation $\alpha = \{\nu J_\nu l j\}$ and $\beta = \{\mu J_\mu l' j'\}$ has been used. The coefficients

$$A_{l\nu}^+(\hat{\boldsymbol{k}}) = \mathrm{i}^l \sum_{\substack{m_l \\ jm}} (s l m_s m_l\,|\,j m)\,(j J_\nu m M_\nu\,|\,J M)\; Y_{l m_l}^*(\hat{\boldsymbol{k}}) \tag{1.14}$$

exhibit the dependence of $\lim \Psi_\nu^+$ on the spin projection m_s, direction $\hat{\boldsymbol{k}}$, and target quantum numbers J_ν, M_ν.

According to (1.13) and (1.14) it is possible to express the asymptotic behavior of Ψ_ν^+ as a superposition of the following N_c limiting functions (N_c may be infinite):

$$\lim \chi_\alpha^+ = \sum_{\beta=1}^{N_c} \frac{\Omega_\beta}{2\mathrm{i}r} \sqrt{\frac{2}{\pi k_\beta}}\,(S_{\beta\alpha}\,\mathrm{e}^{+\mathrm{i}\theta_\beta} - \delta_{\beta\alpha}\,\mathrm{e}^{-\mathrm{i}\theta_\beta}), \tag{1.15}$$

where Ω_β are the symmetry adapted functions $|\beta J M\rangle$. It is understood that the summation over β which defines a *channel* goes over all $\mu J_\mu j' l'$ that give rise to an eigenfunction associated with J and M. Note, however, that $S_{\beta\alpha}$ is independent of M according to rotational symmetry which is implied by (1.12). Whereas wave functions approaching the limiting form (1.13) are needed for the construction of angular distributions, wave functions approaching (1.15) can be used directly to construct total cross sections. In the following sections we assume that the asymptotic form is given by (1.15) or the corresponding ingoing wave modification. Once the wave functions χ_α^+ are known it is possible to obtain the wave fuctions Ψ_ν^+ as linear superpositions.

Let us form the adjoint scattering matrix $\boldsymbol{S}^\dagger$ which together with $\boldsymbol{S}$ satisfies the unitary relation $\boldsymbol{S}\boldsymbol{S}^\dagger = \boldsymbol{S}^\dagger\boldsymbol{S} = \mathbf{1}$. The wave functions

$$\chi_\alpha^- = \sum_{\mu=1}^{N_c} \chi_\mu^+\,(S^\dagger)_{\mu\alpha} \tag{1.16}$$

satisfy, according to (1.15), the ingoing wave boundary condition

$$\lim \chi_\alpha^- = \sum_{\beta=1}^{N_c} \frac{\Omega_\beta}{2\mathrm{i}r} \sqrt{\frac{2}{\pi k_\beta}}\,(\delta_{\beta\alpha}\,\mathrm{e}^{+\mathrm{i}\theta_\beta} - (S^\dagger)_{\beta\alpha}\,\mathrm{e}^{-\mathrm{i}\theta_\beta}). \tag{1.17}$$

This boundary condition is especially needed for the description of the influence of channel interactions on transition probabilities. We shall use it in Sects. 10.β and 20.δ to describe the influence of final-state continuum configuration-interaction effects on nonradiative transition probabilities.

The phase (1.9) pertains to an electron moving in the field of a neutral atom. In inner-shell ionization phenomena one always meets the situation where there is the long-range Coulomb interaction in addition to the short-range interactions. In this case the phase (1.9) must be modified [7] to

$$\theta_\beta = k_\beta r + \frac{Z}{k_\beta} \ln 2k_\beta r - l_\beta \tfrac{\pi}{2} + \sigma_\beta^{\mathrm{C}} + \sigma_\beta, \tag{1.18}$$

where $\sigma_\beta^{\mathrm{C}}$ is the Coulomb phase

$$\sigma_\beta^{\mathrm{C}} = \arg \Gamma \left(l_\beta + 1 - \mathrm{i} \frac{Z}{k_\beta} \right), \tag{1.19}$$

and where σ_β is any additional short-range phase shift. In (1.18) Z is the residual charge of the target ion seen by the scattered electron at large distances. This modification also introduces an extra phase factor $\exp[\pm(\sigma_\beta^{\mathrm{C}} + \sigma_\beta)]$ into the coefficient (1.14). The minus sign corresponds to the ingoing wave boundary condition.

The boundary conditions have been given with respect to the axes of quantization (Fig. 1) which define an atom's (molecule's) fixed coordinate system (body frame). In the theory of angular distributions the boundary conditions must eventually be expressed with respect to a laboratory-fixed coordinate system (lab frame). Since the $\boldsymbol{r}$ dependent part is rotationally invariant in (1.13), the desired boundary conditions are obtained by using the transformation [8]

$$Y_{lm}(\hat{\boldsymbol{k}}) = \sum_{m'} Y_{lm'}(\hat{\boldsymbol{k}}') D^l_{m'm}(\hat{\boldsymbol{R}}) \tag{1.20}$$

in (1.14). In this transformation $D^l_{m'm}(\hat{\boldsymbol{R}})$ are the matrix elements of finite rotations which carry the lab frame into the body frame through the Euler angles $\hat{\boldsymbol{R}} = (\alpha\beta\gamma)$. In (1.20) the direction $\hat{\boldsymbol{k}}'$ is defined with respect to the lab frame. For outgoing wave boundary conditions the direction of the incoming beam $\hat{\boldsymbol{k}}' = (\theta, \phi)$ is usually chosen to be the z' axis. According to (1.20),

$$Y_{lm}(\hat{\boldsymbol{k}}) = Y_{l0}(0,0)\, D^l_{0m}(\hat{\boldsymbol{R}}) = \sqrt{\frac{2l+1}{4\pi}}\, D^l_{0m}(\hat{\boldsymbol{R}})$$

in that case.

2. Nonresonant multichannel scattering. As shown in Sect. 1 the symmetry-adapted stationary scattering wave functions satisfy the outgoing wave boundary condition (1.15) or the ingoing wave boundary condition (1.17) (Fig. 2). In the former case the wave function describes the scattering of an electron which enters into a definite channel but escapes in one of the alternative channels, the probability amplitude being given by the corresponding $\boldsymbol{S}$ matrix element. In the latter case the scattering is viewed from an equivalent point of view of the adjoint scattering matrix $\boldsymbol{S}^\dagger$ which describes what the probability distribution in the entrance channels must be in order to result in an escape into a

---

[7] e.g. Ref. [7], Vol. I, Appendix B
[8] e.g. Ref. [8]

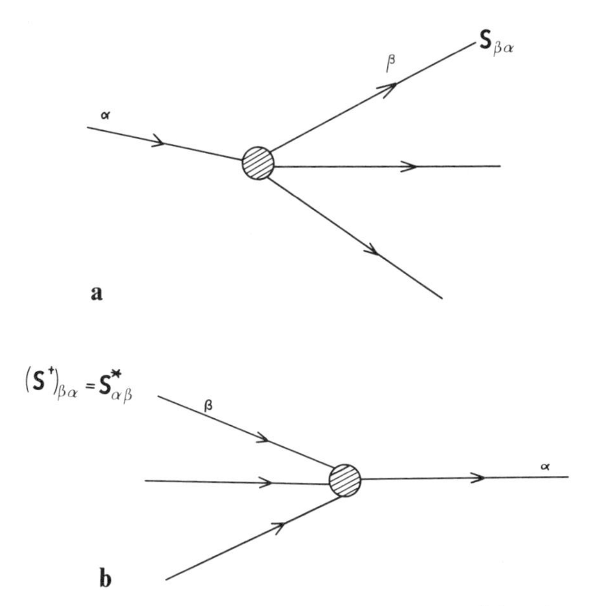

Fig. 2a, b. Outgoing wave (**a**) and ingoing wave (**b**) boundary condition, shown schematically

given channel. According to Sect. 1, the channel is characterized in the JM representation by the total angular momentum and additional quantum numbers of the target atom or ion and the orbital and total angular momentum quantum number of the incoming or outgoing electron.

The multichannel scattering is a consequence of the electron-electron interaction. If the target nucleus and the electrons are described by a central field potential $V(r)$, then the elastic scattering is described by diagonal scattering matrix elements $S_l = e^{2i(\sigma_l^c + \sigma_l)}$.[9] According to (1.15), the corresponding asymptote is

$$\lim \chi_{\alpha\varepsilon} = \frac{1}{r}\sqrt{\frac{2}{\pi k}}\,\Omega_\alpha\, e^{\pm i[\sigma_\alpha^c(\varepsilon) + \delta_\alpha(\varepsilon)]} \sin\theta_\alpha(\varepsilon), \tag{2.1}$$

where $\theta_\alpha(\varepsilon)$ is given by (1.18). In (2.1) $\varepsilon = k^2/2$ is the kinetic energy of the scattered electron which is equal to $k_\alpha^2/2 = E - E_\alpha$ on the energy shell. The phase shift $\sigma_\alpha(\varepsilon)$ characterizes the deviations of the central field from the Coulomb field Z/r of the residual ion. In order to describe the influence of the electron-electron interaction on the scattering the residual interaction $\sum_{\mu>\nu}(1/r_{\mu\nu}) - \sum_\mu V_\mu(r_\mu)$ is considered. As a consequence, states which fulfil the boundary conditions (2.1) will interact. This results in a nondiagonal $\boldsymbol{S}$ matrix which describes the inelastic scattering as well.

In analogy to the ordinary configuration-interaction problem, the solution of the multichannel problem can now be obtained as follows.[10] Given a set of

[9] See any of the Refs. [*1–5*]

[10] We follow the idea of FANO and PRATS (see I.4)

continuum wave functions $\chi_{\alpha\varepsilon}$ which fulfil the boundary condition[11]

$$\lim \chi_{\alpha\varepsilon} = \frac{1}{r}\sqrt{\frac{2}{\pi k_\alpha}}\,\Omega_\alpha \sin\theta_\alpha(\varepsilon), \tag{2.2}$$

a new set $\{\Psi^{\pm}_{\alpha\varepsilon}\}$ is obtained as linear superpositions

$$\Psi^{\pm}_{\alpha\varepsilon} = \sum_{\beta=1}^{N_c} \int_0^\infty \chi_{\beta\tau}\, b^{\pm}_{\beta\alpha}(\tau,\varepsilon)\, d\tau, \tag{2.3}$$

such that

$$\langle \chi_{\beta\varepsilon} | H - E | \Psi^{\pm}_{\alpha\mathcal{E}} \rangle = 0 \qquad \alpha, \beta = 1 \dots N \tag{2.4}$$

for $\mathcal{E} = \mathcal{E}_\alpha = E - E_\alpha$.[12]

The boundary conditions for $\Psi^{\pm}_{\alpha\mathcal{E}}$ are given by

$$\lim \Psi^{\pm}_{\alpha\mathcal{E}} = \sum_{\beta=1}^{N_c} \frac{\Omega_\beta}{2ir}\sqrt{\frac{2}{\pi k_\beta}}\, g^{\pm}_{\beta\alpha}, \tag{2.5}$$

where

$$g^{+}_{\beta\alpha} = S_{\beta\alpha}\, e^{i\theta_\beta} - \delta_{\beta\alpha}\, e^{-i\theta_\beta} \tag{2.6}$$

and

$$g^{-}_{\beta\alpha} = \delta_{\beta\alpha}\, e^{+i\theta_\beta} - (S^\dagger)_{\beta\alpha}\, e^{-i\theta_\beta}. \tag{2.7}$$

Note that the matrix elements $S_{\beta\alpha}$ and the angles θ_β are functions of $\mathcal{E}$. The Hamiltonian matrix is given with respect to the "standing wave" states $\chi_{\alpha\varepsilon}$ by

$$\langle \chi_{\beta\varepsilon'} | H - E | \chi_{\alpha\varepsilon} \rangle = (\varepsilon + E_\alpha - E)\, \delta(\varepsilon' + E_\beta - \varepsilon - E_\alpha)\, \delta_{\beta\alpha} + V_{\beta\alpha}(\varepsilon', \varepsilon, E). \tag{2.8}$$

This definition of the residual-interaction matrix $\boldsymbol{V}(\varepsilon', \varepsilon, E)$ does not imply a partioning of the operator H into the sum of two fixed operators H_0 and V. However, it leads to the usual "post" and "prior" representations of the transition matrix[13] if a partioning of H into $H_\alpha + V_\alpha (\alpha = 1 \dots N_c)$ is defined for each channel.

From (2.3) and (2.4) it follows that (the superscripts $\pm$ are dropped)

$$b_{\beta\alpha}(\varepsilon, \mathcal{E})(\varepsilon + E_\beta - E) + \sum_{\gamma=1}^{N_c} \int_0^\infty V_{\beta\gamma}(\varepsilon, \tau, E)\, b_{\gamma\alpha}(\tau, \mathcal{E})\, d\tau = 0. \tag{2.9}$$

According to Appendix A this system of linear equations can be solved by removing the singularity $E = \varepsilon + E_\beta$ using the transformation

$$b_{\mu\nu}(\varepsilon, \mathcal{E}) = \mathscr{P}\frac{Y_{\mu\nu}(\varepsilon, \mathcal{E})}{E - E_\mu - \varepsilon} + Z_{\mu\nu}(\mathcal{E})\, \delta(E - E_\mu - \varepsilon), \tag{2.10}$$

[11] The phase factor $\exp \pm[\sigma^C_\alpha(\varepsilon) + \sigma_\alpha(\varepsilon)]$ is omitted in the following with the understanding that it has to be considered in (1.12), which is needed for the evaluation of differential cross sections

[12] The subscript α of $\mathcal{E}_\alpha = E - E_\alpha$ is dropped from $\mathcal{E}$ whenever there is no confusion that it refers to the energy shell and is associated with the threshold energy E_α

[13] e.g. Ref. [4], Chap. 18

where $\mathscr{P}$ means the principal Cauchy value of the integral having the factor $Y_{\mu\nu}(\varepsilon,\mathcal{E})(E-E_\mu-\varepsilon)^{-1}$ in the integrand. This leads to

$$-Y_{\beta\alpha}(\varepsilon,\mathcal{E})+\sum_{\gamma=1}^{N_c}\left[\mathscr{P}\int_0^\infty \frac{V_{\beta\gamma}(\varepsilon,\tau,E)\,Y_{\gamma\alpha}(\tau,\mathcal{E})\,d\tau}{E-E_\gamma-\tau}+V_{\beta\gamma}(\varepsilon,\mathcal{E},E)\,Z_{\gamma\alpha}(\mathcal{E})\right]=0. \quad (2.11)$$

The $\boldsymbol{Z}$ matrix can be eliminated from (2.11) by using the boundary conditions (2.5). In order to do this we express the functions (2.3) in terms of $\boldsymbol{Y}$ and $\boldsymbol{Z}$ with the aid of the transformation (2.10). By letting any of the radial distances r approach infinity and using (A.11) of Appendix A we obtain $\lim \Psi^{\pm}_{\alpha\mathcal{E}}$ as a function of $\cos\theta_\gamma$ and $\sin\theta_\gamma$. Since this limit must coincide with (2.5), the following conditions

$$\mathrm{i}\pi\,Y^+_{\beta\alpha}(\mathcal{E},\mathcal{E})+Z^+_{\beta\alpha}(\mathcal{E})=\delta_{\beta\alpha} \qquad -\mathrm{i}\pi\,Y^+_{\beta\alpha}(\mathcal{E},\mathcal{E})+Z^+_{\beta\alpha}(\mathcal{E})=S_{\beta\alpha}(\mathcal{E}) \quad (2.12)$$

$$\mathrm{i}\pi\,Y^-_{\beta\alpha}(\mathcal{E},\mathcal{E})+Z^-_{\beta\alpha}(\mathcal{E})=S^\dagger_{\beta\alpha}(\mathcal{E}) \quad -\mathrm{i}\pi\,Y^-_{\beta\alpha}(\mathcal{E},\mathcal{E})+Z^-_{\beta\alpha}(\mathcal{E})=\delta_{\beta\alpha} \quad (2.13)$$

must prevail on the energy shell. From these equations it follows that $S_{\beta\alpha}=\delta_{\beta\alpha}-2\pi\mathrm{i}\,Y^+_{\beta\alpha}$ and $Y^-_{\beta\alpha}=Y^{+*}_{\alpha\beta}$. This identifies $\boldsymbol{Y}^+$ as the transition matrix $\boldsymbol{T}^+$ and $\boldsymbol{Y}^{-\dagger}$ as the transition matrix $\boldsymbol{T}^-$ and shows that on the energy shell, $\boldsymbol{T}^+=\boldsymbol{T}^-$.[14] According to (1.10), the absolute squares of the transition matrix elements are proportional to the corresponding cross sections. The $\boldsymbol{T}$ matrices satisfy

$$\boldsymbol{T}^+(\varepsilon,\mathcal{E})=\boldsymbol{V}(\varepsilon,\mathcal{E},E)+\lim_{\nu\to 0}\int_0^\infty \boldsymbol{V}(\varepsilon,\tau,E)\,[(E-\tau+\mathrm{i}\nu)\mathbf{1}-\boldsymbol{E}]^{-1}\,\boldsymbol{T}^+(\varepsilon,\tau)\,d\tau \quad (2.14)$$

and

$$\boldsymbol{T}^-(\mathcal{E},\varepsilon)=\boldsymbol{V}(\mathcal{E},\varepsilon,E)+\lim_{\nu\to 0}\int_0^\infty \boldsymbol{T}^-(\mathcal{E},\tau)\,[(E-\tau+\mathrm{i}\nu)\mathbf{1}-\boldsymbol{E}]^{-1}\,\boldsymbol{V}(\tau,\varepsilon,E)\,d\tau, \quad (2.15)$$

which are known as the Lippmann-Schwinger equations. The limiting procedure with respect to ν in these equations is defined in Appendix A.

In spite of the fact that the $\boldsymbol{V}(\varepsilon,\varepsilon',E)$ matrix is real due to the standing wave boundary condition (2.2) the $\boldsymbol{T}$ matrices are not. Hence in practice one usually considers the $\boldsymbol{K}$ matrix which is defined by

$$\boldsymbol{Y}(\varepsilon,\mathcal{E})=\boldsymbol{K}(\varepsilon,\mathcal{E})\,\boldsymbol{Z}(\mathcal{E}). \quad (2.16)$$

This definition together with (2.11) yields the K matrix equation

$$\boldsymbol{K}(\varepsilon,\mathcal{E})=\boldsymbol{V}(\varepsilon,\mathcal{E},E)+\mathscr{P}\int_0^\infty \boldsymbol{V}(\varepsilon,\tau,E)\,[(E-\tau)\mathbf{1}-\boldsymbol{E}]^{-1}\,\boldsymbol{K}(\tau,\mathcal{E})\,d\tau, \quad (2.17)$$

---

[14] e.g. Ref. [4], Chap. 3

which shows that the $\boldsymbol{K}$ matrix is real as long as the $\boldsymbol{V}$ matrix is real. It follows from the boundary conditions (2.12) and (2.13) and the definition (2.16) that

$$\boldsymbol{S}(\mathcal{E})=\frac{\mathbf{1}-\mathrm{i}\pi\boldsymbol{K}(\mathcal{E},\mathcal{E})}{\mathbf{1}+\mathrm{i}\pi\boldsymbol{K}(\mathcal{E},\mathcal{E})}, \tag{2.18}$$

which shows that the $\boldsymbol{K}$ matrix is symmetric on the energy shell.

With regard to the Auger effect the result of this section makes it possible to treat final-state channel interactions. For this purpose we also need the Lippmann-Schwinger equations in a form which exposes the wave function behavior. It follows from (2.3) and (2.10–13) that

$$\Psi_{\alpha\mathcal{E}}^{\pm}=\chi_{\alpha\mathcal{E}}+\lim_{\nu\to 0}\sum_{\beta=1}^{N_c}\int_0^{\infty}\frac{Y_{\beta\alpha}^{\pm}(\tau,\mathcal{E})\,\chi_{\beta\tau}\,d\tau}{E-E_\beta-\tau\pm\mathrm{i}\nu}, \tag{2.19}$$

where $\boldsymbol{Y}^{+}$ is equivalent to the $\boldsymbol{T}^{+}$ matrix and $\boldsymbol{Y}^{-\dagger}$ to $\boldsymbol{T}^{-}$. If it is assumed that the elements of the perturbation matrix $\boldsymbol{V}$ can be expressed with respect to perturbation operators $V_\alpha=H-H_\alpha$, then (2.19) can be cast, according to (2.11), into

$$\Psi_{\alpha\mathcal{E}}^{\pm}=\chi_{\alpha\mathcal{E}}+\lim_{\nu\to 0}\sum_{\beta=1}^{N_c}\int_0^{\infty}\frac{\chi_{\beta\tau}\langle\chi_{\beta\tau}|V_\beta|\Psi_{\alpha\mathcal{E}}^{\pm}\rangle\,d\tau}{E-E_\beta-\tau\pm\mathrm{i}\nu}. \tag{2.20}$$

Note that these equations are equivalent to (2.14) and (2.15) provided that

$$T_{\beta\alpha}^{+}(\varepsilon,\mathcal{E})=\langle\chi_{\beta\varepsilon}|V_\beta|\Psi_{\alpha\mathcal{E}}^{+}\rangle,\qquad T_{\beta\alpha}^{-}(\mathcal{E},\varepsilon)=\langle\Psi_{\beta\mathcal{E}}^{-}|V_\alpha|\chi_{\alpha\varepsilon}\rangle, \tag{2.21}$$

which are the usual post and prior forms of the $\boldsymbol{T}$ matrix elements.

3. Resonant scattering including channel interaction. In this section we examine the origin of an isolated resonance by generalizing the results of Sect. 2 (Fig. 3). It is shown how the internal structure of a channel wave function $\Psi_{\alpha\mathcal{E}}^{\pm}$ can give rise to a resonance behavior in the scattering matrix and the cross section at some energy $\mathcal{E}_r$. Following FANO[3], the idea of a resonance as a "nearly bound" state is invoked in the theory by assuming that it is possible to represent $\Psi_{\alpha\mathcal{E}}^{\pm}$ in the vicinity of $\mathcal{E}_r$ by the "configuration-interaction" (CI) expansion

$$\Psi_{\alpha\mathcal{E}}=a'_\alpha(\mathcal{E})\,\Phi+\sum_{\beta=1}^{N_c}\int_0^{\infty}\chi_{\beta\tau}\,C_{\beta\alpha}(\tau,\mathcal{E})\,d\tau, \tag{3.1}$$

where Φ is square integrable, i.e., $\langle\Phi|\Phi\rangle=1$, and where $\chi_{\beta\tau}$ are the standing wave states which satisfy the boundary conditions (2.2) and define a Hamiltonian matrix which is of the form (2.8). In addition, $\langle\Phi|H-E|\chi_{\alpha\varepsilon}\rangle=M_\alpha(\varepsilon,\mathcal{E})$ is in general nonzero, giving rise to the resonance behavior. The diagonalization of the discrete-continuum Hamiltonian matrix consisting of this matrix element and the matrix (2.8) can be achieved in two steps. First, the matrix (2.8) is prediagonalized using the method described in the previous section, and

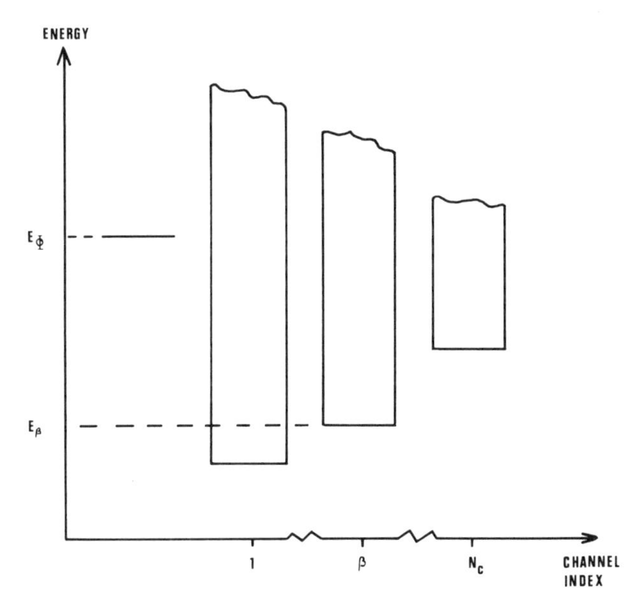

Fig. 3. Energy diagram of an isolated bound state interacting with N_c continua

secondly, a linear combination

$$\Psi^{\pm}_{\alpha\mathcal{E}} = a^{\pm}_{\alpha}(\mathcal{E})\,\Phi + \sum_{\beta=1}^{N_c} \int_0^{\infty} \chi^{\pm}_{\beta\tau}\, C^{\pm}_{\beta\alpha}(\tau, \mathcal{E})\, d\tau \tag{3.2}$$

is formed such that

$$\langle \Phi | H - E | \Psi^{\pm}_{\alpha\mathcal{E}} \rangle = \langle \chi^{\pm}_{\beta\varepsilon} | H - E | \Psi^{\pm}_{\alpha\mathcal{E}} \rangle = 0. \tag{3.3}$$

In (3.2) the continuum wave functions satisfy the boundary conditions (2.5). It is required that $\Psi^{\pm}_{\alpha\mathcal{E}}$ satisfy similar boundary conditions but with the scattering matrix $\boldsymbol{S}$ replaced by another scattering matrix $\boldsymbol{S}'$.

It follows from the requirement (3.3) that

$$(E_{\Phi} - E)\, a^{\pm}_{\alpha}(\mathcal{E}) + \sum_{\beta=1}^{N_c} \int_0^{\infty} M^{\pm}_{\beta}(\tau, \mathcal{E})\, C^{\pm}_{\beta\alpha}(\tau, \mathcal{E})\, d\tau = 0, \tag{3.4}$$

$$a^{\pm}_{\alpha}(\mathcal{E})\, M^{\pm}_{\beta}(\varepsilon, \mathcal{E})^{*} + (\varepsilon + E_{\beta} - E)\, C^{\pm}_{\beta\alpha}(\varepsilon, \mathcal{E}) = 0,$$

where $M^{\pm}_{\beta}(\varepsilon, \mathcal{E}) = \langle \Phi | H - E | \chi^{\pm}_{\beta\varepsilon} \rangle$ and $E_{\Phi} - E = \langle \Phi | H - E | \Phi \rangle$. Note that Φ and $\chi^{\pm}_{\beta\varepsilon}$ are not necessarily orthogonal, which, as we shall indicate in Sects. 10 and 20, is essential for the development of a consistent theory of the Auger effect. This system of linear equations, which has N solutions satisfying the appropriate boundary conditions, becomes singular at $E = \varepsilon_{\beta} + E_{\beta}$. However, this singularity, which appears in the second equation of (3.4), can be removed by the application of the transformation (2.10) to the coefficients $C^{\pm}_{\beta}(\tau, \mathcal{E})$. As in the nonresonant case the $\boldsymbol{Y}$ and $\boldsymbol{Z}$ matrices are determined on the energy shell by the boundary conditions. It follows from the comparison of the wave functions (3.2) with the asymptotic forms (2.5) at large r that these boundary conditions are in the matrix form

$$\begin{aligned} \mathrm{i}\pi \boldsymbol{Y}^{+}(\mathcal{E},\mathcal{E})+\boldsymbol{Z}^{+}(\mathcal{E}) &= \mathbf{1} \\ \boldsymbol{S}(\mathcal{E})[-\mathrm{i}\pi \boldsymbol{Y}^{+}(\mathcal{E},\mathcal{E})+\boldsymbol{Z}^{+}(\mathcal{E})] &= \boldsymbol{S}'(\mathcal{E}) \end{aligned} \tag{3.5}$$

for the outgoing wave $(+)$ case. The ingoing wave $(-)$ boundary conditions are obtained from (3.5) by transposing and taking the complex conjugate. As in the nonresonant case $\boldsymbol{Y}^{-}=\boldsymbol{Y}^{+\dagger}$ on the energy shell. However, $\boldsymbol{Y}^{+}=(\mathbf{1}-\boldsymbol{S}^{+}\boldsymbol{S}')/2\pi\mathrm{i}$ so that the $\boldsymbol{Y}$ matrices cannot directly be identified as $\boldsymbol{T}$ matrices for the resonance process.

Substituting the transformation (2.10) into (3.4) and using the boundary conditions (3.5) yields the solutions of (3.3) in the form

$$\Psi_{\alpha}^{\pm}(\mathcal{E})=\chi_{\alpha}^{\pm}(\mathcal{E})+\frac{M_{\alpha}^{\pm}(\mathcal{E},\mathcal{E})}{E-E_{r}\pm\mathrm{i}\frac{\Gamma}{2}}\left[\Phi+\lim_{\nu\to 0}\sum_{\beta=1}^{N_c}\int\frac{\chi_{\beta\tau}^{\pm}M_{\beta}^{\pm}(\tau,\mathcal{E})^{*}\,d\tau}{E-E_{\beta}-\tau\pm\mathrm{i}\nu}\right], \tag{3.6}$$

where the limiting procedure $\nu\to 0$ is defined by (A.10). In (3.6), where

$$\Gamma=\sum_{\alpha=1}^{N_c}\Gamma_{\alpha}^{\pm}=2\pi\sum_{\alpha=1}^{N_c}|M_{\alpha}^{\pm}(\mathcal{E},\mathcal{E})|^{2}, \tag{3.7}$$

the resonance energy $E_{\mathrm{r}}=E_{\Phi}+\Delta$ deviates from the energy E_{Φ} by the level shift

$$\Delta=\sum_{\alpha=1}^{N_c}\mathscr{P}\int_{0}^{\infty}\frac{|M_{\alpha}^{\pm}(\tau,\mathcal{E})|^{2}\,d\tau}{E-E_{\alpha}-\tau}. \tag{3.8}$$

In addition, it follows that

$$S'_{\beta\alpha}(\mathcal{E})=S_{\beta\alpha}(\mathcal{E})-2\pi\mathrm{i}\frac{M_{\alpha}^{+}(\mathcal{E},\mathcal{E})\,M_{\beta}^{-}(\mathcal{E},\mathcal{E})^{*}}{E-E_{\mathrm{r}}+\mathrm{i}\frac{\Gamma}{2}}, \tag{3.9}$$

from which the $\boldsymbol{T}'$ matrix elements can be obtained by using the relation $\boldsymbol{T}'=(\mathbf{1}-\boldsymbol{S}')/2\pi\mathrm{i}$. This result follows from using the outgoing wave boundary condition whereas the ingoing wave boundary condition yields the adjoint scattering matrix $\boldsymbol{S}^{\dagger}$ as it should be. It can be shown directly from (3.9) that $\boldsymbol{S}'$ is unitary provided that $\boldsymbol{S}$ is unitary. This ensures through the boundary conditions (2.5) that the wave functions $\Psi_{\alpha\mathcal{E}}^{\pm}$ are also normalized per unit energy range. The matrix elements (3.9) describe the influence of a single resonance on the scattering matrix in *presence* of direct scattering. Each element consists of two amplitudes which describe the probability that an electron entering channel α is emitted in channel β. The former amplitude represents the *direct* scattering, whereas the latter represents the scattering via the bound state Φ which is governed by the discrete-continuum interaction matrix elements $M_\mu^\pm(\mathcal{E},\mathcal{E})=\langle\Phi|H-E|\chi_{\beta\mathcal{E}}^\pm\rangle$. The celebrated Breit-Wigner formula[15] is obtained as a special case by putting $S_{\beta\alpha}=0$. In Sect. 4 we shall discuss the consequences of the result (3.9) concerning autoionization and the Auger effect.

The results of this section do not tell anything about how the discrete state Φ should be chosen. It is therefore gratifying to see (Appendix B) that the CI method of FANO[3] as it is developed in this section is equivalent to FESHBACH's[16] projection-operator approach[17] provided the projection operators are chosen as

$$Q=|\Phi\rangle\langle\Phi|, \qquad P=\int_0^\infty \sum_{\beta=1}^{N_c} |\chi_{\beta\tau}^\pm\rangle\langle\chi_{\beta\tau}^\pm|\,d\tau \tag{3.10}$$

and provided the orthogonality relations $\langle\Phi|\chi_{\beta\tau}^\pm\rangle=0$ $(\beta=1\dots N)$ are valid. If the set $\{\Phi,\chi_{\beta\tau}^\pm\}$ would be complete, which it never is in practice, then the CI method would be equivalent to the choice of two Feshbach projection operators P and Q such that

$$P+Q=1, \qquad \lim P\Psi_{\alpha\mathcal{E}}^\pm=\lim \Psi_{\alpha\mathcal{E}}^\pm. \tag{3.11}$$

This choice is not unique. However, as long as no approximations are made in the partitioning of the Schrödinger equation (see Appendix B), the final results concerning the scattering matrix would be the same in the sense that they would be connected by unitary transformations. However, the approximation of identifying the resonances as bound eigenstates of the operator QHQ (i.e., as Φ in our case) brings up the question of the best choice of Q. This question has been discussed extensively in the literature[18] and a few remarks concerning inner-shell vacancy states are made in Sect. 16.

4. Autoionization and the Auger effect as resonance scattering. The concept of nonradiative transitions. The preceding developments of the resonant electron scattering are quite general, the only restriction being that the number of bound states is assumed to be one and that all residual target states are assumed to be discrete. The former restriction can be removed.[19] In FESHBACH's approach (Appendix B) it corresponds to the replacement of the

[15] Breit, G., Wigner, E.P. (1936): Phys. Rev. *49*, 519

[16] Feshbach, H. (1958): Ann. Phys. *5*, 357; (1962): *19*, 287

[17] For previous discussions, see O'Malley, T.F., Geltman, S. (1965): Phys. Rev. *137* A, 1344; Mahaux, C., Weidenmüller, H.A. (1969): Shell-model approach to nuclear reactions, Amsterdam: North-Holland, Chap. 3, Sect. 8; Tweed, R.J. (1976): J. Phys. B *9*, 1725

[18] Without trying to systematize various approaches we would like to mention the following recent references as general guidelines to the subject: Russek, A., Wu, H.C., Owens, J. (1969): Phys. Rev. *180*, 6; Nicolaides, C.A. (1972): Phys. Rev. A *6*, 2078; Froelich, P., Brändas, E. (1975): Phys. Rev. A *12*, 1; Öksüz, I. (1976): Phys. Rev. A *13*, 1507; Brändas, E., Froelich, P. (1977): Phys. Rev. A *16*, 2207; Wendoloski, J.J., Reinhardt, W.P. (1978): Phys. Rev. A *17*, 195

[19] Mies, F.H. (1968): Phys. Rev. *175*, 164; Starace, A.F. (1972): Phys. Rev. B *5*, 1773; Davis, L.C., Feldkamp, L.A. (1977): Phys. Rev. B *15*, 2961; see also the book of MAHAUX and WEIDENMÜLLER, which is closest in spirit to our approach, and Ref. 7 of introduction

projection operator $Q=|\Phi\rangle\langle\Phi|$ by $Q=\sum_{\beta=1}^{N'}|\Phi_\beta\rangle\langle\Phi_\beta|$ in (3.10). This does not essentially alter any results as long as the energy separation between the states Φ_β is large in comparison with Γ_β, which is often true in the case of the Auger effect. The latter restriction is more severe since it leads to the consideration of the boundary conditions in the case of the ejection of two or more electrons.[20] Although a rigorous justification has apparently not been given, we assume that the present theory will not change formally if the set $\{\Omega_\alpha\}$ includes continuum states in connection with the boundary condition (1.1). In a strict theory of the Auger effect we would face this problem since it always involves multiple electron ionization, i.e., the primary ionization process is accompanied by the ejection of at least one Auger electron.

Not only electrons but also other scattering particles can be included in the multichannel scattering theory of the preceeding sections. In particular we shall be concerned about photons as well since inner-shell ionization generally leads to both electron and photon emission. The generalization of the theory to photon-electron systems requires some special considerations since the photon cannot be represented by a wave function in the coordinate space.[21] Another difficulty which is also related to the intrinsic properties of quantized fields is that the level shift (3.8) becomes infinite if the interaction between the discrete and continuum states is governed by the photon-electron interaction.[22] The former obstacle can be handled by relating the boundary conditions to matrix elements of the vector potential or the electric field vector of the photon field. The latter obstacle and related divergencies can be circumvented by the procedures of renormalization[22] or by a truncation of the scattering equations at highest nondivergent order in the photon-electron interaction, which is usually sufficient for practical purposes. Hence it is possible to apply the multichannel theory of the preceeding sections to photon scattering processes without any modifications of the scattering equations themselves. It is only required that the Hamiltonian of the system includes the photon field and the photon-electron interaction so that the wave functions include the photon states.

We shall begin with considerations of autoionization in photoabsorption (neglecting any multiphoton processes) since autoionization is simpler to treat than the Auger effect from the point of view of the boundary conditions. In

[20] For a general discussion see, e.g., Ref. [*4*], Chap. 19. RUDGE has discussed this problem in the context of electron ionization (Rudge, M.R.H. (1968): Rev. Mod. Phys. *40*, 564). His use of hyperspherical coordinates to describe the boundary conditions does not fit the present treatment very well

[21] This is a general statement made in treatises of "Quantum electrodynamics" (see, e.g. Ref. [*17*], Chap. I, Sect. 2.2). This dilemma could evidently be avoided by a suitable generalization of the concept of position for photons (Kraus, K. (1977) in: The uncertainty principle and foundations of quantum mechanics, Price, W.C., Chissick, S.S. (eds.), New York: Wiley, p. 293)

[22] This is an example of divergencies that occur in applications of the scattering theory to systems that are described by the quantum field theory. After renormalization of the electron mass and charge, the divergent quantities turn out to be very small and can usually be neglected. As shown in Sect. 5, the level shift (3.8) is related to the Lamb shift. In addition to the book of AKHIEZER and BERESTETSKII (Ref. [*17*]), see also the article by Fowler, G.N., in Quantum theory III. Radiation and high energy physics, Bates, D.R. (ed.) (1962), New York: Academic Press; Jauch, J.M., Rohrlich, F. (1955): The theory of photons and electrons, New York: Addison-Wesley

autoionization there is an interference between an N-electron bound state and *single* ionization continua, whereas even the single Auger effect requires considerations of the interaction between a single ionization continuum and double ionization continua. The resonance behavior is due to the bound nature of the $(N-1)$-electron state of the residual ion in the single ionization process. If the interaction between the primary ejected electron and the rest of the system can be neglected, then the theory of the Auger effect reduces effectively to that of autoionization. However, then the theory excludes post-collision effects, which are manifestations of interactions between the ionized and autoionized electrons close to ionization thresholds.[23]

For the purpose of introducing the concept of nonradiative transitions the matrix elements of the $\boldsymbol{T}^{+\prime}$ matrix are written in the following form on the energy shell:

$$T_{\beta\alpha}^{+\prime} = T_{\beta\alpha}^{+} + \frac{\langle \chi_{\beta\varepsilon}^{-} | H - E | \Phi \rangle \langle \Phi | H - E | \chi_{\alpha\varepsilon}^{+} \rangle}{E - E_r + \mathrm{i}\frac{\Gamma}{2}}, \tag{4.1}$$

which follows from (3.9) and the general relation $T_{\beta\alpha} = (\delta_{\beta\alpha} - S_{\beta\alpha})/2\pi\mathrm{i}$. According to (3.7), Γ is given in (4.1) by

$$\Gamma(E) = \sum_{\beta=1}^{N_c} \Gamma_\beta = 2\pi \sum_{\beta=1}^{N_c} |\langle \chi_{\beta\varepsilon}^{-} | H - E | \Phi \rangle|^2, \tag{4.2}$$

where we have chosen the ingoing wave channel wave functions for reasons which will become clear later on. Since the cross section for scattering from channel α into channel β is proportional to $|T_{\beta\alpha}^{+\prime}|^2$, this also gives us an opportunity to examine the resonance behavior of the photoabsorption cross section as a function of the incoming photon energy. This is done in detail in Appendix C, where we derive from (4.1) the Fano profile for the two-channel case.

The wave functions in (4.1) and (4.2) describe a system which consists of both photons and electrons. The corresponding Hamiltonian is given by

$$H = H_{\mathrm{el}} + H_{\mathrm{ph}} + H_{\mathrm{int}}, \tag{4.3}$$

where H_{el} refers to the electrons, H_{ph} to the photon field, and H_{int} to the phototon-electron interaction.[24] The zeroth-order one-photon states $|\omega\rangle$

[23] A rather extensive research has been devoted to this subject in the recent years although a comprehensive quantum mechanical approach seems still to be lacking. For references see Read, F.H. (1975): Radiation Research *64*, 23; Morgenstern, R., Niehaus, A., Thielmann, U. (1977): J. Phys. B *10*, 1039; Read, F.H. (1977): J. Phys. B *10*, L 207. See also the articles by Read, F.H. (with Comer, J.), Amusia, M. Ya. (with Kuchev, M. Yu., Sheinerman, S.A.), Heideman, H.G.M. (1980) in: Coherence and correlations in atomic collisions, Kleinpoppen, H., Williams, J.F. (eds.), New York: Plenum Press

[24] The Hamiltonians H_{ph} and H_{int} are defined in Sect. 5. As pointed out above it is possible to develop the theory of photon scattering in very much the same way as for the electrons. This is the only property we need here. Problems associated with the photon emission are treated in Sects. 5 and 6

$=|0, \dots 1, \dots\rangle = a^\dagger_\omega|0, \dots 0, \dots\rangle$ can be characterized by the energy ω, direction $\boldsymbol{k}$, and polarization $\boldsymbol{e}$ of the state in which a photon is created. The corresponding zeroth-order wave functions of the total system consist of product wave functions $|\nu, \omega\rangle$, where $|\nu\rangle$ refers to a bound or standing wave continuum state of the electrons. In addition, there are states with no photons, which we denote by $|\mu\varepsilon, 0\rangle$. Since the photons may either scatter elastically or inelastically or they may be absorbed, there are three types of channels to be considered. The corresponding stationary scattering states $\chi^+_{\alpha\varepsilon}$ and $\chi^-_{\beta\varepsilon}$ in (4.1) fulfil the Lippmann-Schwinger equations (2.20), where we can due to the weakness of the photon-electron interaction, disregard all matrix elements $\langle\chi_{\beta\tau}|V_\beta|\chi^\pm_\alpha\rangle$ with $V_\beta = H_{\text{int}}$. This amounts to neglecting the interaction between and within the elastic and inelastic channels and also between these and the photoabsorption channels. The remaining interaction elements are of the form[25] $\langle\mu\varepsilon, 0|H_{\text{el}} - E|\chi^\pm_{\alpha\varepsilon}\rangle$ that describes electron-electron interactions between and within the photoabsorption channels.

We assume now that a zero-photon discrete state $|\Phi, 0\rangle$ is embedded in the continua defined by the wave functions $|\mu\varepsilon, 0\rangle$[26] and consider the absorption of a photon by the atom in ground state into that region. Consequently, the initial channel α is characterized by a wave function $|i, \omega\rangle$, where $|i\rangle$ is the ground-state wave function of the atom. The wave function of the electrons in the final channel β is given by $\lambda^-_{\beta\varepsilon}$, which is a solution of the Lippmann-Schwinger equation (2.20) with $\langle\chi_{\beta\tau}|V_\beta|\Psi^-_{\alpha\varepsilon}\rangle = \langle\beta\tau|H_{\text{el}} - E|\lambda^-_{\alpha\varepsilon}\rangle$. The corresponding $\boldsymbol{T}$ matrix element is given by

$$T^{+\prime}_{\beta\alpha} = \langle\lambda^-_{\beta\varepsilon}, 0|H_{\text{int}}|i, \omega\rangle + \frac{\langle\lambda^-_{\beta\varepsilon}, 0|H_{\text{el}} - E|\Phi, 0\rangle\langle\Phi, 0|H_{\text{int}}|i, \omega\rangle}{\mathcal{E} - \mathcal{E}_{\text{r}} + i\dfrac{\Gamma}{2}} \tag{4.4}$$

to lowest order in H_{int}. Here $\mathcal{E} = \omega - (E_\beta - E_{\text{i}})$ represents the kinetic energy of the photoelectron and $\mathcal{E}_{\text{r}} = E_{\text{r}} - E_\beta$. Note that according to (4.4) the ingoing wave boundary conditions must be used to treat final-state channel interactions in photoabsorption if the absorption process is described, as is usual, in a perturbative way. Actually, it is clear from (4.1) that these boundary conditions are required to describe final-state interactions in any process, where a projectile excites and ionizes atomic electrons in a resonant fashion.[27]

Let us assume that the second term in (4.4) dominates close to the resonance $\mathcal{E} \cong \mathcal{E}_{\text{r}}$. The cross section for the emission of an electron with the kinetic

---

[25] In the following, E with and without subscripts does not include any photon energies but refers to total energies of states of the atomic electrons

[26] We postpone to Sects. 5 and 6 the discussion of one-photon continua with which the discrete state may interact

[27] A general discussion of the appropriate boundary conditions in the final state has also been given by BREIT and BETHE (Breit, G., Bethe, H.A. (1954): Phys. Rev. *93*, 888) using a time-dependent approach. See also Ref. [*6*], p. 423 and Gell-Mann, M., Goldberger, M.L. (1953): Phys. Rev. *91*, 938

energy in the range $(\mathcal{E}, \mathcal{E}+d\mathcal{E})$ is given under these circumstances by

$$\frac{d\sigma(\mathcal{E})}{d\mathcal{E}}=\frac{\frac{1}{2\pi}\Gamma_\beta(\mathcal{E})\,\sigma'(\omega)}{(\mathcal{E}-\mathcal{E}_r)^2+\Gamma^2/4}, \tag{4.5}$$

where $\sigma'(\omega)$ is the cross section for exciting the atom into the resonance state Φ. Equation (4.5) gives the Breit-Wigner formula,[15] which has a Lorentz form provided $\Gamma_\beta(\mathcal{E})$, $\Gamma(E)$ and $\sigma'(\omega)$ are assumed to be constants.

Suppose $N_0(\omega-\omega_0)$ gives the distribution of the incoming photons per unit area and time centered around some mean energy ω_0. Then

$$\frac{dN_\beta}{d\mathcal{E}}=\frac{\frac{1}{2\pi}\Gamma_\beta(\mathcal{E})\,\sigma'(\omega)\,N_0(\omega-\omega_0)}{(\mathcal{E}-\mathcal{E}_r)^2+\Gamma^2/4} \tag{4.6}$$

gives the number of electrons which are emitted in the energy range $(\mathcal{E}, \mathcal{E}+d\mathcal{E})$ per unit time. Now as in resonance fluorescence[28] this distribution depends on the distribution of the incoming photons. For instance, if $N_0(\omega-\omega_0)$ is a narrow Lorentz distribution such that its halfwidth is much smaller than Γ, $dN_\beta/d\mathcal{E}$ reflects this distribution rather than the distribution whose halfwidth is determined by Γ. On the other hand, if $N_0=\sigma'(\omega)\,N_0(\omega-\omega_0)$ is constant over a range much larger than Γ, the distribution is a Lorentzian with the halfwidth Γ. In that case the total number of ejected electrons per unit time is given approximately by

$$N_\beta=\int_0^\infty \frac{dN_\beta}{d\mathcal{E}}\,d\mathcal{E}=\frac{\Gamma_\beta}{\Gamma}N_0, \tag{4.7}$$

which together with (4.2) shows that the quantities Γ_β and Γ can be interpreted as *nonradiative* decay probabilities.[29] If Γ, in accordance with (4.2), is interpreted as the total probability of nonradiative transitions from the resonance state Φ, then N_0/Γ gives the number of atoms *present* in this state. According to (4.7), this result is consistent with the interpretation of Γ_β as a *partial* nonradiative decay probability since N_β represents the total number of electrons ejected per unit time into channel β.

So far we have seen that in certain cases of weak direct scattering and of broad energy distributions of the incoming radiation, the autoionization can be interpreted as a two-step process, where the excitation of the resonance state is separated from the decay. In the Auger effect this is generally true regardless of the nature of the incoming radiation's distribution. We shall generalize (4.4) in

---

[28] See, e.g. Sect. 20 in Heitler, W. (1960): The quantum theory of radiation. 3rd edn., Oxford: University Press

[29] Actually, Γ_β and Γ should be interpreted as halfwidths rather than probabilities since they are expressed in atomic units of energy. In the following we shall not make any distinctions between these two concepts

a somewhat nonrigorous way to demonstrate this for Auger electron emission following photoionization.

In the Auger process the resonance state Φ actually consists of a distribution of continuum states which represent a discrete $(N-1)$-electron state plus an outgoing photoelectron. The final states $\lambda^-_{\beta\varepsilon}$ are *doubly* ionized states representing the outgoing photo- and Auger electrons coupled to a discrete $(N-2)$-electron state. The $\boldsymbol{T}$ matrix element (4.4) takes the form

$$T^{+\prime}_{\beta\alpha}=\langle\lambda^-_{\beta\varepsilon_1,\varepsilon_2},0|H_{\rm int}|i,\omega\rangle+\int\limits_0^\infty\frac{\langle\lambda^-_{\beta\varepsilon_1,\varepsilon_2},0|H_{\rm el}-E|\varphi_\tau,0\rangle\langle\varphi_\tau,0|H_{\rm int}|i,\omega\rangle\,d\tau}{\varepsilon_1+\varepsilon_2-\varepsilon_{\rm r}-\tau+{\rm i}\Gamma/2}, \quad (4.8)$$

where ε_1 and ε_2 represent the kinetic energies of the outgoing electrons in the final state so that $\varepsilon_1+\varepsilon_2=\omega-(E_\beta-E_{\rm i})$. Consequently $\varepsilon_{\rm r}=E_{\rm r}-E_\beta$, where $E_{\rm r}$ is the resonance energy in the $(N-1)$ electron space and E_β refers to the double-ionization threshold for channel β. If $\omega\gg E_\beta-E_{\rm i}$, it is reasonable to approximate, in analogy to the impulse approximation,[30] the intermediate states φ_τ by product wave functions $|\varphi,0\rangle|\tau\rangle$, where $|\tau\rangle$ represents the continuum electron. If the final state is approximated in the same way, then

$$\langle\lambda^-_{\beta\varepsilon_1,\varepsilon_2},0|H_{\rm el}-E|\varphi_\tau,0\rangle\cong\langle\varepsilon_2|\tau\rangle\langle\lambda^-_{\beta\varepsilon_1},0|H_{\rm el}(N-1)-E(N-1)|\varphi,0\rangle, \quad (4.9)$$

where $\langle\varepsilon_2|\tau\rangle=\delta(\varepsilon_2-\tau)$ provided the wave functions $|\varepsilon_2\rangle$ and $|\tau\rangle$ are orthogonal.[31] Approximation (4.9) is best understood within the independent-electron central field model[32] according to which $\lambda^-_{\beta\varepsilon_1}$ and φ must differ by *two* sets of one-electron quantum numbers. Since $H_{\rm int}$ is a one-electron operator, the matrix element of the direct double ionization would vanish if $\lambda^-_{\beta\varepsilon_1}$ and φ are constructed from *orthogonal* one-electron spin orbitals and $\langle\varphi_\tau,0|H_{\rm int}|i,\omega\rangle$ would reduce to the one-electron matrix element $\langle\tau|H_{\rm int}|\chi\rangle$, where χ is the vacant spin orbital after the photoionization. Hence we obtain

$$\frac{d\sigma(\varepsilon_1)}{d\varepsilon_1}=\frac{\frac{1}{2\pi}\Gamma_\beta(\varepsilon_{\rm r})\cdot\sigma'(\varepsilon_2)}{(\varepsilon_1-\varepsilon_{\rm r})^2+\Gamma^2/4} \quad (4.10)$$

in the weak direct-ionization and high-energy photoionization limit. Here

$$\Gamma_\beta(\varepsilon_{\rm r})=2\pi|\langle\lambda^-_{\beta\varepsilon_{\rm r}}|H_{\rm el}(N-1)-E(N-1)|\varphi\rangle|^2 \quad (4.11)$$

can be identified as the nonradiative transition probability that the resonant inner-shell vacancy state φ decays into the channel β by emission of an

[30] e.g., Ref. [3], Chap. 12, Sect. 3

[31] The allowance for the nonorthogonality between the continuum spin orbitals $|\varepsilon_2\rangle$ and $|\tau\rangle$ accounts for post-collision interactions in lowest order. See Amusia, M.Ya., Kuchev, M.Yu., Sheinerman, S.A. (1980) in: Coherence and correlations in atomic collisions, Kleinpoppen, H., Williams, J.F. (eds.), p. 297, New York: Plenum Press; Åberg, T. (1981) in: Inner-shell and x-ray physics of atoms and solids, Fabian, D.J., Kleinpoppen, H., Watson, L.M. (eds.), New York: Plenum Press, p. 251

[32] See Sect. 7

electron with the kinetic energy $\mathcal{E}_r = E_r - E_\beta \cong E_\varphi - E_\beta$ since $\mathcal{E}_2 = \omega - (E_r - E_i)$. Note that in contrast to (4.5) the differential cross section is separable into a Lorentzian decay part and an excitation part represented by the photoionization cross section $\sigma'(\mathcal{E}_2)$ without further assumptions regarding the distribution of the incoming radiation. Note also that the nonradiative probabilities (4.11) include the effect of final-state channel interactions, which can be taken into account by solving the corresponding Lippmann-Schwinger or $\boldsymbol{K}$ matrix equations in the $(N-1)$-electron space using the ingoing wave boundary conditions

It follows from the general expression (4.1) that (4.10) has a more general validity than anticipated by the present derivation. It provides an adequate approximation of the Auger electron production cross section in inner-shell ionization by a variety of projectiles which have energies not too close to the threshold. The present theory can be generalized to include the subsequent decay of the final state, which assesses the validity of the Weisskopf-Wigner approximation.[33] Effects of closely spaced resonance states and of resonance states close to the double ionization threshold can also be included, in which case it is not adequate to treat the Auger effect as a two-step process.[34] Nevertheless, most of the subsequent developments in this article will be based on the concept of the nonradiative probability as it is given by (4.11).

5. Spontaneous photon emission as resonance scattering. The creation of inner-shell vacancies is followed by both electron and photon emission. In light atoms the nonradiative decay mode usually dominates, whereas the radiative decay mode is also important in heavy atoms. In some cases, such as the K emission around $Z=30$, both decay modes are of equal strength. Hence a comprehensive theory of the Auger effect should also include a study of the influence of these decay modes on each other.

As discussed by most textbooks in quantum mechanics the radiative decay mode is usually treated by the time-dependent perturbation theory, leading to a photon emission rate that is based on Fermi's golden rule. This approach, however, is inadequate for a proper treatment of autoionization including the Auger effect. Hence we present an alternative derivation of the radiative rate by realizing that excitation followed by spontaneous photon emission is an example of resonance scattering. This also allows for the inclusion of effects pertaining to the interaction between the nonradiative and radiative decay modes. These results, which are obtained in the next section, are used to demonstrate under what circumstances the nonradiative and radiative decay probabilities are additive.

Hamiltonian (4.3) forms the basis of the subsequent developments. The electron part H_{el} is assumed to be the nonrelativistic Hamiltonian but may

---

[33] According to this approximation, the total width Γ is the sum of the widths pertaining to the initial and final states, respectively (Weisskopf, V., Wigner, E. (1930): Z. Phys. *63*, 54)

[34] Such cases are expected in particular in connection with so called Super Coster-Kronig transitions. (For example, see Wendin, G., Ohno, M. (1976): Phys. Scr. *14*, 148; Davis, L.C., Feldkamp, L.A. (1978): Phys. Rev. A *17*, 2012)

include the spin-orbit interaction. The photon part H_{ph} is given by[35]

$$H_{ph} = \sum_{\lambda} \hbar\omega_{\lambda} a_{\lambda}^{\dagger} a_{\lambda}, \tag{5.1}$$

where the mode $\lambda = (\boldsymbol{k}, \sigma)$ corresponds to a given direction $\boldsymbol{k}^0$ of propagation and polarization $\boldsymbol{e}_{\sigma}$ $(\sigma = 1, 2)$. The length of the wave vector $\boldsymbol{k}$ is in the λth mode given by ω_{λ}/c. The operators $a_{\lambda}^{\dagger}$ and a_{λ} are the photon creation and annihilation operators, respectively. The interaction Hamiltonian H_{int} is given in the Coulomb gauge[36] by

$$H_{int} = \sum_{\gamma=1}^{N} \left[\frac{e}{m} \boldsymbol{p}_{\gamma} \cdot \boldsymbol{A}(\boldsymbol{r}_{\gamma}) + \frac{e^2}{2m} \boldsymbol{A}(\boldsymbol{r}_{\gamma}) \cdot \boldsymbol{A}(\boldsymbol{r}_{\gamma}) \right], \tag{5.2}$$

where $\boldsymbol{p}_{\gamma} = -i\hbar\nabla_{\gamma}$ in the coordinate space and

$$\boldsymbol{A}(\boldsymbol{r}) = \sum_{\lambda} \sqrt{\frac{\hbar}{2\varepsilon_0 \omega_{\lambda} L^3}} \, \boldsymbol{e}_{\lambda} [a_{\lambda} \, \mathrm{e}^{i\boldsymbol{k}\cdot\boldsymbol{r}} + a_{\lambda}^{\dagger} \, \mathrm{e}^{-i\boldsymbol{k}\cdot\boldsymbol{r}}]. \tag{5.3}$$

The vector potential $\boldsymbol{A}(\boldsymbol{r})$ is a superposition of plane waves that are normalized in a box of volume L^3. This normalization corresponds to a density of one-photon states given by

$$\rho(\omega) = \frac{L^3 \omega^2}{(2\pi c)^3}. \tag{5.4}$$

The plane waves, in analogy with the electron case, could equally well be normalized per unit energy range in (5.3). This would introduce a factor $1/\sqrt{\sigma(\omega_{\lambda})}$, which would make the corresponding one-photon density equal to one.

Suppose that the atom is photoexcited into zero-photon state $|\Phi\rangle|0\rangle = |\Phi, 0\rangle$ that subsequently decays into the state $|f, \omega\rangle = |f\rangle |\boldsymbol{k}^0, \boldsymbol{e}_{\sigma}, \omega\rangle$ for which $\hbar\omega = E_{\Phi} - E_{\mathrm{f}}$. According to Sect. 4, this process is a manifestation of the photon-electron interaction between the discrete state $|\Phi, 0\rangle$ and one of the continua $|v, \boldsymbol{k}^0, \boldsymbol{e}_{\sigma}, \omega\rangle$, where $\hbar\omega$ takes any value between E_v and infinity. The electron continua associated with the photon vacuum state are now disregarded and all electron states v are assumed to be discrete. Each channel β is characterized by the quantum numbers of the electron state v and by the unit vectors $\boldsymbol{k}^0$ and $\boldsymbol{e}_{\sigma}$ (see Fig. 3, where the threshold energy E_{β} should be identified as E_v).

---

[35] We shall use SI units in Sect. 5 and introduce the quantized photon field in the most elementary manner (e.g., Ref. [7], Vol. II, Chap. XXI)

[36] There has been much discussion on the gauge invariance in the conventional time-dependent perturbation theory approach which uses this form of H_{int} as a perturbation (see for example Kobe, D.H., Smirl, A.L. (1978): Am. J. Phys. *46*, 624). Apparently this question has not been related to the possibility to treat the photon-electron interaction by time-independent scattering theory as in our work. However, GRANT (Grant, I.P. (1974): J. Phys. B 7, 1458) has considered the gauge invariance of the matrix elements of the photon-electron interaction from the quantum-electrodynamical point of view

According to (3.1), the resonant scattering via the discrete state $|\Phi, 0\rangle$ is described by the mixing of this and the one-photon states in the wave functions

$$\Psi_\mu(\omega) = a_\mu(\omega)|\Phi, 0\rangle + \sum_{\nu, \boldsymbol{k}^0, \sigma} \int_0^\infty b_\mu(\boldsymbol{k}^0, \sigma, \omega)|\nu, \boldsymbol{k}^0, \sigma, \omega\rangle \rho(\omega)\, d\omega. \tag{5.5}$$

As in Sect. 3 we first consider the interaction between and within the continua. In principle, these interactions are taken into account by the solutions of the Lippmann-Schwinger equations (2.20), where V_β is given by (5.2). From (5.3) it follows that only the A^2 term contributes to V_β, which we neglect as small.[37] Hence the one-photon states in (5.5) are taken to be eigenfunctions of H_{ph}. This also implies that the direct scattering term between the entrance channel $\alpha = |i, \omega'\rangle$ and the escape channel $\beta = |f, \omega\rangle$ is neglected in the $\boldsymbol{T}$ matrix element (4.1), which reduces to

$$T_{\beta\alpha}^{+\prime} = \frac{\langle f, \omega|H_{\text{int}}|\Phi, 0\rangle \langle \Phi, 0|H_{\text{int}}|i, \omega'\rangle}{\hbar\omega - (E_{\text{r}} - E_{\text{f}}) + \mathrm{i}\dfrac{\Gamma}{2}}, \tag{5.6}$$

where only the $\boldsymbol{p} \cdot \boldsymbol{A}$ term contributes to H_{int} and where energy conservation requires that $\hbar\omega = \hbar\omega' - (E_{\text{f}} - E_{\text{i}})$. According to (4.2), Γ is given by

$$\Gamma = \sum_\beta \Gamma_\beta = \frac{2\pi}{\hbar} \sum_{\nu, \boldsymbol{k}^0, \sigma} |\langle \nu, \boldsymbol{k}^0, \sigma, \omega|H_{\text{int}}|\Phi, 0\rangle|^2 \rho(\omega). \tag{5.7}$$

From (5.6) it follows that the number of emitted photons in the frequency range $(\omega, \omega + d\omega)$ and per unit time is given by

$$\frac{dN_f}{d\omega} = \frac{\dfrac{\hbar}{2\pi} \Gamma_{\text{f}} \sigma'(\omega') N_0(\omega' - \omega_0)}{[\hbar\omega - (E_\Phi - E_{\text{f}})]^2 + \dfrac{\Gamma^2}{4}}, \tag{5.8}$$

which is completely analogous to the distribution function (4.6) of autoionized electrons. The photon distribution (5.8) describes the resonance fluorescence which depends on the distribution function $N_0(\omega' - \omega_0)$ of the incoming photons. If $\sigma'(\omega') N_0(\omega' - \omega_0)$ is constant over a wide range of photon energies ω', then $dN_{\text{f}}/d\omega$ is a Lorentz function with the halfwidth Γ and Γ_{f} can be interpreted as the probability that a photon is emitted in the direction $\boldsymbol{k}^0$ with the polarization vector $\boldsymbol{e}_\sigma$ and energy $\hbar\omega = E_\Phi - E_{\text{f}}$. Note that we have omitted the level shift

$$\Delta = \mathscr{P} \sum_{\nu, \boldsymbol{k}^0, \sigma} \int_0^\infty \frac{|\langle \nu, \boldsymbol{k}^0, \sigma, \omega'|H_{\text{int}}|\Phi, 0\rangle|^2 \rho(\omega')\, d\omega'}{E_\Phi - E_\nu - \hbar\omega'} \tag{5.9}$$

---

[37] The iteration of the Lippmann-Schwinger equations would lead to divergencies which may be removed using the technique of renormalization (see I.22). However, as long as the photon-electron interactions are treated to lowest order in perturbation theory, no divergencies occur

from the denominator of (5.8). A straightforward analysis of Δ, which is related to the Lamb shift, would show that it is infinite. However, as is well known, it turns out to be very small after renormalization of the electron mass.[22]

The generalization of (5.8) to the case of photon emission accompanying inner-shell ionization follows exactly the same path as the derivation of (4.10) from (4.6). Hence we shall just give the obvious result, namely, that the photon-emission cross section can be obtained from

$$\frac{d\sigma(\omega)}{d\omega}=\frac{\frac{\hbar}{2\pi}\Gamma_{\mathrm{f}}(\omega_{\mathrm{r}})\cdot\sigma'(\mathcal{E}_2)}{[\hbar\omega-(E_{\varphi}-E_{\mathrm{f}})]^2+\Gamma^2/4} \tag{5.10}$$

under the same circumstances under which (4.10) is valid. Here $\sigma'(\mathcal{E}_2)$ is defined in Sect. 4 and the radiative transition probability $\Gamma_{\mathrm{f}}(\omega_{\mathrm{r}})$ is given explicitly by

$$\Gamma_{\mathrm{f}}(\omega_{\mathrm{r}})=\frac{e^2(E_{\varphi}-E_{\mathrm{f}})}{8\pi^2\varepsilon_0 m^2 c^3\hbar}\left|\langle f|\sum_{\nu=1}^{N-1}\mathrm{e}^{-\mathrm{i}\boldsymbol{k}\cdot\boldsymbol{r}_{\nu}}(\boldsymbol{e}_{\sigma}\cdot\boldsymbol{p}_{\nu})|\varphi\rangle\right|^2 \tag{5.11}$$

in energy units.

Equations (4.10) and (5.10) govern the most common situation in inner-shell ionization. They have one shortcoming however, namely, that whereas Γ in (4.10) refers to nonradiative transitions, it corresponds only to radiative transitions in (5.10). In the next section we shall show that Γ should actually be the *sum* of the nonradiative and radiative widths in both distributions.

6. Final-state photon-electron interactions and additivity of nonradiative and radiative widths. According to the previous sections, the electron and photon emission can be treated in a completely analogous way by interpreting them as manifestations of resonance scattering processes. This indicates that it is also possible to consider their combined effect within this framework. In fact the only essential difference between the considerations in Sects. 4 and 5 was that in the former case no one-photon continua were involved, whereas there were no electron continua in the latter case. We shall drop these assumptions and consider for simplicity such inner-shell processes, where the decay can be separated from the excitation process. This simplification and the neglect of multiphoton and multielectron decay reduce the problem to the one shown by Fig. 4. In the $(N-1)$-electron space a discrete zero-photon state $|i,0\rangle$ interacts with two continua. The first continuum consists of one-photon states, $|f,\boldsymbol{k}^0,\sigma,\omega\rangle$, and the second of singly ionized electron states associated with the photon vacuum, $|\lambda,\varepsilon,0\rangle$. The continuous variables are the photon energy ω and the kinetic energy ε (or τ) of the electron. The generalization to the case where there are several continua of both kinds is straightforward. The two continua may interact in the sense that

$$H_{\mathrm{f}\lambda}(\omega,\varepsilon)=\langle f,\boldsymbol{k}^0,\sigma,\omega|H-E|\lambda,\varepsilon,0\rangle=\langle f,\omega|H_{\mathrm{int}}|\lambda,\tau\rangle \tag{6.1}$$

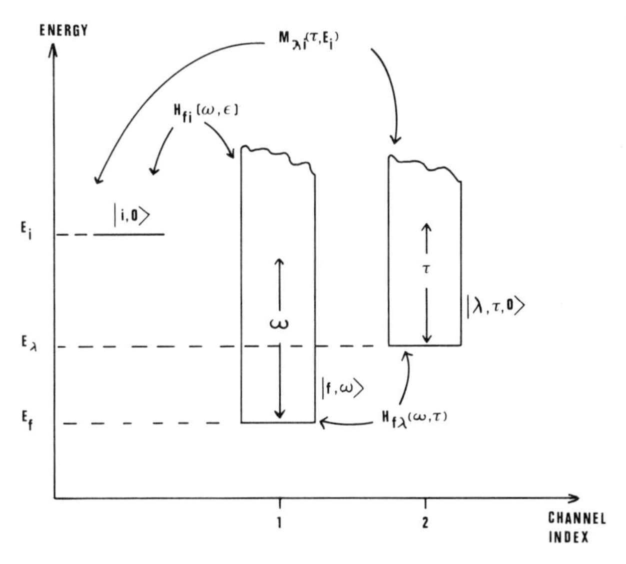

Fig. 4. Photon and electron emission including the photon-electron coupling in the two-channel case

is nonzero due to the $\boldsymbol{p} \cdot \boldsymbol{A}$ part of the interaction Hamiltonian (5.2). The intrachannel interaction can be neglected in both channels so that

$$\begin{aligned} \langle f, \omega' | H - F | f, \omega \rangle &= (E_f + \omega - E)\, \delta(\omega' - \omega) \\ \langle \lambda, \varepsilon' | H - E | \lambda, \varepsilon \rangle &= (E_\lambda + \varepsilon - E)\, \delta(\varepsilon' - \varepsilon). \end{aligned} \tag{6.2}$$

This simplification is justified since any intrachannel interaction effects are due to the weak A^2 term in the photon channel. On the other hand the electron wave functions can always be chosen so that (6.2) is valid.[38] The diagonalization of the continuum matrix, consisting of the matrix elements (6.1) and (6.2), gives rise to two scattering states $\Psi_{\alpha\mathcal{E}}^-$ ($\alpha = 1, 2$) that are superpositions of the zero- and one-photon states. These scattering states, which must fulfil the ingoing wave Lippmann-Schwinger equations (2.20), are represented by

$$\begin{aligned} \Psi_{1\mathcal{E}}^- &= |f, \mathcal{E}\rangle + \lim_{\nu \to 0} \int_0^\infty \frac{H_{f\lambda}^*(\mathcal{E}, \tau)\, |\lambda, \tau\rangle\, d\tau}{E - E_\lambda - \tau - i\nu} \\ \Psi_{2\mathcal{E}}^- &= |\lambda, \mathcal{E}\rangle + \lim_{\nu \to 0} \int_0^\infty \frac{H_{f\lambda}(\omega, \mathcal{E})\, |f, \omega\rangle\, \rho(\omega)\, d\omega}{E - E_f - \omega - i\nu} \end{aligned} \tag{6.3}$$

to first order in the matrix elements $H_{f\lambda}(\omega, \varepsilon)$. Note that the iteration of the Lippmann-Schwinger equations to higher orders would eventually lead to divergences.[22] Nevertheless, the wave functions (6.3) display the interaction between the final photon- and electron-emission channels.

---

[38] An example is given in Sect. 10β

According to definition (3.7), the total width Γ must be the sum of the nonradiative width Γ_{nr} and the radiative width Γ_{r}. This is even true for cases where these quantities cannot be directly interpreted as decay probabilities. This result is ultimately a consequence of the boundary conditions that describe an experimental situation where electrons *or* photons are observed. Hence it follows that the total decay probability of an inner-shell vacancy state is the sum of the nonradiative and radiative decay probabilities. According to (4.2) and (6.3), the corresponding widths are given to first order in the matrix elements $H_{\mathrm{f}\lambda}(\omega, \varepsilon)$ by

$$\Gamma_{\mathrm{r}}(\omega_{\mathrm{r}}) = 2\pi \left| H_{\mathrm{fi}}(\omega_{\mathrm{r}}) + \lim_{\nu \to 0} \int_0^\infty \frac{H_{\mathrm{f}\lambda}(\omega_{\mathrm{r}}, \tau)\, M_{\lambda\mathrm{i}}(\tau, E_{\mathrm{i}})\, d\tau}{E_{\mathrm{i}} - E_\lambda - \tau + i\nu} \right|^2 \rho(\omega_{\mathrm{r}}) \tag{6.4}$$

and

$$\Gamma_{\mathrm{nr}}(\mathcal{E}) = 2\pi \left| M_{\lambda\mathrm{i}}(\mathcal{E}, E_{\mathrm{i}}) + \lim_{\nu \to 0} \int_0^\infty \frac{H_{\lambda\mathrm{f}}(\mathcal{E}, \omega)\, H_{\mathrm{fi}}(\omega, \mathcal{E})\, \rho(\omega)\, d\omega}{E_{\mathrm{i}} - E_{\mathrm{f}} - \omega + i\nu} \right|^2, \tag{6.5}$$

where $\omega_{\mathrm{r}} = E_{\mathrm{i}} - E_{\mathrm{f}}$ and $\mathcal{E} = E_{\mathrm{i}} - E_\lambda$. These equations show how the radiative amplitude $H_{\mathrm{fi}}(\omega_{\mathrm{r}}) = \langle f, \omega_{\mathrm{r}} | H_{\mathrm{int}} | i, 0 \rangle$ and the nonradiative amplitude $M_{\lambda\mathrm{i}}(\mathcal{E}, E_{\mathrm{i}}) = \langle \lambda, \mathcal{E} | H_{\mathrm{el}} - E_{\mathrm{i}} | i, 0 \rangle$ are modified by the interaction between the nonradiative and radiative decay channels. The photon-electron coupling corrections are represented by the $\nu \to 0$ limits of the integrals in (6.4) and (6.5). In $\Gamma_{\mathrm{r}}(\omega_{\mathrm{r}})$ this correction represents a real and virtual process where the state $|i, 0\rangle$ decays by the ejection of an electron which recombines with the atom. As a result the atom ends up in the state $|f\rangle$ after the emission of a photon with the energy ω_{r}. In $\Gamma_{\mathrm{nr}}(\mathcal{E})$ the correction represents a real and virtual process where the state $|i, 0\rangle$ decays by the emission of a photon which is reabsorbed by the atom. As a result an electron with the kinetic energy $\mathcal{E}$ is ejected and the atom is left in a doubly ionized state $|\lambda\rangle$. We shall see in Chap. III that the relativistic theory of the Auger effect automatically incorporates this photon-electron coupling effect. However, there is one important difference, namely, the electron states in (6.4) and (6.5) can be any many-electron state, i.e., they may include correlation, whereas the relativistic theory necessarily is based on the independent-electron model.[39] Hence (6.4) and (6.5) can be used to investigate how the final-state photon-electron interaction affects amplitudes of weak transitions that are solely due to electron correlation effects like the two-electron transitions[40] in x-ray spectra.[41]

The vector potential (5.3) is given in the plane wave representation, which is not very suitable for calculations. However, the theory of the joint electron and photon scattering can be developed very much along the lines of Sects. 1–3

---

[39] Åberg, T. (1977) in: Extended abstracts of V Int. Conf. Vac. Ultrav. Rad. Phys., At. Molec. Phys., Vol. I, Castex, M.C., Pouey, M., Pouey, N. (eds.), p. 73

[40] See Åberg, T., Jamison, K., Richard, P. (1977): Phys. Rev. A, *15*, 172, and references therein

[41] Recently Armstrong et al. (see I.2) have pointed out that the fluorescence yield $\omega_{\mathrm{F}} = \Gamma_{\mathrm{r}}/\Gamma$ may be influenced considerably by electron-photon coupling effects in the final state. It has also been suggested that certain two-electron *x*-ray transitions may be particularly sensitive to such effects (see Åberg, T. (1981) in: Inner-shell and x-ray physics of atoms and solids, Fabian, D.J., Kleinpoppen, H., Watson, L.M. (eds.), New York: Plenum Press, p. 251 and references therein

if the vector spherical harmonics[42] $\mathcal{Y}_{jm}^{(\lambda)}$ are introduced to represent the photon states. This makes it possible to characterize the photon states by a definite energy ω, the angular momentum quantum number j and it's projection m, and by the parity λ. As a consequence, it is still possible to characterize by vector coupling the photon-electron scattering wave functions by a definite total angular momentum quantum number J and the corresponding magnetic quantum number M. We shall not pursue this matter further here since our purpose has been only to demonstrate how the many-electron theory of the Auger effect can be generalized to include the radiative decay and its coupling to the nonradiative decay in a consistent manner.

II. Nonrelativistic theory of nonradiative transitions

The interpretation of the gross features of Auger electron spectra belongs admittedly to one of the successes of the independent-electron central field model of the atom. The distribution of Auger electron energies for a particular atom reflects its shell structure, and the number of Auger electrons ejected in a particular transition can be satisfactorily estimated using central field orbitals in connection with Wentzel's lowest-order probability formula as shown in earlier reviews.[1] Hence we shall begin in Sect. 7 with a brief summary of the central field model of the atom in order to provide the background for the nomenclature and gross classification of nonradiative transitions. Section 8 presents the necessary many-electron theory for the construction of initial and final-state wave functions in a nonradiative transition. This includes considerations of various coupling schemes, procedures for the calculation of spin orbitals, and considerations of the relationship between electron and hole configurations. The frozen-core approximation, which is also introduced in Sect. 8, forms the basis of the derivation of formulae for nonradiative transition probabilities in various coupling schemes in Sect. 9. Improvements of the frozen-core central field model of nonradiative transitions are described in the next section. These involve not only considerations of the intermediate coupling including configuration interaction but also of relaxation and final-state channel interaction effects. We shall conclude Sect. 10 by examining many-body perturbation theory and related approaches to the calculation of nonradiative transition probabilities.

This chapter relies on many aspects of the theory of atomic structure and spectra as well as nuclear structure. For details we would like to refer to the works of CONDON and SHORTLEY [*9*], BETHE and SALPETER [*10*], SOBEL'MAN [*11*], SLATER [*12*], and FROESE-FISCHER [*13*]. The books of SHORE and MENZEL [*14*] and MIZUSHIMA [*15*] are also useful. With regard to the angular momentum coupling we would like to mention the book of the DE SHALIT and TALMI [*16*] besides that of EDMONDS [*8*].

---

[42] e.g., Ref. [*17*], Chap. I, Sect. 4

[1] See reviews listed in the introduction

7. Central field model and classification of nonradiative transitions. If the spin-orbit interaction is ignored and the nucleus is assumed to be infinitely heavy, the Hamiltonian of the atomic electrons is given in atomic units by

$$H=-\tfrac{1}{2}\sum_{\nu=1}^{N}\nabla_{\nu}^{2}-\sum_{\nu=1}^{N}\frac{Z}{r_{\nu}}+\sum_{\nu>\mu}\frac{1}{r_{\nu\mu}}, \qquad (7.1)$$

where Z is the atomic number and N the number of electrons. The spin-independent Hamiltonian (7.1) is invariant with respect to rotations around the nucleus and inversion of the electronic coordinates. Hence the set $\Lambda=\{L^2, L_z, S^2, S_z, P\}$ of commuting operators commutes with H. Here the total angular momentum $\boldsymbol{L}$ is represented by the operators L^2 and L_z, and the total spin $\boldsymbol{S}$ by S^2 and S_z. The operator P is the inversion operator whose eigenvalues ± 1 define the parity of the wave functions. The eigenfunctions of H are simultaneous eigenfunctions of the operators in the set Λ and so are any projections $Q\psi$ and $P\psi$ with the projection operators Q and P defined in Appendix B. Note that although the problem of nonradiative transitions actually requires knowledge of these projections, approximate pure discrete or continuous solutions of the Schrödinger equation $H\psi=E\psi$ can be identified as approximate projected wave functions $Q\psi$ and $P\psi$ provided these solutions are subject to suitable *constraints.*[2] Hence we will make no distinction between ψ and its projections in the following.

The central field model[3] of the atom is based on three postulates which can be stated as follows:

1) Each electron moves in an average field created by the nucleus and all the other electrons.

2) In addition to the orbital angular momentum $\boldsymbol{l}$ each electron has an "intrinsic" angular momentum $\boldsymbol{s}$ that can take only two orientations with respect to a selected direction in space.

3) The occupation number of each state in the central field is either zero or one.

In order to preserve the symmetry of the full Hamiltonian (7.1) in the one-electron space, postulate 1 requires a potential $V(r)$ which is *spherically* symmetric so that the spatial part of the central field wave functions fulfils the Schrödinger equation

$$H_{\mathrm{cf}}\psi=(-\tfrac{1}{2}\nabla^2+V(r))\psi=\mathcal{E}\psi. \qquad (7.2)$$

The Hamiltonian in (7.2) commutes with the operators in the set Λ, which we denote by $\Lambda_1=\{l^2, l_z, s^2, s_z, P\}$ for $N=1$. By expressing l^2 and l_z in spherical coordinates, we obtain their simultaneous eigenfunctions $Y_{lm_l}(\theta,\varphi)$ associated with the eigenvalues $l(l+1)$ and m_l, respectively. For each $l(=0,1,2,\ldots)$, $m_l=+l, l-1, \ldots, -l$. The eigenfunctions Y_{lm_l} are given by

$$Y_{lm_l}(\theta,\varphi)=(-1)^{\frac{m_l+|m_l|}{2}}\left[\frac{(2l+1)(l-|m_l|)!}{4\pi(l+|m_l|)!}\right]^{1/2}P_l^{|m_l|}(\cos\theta)\,\mathrm{e}^{\mathrm{i}m_l\varphi}, \qquad (7.3)$$

---

[2] Such constraints are discussed in Sect. 16. For example, the assumption of a given configuration in the self-consistent-field procedure (Sect. 8) defines a constraint

[3] e.g., Ref. [12]

which are the orthonormal spherical harmonics, where $P_l^{|m_l|}(\cos\theta)$ are the associated Legendre polynomials. Functions (7.3) are also eigenfunctions of the inversion operator P corresponding to the eigenvalues $(-1)^l$. Note that the Condon and Shortley phase convention[4]

$$Y_{l,m_l}=(-1)^{m_l}Y^*_{l,-m_l}, \qquad (m_l>0) \tag{7.4}$$

is valid, which defines the solutions (7.3) uniquely.

Since l^2 and l_z commutes with H_{cf}, the solutions ψ must be of the form

$$\psi=R(r)\,Y_{lm_l}(\theta,\varphi). \tag{7.5}$$

Substitution of this form in (7.2) yields the radial Schrödinger equation

$$\frac{d^2R}{dr^2}+\frac{2}{r}\frac{dR}{dr}+2[\mathcal{E}-U_l(r)]\,R=0, \tag{7.6}$$

where

$$U_l(r)=V(r)+\frac{l(l+1)}{2r^2} \tag{7.7}$$

with the asymptotic behavior $\lim_{r\to\infty} U_l(r)=0$. Hence the nondegenerate discrete spectrum of (7.6) must be in the region $\mathcal{E}<0$. For a given l the possible discrete eigenvalues $\mathcal{E}_1<\mathcal{E}_2<\mathcal{E}_3<\dots$ correspond to solutions $R_1, R_2, R_3\dots$ such that R_1 has no nodes, R_2 has one, R_3 has two, etc.[5] This relationship defines the principal quantum number n by the relation

$$n'=n-l-1, \tag{7.8}$$

where n' is the number of nodes.

In addition to the discrete solutions $R_{nl}(r)$ which fulfil the condition

$$\int_0^\infty R_{n'l'}(r)\,R_{nl}(r)\,r^2\,dr=\delta_{n'n}\delta_{l'l}, \tag{7.9}$$

(7.6) has solutions $R_{kl}(r)$ in the continuous spectrum $\mathcal{E}=k^2/2\geqq 0$. If $Z>N$, which is usually the case in the theory of the Auger effect, then $\lim_{r\to\infty} V(r)=-(Z-N)/r$. The corresponding solutions $R_{kl}(r)$ behave at large distances from the nucleus as

$$\lim_{r\to\infty} R_{kl}(r)=\sqrt{\frac{2}{\pi k}}\frac{1}{r}\sin\left[kr-l\tfrac{\pi}{2}-\eta\ln 2kr+\sigma_l+\delta_l(\mathcal{E})\right], \tag{7.10}$$

where $\eta=-(Z-N)/k$ and $\sigma_l=\arg\Gamma(l+1+i\eta)$ is the Coulomb phase shift. The phase shift $\delta_l(\mathcal{E})$ characterizes the deviations of the central field from the

[4] Ref. [9], Chap. 3, Sect. 4

[5] Morse, P.M., Feshbach, H. (1953): Methods of theoretical physics, New York: McGraw-Hill, Part 1, Chap. 6

Coulomb field $-(Z-N)/r$. The factor $\sqrt{2/\pi k}$ ensures that $R_{kl}(r)$ is normalized per unit energy range, that is,

$$\int_0^\infty R_{k'l'}(r)\, R_{kl}(r)\, r^2\, dr = \delta_{l'l}\, \delta(\mathcal{E}' - \mathcal{E}), \tag{7.11}$$

in accordance with (1.3). Near the origin both $R_{nl}(r)$ and $R_{kl}(r)$ are proportional to r^l.

So far we have explored the consequences of the symmetry and postulate 1 without invoking the spin. According to postulate 2 there is a spatial distribution of the electron corresponding to each of the two possible spin orientations. This property can be described by introducing a spin variable ζ which takes two values. These values are chosen to be the eigenvalues of s_z, namely, $m_s = +1/2$ and $-1/2$. The corresponding eigenfunctions are $\chi_{1/2}(\zeta) = \delta_{\zeta, 1/2}$ and $\chi_{-1/2}(\zeta) = \delta_{\zeta, -1/2}$ which are also eigenfunctions of s^2 pertaining to the eigenvalue $s(s+1)$, where $s = 1/2$.

Combining the consequences of postulates 1 and 2 yields the complete solution of the central field problem. In the absence of the spin-orbit interaction there are $2(2l+1)$ degenerate spin orbitals for each energy $\mathcal{E}_{nl}$ and pair (n, l) or (k, l). These spin orbitals are given by

$$\psi_{nlm_lm_s}(x) = R_{n(k)l}(r)\, Y_{lm_l}(\theta, \varphi)\, \chi_{m_s}(\zeta), \tag{7.12}$$

where x stands for the position-spin coordinate $(\boldsymbol{r}, \zeta)$. The quality of these solutions depends on the choice of the potential $V(r)$. This matter is briefly discussed in Sect. 19 with reference to the Auger effect.

Spin orbitals (7.12) represent the available states (n, l, m_l, m_s) in the central field. Postulate 3 tells us how they are occupied. This leads to the notion of a configuration $(n_1 l_1)^{q_1} (n_2 l_2)^{q_2} \ldots (n_{N_s} l_{N_s})^{q_{N_s}}$. Each shell (nl) is occupied by $q \leqq 2(2l+1)$ electrons which have equal energies given by $\mathcal{E}_{nl}$. For example, the ground-state configuration is determined by the minimum-energy requirement and the maximum occupancy for each shell.

Before we discuss the classification scheme of inner-shell vacancy states we shall introduce the spin-orbit interaction. It is described approximately by the operator [6]

$$H_{so} = \xi(r)(\boldsymbol{l} \cdot \boldsymbol{s}) = \frac{\alpha^2}{2r} \left(\frac{dV}{dr}\right) (\boldsymbol{l} \cdot \boldsymbol{s}), \tag{7.13}$$

where α is the fine-structure constant. The Hamiltonian $H' = H_{cf} + H_{so}$ commutes with l^2, s^2, and P but not with l_z and s_z. Instead, H' as well as H_{cf} commutes with the operators j^2 and j_z which correspond to the total angular momentum $\boldsymbol{j} = \boldsymbol{s} + \boldsymbol{l}$. Since the influence of H_{so} is expected to be small in light atoms even for inner shells, it is natural to solve the Schrödinger equation $H'\psi' = \mathcal{E}'\psi'$ in terms of eigenfunctions of H_{cf} that can also be eigenfunctions of

[6] The origin of this operator is, for example, described in Ref. [*12*], Chap. 23

j^2 and j_z. These functions, given by

$$|n(k)\,s\,l\,j\,m_j\rangle = \sum_{m_l+m_s=m_j} (s\,l\,m_s\,m_l|j\,m_j)\,|n\,s\,l\,m_s\,m_l\rangle, \tag{7.14}$$

where $(s\,l\,m_s\,m_l|j\,m_j)$ are the Clebsch-Gordan coefficients, form a complete set with respect to n and k.[7] Hence each ψ' should be expanded over all n and k. However, if it is assumed that the nondiagonal elements of the spin-orbit parameter matrix $\boldsymbol{\xi}_l = \{\xi_{n'l,\,nl}\}$, where

$$\xi_{n'l,\,nl} = \frac{\alpha^2}{2} \int_0^\infty R_{n'l}(r) \frac{dV}{dr} R_{nl}(r)\,r\,dr, \tag{7.15}$$

are zero, then

$$\mathcal{E}'(nlj) = \mathcal{E}_{nl} + \xi_{nl} \frac{j(j+1) - l(l+1) - s(s+1)}{2}. \tag{7.16}$$

In (7.16), ξ_{nl} has been obtained from (7.15) by putting $n'=n$. Note also that the complete matrix $\boldsymbol{\xi}_l$ would have a continuous and semicontinuous part due to the continuum functions (7.14). Since $j=1/2$ for $l=0$ and $j=l\pm 1/2$ for $l \neq 0$,

$$\mathcal{E}'(nlj) = \mathcal{E}_{nl} + \tfrac{1}{2}\kappa_{jl}\,\xi_{nl}, \tag{7.17}$$

where $\kappa_{1/2,\,0}=0$ and where $\kappa_{jl}=l$ for $j=l+1/2$ and $-(l+1)$ for $j=l-1/2$. The spin-orbit parameter ξ_{nl} is positive since $V(r)$ is an increasing function of r.

Any reasonable choice of $V(r)$ for the ground-state electrons like the Herman-Skillman potential[8] allows us to construct an approximate energy-level diagram of inner-shell vacancy states. Figure 5 shows schematically the level diagram of singly ionized inner-shell vacancy states, denoted by $(nlj)^{-1}$. The energy of each level is given by $-\mathcal{E}'(nlj)$ with respect to the ground-state energy. Figure 5 also shows the nomenclature which we shall use in the description of inner-shell vacancy states. Multiply ionized inner-shell vacancy configurations are denoted by $(n_1 l_1 j_1)^{-q_1} (n_2 l_2 j_2)^{-q_2} \ldots$ or $Q_1^{q_1} Q_2^{q_2} \ldots$, where Q_v stands for the labels defined in Table 1. If the number of missing electrons refers to a shell rather than a subshell, then j is omitted.[9,10] The vacancies are usually specified with respect to the ground-state configuration of the neutral atom.

The usefulness of this nomenclature relies heavily on the success of the central field model of describing the atomic structure. With the aid of an

[7] We couple $\boldsymbol{l}$ to $\boldsymbol{s}$, i.e. $\boldsymbol{j}=\boldsymbol{s}+\boldsymbol{l}$ in accordance with CONDON and SHORTLEY (Ref. [9], Chap. 5, Sect. 4)

[8] Herman, F., Skillman, S. (1963): Atomic structure calculations, Englewood Cliffs: Prentice Hall

[9] We shall use the convention that the quantum numbers n and l defines a shell whereas n, l, and j define a subshell. Each shell has one subshell ($l=0$, $j=1/2$) or two subshells ($l>0$, $j=l\pm 1/2$)

[10] There are many other notations in the literature. LARKINS (Larkins, F.P. (1975) in: Atomic inner-shell processes I, ionization and transition probabilities, Crasemann, B. (ed.), New York: Academic Press, p. 377) has suggested a notation where vacancies are denoted by square brackets: $[(n_1 l_1)^{q_1} (n_2 l_2)^{q_2} \ldots]$. For example, $(1s)^{-1}(2p)^{-2}$ would be denoted by $[1s\,2p^2]$

Table 1. Nomenclature used in labelling energy levels of inner-shell vacancy states in the central field approximation of atoms[a]

l[b]	0	1	1	2	2	3	3
j	1/2	1/2	3/2	3/2	5/2	5/2	7/2
n							
1	K						
2	L_1	L_2	L_3				
3	M_1	M_2	M_3	M_4	M_5		
4	N_1	N_2	N_3	N_4	N_5	N_6	N_7
5	O_1	O_2	O_3	O_4	O_5	O_6	O_7
6	P_1	P_2	P_3	P_4	P_5		
7	Q_1						

[a] For the origin of this nomenclature see the note by M. Inokuti and T. Noguchi (Am. J. Phys. *42*, 1118, 1974). We have only included those symbols which are necessary for the description of energy levels in atoms up to $Z=100$

[b] Usually $l=0, 1, 2, 3 \ldots$ is replaced by $l=s, p, d, f \ldots$

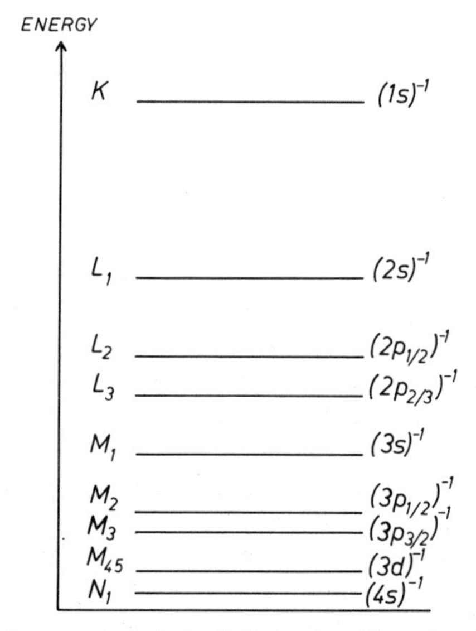

Fig. 5. Schematic energy diagram of one-hole states in zinc. The spin-orbit splitting of the $(3d)^{-1}$ state is not shown

energy analysis that may be more sophisticated than provided by the one-electron energies (7.17), it is usually possible to associate any states that are produced in inner-shell ionization to specific hole configurations. These *principal* configurations may then serve as the zeroth-order approximations in the many-electron theory of inner-shell vacancy states.

The nonradiative transitions can be classified according to Table 1. Suppose that an inner-shell vacancy is initially created in the subshell X_v by some interaction. The nonradiative transition manifests itself through the emission of an electron. Hence the final state must correspond to two subshell vacancies Y_μ and Z_κ. The corresponding double vacancy states must have less energy than the initial state associated with X_ν since the ejected electron carries kinetic energy. Consequently an *Auger transition*[11] is denoted by $X_\nu - Y_\mu Z_\kappa$ or $(n_1 l_1 j_1)^{-1} - (n_2 l_2 j_2)^{-1}(n_3 l_3 j_3)^{-1}$. Any additional initial vacancies that stay passive during the transition can of course be denoted explicitly. The *Coster-Kronig transition*[12] refers to a nonradiative process of the type $X_\nu - X_\mu Z_\kappa$, where X_ν and X_μ pertain to the same shell, i.e., to the same principal quantum number but different subshells. A *Super Coster-Kronig transition*[13] is a nonradiative transition that occurs within a single shell and is denoted by $X_\nu - X_\mu X_\kappa$. Within the central field model the kinetic energy of an Auger electron is given by

$$E_{\text{kin}} = -\mathcal{E}(n_1 l_1 j_1) + \mathcal{E}(n_2 l_2 j_2) + \mathcal{E}(n_3 l_3 j_3) \tag{7.18}$$

for the transition $(n_1 l_1 j_1)^{-1} - (n_2 l_2 j_2)^{-1}(n_3 l_3 j_3)^{-1}$. Naturally, (7.18) is also valid for Coster-Kronig (with $n_1 = n_2$) and Super Coster-Kronig transitions (with $n_1 = n_2 = n_3$). However, we shall often use the word "Auger" to comprise all nonradiative transitions.

8. Many-electron theory and frozen-core approximation. The previous section introduced the building blocks, central field spin orbitals, of the theory of nonradiative transitions. In this section we shall briefly present the necessary many-electron background of this theory.

We start with an orthogonal set of N spin orbitals $\{u_\nu(x)\}$, which we leave unspecified for a moment. For an N-electron system we can construct an antisymmetric wave function from the product of these spin orbitals by forming the Slater determinant

$$\Phi = \frac{1}{\sqrt{N!}} \det |u_1 u_2 \dots u_N| = \frac{1}{\sqrt{N!}} \begin{vmatrix} u_1(x_1) \dots u_1(x_N) \\ u_2(x_1) \dots u_2(x_N) \\ \dots \\ u_N(x_1) \dots u_N(x_N) \end{vmatrix} \tag{8.1}$$

that satisfies the condition

$$\langle \Phi | \Phi \rangle = \prod_{\nu=1}^{N} \langle u_\nu | u_\nu \rangle. \tag{8.2}$$

This formula relates the normalization of Φ to the normalization of the spin orbitals. The Pauli principle is satisfied in the sense that Φ vanishes if any two spin orbitals are identical or whenever $x_\nu = x_\mu$ in (8.1). In other words, two (or more) electrons neither occupy a single spin orbital nor possess the same position in space if their spin projections align.

[11] In recognition of the work of AUGER (Auger, P. (1925): J. Phys. Radium *6*, 205; (1926): Ann. Phys., Paris *6*, 183)

[12] In recognition of the work of COSTER and KRONIG (Coster, D., Kronig, R. de L. (1935): Physica *2*, 13)

[13] This name was introduced by MCGUIRE (McGuire, E.J. (1972): Phys. Rev. A *5*, 1043)

In atomic calculations many-electron wave functions are usually expressed as linear combinations of Slater determinants. This relates any matrix element to those which are given with respect to Slater determinants. In the following we need matrix elements of the many-electron Hamiltonian which is a sum of one- and two-electron operators. For a pair of Slater determinants the necessary formulae of this operator which can be written as

$$H=\sum_{\nu=1}^{N} H_\nu+\sum_{\mu>\nu} H_{\mu\nu}, \tag{8.3}$$

are given by, for example, CONDON and SHORTLEY [9], provided the spin orbitals are members of an orthogonal set. The nonorthogonal case has been treated by LÖWDIN.[14]

We quote the results for two Slater determinants, Φ_1 and Φ_2, which consist of orthogonal spin orbitals. If Φ_1 and Φ_2 have only one pair, v_1 and u_1, of unlike spin orbitals, then

$$\langle\Phi_1|H|\Phi_2\rangle=\pm\left[\langle v_1|H_1|u_1\rangle+\sum_{\mu=1}^{N}\langle v_1 u_\mu\| u_1 u_\mu\rangle\right], \tag{8.4}$$

where the $\mu=1$ term vanishes and where we have introduced the shorthand notation

$$\langle ab\|cd\rangle=\langle ab|H_{12}|cd\rangle-\langle ab|H_{12}|dc\rangle. \tag{8.5}$$

The + sign must be used if an even number of permutations brings the spin orbitals of Φ_1 in the same position as the ones in Φ_2 with v_1 in the place of u_1. The − sign corresponds to an odd number of permutations. With the same sign convention it follows that

$$\langle\Phi_1|H|\Phi_2\rangle=\pm\langle v_1 v_2\|u_1 u_2\rangle \tag{8.6}$$

if Φ_1 has two spin orbitals, v_1 and v_2, that are different from those of Φ_2.

The most important class of matrix elements in the nonrelativistic theory of nonradiative transitions consists of Coulomb matrix elements $\langle ab|1/r_{12}|cd\rangle$, where the spin orbitals are central field spin orbitals $(x|slm_s m_l)$. In this case

$$\langle ab|\frac{1}{r_{12}}|cd\rangle=\delta(m_{s_a}, m_{s_c})\,\delta(m_{s_b}, m_{s_d})\,\delta(m_{l_a}+m_{l_b}, m_{l_c}+m_{l_d}) \cdot\sum_{k=0}^{k_{\max}} c^k(l_a m_a, l_c m_c)\, c^k(l_d m_d, l_b m_b)\, R^k(abcd), \tag{8.7}$$

where the coefficients $c^k(lm, l'm')$ are given by

$$c^k(lm, l'm')=\frac{(l' k m' m'-m|lm)}{\sqrt{2l+1}}\langle l\| C^k \|l'\rangle. \tag{8.8}$$

[14] Löwdin, P.-O. (1955): Phys. Rev. 97, 1474

These coefficients, which are tabulated in most textbooks (e.g., [9], [12]), depend on the Clebsch-Gordan coefficient which couples $\boldsymbol{k}$ to $\boldsymbol{l}'$, producing the resultant $\boldsymbol{l}$ so that $k_{max} \leqq \mathrm{MIN}(l_a+l_c, l_b+l_d)$. In (8.8) the reduced matrix element $\langle l \| C^k \| l' \rangle$ of the Racah tensor $\boldsymbol{C}^k$ of spherical harmonics is given by

$$\langle l \| C^k \| l' \rangle = (-1)^l \sqrt{(2l+1)(2l'+1)} \begin{pmatrix} l & k & l' \\ 0 & 0 & 0 \end{pmatrix}, \tag{8.9}$$

where the symbol $\begin{pmatrix} a & b & c \\ d & e & f \end{pmatrix}$ denotes the $3j$ coefficient. The formulae (8.7) to (8.9) are consistent with the Condon-Shortley phase convention (7.4).

In (8.7) $R^k(abcd)$ are the Slater integrals

$$R^k(abcd) = \int_0^\infty r_1^2\, dr_1 \left[\frac{1}{r_1^{k+1}} \int_0^{r_1} X(r_1, r_2)\, r_2^{k+2}\, dr_2 + r_1^k \int_{r_1}^\infty X(r_1, r_2)\, r_2^{-k+1}\, dr_2 \right], \tag{8.10}$$

where

$$X(r_1, r_2) = R_{n_a l_a}(r_1)\, R_{n_b l_b}(r_2)\, R_{n_c l_c}(r_1)\, R_{n_d l_d}(r_2). \tag{8.11}$$

In the theory of the nonradiative transitions one or two of the radial wave functions $R_{nl}(r)$ usually belongs to the continuum, in which case n should be replaced by the wave number k. The Slater integrals are the ultimate quantities which determine the dynamics of the electrons in the nonradiative process.

An analysis of nonradiative transitions usually starts with the determination of the principal configurations of the initial and final states. Sometimes this concept is quite meaningless, especially when Coster-Kronig transitions are involved. However, even so it is often possible to state what this configuration would be in the absence of many-electron interactions![15] This defines the next step, namely, to examine how many-electron wave functions with proper symmetry and corresponding matrix elements are constructed for a given configuration.

α) LS-coupling. We start by ignoring the spin-orbit interaction. This leads us to consider Hamiltonian (7.1) for which we have already listed some of the commuting operators. The eigenfunctions of these operators, constructed from the central field spin orbitals of the configuration, and the corresponding eigenvalues provide the necessary ingredients for the classification and calculation of the energy *terms*. The energy term is characterized by the quantum numbers L and S, given in the form ${}^{2S+1}L$, where $2S+1$ is the spin multiplicity. If there are several identical pairs (L, S) which correspond to a given configuration, then additional quantum numbers like the seniority (for definition, see, e.g., [11]) are needed for a unique classification of the terms.

If the configuration is given by

$$\gamma = (n_1 l_1)^{q_1} (n_2 l_2)^{q_2} \ldots (n_{N_s} l_{N_s})^{q_{N_s}}, \tag{8.12}$$

[15] For an example, see Gelius, U. (1974): J. Electron. Spectrosc. 5, 985 and Lundqvist, S., Wendin, G. (1974): J. Electron. Spectrosc. 5, 513

then the number of Slater determinants to be considered is

$$M_0 = \prod_{\nu=1}^{N_s} \frac{n_\nu!}{(n_\nu - q_\nu)!\, q_\nu!}, \tag{8.13}$$

where $n_\nu = 2(2l_\nu + 1)$ is the maximum occupation number of νth subshell. These determinants provide the $(nlm_l m_s)$ representation of the configuration, but the $(nljm)$ representation consisting of the spin orbitals (7.14) could equally well be used. It is always possible to find, for example, using the projection technique of LÖWDIN,[16] an $M_0 \times M_0$ unitary transformation $\boldsymbol{D}$ such that the functions

$$|\gamma\omega LM_S M_L\rangle = \sum_\nu D_\nu \Phi_\nu (m_{l_1} m_{s_1} \ldots m_{l_N} m_{s_N}) \tag{8.14}$$

are the possible lineary independent eigenfunctions of L^2, S^2, L_z, and S_z which can be constructed for a given configuration (8.12). Note that each determinant Φ in (8.14) is an eigenfunction of L_z and S_z which belongs to the eigenvalues

$$M_L = \sum_{\nu=1}^{N} m_{l_\nu}, \qquad M_S = \sum_{\nu=1}^{N} m_{s_\nu}. \tag{8.15}$$

The index ω refers to any additional quantum numbers that may be needed to classify linearly independent solutions within a given term. The transformation from the $(SLM_S M_L)$ representation to the $(SLJM)$ representation is achieved by

$$|\gamma\omega SLJM\rangle = \sum_{M_L M_S} (SLM_S M_L | JM) |\gamma\omega SLM_S M_L\rangle, \tag{8.16}$$

where we have, as usual, coupled $\boldsymbol{L}$ to $\boldsymbol{S}$ (i.e., $\boldsymbol{S} + \boldsymbol{L} = \boldsymbol{J}$).[7] In all representations the wave functions are classified in addition to the angular momentum quantum numbers by the parity π, which is given by

$$\pi = (-1)^{\sum_{\nu=1}^{N} l_\nu} \tag{8.17}$$

for the configuration γ.

Since the set $\Lambda = \{L^2, S^2, L_z, S_z, P\}$ commutes with H, the matrix element

$$\begin{aligned} &\langle \gamma' \pi' \omega' L' M_{S'} M_{L'} | H | \gamma \pi \omega SLM_S M_L \rangle \\ &= \delta_{S'S} \delta_{L'L} \delta_{\pi'\pi} \delta_{M_{S'} M_S} \delta_{M_{L'} M_L} E_{\gamma'\gamma\omega'\omega}(SL) \end{aligned} \tag{8.18}$$

for any two configurations γ' and γ. For $\gamma' = \gamma$ the Hamiltonian matrix (8.18) is diagonal only if all the pairs (SL) occur only once. Otherwise the final eigenfunction must be obtained by a diagonalization of each submatrix $\{E_{\gamma\gamma\omega_\mu\omega_\nu}(SL)\}$, where ω_μ and ω_ν pertain to linearly independent wave functions with identical values of S and L.

---

[16] Löwdin, P.-O. (1962): Rev. Mod. Phys. *34*, 520

Whenever the spin-orbit interaction can be neglected, (8.18) provides the energy terms of the initial and final principal configurations, γ_i and γ_f, of any nonradiative transition in terms of one-electron and Slater integrals. This reduction can be achieved using standard techniques (e.g., [*12*]). Equation (8.18) also provides general formulae for the transition amplitude in (4.11) which correspond to a nondiagonal element of H, where γ' and γ differ by two spin orbitals. If we assume that the spin orbitals of γ_i and γ_f belong to the same orthogonal set then the problem of calculating the transition amplitude reduces to the evaluation of the Coulomb matrix element

$$H_{\mathrm{fi}} = \langle \gamma_\mathrm{f} \pi SLM_S M_L \omega_\mathrm{f} | \sum_{i>j} \frac{1}{r_{ij}} | \gamma_\mathrm{i} \pi SLM_S M_L \omega_\mathrm{i} \rangle. \tag{8.19}$$

This matrix element can be reduced to a superposition of Slater integrals for any two configurations γ_f and γ_i.[17] The evaluation starts with a construction of the wave functions in a specific way. First the antisymmetric wave functions for each subshell $(nl)^q$ are constructed using the fractional parentage technique. For incomplete s, p, d, and f subshells the fractional parentage coefficients can be easily found in the literature (e.g., [*11*]). Configuration (8.12) is given in the standard order and the N variables $\{x_\nu\}$ are distributed among the wave functions in an ordered fashion as

$$\boldsymbol{X} = (x_1 x_2 \dots x_{q_1})(x_{q_1+1} \dots x_{q_1+q_2}) \dots (x_{N-q_{N_s}} \dots x_N). \tag{8.20}$$

Each set of coordinates is denoted by $\boldsymbol{X}_\lambda$ so that the wave functions of the subshell $(n_\lambda l_\lambda)^{q_\lambda}$ are given by

$$(\boldsymbol{X}_\lambda | (n_\lambda l_\lambda)^{q_\lambda} \omega_\lambda S_\lambda L_\lambda\} = \{(\boldsymbol{X}_\lambda | (n_\lambda l_\lambda)^{q_\lambda} \omega_\lambda S_\lambda L_\lambda M_{S_\lambda} M_{L_\lambda}), \tag{8.21}$$

where the brackets $(\dots\}$ indicate the whole set. The wave function

$$(\boldsymbol{X} | \omega SLM_S M_L) = [\prod_\lambda (\boldsymbol{X}_\lambda | (n_\lambda l_\lambda)^{q_\lambda} \omega_\lambda S_\lambda L_\lambda\}]^{(\omega SL)}_{M_S M_L}, \tag{8.22}$$

which has been constructed by multiplication and coupling the angular momenta $S_\lambda L_\lambda$ according to some prescription ω with resultants S and L, is not antisymmetric. However, this wave function can be made antisymmetric by the following recipe. An even parity is assigned to the sequence (8.20). Then all possible permutations $P(\boldsymbol{X})$ among the coordinates between *different* sets are carried out but *not* within the sets. An even permutation corresponds to the plus sign and an odd permutation to the minus sign. Consequently the final antisymmetric wave function is

$$\psi(\omega SLM_S M_L) = \mathcal{N}(q_\lambda)^{-1/2} \sum_{\boldsymbol{X}} (-1)^{P(\boldsymbol{X})} (\boldsymbol{X} | \omega SLM_S M_L), \tag{8.23}$$

---

[17] Fano, U. (1965): Phys. Rev. *140*, A 67

where

$$\mathcal{N}(q_\lambda)=\frac{N!}{\prod_\lambda q_\lambda!} \tag{8.24}$$

is the number of distributions which can be formed by the exchange of the coordinates between the sets $\boldsymbol{X}_\lambda$. By representing the wave functions in the matrix element (8.19) by the functions (8.23) a reduction can be achieved with the aid of the Racah algebra or related graphical techniques.[18] The final result can be expressed in terms of Slater integrals (8.10), recoupling coefficients[19] and fractional parentage coefficients. Computer versions of the reduction procedure have been provided and incorporated in atomic programs by HIBBERT.[20] It is important to note that the results given by FANO[17] and later by ARMSTRONG[21] are not consistent with the Condon-Shortley phase convention but utilize a convention defined by FANO and RACAH.[22]

β) Intermediate coupling and configuration interaction. The spin-orbit interaction is represented in the N-electron case by a generalization of operator (7.13).[23] The many-electron operator is given by

$$H_{\mathrm{so}}=\sum_{v=1}^{N}\xi(r_v)(\boldsymbol{l}_v\cdot\boldsymbol{s}_v), \tag{8.25}$$

which is added to Hamiltonian (7.1). As a consequence the new Hamiltonian H' does not even commute with L^2 and S^2 as in the $N=1$ case. However, it still commutes with J^2, J_z, and P, which shows that

$$\langle\gamma'\pi'\omega'S'L'J'M'|H'|\gamma\pi\omega SLJM\rangle=\delta_{J'J}\delta_{M'M}\delta_{\pi'\pi}E^{S'SL'L}_{\gamma'\gamma\omega'\omega}(J) \tag{8.26}$$

for any pair of configurations γ' and γ. Note that due to H_{so} the matrix $\boldsymbol{H}'$ has, in general, nondiagonal elements with respect to wave functions which pertain to the same J and M but which differ in any of the quantum numbers ω, S, and L. Hence the task would be to diagonalize $\boldsymbol{H}'$ using a complete set of eigenfunctions $|\gamma\omega SLJM\rangle$ of the same parity. Using the expansion

$$|\alpha JM\rangle=\sum_{\gamma\omega SL}D(\gamma\omega SL)|\gamma\omega SLJM\rangle \tag{8.27}$$

to transform the Schrödinger equation into a set of linear equations yields

$$\sum_{\gamma\omega SL}D(\gamma\omega SL)[E^{S'SL'L}_{\gamma'\gamma\omega'\omega}(J)-E\delta_{\gamma'\gamma}\delta_{\omega'\omega}\delta_{S'S}\delta_{L'L}]=0, \tag{8.28}$$

---

[18] Briggs, J.S. (1971): Rev. Mod. Phys. *43*, 189

[19] For a definition, see, for example, Fano, U., Racah, G. (1959): Irreducible tensorial sets, New York: Academic Press

[20] Hibbert, A. (1975): Comp. Phys. Comm. *9*, 141 and references therein

[21] Armstrong, Jr., L. (1968): Phys. Rev. *172*, 12

[22] In this convention (see II.18) the spherical harmonics (7.3) are multiplied by an additional factor i^l

[23] e.g., Ref. [*12*], Chap. 23

with the associated secular equation

$$\det|E^{S'SL'L}_{\gamma'\gamma\omega'\omega}(J)-E\,\delta_{\gamma'\gamma}\delta_{\omega'\omega}\delta_{S'S}\delta_{L'L}|=0 \tag{8.29}$$

for the determination of the energy *levels*.

According to the set of equations (8.28), the coefficients $D(\gamma\omega SL)$ are independent of M and there is a separate secular equation for each J. The summation over a truncated set of configurations γ in (8.27) including eventually continuum wave functions defines a configuration-interaction wave function. However, if only *one* configuration is considered and there is no ambiguity due to ω, then the wave function is called an intermediate coupling wave function. In that case $\boldsymbol{H}'$ consists of the M-independent submatrices

$$\begin{array}{c} \\ {}^{2S+1}L_J^{(1)} \\ {}^{2S+1}L_J^{(2)} \\ \vdots \\ {}^{2S+1}L_J^{(t)} \end{array} \begin{array}{c} \quad {}^{2S+1}L_J^{(1)} \quad {}^{2S+1}L_J^{(2)} \ldots {}^{2S+1}L_J^{(t)} \\ \begin{pmatrix} H'_{11} & H'_{12} & \ldots & H'_{1t} \\ H'_{21} & H'_{22} & \ldots & H'_{2t} \\ \vdots & & & \vdots \\ H'_{t1} & \ldots & \ldots & H'_{tt} \end{pmatrix} \end{array}, \tag{8.30}$$

with one for each J such that the nondiagonal matrix elements are only due to the spin-orbit interaction.

Since the spin-orbit interaction is represented by a one-electron operator sum H_{so}, the matrix element $\langle\gamma'\omega'S'L'JM|H_{so}|\gamma\omega SLJM\rangle$ vanishes if γ' and γ differ by two spin orbitals. For $\gamma'=\gamma$ no general formula in terms of spin-orbit parameters ξ_{nl} seems to have been given in the case of nondiagonal elements although tabulations can be found (e.g., [9]). Since $\boldsymbol{S}$ commutes with each $\xi(r_\nu)\boldsymbol{l}_\nu$ and $\boldsymbol{L}$ with each $\boldsymbol{s}_\nu$ in H_{so} (which is a sum of scalar products of irreducible tensors of rank one), the diagonal matrix element is proportional to the $6j$ coefficient $\begin{Bmatrix} S & L & J \\ L & S & 1 \end{Bmatrix}$.[24] Hence it follows that

$$\langle\gamma\omega SLJM|H_{so}|\gamma\omega SLJM\rangle=\lambda_{SL}[J(J+1)-L(L+1)-S(S+1)], \tag{8.31}$$

where the fine-structure splitting constant λ_{SL} depends on γ, ω, S, and L. If the nondiagonal elements of H_{so} are small, then (8.31) gives an idea of the splitting of the term ${}^{2S+1}L$ into levels ${}^{2S+1}L_J$ pertaining to $J=L+S$, $L+S-1$, ... $|L-S|$.

With regard to nonradiative transitions this section gives a clue how to handle spin-orbit coupling and configuration interaction within the nonrelativistic approximation. The energies of the initial and final states are determined, in principle, by secular equation (8.29). The transition probabilities require considerations of (8.19) in connection with expansions (8.16) and (8.27). In Sect. 13 we shall give an example of the intermediate coupling in the relativistic case with considerations of the nonrelativistic limit as well.

---

[24] e.g., Ref. [*11*], Chap. 19

γ) jj coupling. Instead of the general configuration γ which is given by (8.12), we could also consider a configuration

$$\gamma'=(n_1 l_1 j_1)^{s_1}(n_2 l_2 j_2)^{s_2}\dots(n_{N_r} l_{N_r} j_{N_r})^{s_{N_r}}, \tag{8.32}$$

that is given in terms of subshells $(nlj)^s$. For a given γ there are several possibilities to choose γ', for example, p^2 corresponds to $p^2_{3/2}$, $p^2_{1/2}$, and $p_{1/2}p_{3/2}$. The wave functions corresponding to γ' can be constructed analogously to those of γ. First the antisymmetric wave functions $(\boldsymbol{X}_\lambda|(n_\lambda l_\lambda j_\lambda)^{s_\lambda}\omega_\lambda J_\lambda\}$ for each subshell $(n_\lambda l_\lambda j_\lambda)$ are constructed by coupling the individual total angular momenta $\boldsymbol{j}$. Further coupling of the angular momenta $\boldsymbol{j}_\lambda$ and antisymmetrization yields wave functions similar to the *LS*-coupled wave functions (8.23). This procedure can be repeated for each γ' corresponding to γ. This results in the *jj*-coupled representation $(jmJM)$ of γ that would be convenient to use in cases where the spin-orbit interaction dominates. This stems from the fact that the spin-orbit interaction matrix is diagonal in this representation. In the relativistic case this representation must be used since the Dirac central field four-spinors are eigenfunctions of j^2 and j_z. The $(SLJM)$ and $(jmJM)$ representations are connected by a unitary transformation which is unique only for $N=2$. This transformation will be considered in the following since two-vacancy configurations play an important role in the theory of nonradiative transitions.

Using the recipe of Sect. 8α we obtain the antisymmetric two-electron wave functions

$$(x_1x_2|SLJM)=\frac{1}{\sqrt{2\tau}}\left[\Phi^{l_1l_2s_1s_2}_{SLJM}(x_1x_2)+(-1)^{l_1+l_2-S-L}\cdot\Phi^{l_2l_1s_2s_1}_{SLJM}(x_1x_2)\right] \tag{8.33}$$

in *LS* coupling. Here $\tau=2$ for $n_1=n_2$ and $l_1=l_2$, otherwise $\tau=1$. The wave function $\Phi^{ll'ss'}_{SLJM}(x_1x_2)$ is obtained by transformation (8.16) from the wave functions

$$\Phi^{ll'ss'}_{SLM_LM_S}(x_1x_2)=\sum_{\substack{m_l\\m_{l'}}}\sum_{\substack{m_s\\m_{s'}}}(ll'm_lm_{l'}|LM_L)(ss'm_sm_{s'}|SM_S)\cdot(x_1|lsm_lm_s)(x_2|l's'm_{l'}m_{s'}). \tag{8.34}$$

In a similar way we obtain the antisymmetric *jj*-coupled wave functions

$$(x_1x_2|JM)=\frac{1}{\sqrt{2\tau}}\left[\Phi^{j_1j_2}_{JM}(x_1x_2)-(-1)^{j_1+j_2-J}\Phi^{j_2j_1}_{JM}(x_1x_2)\right], \tag{8.35}$$

where $\tau=2$ for $n_1=n_2$, $l_1=l_2$, and $j_1=j_2$. The wave function $\Phi^{jj'}_{JM}(x_1x_2)$ is given by

$$\Phi^{jj'}_{JM}(x_1x_2)=\sum_{mm'}(jj'mm'|JM)(x_1|sljm)(x_2|s'l'j'm'), \tag{8.36}$$

where the spin orbitals $(x|sljm)$ are defined by (7.14) in terms of the set $\{(x|lsm_lm_s)\}$. The difference between the two sets of wave functions, (8.33) and

(8.35), is only due to the different coupling schemes of the four angular momenta, $\boldsymbol{l}_1$, $\boldsymbol{l}_2$, $\boldsymbol{s}_1$, and $\boldsymbol{s}_2$. Therefore wave functions (8.33) and (8.35) are related by the unitary transformation[25]

$$(s_1 s_2[S], l_1 l_2[L] J | s_1 l_1[j_1] s_2 l_2[j_2] J) \\ = \sqrt{(2S+1)(2L+1)(2j_1+1)(2j_2+1)} \begin{Bmatrix} s_1 & s_2 & S \\ l_1 & l_2 & L \\ j_1 & j_2 & J \end{Bmatrix} \tag{8.37}$$

for a given two-electron configuration $\gamma=(n_1 l_1)(n_2 l_2)$. If $n_1=n_2$ and $l_1=l_2$ but $j_1 \neq j_2$, coefficients (8.37) must be multiplied by $\sqrt{2}$.

δ) Variational principle and Hartree-Fock method. So far nothing has been said about the determination of the spin orbitals. We postpone the discussion of calculational methods for the continuum orbitals to Sect. 19 and treat only the calculation of the bound spin orbitals in order to introduce the frozen-core approximation in particular.

Almost all calculations of spin orbitals are based on the variational principle: the eigenvalue equation of a Hermitian operator is equivalent to a variational equation of its expectation value. This implies that the time-independent Schrödinger equation $H\psi=E\psi$ can be replaced by a variational equation

$$\delta\langle\psi|H|\psi\rangle=0, \tag{8.38}$$

with the constraint $\langle\psi|\psi\rangle=1$ for bound-state wave functions. By considering in (8.38) various approximate forms of ψ with free parameters which may be functions themselves, approximate solutions of the Schrödinger equation are obtained.

Whereas equations (8.28) are the result of variations of the coefficients in expansion (8.27) with *fixed* basis functions, the Hartree-Fock equations[26] follow from the variation of the spin orbitals in a single or linear superposition of Slater determinants (8.1). For example, the variation of orthonormal spin orbitals in a single Slater determinant *without* specifying the form of the spin orbitals leads to the *unrestricted Hartree-Fock* (UHF) method. In this case the substitution of Slater determinant (8.1) in (8.38), H is given by (7.1), and the variation of spin orbitals u_ν $(\nu=1 \ldots N)$ under the constraints $\langle u_\nu|u_\mu\rangle=\delta_{\nu\mu}$ leads to a set of equations given by

$$F u_\nu = \sum_{\mu=1}^{N} \lambda_{\mu\nu} u_\mu, \tag{8.39}$$

where F is the nonlocal *Fock operator*

$$F = f_1 + \sum_{\mu=1}^{N} \langle u_\mu \| u_\mu \rangle \tag{8.40}$$

---

[25] Ref. [*11*], Chap. 20. For p^n and d^n $(n>2)$ configurations, see Calvert, J.B., Tuttle, E.R. (1979): Nuovo Cimento B *54*, 413; see also Dyall, K.G., Grant, I.P. (1982): J. Phys. B *15*, L371

[26] e.g., Ref. [*13*]

and $\boldsymbol{\lambda}=\{\lambda_{\mu\nu}\}$is the matrix of the Lagrangian multipliers. Operator f_1 is the one-electron operator $-\nabla_1^2/2-Zr_1^{-1}$. According to (8.5), operator $\langle u\|u\rangle$ acting on an arbitrary function $v(x_1)$ gives

$$\langle u\|u\rangle v=\int\frac{u^*(x_2)\,u(x_2)}{r_{12}}\,dx_2 v(x_1)-\int\frac{u^*(x_2)\,v(x_2)}{r_{12}}\,dx_2 u(x_1), \tag{8.41}$$

where the integration over x_2 involves also the summation over the spin projection ζ_2. The diagonalization of the Hermitian matrix $\boldsymbol{\lambda}$ by a unitary transformation yields

$$F v_\mu=\mathcal{E}_\mu v_\mu, \tag{8.42}$$

which can be solved by iteration of the unrestricted Hartree-Fock equations

$$(f+\sum_{\nu\neq\mu}\langle v'_\nu\|v'_\nu\rangle)\,v_\mu=\mathcal{E}_\mu v_\mu \tag{8.43}$$

until self-consistency is achieved, i.e. $v'_\mu\to v_\mu$ for $\mu=1\ldots N$. Methods which try to find solutions of equations like (8.42) be repeatedly solving equations like (8.43) are called self-consistent-field (SCF) methods.

Without any further constraints like requirements of orthogonality to states lower in energy, the total wave function $\Phi=(N!)^{-1/2}\det|v_1 v_2\ldots v_N|$ provides an estimate of the ground-state wave function and the corresponding energy $E(N)=\langle\Phi|H|\Phi\rangle$ an upper-limit estimate of the ground-state energy. The $(N-1)$-electron wave functions $\Phi^\mu=|(N-1)!|^{-1/2}\det|v_1 v_2\ldots v_{\mu-1}v_{\mu+1}\ldots v_N|$ and the corresponding energies $E_\mu(N-1)=\langle\Phi^\mu|H(N-1)|\Phi^\mu\rangle$ have also an interpretation. Since

$$E_\mu(N-1)-E(N)=-\langle u_\mu|F|u_\mu\rangle \tag{8.44}$$

for *any* set $\{u_\mu\}$, it also holds for $u_\mu=v_\mu$, in which case $\mathcal{E}_\mu=E_\mu(N-1)-E(N)$. Furthermore, according to (8.42) and (8.44), the set $\{v_\mu\}$ makes the variations of *both* energies $E(N)$ and $E_\mu(N-1)$ zero if they are interpreted as energy functionals of a common set of spin orbitals (Koopmans' theorem).[27] Hence the one-electron energies $-\mathcal{E}_\mu$ represent the optimum ionization energies to the singly ionized states Φ^μ if the passive electrons are described by the same spin orbitals in the ground and ionized state. This is the *frozen-core* approximation, which in a generalized form implies the use of a common set of orthogonal spin orbitals for the initial and final state of a transition.

It follows from (8.42) that

$$\langle\Phi_\mu^\nu|H(N)|\Phi\rangle=0, \tag{8.45}$$

where Φ_μ^ν has been obtained from the solution $\Phi=(N!)^{-1/2}\det|v_1\ldots v_\mu\ldots v_N|$ by replacing the orbital v_μ with another orthogonal spin orbital w_ν (Brillouin's

[27] Koopmans, T.A. (1933) (nobel-prize winner in economy in 1975): Physica *1*, 104. Equation (8.44) is sometimes erroneously referred to as Koopmans' theorem

theorem). Related to this theorem, it also follows that

$$\langle \Phi^\nu | H(N-1) | \Phi^\mu \rangle = [\mathcal{E}_\mu + E(N)]\, \delta_{\mu\nu}, \tag{8.46}$$

which implies that the matrix $\boldsymbol{H}(N-1)$ is diagonal with respect to the frozen-core wave functions of the singly ionized states. This reflects their optimum property with respect to variations of both $E(N)$ and $E^\mu(N-1)$ simultaneously. Unfortunately, (8.46) does not imply rigorously that each energy $\mathcal{E}_\mu + E(N)$ provides an upper bound of the true energy as is the case for $E(N)$ with respect to the ground-state energy. The relevance of this question with regard to calculations of energies of nonradiative transitions is discussed in Sect. 16β.

The unrestricted Hartree-Fock equations (8.42) were obtained without imposing any *symmetry and equivalence restrictions* on the spin orbitals. For an example of these restrictions we consider the ground state of the lithium atom. By requiring that there are three spin orbitals with pure s character corresponding to the configuration $1s^2 2s$, a symmetry restriction is imposed. By requiring that the 1s spin orbitals have identical spatial dependence, an equivalence restriction is imposed.

In the *restricted Hartree-Fock* (RHF) methods the spin orbitals are restricted to be central field spin orbitals (7.12). The energy functional $\langle \psi | H(N) | \psi \rangle$ is calculated with respect to eigenfunctions of L^2, S^2, L_z, S_z, and P for a single or a superposition of several configurations. According to Sect. 8α, this energy functional can be expressed as a function of the radial wave functions $R_{nl}(r)$ and unknown expansion coefficients if several configurations are involved as in (8.27). In the case of a single configuration the variation of the functions $R_{nl}(r)$ results in Hartree-Fock equations that can be written for each term and solved numerically or analytically.[28] In the case of several configurations the simultaneous variation of the functions $R_{nl}(r)$ and the expansion coefficients lead to Hartree-Fock-like equations, which can also be solved using the SCF technique. In this case the method is called the multiconfiguration Hartree-Fock (MCHF) method. In these RHF methods the nondiagonal Lagrangian multipliers cannot be eliminated in general by a unitary transformation and the diagonal Lagrangian multipliers may *not* satisfy Koopmans' theorem.[29]

A method frequently used in calculations of nonradiative transition energies and in some cases also probabilities is the term-averaged HF method. It is based on the average energy of the configuration given by

$$E_{\mathrm{av}} = \frac{\sum\limits_{\mu} (2L_\mu + 1)(2S_\mu + 1)\, E(S_\mu L_\mu)}{\sum\limits_{\mu} (2L_\mu + 1)(2S_\mu + 1)}, \tag{8.47}$$

where the summation over μ includes all possible terms (note that the same $(S_\mu L_\mu)$ combination may sometimes appear more the once) for a given con-

[28] Numerical techniques are described in Ref. [*13*]. For analytical techniques, see Roothaan, C.C., Bagus, P.S. (1963) in: Methods of computational physics, Vol. II, New York: Academic Press

[29] See Ref. [*13*], Chap. 2 for details

figuration γ. In this case E_{av} can be written in a general form that is a functional of the radial functions and the occupation numbers q_v for each subshell of γ. The variation of the radial wave functions leads to a set of Hartree-Fock equations that can be written in a general form involving the occupation numbers q_v as parameters. The diagonal Lagrangian multipliers $\varepsilon_{\mu\mu}$ satisfy for a given set of radial wave functions the equation

$$\varepsilon_{\mu\mu} = E_{av}(q_\mu) - E_{av}(q_\mu - 1), \tag{8.48}$$

which is analogous to Koopmans' theorem. Equations for the nondiagonal Lagrangian multipliers $\varepsilon_{\mu v}$ ($u \neq v$) can also be written. These equations show that $\varepsilon_{\mu v}$ is determined uniquely unless $q_\mu = q_v$ for two incomplete subshells of same symmetry, i.e., l value. On the other hand, if $\varepsilon_{\mu v}$ is omitted, it leads to nonorthogonal radial wave functions. An example of this symmetry collapse is given in Sect. 18α since it can greatly affect nonradiative transition energies.

Within the framework of RHF methods it is evident that we can perform a variation of the spin orbitals with respect to any *fixed* configuration $\bar{\gamma}$ that may contain inner-shell vacancies. The resulting radial wave functions $\bar{R}_{nl}(r)$ are not orthogonal to the optimized ground-state radial functions $R_{n'l}(r)$ of the same symmetry so that

$$\langle nl | n'l \rangle = \int_0^\infty \bar{R}_{nl}(r) R_{n'l}(r) r^2 \, dr \neq \delta_{nn'}. \tag{8.49}$$

The separate optimization of initial- and final-state spin orbitals by self-consistent-field procedures for a given transition is called the ΔE_{SCF} method, and the difference between the ΔE_{SCF} and frozen-core transition energy is called the relaxation energy. This method has not been fully justified from the point of view of the variational principle of excited states. However, as pointed out in Sect. 16γ, it can be shown using perturbation theory and the unrestricted formalism that the relaxation correction must be included together with other correlation energy contributions to the ionization energy. It has also been shown that the overlap elements (8.49) have physical significance with regard to multiple excitation probabilities in various inner-shell processes.[30] The influence of the overlap elements (8.49) on nonradiative transition probabilities will be discussed in Sects. 10α and 20γ.

The spin-orbit interaction (8.25) is treated as a perturbation in the nonrelativistic HF schemes. It is rather difficult to combine the intermediate-coupling analysis with the ΔE_{SCF} procedure although some obstacles may be circumvented by using the transition operator method of GOSCINSKI et al.[31] However, within the relativistic Hartree-Fock (Dirac-Fock) framework (see Chap. III) the spin-orbit interaction is automatically included in the ΔE_{SCF} method.

---

[30] For a review and further references, see Åberg, T. (1976) in: Photoionization and other probes of many-electron interactions, Wuilleumier, F.J. (ed.), New York: Plenum Press, p. 49

[31] Goscinski, O., Howat, G., Åberg, T. (1975): J. Phys. B *8*, 11

ε) Electrons and holes. Configurations having almost complete shells are characteristic of inner-shell vacancy states. Hence it is useful to know whether any information can be obtained about the wave functions and matrix elements of a hole configuration $\mathscr{H}$ if the corresponding quantities are given for a conjugate electron configuration $\mathscr{E}$. With mutually conjugate configurations we mean two configurations whose occupation numbers, $q_\mu(\mathscr{H})$ and $q_\mu(\mathscr{E})$, fulfill the relation

$$q_\mu(\mathscr{H})+q_\mu(\mathscr{E})=2(2l_\mu+1) \tag{8.50}$$

for each incomplete shell μ.[32] If the spin-orbit interaction is included and the $nljm$ representation is used, then

$$q'_\mu(\mathscr{H})+q'_\mu(\mathscr{E})=2j_\mu+1 \tag{8.51}$$

for each subshell in an incomplete shell μ. In (8.51) $j=l-1/2$ $(l>0)$ and $j=l+1/2$ for each μ.

We shall begin by examining the relationship between $\mathscr{H}$ and $\mathscr{E}$ in the *LS*-coupling case and consider the transformation from the (nlm_lm_s) representation to the (SLM_SM_L) representation. Consider a wave function $|SLM_SM_L\rangle$ of $\mathscr{E}$ given by

$$|\mathscr{E}, SLM_SM_L\rangle=\sum_{\alpha=1}^{Q} C_\alpha(\mathscr{E})D_\alpha(\ldots,(nlm_lm_s),\ldots), \tag{8.52}$$

where D_α are the Slater determinants consisting of spin orbitals $(x|nlm_lm_s)$ in standard order, which is defined as

$$1s1\bar{s}2s2\bar{s}2p_12\bar{p}_12p_02\bar{p}_02p_{-1}2\bar{p}_{-1}\ldots(nlm_l+1/2),(nlm_l-1/2)\ldots \tag{8.53}$$

From these Slater determinants new Slater determinants $\bar{D}_\alpha$ are obtained that include for each incomplete shell those spin orbitals that are missing in D_α. The resulting wave function

$$\psi=\sum_{\alpha=1}^{Q} \bar{C}_\alpha(\mathscr{H})\bar{D}_\alpha(\ldots,(nlm_lm_s)^{-1},\ldots) \tag{8.54}$$

is obviously an eigenfunction of L_z and S_z that belongs to the eigenvalues $-M_L$ and $-M_S$, respectively. According to Appendix D, it is also an eigenfunction of L^2 and S^2 that belongs to the same L and S as function (8.52). So also is then the conjugate wave function obtained from function (8.54) by replacing each m_l and m_s by $-m_l$ and $-m_s$ and by restoring the standard order in the determinants $\bar{D}_\alpha$. Consequently each coefficient $\bar{C}_\alpha(\mathscr{H})$ in

$$|\mathscr{H}, SLM_SM_L\rangle=\sum_{\alpha=1}^{Q} \bar{C}_\alpha(\mathscr{H})\bar{D}_\alpha(\ldots,(nl-m_l-m_s)^{-1},\ldots) \tag{8.55}$$

[32] Condon and Shortley (Ref. [9], Chap. XII) consider this complementarity only within one incomplete shell. All spin orbitals outside this shell are assumed to be identical in their work

is related to the coefficients $C_\alpha(\mathscr{E})$ by

$$\bar{C}_\alpha(\mathscr{H}) = \Gamma(M_L, M_S)\, C_\alpha(\mathscr{E}), \tag{8.56}$$

where $\Gamma(M_L, M_S)$ is a phase factor.

According to (D.5) of Appendix D, each nonvanishing determinant produced by the step-down operator $L_x - iL_y$ contains either (*i*) a sequence $\ldots, m_l - 1, \bar{m}_l, \ldots$ or $\ldots, \bar{m}_l - 1, m_l - 1, \ldots$ that are brought to standard order by *one* permutation or (*ii*) a sequence that is in standard order.[33] In the latter case the original determinant has two subsequent m_l values which differ by two. The transformation from $\mathscr{E}$ to $\mathscr{H}$ maps (*i*) on (*ii*) and vice versa. Hence the simplest choice is $\Gamma(M_L, M_S) = (-1)^{M_L}$, which is consistent with the sequence of spin orbitals in standard order (8.53) in both wave functions (8.52) and (8.55). However, this choice of CONDON and SHORTLEY is not necessarily consistent with the phase obtained by constructing the hole wave functions according to the general recipe of Sect. 8α except *within* each individual term (SL). In the case of two-hole configurations $\mathscr{H} = (n_a l_a)^{-1}(n_b l_b)^{-1}$ a consistent overall phase of wave functions (8.55) can be obtained as follows. If it is required that the application of the coupling technique of Sect. 8α to the two-hole wave functions should lead to a single determinant with the spin orbitals in the standard order (8.53) for the corresponding closed shells, then the complete phase factor should be

$$\Gamma(M_L, M_S) = (-1)^{l_a + l_b + m_{l_a} + m_{l_b}}, \tag{8.57}$$

which reduces to the Condon-Shortley phase factor for $l_a + l_b$ even.

The most important result of the equivalence of the wave functions (8.52) and (8.55) is the relationship

$$\begin{aligned} &\langle \mathscr{H}' SLM_S M_L | H(N_{\mathscr{H}}) | \mathscr{H} SLM_S M_L \rangle \\ &\quad = \langle \mathscr{E}' SLM_S M_L | H(N_{\mathscr{E}}) | \mathscr{E} SLM_S M_L \rangle, \end{aligned} \tag{8.58}$$

where $\mathscr{E}'$ and $\mathscr{E}$ are assumed to differ either by one in each occupation number of two separate shells or by two in the occupation number of one shell. In (8.58) $N_{\mathscr{H}}$ is the number of electrons in $\mathscr{H}$ and $N_{\mathscr{E}}$ in $\mathscr{E}$. Another consequence of this equivalence is that although the average energy (8.47) of $\mathscr{E}$ is different from that of $\mathscr{H}$, the differences $E_{\mathrm{av}} - E(LS)$ are identical. In other words the electron and hole configurations have the same energy separations between the terms.

It follows from (8.16) and the correspondence between the wave functions (8.52) and (8.55) that the wave functions $|\mathscr{E} SLJM\rangle$ and $|\mathscr{H} SLJM\rangle$ are eigenfunctions of J^2 and J_z that pertain to the same eigenvalues. However, with regard to the spin-orbit interaction it must be realized that

$$\langle \mathscr{H} S' L' JM | H_{\mathrm{so}} | \mathscr{H} SLJM \rangle = -\langle \mathscr{E} S' L' JM | H_{\mathrm{so}} | \mathscr{E} SLJM \rangle, \tag{8.59}$$

[33] The notation $\bar{m}$ means that the spin orbital is associated with $m_s = -1/2$

which implies that the sign of each spin-orbit parameter ξ_{nl} must be changed in each element of the "intermediate-coupling matrices" (8.30) when going from an electron to a hole configuration. This is most easily seen by transforming both sides of (8.59) into a matrix which is given with respect to the $(jjJM)$ representation. As a result there are only diagonal elements which according to (7.17) are given by

$$\langle \mathscr{E} jjJM | H_{\mathrm{so}} | \mathscr{E} jjJM \rangle = \tfrac{1}{2} \sum_{\mu, j_\mu} q_{j_\mu} \kappa_{j_\mu l_\mu} \xi_{n_\mu l_\mu}, \tag{8.60}$$

where the summation runs only over the incomplete subshells $(n_\mu l_\mu j_\mu)$. Since the sum in (8.60) would be zero for each complete shell, it follows that

$$\langle \mathscr{E} jjJM | H_{\mathrm{so}} | \mathscr{E} jjJM \rangle + \langle \mathscr{H} jjJM | H_{\mathrm{so}} | \mathscr{H} jjJM \rangle = 0, \tag{8.61}$$

which shows that (8.59) is valid.

In order to establish a unique correlation between *jj*-coupled wave functions of $\mathscr{E}$ and $\mathscr{H}$, we first define a standard order, which is given by

$$\begin{aligned} &1s_{1/2},\ 1s_{-1/2},\ 2s_{1/2},\ 2s_{-1/2},\ 2p^*_{1/2},\ 2p^*_{-1/2},\ 2p_{3/2},\\ &2p_{1/2},\ 2p_{-1/2},\ 2p_{-3/2},\ \ldots,\ nl^*_{m_j},\ \ldots,\ nl_{m_j},\ \ldots, \end{aligned} \tag{8.62}$$

where the notation $nl^*_{m_j}$ refers to $j=l-1/2$ and nl_{m_j} to $j=l+1/2$. We consider a wave function $|\mathscr{E}, JM\rangle$ that is given as a function of Slater determinants, where the spin orbitals $|nljm\rangle$ are listed in the standard order (8.62). By replacing in each determinant those spin orbitals which are present in incomplete shells with the missing ones, and by changing each m to $-m$ with the standard order (8.62) restored, the conjugate wave function $|\mathscr{H}, JM\rangle$ is obtained. This can be shown by applying the technique of Appendix D to eigenfunctions of the total angular momentum. It also follows that an application of the technique of Sect. 8γ reduces an arbitrary conjugate two-hole wave function $|(n_a l_a j_a)^{-1} (n_b l_b j_b)^{-1} JM\rangle$ to a closed-shell single determinant with the spin orbitals in the standard order (8.62). Hence the two-hole wave functions form a consistent set with respect to the phase in *jj* coupling.

There is also the question of the transformation from the $(SLJM)$ representation to the $(jjJM)$ representation and vice versa. For that purpose we consider the transformation between the $(nlm_l m_s)$ and $(nljm)$ representations in both $\mathscr{E}$ and $\mathscr{H}$. It follows from the work of Shortley[34] that this transformation would be the same in $\mathscr{E}$ and $\mathscr{H}$ provided a phase factor

$$\Gamma(r, p, l, M_L) = (-1)^{M_L + (r+1)(p+l)} \tag{8.63}$$

is introduced for each incomplete shell. Note that Shortley's standard order is different from that of (8.62), which is commonly used in the relativistic central field approach.[35] Nevertheless, there would be no difference in phase factor (8.63), where r is the number of the electrons in the shell (nl) and p the number

[34] Shortley, G.H. (1933): Phys. Rev. *43*, 451 and Ref. [9], Chap. XII

[35] Grant, I.P. (1970): Adv. Phys. *19*, 747

of $j=l-1/2$ electrons in that shell. It follows from (8.57) and (8.63) that the adoption of the phase factor $(-1)^{M_L}$ for conjugate hole states in the $(nlm_l m_s)$ scheme and of $(-1)^{(r+1)(p+l)}$ for conjugate hole states in the $(nljm)$ scheme leads to the *same* set of transformations between the four representations $(nlm_l m_s)$, $(nljm)$, $(SLJM)$, and $(jjJM)$ in $\mathscr{E}$ and $\mathscr{H}$. Note that (8.63) must be taken into account in applications of transformation (8.37) to two-*hole* wave functions.

9. Probability of nonradiative transitions. To the extent the Auger process can be treated as a two-step process, the decay part is represented by the width

$$\Gamma=2\pi|\langle f|H-E|i\rangle|^2 \tag{9.1}$$

for a nonradiative transition between the initial bound state $|i\rangle$ and the final continuum state $|f\rangle$, which is assumed to be normalized per unit energy range. In (9.1), which is given in atomic units, H is the total Hamiltonian of the initially ionized atom and E the total energy of the initial state. As is explained in Sect. 3 and Appendix B, neither $|i\rangle$ nor $|f\rangle$ should be identified as eigenfunctions H but rather as projections of the total eigenfunction $\psi(E)$.

According to Sect. 4 the final state is represented by a wave function which fulfils the ingoing spherical wave boundary condition (2.7) in the case of N_c channels. For $N_c=1$ this boundary condition is equivalent to the standing wave boundary condition (2.2). This is also true if the $\boldsymbol{S}$ matrix is assumed to be diagonal, which implies that there is no final-state channel mixing. This assumption is made in the following. It is also assumed that the frozen-core approximation is valid so that $\langle f|i\rangle=0$, in which case the E term can be dropped from (9.1).

Suppose the wave functions $|i\rangle$ and $|f\rangle$ are given by Slater determinants (8.1), where the form of the spin orbitals is unspecified. The initial-state Slater determinant contains two spin orbitals u_2 and u_3 which are vacant in the final state. The final-state Slater determinant contains the spin orbital u_1 which is vacant in the initial state and the continuum spin orbital $v=v(\mathcal{E})$. Since all other spin orbitals are identical, the width (9.1) reduces, according to (8.6), to

$$\Gamma=2\pi|\langle u_1 v\|u_2 u_3\rangle|^2, \tag{9.2}$$

where the matrix element

$$\langle u_1 v\|u_2 u_3\rangle=\langle u_1 v|\frac{1}{r_{12}}|u_2 u_3\rangle-\langle u_1 v|\frac{1}{r_{12}}|u_3 u_2\rangle \tag{9.3}$$

consists of a direct and exchange amplitude. Equations (9.2) and (9.3) are equivalent to WENTZEL's ansatz provided the spin orbitals u_1, v, u_2, and u_3 are chosen to be hydrogenic ones. In the literature[36] (9.2) is usually taken as the starting point for the theory of the Auger effect. Normalizations of v, different from per unit energy range, are used which introduce a final-state density

[36] See the reviews listed in the introduction

function $\rho(\mathcal{E})$ different from unity in (9.2). In Appendix E various normalization conventions with associated values of $\rho(\mathcal{E})$ are presented. Some conversion factors are also given which allows a transfer from the concept of width to the concept of probability or rate. Keeping this in mind, we shall give all formulae in terms of widths and in atomic energy units but interpret them as probabilities or rates. The concepts of probability and rate we shall use synonymously.

In the derivation of (9.2) symmetry requirements were ignored. The symmetry of the initial and final many-electron wave functions is taken into account in the following by considering separately the *LS*-, *jj*-, and intermediate-coupling case. The mixed-coupling case including configuration interaction is also considered. It is assumed that the atom has initially a closed-shell structure in its ground state. The section is concluded with a discussion of the open-shell case.

α) LS- and jj-coupling limit. The initial state is assumed to be singly ionized so that in *LS* coupling the corresponding term is a doublet that is given by $(n_1 l_1)^{-1}\,{}^2L'$, where $L'=l_1$. The final states are given by $(n_2 l_2)^{-1}(n_3 l_3)^{-1}\,{}^{2S+1}L\mathcal{E}\,l\,{}^2L'$, where the momenta $\boldsymbol{s}$ and $\boldsymbol{l}$ are coupled to the momenta $\boldsymbol{S}$ and $\boldsymbol{L}$ of the doubly ionized core to give ${}^2L'$. Clearly, this coupling defines several channels for a given distribution of initial and final vacancies, denoted by $(n_1 l_1)^{-1}$ and $(n_2 l_2)^{-1}(n_3 l_3)^{-1}$, respectively. The values of the orbital quantum number l are restricted by the requirement of equal parity for the initial and final state, which gives

$$(-1)^{l_1+l_2+l_3+l}=1. \tag{9.4}$$

We consider the initial-state wave function $|iS'L'M_{S'}M_{L'}\rangle$ which is a single Slater determinant and the corresponding final-state wave function $|f(SL)\,S'L'M_{S'}M_{L'}\rangle$ which is usually a linear combination of several Slater determinants. In each of these determinants the spin orbitals are given in the standard order (8.53) and the phase convention (8.57) is adopted for the wave functions. The matrix element in (9.1) reduces to

$$A(SL)=\langle f(SL)S'L'M_{S'}M_{L'}|\sum_{\mu>\nu} r_{\mu\nu}^{-1}|iS'L'M_{S'}M_{L'}\rangle,$$

which can be expressed in terms of the two-electron wave functions $(x_1 x_2|n_1 l_1,\,\mathcal{E}l,\,SLM_SM_L)$ and $(x_1 x_2|n_2 l_2, n_3 l_3, SLM_SM_L)$. Both functions are defined according to (8.33) and (8.34) except that JM is replaced by M_S and M_L. Note that there is also the convention of standard order in the two-electron wave functions. The result, which is a consequence of the property

$$(j_1 j_2 m_1 m_2|JM)=(-1)^{j_2+m_2}\sqrt{\frac{2J+1}{2j_1+1}}\,(j_2 J-m_2 M|j_1 m_1) \tag{9.5}$$

of the Clebsch-Gordan coefficients, is given by

$$A(SL)=(-1)^{\sigma}\sqrt{\frac{(2S+1)(2L+1)}{2(2l_1+1)}} \cdot \langle n_1 l_1 \mathcal{E} l SLM_S M_L | \frac{1}{r_{12}} | n_2 l_2 n_3 l_3 SLM_S M_L \rangle, \tag{9.6}$$

where $\sigma = l + l_1 - L - S$. Using techniques outlined in Sect. 8, the two-electron matrix element in the amplitude $A(SL)$ can be expressed in terms of the Slater integrals (8.10). The result is given by

$$A(SL)=(-1)^{l+l_2+L+\sigma}\sqrt{\frac{(2S+1)(2L+1)}{2\tau(2l_1+1)}} \cdot \sum_{\mu} [D(\mu l_1 l l_2 l_3 L)+(-1)^{L+S} E(\mu l_1 l l_2 l_3 L)], \tag{9.7}$$

where the direct amplitude $D(\mu l_1 l l_2 l_3 L)$ is expressed by

$$D(\mu l_1 l l_2 l_3 L)=\langle l_1 \| C^{\mu} \| l_2 \rangle \langle l \| C^{\mu} \| l_3 \rangle \begin{Bmatrix} l_1 & l & L \\ l_3 & l_2 & \mu \end{Bmatrix} R^{\mu}(n_1 l_1 \mathcal{E} l n_2 l_2 n_3 l_3). \tag{9.8}$$

The exchange amplitude E is obtained from D by changing the indexes 2 and 3 everywhere in (9.8). Consequently, the μ values, which are restricted by triangular conditions, may be different for the D and E terms. The reduced matrix element $\langle l \| C^{\mu} \| l' \rangle$ is defined by (8.9) and τ is two for $n_2 = n_3$ and $l_2 = l_3$, otherwise one. Equations (9.7) and (9.8) agree with those given by WALTERS and BHALLA[37] except for the phase factor $(-1)^{\sigma}$ in (9.6).

In jj coupling we can derive for the amplitude of a transition between the initial hole states $(n_1 l_1 j_1)^{-1} J'$ and the final states $(n_2 l_2 j_2)^{-1} (n_3 l_3 j_3)^{-1} J (\mathcal{E} l j) J'$ formulae which are equivalent to (9.6), (9.7), and (9.8). The result is

$$A(J)=(-1)^{\sigma'}\sqrt{\frac{2J+1}{2j_1+1}} \langle n_1 l_1 j_1 \mathcal{E} l j JM | \frac{1}{r_{12}} | n_2 l_2 j_2 n_3 l_3 j_3 JM \rangle \tag{9.9}$$

in terms of two-electron wave functions (8.35). In analogy to (9.7) $\sigma' = j - j_1 + J$ in the phase factor. The reduction of the matrix element in (9.9) is similar to the procedure which yields (9.7) and (9.8). The result is

$$A(J)=(-1)^{\sigma'+j+j_2+J}\sqrt{\frac{2J+1}{\tau(2j_1+1)}} \cdot \sum_{\mu} [D(\mu j_1 j j_2 j_3 J)+(-1)^{J} E(\mu j_1 j j_2 j_3 J)], \tag{9.10}$$

[37] Walters, D.L., Bhalla, C.P. (1971): Phys. Rev. A 3, 1919

where the direct amplitude is given by

$$D(\mu j_1 j j_2 j_3 J) = \langle j_1 \| C^\mu \| j_2 \rangle \langle j \| C^\mu \| j_3 \rangle \begin{Bmatrix} j_2 & j_3 & J \\ j & j_1 & \mu \end{Bmatrix} \cdot R^\mu(n_1 l_1 j_1 \varepsilon l j n_2 l_2 j_2 n_3 l_3 j_3), \tag{9.11}$$

where the reduced matrix elements $\langle j \| C^\mu \| j' \rangle$ are given by (12.13), and where the exchange amplitude is obtained from (9.11) by changing the indexes 2 and 3 everywhere.[38] Equations (9.10) and (9.11) agree with those of ASAAD[39] except for σ'.

Several probability (actually width, see Appendix E) concepts emerge from considerations of the amplitudes (9.6) and (9.10). By using the $(SLJM)$ representation in (9.6), we obtain, for instance,

$$w_{n_1 l_1}(\varepsilon l, n_2 l_2, n_3 l_3, SL) = 2\pi \frac{(2S+1)(2L+1)}{2(2l_1+1)} \cdot \left| \langle n_1 l_1 \varepsilon l SLJM | \frac{1}{r_{12}} | n_2 l_2 n_3 l_3 SLJM \rangle \right|^2, \tag{9.12}$$

where any J and M values that are consistent with S and L can be used. Equation (9.12) represents the probability that a given initial $(n_1 l_1)^{-1}$ hole state decays into any of the final $(n_2 l_2)^{-1} (n_3 l_3)^{-1}$ two-hole states with the emission of an electron that carries the angular momentum l. The summation over l in (9.12) yields the *component*, *term*, or *multiplet* transition rate

$$w_{n_1 l_1}(n_2 l_2, n_3 l_3, SL) = \sum_l w_{n_1 l_1}(\varepsilon l, n_2 l_2, n_3 l_3, SL). \tag{9.13}$$

The summation over all S and L of the two-electron configuration $(n_2 l_2, n_3 l_3)$ gives the *group* transition rate

$$w_{n_1 l_1}(n_2 l_2, n_3 l_3) = \sum_{SL} w_{n_1 l_1}(n_2 l_2, n_3 l_3, SL). \tag{9.14}$$

This probability concept is independent of the coupling scheme provided the vacancies $(n_1 l_1)^{-1}$, $(n_2 l_2)^{-1}$, and $(n_3 l_3)^{-1}$ and the outgoing electron are represented by the same radial wave functions in all schemes. By using the transformation (8.37) in (9.14) or directly the jj coupling amplitude (9.9), we obtain

$$w_{n_1 l_1}(n_2 l_2, n_3 l_3) = \sum_{l j j_2 j_3 J} w_{n_1 l_1 j_1}(\varepsilon l j, n_2 l_2 j_2, n_3 l_3 j_3, J), \tag{9.15}$$

where

$$w_{n_1 l_1 j_1}(\varepsilon l j, n_2 l_2 j_2, n_3 l_3 j_3, J) = 2\pi \frac{(2J+1)}{2j_1+1} \cdot \left| \langle n_1 l_1 j_1 \varepsilon l j J M | \frac{1}{r_{12}} | n_2 l_2 j_2 n_3 l_3 j_3 J M \rangle \right|^2 \tag{9.16}$$

[38] Note that the definition (12.13) corresponds to the coupling of $\boldsymbol{s}$ to $\boldsymbol{l}$, i.e. $\boldsymbol{j} = \boldsymbol{l} + \boldsymbol{s}$, which is opposite to what is used by CONDON and SHORTLEY [9] (see also II.7 and III.12).

[39] Asaad, W.N. (1963): Nucl. Phys. *44*, 415

represents the probability that a given $(n_1 l_1 j_1)^{-1}$ hole state decays into any of the $(n_2 l_2 j_2)^{-1}(n_3 l_3 j_3)^{-1}$ two-hole states with the emission of an electron that carries an orbital angular momentum l and a total angular momentum j. Finally, we may sum $w_{n_1 l_1}(n_2 l_2, n_3 l_3)$ over all possible final vacancies $(n_2 l_2)^{-1}$ and $(n_3 l_3)^{-1}$ of a shell or several shells in order to obtain the *total* transition rate $w_{n_1 l_1}$ of a given initial vacancy $(n_1 l_1)^{-1}$.

Provided the spin-orbit interaction can be neglected, the formulae given above are valid for any Auger or Coster-Kronig process which occurs in an atom or ion with initially one vacancy in a closed-shell structure. If the spin-orbit splitting cannot be ignored, intermediate coupling should be used, in which case the probability is redistributed among transitions between different energy levels although the total probability (9.14) remains invariant. However, in the case of Coster-Kronig transitions of the type $(n_1 l_1 j_1)^{-1} - (n_2 l_2 j_2)^{-1}(n_3 l_3 j_3)^{-1}$, where $n_1 = n_2$ and $l_1 = l_2$, e.g., $L_2 - L_3 X$, the total probability cannot be evaluated from (9.14).[40] Instead the jj-coupling formula

$$w_{n_1 l_1 j_1}(n_2 l_2 j_2, n_3 l_3) = \sum_{l j j_3 J} w_{n_1 l_1 j_1}(\mathcal{E} l j, n_2 l_2 j_2, n_3 l_3 j_3, J) \tag{9.17}$$

may be used.

β) Intermediate coupling and configuration interaction. In the closed-shell case the intermediate coupling description of the initial state reduces to that of the jj-coupling limit. According to (8.27) the final-state wave funcion becomes

$$|\alpha J' M'\rangle = \sum_{SL} \sum_{Mm} D_{\alpha J}(SL)(J j M m | J' M')\,|(n_2 l_2)^{-1}(n_3 l_3)^{-1} SLJM\rangle\,|\mathcal{E} l j m\rangle, \tag{9.18}$$

where $J' = j_1$ $M' = -m_1$ due to the missing initial spin orbital $|n_1 l_1 j_1 m_1\rangle$. There is no summation over ω since the two-hole system is fully specified by the set $\{n_2 l_2 n_3 l_3 SLJM\}$. The evaluation of the matrix element $\langle \alpha J' M'| \cdot \sum_{\mu>\nu} r_{\mu\nu}^{-1} |(n_1 l_1 j_1)^{-1} J' M'\rangle$ follows closely the procedure which was established in Sect. 9α. For that purpose the two-hole wave function $|(n_2 l_2)^{-1}(n_3 l_3)^{-1} SLJM\rangle$ is expressed as a linear combination of the jj-coupled two-hole wave functions $|(n_2 l_2 j_2)^{-1}(n_3 l_3 j_3)^{-1} JM\rangle$. The expansion coefficients are denoted by $C_{JSL}(j_2 j_3)$. The probability corresponding to (9.16) becomes

$$w_{n_1 l_1 j_1}(\mathcal{E} l j, n_2 l_2, n_3 l_3, J) = 2\pi \frac{2J+1}{2j_1+1} \left| \sum_{SL} \sum_{j_2 j_3} D_{\alpha J}(SL)\, C_{JSL}(j_2 j_3) \cdot \langle n_1 l_1 j_1 \mathcal{E} l j J M | \frac{1}{r_{12}} | n_2 l_2 j_2 n_3 l_3 j_3 J M\rangle \right|^2, \tag{9.19}$$

where the coefficients $D_{\alpha J}(SL)$ may be otained by diagonalizing for each J the intermediate-coupling matrices (8.30). If the matrix elements in these matrices are given with respect to the two-electron wave functions $|n_2 l_2 n_3 l_3 SLJM\rangle$, the sign of each spin-orbit parameter ξ_{nl} must be reversed in each element (see Sect. 8ε).

[40] Kostroun, V.O., Chen, M.H., Crasemann, B. (1971): Phys. Rev. A *3*, 533

With regard to the phase factors two things must be observed. Since matrix elements are given with respect to one- and two-electron wave functions, the $(n_1 l_1 j_1)^{-1}$ and $(n_2 l_2 j_2)^{-1}(n_3 l_3 j_3)^{-1}$ wave functions *must* be interpreted as *hole* wave functions for which the phase conventions (8.57) and (8.63) apply. Changing the sign of the spin-orbit parameters in the intermediate-coupling matrices assures that the coefficients $D_{\alpha J}(SL)$ form a consistent set with respect to the change from the hole to electron picture except that the diagonalization procedure does not determine the absolute phases of the resulting wave functions. Matrix (8.37) relates the two-electron wave function $(x_1 x_2 | n_2 l_2 n_3 l_3 SLJM)$ and $(x_1 x_2 | n_2 l_2 j_2 n_3 l_3 j_3 JM)$ to each other so that the elements of this matrix are closely related to the coefficients $C_{JSL}(j_2 j_3)$ but not identical due to the phase factor (8.63). In the case where $n_2 l_2$ and $n_3 l_3$ pertain to two different shells, there is no ambiguity so that $C_{JSL}(j_2 j_3)$ can be put equal to the coefficients (8.37), whereby the phase factor $(-1)^{l_2+l_3}$ may be dropped from (8.57). In the case of $n_2 = n_3$ and $l_2 = l_3$ the corresponding coefficients $C_{JSL}(j_2 j_3)$ should be obtained by multiplying the coefficients (8.37), in addition to $\sqrt{2}$, by $(-1)^{l_2}$ for $p=0{,}2$ and by $(-1)^{l_2+1}$ for $p=1$. This change of sign seems, however, to lead to a controversy with regard to the *LS*-coupling limit if adopted. Whereas the mixed-coupling formula of El Ibyari et al.[41] [which follows from (9.19) if the coefficients $D_{\alpha J}(SL)$ are set equal to $\delta_{SL,S'L'}$ and $C_{JSL}(j_2 j_3)$ are set equal to the two-electron transformation coefficients (8.37)] correctly predicts the allowed final terms of the double-hole states, the many-electron version of (9.19) does not. For example, the amplitude of the $1s_{1/2}^{-1} \rightarrow 2p^{-2}\,{}^3P_2 \varepsilon dj$ $(j=3/2, 5/2)$ transitions would not vanish in the absence of the spin-orbit interaction if calculated with the many-electron transformation coefficients $C_{JSL}(j_2 j_3)$.

The coupling scheme adopted in the probability concept (9.19), namely, the *jj* coupling for the initial state and the intermediate coupling for the final state, is often called *mixed*.[42] The generalization to the case of configuration mixing in the final two-hole state is straightforward except for the phase. The summation over S and L in (9.18) is replaced by a more general summation which includes alternative $n_2 l_2$ and $n_3 l_3$ values in accordance with parity rule (9.4). This results in a summation, not only over S and L in the probability (9.19), but also over several two-electron configurations $\gamma = (n_2 l_2, n_3 l_3)$. Each γ gives rise to a number of possible S and L values which together with j_2, j_3, and J define the coefficients $C_{JSL}(j_2 j_3)$.

From (9.19) the various probability concepts listed at the end of the previous section can be obtained. The summation over l, j, and J gives the total probability $w_{n_1 l_1}(n_2 l_2, n_3 l_3)$. In the case of final-state core mixing, the two-electron label $(n_2 l_2, n_3 l_3)$ indicates only the principal configuration.

γ) The open-shell case. In Sects. 8α and 8γ the general method to calculate Coulomb matrix elements with respect to *LS*- and *jj*-coupled wave functions of two arbitrary configurations was described. This method has been utilized by

[41] El Ibyari, S.N., Asaad, W.N., McGuire, E.J. (1972): Phys. Rev. A 5, 1048

[42] This scheme was introduced by Asaad and Mehlhorn [Asaad, W.N., Mehlhorn, W. (1968): Z. Phys. *217*, 304] originally for pure *LS* coupling in the final state

BHALLA[43] to calculate nonradiative K transition rates in the neon atom. The generalization to include configuration interaction and intermediate coupling is straightforward but tedious. However, any phase ambiguities are avoided by the judicious use of the many-electron picture for both the initial and final state.

MCGUIRE[44] has described an elegant method which leads to explicit formulae for transition rates in any nonclosed-shell atom. It is based on the observation that the amplitudes (9.6) and (9.9) could equally well have been obtained by considering transitions from initial two-hole states in $(n_1 l_1)^{-1}(\varepsilon l)^{-1}\,{}^{2S+1}L$ to final two-hole states in $(n_2 l_2)^{-1}(n_3 l_3)^{-1}\,{}^{2S+1}L$ and their counterpart in the jj coupling. In light of (8.58) the only difference in the results with respect to the amplitudes (9.6) and (9.9) would be in the phase factor. Amplitude (9.6) would be without the phase factor $(-1)^{\sigma}$ and amplitude (9.9) without $(-1)^{\sigma'}$. The statistical factors do not follow automatically from this approach, they have to be included separately.

MCGUIRE's formulae cover three types of nonradiative transitions in nonclosed-shell atoms. These include

$$(n_1 l_1 j_1)^{-r_1}(n_2 l_2)^{-r_2}(n_3 l_3)^{-r_3}X^{-r} - (n_1 l_1 j_1)^{-r_1+1}(n_2 l_2)^{-r_2-1}(n_3 l_3)^{-r_3-1}X^{-r} \quad (9.20)$$

transitions, where $n_2 l_2$ and $n_3 l_3$ may or may not be identical. In addition, he considers Coster-Kronig transitions with equivalent initial- and final-state holes which correspond to the replacement of $n_2 l_2$ by $n_1 l_1 j_1'$ in (9.20), where j_1' is different from j_1. The symbol X^{-r} refers to any number of nonclosed shells with electrons which do not participate in the transitions. Since MCGUIRE's formulae are given in the mixed-coupling scheme involving jj coupling in the initial state and LS coupling in the final state, there may be some concern about the relative phases. According to Sect. 9β, there could be a change of the relative signs of the various jj amplitudes in the mixed-coupling probability (9.19) due to the transformation from the many-electron picture in the case of equivalent final holes ($n_2 = n_3$, $l_2 = l_3$).

We shall not reproduce the general probability formulae, which involve fractional parentage coefficients and nj symbols in addition to the basic elements (9.8) and (9.11), but refer to the literature for various component and group transition rates.[45] Applications to specific open-shell configurations can be found in the works of MCGUIRE, CHEN, and CRASEMANN.[46] Here we only consider the total transition rate for a given initial vacancy in order to give a crude idea how the non-radiative probability behaves as a function of the distribution of the vacancies. In the case of nonequivalent holes in the final state the total probability is given by

$$P(r_1, r_2, r_3) = r_1 \frac{(4l_2+2-r_2)(4l_3+2-r_3)}{(4l_2+2)(4l_3+2)} P(1,0,0) \quad (9.21)$$

[43] Bhalla, C.P. (1975): Phys. Rev. A *12*, 122

[44] McGuire, E.J. (1975) in: Atomic inner-shell processes ionization and transition probabilities, Vol. I, Crasemann, B. (ed.), New York: Academic Press, p. 293

[45] McGuire, E.J. (1975): Phys. Rev. A *12*, 330. See also II.44

[46] See, for example, McGuire, E.J. (1975): Phys. Rev. A *11*, 1889; (1978): Phys. Rev. A *17*, 182; Chen, M.H., Crasemann, B. (1974): Phys. Rev. A *10*, 2232; (1977): Phys. Rev. A *16*, 1495

as a function of the number of the initial vacancies in the subshells $(n_1 l_1)$, $(n_2 l_2)$, and $(n_3 l_3)$, respectively. Equation (9.21) is based on the crude approximation of using the same radial wave functions for the evaluation of each probability $P(r_1, r_2, r_3)$ and of assuming that the Auger electron energies are unaffected by the passive electrons. For equivalent holes in the final state $(n_2 = n_3, l_2 = l_3)$ the total probability behaves in the same approximation as

$$P(r_1, r_2) = r_1 \frac{(4l_2+2-r_2)(4l_2+1-r_1)}{(4l_2+2)(4l_2+1)} P(1,0), \tag{9.22}$$

but this is only valid provided it is obtained as an average over the (ω, S, L) quantum numbers of the final $(n_2 l_2)^{-r_2}$ configuration.[47]

δ) Relationship between the intensity and probability. According to Sect. 4, the separation of the excitation and decay processes makes it possible to express the integrated intensity of an Auger line in terms of the initial-state population and transition rate. We assume as in Chap. IV the usual experimental situation in which a collimated monoenergetic beam of particles which has the flux density I_0 hits a thin target of randomly oriented atoms. If the cross section of the excitation to a particular final state $|\gamma JM\rangle$ is $\sigma(\gamma JM)$, then $N'(\gamma JM) = \sigma(\gamma JM) I_0$ is the number of states which are produced in unit time per target atom. The initial-state population which gives the relative number of γJM states present in the target is

$$N(\gamma JM) = \frac{N'(\gamma JM)}{P_t(\gamma JM)}, \tag{9.23}$$

where $P_t(\gamma JM)$ is the total probability of decay from the state γJM to all possible final states. Note that in general $P_t(\gamma JM)$ is a sum of nonradiative and radiative transition probabilities (see Sect. 6). The "cross section" $\sigma(\gamma JM)$ may also contain contributions from excitations to states higher in energy than $|\gamma JM\rangle$ via various decay processes.

We denote the probability of nonradiative transition from the state γJM to states $\gamma' JM$ by $P_{\gamma\gamma'}(JM)$. For example, if (9.19) is valid, then $J = j_1$. The total number of electrons I emitted per unit time and target atom in these transitions is given by

$$I = (2J+1) P_{\gamma\gamma'}(JM) \overline{N(\gamma JM)}, \tag{9.24}$$

where $\overline{N(\gamma JM)}$ is the averaged initial-state population defined by

$$\overline{N(\gamma JM)} = \frac{1}{2J+1} \sum_M N(\gamma JM). \tag{9.25}$$

[47] The general proof of the formulae (9.21) and (9.22) is due to McGuire (see II.44 and Sandia Lab. Res. Rep. SC-RR-69-137, 1969). They were introduced in a heuristic manner by Åberg (Åberg, T. (1968): Phys. Lett. *26* A, 515) for $KL-L^3$ transitions and by Larkins, F.P. (1971): J. Phys. B *4*, L29 for a more general case

As we shall see in Chap. IV $\overline{N(\gamma JM)}$ is equal to $N(\gamma JM)$ (natural excitation) only under very special circumstances since in general $N(\gamma JM)$ depends on M. According to (9.24), the intensity of a line between two levels is related to the *strength* $(2J+1)P_{\gamma\gamma'}(JM)$. It determines the intensity ratio of transitions between various levels of the configuraions γ and γ' provided $\overline{N(\gamma JM)}$ is level independent, i.e., independent of J.

If the spin-orbit splitting can be ignored, then it is possible to consider the transition probability $P_{\gamma\gamma'}(SLJM)$ of transitions from a state $|\gamma SLJM\rangle$ to various states $|\gamma' SLJM\rangle$. In the case that (9.12) is valid $L=l_1$ and $S=1/2$. The intensity of a *multiplet* which is related to the number of electrons emitted in transitions between two *terms* is given by

$$I=(2L+1)2S+1)\,P_{\gamma\gamma'}(SLJM)\,\overline{N(\gamma SLJM)}, \tag{9.26}$$

where the average initial-state population is defined by

$$\overline{N(\gamma SLJM)}=\frac{1}{(2L+1)(2S+1)}\sum_{JM} N(\gamma SLJM). \tag{9.27}$$

According to (9.26) the multiplet intensity ratios are determined solely by the strengths $(2L+1)(2S+1)P_{\gamma\gamma'}(SLJM)$ provided $\overline{N(\gamma SLJM)}$ is term independent, i.e., independent of the quantum numbers S and L.

According to Chap. IV, the experimental situation for which (9.24) and (9.26) can be used to represent line intensities would in general requre flux measurements as a function of the emission angle of the Auger electrons. However the anisotropy of the angular distribution of the Auger electrons is usually small, which makes it reasonable to also use the intensities (9.24) and (9.26) to represent line intensities for the Auger electron emission in specific directions.

10. Beyond the independent-electron frozen-core approximation

α) Relaxation. In Sect. 3 the theory of resonance scattering including channel interaction has been developed without assuming orthogonality between the discrete state Φ and the final-state channel wave functions $\chi_{\bar{\beta}\varepsilon}$. In the following we shall assume that there is no channel mixing and consider the transition rate $2\pi|\langle\lambda_\varepsilon|H-E|\varphi\rangle|^2$ where both $|\varphi\rangle$ and $|\lambda\rangle$ are constructed from one-electron spin orbitals in the $(N-1)$ electron space.

In the ΔE_{SCF} method (see Sect. 8δ) the initial and final states must be optimized separately and the continuum spin orbital solved in the field of the final doubly ionized ion. As a consequence there is a nonorthogonality problem due to the overlap elements (8.49).

In order to explore the form of the terms introduced by using nonorthogonal spin orbitals, we examine, following HOWAT et al.[48], the rate (9.1)

[48] Howat, G., Åberg, T., Goscinski, O. (1978): J. Phys. B *11*, 1575

in the UHF method which was described in Sect. 8δ. The total wave functions are represented by the single Slater determinants

$$\begin{aligned} U_{\mathrm{i}} &= \frac{1}{[(N-1)!]^{1/2}} \det |u_2 u_3 \ldots u_\mu \ldots u_N|, \\ V_{\mathrm{f}} &= \frac{1}{[(N-1)!]^{1/2}} \det |v_1 v_\varepsilon \ldots v_\mu \ldots v_N|, \end{aligned} \tag{10.1}$$

where the initial-state spin orbitals u_2 and u_3 are replaced by a bound-state spin orbital v_1 and a continuum spin orbital v_ε in the final state. The notation $u \to v$ denotes the alteration in the passive spin orbitals due to the change of the average field on passing from initial to final state. Using LÖWDIN's[2] expressions and retaining all terms linear in the nondiagonal overlap elements $\langle v_{\mathrm{f}} | u_{\mathrm{i}} \rangle = \langle f' | i \rangle$ and the quadratic overlap term associated with E, the amplitude in (9.1) becomes

$$\begin{aligned} &\langle V_{\mathrm{f}} | H - E | U_{\mathrm{i}} \rangle \\ &= \prod_{\nu \neq 2,3} \langle \nu' | \nu \rangle \left[\langle 1\varepsilon \| 23 \rangle + \begin{vmatrix} \langle 1|F|2 \rangle & \langle 1|F|3 \rangle \\ \langle \varepsilon|2 \rangle & \langle \varepsilon|3 \rangle \end{vmatrix} + \begin{vmatrix} \langle 1|2 \rangle & \langle 1|3 \rangle \\ \langle \varepsilon|F|2 \rangle & \langle \varepsilon|F|3 \rangle \end{vmatrix} \right. \\ &\quad - \mathscr{E} \begin{vmatrix} \langle 1|2 \rangle & \langle 1|3 \rangle \\ \langle \varepsilon|2 \rangle & \langle \varepsilon|3 \rangle \end{vmatrix} + \sum_{\mu \neq 2,3} \begin{vmatrix} \langle \mu'|2 \rangle & \langle \mu'|3 \rangle \\ \langle 1\varepsilon \| 2\mu \rangle & \langle 1\varepsilon \| 3\mu \rangle \end{vmatrix} \\ &\quad \left. + \begin{vmatrix} \langle 1|\mu \rangle & \langle \varepsilon|\mu \rangle \\ \langle 1\mu' \| 23 \rangle & \langle \varepsilon\mu' \| 23 \rangle \end{vmatrix} \right], \end{aligned} \tag{10.2}$$

where F is the "Fock-like" operator [cf. operator (8.40)]

$$F = f + \sum_{\mu \neq 2,3} \langle \mu' \| \mu \rangle. \tag{10.3}$$

The coefficient $\mathscr{E}$ of the quadratic overlap term is given by

$$\mathscr{E} = E - \tfrac{1}{2} \sum_{\mu \neq 2,3} (\langle \mu' | f | \mu \rangle + \langle \mu' | F | \mu \rangle), \tag{10.4}$$

which involves a partial cancellation of two large terms.

It is seen that the corrections to the lowest-order Auger amplitude, $\langle 1\varepsilon \| 23 \rangle$, involve both one- and two-electron interaction matrix elements in addition to a scaling factor. For the purpose of estimation we can put $\mu' = \mu$ in the coefficients of the overlap elements. This permits us to use the UHF equations for the initial one-hole state, namely,

$$[F + \langle 2 \| 2 \rangle + \langle 3 \| 3 \rangle]\, u_\mu = \varepsilon_\mu u_\mu, \tag{10.5}$$

to extract from (10.2) the terms linear in overlap elements. The result is

$$\begin{aligned}\langle V_{\rm f}|H-E|U_{\rm i}\rangle \cong &\langle 1\mathcal{E}\|23\rangle - \langle\mathcal{E}|3\rangle\langle 13\|23\rangle \\ &-\langle\mathcal{E}|2\rangle\langle 12\|23\rangle - \langle 1|2\rangle\langle 2\mathcal{E}\|23\rangle - \langle 1|3\rangle|\langle 3\mathcal{E}\|23\rangle \\ &- \sum_{\mu\neq 2,3}[\langle\mu'|2\rangle\langle 1\mathcal{E}\|\mu 3\rangle + \langle\mu'|3\rangle\langle 1\mathcal{E}\|2\mu\rangle \\ &+\langle 1|\mu\rangle\langle\mu'\mathcal{E}\|23\rangle + \langle\mathcal{E}|\mu\rangle\langle 1\mu'\|23\rangle],\end{aligned} \tag{10.6}$$

which reveals the effect of partially removing processes other than $2, 3 \to 1, \mathcal{E}$ from the lowest-order amplitude. Note that the use of the erroneous amplitude $\langle V_{\rm f}|\sum_{\alpha>\beta} r_{\alpha\beta}^{-1}|U_{\rm i}\rangle$, which does not yield one-electron integrals, misses certain cancellations that are correctly treated by (10.2).

In Sect. 20γ, (10.2) is used in the RHF case determinant by determinant for the evaluation of a few $\Delta E_{\rm SCF}$ rates. These rates are expressed in terms of radial matrix elements and Slater integrals, which makes it possible to estimate the effect of relaxation in particular cases.

β) Interchannel interaction. The fundamental result of Sect. 4 is that the calculation of the partial rate of nonradiative transitions involves a final-state many-electron continuum wave function that includes the channel interaction and satisfies the ingoing spherical wave boundary condition (1.17). According to (2.19), the partial rate

$$\Gamma_\alpha = 2\pi|\langle\lambda_{\alpha\mathcal{E}}^-|H-E|\varphi\rangle|^2 \tag{10.7}$$

becomes

$$\Gamma_\alpha = 2\pi\left|\langle\lambda_{\alpha\mathcal{E}}|H-E|\varphi\rangle + \lim_{\nu\to 0}\sum_{\beta=1}^{N_{\rm c}}\int_0^\infty \frac{T_{\alpha\beta}^-(\tau,\mathcal{E})\langle\lambda_{\beta\tau}|H-E|\varphi\rangle\,d\tau}{E-E_\beta-\tau+{\rm i}\nu}\right|^2, \tag{10.8}$$

where the $\boldsymbol{T}^-$ matrix is obtained from (2.15). It follows from this equation and (10.8) that the partial rate is given to the lowest order in the interaction matrix $\boldsymbol{V}(\varepsilon, \mathcal{E}, E)$ by

$$\begin{aligned}\Gamma_\alpha = 2\pi\Bigg|&\langle\lambda_{\alpha\mathcal{E}}|H-E|\varphi\rangle + \mathcal{P}\sum_{\beta=1}^{N_{\rm c}}\int_0^\infty \frac{V_{\alpha\beta}(\tau,\mathcal{E},E)\langle\lambda_{\beta\tau}|H-E|\varphi\rangle\,d\tau}{E-E_\beta-\tau} \\ &-{\rm i}\pi\sum_{\beta=1}^{N_{\rm c}} V_{\alpha\beta}(\mathcal{E},\mathcal{E},E)\langle\lambda_{\beta\mathcal{E}}|H-E|\varphi\rangle\Bigg|^2,\end{aligned} \tag{10.9}$$

which may be a reasonable approximation provided the matrix elements $V_{\alpha\beta}(\mathcal{E},\mathcal{E},E)$ are considerably less than unity. Note that there is no restriction with regard to the generality of the bound-state wave function φ and of that of the residual ion in the final state in (10.8) and (10.9).

A modification of Brillouin's theorem, described in Sect. 8δ, proves that $\beta\neq\alpha$ in (10.9) if UHF wave functions are used. Suppose that the continuum spin orbital $v_{\mathcal{E}}$ has been solved for various $\mathcal{E}$ values in the field of the doubly ionised ion with *fixed* bound-state orbitals by the application of SEATON's[49] method. Using these orbitals, we can construct the channel interaction matrix

[49] Seaton, M.J. (1953): Phil. Trans. (London) A *245*, 469

elements

$$\langle \lambda_{\alpha\mathcal{E}'}|H-E|\lambda_{\alpha\mathcal{E}}\rangle = \langle v_{\mathcal{E}'}|F|v_{\mathcal{E}}\rangle, \tag{10.10}$$

where F is now the Fock operator. Since F is a Hermitian operator and since $Fv_{\mathcal{E}}=\mathcal{E}v_{\mathcal{E}}$, the matrix elements (10.10) are zero if $\mathcal{E}'\neq\mathcal{E}$. If the continuum spin orbital is solved in an analogous way using RHF spin orbitals, the intrachannel interaction also vanishes, which is not true for the local-potential approximations. Calculations based on (10.9) and RHF spin orbitals are discussed in Sect. 20δ, where also a comparison with the many-body-perturbation-theory method (see the next section) is made.

γ) Perturbation approaches to the Auger effect. It is often forgotten that WENTZEL's derivation of the lowest-order probability of nonradiative transitions in 1927 apparently was one of the first applications of the scattering approach to a quantum-mechanical resonance process. For the Coulomb interaction it lead to an approximate probability formula which later became known as "Fermi's Golden Rule". We shall reproduce WENTZEL's perturbation approach but by using the formalism of Sects. 3 and 4.

WENTZEL considered two electrons in hydrogenic orbitals $(\boldsymbol{r}|n_2 l_2 m_2)$ and $(\boldsymbol{r}|n_3 l_3 m_3)$ and constructed the two-electron resonance state $\psi^+(\boldsymbol{r}_1,\boldsymbol{r}_2)$ by using perturbation theory up to the first order. This wave function is given by

$$\begin{aligned}\psi^+(\boldsymbol{r}_1,\boldsymbol{r}_2) &= (\boldsymbol{r}_1,\boldsymbol{r}_2|n_2 l_2 m_2, n_3 l_3 m_3)\\ &\quad+\lim_{\nu\to 0}\sum_{\substack{l_1 m_1\\ l\ m}}\ \sum_{\substack{n_1\neq\\ n_2,n_3}}\int\frac{M^{n_1 l_1 m_1,\tau lm}_{n_2 l_2 m_2, n_3 l_3 m_3}(\boldsymbol{r}_1,\boldsymbol{r}_2|n_1 l_1 m_1,\tau lm)\,d\tau}{\mathcal{E}_{n_2}+\mathcal{E}_{n_3}-\mathcal{E}_{n_1}-\tau+\mathrm{i}\nu},\end{aligned} \tag{10.11}$$

where the two-electron wave functions represent the symmetric or antisymmetric space parts of the total antisymmetric wave functions $(x_1 x_2|nlm_l m_s, n'l'm_{l'}m_{s'})$. The integration over τ is assumed to include also the summation over the discrete quantum numbers $n\,(\neq n_2, n_3)$. Since WENTZEL did not consider the primary excitation process, he interpreted his scattering wave function as the initial-state wave function of the nonradiative process and consequently uses the outgoing spherical wave boundary condition as indicated by the $+\mathrm{i}\nu$ term in (10.11). According to (2.1), $(\boldsymbol{r}|\mathcal{E}lm)$ behaves at large r as

$$(\boldsymbol{r}|\mathcal{E}lm)\sim\sqrt{\frac{2}{\pi kr}}\,Y_{lm}(\theta,\phi)\sin\theta(\mathcal{E},r), \tag{10.12}$$

where $\theta(\mathcal{E},r)$ is given by (1.18). Hence $\psi^+(\boldsymbol{r}_1,\boldsymbol{r}_2)$ has at large r_1 or r_2 the limiting form

$$\begin{aligned}\Phi(\boldsymbol{r}_1,\boldsymbol{r}_2) &= \lim_{r_1(r_2)\to\infty}\psi^+(\boldsymbol{r}_1,\boldsymbol{r}_2)\\ &= -\pi\sum_{\substack{n_1 l_1 m_1\\ lm}} M^{n_1 l_1 m_1,\mathcal{E}lm}_{n_2 l_2 m_2, n_3 l_3 m_3}\sqrt{\frac{2}{\pi k}}\\ &\quad\cdot\frac{1}{\sqrt{2}}\Big[(\boldsymbol{r}_1|n_1 l_1 m_1)\frac{\mathrm{e}^{ikr_2}}{r_2}Y_{lm}(\theta_2,\phi_2)\\ &\quad\pm(\boldsymbol{r}_2|n_1 l_1 m_1)\frac{\mathrm{e}^{ikr_1}}{r_1}Y_{lm}(\theta_1,\phi_1)\Big],\end{aligned} \tag{10.13}$$

where the kinetic energy $\mathcal{E}=k^2/2=-\mathcal{E}_{n_1}+\mathcal{E}_{n_2}+\mathcal{E}_{n_3}$ is the resonance energy. Equation (10.13) follows from the application of formulae (A.10) and (A.11) of Appendix A to (10.11) after the substitution of the limit (10.12). Using (10.13) it is possible to calculate the current density vector

$$j=\mathrm{Re}[-\mathrm{i}\Phi(\boldsymbol{r}_1,\boldsymbol{r}_2)^*(\nabla_1+\nabla_2)\Phi(\boldsymbol{r}_1,\boldsymbol{r}_2)] \tag{10.14}$$

in the six-dimensional space. From this density, the total flux w of emitted electrons is obtained by integrating over the surface of a large sphere in the $\boldsymbol{r}_1$ and $\boldsymbol{r}_2$ spaces, respectively. The result is

$$w=2\pi\sum_{\substack{n_1 l_1 m_1\\ lm}}|M^{n_1 l_1 m_1,\mathcal{E}lm}_{n_2 l_2 m_2,n_3 l_3 m_3}|^2, \tag{10.15}$$

which corresponds to the group probability $w_{n_1 l_1}$ in Sect. 9 except that there is also a summation over $n_1 l_1$. A comparison of (10.11) and (3.6) shows that $\psi^+(\boldsymbol{r}_1,\boldsymbol{r}_2)$ represents an approximation of the total scattering wave function in the vicinity of the resonance apart from the normalization factor $a^+(\mathcal{E})$. Final-state channel mixing is not included and hydrogenic wave functions are used. With regard to the former approximation the nondimensional interaction matrix elements $V_{\beta\alpha}(\varepsilon',\varepsilon,E)$ defined by (2.8) serve, in comparison to unity, as approximation indices.

The many-body-perturbation-theory (MBPT) method of Chase et al.[50] and Kelly[51] is closely related to Wentzel's original method although it is based on the analysis of the complex energy rather than the scattering wave function of the initial resonance state. According to Sect. 4, the parameter Γ in the complex energy $E=E_r-\mathrm{i}\Gamma/2$ can be interpreted as the total width or rate of the resonance state provided the decay process can be separated from the excitation process. Hence, if it is possible to determine the complex part of the resonance energy using perturbation theory, the nonradiative width is obtained.

We shall demonstrate this fact by using a basis set of UHF spin orbitals v_μ which are eigenfunctions of (8.42). The first occupied N spin orbitals v_μ $(\mu=1\ldots N)$ are completed with the virtual spin orbitals v_ν $(\nu=N+1\ldots)$, which are eigenfunctions of the fixed Fock operator F, which is constructed from the first N spin orbitals. As a consequence of Brillouin's theorem (8.45), the total energy with respect to the many-electron wave function that is expanded in the complete set of the Slater determinants $(N!)^{-1/2}\det|v_{\kappa_1},v_{\kappa_2}\ldots v_{\kappa_n}|$ becomes

$$E=E_0+\sum_{\substack{\nu_1\nu_2\\ \mu_1\mu_2}}\langle\nu_1\nu_2\|\mu_1\mu_2\rangle C^{\nu_1\nu_2}_{\mu_1\mu_2}, \tag{10.16}$$

where E_0 is the energy with respect to the Slater determinant

$$\Phi_0=(N!)^{-1/2}\det|v_1,v_2\ldots v_N| \tag{10.17}$$

---

[50] Chase, R.L., Kelly, H.P., Köhler, H.S. (1971): Phys. Rev. A *3*, 1550

[51] Kelly, H.P. (1975): Phys. Rev. A *11*, 556

and where $C^{\nu_1\nu_2}_{\mu_1\mu_2}$ are expansion coefficients of Slater determinants in which any two occupied orbitals (v_{μ_1}, v_{μ_2}) in Φ_0 have been replaced by any two virtual orbitals (v_{ν_1}, v_{ν_2}) in an ordered fashion. First-order perturbation theory gives

$$C^{\nu_1\nu_2}_{\mu_1\mu_2} = \frac{\langle \nu_1 \nu_2 \| \mu_1 \mu_2 \rangle}{\mathcal{E}_{\mu_1} + \mathcal{E}_{\mu_2} - \mathcal{E}_{\nu_1} - \mathcal{E}_{\nu_2}}, \tag{10.18}$$

so that

$$E - E_0 = \sum_{\substack{\nu_1\nu_2 \\ \mu_1\mu_2}} \frac{|\langle \nu_1 \nu_2 \| \mu_1 \mu_2 \rangle|^2}{\mathcal{E}_{\mu_1} + \mathcal{E}_{\mu_2} - \mathcal{E}_{\nu_1} - \mathcal{E}_{\nu_2}} \tag{10.19}$$

in the second order. Since $\mathcal{E}_\nu > \mathcal{E}_\mu$ in the denominators of (10.19), they are all nonzero. However, suppose we shall calculate the energy $E(k^{-1})$ with respect to a set $v_1 \dots v_{k-1}, v_{k+1} \dots v_N$, where the inner-shell spin orbital v_k is missing. Then there is a corresponding spin orbital v_{ν_k} in the set of virtual spin orbitals, and the denominator $\mathcal{E}_{\mu_1} + \mathcal{E}_{\mu_2} - \mathcal{E}_{\nu_k} - \mathcal{E}_{\nu_2}$ may vanish for positive $\mathcal{E}_{\nu_2}$ values (a discrete negative singular energy would be possible but rare). The corresponding part of the energy difference $E(k^{-1}) - E_0(k^{-1})$ is given in accordance with (A.10) of Appendix A by

$$E_{\mu_1\mu_2}(k^{-1}) - E_0(k^{-1}) = \lim_{\nu \to 0} \int \frac{|\langle \nu_k \nu_\tau \| \mu_1 \mu_2 \rangle|^2 \, d\tau}{\mathcal{E}_{\mu_1} + \mathcal{E}_{\mu_2} - \mathcal{E}_{\nu_k} - \tau + i\nu}, \tag{10.20}$$

having an imaginary part $-i\pi |\langle \nu_k \nu_\varepsilon \| \mu_1 \mu_2 \rangle|^2$ that can be identified as $-i\Gamma/2$. Hence the width is given in accordance with (9.2) by

$$\Gamma = 2\pi |\langle \nu_k \nu_\varepsilon \| \mu_1 \mu_2 \rangle|^2, \tag{10.21}$$

where $\mathcal{E} = \mathcal{E}_{\mu_1} + \mathcal{E}_{\mu_2} - \mathcal{E}_{\nu_k}$.

The procedure leading to the lowest-order formula (10.21) can be made more accurate by including second- and higher-order perturbation corrections to the perturbation expansion of the coefficients $C^{\nu_1\nu_2}_{\mu_1\mu_2}$, which then gives third- and higher-order corrections to the energy difference $E(k^{-1}) - E_0(k^{-1})$. These corrections may contain singularities, which are treated in the same way as in (10.20). This results, according to (10.16), in a sum of imaginary amplitudes of various order that can be put equal to $-i\Gamma/2$. This procedure can be made systematic by using the linked cluster diagrammatic many-body perturbation expansion of Brueckner and Goldstone[52] in the analysis of the correlation energy $E - E_0$. This technique has been used by Kelly[51] in his calculation of the neon K Auger rates. His calculation and the diagrams involved are discussed in Sect. 20δ. In comparison with Wentzel's method the only difference is that whereas Wentzel treats the coefficients $C^{\nu_1\nu_2}_{\mu_1\mu_2}$ to lowest order in the perturbation theory, Kelly considers the imaginary part of the corresponding correlation energy (10.16) also to higher orders.

A third many-body theory techniques which has found some application in calculations of nonradiative rates is the random-phase-with-exchange (RPAE)

[52] Goldstone, J. (1957): Proc. R. Soc. London. A *239*, 267

method of AMUSIA et al.[53] In the unrestricted formalism the effective nonradiative direct amplitude is given by

$$\langle u_1 v|L(\omega)|u_2 u_3\rangle = \langle u_1 v|u_2 u_3\rangle + \lim_{\delta\to 0}\sum_{\nu,\mu}\left[\frac{\langle u_1 v_\nu \| u_2 u_\mu\rangle\langle v u_\mu|\Gamma(\omega)|u_3 v_\nu\rangle}{\omega-\mathcal{E}_\nu+\mathcal{E}_\mu+\mathrm{i}\delta} - \frac{\langle u_1 u_\mu \| u_2 v_\nu\rangle\langle v v_\nu|\Gamma(\omega)|u_3 u_\mu\rangle}{\omega+\mathcal{E}_\nu-\mathcal{E}_\mu-\mathrm{i}\delta}\right], \tag{10.22}$$

where the effective Coulomb interaction satisfies a similar equation, namely,

$$\langle ab|\Gamma(\omega)|cd\rangle = \langle ab\|cd\rangle + \lim_{\delta\to 0}\sum_{\mu,\nu}\left[\frac{\langle a\nu\|c\mu\rangle\langle b\mu|\Gamma(\omega)|d\nu\rangle}{\omega-\mathcal{E}_\nu+\mathcal{E}_\mu+\mathrm{i}\delta} - \frac{\langle a\mu\|c\nu\rangle\langle b\nu|\Gamma(\omega)|d\mu\rangle}{\omega+\mathcal{E}_\nu-\mathcal{E}_\mu-\mathrm{i}\delta}\right], \tag{10.23}$$

where $\omega=\mathcal{E}_2-\mathcal{E}_1$. By iteration of these equations it is possible to recognize some correlation with MBPT diagrams which appear in the analysis of the coefficients $C^{\nu_1\nu_2}_{\mu_1\mu_2}$ of the hole-state wave function. However, there is no detailed comparison of the MBPT and RPAE methods (see CHANG and FANO[54] for a general discussion). The RPAE method in its present form, strictly applicable only to closed-shell atoms, can only be used for the calculation of the total rates in the $(nlm_l m_s)$ representation. With the MBPT method, on the other hand, it is possible to calculate multiplet transition rates.[51] The relativistic RPAE method of JOHNSON et al.[55], especially if it is possible to generalize this method for the open-shell case, may be proven useful in the analysis of nonradiative transitions.

III. Relativistic theory of nonradiative transitions

As shown in the previous chapters the Auger effect is a consequence of the electron-electron interaction. Since a fully Lorentz-invariant quantum theory of many-electron systems does not exist, a formulation of the relativistic theory of the Auger effect is necessarily based on quantum electrodynamical considerations (e.g., AKHIEZER and BERESTETSKII [*17*]). In quantum electrodynamics the interaction between a pair of electrons is described in terms of the retarded exchange of virtual photons. If this exchange is taken into account to the first nonvanishing order in the fine-structure constant $\alpha=e^2/4\pi\varepsilon_0\hbar c$, it results in MØLLER's famous formula for two interacting electrons.[1] This formula, since

[53] Amusia, M. Ya., Cherepkov, N.A. (1975): Case Stud. At. Phys. 5, 47 and references therein. See also Wendin, G., Ohno, M. (1976): Phys. Scr. *14*, 148 for a related approach

[54] Chang, T.N., Fano, U. (1976): Phys. Rev. A *13*, 263

[55] Johnson, W.R., Cheng, K.T. (1978): Phys. Rev. Lett. *40*, 1167 and references therein

[1] Møller, C. (1931): Z. Phys. *70*, 786; (1932): Ann. Phys. *14*, 531; see also Rose, M.E. (1961) Relativistic electron theory, New York: Wiley, Appendix E

the fundamental work of MASSEY and BURHOP,[2] has formed the basis for almost all calculations of relativistic Auger transition probabilities,[3] which have mostly been done in the *jj* coupling. Recently, however, BRIANÇON and DESCLAUX[4] have presented some Dirac-Fock multiconfiguration calculations of Auger electron energies and ASAAD and PETRINI[5] have made an intermediate-coupling analysis of component transition rates.

In Sect. 11 we shall introduce the covariant Hamiltonian of a pair of electrons that interact only through the free electromagnetic field. This coupling is treated to the lowest-nonvanishing order, which leads to MØLLER's formula for the energy of two interacting electrons. The connection to the nonrelativistic photon-electron coupling theory of Sect. 6 is also established. Section 12 presents a derivation of the relativistic nonradiative probability in the *jj*-coupling limit. In Sect. 13 we discuss the relativistic intermediate-coupling approach and its relationship to the Dirac-Fock multiconfiguration method by treating the mixing within the $n'snp^5$ configuration as an example.

11. Basic Hamiltonian and the electron-electron interaction. In order to describe the electron-electron interaction in terms of virtual photons we need Maxwell's equations in their Lorentz-invariant form. For that purpose we introduce the four-dimensional space defined by $x=(x_1, x_2, x_3, x_4)=(x, y, z, \mathrm{i}ct)$. We define two four-vectors in that space, namely, $A=(A_x, A_y, A_z, \mathrm{i}\phi/c)$ and $j=(j_x, j_y, j_z, \mathrm{i}c\rho)$, as combinations of the vector and scalar potentials and of the current and the charge density, respectively. The components of A fulfil the wave equation

$$\Box A_\mu = \sum_{\nu=1}^{4} \frac{\partial^2 A_\mu}{\partial x_\nu^2} = -\mu_0 j_\mu \quad (\mu=1, 2, 3, 4), \tag{11.1}$$

which has this form provided that the Lorentz condition

$$\sum_{\mu=1}^{4} \frac{\partial A_\mu}{\partial x_\mu} = 0 \tag{11.2}$$

is fulfilled. Equations (11.1) and (11.2) are invariant and the field vectors remain unaltered in any *gauge* transformation

$$A'_\mu = A_\mu + \frac{\partial \eta}{\partial x_\mu}, \tag{11.3}$$

where η fulfils the equation

$$\Box \eta = 0 \tag{11.4}$$

---

[2] Massey, H.S.W., Burhop, E.H.S. (1936): Proc. R. Soc. Lond. A *153*, 661

[3] Asaad, W.N. (1959): Proc. R. Soc. Lond. A *249*, 555; Listengarten, M.A. (1961): Bull. Acad. Sci. USSR Phys. Ser. *25*, 803; (1962): *26*, 182; Bhalla, C.P., Ramsdale, D.J. (1970): Z. Phys. *239*, 95; Bhalla, C.P., Rosner, H.R., Ramsdale, D.J. (1970): J. Phys. B *3*, 1232; Chen, M.H., Laiman, E., Crasemann, B., Aoyagi, M., Mark, H. (1979): Phys. Rev. A *19*, 2253

[4] Briançon, Ch., Desclaux, J.P. (1976): Phys. Rev. A *13*, 2157

[5] Asaad, W.N., Petrini, D. (1976): Proc. R. Soc. Lond. A *350*, 381

but is otherwise arbitrary. In the free-field space ($j_\mu=0$) the solution of (11.1) can be obtained in terms of plane waves. If the electromagnetic field is closed in a normalization box of volume Ω and the quantization rules are applied to the plane wave expansion coefficients after an appropriate generalization[6] of the condition (11.2), then the Lorentz-invariant vector potential operators

$$A_\mu(x)=\frac{1}{\sqrt{\Omega}}\sum_{k,\lambda}\sqrt{\frac{\hbar}{2\varepsilon_0\omega}}\, e_\mu^{(\lambda)}(c_{k,\lambda}\,\mathrm{e}^{\mathrm{i}kx}+c_{k,\lambda}^\dagger\,\mathrm{e}^{-\mathrm{i}kx}) \qquad \mu=1,2,3,4 \tag{11.5}$$

are obtained. Here kx is the scalar product of the four-vector x and the four-dimensional propagation vector $k=(\boldsymbol{k},\mathrm{i}\omega/c)$ which satisfies the condition $k^2=|\boldsymbol{k}|^2-\omega^2/c^2=0$. The four-vectors $e^{(\lambda)}$ are the polarization unit vectors fulfilling the orthogonality relations $e^{(\lambda)}e^{(\lambda')}=\delta_{\lambda\lambda'}$ and $e_\mu^{(\lambda)}e_\nu^{(\lambda)}=\delta_{\mu\nu}$. Two of them ($\lambda=1,2$) can be chosen orthogonal to $\boldsymbol{k}$ and one parallel to $\boldsymbol{k}$ ($\lambda=3$). It follows in accordance with the orthogonality relations that $e^{(\lambda)}k=0$ ($\lambda=1,2$; transverse photons) and $e^{(3)}k=\omega/c$ (longitudinal photons). The fourth vector for which $e^{(4)}k=\mathrm{i}\omega/c$ corresponds to the scalar photons which are responsible for the Coulomb interaction. The operators $c_{k,\lambda}$ and $c_{k,\lambda}^\dagger$ are the photon annihilation and creation operators, respectively.

Having briefly established the properties of the electromagnetic field we also need to consider Dirac's equation of an electron moving in an external field. Suppose that this field is represented by the vector potential $A=(A_x,A_y,A_z,\mathrm{i}\phi/c)$. Then Dirac's equation, which is satisfied by the four-spinor

$$\psi=\begin{pmatrix}\psi_1\\ \psi_2\\ \psi_3\\ \psi_4\end{pmatrix}, \tag{11.6}$$

is given by

$$[\mathrm{i}c\gamma\boldsymbol{p}+\mathrm{i}ce\gamma\boldsymbol{A}+mc^2\mathbf{1}]\,\psi=0 \tag{11.7}$$

where the components of $\boldsymbol{p}$ are the operators $-\mathrm{i}\hbar(\partial/\partial x_\nu)\mathbf{1}$ ($\nu=1,\ldots 4$) and where e is the absolute value of the charge of the electron. In (11.7) each component of $\gamma=(\gamma_1,\gamma_2,\gamma_3,\gamma_4)$ is a 4×4 matrix defined below and $\mathbf{1}$ is the 4×4 unit matrix. The matrices γ are related to the 2×2 Pauli matrices $\sigma_x,\sigma_y,\sigma_z$ and the unit 2×2 matrix I by

$$\gamma_\mu=-\mathrm{i}\beta\alpha_\mu \quad (\mu=1,2,3), \qquad \gamma_4=\beta, \tag{11.8}$$

where

$$\alpha_\mu=\begin{pmatrix}0 & \sigma_\mu\\ \sigma_\mu & 0\end{pmatrix} \quad (\mu=1,2,3) \quad \text{and} \quad \beta=\begin{pmatrix}I & 0\\ 0 & -I\end{pmatrix}. \tag{11.9}$$

The simplest covariant Hamiltonian which we can construct for a pair of electrons moving in a central field and interacting with each other is

$$\kappa=\mathrm{i}c\gamma_1\boldsymbol{p}_1+\mathrm{i}c\gamma_2\boldsymbol{p}_2+2mc^2\mathbf{1}+\mathrm{i}ec\gamma_1\boldsymbol{A}_1'+\mathrm{i}ec\gamma_2\boldsymbol{A}_2', \tag{11.10}$$

[6] Ref. [17], Sect. 16

where $\boldsymbol{A}'_\nu$ $(\nu=1,2)$ is the sum of the free-field operator (11.5) and a central field potential $(0,0,0,\mathrm{i}\,V(r)/c)$. Hence in this model for which $\kappa\psi=0$ the electrons are coupled to the *same* photon field and interact through the exchange of photons. The central field potential is assumed to include the potential of the infinite heavy nucleus and to account for screening effects due to electrons not participating in the nonradiative transition. According to (11.7) each electron's stationary-state four-component wave function θ satisfies, in the absence of the photon field, the equation

$$(c\bar{\boldsymbol{\alpha}}\cdot\bar{\boldsymbol{p}}+mc^2\boldsymbol{\beta}-eV(r)\mathbf{1})\,\theta=\mathcal{E}\theta. \tag{11.11}$$

Since the interaction between the two electrons is described by the exchange of photons via the operator (11.5), the potential $V(r)$ should exclude any effects pertaining to this interaction.

According to (11.10) and (11.11), nonradiative transitions of the two electrons moving in the central field would be a consequence of the perturbing potential $H'=\mathrm{i}ec(\gamma_1\boldsymbol{A}_1+\gamma_2\boldsymbol{A}_2)$ which describes the exchange of virtual photons. Since there are no photons in the initial and final states and since A is linear in photon annihilation and creation operators, the transition probability must be at least of second order in H'. Consequently, our considerations are based on the transition amplitude

$$M_{\mathrm{fi}}=\sum_\gamma \frac{\langle f|H'|\gamma\rangle\langle\gamma|H'|i\rangle}{E_{\mathrm{i}}-E_\gamma}, \tag{11.12}$$

where the Dirac brackets refer to integration over all x_ν $(\nu=1\ldots4)$ and where the wave functions $|i\rangle$ and $|f\rangle$ are antisymmetric stationary-state Dirac spinors of the form

$$|u\rangle=\frac{1}{\sqrt{2}}\begin{vmatrix}u_1 & u_2\\ u_1 & u_2\end{vmatrix}|0\rangle. \tag{11.13}$$

In this wave function $|0\rangle$ represents the photon vacuum and

$$u_\mu=\frac{1}{\sqrt{2\pi c}}\,\theta_\mu \mathrm{e}^{-\mathrm{i}\omega_\mu t}, \tag{11.14}$$

where the normalization factor is chosen so that $\langle u'_\mu|u_\mu\rangle=\mathrm{i}\delta(\omega'_\mu-\omega_\mu)$ with respect to $x_4=\mathrm{i}ct$. The time-independent four-spinors θ_μ are assumed to satisfy (11.11). Since $|i\rangle$ represents a positive-energy electron state $E_{\mathrm{i}}-E_\gamma\gtrsim 2mc^2$ for intermediate negative-energy positron states in (11.12). Hence it is assumed that the states $|\gamma\rangle$ include only electron states of the form (11.13) but with $|0\rangle$ replaced by one-photon states $|0\ldots k_\lambda\ldots\rangle$.

Since H' is a one-electron operator, the intermediate states differ by one four-spinor in addition to one photon from the initial and final wave functions $|i\rangle$ and $|f\rangle$ which are associated with the zero-photon states. In Fig. 6 the routes via four possible intermediate continua are pictured. The initial state is embedded in these continua so that the perturbation matrix element M_{fi} has to be evaluated from (11.12) by replacing E_μ by $E_\mu-\mathrm{i}\nu$ and by going to the limit

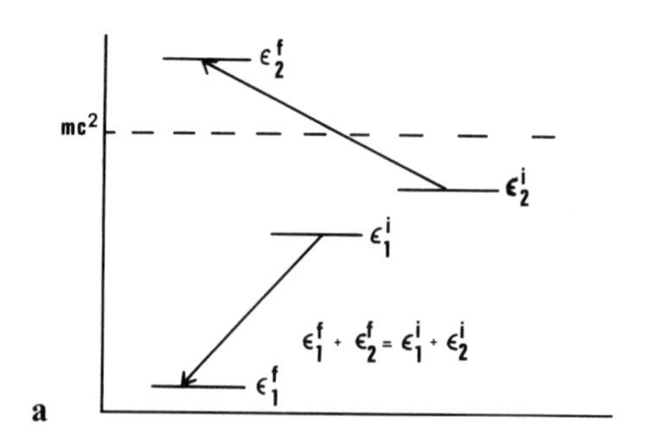

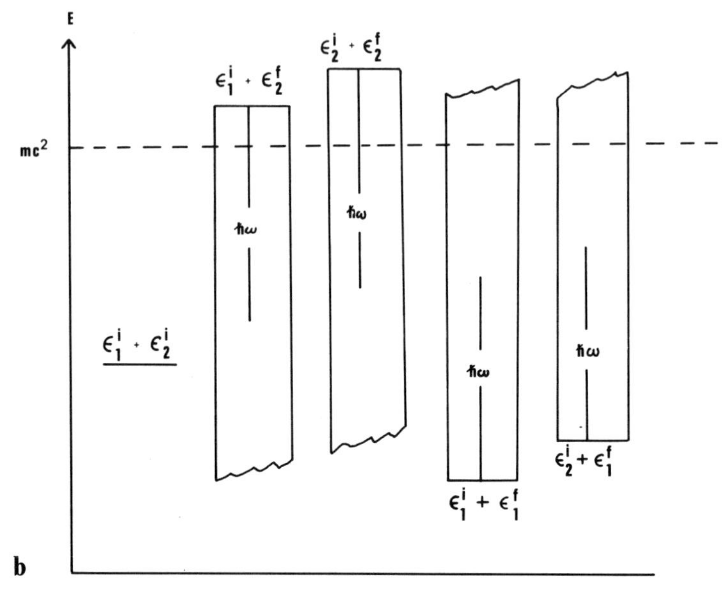

Fig. 6. (**a**) Auger process is pictured within the one-electron model. The electrons in the states $|i_1\rangle$ and $|i_2\rangle$ of the initial ionized many-electron state are transferred to the initially vacant core state $|f_1\rangle$ and to the continuum state $|f_2\rangle$. (**b**) Four groups of possible intermediate states are shown. Either the photon emission is associated with the electron transfer to the continuum state $|f_2\rangle$ or with the transition to the core state $|f_1\rangle$. The initial state is in resonance with the corresponding continua

$\lim_{\nu\to 0} M_{\mathrm{fi}}(\nu)$. This procedure is consistent with the nonrelativistic perturbation approach of Sect. 10γ. Hence

$$M_{\mathrm{fi}} = -\lim_{\nu\to 0} \frac{e^2\hbar c^2}{2\varepsilon_0(2\pi)^3} \sum_{\lambda\mu\mu'} \int \frac{d^3k\,d\omega}{|\omega|}\,\delta(\omega^2 - k^2c^2)$$

$$\cdot\left[\frac{\langle f_1 f_2|\gamma^1_\mu \gamma^2_{\mu'} e^{(\lambda)}_\mu e^{(\lambda)}_{\mu'} \mathrm{e}^{-ik(x_1-x_2)}|i_1 i_2\rangle}{\Delta_1 - \hbar\omega + \mathrm{i}\nu} + \frac{\langle f_1 f_2|\gamma^1_\mu \gamma^2_{\mu'} e^{(\lambda)}_\mu e^{(\lambda)}_{\mu'} \mathrm{e}^{ik(x_1-x_2)}|i_1 i_2\rangle}{-\Delta_1 + \hbar\omega + \mathrm{i}\nu} - X\right], \tag{11.15}$$

where the summation over the four-dimensional propagation vector k ($k^2 - \omega^2/c^2 = 0$) has been replaced by the integration over k and ω. The initial and final states are now in product forms. Hence X is the exchange term which is similar to the sum of the first two terms in (11.15) except that $|f_1 f_2\rangle$ should be replaced by $|f_2 f_1\rangle$ and $\Delta_1 = \mathcal{E}_1^{\mathrm{i}} - \mathcal{E}_1^{\mathrm{f}}$ by $\Delta_2 = \mathcal{E}_2^{\mathrm{i}} - \mathcal{E}_1^{\mathrm{f}}$. Using the orthogonality properties of the polarization vectors $e_\mu^{(\lambda)}$ and the fact that

$$\sum_{\mu=1}^{4} \gamma_\mu^1 \gamma_\mu^2 = (1 - \bar{\boldsymbol{\alpha}}^1 \cdot \bar{\boldsymbol{\alpha}}^2) = \alpha(12), \tag{11.16}$$

one obtains the direct part of M_{fi} after integrating over ω, the time-dependent factors in the matrix elements, and over the direction of the space part of the four-vector k as

$$M_{\mathrm{fi}}^{\mathrm{d}} = -\lim_{\nu \to 0} \frac{e^2}{4\pi^2 \varepsilon_0} \int_0^\infty dk \langle f_1 f_2 | \alpha(12) \frac{\sin k r_{12}}{r_{12}} | i_1 i_2 \rangle \cdot \left[\frac{1}{k_1 - k + \mathrm{i}\nu} + \frac{1}{-k_1 - k + \mathrm{i}\nu} \right], \tag{11.17}$$

where $k_1 = \Delta_1/\hbar c$. The exchange part $M_{\mathrm{fi}}^{\mathrm{e}}$ is similar to $M_{\mathrm{fi}}^{\mathrm{d}}$ except that $|f_1 f_2\rangle$ is replaced by $|f_2 f_1\rangle$ and k_1 by $k_2 = \Delta_2/\hbar c$. With the aid of (A.12) we obtain

$$M_{\mathrm{fi}} = M_{\mathrm{fi}}^{\mathrm{d}} - M_{\mathrm{fi}}^{\mathrm{e}} = \frac{e^2}{4\pi\varepsilon_0} [\langle f_1 f_2 | O_{12}(k_1) | i_1 i_2 \rangle - \langle f_2 f_1 | O_{12}(k_2) | i_1 i_2 \rangle], \tag{11.18}$$

where

$$O_{12}(k) = (\mathbf{1} - \bar{\boldsymbol{\alpha}}_1 \cdot \bar{\boldsymbol{\alpha}}_2) \frac{\mathrm{e}^{\mathrm{i}k r_{12}}}{r_{12}} \tag{11.19}$$

can be interpreted as an energy-dependent operator. The electron-electron interaction amplitude (11.18) was derived by Møller[1] in the context of the relativistic scattering of free electrons and rederived by Hulme[7] for the case of bound electrons. This result can also readily be obtained by using the S matrix formalism of quantum electrodynamics.[8] However, the present derivation shows clearly the relationship between (11.18) and the corresponding nonrelativistic amplitude (6.5) which includes photon-electron coupling correlations.

From the definition of γ_4 and (11.15) and (11.16) it can be seen that the scalar photons are responsible for the Coulomb interaction which is represented by the first term of the operator $\lim_{k \to 0} O_{12}(k)$. The expansion of retardation factor to second order in k multiplied by the r_{12}^{-1} term plus the $(\bar{\boldsymbol{\alpha}}_1 \cdot \bar{\boldsymbol{\alpha}}_2)/r_{12}$ term yields the k-independent operator

$$O_{12} = \frac{1}{r_{12}} \mathbf{1} - \frac{\bar{\boldsymbol{\alpha}}_1 \cdot \bar{\boldsymbol{\alpha}}_2}{r_{12}} + \frac{(\bar{\boldsymbol{\alpha}}_1 \cdot \boldsymbol{\nabla}_2)(\bar{\boldsymbol{\alpha}}_1 \cdot \boldsymbol{\nabla}_2) r_{12}}{2}. \tag{11.20}$$

---

[7] Hulme, H.R. (1936): Proc. R. Soc. Lond. A *154*, 487

[8] Ref. [*17*], Sect. 37

where the sum of the second and third term represents the Breit interaction.[9] The derivation of (11.20), which represents the interaction of two electrons up to terms of order $(v/c)^2$ in the velocities, is based on the assumption that the stationary-state four-spinors appearing in the amplitude (11.18) are solutions of the Dirac equation (11.11). Note that this assumption is not a necessary one for the derivation of (11.18) from (11.12) although it was made for the purpose of identifying the one-electron energies $\mathcal{E}_\nu^{\mathrm{i}}$ and $\mathcal{E}_\nu^{\mathrm{f}}$ $(\nu = 1, 2)$.

The relativistic theory, presented in this section, is equivalent to the nonrelativistic one which was developed in Sects. 5 and 6 but in the independent-electron limit. This is most easily seen by comparing (6.5) with (11.15). If we use (11.16) to separate the Coulomb part, then the rest, which is governed by the operator $\bar{\boldsymbol{\alpha}} \cdot \bar{\boldsymbol{A}}$, corresponds to the photon-electron coupling correction in the nonradiative decay probability (6.5). The nonrelativistic limit is established by the correspondence $\bar{\boldsymbol{\alpha}} \cdot \bar{\boldsymbol{A}} \rightarrow \boldsymbol{A} \cdot \boldsymbol{p}/mc$, where the former operator is operating on Dirac functions and the latter operator on Pauli functions. This correspondence implies that the Coulomb gauge has been chosen in the latter case. The vector-potential operator is then given by (5.3), which contains only terms corresponding to transverse photons. Hence the calculation of nonradiative probabilities which is based on MØLLER's formula can be understood as the relativistic counterpart to WENTZEL's ansatz which includes photon-electron coupling effects but no electron correlation and relaxation effects.

It would also be possible to use relativistic wave functions in (6.5), in which case mass variation and spin-orbit effects would be incorporated in the electron-electron interaction amplitude, whereas the corrections due to the interaction between the electrons and the transverse photons would be represented by the integral expression. The latter corrections are usually very small, as was suggested long ago by TAYLOR and MOTT[10] in connection with the question of whether the Auger effect and internal conversion should be treated as internal absorption of x and gamma rays or as a rearrangement process of charged particles.

12. A relativistic nonradiative transition probability. In the previous section we have discussed the origin of the relativistic theory of the Auger effect by using the unrestricted form of the wave functions. In this section we impose the appropriate symmetry constraints on the wave functions which leads to a classification of the various Auger transitions according to quantum numbers and configurations of the initial and final vacancies.

The use of MØLLER's electron-electron interaction amplitude (11.18) is strictly consistent only in connection with eigensolutions of Dirac's central field equation (11.11). The corresponding Hamiltonian commutes with $\boldsymbol{j}^2 = (\bar{\boldsymbol{l}} + \bar{\boldsymbol{s}})^2$, $\boldsymbol{j}_z$, and $\boldsymbol{P}$, which are all represented by 4×4 operator matrices. In particular,

---

[9] Ref. [17], Sect. 38

[10] Taylor, H.M., Mott, N.F. (1933): Proc. R. Soc. Lond. A *142*, 215; see also Tralli, N., Goertzel, G. (1951): Phys. Rev. *83*, 399. BURHOP and CHATTARJI (see introduction) also discuss this question in their books

the spin is represented by the matrix operator

$$\bar{\boldsymbol{s}}=\frac{\hbar}{2}\bar{\boldsymbol{\sigma}}'=\frac{\hbar}{2}\begin{pmatrix}\bar{\boldsymbol{\sigma}} & 0\\ 0 & \bar{\boldsymbol{\sigma}}\end{pmatrix}, \tag{12.1}$$

where $\bar{\boldsymbol{\sigma}}=(\sigma_x, \sigma_y, \sigma_z)$ is the Pauli matrix in the vector form, and the permutation operator $\boldsymbol{P}$ by the matrix operator

$$\boldsymbol{P}=\boldsymbol{\beta}\,P, \tag{12.2}$$

where P is the ordinary permutation operator. However, instead of operator (12.2), the eigenvalues of another commuting operator, namely, $\boldsymbol{K}=\boldsymbol{\beta}(\mathbf{1}+\bar{\boldsymbol{\sigma}}'\cdot\bar{\boldsymbol{l}})$, are used together with $\boldsymbol{j}^2$ and $\boldsymbol{j}_z$ to classify the solutions of Dirac's central field equation. As shown in many textbooks[11] the solutions can be written in terms of the Dirac central field four-spinors

$$\psi_{\kappa m}=\begin{bmatrix} R_{+1}(r)\langle\theta\varphi;\,\zeta|j-\tfrac{1}{2}a, j, m\rangle\\ iR_{-1}(r)\langle\theta\varphi;\,\zeta|j+\tfrac{1}{2}a, j, m\rangle\end{bmatrix}, \tag{12.3}$$

where R_{+1} and R_{-1} are the large and small radial components, respectively, and where the spin-orbit eigenfunctions[12]

$$\langle\theta\varphi;\,\zeta|\lambda, j, m\rangle=\sum_{\mu} Y_l^{m-\mu}(\theta,\varphi)\,\phi^{\mu}(\zeta)\,(l, 1/2, m-\mu, \mu|j, m) \tag{12.4}$$

are superpositions of the Pauli two-spinors $\phi^{1/2}=\begin{pmatrix}\delta_{\zeta,1/2}\\ 0\end{pmatrix}$ and $\phi^{-1/2}=\begin{pmatrix}0\\ \delta_{\zeta,-1/2}\end{pmatrix}$. In the eigenfunction (12.4) $\lambda=j-a\beta/2$, where $\beta=\pm 1$ labels the row in wave function (12.3) and $a=\pm 1$ distinguishes between the two possible couplings of spin and angular momenta in $j=l+a/2$. Since $\boldsymbol{K}\psi_{\kappa m}=-\kappa\psi_{\kappa m}$, where $\kappa=l$ for $j=l-1/2$ and $-(l+1)$ for $j=l+1/2$, κ is connected to a by $\kappa=-(j+1/2)a$. It can also be shown that

$$\boldsymbol{P}\psi_{\kappa m}=(-1)^l\psi_{\kappa m}, \tag{12.5}$$

so that the four-spinor $\psi_{\kappa m}$ has the same parity as the large component which corresponds to $l=j-a/2$ in the nonrelativistic limit.

As in the nonrelativistic case it is possible to construct symmetry-adapted many-electron wave functions $|\gamma JM\rangle$ from Dirac central-field four-spinors representing a given configuration.[13] Each subshell is considered separately

---

[11] e.g. Ref. [17], Sect. 12; Ref. [7], Vol. II, Chap. XX; see also the review by Grant (Grant, I.P. (1970): Adv. Phys. *19*, 747)

[12] Note that it is customary to couple $\boldsymbol{s}$ to $\boldsymbol{l}$ (i.e. $\boldsymbol{j}=\boldsymbol{l}+\boldsymbol{s}$) in relativistic applications in contrast to the nonrelativistic case (7.14). The difference in the phase factor is $(-1)^{l+1/2-j}$. Note also that the phase factor in (12.13) corresponds to the relativistic convention

[13] Grant, I.P., Pyper, N.C. (1976): J. Phys. B *9*, 761

and the corresponding antisymmetric wave functions $|j^s, v, \omega, J', M'\rangle$ are constructed in the jj coupling. Here s denotes the occupation number of the subshell (nlj). A wave function which corresponds to the total angular momentum quantum number J' may be further characterized by the seniority number v and by ω, which resolves any further degeneracy. The final wave function is obtained by successive angular momentum coupling of the wave functions of each shell in a standard order and by proper antisymmetrization (see Sect. 8α). Hence it is possible to express matrix elements of both the Coulomb and Breit interaction as a superposition of radial integrals for any pair of antisymmetric many-electron wave functions $|\gamma JM\rangle$ and $|\gamma' JM\rangle$.[13] This procedure makes it also possible to construct relativistic Auger amplitudes for any initial and final configurations by using the appropriate Breit interaction operator. For many-electron systems[14] this operator is given in accordance with (11.20) by

$$B = \frac{-e^2}{4\pi\varepsilon_0} \sum_{\mu>\nu} \left[\frac{\bar{\boldsymbol{\alpha}}_\mu \cdot \bar{\boldsymbol{\alpha}}_\nu}{r_{\mu\nu}} + \frac{(\bar{\boldsymbol{\alpha}}_\mu \cdot \boldsymbol{\nabla}_\mu)(\bar{\boldsymbol{\alpha}}_\nu \cdot \boldsymbol{\nabla}_\nu)\, r_{\mu\nu}}{2}\right]. \tag{12.6}$$

It may also be possible to generalize this procedure by considering directly the second-order perturbation formula (11.12) with many-electron wave functions and the many-electron operator $H' = iec \sum_\mu \boldsymbol{A}_\mu$. Attempts in this direction have apparently been made.[15, 16]

We shall now consider the case of Auger transitions in an initially closed-shell atom. Both the initial single-vacancy state $|(n_1 l_1 j_1)^{-1} J_1 M_1\rangle$ and the final double-vacancy state $|(n_2 l_2 j_2)^{-1} (n_3 l_3 j_3)^{-1} J(\varepsilon l j) J_1 M_1\rangle$ are constructed in jj coupling from orthogonal Dirac central field four-spinors. Analogously with the nonrelativistic case (we omit the phase factor which is due to the transformation from the many- to the two-electron picture; see Sect. 9α) the Auger amplitude is given by

$$\begin{aligned}\langle f|H-E|i\rangle = \sqrt{\frac{2J+1}{(2j_1+1)\tau}}\, [\langle j_1 j JM|O_{12}(k_2)|j_2 j_3 JM\rangle \\ -(-1)^{j_2+j_3-J} \langle j_1 j JM|O_{12}(k_3)|j_3 j_2 JM\rangle],\end{aligned} \tag{12.7}$$

where $\tau=1$ in the case of inequivalent final vacancies and $\tau=2$ in the case of equivalent vacancies. The operator $O_{12}(k)$ is the Møller operator (11.19) with $k_2 = \mathcal{E}_2 - \mathcal{E}_1$ and $k_3 = \mathcal{E}_3 - \mathcal{E}_1$. The wave functions $|j_1 j JM\rangle$ and $|j_2 j_3 JM\rangle$ are the two-electron wave functions of product form which are constructed from the Dirac's central field four-spinors by coupling $\boldsymbol{j}$ to $\boldsymbol{j}_1$ and $\boldsymbol{j}_3$ to $\boldsymbol{j}_2$, respectively. Note that $\boldsymbol{j}_2$ must be coupled to $\boldsymbol{j}_3$ in the latter matrix element of (12.7). Further reduction of the amplitude (12.7) is based on the expansion[2]

$$\frac{e^{ikr_{12}}}{r_{12}} = \sum_{\nu=0}^{\infty} \gamma_\nu(k; r_1, r_2)\, P_\nu(\cos\theta), \tag{12.8}$$

---

[14] Kim, Y.-K. (1967): Phys. Rev. *154*, 17

[15] Mann, J.B., Johnson, W.R. (1971): Phys. Rev. A *4*, 41

[16] Huang, K.N. (1978): J. Phys. B *11*, 787

where

$$\gamma_\nu(k; r_1, r_2) = \frac{2\nu+1}{k r_1 r_2} \cdot \begin{cases} \eta_\nu(kr_1)\,\zeta_\nu(kr_2) & r_1 < r_2 \\ \eta_\nu(kr_2)\,\zeta_\nu(kr_1) & r_2 < r_1 \end{cases}. \tag{12.9}$$

The functions $\eta_\nu(kr)$ and $\zeta_\nu(kr)$ are related to the spherical Bessel functions $J_{\nu+1/2}(kr)$ of half-odd order by

$$\begin{aligned} \eta_\nu(kr) &= \sqrt{\tfrac{1}{2}\pi k r}\, J_{\nu+1/2}(kr) \\ \zeta_\nu(kr) &= \sqrt{\tfrac{1}{2}\pi k r}\, [(-1)^\nu J_{-\nu-1/2}(kr) + \mathrm{i} J_{\nu+1/2}(kr)]. \end{aligned} \tag{12.10}$$

The Racah technique of spherical tensor operators can readily be used for the evaluation of the Auger amplitude (12.7) in terms of the radial integrals

$$\begin{aligned} R_k^\nu(1\beta_1, \varepsilon\beta_\varepsilon, 2\beta_2, 3\beta_3) = \int_0^\infty \int_0^\infty & R_{1\beta_1}(r_1) R_{\varepsilon\beta_\varepsilon}(r_2) \gamma_\nu(k; r_1, r_2) \\ & \cdot R_{2\beta_2}(r_1) R_{3\beta_3}(r_2) r_1^2 r_2^2 \, dr_1 \, dr_2. \end{aligned} \tag{12.11}$$

This procedure yields

$$\begin{aligned} \langle j_1 j J M | O_{12}(k_2) | j_2 j_3 J M \rangle = & (-1)^{j+j_2+J} \sum_\nu \sum_{\beta\beta'} \begin{Bmatrix} j_2 & j_3 & J \\ j & j_1 & \nu \end{Bmatrix} \\ & \cdot [\langle j_1 \| C^\nu \| j_2 \rangle \langle j \| C^\nu \| j_3 \rangle R_{k_2}^\nu(1\beta, \varepsilon\beta', 2\beta, 3\beta') \pi'(\kappa_1 \kappa_2 \nu) \pi'(\kappa\kappa_3\nu) \\ & - \sum_L (-\beta\beta') \langle j_1 \| C^L \| j_2 \rangle \langle j \| C^L \| j_3 \rangle R_{k_2}^\nu(1\beta, \varepsilon\beta', 2(-\beta), 3(-\beta')) \\ & \cdot E_{-\beta}^\nu(\kappa_1\kappa_2 L) E_{-\beta'}^\nu(\kappa\kappa_3 L) \pi(\kappa_1\kappa_2\nu) \pi(\kappa\kappa_3\nu)]. \end{aligned} \tag{12.12}$$

The exchange amplitude can be obtained from the direct amplitude (12.12) by interchanging the indexes 2 and 3 everywhere.

In (12.12)

$$\langle j \| C^\mu \| j' \rangle = (-1)^{j+1/2} \sqrt{(2j+1)(2j'+1)} \begin{pmatrix} j & \mu & j' \\ 1/2 & 0 & -1/2 \end{pmatrix} \tag{12.13}$$

is the reduced matrix element of the tensor operator $\boldsymbol{C}^\mu$.[12] The coefficients $\pi(\kappa\kappa'\mu)$, given by

$$\pi(\kappa\kappa'\mu) = \tfrac{1}{2}[1 - (-1)^{l+l'+\mu}], \tag{12.14}$$

incorporate parity selection rules explicitly. The coefficients $\pi'(\kappa\kappa'\mu)$ are defined by

$$\pi'(\kappa\kappa'\mu) = 1 - \pi(\kappa\kappa'\mu). \tag{12.15}$$

Finally the coefficients $E_\beta^\mu(\kappa\kappa' L)$ are given by[13]

$$E_\beta^\mu(\kappa\kappa' L) = \begin{cases} \dfrac{L - \beta(\kappa' - \kappa)}{[L(2L-1)]^{1/2}} & \mu = L-1 \\ \dfrac{-\beta(\kappa + \kappa')}{[L(L+1)]^{1/2}} & \mu = L \\ \dfrac{-(L+1) - \beta(\kappa' - \kappa)}{[(L+1)(2L+3)]^{1/2}} & \mu = L+1. \end{cases} \tag{12.16}$$

In (12.12) the radial integrals $R_{k_2}^{\nu}(1\beta, \varepsilon\beta', 2\beta, 3\beta')$ correspond to the Coulomb term in the Møller operator. The exchange of the transverse and longitudinal photons is represented by the sum containing the radial integrals $R_{k_2}^{\nu}(1\beta, \varepsilon\beta', 2(-\beta), 3(-\beta'))$ in which the large and small components are mixed. Hence this part, which is related to the Breit interaction, vanishes in the nonrelativistic limit. The radial integrals $R_{k_2}^{\nu}(1\beta, \varepsilon\beta', 2\beta, 3\beta')$ approach the ordinary Slater integrals, and amplitude (12.12) becomes identical with the nonrelativistic amplitude (9.10) in the *jj*-coupling limit apart from the phase factor.

13. Intermediate coupling. In the nonrelativistic case the intermediate coupling was defined as a kind of limited configuration interaction due to the spin-orbit interaction. In this coupling scheme mixing between levels pertaining to the same J value but different terms of a given configuration is considered (Sect. 8β). According to DESCLAUX,[17] the relativistic analogue would then be to consider the interaction between various *jj*-coupled levels which correspond to the same J value and configuration.

In order to illustrate the relativistic intermediate-coupling method and its relationship to the nonrelativistic counterpart, we consider the mixing within the $n'snp^5$ configuration in detail. There are two levels corresponding to $J=1$, namely, $^1P_1$ and $^3P_1$ in the nonrelativistic case and those pertaining to the subconfigurations $s_{1/2}^{-1}p_{1/2}^{-1}$ and $s_{1/2}^{-1}p_{3/2}^{-1}$ in the relativistic case. Consequently, for a given M (1, 0, or -1), the relativistic intermediate-coupling wave functions ψ_1 and ψ_2 are linear superpositions of the wave functions $|1/2\ 1/2\rangle = |s_{1/2}^{-1}p_{1/2}^{-1}JM\rangle$ and $|1/2\ 3/2\rangle = |s_{1/2}^{-1}p_{3/2}^{-1}JM\rangle$ so that

$$\begin{aligned}\psi_1 &= -\sin\phi\,|1/2\ 1/2\rangle + \cos\phi\,|1/2\ 3/2\rangle\\ \psi_2 &= \ \ \cos\phi\,|1/2\ 1/2\rangle + \sin\phi\,|1/2\ 3/2\rangle.\end{aligned} \tag{13.1}$$

In the nonrelativistic limit with vanishing spin-orbit interaction the coefficients $\sin\phi$ and $\cos\phi$ would be given directly by the unitary transformation (8.37) which gives the relationship between the two-particle *LS*- and *jj*-coupled wave functions. As a result $\sin\phi = \sqrt{1/3}$ and $\cos\phi = \sqrt{2/3}$, which relates ψ_1 to $^1P_1$ and ψ_2 to $^3P_1$, respectively.

The relativistic Hamiltonian submatrix corresponding to $J=1$ can be evaluated using known techniques[18] for the Hamiltonian (16.10) which consists of a sum of Dirac one-electron Hamiltonians plus the unretarded Coulomb interactions between the electrons. The matrix is given by

$$\begin{array}{c} \\ |1/2\ 1/2\rangle \\ |1/2\ 3/2\rangle \end{array}\!\!\begin{array}{c} \quad|1/2\ 1/2\rangle \qquad\qquad\qquad |1/2\ 3/2\rangle \\ \begin{pmatrix} E_{\text{av}}^{1/2\,1/2} + \frac{1}{18}G^1(sp^*) & -\frac{2}{9}\sqrt{2}R^1(spp^*s) \\ -\frac{2}{9}\sqrt{2}R^1(spp^*s) & E_{\text{av}}^{1/2\,3/2} + \frac{5}{18}G^1(sp) \end{pmatrix}, \end{array} \tag{13.2}$$

[17] Desclaux, J.P. (1972): Int. J. Quantum Chem. 6, 25; Desclaux, J.P. (1976): In: Photoionization and other probes of many-electron interactions, Wuilleumier, F.J. (ed.), New York: Plenum Press, p. 367

[18] See, e.g., the review by GRANT (III.11)

where we have used the notation $p_{1/2}=p^*$, $p_{3/2}=p$. The single-configuration jj-average energies $E_{\mathrm{av}}^{j'j}$ are given by

$$E_{\mathrm{av}}^{j'j}=\frac{\sum_{\alpha}(2J_{\alpha}+1)\,E_{J_{\alpha}}^{j'j}}{\sum_{\alpha}(2J_{\alpha}+1)}, \tag{13.3}$$

with $J=1,\,0$ for $j'=1/2$, $j=1/2$ and $J=2,1$ for $j'=1/2$, $j=3/2$. In the nonrelativistic limit matrix (13.2) becomes

$$\begin{array}{c} \\ |1/2\;1/2\rangle \\ |1/2\;3/2\rangle \end{array}\begin{array}{c}\begin{array}{cc} |1/2\;1/2\rangle & \qquad |1/2\;3/2\rangle \end{array}\\ \begin{pmatrix} E_{\mathrm{av}}+\xi+\frac{1}{18}G^1 & -\frac{2}{9}\sqrt{2}\,G^1 \\ -\frac{2}{9}\sqrt{2}\,G^1 & E_{\mathrm{av}}-\frac{1}{2}\xi+\frac{5}{18}G^1 \end{pmatrix}\end{array}, \tag{13.4}$$

where $G^1=G^1(sp)$ is the Slater exchange integral and $\xi=\xi_{np}$ the spin-orbit interaction parameter for the np shell. The energy E_{av} which is obtained from (13.3) by forming the many-configuration jj average

$$E_{\mathrm{av}}=\tfrac{1}{3}E_{\mathrm{av}}^{1/2\;1/2}+\tfrac{2}{3}E_{\mathrm{av}}^{1/2\;3/2} \tag{13.5}$$

coincides in the nonrelativistic limit with the LS average energy

$$E_{\mathrm{av}}=\tfrac{3}{4}E(^3P)+\tfrac{1}{4}E(^1P). \tag{13.6}$$

Note that in matrix (13.4) ξ is the only remainder of the relativistic effects. If we express the spin-orbit interaction contributions in matrix (13.4) in terms of the differences $E_{\mathrm{av}}^{j'j}-E_{\mathrm{av}}$, matrices (13.2) and (13.4) are formally equivalent and can readily be diagonalized using transformation (13.1).

However, in the nonrelativistic case one usually starts from a Hamiltonian matrix which is given with respect to the LS-coupled wave functions. This matrix can be obtained from the matrix (13.4) by performing the unitary transformation (13.1) with $\sin\phi=\sqrt{1/3}$ and $\cos\phi=\sqrt{2/3}$ as given by (8.37). The relativistic counterpart can be obtained in the same way from matrix (13.2). The diagonalization of the resulting nonrelativistic matrix

$$\begin{array}{c} \\ ^1P_1 \\ \\ ^3P_1 \end{array}\begin{array}{c}\begin{array}{cc} ^1P_1 & \qquad\qquad ^3P_1 \end{array}\\ \begin{pmatrix} E_{\mathrm{av}}+\dfrac{1}{2}G^1 & -\dfrac{\sqrt{2}}{2}\xi \\ -\dfrac{\sqrt{2}}{2}\xi & E_{\mathrm{av}}+\dfrac{1}{2}\xi-\dfrac{1}{6}G^1 \end{pmatrix}\end{array} \tag{13.7}$$

using the transformation

$$\begin{aligned} \psi_1&=-\sin\phi'|^3P_1\rangle+\cos\phi'|^1P_1\rangle \\ \psi_2&=\cos\phi'|^3P_1\rangle+\sin\phi'|^1P_1\rangle \end{aligned} \tag{13.8}$$

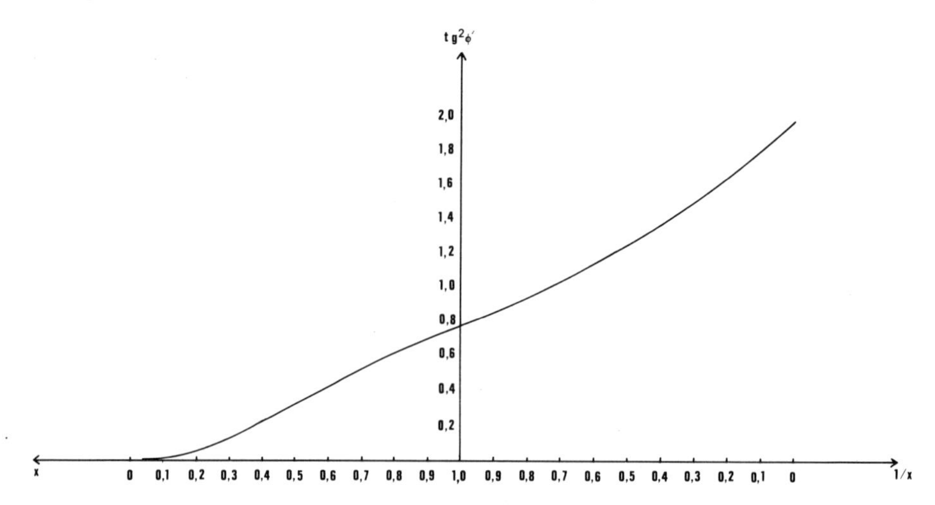

Fig. 7. Square of the intermediate-coupling parameter tg ϕ' as a function of the strength parameter $x=\xi/G'$

yields the energies

$$E_{1,2}=E_{\text{av}}+\tfrac{1}{6}G^1+\tfrac{1}{4}\xi\pm\sqrt{\tfrac{1}{9}G^{1^2}+\tfrac{9}{16}\xi^2-\tfrac{1}{6}G^1\xi} \tag{13.9}$$

and the ratio

$$\operatorname{tg}\phi'=\frac{\sqrt{2}\sin\phi-\cos\phi}{\sin\phi+\sqrt{2}\cos\phi}=\frac{\frac{\sqrt{2}}{2}x}{\frac{1}{3}-\frac{1}{4}x+\sqrt{\frac{9}{16}x^2-\frac{1}{6}x+\frac{1}{9}}}, \tag{13.10}$$

where $x=\xi/G^1$. In the relativistic case a reasonable approximation would be to replace $G^1(sp^*)$ and $G^1(sp)$ by $R^1(spp^*s)$ in the corresponding matrix. If we define the relativistic analogue of ξ as

$$\xi^*=\tfrac{2}{3}(E_{\text{av}}^{1/2\ 1/2}-E_{\text{av}}^{1/2\ 3/2}), \tag{13.11}$$

then (13.9) and (13.10) remain valid also in the relativistic case with R^1 instead of G^1 and $x^*=\xi^*/R^1$ instead of x. Figure 7 shows $\operatorname{tg}^2\phi'$ as a function of x. The typical rapid variation of this function around $x=1$ reveals the important intermediate-coupling regime for numerous nonradiative and radiative transition probabilities involving s and p electrons. The relationship $x=x(Z)$ establishes the connection to the periodic table. Note that the relativistic effects influence transition probability ratios mainly through eventual differences between x^* and x.

The preceding example demonstrates that a relativistic analogue of the intermediate coupling can be obtained by transforming the relativistic Hamiltonian submatrix which is given with respect to the jj-coupled wave functions into one which is given with respect to *LS*-coupled wave functions. This transformation is purely geometrical, i.e., independent of the nature of the basis set. The diagonalization of the resulting matrix yields results which are analogous to the nonrelativistic ones but which can be expressed in terms of relativistic Slater integrals and generalized spin-orbit parameters. This procedure implies a fixed set of four-spinors without any optimization of the total energy. However, since intermediate coupling implies mixing between *jj* subconfigurations in the relativistic case, the Dirac-Fock multiconfiguration-interaction method (see also Sect. 18α) can be used to perform intermediate-coupling calculations in certain cases.[4,17] In our example this would imply the optimization of the radial components of the four-spinors with respect to vatiations of either energy

$$\begin{aligned} E_v = \langle \psi_v | H | \psi_v \rangle = & \sin^2\phi\, E_{\mathrm{av}}^{1/2\ 1/2} + \sin^2\phi\, \tfrac{1}{18} G^1(sp^*) \\ & + \cos^2\phi\, E_{\mathrm{av}}^{1/2\ 3/2} + \cos^2\phi\, \tfrac{5}{18} G^1(sp) \\ & \pm \tfrac{4}{9}\sqrt{2} \sin\phi \cos\phi\, R^1(spp^*s), \end{aligned} \tag{13.12}$$

where the plus sign corresponds to $v=1$ and the minus to $v=2$. The Hamiltonian operator H in (13.12) is the same as the one which has been used to construct matrix (13.2) so that the variation of energy (13.12) corresponds to the diagonalization of matrix (13.2) with the simultaneous optimization of the radial components of the four-spinors. The wave functions are expressed in terms of optimized coefficients $\sin\phi$ and $\cos\phi$ which yield, in addition to improved intermediate-coupling energies E_1 and E_2, an optimized ratio $\mathrm{tg}\,\phi'$ through (13.10). Since the multiconfiguration procedure accounts for the effect of relaxation on the spin-orbit interaction, $\mathrm{tg}\,\phi'$ may be especially sensitive to the optimization close to the nonrelativistic limit.[19]

With regard to absolute intensities we note that the *jj*-coupling amplitude formula (12.12) can be used to generate intermediate-coupling transition amplitudes since these can be expressed in accordance with (9.19) as linear superpositions of *jj*-coupling amplitudes. However, the phase factors introduced in Sect. 9α must eventually be taken into account in the transformation from the many-electron to two-electron picture.

IV. Angular distribution of Auger electrons

Mehlhorn's[1] prediction that Auger electrons emitted by decay of hole states having total angular momentum quantum number $J > 1/2$ can exhibit a nonisotropic distribution has excited considerable interest in this area, both

[19] The $1s2p^5$ case has been studied in detail by Åberg, T., Suvanen, M. (1982) in: Advances in *X*-ray spectroscopy, Bonnelle, C., Mande, C. (eds.), Oxford: Pergamon Press

[1] Mehlhorn, W. (1968): Phys. Lett. A *26*, 166

theoretical and experimental.[2–13] In at least two connections can studies of this anisotropy be of importance. Firstly, as we shall see below, it may be possible to extract features of both the Auger decay process and the primary event which created the hole. For example, applications have been directed towards calculating the parameters directing the nonisotropic Auger emission and then comparing with experimental observations. The sensitivity of these parameters to the wave functions and approximations employed provides a good test of the theoretical approach. Secondly, as will become more apparent when the theoretical description of angular distributions is discussed, future applications can be envisaged which will probe the two-step process (creation of the hole and *then* independent decay) conventionally used to describe the Auger effect (see Sect. 4).

In general terms, the fundamental feature underlying the phenomenon can be understood as follows. The production amplitudes between the initial state and the JM substates of the intermediate hole state created by some incoming beam of photons or particles are, in general, unequal, in which case the hole state is said to become either aligned or polarized with respect to the axis defined by the beam direction. The nonisotropic Auger distribution reflects this inequality. When the mechanism by which the hole is formed does not define any unique quantization axis (for example, by some radioactive decay process), we may expect the distribution to be isotropic.

In Sects. 15 and 16, we discuss briefly the features of Auger electron angular distributions obtained using high-energy electron impact and photon sources to create the hole state. In so doing, we try to emphasise the relationship between the theoretical formulation and the experimentally observed quantities. Although we could in principle develop the theory using the scattering formalism of Chap. I, we shall, as is customary, assume from the beginning a two-step process and use density matrices and their symmetry properties to describe the angular distribution.[14]

14. General considerations. We consider the following experiment: an incoming beam of unpolarized photons or high-energy electrons is employed to produce an inner-shell hole state which decays by Auger electron emission. We

---

[2] Cleff, B., Mehlhorn, W. (1971): Phys. Lett. *37* A, 3
[3] Flügge, S., Mehlhorn, W., Schmidt, V. (1972): Phys. Rev. Lett. *29*, 7
[4] Cleff, B., Mehlhorn, W. (1974): J. Phys. B *7*, 605
[5] Cleff, B., Mehlhorn, W. (1974): J. Phys. B *7*, 593
[6] Döbelin, E., Sandner, W., Mehlhorn, W. (1974): Phys. Lett. *49* A, 7
[7] McFarlane, S.C. (1975): J. Phys. B *8*, 895
[8] Eichler, J., Fritsch, W. (1976): J. Phys. B *9*, 1477
[9] Omidvar, K. (1977): J. Phys. B *10*, L55
[10] Berezhko, E.G., Kabachnik, N.M. (1977): J. Phys. B *10*, 2467
[11] Berezhko, E.G., Kabachnik, N.M., Rostovsky, V.S. (1978): J. Phys. B *11*, 1749
[12] Berezhko, E.G., Kabachnik, N.M., Sizov, V.V. (1978): J. Phys. B *11*, 1819
[13] Sandner, W., Schmitt, W. (1978): J. Phys. B *11*, 1833
[14] See, e.g., Devons, S., Goldfarb, J.B. (1957) in: Handbuch der Physik, Vol. XLII, Flügge, S. (ed.), Heidelberg: Springer, p. 362; Frauenfelder, H., Steffen, M. (1965) in: Alpha-, beta- and gamma-ray spectroscopy, Vol. II, Siegbahn, K. (ed.), Amsterdam: North-Holland, p. 997

shall not discuss the anisotropy or polarization properties of the possible x-ray emission processes although it could be done with minor modifications of the following treatment.[10, 12, 15] The characteristic quantity in which we are interested is the angular distribution of the Auger electrons with respect to the quantization axis defined by the direction of the incoming beam. Two points are to be noted about this experiment. Firstly, the ionized (primary) electron (and scattering plane for particle impact) is not observed, and secondly, the detectors are assumed to be insensitive to the spin projection of the Auger electron. The first restriction will be removed in Sect. 15γ while the second can also be removed.[14]

The theoretical description treats our experiment as a two-step process in analogy to a two-stage cascade of γ rays.[14] Omitting multiplicative constants, the angular function which we seek is given by

$$W(\Omega_a)=\mathscr{S}\sum_{\alpha\gamma}|\sum_{\beta}\langle c\gamma|H_2|b\beta\rangle\langle b\beta|H_1|a\alpha\rangle|^2, \tag{14.1}$$

where H_1 and H_2 are the interaction operators responsible for the various transitions and where Ω_a denotes the solid angle in which the Auger electron is emitted. In (14.1) a and α, etc., denote respectively the various quantum numbers and their projections with respect to the beam axis. Here a represents the initial state, b the intermediate hole state with the wave function of the primary electron included in the matrix element of H_1, and c the final state and Auger electron. The symbol $\mathscr{S}$ represents a sum (or average) over *all* the experimentally unobserved variables. This feature of (14.1) lies at the heart of the link between experiment and theory.

In writing (14.1) we have already made some assumptions: (i) the initial state is randomly oriented, (ii) the intermediate hole state b is well separated from any other; otherwise a sum over b enters in addition to β. In Auger emission this case occurs when the spin-orbit splitting is much larger than the width of the hole state. If the spin-orbit splitting is small and the LS coupling is valid, there is a summation over b. (iii) The restriction to a specific final state c implies that the transition to c does not overlap with another (in particular, we do not consider degenerate transitions). The sum over γ asserts that we do not observe the residual ion. (iv) No process occurs by which the intermediate states have their alignment disturbed (for example, by collisions) before the decay takes place. In order to be more precise, we define the amplitude integrals in (14.1) by

$$\begin{aligned}\langle b\beta|H_1|a\alpha\rangle &= \langle J_2 M_2 \boldsymbol{k}_p \sigma_p|H_1|J_1 M_1\rangle\\ \langle c\gamma|H_2|b\beta\rangle &= \langle J_3 M_3 \boldsymbol{k}_a \sigma_a|H_2|J_2 M_2\rangle,\end{aligned} \tag{14.2}$$

where σ_ν is the spin projection and $k_\nu^2/2$ the energy of the electron ($\nu=p$ or a) ejected in the direction $\hat{\boldsymbol{k}}_\nu$. J and M denote the total angular momentum quantum numbers and their projections of the initial atom and intermediate-

[15] McFarlane, S.C. (1972): J. Phys. B *5*, 1906; Scofield, J.H. (1976): Phys. Rev. A *14*, 1418

and final-state ions. Note also that, for example, $|J_2 M_2 \boldsymbol{k}\sigma\rangle$ is to be taken as representing an antisymmetric wave function which does not correspond to any definite angular momentum. It follows from Sect. 1 that when final-state interchannel interactions are important (see also Sect. 10β), the correct form for the wave function is

$$|J_2 M_2 \boldsymbol{k}\sigma\rangle = \sum_K \mathrm{i}^l \mathrm{e}^{-\mathrm{i}\delta_l} Y^*_{lm_l}(\hat{\boldsymbol{k}})(s l \sigma m_l|jm)(jJ_2 m M_2|JM)|(J_2 l j)JM\rangle, \quad (14.3)$$

where $(j_1 j_2 m_1 m_2|j_3 m_3)$ is a Clebsch-Gordan coefficient and δ_l is the Coulomb plus short-range potential phase shift of the l'th partial wave. The symbol K stands for the sextet of summation indices J, M, j, m, l, and m_l. Recall that, asymptotically, wave function (14.3) is comprised of an outgoing Coulomb wave in direction $\hat{\boldsymbol{k}}$ in channel $J_2 M_2$ and incoming spherical Coulomb waves in all channels which can yield a state of total angular momentum J. When interchannel interactions can be neglected, as is usually assumed,[5] the simplification lies solely in the structure of the function $|(J_2 l j)JM\rangle$, which then asymptotically contains an outgoing and incoming spherical Coulomb wave in channel $J_2 M_2$ *only*. It can therefore be appreciated that the theoretical formulation of the angular distribution is independent of whether or not interchannel interaction is incorporated, although we stress that the *numerical* value of the quantities which direct the emission may depend strongly on this feature.

Recalling the nature of our experiment, function (14.1) can be written

$$W(\Omega_a) = \mathscr{S} \sum_{M_1 M_3} \sum_{\sigma_a \sigma_p} \sum_{M_2 M_2'} \langle J_3 M_3 \boldsymbol{k}_a \sigma_a|H_2|J_2 M_2\rangle \langle J_2 M_2 \boldsymbol{k}_p \sigma_p|H_1|J_1 M_1\rangle$$
$$\cdot \langle J_2 M_2' \boldsymbol{k}_p \sigma_p|H_1|J_1 M_1\rangle^* \langle J_3 M_3 \boldsymbol{k}_a \sigma_a|H_2|J_2 M_2'\rangle^*. \quad (14.4)$$

This "master equation" can be expressed as the trace of a product of two density matrices, each (Hermitian) matrix being concerned with either the primary or the decay event:

$$W(\Omega_a) = \sum_{M_2 M_2'} \rho_a(M_2, M_2')\, \rho_p(M_2', M_2), \quad (14.5)$$

where

$$\rho_p(M_2', M_2) = \mathscr{S}_p \sum_{M_1 \sigma_p} \langle J_2 M_2' \boldsymbol{k}_p \sigma_p|H_1|J_1 M_1\rangle^* \langle J_2 M_2 \boldsymbol{k}_p \sigma_p|H_1|J_1 M_1\rangle \quad (14.6)$$

$$\rho_a(M_2, M_2') = \mathscr{S}_a \sum_{M_3 \sigma_a} \langle J_3 M_3 \boldsymbol{k}_a \sigma_a|H_2|J_2 M_2\rangle \langle J_3 M_3 \boldsymbol{k}_a \sigma_a|H_2|J_2 M_2'\rangle^*. \quad (14.7)$$

This partitioning, which corresponds in effect to (4.10), is a very convenient one since we can discuss each matrix separately, making the necessary changes in H_1, etc. Note that at this stage, ρ_p is nondiagonal and the angular correlation does not depend solely on production amplitudes for the $J_2 M_2$ states. We should also remark that this partitioning is implied by the assumption of a two-step process in which post-collision effects[16] are neglected. While this is probably true for excitation far from threshold, it does suggest that angular distribution measurements may provide an avenue to detect the energy region in which the two-step model becomes inapplicable. The theory should then be recast in terms of a double ionization process as discussed in Sect. 4.

In the following subsections we discuss photoionization and fast-electron-impact experiments. We explore the form of the ρ_p matrices appropriate to these sources and discuss later the ρ_a matrix in Sect. 15.

α) Photoionization. Assuming that the photon beam is both monochromatic and unpolarized, and using the dipole-length formula for the interaction operator H_1, (14.6) can be written (again ignoring multiplicative constants)

$$\rho_p(M_2', M_2) = \mathscr{S} \sum_{M_1\sigma} \sum_{\mu} \langle J_2 M_2' \boldsymbol{k}\sigma | K_\mu^1 | J_1 M_1\rangle^* \langle J_2 M_2 \boldsymbol{k}\sigma | K_\mu^1 | J_1 M_1\rangle. \quad (14.8)$$

The summation over μ runs over the two states of circular polarization of the photon beam, and K_μ^1 $(\mu = \pm 1)$ is a standard component of an irreducible tensor operator of order one,[17]

$$K_1^1 = -\frac{1}{\sqrt{2}}(X + \mathrm{i}Y) \qquad K_{-1}^1 = \frac{1}{\sqrt{2}}(X - \mathrm{i}Y), \quad (14.9)$$

where for example $X = \sum_\nu x_\nu$ is the sum of the x coordinates of the electrons in the atom.

Introducing expansion (14.3) into (14.8) we obtain

$$\begin{aligned}\rho_p(M_2', M_2) = \mathscr{S} \sum_{M_1\sigma} \sum_{\mu} \sum_{KK'} a_{K'M_2'\sigma}\, a^*_{KM_2\sigma} \langle (J_2 l' j') J' M' | K_\mu^1 | J_1 M_1\rangle^* \\ \cdot \langle (J_2 l j) J M | K_\mu^1 | J_1 M_1\rangle,\end{aligned} \quad (14.10)$$

where

$$a_{KM_2\sigma} = \mathrm{i}^l\, \mathrm{e}^{-\mathrm{i}\delta_l}\, Y^*_{lm_l}(\hat{\boldsymbol{k}})\, (s l \sigma m_l | j m)(j J_2 m M_2 | J M). \quad (14.11)$$

Use of the Wigner-Eckart theorem[18] yields

$$M_2' + m_l' = M_2 + m_l. \quad (14.12)$$

Since we specified that the beam is unpolarized, ρ_p is symmetrical about the beam axis and a simple application of angular momentum theory shows that

$$\rho_p(-M_2', -M_2) = (-1)^{M_2' - M_2} \rho_p(M_2', M_2). \quad (14.13)$$

A further consequence of the unpolarized nature of the photon beam is that ρ_p should be independent of any choice of orientation of the X and Y axes.

[16] See references in I.23
[17] See, e.g., Ref. [7], Vol. II, p. 573
[18] See, e.g., Ref. [8]

Treating our axis system as having been derived from an alternative choice by rotation through an angle α about the beam axis, one obtains the same quantities as in (14.10) multiplied by a phase factor $\exp i(m_l'-m_l)\alpha$. Averaging over all the possible orientations α shows that $m_l=m_l'$ and ρ_p is *diagonal*. Using (14.13), which is a consequence of the reflection symmetry of the density matrix, shows also that these diagonal matrix elements depend only on the *magnitude* of M_2 and not on its sign. The fact that the primary electron is not observed means that we can finally specify $\mathscr{S}$ to be an integration over the angular coordinates of the electron. Thus, we can see that the non-zero elements of ρ_p are proportional to the production probabilities for creating the state J_2M_2 from the randomly oriented initial states J_1M_1. The trace of ρ_p is simply related to the *strength* of a transition as defined by CONDON and SHORTLEY [9]. As a contrast, consider the case when the light beam is polarized in the x direction thus defining a unique orientation of the coordinate system. Equation (14.8) becomes

$$\begin{aligned}\rho_p(M_2',M_2)=\mathscr{S}\sum_{M_1\sigma}\langle J_2M_2'\boldsymbol{k}\sigma|K_{-1}^1-K_1^1|J_1M_1\rangle^*\\ \cdot\langle J_2M_2\boldsymbol{k}\sigma|K_{-1}^1-K_1^1|J_1M_1\rangle.\end{aligned} \tag{14.14}$$

Equation (14.13) still holds due to the symmetry with respect to reflections in the xz plane, but ρ_p is no longer diagonal in general.

β) Ionization by high-energy electron impact. We shall restrict our discussion to electron impact, in which case the center of mass coordinate system coincides with the autoionizing system to a good approximation. For a discussion of other projectiles, see for example the work of EICHLER and FRITSCH.[8] We also consider only the high-energy region in which the first Born approximation for excitation amplitudes is a good approximation. Differences between this and the previous case of photoionization are to be noted. Firstly, a range of energies of the primary electron is obtained which depends on the energy and momentum transfer between atom and projectile. Secondly, since the scattering direction and energy of the projectile are not to be observed in our experiment, one must integrate over these quantities in $\mathscr{S}_p$.

At large incident energies, the matrix elements in (14.6) are proportional to the so-called form factor[19]

$$F(\boldsymbol{p})=\langle J_2M_2\boldsymbol{k}\sigma|\sum_{\nu=1}^{N}e^{i\boldsymbol{p}\cdot\boldsymbol{r}_\nu}|J_1M_1\rangle, \tag{14.15}$$

in which $\boldsymbol{p}=\boldsymbol{p}_i-\boldsymbol{p}_f$ is the momentum transfer during the collision. Using the Rayleigh expansion (1.4) one can write this matrix element as

$$F(\boldsymbol{p})=4\pi\sum_{\lambda\kappa}i^\lambda Y_{\lambda\kappa}^*(\hat{\boldsymbol{p}})\langle J_2M_2\boldsymbol{k}\sigma|T_\kappa^{(\lambda)}|J_1M_1\rangle, \tag{14.16}$$

in which

$$T_\kappa^{(\lambda)}=\sum_\nu j_\lambda(pr_\nu)\,Y_{\lambda\kappa}(\hat{\boldsymbol{r}}_\nu) \tag{14.17}$$

[19] See, e.g., Inokuti, M. (1971): Rev. Mod. Phys. *43*, 297

is an irreducible tensor operator of order λ. After integrating over the angles defining the direction $\hat{\boldsymbol{p}}$ of the vector $\boldsymbol{p}$, and again recognizing the rotational symmetry about the beam axis, it can be seen that ρ_p is also diagonal in this case. Equation (14.13) also holds. After integrating over the angle variables of the primary electron, $\rho_p(M_2, M_2)$ becomes proportional to the cross section for ionization giving state $J_2 M_2$. For explicit formulae see BEREZHKO and KABACHNIK.[10] The similarity to the photoionization case is enhanced by recalling that at very high impact energies and small momentum transfer, the Born approximation becomes the Bethe approximation.[19]

To summarize, the symmetry of our experiment about the beam axis for both (unpolarized) photoionization and electron impact dictates that ρ_p is diagonal and the matrix elements depend only on the magnitude of M_2. Each diagonal matrix element is related to the production probability for each substate $J_2 M_2$ referred to in the following as $P(J_2 M_2)$. Using the definition of the statistical tensor[14]

$$\rho_{\lambda\kappa}(J_2, J_2) = \sum_{M_2 M_2'} (-1)^{J_2 - M_2'} (J_2 J_2 M_2' - M_2 | \lambda\kappa)\, \rho_p(M_2', M_2), \qquad (14.18)$$

we can see that the diagonal nature of ρ_p and (14.13) allows only for $\kappa = 0$ and λ even. Thus the states $J_2 M_2$ are said to be aligned.[14] Polarization of the states (λ also odd, $\kappa \neq 0$) occurs by either using plane-polarized light or for electron impact by detecting the scattering plane of the incident particle.

15. Auger electron angular distributions

α) Noncoincidence experiments. Since we have already noted that ρ_p is diagonal, we shall restrict our discussion to the case $M_2 = M_2'$ in (14.7). According to (9.1), $H_2 = H - E$ so that the diagonal matrix elements of ρ_a are given by

$$\rho_a(M_2, M_2) = \mathscr{S} \sum_{M_3 \sigma} \left| \sum_K a^*_{K M_3 \sigma} \langle (J_3 l j) J M | H - E | J_2 M_2 \rangle \right|^2, \qquad (15.1)$$

where the coefficients $a_{K M_3 \sigma}$ are defined by (14.11) after replacement of $J_2 M_2$ by $J_3 M_3$. This expression may be simplified by using the selection rules for the Auger transition,

$$\Delta J = \Delta M = 0,$$

while parity is conserved. Thus, the sum over K in (15.1) becomes

$$\sum_{\substack{jm \\ l m_l}} (-i)^l e^{i\delta_l} Y_{l m_l}(\hat{\boldsymbol{k}}) (s l \sigma m_l | j m)(j J_3 m M | J_2 M_2) \langle (J_3 l j) J_2 M_2 | H - E | J_2 M_2 \rangle. \qquad (15.2)$$

The nonradiative transition amplitude in (15.2) is independent of M_2 and is in effect a reduced matrix element $A(J_2 \to J_3 l j)$, the evaluation of which has been discussed in Sect. 9. After some manipulation, the angular distribution function

(14.5) becomes

$$W(\theta)=\sum_{M_2M_3\sigma} P(J_2M_2)|\sum_{jl}(-\mathrm{i})^l \mathrm{e}^{\mathrm{i}\delta_l}\Theta_{l,M_2-M_3-\sigma}A(J_2\to J_3lj)$$
$$\cdot(sl\sigma M_2-M_3-\sigma|jM_2-M_3)(jJ_3M_2-M_3M_3|J_2M_2)|^2, \tag{15.3}$$

where $P(J_2M_2)$ is the production probability of the state J_2M_2 according to Sect. 14β. We have also taken into consideration that the Auger distribution is independent of the azimuthal angle and extracted the φ dependence in (15.3) from

$$Y_{lm_l}(\hat{\boldsymbol{k}})=\frac{1}{\sqrt{2\pi}}\Theta_{lm_l}(\theta)\,\mathrm{e}^{\mathrm{i}m_l\varphi}. \tag{15.4}$$

Since the θ dependence of W is constructed from products of the form

$$\Theta_{l,m_l}\Theta_{l',m_l}=\sum_{\nu}\frac{2\nu+1}{2}(-1)^l\langle l\|C^\nu\|l'\rangle\begin{pmatrix} l & l' & \nu\\ m_l & -m_l & 0\end{pmatrix}P_\nu(\cos\theta), \tag{15.5}$$

in which both functions on the right-hand side have the same parity ($l+l'$ $=$even), the distribution of Auger electrons always involves the Legendre polynomials $P_\nu(\cos\theta)$ of even order and is therefore symmetrical about $\theta=\pi/2$. Note that when $P(J_2M_2)$ is independent of M_2 (e.g., $J=1/2$), the distribution is isotropic. In general, however, $W(\theta)$ involves both production and Auger amplitudes as well as the Auger partial wave phase shifts. When only one partial wave is emitted, i.e., when $S_3=L_3$ and $J=0$, the distribution is independent of A and δ_l. In this case, (15.3) yields

$$W(\theta)=\sum_{M_2\sigma} P(J_2M_2)|\Theta_{l,M_2-\sigma}(sl\sigma M_2-\sigma|J_2M_2)|^2\,|A(J_2\to 0lJ_2)|^2. \tag{15.6}$$

CLEFF and MEHLHORN [5] point out that the different J states arising from the term S_3L_3 usually overlap and the most favorable case for experimental observations is when the final state is $^1S_0$. These authors neglected the coupling between the Auger channels and made use of pure $SLJM$ coupling in their original derivation thus permitting a decomposition of the function $|(J_3lj)J_2M_2\rangle$ into its LS-coupled constituents. This yields a very simple form for (15.6) in the case $S_3=L_3=J_3=0$, namely,

$$W(\theta)=|A(S_2L_2\to{}^1S_0l_2)|^2\sum_{M_2\sigma}P(J_2M_2)|\Theta_{L_2,M_2-\sigma}|^2(S_2L_2\sigma M_2-\sigma|J_2M_2)^2. \tag{15.7}$$

If it is preferred to retain the more general form of (14.3) in which the interchannel interaction is included, note that (15.7) does not apply in this case.

Following CLEFF and MEHLHORN,[5] we define a relative angular intensity as

$$W_{\mathrm{rel}}(\theta)=4\pi W(\theta)/W_{\mathrm{tot}}, \tag{15.8}$$

where W_{tot} is the total intensity of Auger electrons ejected into a solid angle of 4π. According to (15.3) and (15.5), this relative distribution can be written

$$W_{rel}(\theta)=\sum_{\mu} A_{2\mu} P_{2\mu}(\cos\theta); \qquad A_0=1, \tag{15.9}$$

where $\mu \leqq J_2$ and where the anisotropy coefficients $A_{2\mu}$ depend in general on the production probabilities, nonradiative transition amplitudes, and the phase shifts. For the special case (15.7) the anisotropy coefficients have been evaluated by CLEFF and MEHLHORN for the important cases when the hole is introduced into a closed shell:

$$\begin{aligned}
&\left.\begin{aligned} &S_2=\tfrac{1}{2},\ L_2=1,\ J_2=\tfrac{3}{2}\\ &S_2=\tfrac{1}{2},\ L_2=2,\ J_2=\tfrac{3}{2}\end{aligned}\right\} \quad A_2=\frac{P(\frac{3}{2}\frac{1}{2})-P(\frac{3}{2}\frac{3}{2})}{P(\frac{3}{2}\frac{1}{2})+P(\frac{3}{2}\frac{3}{2})}\\
&S_2=\tfrac{1}{2},\ L_2=2,\ J_2=\tfrac{5}{2} \qquad A_2=\tfrac{2}{7}\,\frac{4P(\frac{5}{2}\frac{1}{2})+P(\frac{5}{2}\frac{3}{2})-5P(\frac{5}{2}\frac{5}{2})}{P(\frac{5}{2}\frac{1}{2})+P(\frac{5}{2}\frac{3}{2})+P(\frac{5}{2}\frac{5}{2})}\\
&\qquad\qquad\qquad\qquad\qquad\quad A_4=\tfrac{3}{7}\,\frac{2P(\frac{5}{2}\frac{1}{2})-3P(\frac{5}{2}\frac{3}{2})+P(\frac{5}{2}\frac{5}{2})}{P(\frac{5}{2}\frac{1}{2})+P(\frac{5}{2}\frac{3}{2})+P(\frac{5}{2}\frac{5}{2})},
\end{aligned} \tag{15.10}$$

where $P(J_2 M_2)$ can be related to the production probabilities $P(S_2 L_2 M_{S_2} M_{L_2})$ in LS coupling through

$$P(J_2 M_2)=\sum_{M_{S_2} M_{L_2}} (S_2 L_2 M_{S_2} M_{L_2}|J_2 M_2)^2\, P(S_2 L_2 M_{S_2} M_{L_2}) \tag{15.11}$$

provided the spin-orbit interaction is neglected in the description of the ionization process.

An examination of (15.10) shows that the directing factor in the angular distribution for transitions to $^1S_0$ states is the ratio of production probabilities. Departures of this quantity from unity will result in an anisotropy in the angular distribution. These ratios depend on the excitation energy and on the mode as do the anisotropy coefficients.

Until now it has been assumed that the intermediate hole state b in (14.1) is nondegenerate with respect to J_2. In the LS-coupling limit an incoherent sum over $b=J_2$ enters in addition to $\beta=M_2$ in (14.1). After reexpressing (14.3) in LS coupling and after some summations[20] the angular distribution function (14.5) becomes

$$\begin{aligned} W(\theta)=\sum_{M_{L_2} M_{L_3}} & P(S_2 L_2 M_{L_2}) \Big|\sum_l (-\mathrm{i})^l \mathrm{e}^{\mathrm{i}\delta_l}\, \Theta_{l, M_{L_2}-M_{L_3}}(\theta)\\ &\cdot A(S_2 L_2 \to (S_3 L_3) s l)(L_3 l M_{L_3} M_{L_2}-M_{L_3}|L_2 M_{L_2})\Big|^2, \end{aligned} \tag{15.12}$$

where it has been assumed that the production probability $P(S_2 L_2 M_{L_2})$ is independent of M_{S_2}. The result (15.12), which is analogous to (15.3), may be

[20] Mehlhorn, W. (1978): Lecture notes on electron spectroscopy of Auger and autoionizing states: experiment and theory (University of Aarhus, unpublished)

subject to the same type of simplifications that resulted in (15.6) and (15.7). For instance, if $L_3=0$, in which case $l=L_2$, and if the interchannel interaction is neglected, (15.12) becomes

$$W(\theta)=|A(S_2L_2\to(S0)sl)|^2\sum_{M_L}P(S_2L_2M_{L_2})|\Theta_{lM_L}(\theta)|^2. \tag{15.13}$$

Written in the form (15.9), the anisotropy coefficients $A_{2\mu}$ of the relative distribution (15.8) are, in some important cases, where S_2 is arbitrary ($S_3=S_2\pm 1/2$) but L_2 fixed,

$$\begin{aligned} L_2=1 \quad & A_2=2\frac{P(10)-P(11)}{P(10)+2P(11)} \\ L_2=2 \quad & A_2=\frac{10}{7}\frac{P(20)+P(21)-2P(22)}{P(20)+2P(21)+2P(22)} \\ & A_4=\frac{6}{7}\frac{3P(20)-4P(21)+P(22)}{P(20)+2P(21)+2P(22)}. \end{aligned} \tag{15.14}$$

Here the production probabilities $P(S_2L_2M_{L_2})$ have been denoted by $P(L_2M_{L_2})$. Note that the only difference between (15.14) and (15.10) after the substitution of (15.11) is in the coefficients of the ratios given above.

Equation (15.3) is valid provided the hole-state level $S_2L_2J_2$ is well separated from any other level of the same symmetry. This may even be true in the intermediate coupling regime, in which case the general structure of wave function (14.3) assures that (15.3) is still valid. Equation (15.12) is valid in the extreme case where all levels pertaining to S_2L_2 are degenerate but well separated from any other levels. However, as stressed by MEHLHORN,[20] none of these equations comprises the case where the levels overlap, i.e., the separations are comparable to the halfwidths of the corresponding states. Within the framework of the scattering approach of Chapt. II, a proper treatment would require a generalization of the theory to the case[21] of overlapping levels interacting with several continua. Another possibility would be to use a time-dependent perturbation approach in analogy to the work of PERCIVAL and SEATON[22] on polarization of light emitted by collisionally aligned atoms.[20] Recently MEHLHORN and TAULBJERG[23] have formulated the problem along these lines using the density matrix formulation and derived a formula for the angular distribution of autoionizing electrons in the case of an arbitrary ratio between the fine-structure splitting and halfwidth of the autoionizing levels. Their work has been generalized by BRUCH and KLAR[23] who also considered the influence of a partially unresolved hyperfine structure on the angular distribution pattern.

Coupling to outer open shells may also cause splitting. Provided this or any other splitting is small compared to the halfwidths of the states involved, it

[21] See references in I.19

[22] Percival, I.C., Seaton, M.J. (1958): Phil. Trans. Soc. A *251*, 113

[23] Mehlhorn, W., Taulbjerg, K. (1980): J. Phys. B *13*, 445; Bruch, R., Klar, H. (1980): J. Phys. B *13*, 1363 and 2885

will have no influence on the angular dependence since any extra quantum numbers will automatically be included in the summation $\mathscr{S}$ in (14.1) and averaged out.

β) Some features of noncoincidence angular distributions. Angular distributions of Auger electrons have been measured as a function of energy following electron impact.[2, 4, 6, 13, 24] Also heavier projectiles[24, 25] have been used in a few experiments which include some studies of autoionizing transitions as a function of angle.[26, 27] In all experiments the final state has been a $^1S_0$ state, whereas the initial hole has been either a $2p_{3/2}$(Ar, Mg) or a $3d_{3/2,5/2}$(Kr) hole in Auger experiments. The alignment parameter of interest is A_2 in these experiments, A_4 enters only in the case of $3d_{5/2}$.

The common characteristics of A_2, which have also been obtained from (15.10) and (15.11) using hydrogenic or Hartree-Slater wave functions in the first Born approximation,[9, 10] can be summarized as follows. (i) A_2 is rather small (<0.1) except possibly close to the threshold. This differs from what is found in autoionization. For example, the transition $2p^5 3s^2\,{}^2P - 2p^6\,{}^1S_0 \varepsilon p\,{}^2P$ is strongly anisotropic in sodium.[26, 29] (ii) There is a sign reversal of A_2 at about 10 to 100 times the threshold energy which can be expected on general grounds.[28] Additional reversals of sign at higher and lower energies should also occur. (iii) The behavior of A_2 as a function of energy seems to be predicted reasonably well by Hartree-Slater calculations in the region, where the first Born approximation is expected to be valid.[13, 24] The hydrogenic approximation is in general very poor. Besides that, recall that the angular distribution for transitions to $J_3>0$ depends on partial wave phase shifts which the hydrogenic approximation is not able to yield.

There are no measurements of the angular distribution of Auger electrons in photoionization although the alignment of Cd atoms has been observed in one case by measuring the polarization of the resulting photon emission.[30] The interpretation of the results would be more clear-cut, in particular close to the threshold, since the photon-electron interaction is well understood and the alignment parameters can be related to the wave functions more easily than in the case of electron impact. There exist some model calculations which show effects of Cooper minima and potential barriers on the alignment parame-

---

[24] Rødbro, M., DuBois, R., Schmidt, V. (1978): J. Phys. B *11*, L 551

[25] Polarization of x-rays has been studied in transitions from $1s^{-1}2p^{-1}\,{}^1P_1$ states to $2p^{-2}\,{}^1S_0$ and $2s^{-2}\,{}^1S_0$ states in ion-atom collisions. See, Jamison, K., Richard, P. (1977): Phys. Rev. Lett. *38*, 484; Jamison, K., Richard, P., Hopkins, F., Matthews, D.L. (1978): Phys. Rev. A *17*, 1642; Jamison, K., Newcomb, J., Hall, J.M., Schmiedekamp, C., Richard, P. (1978): Phys. Rev. Lett. *41*, 1112

[26] Breuckmann, E., Breuckmann, B., Mehlhorn, W., Schmitz, W. (1978): J. Phys. B *10*, 3135

[27] Bordenave-Montesquieu, A., Benoit-Cattin, P., Gleizes, A., Merchez, H. (1975): J. Phys. B *8*, L 350; Edwards, A.K. (1975): Phys. Rev. A *12*, 1830; (1976): A *13*, 1654; Ziem, P., Morgenstern, R., Stolterfoht, N. (1978): Abstr. of X Int. conf. phys. electr. atom. collisions, Paris, Vol. 2, p. 1002; Kessel, Q.C., Morgenstern, R., Müller, B., Niehaus, A., Thielmann, U. (1978): Phys. Rev. Lett. *40*, 645

[28] Fano, U., Macek, J.H. (1973): Rev. Mod. Phys. *45*, 553

[29] Thedosiou, C.E. (1977): Phys. Rev. A *16*, 2232

[30] Caldwell, C.D., Zare, R.N. (1977): Phys. Rev. A *16*, 255

ters.[3, 11] However, no explicit connection seems to have been established to the Fano-Dill theory[31] of angular distributions of photoejected electrons in terms of an angular momentum transfer basis. Nor have relativistic effects[32] been discussed very much, although they influence in certain cases the angular distribution of photoelectrons drastically.[33] Such investigations would be of value in light of the experimental possibilities offered by the synchrotron radiation of variable energy.

γ) Coincidence experiments. Overall, it appears that the anisotropy coefficient (or equivalently, the alignment of the intermediate hole states) seems to be rather small in noncoincidence experiments, although in the case of electron excitation the following autoionization transition may be strongly anisotropic.[26] Hence MCFARLANE[7] and BEREZHKO et al.[12] have suggested coincidence experiments which could be employed to enhance the anisotropic behavior of the Auger electron distribution. Similar ideas have been put forward by BALASHOV et al.[34] in the case of autoionization.

In a photoionization experiment, the angular distributions of the primary ejected electron and the Auger electron could be detected in coincidence.[7] Using the formalism which we have described previously, we can analyse this situation. When the photons are unpolarized, cylindrical symmetry is retained and ρ_p is still diagonal. Since θ_p is now to be observed, we do not integrate over θ as we did in Sect. 14α. One obtains in this case

$$\rho_p(M_2, M_2; \theta_p) = \sum_{M_1\sigma} \sum_{\mu} |\sum_{K} a^*_{KM_2\sigma} \langle (J_2 lj) JM | K^1_\mu | J_1 M_1 \rangle|^2. \tag{15.15}$$

This quantity is proportional to the differential photoionization cross section for ejection of an electron at an angle θ_p to the beam on making a transition to $J_2 M_2$ from a randomly oriented initial state. The trace $\rho_p(\theta_p)$ gives the angular distribution of photoelectrons as discussed by, e.g., FANO and DILL[31]. Note that since $l+l'$ is even the distribution is symmetrical about $\theta_p = \pi/2$. The correlation function in which we are interested, $W(\theta_p, \theta_a)$, takes exactly the same form as (15.3) after substituting $P(J_2 M_2; \theta_p)$ for $P(J_2 M_2)$. Further, the anisotropy coefficients appearing in (15.9) and (15.10) are of the same form but become θ-dependent quantities.

In an electron impact experiment the inelastically scattered electron could be detected at a fixed angle in coincidence with an Auger electron.[12] Although ρ_p is in general nondiagonal in a frame in which the beam axis is chosen to be the quantization axis, some simplifications occur for ionization by high-energy electron impact for which (14.15) is valid. It can be shown that the statistical

---

[31] Fano, U., Dill, D. (1972): Phys. Rev. A *6*, 185; Dehmer, J.L., Dill, D. (1979) in: Electron and photon molecule collisions, McKoy, V., Rescigno, T., Schneider, B. (eds.), New York: Plenum Press

[32] Oh, S.D., Pratt, R.H. (1974): Phys. Rev. A *10*, 1198

[33] Tseng, H.K., Pratt, R.H., Yu, S., Ron, A. (1978): Phys. Rev. A *17*, 1061; Johnson, W.R., Cheng, K.T. (1978): Phys. Rev. Lett. *40*, 1167

[34] Balashov, V.V., Lipovetsky, S.S., Senashenko, V.S. (1972): Phys. Lett. *39* A, 103; Balashov, V.V., Grum-Grzhimailo, A.N., Kabachnik, N.M., Magunov, A.I., Strakhova, S.I. (1978): Phys. Lett. *67* A, 266

tensors (14.18) are diagonal (i.e., $\kappa=0$) with respect to a frame in which the axis, of quantization coincides with the momentum transfer $\boldsymbol{p}$.[12] According to (1.20), the relationship between the two sets of statistical tensors is given by

$$\rho_{\lambda\kappa}^{\boldsymbol{p}_i}(J_2,J_2)=\rho_{\lambda 0}^{\boldsymbol{p}}(J_2,J_2)\,Y_{\lambda\kappa}^{*}(\theta_p,0)\sqrt{\frac{4\pi}{2\lambda+1}}, \qquad (15.16)$$

where θ_p is the angle between $\boldsymbol{p}$ and $\boldsymbol{p}_i$ and where λ is even. Thus the angular distribution of the Auger electrons is rotational symmetric with respect to the direction of $\boldsymbol{p}$ and symmetric to a plane perpendicular to $\boldsymbol{p}$. Any deviations from this symmetry would indicate a failure of the first Born approximation. Instead of the distribution function (15.9), one obtains

$$W_{\text{rel}}(\theta_p,\theta_a)=\sum_{\mu}\sum_{\kappa=-2\mu}^{+2\mu} B_{2\mu}(p,\mathcal{E})\,P_{2\mu}^{|\kappa|}(\cos\theta_p)\,Y_{2\mu,\kappa}(\theta_a,\varphi_a), \qquad B_0=1, \quad (15.17)$$

where θ_a is the angle between the wave vector $\boldsymbol{p}_a$ of the Auger electron and $\boldsymbol{p}_i$, and φ_a is the angle between the plane of reaction (defined by $\boldsymbol{p}_i$ and $\boldsymbol{p}_f$) and the plane of emission (defined by $\boldsymbol{p}_i$ and $\boldsymbol{p}_a$). The coefficients $B_{2\mu}(p,\mathcal{E})$ depend in general, in addition to the nonradiative transition matrix elements, on products of reduced matrix elements of operator (14.17) which are functions of both the magnitude of the momentum transfer p and the kinetic energy $\mathcal{E}$ of the ejected electron in the primary process. In (15.17) $\mu \leqq J_2$. For explicit formulae, see the work of BEREZHKO et al.[12]

For zero-energy photoelectrons MCFARLANE[7] showed using the hydrogenic approximation that $A_2(\theta)$ for a $2p_{3/2}$ hole state in a closed shell has two features. (i) $A_2(\theta)$ varies rapidly with θ, the angle of photoejection, and could attain large values ($\simeq 0.5$). The small values for the anisotropy coefficient in the decay of $2p_{3/2}$ holes at certain θ values are suggested to arise from the cancellation of regions where $P(2p_0,\theta)<P(2p_{\pm 1},\theta)$ and $P(2p_0,\theta)>P(2p_{\pm 1},\theta)$. (ii) The variation of $P(J_2M_2,\theta)$ is found to be sensitive to the phase shift difference $\delta_{l+1}-\delta_{l-1}$ of the two partial waves of the photoelectron. For electron impact BEREZHKO et al.[12] showed that the anisotropy is considerable for the decay of $2p$ and $3d$ holes if the Auger electron is detected in coincidence with the scattered electron. They found also large differences between Hartree-Slater and hydrogenic results.

V. The calculation of nonradiative transition energies

As pointed out in Chap. II, an inherent faith in the shell structure of the atom is the prominent feature of Auger electron spectroscopy. Indeed, the nonrelativistic frozen-core central field model is heavily embedded in the theory and is kept alive in the nomenclature for such transitions as was discussed in Sect. 7. The following three sections present a discussion of the problems and successes in the calculation of nonradiative transition energies which goes beyond this model. We shall concentrate our attention on aspects that are not included in the earlier reviews[1] while still providing an overview of the available techniques.

The probability concepts introduced in Sects. 9–12 are based on the notion of initial and final hole states. In Sect. 16 we shall examine hole-state and binding energies in general. Factors not considered by the nonrelativistic frozen-core central field model but which must be incorporated in rigorous calculations of nonradiative transition energies are introduced. In Sects. 17 and 18 we shall trace the development of the calculation of nonradiative transition energies in free atoms and ions with respect to the various improvements discussed in Sect. 16. We shall follow the two major avenues in this development, namely, the semiempirical and the "ab initio" approach.

16. General aspects of hole-state and binding energies

α) The level shift. As pointed out in Sect. 3 the autoionizing state or hole state is approximately described within Feshbach's projection operator formalism as a discrete eigenfunction of the projected Hamiltonian operator $QHQ\,(=H_Q)$:

$$H_Q(Q\psi)=E_0(Q\psi). \tag{16.1}$$

The projection operator Q is defined in Appendix B. In the FANO formalism the wave function $Q\psi$ should be identified as the discrete state Φ interacting with the continuum. Although the choice of the projection operator Q in (3.11) and consequently in (16.1) is not unique, the location of the resonance energy

$$E_r=E_0+\Delta \tag{16.2}$$

is unique despite the arbitrariness involved in the calculation of E_0 and Δ. The energy-dependent level shift Δ which is given by (3.8) and accounts for the interaction between the discrete state and continuum may be positive or negative. Since the nature and choice of the channels are determined by the scattering process, Δ depends in principle also on the type of scattering. In the two-step approximation (Sect. 4) this dependence, as well as the dependence on energy, is neglected. According to Sect. 5, the shift due to the photon continuum is included in the quantum-electrodynamical effects.

As indicated by (4.10) Δ is interlinked with the partial widths (4.11) which describe the decay of the ionized atom. Hence, it may very well be that the optimum choice of Q with respect to the energy (for methods, see footnote I.18) which results in a good estimate of E_{res} by E_0 does not necessarily provide accurate partial widths. This must be kept in mind in the discussion of the level shift.

It is well known that if the hole state is well isolated from other hole states of the same symmetry, then the ΔE_{SCF} method (Sect. 8δ) generally provides a

[1] See Introduction

remarkably good estimate of the energy after the inclusion of relativistic and quantum-electrodynamical effects. Yet, this method corresponds to an approximate Q space which consists only of a single basis element.[2] With regard to energy, improvements of this trial Q space would amount to estimates of the correlation energy using, for example, configuration-interaction calculations which are limited to configurations containing the same hole as the Hartree-Fock configuration. Any residual discrepancy between the results of such an analysis and experiments includes a level shift. On the basis of this procedure an estimate of $|\Delta| < 0.2\,\mathrm{eV}$ has been suggested for the Ne $1s^{-1}$ hole state.[3] According to DESLATTES et al.,[4] the difference between calculated $K\alpha_1$ transition energies and experimental data increases with the atomic number in a linear fashion, being still less than 10 eV for heavy elements. The calculated values do not include the correlation energy. Nevertheless, the $1s^{-1}$ and $2p^{-1}$ level shifts could also contribute to the systematic deviation between theory and experiment. It would also be possible to obtain an estimate of Δ from (3.8) by using the technique[5] developed for the evaluation of final-state channel interaction matrix elements.

According to (3.8), the absolute value of the level shift is expected to increase as the width of the resonance state increases, although strict proportionality[6] is not evident. If in addition the hole state is not well isolated from other hole states of the same symmetry, peculiar problems arise and the ΔE_{SCF} energy ceases from being a good zeroth-order approximation. For example, as shown by WENDIN and OHNO[7] the $4p_{1/2}$ hole state in xenon is strongly coupled to nearly degenerate hole states and continua which are associated with two $4d$ vacancies. As a consequence the $4p_{1/2}$ line in the photoelectron spectrum of xenon is completely smeared out and in this sense the concept of the $4p_{1/2}$ hole becomes meaningless. Additional complications occur if there is also a strong interference between the indirect and direct ionization channels as in the case of the manganese $3p$ photoabsorption. As shown by DAVIS and FELDKAMP[8] there is an interference between two processes, namely, $3p^6\,3d^5\,4s^2 + \hbar\omega \rightarrow 3p^5\,3d^6\,4s^2 \rightarrow 3p^6\,3d^4\,4s^2\,\varepsilon f$ and $3p^6\,3d^5\,4s^2 + \hbar\omega \rightarrow 3p^6\,3d^4\,4s^2\,\varepsilon f$. In addition the $3p$ spin-orbit coupling between the adjacent ^6P and ^6D states of the $3p^5\,3d^6\,4s^2$ configuration must be taken into account. In both examples the origin of the large broadening can be associated with Super Coster-Kronig transitions, namely, with N_2-$N_{45}N_{45}$ and M_{23}-$M_{45}M_{45}$ transitions, respectively.

β) Variational collapse. The procedures used to determine E_0 in (16.2) and the energies of the residual ions necessitate some realization of H_Q. While both

---

[2] Russek, A. et al. (I.18)

[3] Moser, C.M., Nesbet, R.K., Verhaegen, G. (1971): Chem. Phys. Lett. *12*, 230

[4] Deslattes, R.D., Kessler, Jr., E.G., Jacobs, L., Schwitz, W. (1979): Phys. Lett. *71* A, 411

[5] See II.48. This procedure was recently used by CHEN, CRASEMANN, and MARK (1981): (Phys. Rev. A *24*, 1158) in their evaluation of $2s$ level shifts for $30 \leqq Z \leqq 47$

[6] Nicolaides, C.A. (1972): Phys. Rev. A *6*, 2078

[7] Wendin, G., Ohno, M. (1976): Phys. Scripta *14*, 148

[8] Davis, L.C., Feldkamp, L.A. (1978): Phys. Rev. A *17*, 2012

variational and perturbation techniques[9] are employed without an explicit construction of Q, it is important to recognize that such highly excited species as hole states present a formal difficulty well known in connection with excited but bound states, that of variational collapse.

For a state lowest in energy of its symmetry type, the expectation value of the Hamiltonian provides an upper bound to the exact energy. Should one desire that a wave function describing an excited state also yield an upper bound, it is necessary that the trial function is orthogonal to *all* exact eigenfunctions of lower energy and same symmetry type. HYLLERAAS and UNDHEIM[10] and McDONALD[11] showed that the roots of the secular equations involving trial functions for all these states also provide upper bounds, i.e., one requires no knowledge of the *exact* wave functions for the lower states. In practice this means that a set $\{\Phi_k\}$ of wave functions which approximate the M *first* eigenfunctions has the upper-bound property if the condition of noninteracting matrix elements, i.e., $\langle\Phi_k|\Phi_l\rangle=\delta_{kl}$, $\langle\Phi_k|H|\Phi_l\rangle=E_k\delta_{kl}$, is satisfied. Since each energy eigenvalue is usually degenerate, the corresponding approximate wave functions Φ_k must have the same symmetry as the corresponding exact wave fuctions.

Without the upper-bound property there is a danger of crossing to an excitation from a configuration of lower energy and so obtain an energy *below* the true position. For example, the $1s^2\,2s\,2p^5\,{}^3P_0$ state of oxygen lies above two other ${}^3P_0$ states, viz., $1s^2\,2s^2\,2p^3\,3s$ and $1s^2\,2s^2\,2p^3\,3d$. In this case it is easy to obtain a strict upper bound by performing a configuration-interaction (CI) calculation among these three states (at least).[12] However, for the $1s^2\,2p^2\,{}^1S$ state of Be, an *infinite* number of $1s^2\,2sns\,{}^1S$ states lie below it, as well as the quasi-degenerate ground state, and one is forced to neglect some of these lower states in a CI calculation.[12] This is typical of the case in which we are interested, a deep hole state lies above an infinite number of states of the same symmetry as well as the continuum threshold. Thus, one can see that the problem becomes acute since the calculation of hole-state energies is necessarily based on the choice of some approximate Q. For example, the use of the single-configuration ΔE_{SCF} procedure is now common in such calculations.

In order to explore this problem further, consider the set of hole-state wave functions $\{\Phi_k\}$ of the same symmetry which can be constructed from a given set of orthonormal orbitals. To obtain an approximate upper bound to any hole-state energy, one solves the secular equation of the linear equations,

$$\begin{vmatrix} H_{11} & \boldsymbol{H}_{1A} \\ \boldsymbol{H}_{A1} & \boldsymbol{H}_{AA} \end{vmatrix} \begin{vmatrix} C_1 \\ \boldsymbol{C}_A \end{vmatrix} = E \begin{vmatrix} C_1 \\ \boldsymbol{C}_A \end{vmatrix}, \tag{16.3}$$

where the hole state of interest is labelled by subscript 1. Solving for the column vector $\boldsymbol{C}_A$ of coefficients, one obtains

$$\boldsymbol{C}_A = -(\boldsymbol{H}_{AA} - E)^{-1}\,\boldsymbol{H}_{A1}\,C_1. \tag{16.4}$$

---

[9] See references listed in I.18

[10] Hylleraas, E.A., Undheim, B. (1930): Z. Phys. *65*, 759

[11] MacDonald, J.K.L. (1933): Phys. Rev. *43*, 830

[12] Luken, W.L., Sinanoğlu, O. (1976): A. Data. Nucl. Data. Tables *18*, 525

In lowest order the coefficients associated with function k and the energy are

$$C_k \simeq \frac{H_{k1}}{H_{11} - H_{kk}} C_1 \tag{16.5}$$

$$E \simeq H_{11} - \sum_k \frac{|H_{k1}|^2}{H_{kk} - H_{11}}. \tag{16.6}$$

We conclude that the degree of mixing among the hole states depends on the *separation* between the hole states as well as the matrix element H_{k1}. Two states disparate in energy are not expected to mix strongly in lowest order, and one can assume that the first approximation to the energy, H_{11}, will be altered but little. The near degeneracy would on the other hand produce variational problems as pointed out by NICOLAIDES.[6]

If the orbital basis is obtained from a UHF calculation on the hole state 1, then Brillouin's theorem shows according to Sect. 8δ that H_{k1} is zero for single excitations. Among others, HSU et al.[13] have shown that this is also true for doublet hole states obtained from closed-shell configurations when each determinant Φ_k, is constructed as a single excitation from the optimal occupied orbitals obtained by way of Roothaan's analytical SCF procedure for Φ_1. For instance, in argon the $(1s)^{-1}$ hole state does not interact with $(2s)^{-1}$ or $(3s)^{-1}$ when optimal $(1s)^{-1}$ orbitals are employed. However, there will be matrix elements with doubly excited configurations so that the upper-bound property is not a strict one.

When the set $\{\Phi_k\}$ is not mutually orthogonal, consisting, for example, of separately optimized functions, one observes similar behavior to (16.5):

$$C_k \simeq \left[\frac{H_{k1} - H_{kk} S_{k1}}{E - H_{kk}} - S_{k1}\right] C_1, \tag{16.7}$$

where $S_{k1} = \langle \Phi_k | \Phi_1 \rangle$. Note that the second term projects out part of the contribution of Φ_k to Φ_1 in lowest order. From calculations on binding energies (BE) for neon and argon and isoelectronic ions BAGUS[14] found that the overlap between the various hole states was sufficiently small ($|S_{k1}| < 0.01$) to suggest that any change in the binding energy by diagonalizing the Hamiltonian interaction matrix is also likely to be small.

The condition of noninteracting matrix elements with *all* lower states is a strong one and difficult to implement in practice, even for two-electron systems. Hence a weaker condition, that of approximate orthogonality to lower states, is imposed.[6] Such constraints will raise the calculated energy of the state and so, in a sense, "buoy" up the state. For example, the calculations of O'MALLEY and GELTMAN[15] on two-electron systems may be looked upon from this point of view. They evaluated energies of autoionizing states below the $n = 2$ hydrogenic level by a variational treatment of the expectation value of the

[13] Hsu, H., Davidson, E.R., Pitzer, R.M. (1976): J. Chem. Phys. *65*, 609
[14] Bagus, P.S. (1965): Phys. Rev. *139*, A 619
[15] O'Malley, T.F., Geltman, S. (1965): Phys. Rev. *137*, A1344

Hamiltonian QHQ. Here $Q=Q_1Q_2$ with $Q_\nu=1-|1s(\nu)\rangle\langle 1s(\nu)|$ ($\nu=1,2$ refer to the radius vectors $\boldsymbol{r}_1$ and $\boldsymbol{r}_2$, respectively) so that the excited states contain no contribution from configurations with a $1s$ hydrogenic orbital. This choice of Q also ensures that $P\psi$ ($P=1-Q$) has the correct asymptotic behavior given by (B.7) in Appendix B.

The phenomenon of variational collapse does not seem to have arisen in theoretical studies of spectra associated with deep inner-shell hole states although in a relativistic treatment not even the ground-state energy has a lower bound due to the negative-energy continuum. For example, relativistic ΔE_{SCF} calculations including quantum-electrodynamical corrections can predict K and L_3 BE's within 10–15 eV and 5 eV, respectively, for $90 \lesssim Z \lesssim 100$.[4,16] Only for outer shells does the relative accuracy fall, which is attributed to correlation effects. In fact, the relativistic calculations of HUANG et al.[17] agree well throughout the periodic table except for some cases (a few binding energies are shown in Table 2). These observations may perhaps be attributed to the strong central field character of inner domains of the atom and by a judicious choice of orbitals. That is, one imposes strong restrictions on the nature of the orbitals to be selected: orbital occupany defining the configuration, angular momenta, and the nodal structure of the orbitals.[6] Moreover, autoionizing hole states of the same symmetry are usually well separated while we expect a significant danger of variational collapse only when near degeneracy occurs. Thus, in particular for the energies of multiple-vacancy atoms one should keep this in mind.

γ) Perturbation analysis of binding energies. As PICKUP and GOSCINSKI[18] have forcibly demonstrated, the binding energy of an electron in orbital k of a closed-shell atom is given at lowest order with respect to a UHF basis by

$$-\mathrm{BE}_k=\varepsilon_k+\sum_{i,b}\frac{|\langle bk\|ik\rangle|^2}{\varepsilon_i-\varepsilon_b}+\frac{1}{2}\sum_{\substack{i\neq k\\ a,b}}\frac{|\langle ki\|ab\rangle|^2}{\varepsilon_k+\varepsilon_i-\varepsilon_a-\varepsilon_b}+\frac{1}{2}\sum_{\substack{j,i\neq k\\ a}}\frac{|\langle ka\|ij\rangle|^2}{\varepsilon_k+\varepsilon_a-\varepsilon_j-\varepsilon_i},\tag{16.8}$$

where ε_k is the UHF eigenvalue in (8.42). Hence $-\varepsilon_k$ is the binding energy according to Koopmans' theorem (see Sect. 8δ). In the correction terms to ε_k the notation $\langle ab\|cd\rangle$ is defined by (8.5) and i, j, k indicate occupied spin orbitals whereas a, b refer to the unoccupied (virtual) spin orbitals. The first summation represents the response of the passive electrons to the removal of the electron in orbital k, i.e. orbital reorganization or relaxation energy. The second and third terms are to be associated with the pair correlations lost upon ionization and the change in passive pair correlations, respectively. The first two summations have opposite signs and tend to cancel when k is an outer-shell electron, i.e., the region in which Koopmans' theorem is a good approximation. As the hole becomes progressively deeper, the negative relax-

---

[16] Krause, M.O., Nestor, C.W. (1977): Phys. Scr. *16*, 285

[17] Huang, K., Aoyagi, M., Chen, M.H., Crasemann, B., Mark, H. (1976): A. Data. Nucl. Tables *18*, 243

[18] Pickup, B.T., Goscinski, O. (1973): Mol. Phys. *26*, 1013

Table 2. Contributions to the relativistically calculated binding energy (eV) for $1s_{1/2}$, $2s_{1/2}$ and $2p_{1/2}$ hole states as calculated by HUANG et al.[a] E_f: Koopmans' binding energy[b]; E_R: relaxation energy[c]; $E_{M,ret}$: magnetic and retardation; E_{VP}: vacuum polarization; E_{self}: screened self energy; E: total binding energy; E_{exp}: experimental binding energy[d]

		$1s_{1/2}$	$2s_{1/2}$	$2p_{1/2}$
Ar	E_f	3,241.6	337.74	262.10
	E_R	−32.9	−10.89	−11.96
	E_M	− 2.4	− 0.10	− 0.21
	E_{ret}	0.1	< 0.01	0.02
	E_{VP}	0.1	0.01	< 0.01
	E_{self}	− 1.1	− 0.09	< 0.01
	E	3,205.4	326.66	249.95
	E_{exp}	3,205.9 (5)	326.3 (1)	250.56 (7)
Kr	E_f	14,413.6	1,961.4	1,765.3
	E_R	−54.9	−28.0	−30.5
	E_M	−23.7	− 1.7	− 3.3
	E_{ret}	1.6	0.1	0.4
	E_{VP}	1.3	0.1	< 0.1
	E_{self}	−12.1	− 1.3	< 0.1
	E	14,325.8	1,930.5	1,731.9
	E_{exp}	14,327.2 (5)	1,924.6 (8)	1,730.9 (5)
Xe	E_f	34,756.3	5,509.4	5,161.5
	E_R	−69.6	−37.7	−41.5
	E_M	−88.9	− 8.3	−14.9
	E_{ret}	6.5	0.6	1.7
	E_{VP}	6.9	0.8	< 0.1
	E_{self}	−48.6	− 6.1	− 0.1
	E	34,562.5	5,458.7	5,106.7
	E_{exp}		5,453.2 (4)	5,107.2 (4)
Rn	E_f	99,082	18,199	17,479
	E_R	−100	−60	−66
	E_M	−420	−49	−84
	E_{ret}	31	4	9
	E_{VP}	60	9	1
	E_{self}	−257	−40	− 5
	E	98,395	18,063	17,334

[a] See V. 17. According to a recent refinement of CHEN et al. (M.H. CHEN, B. CRASEMANN, M. AOYAGI, K.N. HUANG, and H. MARK, to be published) the $1s_{1/2}$ binding energy of radon should be increased by 6 eV and the $2s_{1/2}$ energy by 1 eV. These shifts are due to the employment of a Fermi (rather than uniform) distribution of nuclear charge, of an orbital-energy dependent Breit interaction, and of a screened self-energy correction

[b] From Desclaux, J.P. (1973): At. Data Nucl. Tables *12*, 311

[c] Obtained by combining the results of HUANG et al. (see V. 17) and DESCLAUX

[d] The experimental binding energies are from V. 22 except the krypton 1*s* value which is from M. BREINIG et al. (Stanford Synchr. Rad. Lab. Rep. 79/03, 1979)

ation term dominates. Indeed, it has long been observed that the orbital reorganization effects play a major role on inner-shell ionization phenomena.[19] Working within the UHF formalism, HEDIN and JOHANSSON[20] demonstrated

[19] Kennard, E.H., Ramberg, E. (1934): Phys. Rev. *46*, 1034

[20] Hedin, L., Johansson, A. (1969); J. Phys. B *2*, 1336

that the relaxation energy E_{R} for ionization from orbital k is given to a good approximation by the expectation value of a "polarization operator" $V_{\mathrm{p}}(k)$:

$$E_{\mathrm{R}}^{k}=-\tfrac{1}{2}\langle k|\,V_{\mathrm{p}}(k)\,|k\rangle=-\tfrac{1}{2}\sum_{k\neq i}(\langle i\tilde{k}\|i\tilde{k}\rangle-\langle ik\|ik\rangle), \tag{16.9}$$

a relationship which has also been investigated by PICKUP and GOSCINSKI.[18] The tilde denotes the relaxed orbitals of the final state. From pilot calculations on Na, Na^{+}, K, and K^{+} HEDIN and JOHANSSON made the important observation that for closed-shell ions the shell-by-shell contribution to the relaxation energy goes in the order

$$\text{inner} \ll \text{intra shell} < \text{outer}.$$

Physical intuition also promptes one to expect the more loosely bound outer orbitals to respond more readily to the deshielding of the nucleus by removal of an inner electron, but (16.9) quantifies this expectation. As an example, the inner, intra and outer-shell contributions to the relaxation energy in K^{+} for $(2s)^{-1}$ and $(2p)^{-1}$ are respectively 0, 0.26, 9.5 eV and 0.53, 2.2, 8.3 eV. Note that the intra- and outer-shell contributions increase markedly with the number of electrons. For numerical purposes (16.9) is more accurate than the corresponding term in (16.8) which overestimates the relaxation energy substantially.[21]

δ) Relativistic and quantum-electrodynamical effects. As the atomic number increases, the nonrelativistic treatment of atomic structure breaks down. The total relativistic energy can be partitioned into three parts.

(i) The electrostatic energy in which spin-orbit coupling and mass variation are explicitly incorporated by the noncovariant operator

$$\sum_{\nu=1}^{N}\left[c\,\bar{\boldsymbol{\alpha}}_{\nu}\cdot\bar{\boldsymbol{p}}_{\nu}+mc^{2}\boldsymbol{\beta}_{\nu}-eV_{\mathrm{n}}(r_{\nu})\mathbf{1}\right]+\sum_{\mu<\nu}^{N}\frac{e^{2}}{4\pi\varepsilon_{0}r_{\mu\nu}}\mathbf{1}. \tag{16.10}$$

The matrices $\boldsymbol{\alpha}$ and $\boldsymbol{\beta}$ are taken to have their usual form (11.9). $V_{\mathrm{n}}(r_{\nu})$ denotes the electron-nuclear interaction and depends on the model adopted for the nucleus. Spin-orbit coupling lifts the degeneracy of the two states of total angular momentum $j=l\pm 1/2$ formed by creating a hole in the (closed) nl shell. This splitting increases rapidly with Z (roughly as Z^{4}) and (usually) falls rapidly as n and l increase, although for heavy atoms even outer-shell splittings are substantial. As an example, the splitting $\delta_{2p}=E_{2p_{3/2}}-E_{2p_{1/2}}$ is 2.1 eV at $Z=18$ and 320 eV at $Z=54$ while for $Z=36$ we have $\delta_{2p}=52.5$ eV, $\delta_{3p}=7.8$ eV, and $\delta_{3d}=1.2$ eV.[22] The relationship between electron mass and velocity leads to a contraction of 1s orbitals towards the nucleus and, by self-consistency and orthogonality requirements, also of the outer s orbitals but in a less striking

[21] Goscinski, O., Hehenberger, M., Roos, B., Siegbahn, P. (1975): Chem. Phys. Lett. *33*, 427

[22] Siegbahn, K., Nordling, C., Johansson, G., Hedman, J., Hedén, P.F., Hamrin, K., Gelius, U., Bergmark, T., Werme, L.O., Manne, R., Baer, Y. (1969): ESCA applied to free molecules, Amsterdam: North-Holland

manner. The mass-velocity contraction competes with a tendency towards expansion due to increased nuclear shielding of the outer orbitals. DESCLAUX[23] has highlighted the general trends.

The many-electron operator (16.10) forms the basis of the multiconfiguration Dirac-Fock (MCDF) method.[24] The steps to obtain approximate solutions of the corresponding time-independent Schrödinger equation are very similar to those of the nonrelativistic MCHF method[25] except that the variational basis functions are Dirac central field four-spinors (12.3) with undetermined radial parts. In Sects. 17 and 18 we shall frequently refer to configuration-average or multiconfiguration DF calculations of Auger electron energies. The use of the MCDF method in the analysis of x-ray spectra has been discussed by ÅBERG and SUVANEN.[26]

(ii) The Breit interaction is the leading quantum-electrodynamical (QED) correction to the instantaneous Coulomb interaction in (16.10), corresponding to the exchange of a single transverse photon in the Coulomb gauge. The general form of the Breit operator is (see also Sect. 11)

$$B(r_{12}) = \frac{-e^2}{4\pi\varepsilon_0}\left[\bar{\boldsymbol{\alpha}}_1 \cdot \bar{\boldsymbol{\alpha}}_2 \frac{1}{r_{12}} + \frac{1}{2}(\bar{\boldsymbol{\alpha}}_1 \cdot \nabla_1)(\bar{\boldsymbol{\alpha}}_2 \cdot \nabla_2)\, r_{12}\right] \tag{16.11}$$

for direct two-electron matrix elements. For exchange matrix elements a more general form involving the orbital-energy difference has been suggested.[27,28] The first and second operators in (16.11) [which is correct to order $(Z\alpha)^2$] are known as the magnetic and retardation terms, respectively. In order of magnitude, the Breit energy is of the order $\alpha(Z\alpha)$ relative to the Coulomb energy and grows as $Z^{3.6}$ for the ground state of atoms at large Z.[27]

The Breit-energy correction to transition and ionization energies would in principle involve the evaluation of the matrix element

$$E^B(J) = \langle JM | \sum_{\mu > \nu} B(r_{\mu\nu}) | JM \rangle \tag{16.12}$$

for initial and final levels. Here the wave function $|JM\rangle$ could be obtained by the DF or MCDF method, utilizing operator (16.10). The operator $B(r_{\mu\nu})$ in (16.12) is defined by (16.11) and has been previously given by (12.6) in connection with the discussion of relativistic calculations of nonradiative tran-

---

[23] Desclaux, J.P. (1973): A. Data. Nucl. Data. Tables *12*, 311; (1976) in: Photoionization and other probes of many-electron interactions, Wuilleumier, F.J. (ed.), New York: Plenum, p. 367

[24] Desclaux, J.P. (1975): Comp. Phys. Comm. *9*, 31; Grant, I.P., McKenzie, B.J., Norrington, P.H., Mayers, D.F., Pyper, N.C. (1980): Comp. Phys. Comm. *21*, 207; McKenzie, B.J., Grant, I.P., Norrington, P.H. (1980): Comp. Phys. Comm. *21*, 233

[25] e.g. Ref. [*13*]

[26] See III.19

[27] Mann, J.B., Johnson, W.R. (1971): Phys. Rev. A *4*, 41

[28] Grant, I.P., Pyper, N.C. (1976): J. Phys. B *9*, 761 and references therein

sition probabilities. Usually the energy correction (16.12) is not evaluated for each individual level but in an average fashion.[29]

(iii) The Lamb shift contains further QED corrections, the electron self-energy (emission and absorption of a virtual photon by a single electron) and the vacuum polarization (creation and annihilation of virtual electron-positron pairs by the central field acting on the electron) being the most important ones.[30] The Lamb shift is of order $Z\alpha$ with respect to the Breit energy and grows as Z^4 at large Z.

The self-energy of an electron in a Coulomb field has been considered up to all orders in $Z\alpha$ by MOHR.[31] Together with the vacuum-polarization corrections of WICHMANN and KROLL[32] as well as other higher-order QED corrections, these results have yielded very accurate Lamb shifts for hydrogen-like ions.[33] Since the self-energy and the vacuum polarization are basically single-electron properties, these results can be used as guidance for obtaining QED corrections of binding energies of inner-shell electrons in many-electron atoms. The binding-energy corrections can in turn used according to (7.18) for the evaluation of QED corrections of nonradiative transition energies. CHENG and JOHNSON[34] have used a more realistic central field potential[35] than the Coulomb potential in their evaluation of Lamb shifts of $1s$ electrons in atoms with Z in the range 70 to 160.

In the recent compilation of relativistically calculated BE's for atoms with Z in the range 2 to 106, HUANG et al.[17] used the Hartree-Slater local exchange potential (see Sect. 19γ) to generate the appropriate central field orbitals for the initial ground state and final hole states and evaluated all the QED corrections discussed above. Table 2 shows a selection of their results, including an estimate of relaxation energies for the innermost shells of noble gas atoms. There, one can see the rapid growth of all terms as Z increases. For $1s_{1/2}$ and $2s_{1/2}$ holes, the Lamb shift contribution increases rapidly until even at moderate atomic number $Z \sim 36$ it is of the same order of magnitude as the magnetic and retardation term. The Breit energy calculated from (16.11) can be seen to outstrip the relaxation energy as Z increases. For deep hole states, especially $1s_{1/2}$, the model assumed for the nucleus has a major effect on the calculated binding energy.[29, 35] In the calculations of HUANG et al.[17] the homogeneously charged sphere model was used. DESCLAUX's rigorous DF calculations showed that the innermost ground-state orbital eigenvalues were

[29] Techniques which avoid this approximation have recently been developed: Cheng, K.T., Desclaux, J.P., Kim, Y.-K. (1978): J. Phys. B *11*, L359; Grant, I.P., McKenzie, B.J. (1980): J. Phys. B *13*, 2671. The latter work considers the general expression for non-diagonal elements involving the orbital-energy differences. It has been used in a program (McKenzie, B.J., Grant, I.P., Norrington, P.H. (1980): Comp. Phys. Comm. *21*, 233) which calculates Breit and other QED corrections to energy levels in the MCDF scheme

[30] For a general discussion, see the references in I.22

[31] Mohr, P.J. (1974): Ann. Phys. (New York) *88*, 26 and 52; (1975): Phys. Rev. Lett. *34*, 1050

[32] Wichmann, E.H.W., Kroll, N.M. (1954): Phys. Rev. *96*, 232; (1956): *101*, 843

[33] Brodsky, S.J., Mohr, P.J. (1978) in: Structure and collisions of ions and atoms, Topics in Current Physics, Vol. 5, Sellin, I.A. (ed.), Berlin, Heidelberg, New York: Springer, p. 3

[34] Cheng, K.T., Johnson, W.R. (1976): Phys. Rev. A *14*, 1943

[35] Desiderio, A.M., Johnson, W.R. (1971): Phys. Rev. A *3*, 1267

increased (became less negative) on changing from a point to a finite nucleus. For example, for radon the shifts were $1s_{1/2}$ 99.3 eV, $2s_{1/2}$ 15.8 eV, and $3s_{1/2}$ 3.8 eV.[23] The $l \neq 0$ subshells were less affected. For the Fermi distribution CARLSON and NESTOR[36] have shown that the binding energies of heavy atoms ($95 \leqq Z \leqq 130$) are quite sensitive to choice of the nuclear radius and diffuseness parameter. Below $Z \lesssim 90$ the uncertainty coming from the choice of these parameters seems to be less than 20 eV for K binding energies and substantially less for other binding energies. Nevertheless, the implication is that both the size and model for the nucleus is a nontrivial factor in heavy atoms.

ε) Correlation. For an isolated inner-shell hole state the correlation energy contribution to the corresponding BE may be roughly defined as the difference between an accurate QHQ and a ΔE_{SCF} calculation. Since the relativistic atomic theory presents the problem of both positive and negative (positron) continua,[37] leading to difficulties in the perturbation treatment of the correlation energy, it is usually defined with respect to a nonrelativistic Hamiltonian H. This definition excludes the level shift from the correlation energy which would in principle be determined from a variationally well-defined calculation if the projection operator Q could be properly defined. Since Q is not usually known, one has to rely in practice on approximate configuration-interaction calculations of the hole-state and ground-state energies, or one may try to adopt the propagator technique[18, 38] for the *direct* evaluation of the BE's. Alternatively, one may use various configuration-interaction[39] and many-body-perturbation-theory[40] techniques, but such studies have been conducted only for a few hole states in light atoms.[3, 41]

This lack of quantitative information about the correlation energy has fostered a tendency to compare relativistic ΔE_{SCF} results which include the Lamb shift with experimental BE's, which have, if possible, been corrected for solid-state effects whenever necessary. A substantial discrepancy may be taken to indicate contributions from the electron correlation including the level shift discussed in Sect. 16α. In cases where the hole state is not well separated from other hole states, it is very difficult and may be meaningless to extract the correlation controlled factors from the interactions which are responsible for the nonradiative decay. The hole states involving the decay by Super Coster-Kronig transitions mentioned in Sect. 16α serve as examples. With these reservations in mind we shall give below an idea of the order of magnitude of the correlation-energy contribution to a few BE's.

---

[36] Carlson, T.A., Nestor, Jr., C.W. (1977): A. Data. Nucl. Data. Tables *19*, 153

[37] e.g. Armstrong, Jr., L. (1978) in: Structure and collisions of ions and atoms, Topics in Current Physics, Vol. 5, Sellin, I.A. (ed.), Berlin, Heidelberg, New York: Springer, p. 69; see also Sucher, J. (1980): Phys. Rev. A *22*, 348

[38] Cederbaum, L.S., Domcke, W. (1977) in: Adv. Chem. Phys. *36*, Prigogine, I., Rice, S.A. (eds.), New York: Wiley, p. 205; Schirmer, J., Cederbaum, L.S. (1978): J. Phys. B *11*, 1889

[39] Beck, D.R., Nicolaides, C.A. (1978) in: Excited states in quantum chemistry, Nicolaides, C.A., Beck, D.R. (eds.), Dordrecht: Reidel, p. 329; Luken, W.L., Sinanoğlu, O. (1976): Phys. Rev. A *13*, 1293; see also Condon, E.U. (1968): Rev. Mod. Phys. *40*, 872

[40] Kelly, H.P. (1969) in: Adv. Chem. Phys. *14*, Prigogine, I. (ed.), New York: Wiley, p. 129

[41] Chase, R.L., Kelly, H.P., Köhler, H.S. (1971): Phys. Rev. A *3*, 1550

The pair correlation energy of K-shell electrons, which is fairly independent[42] of Z, is about -1 eV.[43] Since this energy disappears in the ionization, it is not surprising to find a correlation correction of K BE's which is about $+1$ eV throughout the periodic table. In combination with the recent analysis of DESLATTES et al.[4] of K_α energies this result suggests that in the L_{23} BE's less than 10 eV is due to correlation. For $Z \lesssim 36$ the L_{23} correlation energy is probably less than 2 eV, which should be true also for the L_1 correlation energy. For outer shells larger discrepancies are found near regions where strong quasidegeneracy occurs. The work of KRAUSE and NESTOR[16] on binding and transition energies in the actinides illustrates this point. The DF theory is found to overestimate the N_2, N_3, N_4, and N_5 BE's by 10–16 eV, which is more than the discrepancy of other N or M BE's. Configuration interaction with nearby states of the same type that perturb the $4p$ levels in the region $50 \lesssim Z \lesssim 70$, namely, with the $4d^8\, nf, \varepsilon f$ states, is thought to be important although there is not an obliteration of the $4p_{1/2}$ hole like in xenon (see Sect. 16α). All estimates given above are subject to an uncertainty that involves the level shift.

17. Semiempirical treatments. The energy of an emitted Auger electron where the initial state has a hole in the U shell and the final state has two in the V and W subshells (which may be equivalent, see Sect. 7 for the nomenclature) can be assertained as

$$\begin{gathered} X(U)^+ \to X(VW)^{2+} + e^- \\ E(U-VW) = E[X(U)^+] - E[X(VW)^{2+}]. \end{gathered} \tag{17.1}$$

Two approaches are commonly applied to evaluate this energy, and, as we shall see in Sect. 20β, the choice is critical for transition probability calculations. First, we have the direct procedure: compute separately the energy of each state involved in the transition. This will be discussed in the next section. Alternatively, estimate the energy difference by resorting to experimental data from related spectroscopies. The latter method was for some time the major avenue to the transition energies, in particular, for the $K-LL$ process.

In order to relate the energy of the emitted electron to known quantities, (17.1) is conventionally broken up into three intermediate steps:[44]

$$\begin{gathered} X \to X(U)^+ + e^- \\ X \to X(V)^+ + e^- \\ X(V)^+ \to X(VW)^{2+} + e^-. \end{gathered} \tag{17.2}$$

This procedure yields

$$E(U-VW) = \mathscr{E}(U) - \mathscr{E}(V) - \mathscr{E}(W;V), \tag{17.3}$$

[42] Das, G. (1970): J. Chem. Phys. *52*, 1004
[43] Åberg, T. (1967): Phys. Rev. *162*, 5
[44] Asaad, W.N., Burhop, E.H.S. (1958): Proc. Phys. Soc. Lond. *71*, 369

where the quantities on the right-hand side of (17.3) are the appropriate binding energies. The values of $\mathscr{E}(U)$ and $\mathscr{E}(V)$ may be obtained readily from tabulations of BE's, while the third quantity $\mathscr{E}(W; V)$, the binding energy of an electron in the W shell in the presence of a V-shell vacancy, is not usually known. In the absence of a more reliable procedure, this energy was estimated as $\mathscr{E}(W)$ for an atom of nuclear charge one unit larger than the atom being considered.[45] BERGSTRÖM and HILL[46] recognized that the nucleus is only partially deshielded by the removal of the V electron and modified this to

$$\mathscr{E}(W; V)=\Delta z[\mathscr{E}(W)_{Z+1}-\mathscr{E}(W)_Z]+\mathscr{E}(W)_Z, \tag{17.4}$$

where Δz is to be fitted to experiment. Δz is not unique and depends on the order of ionization in (17.2).[47] ALBRIDGE and HOLLANDER[48] later averaged the processes in (17.4) to yield

$$\begin{aligned} E(U-VW)=&\mathscr{E}(U)-\mathscr{E}(V)-\mathscr{E}(W)\\ &-\tfrac{1}{2}\Delta z[\mathscr{E}(V)_{Z+1}-\mathscr{E}(V)_Z+\mathscr{E}(W)_{Z+1}-\mathscr{E}(W)_Z]. \end{aligned} \tag{17.5}$$

A catalogue of Auger energies has been compiled by COGHLAN and CLAUSING[49] based on this formula with $\Delta z=1$, and the application to L- and M-shell spectra has been reviewed by HAYNES.[50] Although such formulae are useful for indicating the general region in which an Auger transition occurs, they are too crude for accurate work since, as well as oversimplifying the dynamics, measured BE's and Auger energies are dependent on the phase of the target. Nor is the multiplet structure of the final state incorporated. It is also found that such empirical equations lead to significant errors when applied to low-energy Coster-Kronig (CK) or Super CK transitions.[51] However, in common with other techniques which employ experimental data, they do have the advantage that some important quantities such as the Lamb shift are automatically built in.

A more satisfactory scheme was constructed by ASAAD and BURHOP[44] when the quantities in (17.3) were evaluated with reference to the nonrelativistic Hartree-Fock (HF) expressions for the energies. *Assuming* that some set of orthonormal state-independent one-electron orbitals is available, $E(U-VW)$ can be written explicitly in terms of BE's, spin-orbit coupling parameters, and Slater integrals, all of which are, in principle, calculable quantities. Within the intermediate coupling scheme (see Sect. 8β), one obtains for $K-LL$

[45] Burhop, E.H.S. (1952): The Auger effect and other radiationless transitions, Cambridge: Cambridge University Press

[46] Bergström, I., Hill, R.D. (1954): Ark. Fys. *8*, 21

[47] Erman, P., Sujkowski, Z. (1961): Ark. Fys. *20*, 209

[48] Albridge, R.G., Hollander, J.M. (1961): Nucl. Phys. *27*, 554; see also Chung, M.F., Jenkins, L.H. (1970): Surf. Sci. *22*, 479

[49] Coghlan, W.A., Clausing, R.E. (1973): At. Data *5*, 317

[50] Haynes, S.K. (1973) in: Proc. Int. Conf. Inn. Shell Ioniz. Phenomena future appl. Atlanta, Fink, F.W., Manson, S.T., Palms, J.M., Venugopala Rao, P. (eds.), US At. Energy Comm. Rep. No CONF-720404, Oak Ridge, Vol. I, p. 559

[51] e.g. McGuire, E.J. (1974): Phys. Rev. A *9*, 1840; Larkins, F.P. (1974): J. Phys. B *7*, 37

transitions by a diagonalization of the appropriate matrix (8.30) the well-known equations

$$
\begin{aligned}
E(K-L_1L_1) &= \mathscr{E}(K)-2\mathscr{E}(L_1)-F^0(2s2s) \\
E(K-L_1L_2;\,{}^1P_1) &= \mathscr{E}(K)-\mathscr{E}(L_1)-\mathscr{E}(L_2)-F^0(2s2p)+\tfrac{3}{4}\xi_{2p} \\
&\quad -\{[\tfrac{1}{3}G^1(2s2p)-\tfrac{1}{4}\xi_{2p}]^2+\tfrac{1}{2}\xi_{2p}^2\}^{1/2}
\end{aligned}
\tag{17.6}
$$

and seven other equations. In (17.6) $F^k(ab)=R^k(ab, ab)$ and $G^k(ab)=R^k(ab, ba)$ with R^k given by (8.10) and ξ_{2p} being the spin-orbit parameter (7.15). Later, the configuration interaction between the $(2s)^{-2}$ and $(2p)^{-2}$ ${}^1S_0$ states was included.[52] Inserting calculated values of the Slater integrals F^k and G^k (usually neutral atom values since they are most readily available) and observed BE's into these equations gives poor agreement with experiment, although satisfactory *fits* of the integrals to polynomials in Z can be made after including corrections which simulate the relativistic values of these integrals.[53]

The major importance of formulae like (17.6) lies in the recognition that "universal" Auger energy formula should, as a minimum requirement, reproduce the variation in coupling scheme throughout the periodic table. That is, as the relative importance of the spin-orbit coupling to electrostatic interaction increases from low to high Z, the $K-LL$ spectrum shows a five-line spectrum splitting to nine lines at intermediate Z and collapsing to six for the heaviest atoms. Moreover, the equations are firmly based on a well-understood scheme.

One defect of (17.6) is that the response of the passive spectator orbitals to the loss of one or more electrons is partially ignored,[54] while we have already demonstrated in Sect. 16 that the "relaxation effect" on inner-shell BE's is by no means negligible. SHIRLEY's[55] recent study of the $K-LL$ energies employs the HEDIN-JOHANSSON[20] formalism (see Sect. 16γ) to produce a "static relaxation term" which is added to each of equations (17.6). The improvement over the previous calculations is dramatic, thus rectifying much of the discrepancy between calculated integrals and those derived from experiment.

This remarkable accord with experiment justifies a thorough examination of the approximations involved in the relaxation corrections since there are some implicit assumptions built into SHIRLEY's derivation which should be revealed before applying the model. Detailed analysis of relaxation effects on the $L_3-M_{45}M_{45}$ ${}^1G_4$ transition in zinc by HOOGEWIJS et al.[56] have also highlighted some discrepancies as has the work of KIM et al.[57] on $L_{23}-M_{45}M_{45}$ spectra in nickel, copper, and zinc. A general discussion of semi-empirical models has been presented by LARKINS.[58] We shall derive the relax-

[52] Asaad, W.N. (1965): Nucl. Phys. *66*, 494

[53] Briançon, L. (1970): Ann. Phys. (Paris) *5*, 151; Asaad, W.N., in Fink et al. (see footnote V.50), p. 455

[54] Shirley, D.A. (1972): Chem. Phys. Lett. *17*, 312

[55] Shirley, D.A. (1973): Phys. Rev. A *7*, 1520

[56] Hoogewijs, R., Fiermans, L., Vennik, J. (1976): Chem. Phys. Lett. *38*, 192

[57] Kim, K.S., Gaarenstroom, S.W., Winograd, N. (1976): Chem. Phys. Lett. *41*, 503; (1976): Phys. Rev. B *14*, 2281

[58] Larkins, F.P. (1978): Chem. Phys. Lett. *55*, 335

ation correction within the UHF scheme (for the definition, see Sect. 8δ), denoting the spin orbitals involved in the nonradiative transition by V, W and the passive set by k. A subscript on these orbitals, o, v, w, or vw, denotes the ground or hole state of the configuration to which they belong. Following HEDIN and JOHANSSON[20] and SHIRLEY[55], the BE of the spin orbital W in the initial and V-hole states is given according to (16.9) as

$$\mathscr{E}(W) = -\varepsilon(W_o) - \tfrac{1}{2} \sum_{k \neq W} [\langle W_o k_w \| W_o k_w \rangle - \langle W_o k_o \| W_o k_o \rangle] \tag{17.7}$$

$$\mathscr{E}(W; V) = -\varepsilon(W_v) - \tfrac{1}{2} \sum_{k \neq V, W} [\langle W_v k_{vw} \| W_v k_{vw} \rangle - \langle W_v k_v \| W_v k_v \rangle], \tag{17.8}$$

respectively. $\varepsilon(W)$ is the UHF eigenvalue of orbital W. The difference between (17.8) and (17.7), Δ, gives the amount by which the binding energy $\mathscr{E}(W)$ is to be supplemented due to the presence of the hole in V. Expressing the eigenvalues in terms of their constituent one-electron and two-electron integrals, one finds according to (8.40) and (8.44) that

$$\begin{aligned} \Delta = & -\langle W_v | f | W_v \rangle + \langle W_o | f | W_o \rangle - \tfrac{1}{2} \sum_{k \neq V, W} \{\langle W_v k_{vw} \| W_v k_{vw} \rangle - \langle W_o k_w \| W_o k_w \rangle \\ & + \langle W_v k_v \| W_v k_v \rangle - \langle W_o k_o \| W_o k_o \rangle\} \\ & + \tfrac{1}{2}\{\langle W_o V_w \| W_o V_w \rangle + \langle W_o V_o \| W_o V_o \rangle\}. \end{aligned} \tag{17.9}$$

The last term of this expression is the interaction between W and V, which involves integrals from both ground and W-hole states. This interaction is approximately taken into account by the Asaad-Burhop formulae. To obtain SHIRLEY's formula for the so-called static relaxation term (Eq. 10 of footnote 55) one must assume that the one-electron integrals and the second and third terms of the summation (both being denoted by $V^{*\prime}$ in footnote 55) cancel. This is tantamount to the conditions that $W_v \cong W_o$ and $k_v \cong k_w$. The latter condition asserts that the passive orbitals respond in a similar fashion *irrespective* of whether the hole is in V or W, i.e., V is equivalent to W, or that they are in the same shell, arguing in accordance with Sect. 16γ that outer-shell relaxation is of major importance. While this may be approximately true for $K-LL$ transitions, it is more difficult to justify for $K-LX$ transitions, where X is a shell "outside" L. The assumption, $W_v \simeq W_o$, carries a similar rider: the hole should not be disparate from W – e.g., if V is L and W is X, the response of W to the creation of a hole in V may be large. Alternatively, V should be "outside" W, but this may interfere with the assumptions concerning k. The assumptions above also force the neglect of some of the intrashell relaxation energy that corresponds in (17.9) to those spin orbitals k that occupy the same shells as V and W. This error is likely to increase with the principal quantum number and occupation number of the shell.

Accepting these assumptions, the "static" relaxation term to be added to the Asaad-Burhop formulae takes the form

$$R = \tfrac{1}{2} \sum_{k \neq V, W} [\langle W_o k_{vw} \| W_o k_{vw} \rangle - \langle W_o k_o \| W_o k_o \rangle]. \tag{17.10}$$

The addition and subtraction of the matrix elements $\langle W_v k_w \| W_v k_w \rangle$ in (17.10) and the assumption that $\langle W_v V_w \| W_v V_w \rangle \cong \langle W_o V_o \| W_o V_o \rangle$ yields approximately

$$R \cong \sum_{k \neq w} [\langle W_v k_w \| W_v k_w \rangle - \langle W_o k_o \| W_o k_o \rangle]. \tag{17.11}$$

According to this result, R is approximately twice the so-called dynamic relaxation energy [the summation term in (17.7)]. The equivalent-cores approximation is employed to evaluate R for V and W in the same shell as a quantity dominated by outer-shell relaxation. Returning to the RHF formalism (see Sect. 8δ), one obtains from (17.11)

$$R(nl) \cong r + \sum_{n' > n} \sum_{l'} N(n'l')[M(nl, n'l')_{Z+1} - M(nl, n'l')_Z], \tag{17.12}$$

where r is the intrashell relaxation and $N(n'l')$ the occupation number of the outer shell. The subscript of the average interaction $M(nl, n'l')$ denotes the atomic number of the atom whose ground-state orbitals are to be employed in[59]

$$M(nl, n'l') = F^0(nl\,n'l') - \sum_k \frac{c^k(l0, l'0)}{[(4l+2)(4l'+2)]^{1/2}} G^k(nl\,n'l'). \tag{17.13}$$

Thus, R is independent of the multiplet structure of the transitions and no recipe for the evaluation of the intrashell relaxation contribution r is given.

In order for the integrals $F^0(2s\,2s)$, $F^0(2s\,2p)$ and $F^0(2p\,2p)$ to reflect relativistic effects, the theory was fitted to tabulated experimental Auger energies[60] by applying a correction factor $(1 + aZ^2)$ to each. The integrals $F^2(2p\,2p)$ and $G^1(2s\,2p)$ were retained at their nonrelativistic values[61]. The r values were estimated from the work of HEDIN and JOHANSSON.[20] Above $Z = 18$, r was put equal to zero. As mentioned above, the method works well for $K-LL$ transitions despite being applied to solid-state spectra using binding energies from solid-state experiments. SHIRLEY has also applied his relaxation model to the study of Auger energy shifts of $L_3 - M_{45}M_{45}$ transitions in the presence of an M_{45} spectator vacancy[62] for $70 \leq Z \leq 100$ and finds good agreement with available experimental data.

In an application of SHIRLEY's model to the $L_3 - M_{45}M_{45}$ spectra of metallic copper and zinc, KOWALCZYK et al.[63] introduced the extra-atomic relaxation contributed by the host of the atom. They considered the screening of a localized final $3d$ hole by the valence electrons from an atomic point of view and calculated presumably an upper limit of the extra-atomic relaxation contribution from (17.13) as $M(3d, 4s)$ for zinc and as $M(3d, 4p)$ for copper. Alter-

[59] Slater, J.C., Mann, J.B., Wilson, T.M., Wood, J.H. (1969): Phys. Rev. *184*, 672

[60] Sevier, K.D. (1972): Low energy electron spectrometry, New York: Wiley

[61] Mann, J.B. (1967): Los Alamos Sci. Lab. Rept. No LASL-3690

[62] Shirley, D.A. (1974): Phys. Rev. A *9*, 1549

[63] Ley, L., Kowalczyk, S.P., McFeely, F.R., Pollak, R.A., Shirley, D.A. (1973): Phys. Rev. B *8*, 2387; Kowalczyk, S.P., Ley, L., McFeely, F.R., Pollak, R.A., Shirley, D.A. (1974): Phys. Rev. B *9*, 381

natives[64] and extensions[65] of this model have been considered in the literature. Although an analysis of this and other solid-state corrections is outside the scope of this article, we note that in many cases the major solid-state effect is a uniform shift of the nonradiative spectrum towards higher kinetic energies when going from vapor to solid phase.

We have seen that the application of SHIRLEY's model is likely to be limited by the several approximations involved. Recently, LARKINS[66] has proposed and implemented a more rigorous semiempirical treatment of the relaxation and relativistic corrections to $K-LL$ Auger energies which seems to be of wider applicability.[67,68] Similar ideas have been put forward by others[56,57] but in a more limited context. The transition energy (17.3) with the notation generalized to include multiplet splitting is rewritten in the form

$$E(U-VW;\, ^{2S+1}L_J) = \mathscr{E}(U) - \mathscr{E}(V) - \mathscr{E}(W) - \Delta(^{2S+1}L_J), \tag{17.14}$$

where

$$\Delta(^{2S+1}L_J) = E_0 + E[VW;\, ^{2S+1}L_J] - E[V] - E[W]. \tag{17.15}$$

In Δ, E_0 denotes the energy of the ground state and $E[X]$ that of the hole states. The energies in (17.14) and (17.15) are in principle exact, but in practice LARKINS resorts to the quantities calculated by the self-consistent relativistic Dirac-Fock (DF) method. Casting (17.14) into a form reminiscent of SHIRLEY's equations,

$$\Delta(^{2S+1}L_J) = \langle VW;\, ^{2S+1}L_J\rangle - R[VW], \tag{17.16}$$

where (minus) $\langle VW;\, ^{2S+1}L_J\rangle$ represents the interaction between holes V and W as given in (17.6). The correction term R reflects the orbital reorganisation and corresponds to the "static" relaxation energy (17.10). Note that its value depends on the orbital basis used to ascertain the multiplet splitting through

$$R[VW] = \langle VW;\, ^{2S+1}L_J\rangle - E_0 - E[VW;\, ^{2S+1}L_J] + E[V] + E[W]. \tag{17.17}$$

No approximations have been invoked. A rigorous evaluation of R, two-electron integrals, and BE's using the DF Hamiltonian (16.10) leads directly to the difference between two optimized calculations. However, R is assumed to

---

[64] From a solid-state point of view the extra-atomic relaxation is related to the net polarization of the surrounding medium of an ionic core in which there is a change of the charge from one extra unit to two extra units as a result of the nonradiative transition. The energy of radiative Auger transitions in solids (for the general relationship to the Auger transitions, see e.g. Åberg, T. (1975) in: Atomic inner-shell processes, I. Ionization and transition probabilities, Crasemann, B. (ed.), New York: Academic Press, p. 353) has been analyzed from this point of view: Utriainen, J., Linkoaho, M., Åberg, T. (1973) in: Proc. int. symp. X-ray spectra electronic struct. matter, Faessler, A., Wiech, G. (eds.), München, Vol. I, p. 382; Åberg, T., Utriainen, J. (1975): Solid State Comm. *16*, 571; Åberg, T. (1977) in: X-ray photoelectron spectroscopy, Nemoshkalenko, V.V. (ed.) (Kiev: Naukova Dumka), p. 19; see also Almbladh, C.O., Barth, U. von (1976): Phys. Rev. B *13*, 3307

[65] Larkins, F.P. (1977): J. Phys. C *10*, 2453 and 2461; see also Hoogewijs, R., Fiermans L., Vennik, J. (1976): Chem. Phys. Lett. *38*, 471; Kim et al. (V.57)

[66] Larkins, F.P. (1976): J. Phys. B *9*, 47

[67] Crotty, J.M., Larkins, F.P. (1976): J. Phys. B *9*, 881

[68] Larkins, F.P. (1977): A. Data Nucl. Data Tables, *20*, 311

depend mainly on the principal quantum number of the two holes V and W and adequately represented by the final s-hole term $R[V_1 W_1]$. The error introduced by this approximation is ~ 1–4 eV for $Z<80$ and ~ 6 eV at $Z=102$ for $K-LL$ transitions.[66] The multiplet splitting integrals in the intermediate-coupling ansatz (17.6) are replaced by averaged ground-state relativistic values obtained from a weighted-average correspondence suggested by LARKINS[69] between nonrelativistic and relativistic 'LS multiplet' energies. For example,

$$\begin{aligned} &F^0(2s\,2s) \to F^0(2s_+\,2s_+) \\ &F^0(2s\,2p) \to \tfrac{1}{3}[F^0(2s_+\,2p_-) + 2F^0(2s_+\,2p_+)] \\ &G^1(2s\,2p) \to \tfrac{1}{3}[G^1(2s_+\,2p_-) + 2G^1(2s_+\,2p_+)], \end{aligned} \tag{17.18}$$

where $+(-)$ denote the $j=l\pm 1/2$ subshells. The energies in (17.17) are replaced by the relativistic "LS average" energy[70] of the $ns^{-1}n's^{-1}$ configurations involved. In particular

$$R[V_1 W_1] = \langle V_1 W_1 \rangle_{\text{av}} - E_0^{\text{av}} - E^{\text{av}}[X_1 Y_1] + E^{\text{av}}[X_1] + E^{\text{av}}[Y_1], \tag{17.19}$$

where

$$\langle V_1 W_1 \rangle_{\text{av}} = F^0(ns_+\,n's_+) - \frac{(1-\delta_{nn'})}{2} G^0(ns_+\,n's_+). \tag{17.20}$$

Polynomial fits to R and Slater integrals were obtained using nine closed-shell species from the range $10 \leqq Z \leqq 102$. Combining these quantities with observed BE's and experimental spin-orbit parameters ξ_{nl}, based on

$$\xi_{nl} = \frac{2}{(2l+1)} [\mathscr{E}(X_{l-1/2}) - \mathscr{E}(X_{l+1/2})], \tag{17.21}$$

$K-LL$ energies were evaluated with and without CI between the $(2s)^{-2}$ and $(2p)^{-2}$ $^1S_0$ states. Despite the inevitable intrusion of solid-state effects, the match between theory and experiment is very good. CROTTY and LARKINS[67] also investigated the K and L spectra of Ar and obtained excellent agreement between the semiempirical theory and experiment. The theory has also been used for the preparation of a comprehensive table of semiempirical Auger electron energies[68] which includes a wide range of nonradiative transitions in elements $10 \leqq Z \leqq 100$. Except for the rare-gas atoms the energies are given for the solid phase by including a correction[65] due to the extraatomic relaxation. The accuracy is estimated to be 1 to 5 eV for those elements for which the binding energies are known to a much higher accuracy. In heavy elements there are uncertainties of 10 to 20 eV in the binding energies (see Sect. 16ε) which induces an unreliability of the same order of magnitude in the Auger electron energies. In Tables 3 and 4 we compare a few semiempirical $K-LL$ energies with energies, calculated ab initio, and with experimental results. The

---

[69] Larkins, F.P. (1976): J. Phys. B 9, 37

[70] This energy corresponds to (18.1); for reference, see V.69

Table 3. $K-LL$ Auger energies for neon and argon (in eV)

Transition	Ne			Ar		
	Semi-empirical[a]	MCDF[b]	Experiment[c]	Semi-empirical[a]	MCDF[b]	Experiment[d]
KL_1L_1	747.0	747.5	748.0±0.5	2,510.6	2,511.6	2,508.9±0.6
KL_1L_2 $^1P_1$	770.9	771.6	771.4±0.5	2,576.0	2,575.6	2,575.8±0.4
$^3P_0$	781.7	783.7		2,599.3	2,599.2	
KL_1L_3 $^3P_1$	781.8	783.7	781.9±0.5	2,600.1	2,600.0	2,599.4±0.4
$^3P_2$	781.8	783.8		2,601.4	2,601.4	
KL_2L_2 $^1S_0$	800.9	801.8	800.4±0.5	2,652.2	2,649.9	2,650.6
KL_2L_3 $^1D_2$	803.4	806.9	804.1±0.4	2,660.2	2,661.8	2,660.6±0.2
KL_3L_3 $^3P_0$	806.1	810.3		2,666.7	2,668.4	2,666.8±0.5
$^3P_2$	806.2	810.4		2,668.6	2,670.6	2,669.1±0.4

[a] For reference, see V. 68
[b] For reference, see V. 78
[c] Körber, H., Mehlhorn, W. (1966): Z. Phys. *191*, 217
[d] Krause, M.O., unpublished (see Krause, M.O., in: Proc. Sec. Int. Conf. Inner Shell Ioniz. Phen., Mehlhorn, W., Brenn, R. (eds.), Freiburg 1976, Invited Papers, p. 184). The KL_2L_2 $^1S_0$ value is affected by satellite interference

Table 4. $K-LL$ Auger energies for uranium and americium (in eV)

Transition	U			Am		
	Semi-empirical[a]	MCDF[b]	Experiment[c]	Semi-empirical[a]	MCDF[b]	Experiment[d]
KL_1L_1	71,776	71,749	71,745±20	77,040	77,021	77,012±23
KL_1L_2 $^1P_1$	72,606	72,560	72,560±20	77,919	77,878	77,895±12
$^3P_0$	72,662	72,614	72,620±20	77,977	77,942	
KL_1L_3 $^3P_1$	76,334	76,332	76,320±20	82,308	82,300	82.320±500
$^3P_2$	76,442	76,430	76,430±20	82,419	82,414	82,470±500
KL_2L_2 $^1S_0$	73,373	73,272	73,320±40	78,733	78,670	
KL_2L_3 $^1D_2$	77,157	77,125	77,130±20	83,179	83,152	83,169±12
KL_3L_3 $^3P_0$	80,891	80,889		87,573	87,575	87,360±50
$^3P_2$	80,954	80,955		87,638	87,642	87,602±24

[a] For reference, see V. 68. The values include a solid state correction
[b] For reference, see V. 78. The values include the Lamb shift
[c] Briançon, Ch. (1970): Ann. Phys. (Paris) *5*, 151
[d] Porter, F.T., Ahmad, I., Freedman, M.S., Milsted, J., Friedman, A.M. (1974): Phys. Rev. C *10*, 803

semiempirical method has been extended to an analysis of $L-LM$ Coster-Kronig processes in solids.[71]

A major difference between the approaches of LARKINS and SHIRLEY is that

[71] Larkins, F.P. (1978): J. Phys. C *11*, 1965

the former includes from the outset the possibility of including both inter- and intra-shell relaxation (to an uncertain extent although the approximations invoked seem to be justified) as well as an ab initio treatment of the relativistic correction[72] to the "nonrelativistic" Slater integrals. A comparison[73] between the methods may therefore cast some light on the effect of the approximations involved in SHIRLEY's static relaxation correction model. Firstly, the neglect of relativistic corrections to $F^2(2p\,2p)$ by SHIRLEY is not critical, but the nonrelativistic $G^1(2s\,2p)$ seriously underestimates the relativistic value at large Z. Moreover, SHIRLEY's scaling of the F^0 integrals by $(1+aZ^2)$ is also unrealistic, the a values obtained by LARKINS being approximately one half of those of SHIRLEY, while a form $(1+aZ^2+bZ^4)$ is superior. These errors are found to be partially compensated by an overestimate of the relaxation term. Overall, LARKINS's $K-LL$ energies tend to be smaller for $Z<60$, the converse obtains at higher Z.

The success of LARKINS's method, which is strongly based on the DF independent-particle method, hints at the advantage which may be gained by performing ab initio calculations provided all important relativistic factors are included. In this connection, it is worthwhile to repeat that these semiempirical models, via experimental BE's, include the Lamb shift and some solid-state corrections. It is not clear where the Breit-energy corrections enter since we do not expect them to be strictly additive in the final two hole state. Another area of uncertainty lies in the multiplet splitting, which may also result in changes in this interaction and relaxation.

In some cases it has been possible to combine accurately calculated K binding energies with optical energies of the doubly ionized final state. In this way NICOLAIDES and BECK[74] have determined the $K-L_1L_1$ ^1S transition energies for atomic carbon and oxygen and the $K-L_{23}L_{23}$ ^1D energy of neon. With our present knowledge of correlation corrections of binding energies this technique may provide reliable information about the accuracy of Auger energies obtained by other methods and in a wider range.

18. Ab initio calculations

α) Self-consistent-field methods. Available procedures for the calculation of nonradiative transition energies can be divided into two categories.

(i) Nonrelativistic Hartree-Fock methods including the multiconfiguration mode (MCHF). The basis of these self-consistent-field (SCF) methods was discussed in Sect. 8δ. Intermediate coupling is not automatically included, but the diagonal spin-orbit parameter, given by (7.15), can be calculated for an arbitrary subshell using HF radial wave functions. A few computer programs are available.[75] A number of methods can be considered to be approximate

---

[72] For an analysis of this correction see III.19

[73] See V.58 and V.66

[74] Nicolaides, C.A., Beck, D.R. (1974): Chem. Phys. Lett. *27*, 269; (1976): J. Elect. Spec. *8*, 249

[75] Froese Fischer, C. (1969): Comput. Phys. Commun. *1*, 151; (1972): *4*, 107; (1978): *14*, 145. The corresponding programs will be identified as MCHF, MCHF72, and MCHF77 in the following. See also Ref. [*13*] for computational details

Hartree-Fock methods, like the Hartree-Slater (HS) local $X_{\alpha\beta}$ exchange method (see Sect. 19β). These methods can provide nonradiative transition energies of HF accuracy provided correct HF expressions are used for the energy calculations.

Usually the initial and final states are separately optimized (ΔE_{SCF} method). The transition-operator method,[76] which is based on an optimization of the initial and final states simultaneously and provides a common set of orthogonal spin orbitals for both states, has been found to mimic the ΔE_{SCF} method very closely. It seems to yield nonradiative transition energies of HF accuracy but avoids the nonorthogonality problem.[77]

(ii) Relativistic Dirac-Fock methods including the multiconfiguration mode (MCDF). As explained in Sect. 13 the configuration-interaction option can be used for intermediate-coupling calculations. The points which were mentioned in Sect. 16δ regarding the Breit interaction, nuclear potential, and QED corrections are essential for the calculation of nonradiative transition energies. Note also that the MCDF method accounts for the influence of relaxation on the spin-orbit interaction, which may be appreciable in multiple-vacancy states.[26] Computer programs are available for MCDF calculations.[24] As in the nonrelativistic case Dirac-Fock-Slater (DFS) local-exchange methods may provide good zeroth-order approximations of the DF radial wave functions.

As an illustration of the power of the ΔE_{SCF} calculations when applied to $K-LL$ Auger energies, Tables 3 and 4 compare the MCDF calculations of BRIANÇON and DESCLAUX,[78] with LARKINS's[68] semiempirical technique, and with experiments for Ne, Ar, U, and Am. Despite the neglect of any solid-state corrections and the use of a uniform charge distribution of the nucleus of the two latter atoms, the agreement between the MCDF results and the other columns is excellent. The MCDF calculations include the Lamb shift, estimated for U to be $1s_{1/2}$ 245 eV, $2s_{1/2}$ 50 eV, $2p_{1/2}$ 3 eV.[78] Note that the $K-L_1L_1$ transition provides a good test for the $2s_{1/2}$ Lamb shift since it enters twice. According to Sect. 16ε, the relativistic corrections already dominate over the correlation corrections at Ar since the relativistic contribution to K BE in Ar is about 10 eV.

The MCDF calculations of BRIANÇON and DESCLAUX[78] were done in the intermediate-coupling scheme in which the radial components of the final-state four-spinor were optimized simultaneously with the diagonalization of the Hamiltonian matrix with respect to jj-coupled wave functions. This iterative procedure was carried out until self-consistency was achieved. In Sect. 13 the intermediate coupling in the relativistic regime was discussed in detail for the $2s^{-1}2p^{-1}$ configuration. The p^{-2} configuration has four jj energy levels: $p_{1/2}^{-2}$ $J=0$, $p_{3/2}^{-2}$ $J=0, 2$, $p_{3/2}^{-1}p_{1/2}^{-1}$ $J=2$. Hence we see that the two $J=0$ and the two $J=2$ states are mixed in this case. The variation was carried out for each J state *separately*. A cruder procedure would have been to make the many-configuration jj average

$$E_{\mathrm{av}}^{nl}(\Gamma)=\sum_{\gamma} C_{\gamma} E_{\mathrm{av}}^{j'j}(\Gamma)/\sum_{\gamma} C_{\gamma} \tag{18.1}$$

---

[76] Goscinski, O., Howat, G., Åberg, T. (1975): J. Phys. B *8*, 11

[77] Howat, G., Goscinski, O., Åberg, T. (1974): Phys. Fenn. *9*, suppl. S 1, 241

[78] Briançon, C., Desclaux, J.P. (1976): Phys. Rev. A *13*, 2157

stationary. In (18.1) $C_\gamma = \sum_\alpha (2J_\alpha + 1)$ and $E_{av}^{j'j}$ is the single-configuration jj-average (13.3). The summation over α pertains to all J of the subconfiguration γ. The resulting jj-coupled wave functions could have then been used to diagonalize the appropriate Hamiltonian matrices like the one given by (13.2) which corresponds to $J=1$ and $2s^{-1}2p^{-1}$. An inspection of the mixing coefficients[78] shows that Ne is an almost pure LS case, whereas U and Am are very close to the jj limit. Note, however, that according to Table 3, seven lines instead of six have been observed.

Studies employing ΔE_{SCF} techniques, in particular by LARKINS,[79] have proved to be useful for the investigation of relaxation and valence-electron defect configurations on nonradiative transition energies. Nonrelativistic SCF calculations on Ne $K-LL$, Ar $L_1-L_{23}M_{123}$, $L-MM$ were later extended to encompass the M-shell Coster-Kronig (CK), Super CK, and Auger spectra of Kr.[79] Although the match between theory and experiment was judged to be good, significant discrepancies persisted, particularly for the transitions $L-M_1M_1$ in Ar and $M-N_1N_1$ in Kr. A reassignment of these spectra by MCGUIRE[80] has reduced this disagreement to some extent. Especially in studies of the decay of multiply ionized states which result from ion-atom collisions, BHALLA[81] has used the $X_{\alpha\beta}$ local exchange method (see Sect. 19β). Energies were computed using the correct HF expression and yielded results virtually identical to HF results for Ne $K-LL$ transitions.[82]

The versatility of the ΔE_{SCF} method makes it an important tool for the investigation of phenomena for which the experimental data is incomplete and for the critical assessment of the semiempirical model formulae. To this end, in a study of the $L_1-L_{23}M$ CK energies of atoms $11 \leqq Z \leqq 50$ with multiple ionization in the valence shell, LARKINS demonstrated that the cut-off point for the transition is strongly dependent on the defect configurations of the ions.[79] CHEN et al.[83] have calculated relativistic $L-LX$ CK energies for neutral atoms in the range of Z from 11 to 103. They have used the DFS method for the production of zeroth-order wave functions which were used in *average* DF expressions of the total initial- and final-state energy. The QED corrections[17] of calculated binding energies were adapted for the CK transition energies. Their results clearly demonstrate the inadequacy of using (17.5) and related semiempirical formulae.

Nonradiative transition energies are often calculated as in the example above, namely, as differences between average total energies which are given in the nonrelativistic limit by the LS-average energy (8.47) and in the relativistic limit by the many-configuration jj average (18.1). This procedure gives a transition energy between two configurations that should not be confused with an *average transition* energy, i.e., an average of transition energies

[79] Larkins, F.P. (1975) in: Atomic inner-shell processes I, ionization and transition probabilities, Crasemann, B. (ed.), New York: Academic Press, Chap. 10, p. 377 and references therein

[80] McGuire, E.J. (1975): Phys. Rev. A *11*, 17, 1880

[81] Bhalla, C.P. (1975): Phys. Rev. A *12*, 122

[82] Bhalla, C.P., Folland, N.O., Hein, M.A. (1973): Phys. Rev. A *8*, 649

[83] Chen, M.H., Crasemann, B., Huang, K.N., Aoyagi, M., Mark, H. (1977): A. Data Nucl. Data Tables *19*, 97

obtained between energy terms or levels and weighted by the intensity distribution of the transitions involved. The main reason why the two transition energies may be different is that the former procedure also gives weight to multiplets or components whose access may be forbidden by the selection rules.

The next step of sophistication would be to generate the orbital basis set by a variation of an energy functional that is based neither on (8.47) nor (18.1) but to calculate term or level energies using proper expressions. Depending on the open-shell configuration, the results may differ from those which are based on separate optimization of each term or level. A striking example is provided by the $2p$ radial functions in $1s^2\,2s\,2p\ {}^{1,3}P$ state of beryllium.[84] The expectation value $\langle r\rangle_{2p}$ for the triplet, 2.88 a.u., is increased to 5.03 a.u. in the singlet due to the unusually large effect of the exchange term $G^1(2s\,2p)$. An average-energy calculation, based on optimization of $E_{av}=[E({}^1P)+3E({}^3P)]/4$ and, strongly weighted towards the triplet, yields $\langle r\rangle_{2p}=3.06$ a.u. The calculated splitting $E({}^1P)-E({}^3P)$ is increased from 0.117 a.u. to 0.135 a.u. by the average basis set. This case and other less extreme examples[85] indicate that some care should be exercised when nonradiative transition energies are approximated by averaging procedures. This is also true for relativistic calculations, where the need of averaging procedures is greater than in the nonrelativistic case due to longer computer times. For example, DESCLAUX[23] has pointed out that the use of the many-configuration jj average (18.1) forces the radial components of the $j=l\pm 1/2$ four-spinors to behave in a more similar manner than they do in the single-configuration jj-average energy (13.3), i.e., level-dependent features are diluted. The Breit interaction energy may also be sensitive to the choice between the average or level form.[29]

Especially in the calculation of nonradiative transition energies between multiply ionized states, one frequently meets problem configurations for which the ordinary HF or DF SCF procedures break down. We shall illustrate such cases before concluding this section.

(i) The MCDF program of DESCLAUX[24] assumes that the nondiagonal interaction matrix elements are comprised of two-electron integrals only. It is therefore inapplicable for certain intermediate-coupling calculations. An example is provided by the $(np)^2(n's)$ configuration. The two parts of the wave function which concerns us stem from the coupling of the p^2 holes ($j_1=3/2$, $j_2=1/2$) $J=1,2$ and $j=1/2$ for the s hole to give a total $J=3/2$. Specifically, for $M_J=3/2$, these combinations are given by

$$\begin{aligned}\Phi_1(P^2\!:J=2;\ s\!:j=\tfrac{1}{2})=&\sqrt{\tfrac{4}{5}}\,|\tfrac{3}{2}\tfrac{3}{2};\ \tfrac{1}{2}\tfrac{1}{2};\ \tfrac{1}{2}-\tfrac{1}{2}|-\frac{1}{2\sqrt{5}}\,|\tfrac{3}{2}\tfrac{3}{2};\ \tfrac{1}{2}-\tfrac{1}{2};\ \tfrac{1}{2}\tfrac{1}{2}|\\&-\tfrac{1}{2}\sqrt{\tfrac{3}{5}}\,|\tfrac{3}{2}\tfrac{1}{2};\ \tfrac{1}{2}\tfrac{1}{2};\ \tfrac{1}{2}\tfrac{1}{2}|,\\\Phi_2(P^2\!:J=1;\ s\!:j=\tfrac{1}{2})=&-\tfrac{1}{2}\,|\tfrac{3}{2}\tfrac{1}{2};\ \tfrac{1}{2}\tfrac{1}{2};\ \tfrac{1}{2}\tfrac{1}{2}|+\frac{\sqrt{3}}{2}\,|\tfrac{3}{2}\tfrac{3}{2};\ \tfrac{1}{2}-\tfrac{1}{2};\tfrac{1}{2}\tfrac{1}{2}|.\end{aligned}\tag{18.2}$$

The symbols $|j_1m_1;j_2m_2;j_3m_3|$ stand for Slater determinants (8.1). Matrix

[84] Froese Fischer, C. (1967): J. Chem. Phys. *47*, 4010

[85] See e.g. Ref. [*13*].

elements between Φ_1 and Φ_2 involve *one-electron* integrals from the two matching determinants, and the current versions cannot fully handle this situation. Of course, *given* a basis set, there is no problem. One might expect to find similar difficulties when a particular value of J is reached by more than one route. One-electron integrals also appear generally in MCDF or MCHF calculations when two states differ by only a single excitation from one shell to another, e.g., $(2s)^{-2} \rightarrow (2s)^{-1}(3s)^{-1}\,{}^1S$.

(ii) Another type of configuration for which the normal DF or HF procedure breaks down is when two open (sub)shells of the same symmetry have equal occupation numbers. A typical case discussed by SHARMA and COULSON[86] is $1s\,2s\,{}^1S$. The usual procedure of imposing orthogonality between $1s$ and $2s$ orbitals via Lagrangian multipliers (see Sect. 8 δ) leads to a consistency condition between two-electron integrals: $R^0(1s\,1s\,1s\,2s) = R^0(1s\,2s\,2s\,2s)$. As Z increases, orthonormal $1s$ and $2s$ orbitals tend asymptotically to become hydrogenic and do not satisfy this criterion. COHEN and KELLY[87] showed that the $1s$ orbital, in order to satisfy this consistency criterion, acquired a node and a tail, while the calculated energy lay below the observed energy. The reason is clear from SHARMA and COULSON's analysis. To first order in $1/Z$, the excited ${}^1S$ state is not orthogonal to the ground state when $1s$ and $2s$ are constrained to be orthogonal – variational collapse has occurred. Should one constraint orthogonality to the $1s_g^2$ ground state by forcing $2s$ to be orthogonal to $1s_g$ as well as to the excited-state $1s$ orbital, an upper bound is obtained. These constraints are severe enough to raise the energy appreciably. However, COHEN and KELLY showed that when $\langle 1s|2s\rangle \neq 0$, the constraint $\langle 1s_g|2s\rangle = 0$ has a less marked effect on the energy and a better energy is obtained in the nonorthogonal case.

An earlier version[75] of FROESE-FISCHER's MCHF program as well as DESCLAUX's MCDF program[24] assume zero Lagrangian nondiagonal multipliers between radial wave functions that belong to two incomplete shells (subshells) of same symmetry and occupation. This does not always lead to orthogonality between the wave functions, and consequently the correct HF (or DF) solution is not obtained. The MCHF72 and MCHF77 programs avoid this problem by using a special procedure to estimate the nondiagonal parameters whenever they *should* be nonzero in such a case.[75] In addition, the latter version also allows for mixing between configurations that differ by one electron provided that the resulting MCHF problem has a unique solution. Similar options are provided by the MCDF program of GRANT et al.[24, 88]

β) Correlation effects. As outlined in Sect. 16 ε any major discrepancy between binding energies calculated by independent-particle models (in general, relativistic) and experiment is to be attributed to the omission of electron correlation effects which are, by definition, beyond the scope of these schemes. However, it should be made clear that the apparent success of any inde-

[86] Sharma, C.S., Coulson, C.A. (1962): Proc. Phys. Soc. *80*, 81

[87] Cohen, M., Kelly, P.S. (1965): Can. J. Phys. *43*, 1867

[88] Note, however, that the program of GRANT et al. does not allow for the one-electron integrals in the energy-functional which is used to obtain the equations for the radial functions

pendent-particle scheme is *always* indicative of a certain degree of cancellation between neglected contributions to the energy of initial and/or final states. For example, an investigation of the *absolute* energies of the nonradiative transitions arising from a well-defined hole state should treat the complicated mixture of correlations both broken and altered in passing from initial to final state. The perturbation analysis of binding energies in Sect. 16γ serves to illustrate this point. Although *relative* positions depend only upon the differences in final states for an initial single-hole state, it is not guaranteed that such cancellations will take place even within the same configuration. Electron correlation is besides other things a spin-dependent phenomenon so that two multiplets of the same L but different S can have different correlation energies. In such cases the splitting cannot be represented solely by the usual combination of Slater integrals. Fortunately, strong correlation effects can sometimes be seen to operate through the appearance of the Auger spectrum, e.g., satellites forbidden in the independent-particle model acquire intensity by configuration mixing.[89] Also, in many applications the correlation energies can be neglected, being typically a few electron volts.

In practice, a recognition of the role played by correlation on nonradiative transition energies has been incorporated at two levels. The first, typified by the CI invoked by ASAAD[52] between $(2s)^{-2}$ and $(2p)^{-2}\,{}^1S_0$ for $K-LL$ transitions, focusses attention on the residual-ion hole states of the same symmetry which lie close together and can therefore perturb each other strongly. By its very nature, this type of approach is more suited to the discussion of large shifts of relative positions of Auger lines. Secondly, one has the possibility of calculating the total correlation energy of a particular state explicitly. Due to the possibility of cancellation at the lowest order of approximation, it is imperative that the initial and final states are treated in the same manner before predicting absolute energies. Between these two limits one can envisage schemes in which some features, such as correlation of passive inner-shell electrons, are assumed to vary but little from state to state. The calculations are then tailored to ascertain "the rest." The relative energy calculations of Auger transitions in small molecules by HILLIER and KENDRICK[90] can be classified in this way: CI between all possible double hole states of the same symmetry constructed from the valence-shell orbitals were supplemented by single excitations from these configurations into the lower virtual orbitals.

The first approach described above is well illustrated by the work of AKSELA and AKSELA[91] on the $M_{45}-N_{45}N_{45}$ Auger spectrum of the free cadmium atom. Figure 8 shows the calculated (HF) and observed position of each transition relative to ${}^1G_4$ at three levels of approximation: extreme LS coupling, intermediate coupling (IC), and IC with CI (IC$-$CI) between all the hole states of the same symmetry which can be formed from the configurations $(4s)^{-2}$, $(4p)^{-2}$, $(4s)^{-1}(4d)^{-1}$, $(4d)^{-2}$, $(5s)^{-2}$, $(4s)^{-1}(5s)^{-1}$, and $(4d)^{-1}(5s)^{-1}$. In

[89] See in particular reference 3 in the introduction and Mehlhorn W. (1976) in: Photoionization and other probes of many-electron interactions, Wuilleumier, F.J. (ed.), New York: Plenum Press, p. 309

[90] Hillier, I.H., Kendrick, J. (1976): Mol. Phys. *31*, 849

[91] Aksela, H., Aksela, S. (1974): J. Phys. B 7, 1262

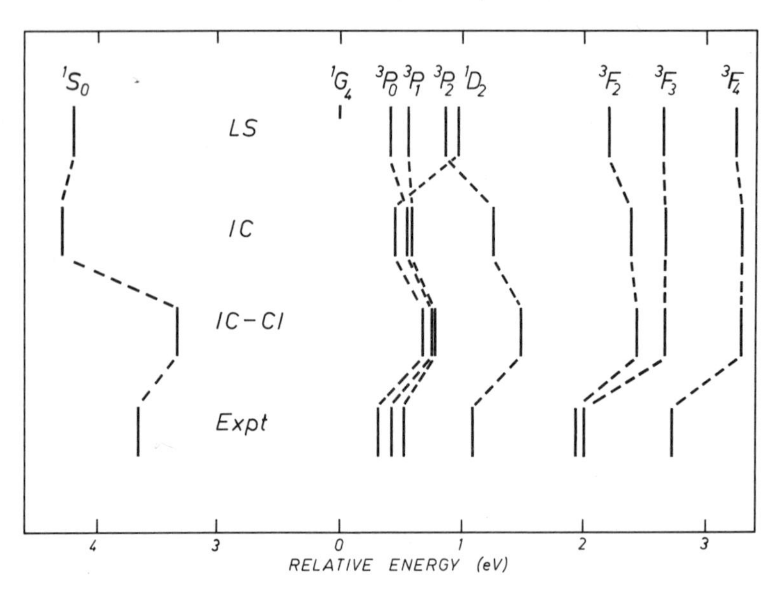

Fig. 8. Multiplet splitting of $M_{45}-N_{45}N_{45}$ Auger transitions in the cadmium atom according to AKSELA and AKSELA[91]. The CI configurations are given in the text

the actual calculation, (8.58) and (8.59) were taken into account and the CI was performed between the two-electron configurations. The effect of CI is most marked for $J=0, 1$, and 2 states and appears to be as important as IC. Note that the splittings between the lines are not too well reproduced by any model. WEIGHTMAN[92] has ascribed this to a reduction of the $F^2(4d\,4d)$ and $F^4(4d\,4d)$ Slater integrals due to the increased shielding of the nucleus by the relativistic contraction of inner orbitals. However, recent relativistic Dirac-Fock calculations[93] of these electron interaction integrals have not borne out WEIGHTMAN's suggestion, since the relativistic and nonrelativistic integrals were found to be very close in magnitude. He has also shown that the calculated splitting may be improved by treating the F^2 and F^4 integrals as parameters within IC only. Due to the apparent importance of the CI, it is to be expected that the semi-empirical fit may be modified when incorporated. For $K-LL$ transitions, the IC$-$CI calculation is found, for $Z>30$, to decrease the $K-L_1L_1\,{}^1S_0$ transition energy by 5–6 eV and increase $K-L_{23}L_{23}\,{}^1S_0$, $K-L_{23}L_{23}\,{}^3P_0$ by 3–4 and 1–3 eV, respectively.[66] Usually, agreement with experiment is increased although the effect is quite small here. Intuition might lead one to expect that correlation contributions tend to increase their significance as both the transition energy decreases and the number of hole states of similar energy rises on approaching the outer regions of the atom.

Of the methods which can be applied for calculating the entire correlation energy, many-body perturbation theory is one of the most powerful. Although

---

[92] Weightman, P. (1976): J. Phys. C 9, 1117

[93] Suvanen, M., Åberg. T. (unpublished)

KELLY[94] has described the procedure for dealing with open shells in LS coupling, the MBPT formalism is most easily applied to cases which have their first-approximation wave function as a single determinant. Moreover, the method suffers from the defect that the relativistic formulation introduces the positron states of negative energy, leading to difficulties. Little work has been reported even for the nonrelativistic case - although KELLY, by estimating the contribution of relativistic and correlation corrections from related calculations, has reproduced the experimentally observed Ne $K-LL$ energies with a very good accuracy.[95] The alternative to MBPT, extensive CI, also presents the problem of dealing with the positron states in a complete expansion of the total wave function, so it would seem that a complete account of correlation is beyond us at present. However, it should be possible to treat, by MCDF, the important features which dominate the correlation energy, at least with regard to final-state level splittings.[96]

VI. The calculation of nonradiative transition amplitudes

The computation of nonradiative transition rates via the two-electron integrals in either the WENTZEL or relativistic MØLLER ansatz has advanced hand-in-hand with the refinement of theoretical tools for atomic-structure investigation. The early pioneering studies were forced to ignore details of atomic structure and to utilize screened hydrogenic functions for both bound and continuum orbitals. Multiplet structure appeared only in the essentially geometrical factors in the coefficients of the matrix elements. However, with the availability of fast electronic computers, it is now possible to reflect the variations in atomic structure of initial and final states and, ultimately, in the decay rates. The discussion in Sect. 19 outlines the main features involved in the actual calculation of the one-electron functions, which constitute the building blocks of the nonradiative transition amplitudes. We shall describe the equations for the continuum orbitals, relativistic and nonrelativistic, as well as the various potentials which have been employed therein. Some aspects of the numerical procedures used to solve the radial equations are also discussed. In Sect. 20 we shall focus our attention on various factors whose influence on nonradiative transition amplitudes has recently been studied but not fully explored. These effects include the relativistic analogue of intermediate coupling, the choice of the potential, basis set, and energy. Relaxation and correlation effects are also discussed.

---

[94] Kelly, H.P. (1966): Phys. Rev. *144*, 39

[95] See II.51

[96] See III.19 and the following works which analyze the level structure of transitions in open-shell systems using the MCDF method: Desclaux, J.P., Kim, Y.-K. (1975): J. Phys. B *8*, 1177; Kim, Y.-K., Desclaux, J.P. (1976): Phys. Rev. Lett. *36*, 139; Armstrong, Jr., L., Fielder, W.R., Lin, D.L. (1976): Phys. Rev. A *14*, 1114; Grant, I.P., Mayers, D.F., Pyper, N.C. (1976): J. Phys. B *9*, 2777; Rose, S.J., Pyper, N.C., Grant, I.P. (1978): J. Phys. B *11*, 755; Rose, S.J., Grant, I.P., Pyper, N.C. (1978): J. Phys. B *11*, 3499

19. Computation of radial wave functions

α) The nonrelativistic treatment. The Hartree-Fock (HF) single-particle picture provides the main practical avenue for rate calculations on complex atoms, and we shall restrict our discussion to this model although some generalizations are indicated below. In Sect. 8δ the bound-state HF method was discussed. Here we consider the problem presented by the calculation of the wave function for the ejected electron moving in the field of a residual N-electron ion. In accordance with (8.23) the wave function

$$\Psi_{\mathcal{E}}=(N+1)^{-1/2}\sum_{\nu=1}^{N+1}(-1)^{N+1-\nu}\hat{P}_{N+1,\nu}\,\varphi_{\mathcal{E}}(X_{N+1})\,\Phi(\boldsymbol{X}_N) \tag{19.1}$$

represents an antisymmetric $N+1$ electron function which describes such a situation. The operator $\hat{P}$ permutes the electron labels $N+1$ and ν; the notation $\boldsymbol{X}_N$ represents the space-spin coordinates of all N electrons; the function Φ is normalized to unity. For simplicity, we have written only the spin orbital $\varphi_{\mathcal{E}}$ for the continuum electron of energy $\mathcal{E}$ and ignored angular momentum coupling since the extension is obvious.[1]

At this stage, the representation of the ionic wave function is unspecified. It could conceivably be fashioned from a superposition of configurations, i.e., incorporating some electron correlation features in the residual ion, or even a function with explicit r_{12} dependence. Restricting ourselves to the usual case when Φ is a combination of determinants representing a single configuration, one may ask which orbital basis set is appropriate – those for the initial hole state or those for the final state? It has been common to take the former choice although this is by no means universal (see Sect. 20γ). Employing orbitals from a calculation on the residual ion poses a nonorthogonality problem which we shall explore in Sect. 20γ.

Subsequent to a choice of bound-orbital basis set, the best $\varphi_{\mathcal{E}}$ can be obtained by employing the Kohn variational method, which makes the functional

$$L=\langle\Psi_{\mathcal{E}}|H-\mathcal{E}-E_{\Phi}|\Psi_{\mathcal{E}}\rangle \tag{19.2}$$

stationary under arbitrary variations of the radial part of $\varphi_{\mathcal{E}}$.[2] Since

$$r\,\varphi_{\mathcal{E}}=P_{\mathcal{E}l}(r)\,Y_{lm_l}(\hat{r})\,\chi_{m_s}(\zeta) \tag{19.3}$$

according to (7.12), we obtain immediately

$$\int d\boldsymbol{X}_N\,d\hat{r}_{N+1}\,d\zeta_{N+1}\,\Phi^*(\boldsymbol{X}_N)\,Y^*_{lm_l}(\hat{r}_{N+1})\,\chi_{m_s}(N+1)\,[H-\mathcal{E}-E_{\Phi}]\,\Psi_{\mathcal{E}}=0. \tag{19.4}$$

This is an inhomogeneous second-order integro-differential equation of the general form

$$\left[-\frac{1}{2}\frac{d^2}{dr^2}-\frac{Z}{r}+\frac{l(l+1)}{2r^2}-\mathcal{E}\right]P_{\mathcal{E}l}(r)+V(r)\,P_{\mathcal{E}l}(r)=X_{\mathcal{E}l}(r), \tag{19.5}$$

[1] See, e.g. Stewart, A.L. (1967) in: Adv. At. Mol. Phys., Vol. 3, Bates, D.R., Esterman, I. (eds.), New York: Academic Press, p. 1

[2] Kohn, W. (1948): Phys. Rev. *74*, 1763; Seaton, M.J. (1953): Philos. Trans. (London) *245*, 469

in which the "direct" potential is given by

$$V(r)=\sum_{knl} a_{nl}^{k} \frac{Y^{k}(nl, nl; r)}{r} \qquad (0 \leqq k \leqq 2l \text{ and even}), \tag{19.6}$$

where $Y^{k}(nl, nl; r)$ corresponds to $n=n'$ and $l=l'$ in

$$Y^{k}(nl, n'l'; r)=\int_{0}^{r} ds \left(\frac{s}{r}\right)^{k} P_{nl}(s) P_{n'l'}(s)+\int_{r}^{\infty} ds \left(\frac{r}{s}\right)^{k+1} P_{nl}(s) P_{n'l'}(s). \tag{19.7}$$

The term $X_{\mathcal{E}l}(r)$ contains the nonlocal exchange potential and, when necessary, the nondiagonal Lagrangian multipliers λ_{nl}, ensuring orthogonality between bound and continuum radial functions of the same orbital angular momentum quantum number:

$$X_{\mathcal{E}l}(r)=\sum_{knl'} b_{nl'}^{k} \frac{Y^{k}(nl', \mathcal{E}l; r)}{r} P_{nl'}(r)+\sum_{n} \lambda_{nl} P_{nl}(r) \tag{19.8}$$

$$|l-l'| \leqq k \leqq l+l'.$$

The summations run over all the occupied shells n, l of the residual ion, and the coefficients a^{k}, b^{k} depend on the angular momentum quantum numbers of the ion core and of the final state $\Psi_{\mathcal{E}}$. The boundary conditions satisfied by the solution $P_{\mathcal{E}l}(r)$ are

$$\begin{aligned} &P_{\mathcal{E}l}(0)=0, \\ &P_{\mathcal{E}l}(r) \underset{r \to \infty}{\sim} \sqrt{\frac{2}{\pi k}} \sin\left(kr+\frac{\bar{Z}}{k} \ln 2kr-\frac{l\pi}{2}+\sigma_{l}+\delta_{l}\right), \end{aligned} \tag{19.9}$$

where $k^{2}=2\mathcal{E}$, $\bar{Z}=Z-N$ and

$$\sigma_{l}=\arg \Gamma(l+1-i\bar{Z}/k) \tag{19.10}$$

in accordance with (7.10). Note that $P_{\mathcal{E}l}(r)$ is normalized in the energy scale according to (7.11). The phase shift δ_{l} is due to non-Coulombic terms in the direct potential. Note that we have chosen the standing wave representation for the continuum function. The transformation to incoming and outging waves has already been seen in Sect. 1. Equation (19.5) provides a foundation for describing the potentials employed for calculating nonradiative transition amplitudes within the nonrelativistic scheme. We may emphasise that *only* when the continuum function is calculated by the rigorous HF method, (19.5), is the condition

$$\langle \Psi_{\mathcal{E}'} | H-E | \Psi_{\mathcal{E}} \rangle=\delta(\mathcal{E}'-\mathcal{E})(\mathcal{E}+E_{\Phi}-E) \tag{19.11}$$

satisfied.

β) Some approximate potential methods. KELLY and co-workers[3], in a series of calculations on the neon $K-LL$ Auger process, employed the Silverstone-

[3] See II.50 and II.51

Huzinaga pseudopotential method. This is a convenient scheme for the generation of one-electron functions suitable for many-body-perturbation-theory (MBPT) calculations[4] since all orbitals of the same orbital angular momentum quantum number are computed as eigenfunctions of a common Hermitian operator. This tailoring for the purposes of MBPT makes it possible to choose the potential to mimic the situation in the final state of the nonradiative transition without an alteration of the initial bound-state orbitals. As an example, a bound-state basis set of s orbitals ($1s$ and $2s$) for neon can be calculated with the potentials [cf. (19.6) and (19.8)]

$$V(r)=[Y^0(1s,1s;r)+2Y^0(2s,2s;r)+6Y^0(2p,2p;r)]/r \tag{19.12}$$

$$X_{ns}(r)=[Y^0(1s,ns;r)P_{1s}+Y^0(2s,ns;r)P_{2s}+Y^1(2p,ns;r)P_{2p}]/r, \tag{19.13}$$

which are closely related to the RHF potential obtained from the variation of the $(1s)^{-1}$ hole state, differing only by $Y^0(1s,2s;r)P_{1/s}/2r$ in $X_{2s}(r)$. It is apparent that one can add to this potential any term of the form

$$V^*(r)=QRQ, \tag{19.14}$$

without altering eigenvalues and eigenfunctions of the occupied orbitals when the operator Q projects onto the orthogonal complement to the manifold of the occupied orbitals, i.e., $Q=1-|P_{1s}\rangle\langle P_{1s}|-|P_{2s}\rangle\langle P_{2s}|$. One is at liberty to select the Hermitian operator R that approximates the desired physical situation. For example, when

$$R=J^0(1s)-[2J^0(2s)-K^0(2s)], \tag{19.15}$$

where

$$J^0(1s)P_{ns}=\frac{Y^0(1s,1s;r)}{r}P_{ns}; \qquad K^0(2s)=\frac{Y^0(2s,ns;r)}{r}P_{2s} \tag{19.16}$$

for excited-state s functions, there is a cancellation of the $2s$ Coulomb potential terms and an extra $1s$ term is introduced, i.e., the situation is typical of the $1s^2\,2p^6$ residual ion in the $K-L_1L_1$ transition. Similarly,

$$R=J^0(1s)-\tfrac{1}{6}K^1(1s)-J^0(2s)+\tfrac{1}{6}K^1(2s) \tag{19.17}$$

simulates the $1s^2\,2s\,2p^5$ core for $l=1$ functions. There is obviously some arbitrariness in the choice of R. Note that one does not have to introduce Lagrangian multipliers to ensure orthogonality.

A useful and popular method used to approximate the solution of the *HF* equation is provided by the local-exchange-potential approximations, the first in our hierarchy of methods which have been used. All have their roots in the treatment of the exchange potential of the electron gas and have been in-

[4] Kelly, H.P. (1975) in: Atomic inner-shell processes ionization and transition probabilities, Vol. I, Crasemann, B. (ed.), New York: Academic Press, p. 331

vestigated intensively for atoms, molecules, and solids.[5] The primary motivation behind the upsurge in interest in these methods lies in the complication caused by the presence of the nonlocal exchange potential in (19.5), which depends itself on the radial function being calculated. This difficulty is side-stepped by introducing a statistical exchange approximation. In its simplest form[6] it is

$$X_{nl}(r) \to X_\alpha(r)\, P_{nl}(r); \qquad X_\alpha(r) = 3\alpha[(3/8\pi)\,\rho(r)]^{1/3}, \tag{19.18}$$

to be used in conjunction with an average of the direct potential (19.6) so that the total electronic potential becomes

$$V(r) = \sum_{nl} w_{nl} \frac{Y^0(nl, nl; r)}{r} - X_\alpha(r), \tag{19.19}$$

where w_{nl} is the occupation number of the n, l shell. In (19.18) α is the so-called exchange parameter and $\rho(r)$ is the spherically averaged electron density

$$\rho(r) = \frac{1}{4\pi r^2} \sum_{nl} w_{nl} P_{nl}(r)^2. \tag{19.20}$$

GASPAR and KOHN and SHAM, in a variational treatment of a nearly homogeneous gas, showed that a value of $\alpha_{\mathrm{KSG}} = 2/3$ is to be preferred over Slater exchange, $\alpha_{\mathrm{S}} = 1$.[7] However, treating α as an adjustable parameter and minimizing the expectation value of the atomic Hamiltonian leads to a value slightly larger than 2/3, varying from atom to atom. We should note here that ROSÉN and LINDGREN[8] have proposed an optimized potential taking the form

$$X_\alpha^{\mathrm{opt}}(r) = \frac{3\alpha}{r}[(3/8\pi)\, r^{k+2} \rho(r)^m]^{1/3}, \tag{19.21}$$

in which m, k and α are to be optimized. It is found that m is nearly one, while there is some correlation between k and α values. SLATER et al. have also discussed modified exchange potentials.[9] Since applications to calculations of nonradiative transition amplitudes have been confined to the X_α potential, (19.18) and (19.19), we shall focus attention on this potential.

One obvious objection to (19.19) is that it does not have the correct asymptotic behavior at large r, approaching N/r instead of $(N-1)/r$. This is due to an extra self-interaction which is automatically absent in the HF treatment. To correct for this feature, the so-called LATTER tail correction[10] is

[5] Slater, J.C. (1974): The self-consistent fields for molecules and solids, New York: McGraw-Hill

[6] See II.8

[7] Gaspar, R. (1954): Acta Phys. Acad. Sci. Hung. *3*, 263; Kohn, W., Sham, L.J. (1965): Phys. Rev. *140*, A 1133

[8] Rosén, A., Lindgren, I. (1968): Phys. Rev. *176*, 114

[9] Wilson, T.M., Wood, J.H., Slater, J.C. (1969): Phys. Rev. *179*, 28; Slater, J.C., Wilson, T.M., Wood, J.H. (1970): Phys. Rev. A *2*, 620

[10] Latter, R. (1955): Phys. Rev. *99*, 510

applied to *force* the proper behavior, viz., at radii greater than r_0, where

$$V(r_0)=(N-1)/r_0, \tag{19.22}$$

the Coulomb-exchange potential (19.19) is replaced by $(N-1)/r$.[11] Thus we have the basis of the X_α Hartree-Slater technique: for a given set of bound orbital occupation numbers, the discrete radial functions are determined self-consistently using potential (19.19). For continuum functions, the *same* potential is employed with the appropriate ejected electron energy. Orthogonality between bound and continuum functions is assured automatically.

When higher-order corrections for the inhomogeneous nature of the atomic electron distribution are incorporated,[12] one obtains the $X_{\alpha\beta}$ method. All main features, including the tail correction, remain identical to the X_α method except for the exchange potential, which becomes modified to

$$X_{\alpha\beta}(r)=X_\alpha(r)\left[1+\tanh\frac{\beta}{\alpha}G(r)\right] \tag{19.23}$$

$$G(r)=\left[\frac{4}{3}\left(\frac{\nabla\rho}{\rho}\right)^2-2\frac{\nabla^2\rho}{\rho}\right]\rho^{-2/3}. \tag{19.24}$$

The hyperbolic tangent was chosen to ensure that the divergences of $G(r)$ at very small and large r are removed. Test calculations showed that the optimum α was exactly α_{KSG} while β differed only slightly from atom to atom. With a value $\beta\sim 0.003$, SCHWARTZ and HERMAN[12] showed that the virial theorem, $2\langle T\rangle/\langle V\rangle=-1$, is nearly satisfied for representative atoms. The inhomogeneity correction is found to modulate the X_α potential by up to 10% at intermediate r and yields good bound state functions – as measured by comparison with HF quantities.

On a more empirical footing than the X_α technique and its variants are the potentials devised and employed by GREEN, SELLIN, and ZACHOR.[13] Both exchange and direct potential are replaced by an analytical expression,

$$\begin{aligned} V(r)&=-\frac{(Z-1)}{r}[\Omega(r)-1]\\ \Omega(r)&=[H(e^{r/d}-1)+1]^{-1}, \end{aligned} \tag{19.25}$$

in which H and d are adjustable parameters chosen such that the eigenvalues produced match the HF eigenvalues of the *ground-state* neutral atom. Thus, rigorously speaking, this potential is not applicable to the nonradiative case of a residual ion of charge two. Nevertheless, applications to calculation of nonradiative transition amplitudes have been made despite this inconsis-

[11] Liberman, D.A. (1970): (Phys. Rev. B *2*, 244) has devised a more sophisticated treatment of the asymptotic region

[12] Schwartz, K., Herman, F. (1972): J. Phys. (Paris) *33*, C 3, 277 and references therein. See also Folland, N.O. (1971): Phys. Rev. A *3*, 1535

[13] Green, A.E.S., Sellin, D.L., Zachor, A.S. (1969): Phys. Rev. *184*, 1

tency.[14] SZYDLIK and GREEN[15] have extended their parameters tabulations to ions and atoms for $Z \leqq 18$.

This brief résumé shows that the means by which nonradiative transition matrix elements can be calculated present a spectrum ranging in complexity from HF to semiempirical potential methods. The disadvantage of the approximate techniques described above is that the direct connection with wave function (19.1), angular momentum coupling, and orbital structure disappears from the radial equations. The X_α potentials formed from the initial state disconnect $P_{\varepsilon l}$ even from the *configuration* of the residual ion. Therefore the dependence of the radial continuum function on multiplet structure of the ion (and hence the amplitude) is only partially accounted for by the energy adopted for the ejected electron. At the WENTZEL ansatz level of approximation, the convenience of having local exchange is a distinct advantage from the practical point of view. But for higher accuracy and consistency the deficiencies of the potential must be rectified and the correction may not be small, so that apparent agreement with experiment should be viewed with caution.

γ) The relativistic treatment. The total wave function for the final continuum state may be written exactly as in the nonrelativistic case (19.1) with fundamental differences peculiar to the relativistic formulation built into the orbital basis. Firstly, the one-electron function φ_ε has four components according to (12.3). We shall denote the large component $rR_{+1}(r)$ by P_κ and the small component $rR_{-1}(r)$ by Q_κ in the following. The phase in (12.3) has been chosen such that P and Q are both real. The quantum number κ is defined in Sect. 12. Secondly, the ion-core function Φ is a linear combination of determinants constructed from four-spinors (12.3). As before, we leave the coupling scheme, *jj* or intermediate, unspecified. The theory leading to the equations for the radial functions proceeds exactly as before with the electrostatic Hamiltonian (16.10) taking the place of the nonrelativistic Hamiltonian. One obtains in atomic units $(c=\alpha^{-1})$[16]

$$\begin{aligned}\frac{dP_\kappa}{dr}+\frac{\kappa}{r}P_\kappa+\frac{1}{c}\left[-(\mathcal{E}+2c^2)+V_n(r)+V(r)\right]Q_\kappa+X_Q(r)&=0\\ \frac{dQ_\kappa}{dr}-\frac{\kappa}{r}Q_\kappa-\frac{1}{c}\left[-\mathcal{E}+V_n(r)+V(r)\right]P_\kappa-X_P(r)&=0.\end{aligned} \tag{19.26}$$

The inhomogeneous term has the form

$$\begin{aligned}X_P(r)&=-\sum_{kn\kappa'} b^k_{n\kappa'}\frac{Y^k(n\kappa',\mathcal{E}\kappa;r)}{cr}P_{n\kappa'}+\frac{1}{c}\sum_n \lambda_{n\kappa}P_{n\kappa}\\ X_Q(r)&=-\sum_{kn\kappa'} b^k_{n\kappa'}\frac{Y^k(n\kappa',\mathcal{E}\kappa;r)}{cr}Q_{n\kappa'}+\frac{1}{c}\sum_n \lambda_{n\kappa}Q_{n\kappa},\end{aligned} \tag{19.27}$$

[14] Chen, M.H., Crasemann, B. (1973) in: Proc. Int. Conf. Inn. Shell Ioniz. Phenomena Future Appl. Atlanta, Fink, F.W., Manson, S.T., Palms, J.M., Venugopala Rao, P. (eds.), Us. At. Energy Comm. Rep. No CONF-720404, Oak Ridge, Vol. I, p. 43

[15] Szydlik, P.P., Green, A.E.S. (1974): Phys. Rev. A 9, 1885

[16] See II.35

while the direct potential has the same form as in (19.6). In the definition of the Y^k functions (19.7), the nonrelativistic product $P_{nl}P_{n'l'}$ is replaced by $P_{n\kappa}P_{n'\kappa'} + Q_{n\kappa}Q_{n'\kappa'}$. Sums over l are replaced by sums over κ. For a homogeneously charged sphere, radius $R = 2.2677 \times 10^{-5} A^{1/3}$ a.u., where A is the atomic mass number, the nuclear potential $V_n(r)$ is given by

$$V_{\mathrm{n}}(r) = \begin{cases} -\dfrac{Z}{2R}\,[3-(r/R)^2] & r<R \\ -\dfrac{Z}{r} & r>R. \end{cases} \tag{19.28}$$

Other shapes, like the Fermi distribution, have also been employed.[16,17] In (19.26) the rest mass energy, c^2 a.u., has been subtracted from the energy of the ejected electron $\mathcal{E}(\geqq 0)$. Comparison of (19.26) with (19.5) shows that the DF treatment replaces the second-order integro-differential HF equation by two coupled first-order equations. This comparison may be completed by a specification of the boundary conditions to be satisfied by the (energy normalized) radial continuum functions[18]:

$$\begin{aligned} &P_{\mathcal{E}\kappa}(0) = Q_{\mathcal{E}\kappa}(0) = 0 \\ &P_{\mathcal{E}\kappa}(r) \underset{r\to\infty}{\sim} \sqrt{\frac{\mathcal{E}+2c^2}{\pi c^2 k}} \cos[\theta_\kappa(\mathcal{E}) + \delta_\kappa(\mathcal{E})] \\ &Q_{\mathcal{E}\kappa}(r) \underset{r\to\infty}{\sim} -\sqrt{\frac{\mathcal{E}}{\pi c^2 k}} \sin[\theta_\kappa(\mathcal{E}) + \delta_\kappa(\mathcal{E})], \end{aligned} \tag{19.29}$$

where

$$\begin{aligned} &\theta_\kappa = kr + y \ln 2kr - \arg\Gamma(\bar{\gamma} + \mathrm{i}y) - \tfrac{1}{2}\pi\bar{\gamma} + \eta, \\ &k^2 = 2\mathcal{E} + \mathcal{E}/c^2; \qquad \bar{\gamma} = +(\kappa^2 - \bar{Z}^2/c^2)^{1/2}, \\ &y = \bar{Z}(\mathcal{E}+c^2)/c^2 k; \quad \mathrm{e}^{2\mathrm{i}\eta} = -\frac{(\kappa - \mathrm{i}y/(\mathcal{E}+c^2))}{\bar{\gamma} + \mathrm{i}y}, \end{aligned} \tag{19.30}$$

and δ_κ is the phase shift due to non-Coulombic terms. In the extreme nonrelativistic limit, $\bar{Z}/c \ll 1$, $\mathcal{E}/c^2 \ll 1$, the large component becomes identical to (19.9) and the small becomes negligible.

In the same way as described in Sect. 19β for the nonrelativistic case, one can envisage a whole series of approximate potentials which could be employed to simulate the more rigorous DF potential in (19.26). We have already seen that the statistical potential methods are useful for bound-state problems,[19] and indeed, most of the available calculations of nonradiative tran-

---

[17] See also V.23 and V.36

[18] Rose, M.E. (1961): Relativistic electron theory, New York: Wiley

[19] See V.17

sition rates use the relativistic analogue of the X_α method (see Sect. 20α):

$$X_{P(\mathrm{or}\,Q)} \rightarrow X^{\alpha}_{P(\mathrm{or}\,Q)} = -\frac{3\alpha}{c}\left[\frac{3}{8\pi}\,\rho(r)\right]^{1/3} P(\mathrm{or}\,Q), \tag{19.31}$$

where

$$\rho(r)=\frac{1}{4\pi r^2}\sum_{n\kappa} w_{n\kappa}(P_{n\kappa}^2+Q_{n\kappa}^2). \tag{19.32}$$

The Latter tail correction is to be applied as before.

δ) Some practical considerations. There are too few details in the literature concerning the intimate details of the numerical techniques, which have been employed to obtain continuum orbitals, for the evaluation of the nonradiative transition amplitudes, especially pertaining to the radial mesh. However, the intrinsic differences between the nature of bound and continuum functions allow one to make certain statements regarding the overall strategy. Basically, the continuum function calculation may be divided into three phases: (i) choice of mesh and interpolation; (ii) numerical integration of differential equations; (iii) normalization.

(i) The first follows from the oscillatory nature of the radial continuum functions at large r, which contrasts with the exponential decay of bound functions. For the latter orbitals, in order to accomodate the rapidly changing nuclear potential at small r, the mesh points for numerical integration must lie close together in this region, while there is no major loss of accuracy in permitting an increase at large r. For example, a logarithmic mesh of the form $r=\mathrm{e}^x$ can be employed.[16,20] Obviously, this type of mesh is prohibited for the continuum function, being inadequate to represent the oscillating function at large r. One suitable mesh has been suggested by Chernysheva et al.[21] in which the independent variable is given by $x=ar+b\ln r$ with $b\sim 1$. At large and small r, the radial step sizes are

$$\Delta r \sim \Delta x/a \quad \text{and} \quad \Delta r \sim r(\mathrm{e}^{\Delta x/b}-1), \tag{19.33}$$

respectively, i.e., equidistant steps at large r and decreasing with r near the nucleus. (As a rough guide, Bhalla and Ramsdale[22] have suggested that the radial mesh be chosen such that about 20 points are contained in a single wavelength, $\lambda \sim 2\pi/k$.) Thus, either the bound-state orbitals are also solved on such a mesh, or an interpolation stage is necessary.

(ii) The degree of complexity involved in the second phase, solution of the differential equations, depends heavily on whether the exchange potential is local or nonlocal. It is worth noting that numerical integration techniques are not universally employed. McGuire[23] replaced the HS potential by seven

---

[20] See Ref. [*13*], Sect. 6.2

[21] Chernysheva, L.V., Cherepkov, N.A., Radojević, V. (1976): Comput. Phys. Commun. *11*, 57

[22] See III.3

[23] McGuire, E.J. (1969): Phys. Rev. *185*, 1; (1970): Phys. Rev. A *2*, 273

straight lines and solved the radial equation (for bound and continuum) in each region exactly in terms of Whittaker functions with matching at the boundaries. However, WALTERS and BHALLA[24], using numerical methods, pointed out that this procedure leads to inaccuracies that produce spurious local structure in the Z dependence of some rates. It would seem that seven straight lines are not enough.

The most popular method for solving second-order differential equations is the Numerov technique. The derivative in (19.5) is replaced by a finite difference expression

$$\frac{P_{\nu+1}-2P_\nu+P_{\nu-1}}{\Delta r^2}=\left[-2\mathcal{E}+\frac{l(l+1)}{r_\nu^2}-\frac{2Z}{r_\nu}+2V(r_\nu)\right]P_\nu-2X(r_\nu)$$
$$=f(r_\nu)\,P_\nu-2X(r_\nu). \tag{19.34}$$

Knowing the values of P at the points ν, $\nu-1$, one can solve for $P_{\nu+1}$ and repeat the process. Integration is usually started by using series expansions for P near the origin,[20]

$$P_{\varepsilon l}\sim r^{l+1}\left(1-\frac{Zr}{l+1}+\dots\right). \tag{19.35}$$

FROESE-FISCHER has described a "deferred difference correction"[25] to be used in conjunction with the Numerov procedure in order to increase the accuracy for a given mesh. A correction term,

$$CP_\nu=\Delta r^2\left(-\frac{1}{240}\,\delta^4+\dots\right)P_\nu'', \tag{19.36}$$

calculated from a *previous* application of the Numerov method is added to the right-hand side of (19.34). On iterating this process, one converges to a solution of order Δr^6 which may be compared with an error $O(\Delta r^4)$ obtained with (19.34). It should also be noted that the Numerov method is unstable when $f(r)<0$ (i.e., at small r for the continuum function). Alternative methods have been devised to overcome this disadvantage, e.g., the Gel'fand-Lokutsiyevski direct chasing method.[26] However, no problems arising from the straightforward application of (19.34) seem to have been noted in calculations of nonradiative transition amplitudes.

Less information is available on numerical techniques used in the relativistic problem. However, for the first-order coupled equations, BHALLA and RAMSDALE[22] employed the (fourth-order) Runge-Kutta technique. This method suffers from the disadvantage of too many function evaluations at each step. We refer also to the works of MATESE and JOHNSON and PRATT et al. for some alternative numerical techniques.[27]

---

[24] Walters, D.L., Bhalla, C.P. (1971): Phys. Rev. A *3*, 1919 and A *4*, 2164

[25] Froese Fischer, C. (1971): Comp. Phys. Commun. *2*, 124

[26] Berezin, I.S., Zhidkov, N.P. (1964): Computing methods, Vol. II, Oxford: Pergamon Press

[27] Matese, J.J., Johnson, W.R. (1965): Phys. Rev. *140* A 1; Pratt, R.H., Levee, R.D., Pexton, R.L., Ron, A. (1964): Phys. Rev. *134* A 898, 916; Pratt, R.H., Ron, A., Tseng, H.K. (1973): Rev. Mod. Phys. *45*, 273

Whatever numerical procedure is employed, one has to start the integration by some means for both P and Q. As before, series expansions near $r=0$ provide a convenient route. The nature of these expressions depends on the nuclear model and the potential. Unlike the nonrelativistic case, the radial wave functions $(P/r, Q/r)$ for s and $p_{1/2}$ ($|\kappa|=1$, $j=1/2$) are weakly singular at the origin for a point nucleus,

$$\begin{aligned} P &\sim r^{\gamma}[1+a_1 r+\ldots] \\ Q &\sim r^{\gamma}[b_0+b_1 r+\ldots], \end{aligned} \tag{19.37}$$

since $\gamma=(1-Z^2/c^2)^{1/2}<1$ for $Z<137$. This singularity is removed for a finite nucleus:

$$\begin{aligned} P &\sim r^{l+1}[1+a'_2 r^2+\ldots] \\ Q &\sim r^{l}[b'_0+b'_2 r^2+\ldots] \quad (b'_0=0,\ l=0). \end{aligned} \tag{19.38}$$

Alternatively, one can arbitrarily choose a value for P near $r=0$ and approximate the ratio P/Q using series expansions and proceed to integrate.

Due to the nonlocal potential and Lagrangian multipliers, there is also the added complexity of solving the differential equations for the Y^k potentials.[28] One can envisage an iterative scheme starting with $X=0$ in which each solution is repeatedly fed back into the equations until self-consistency is achieved. A combination of the type

$$P_{\nu+1}(\mathrm{in})=(1-c)\,P_\nu(\mathrm{out})+c\,P_\nu(\mathrm{in}) \tag{19.39}$$

generally gives better convergence. The accelerating parameter c is given a value in the range from zero to one. Iterative methods are not the only possibility for the nonrelativistic equations, and MARRIOTT has described a noniterative scheme which has been used extensively in close-coupling calculations.[29] SMITH has also summarized other numerical techniques which have been applied.[30]

(iii) Once the solutions are obtained, it is necessary to normalize, i.e., apply boundary conditions (19.9) and (19.29). Normalization procedures for the nonrelativistic continuum functions abound.[1,31] We shall therefore emphasize here that algorithms based on the asymptotic forms (19.9) and (19.29) do not provide a good method since one needs to integrate out to excessively large radii before these forms become even approximately valid. It is better to circumvent unnecessary computation. For example, SEATON and PEACH[32] described the so-called Strömgren method, useful in the (practical) asymptotic

---

[28] See Ref. [*13*] and II.35

[29] Marriott, R. (1958): Proc. Phys. Soc. Lond. *72*, 121; Smith, K., Burke, P.G. (1961): Phys. Rev. *123*, 174

[30] Smith, K. (1971): The calculation of atomic collision processes, New York: Wiley

[31] Burgess, A. (1963): Proc. Phys. Soc. *81*, 442; Martins, P. de A.P. (1968): J. Phys. B (Proc. Phys. Soc.) *1*, 154; Norcross, D.W. (1969): Comput. Phys. Commun. *1*, 88

[32] Seaton, M.J., Peach, G. (1962): Proc. Phys. Soc. *79*, 1296

region where (19.5) becomes

$$P'' + f(r)\, P = 0$$
$$f(r) = k^2 + 2\bar{Z}/r - l(l+1)/r^2. \tag{19.40}$$

The solution P is

$$P = \sqrt{\frac{2}{\pi}}\, \xi^{-1/2} \sin[\phi(r) + \delta]$$
$$\phi(r) = \int_a^r \xi\, dr \tag{19.41}$$
$$\xi^2 = f + \xi^{1/2} \frac{d^2}{dr^2} \xi^{-1/2}$$

and the lower limit a is chosen to satisfy the boundary condition (19.9). One may solve for ξ by iteration (usually one is sufficient):

$$\xi^{(0)} = f^{1/2}$$
$$\xi^{(1)} = \left[f + f^{1/2} \frac{d^2}{dr^2} f^{-1/2}\right]. \tag{19.42}$$

BURGESS provides analytic expressions for ϕ for the first iterated formula, while MARTINS describes the general procedure.[31] Collecting all terms up to order $1/r^4$, $\phi^{(1)}$ becomes $(k>0)$

$$\phi^{(1)} = x + m^{-1} \ln(1 + m^2\rho + mx) + \sigma_l - m^{-1} - m^{-1} \ln m - \tfrac{1}{2}\pi l + y$$
$$- \frac{x(3m^2 t + 4) + m\rho(3m^2 t + 2) + mt}{24(1 + m^2 t)\, x(x + m\rho)} + \frac{5(\rho - t)}{24 x^3}, \tag{19.43}$$

where

$$m = k/\bar{Z}; \quad \rho = \bar{Z} r; \quad x = (m^2\rho^2 - 2\rho - t)^{1/2}; \quad t = l(l+1)$$
$$y = \begin{cases} \left(\dfrac{t + 1/8}{\sqrt{t}}\right) \operatorname{arc\,cos}\left[\dfrac{\rho - t + mtx}{(1 + m^2 t)\rho}\right] & l>0 \\[2ex] \dfrac{1}{4(x + m\rho)} & l=0. \end{cases} \tag{19.44}$$

The constant N, which relates normalized and calculated functions $P = N P_{\text{calc}}$, is given by

$$N^{-1} = \tfrac{1}{2}\left[(a_1 - a_2)^2 \operatorname{cosec}^2\theta + (a_1 + a_2)^2 \sec^2\theta\right]^{1/2}, \tag{19.45}$$

in which

$$\theta = \phi^{(1)}(r_2) - \phi^{(1)}(r_1);$$
$$a_\mu = \sqrt{\frac{\pi\, \xi^{(1)}(r_\mu)}{2}}\, P_{\text{calc}}(r_\mu) \quad (\mu = 1, 2), \tag{19.46}$$

can be obtained from (19.42) and (19.43). After normalization, the phase shift may be obtained by comparing the calculated function with the regular and irregular solutions:

$$\begin{aligned} NP_{\varepsilon l} &= F\cos\delta_l + G\sin\delta_l \\ F = \sqrt{\frac{2}{\pi}}\,\xi^{-1/2}\sin\phi(r); &\qquad G = \sqrt{\frac{2}{\pi}}\,\xi^{-1/2}\cos\phi(r). \end{aligned} \tag{19.47}$$

Normalization has been discussed at some length since the HF equations also present the possibility that the asymptotic potential can be more complicated than that in (19.40). Any term in the direct potential with $k>0$ behaves at large r as $1/r^{k+1}$. For example, if the core is $sp^5\ {}^{1,3}P$ coupled to $l=1$ giving ${}^2S$, one has such a term in $Y^2(2p,2p;r)$. As BURGESS[31] has shown, the procedure described above can easily be generalized to such cases (see also the alternative method described by NORCROSS[31]). The state of the art is apparently not so well developed for relativistic wave functions. For example, BHALLA and RAMSDALE[22] write the asymptotic radial functions as

$$\begin{aligned} P_\kappa &\sim A(r)\cos(kr+\delta) \\ Q_\kappa &\sim -A(r)\sin(kr+\delta), \end{aligned} \tag{19.48}$$

where $A(r)$ is an oscillating function of r converging to unity. The calculated functions were normalized by dividing P and Q by the average of A over a few cycles. An alternative method has recently been described by ONG and RUSSEK.[33]

20. On factors affecting the nonradiative transition amplitudes

α) Relativistic effects. At the present time, relativistic calculations based on realistic model wave functions have been performed for K-. L-, and M-shell processes using the relativistic HS (Dirac-Fock-Slater, DFS) model.[34] In early exploratory studies, relativistic-screened-hydrogenic, Thomas-Fermi, and Hartree wave functions have been utilized.[35] In all cases, the calculations, based on MØLLER's formula (11.8), were performed within the jj-coupling scheme, which we have seen is strictly applicable only to the very highest atomic numbers. For $Z \geqq 55$, BHALLA and RAMSDALE's[22] calculations on $K-LL$ transitions agree well with experiment but are seriously in error at lower atomic

[33] Ong, W., Russek, A. (1978): Phys. Rev. A *17*, 120

[34] See Bhalla et al. and Chen et al. in III.3 and the recent works by Chen et al.: K-shell rates (Chen, M.H., Crasemann, B., Mark, H. (1980): Phys. Rev. A *21*, 436), K- and L-shell rates (Chen, M.H., Crasemann, B., Mark, H. (1979): At. Data Nucl. Data Tables *24*, 13), L-shell rates (Chen, M.H., Crasemann, B., Aoyagi, M., Mark, H. (1979): Phys. Rev. A *20*, 385; Chen, M.H., Crasemann, B., Mark, H. (1981): Phys. Rev. A *24*, 177), M-shell rates (Chen, M.H., Crasemann, B., Mark, H. (1980): Phys. Rev. A *21*, 449). For a summary, see Crasemann, B. (1981) in: Inner-Shell and X-ray Physics of Atoms and Solids, Fabian, D.J., Kleinpoppen, H., Watson, L.M. (eds.), New York: Plenum Press, p. 97

[35] E.g. the review by BURHOP and ASAAD (see Introduction)

numbers.[36] In the attempt to remedy this situation, ASAAD and PETRINI[37] formulated the relativistic analogue of intermediate coupling (IC) with CI for the $K-LL$ process.

ASAAD and PETRINI made use of the procedure described in Sect. 13 without including the optimization of the wave functions. The problem was reduced to a two-electron problem in which the outgoing electron's wave function $|\varepsilon l j\rangle$ was coupled to an $1s_{1/2}$ electron for a given J, M and the corresponding relativistic two-electron wave function was formed by a diagonalization of the relativistic IC matrix which was given with respect to a jj-coupled basis. The $J=0$ matrix included the CI between the $2s_{1/2}^{-2}$, $2p_{1/2}^{-2}$, and $2p_{3/2}^{-2}$ configurations. The $J=1$ matrix is given by (13.2), except that ASAAD and PETRINI did not use the $E_{\mathrm{av}}^{j'j}$ energies in the diagonal elements of their matrices but subtracted the ground-state energy. The difference was expressed in terms of the L binding energies, which were taken from the compilation of SEVIER.[38] The relaxation correction was introduced after the diagonalization using LARKINS's procedure (see Sect. 17). The SLATER integrals in both the diagonal and nondiagonal elements were approximated by LARKINS's relativistic "average" integrals.[39] The nine $K-LL$ IC-CI rates were calculated using BHALLA and RAMSDALE's HS values of the relativistic radial integrals (12.11). It was found that the total nonrelativistic $K-LL$ rate is only slightly less than the relativistic value except at the highest $Z(=48)$ studied but that the individual rates are rather disparate.

The absolute rates of the jj sextet (for the sake of comparison, we have combined the nine lines in ASAAD and PETRINI's calculations into six) calculated in this manner are shown in Fig. 9. One can see that the (IC-CI) effects on these rates are most marked for $Z \lesssim 55$, resulting in significant redistributions of intensity among the six groups: L_2L_3, L_1L_2 and L_2L_2 increase at the expense of the other three. Note also the convergence on the pure "jj group" limiting intensities at higher values of Z except for L_1L_1 and L_2L_2. These two transitions are always different from jj coupling for all Z due to the important CI between the residual ionic states.[40]

To set these results in perspective, shown in Fig. 10 is a comparison of the jj-group rates relative to $K-L_1L_1$ as predicted by both ASAAD and PETRINI and BHALLA and RAMSDALE with the experimental data compiled by BURHOP and ASAAD.[35] While the agreement at large Z is very good, the scatter of the

---

[36] Geiger, J.S. (1973) in: Proc. Int. Conf. Inn. Shell Ioniz. Phenomena Future Appl. Atlanta, Fink, F.W., Manson, S.T., Palms, J.M., Venugopala Rao, P. (eds.): Us. At. Energy Comm. Rep. No CONF-720404, Oak Ridge, Vol. I, p. 523

[37] See III.5. Recently, CHEN et al. (Chen, M.H., Crasemann, B., Mark, H. (1980): Phys. Rev. A *21*, 442) have used their DFS Auger matrix elements in a similar calculation. Their results are in gratifying agreement with those of Asaad and Petrini, thus confirming the trends shown by Figs. 9 and 10

[38] See V.60

[39] See V.69

[40] Note that there is an arithmetic error in the L_2L_3 rates of BHALLA and RAMSDALE (Asaad, W.N., private communication). These rates should be for each Z: 21, 6.98; 30, 8.83; 35, 9.44; 41, 10.03; 48, 10.56; 55, 10.97; 63, 11.37; 70, 11.67; 80, 12.05; 81, 12.07; 93, 12.51. These corrected values were employed in the construction of Figs. 9 and 10

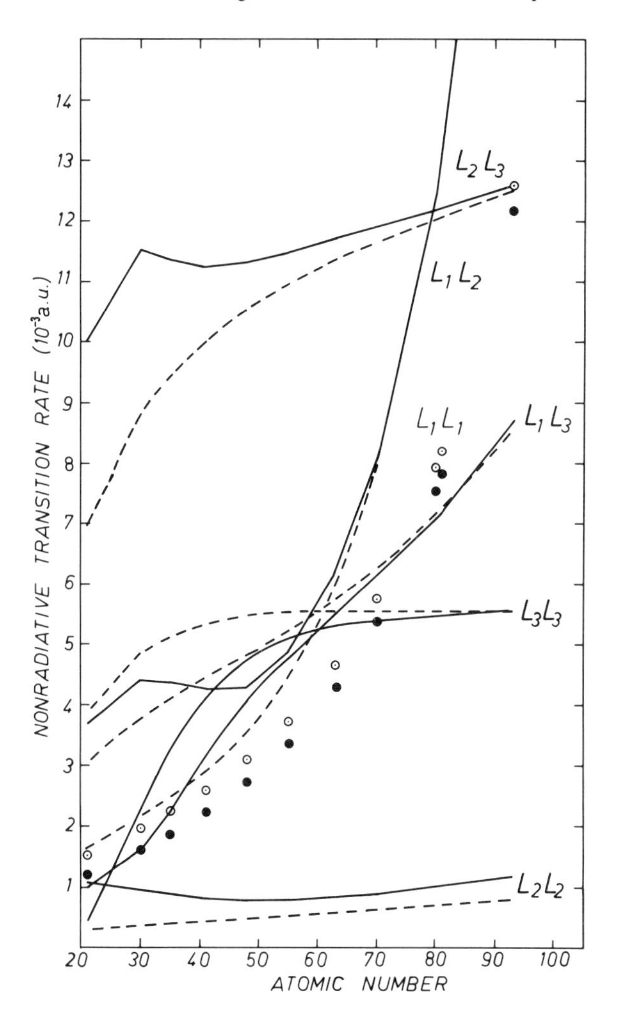

Fig. 9. Nonradiative $K-LL$ transition rates as a function of the atomic number in *jj* and intermediate coupling. The *jj* coupling values of BHALLA and RAMSDALE (III.3, see also footnote VI. 40) are shown as dashed lines and open circles ($K-L_1L_1$). The solid lines and the circles correspond to the intermediate-coupling values of ASAAD and PETRINI (III.5)

experimental observations at low Z makes any definite statement rather difficult, although it does seem that the general trend is better described by the IC-CI calculations, especially for L_1L_2. Note the differing behavior of the L_1L_2 and L_1L_3 transitions at low Z due to the superiority of intermediate coupling to the *jj*-coupling description at these atomic numbers. This trend is in accordance with the universal result regarding the intermediate coupling between $(n's)(np)^5$ states shown in Fig. 7. GEIGER's conclusion that the experimental data are inadequate in the region of Z ranging from 20 to 50 is still applicable.[36]

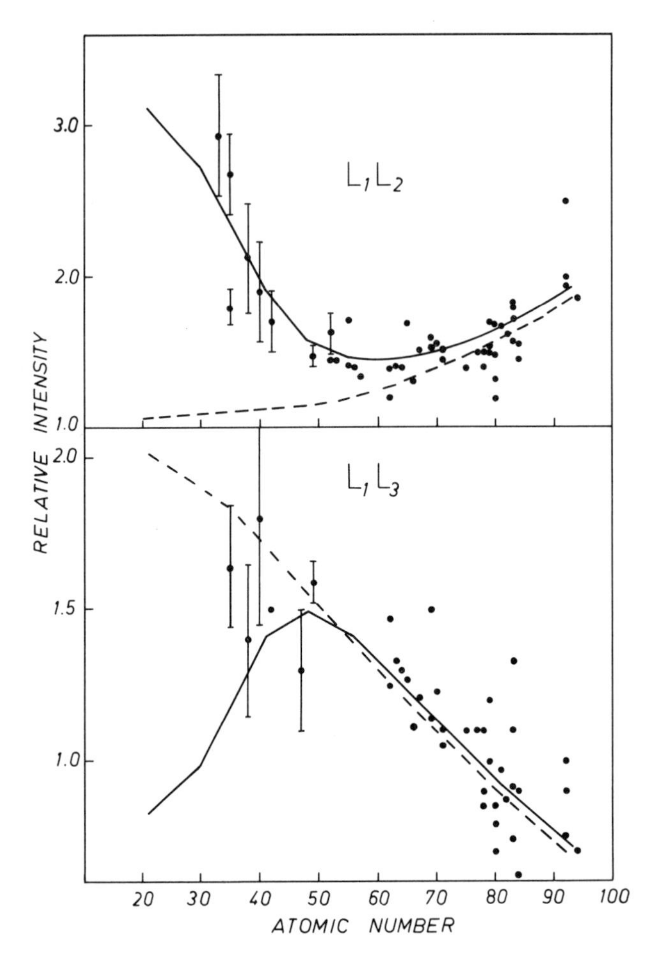

Fig. 10. A comparison between calculated and experimental relative rates. The rates are given with respect to the $K-L_1L_1$ rate. The solid lines correspond to the intermediate-coupling values of ASAAD and PETRINI (III.5) and the dashed lines to the *jj*-coupling values of BHALLA and RAMSDALE (III.3, see also footnote VI. 40). The experimental points are from a compilation by BURHOP and ASAAD (see the Introduction) (Fig. 10 continued on opposite page)

For the $L_1-L_2M_{45}$ CK transitions, $Z=32$–41, TALUKDAR and CHATTARJI[41] presented rates based on screened analytic solutions to the Biedenharn symmetric Dirac-Coulomb Hamiltonian. This suggests that relativistic effects are pronounced especially close to the cutoff point of the transition. While agreeing with MCGUIRE's criticism that relativistic effects should be ascertained with a more realistic model, the effects are large enough to warrant further studies.[42] CHEN et al.[34] have used the relativistic HS (DFS) model for

[41] Talukdar, B., Chattarji, D. (1970): Phys. Rev. A *1*, 33

[42] McGuire, E.J. (1971): Phys. Rev. A *3*, 587; see also the review by Bambynek et al. in the Introduction

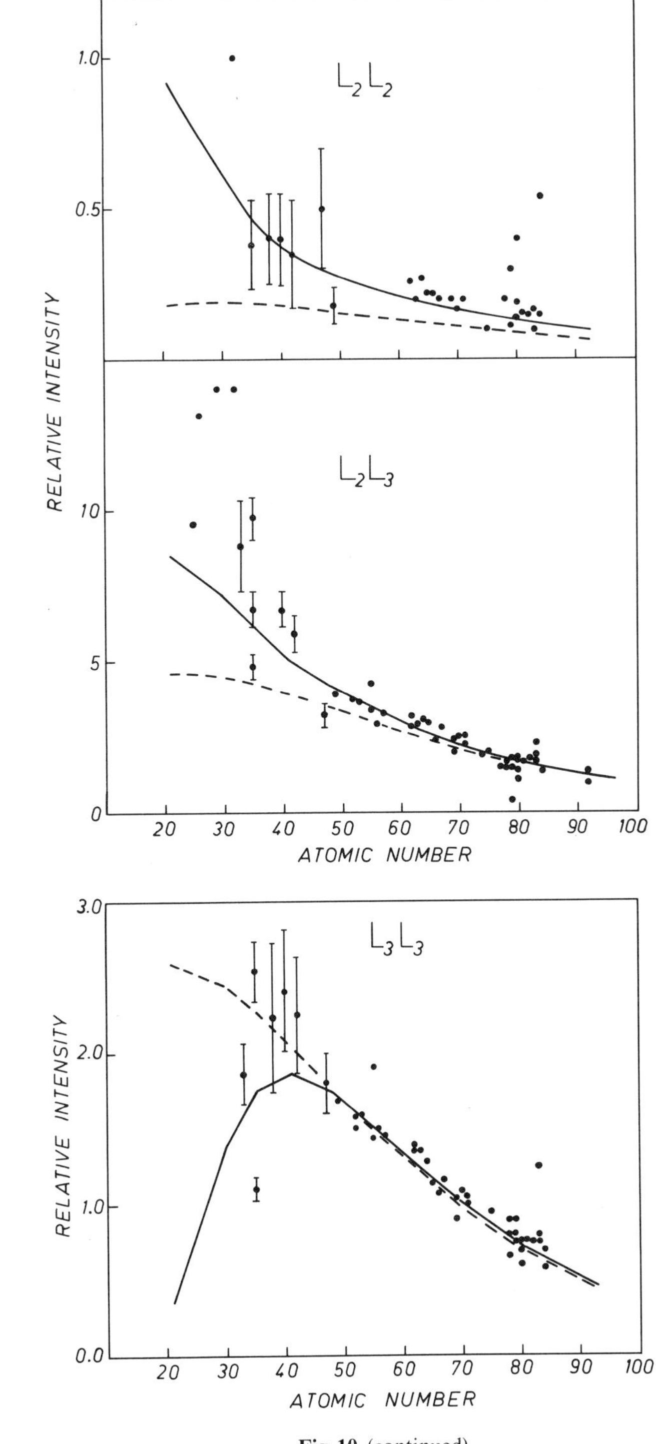

Fig. 10 (continued)

extensive calculations of L-shell nonradiative rates in the range $Z=70$–96. The relativistic effects are found to affect the L_v ($v=1, 2, 3$) total rates by 10 to 25% and individual transition rates to various final jj configurations by 40% at $Z=80$. For some weak M-shell transitions the effect seems to be even larger.[34]

According to the theory in Sects. 6 and 11, the relativistic amplitude includes final-state photon-electron interaction effects. Unfortunately, it is impossible to disentangle these effects from the mass variation and spin-orbit coupling on the basis of available relativistic calculations.

β) The choice of the local potential and kinetic energy. The sensitivity of rate calculations to the exchange potential parameter (see Sect. 19β) has been discussed by Walters and Bhalla[24] for the K and L_{23} shells. As a general trend, for $K-LL$ transitions, Slater exchange, $\alpha=1$, leads to total rates some 15 to 25% larger than does α_{KSG} for $Z<10$ but has progressively less influence at higher Z, being about 2% at $Z=50$. The effect of using the $X_{\alpha\beta}$ potential is less than that of the exchange parameter but still about 5% for $Z<10$. For the total L_{23} rate, the difference becomes more significant. At $Z=17$, the difference between Slater exchange and $X_{\alpha\beta}$ is about 30% in absolute value, falling to 5% at $Z=50$. This sensitivity to exchange approximation apparently increases with the principal quantum number of the shell considered (concomitant, of course, with the decrease in ejected electron energy *vide infra*). The investigations by Chen et al.[43] demonstrated the effect of exchange potential on total N_1 and N_3 rates for the elements $Z=46$–50. In particular, for the N_3 rate at $Z=50$, dominated by the $N_3-N_{45}N_{45}$ Super CK transitions, $\alpha=1$ leads to a rate which is about 1.7 larger than that based on α_{KSG}. This variability is of some concern, since McGuire[23] employs Slater exchange while Bhalla and co-workers favor the KSG value - note that each choice defines a different set of bound orbitals.

We have seen in Sect. 19β that the Green-Sellin-Zachor potential does not approach the correct limit at large r and consequently affects the normalization and behavior of the continuum function. With this in mind, it is therefore somewhat surprising to find that for $K-LL$ transitions[44] the agreement with $X_{\alpha\beta}$ matrix elements[45] was better than 3%. Total L_2 rates based on this model[14] agree with those[24] based on $X_{\alpha\beta}$ to about 5% at low and intermediate atomic numbers and to about 10% at high atomic numbers. It is apparently not known whether this agreement is universal and extends to component rates in general.

The effect of including the exchange correctly in the potential can be seen when HF calculations are compared with their HS counterparts. In Sect. 18α we remarked that the Ne $K-LL$ energies predicted by the $X_{\alpha\beta}$ method were all but identical to the HF values. This suggests that the HF and HS bound orbitals of the $(1s)^{-1}$ state are rather similar and that one may gain some insight into the effect of potential on rate by a comparison of the results of both procedures. A sample of such calculations for neon, magnesium, and

[43] Chen, M.H., Crasemann, B., Yin, L.I., Tsang, T., Adler, I. (1976): Phys. Rev. A *13*, 1435
[44] Chen, M.H., Crasemann, B. (1973): Phys. Rev. A *8*, 7
[45] Walters, D.L., Bhalla, C.P. (1971): At. Data *3*, 301

Table 5. A comparison of $K-LL$ Auger decay rates calculated for neon, magnesium, and argon by the X_α, $X_{\alpha\beta}$, and HF methods employing initial $(1s)^{-1}$ bound-state orbitals and LS coupling.[a] No configuration interaction is included. Units are 10^{-3} a.u.

Transition	Ne			Mg		Ar	
	X_α	$X_{\alpha\beta}$	HF	X_α	HF	X_α	HF
$K-L_1L_1$ ^1S	0.82	0.83	1.14	1.01	1.33	1.35	1.63
$K-L_1L_{23}$ ^1P	1.86	1.83	2.33	2.46	2.91	3.56	4.04
$K-L_1L_{23}$ ^3P	0.56	0.61	1.07	0.72	1.17	0.95	1.30
$K-L_{23}L_{23}$ ^1S	0.40	0.39	0.44	0.56	0.58	0.86	0.89
$K-L_{23}L_{23}$ ^1D	5.19	5.14	5.44	7.28	7.27	10.81	11.03
Total	8.83	8.81	10.41	12.03	13.26	17.53	18.89

[a] For references, see VI.45 (X_α), V.82 ($X_{\alpha\beta}$), and VI.46 (HF)

argon is shown in Table 5. The HF continuum functions were calculated by solving the appropriate equations with the correct nonlocal exchange potential for each final ^2S state constructed with the $(1s)^{-1}$ bound-state orbitals.[46] No configuration interaction or intermediate coupling is included. The results are seen to be markedly different, the HF values being, in general, larger than those of HS. Although the ratio $(\mathrm{HF}-X_\alpha)/X_\alpha$ calculated for each rate tends to fall with Z, it is still substantial for Ar. For the total rate this ratio is 18% (Ne), 10% (Mg), and 7.7% (Ar). If our assumption regarding the bound-state HS and HF orbitals holds good, this sensitivity is due *only* to the form of the continuum orbital as a result of the potential. As we shall see below, such results have some bearing on those calculations which include higher-order corrections to the WENTZEL ansatz.

The kinetic energy employed to calculate the nonradiative transition amplitude is a critical factor, especially for low-energy CK[47,48] and Super CK[43,49,50,51] transitions. A dramatic illustration of this dependence is shown in Table 6 for calculated total N_1 rates of silver and tin. Two energy estimates of the dominating $N_1-N_{23}N_{45}$ and $N_1-N_{45}N_{45}$ Super CK transitions are shown - as derived from (17.4) ($\Delta z=1$) and from relativistic HS calculations using the statistical energy.[43] The former procedure consistently overestimates the transition energy by 10 to 15 eV[48] and also predicts that the $N_1-N_{23}N_{45}$ transition in tin is allowed. Consequently, not only is $\Gamma(N_1)$ overestimated by a factor of 10 to 25, it is predicted to increase with atomic number in this region, contrary to experiment. A similar sensitivity to energy employed is exhibited by L- and M-shell CK transitions, especially near to the cutoff point of the transition where the energy may be very low.[48]

[46] See II.48 (neon) and Howat, G. (1978): J. Phys. B *11*, 1589. The argon results are unpublished

[47] Callan, E.J. (1963): Rev. Mod. Phys. *35*, 524; McGuire, E.J. (1971): Phys. Rev. A *3*, 1801

[48] Yin, L.I., Adler, I., Chen, M.H., Crasemann, B. (1973): Phys. Rev. A *7*, 897

[49] McGuire, E.J. (1974): Phys. Rev. A *9*, 1840

[50] Yin, L.I., Adler, I., Tsang, T., Chen, M.H., Ringers, D.A., Crasemann, B. (1974): Phys. Rev. A *9*, 1070

[51] Chen, M.H., Crasemann, B., Aoyagi, M., Mark, H. (1978): Phys. Rev. A *18*, 802

Table 6. A comparison of Super CK energies and widths for the N_1 level of silver and tin as calculated by McGUIRE and by CHEN et al. All quantities in eV

	McGUIRE[a]			CHEN et al.[b]			
	$N_1-N_{23}N_{45}$	$N_1-N_{45}N_{45}$	$\Gamma(N_1)$	$N_1-N_2N_{45}$	$N_1-N_3N_{45}$	$N_1-N_{45}N_{45}$	$\Gamma(N_1)$
Ag	26	83	41.9 (7.0)[c]	7.82	11.57	64.27	4.13
Sn	15	81	78.2 (3.1)[c]	forbidden		63.1	2.89

[a] McGuire, E.J. (1974): Phys. Rev. A *9*, 1840. Equation (17.5), $\Delta z=1$, using experimental BE's. $4p$ hole-state HS orbitals

[b] Reference VI. 43. Energies from relativistic HS calculations. Neutral atom HS orbitals

[c] Using Asaad-Burhop formula with ionization energies corrected for work function

γ) The effect of bound orbital basis set and relaxation. Apart from a brief comment in the early work of McGUIRE,[23] there have been few systematic investigations of the variation in calculated rate with discrete orbital basis. This is perhaps due to a predilection for the HS method, in which the orbital basis and potential for the continuum function are inextricably linked. In application of this method it has been customary to use initial hole-state wave functions throughout the calculation, whereas neutral-atom wave functions have been used in connection with the Green-Sellin-Zachor potential.[14] Note that, due to the tail correction (19.22), the HS potential coresonding to the initial state approaches $(Z-N-2)/r$, i.e., that of the doubly ionized core, for large r.

In the HF method the nonradiative transition amplitudes are solely determined by the bound-orbital basis set. Each choice of configuration and specification of coupling lead to a variationally optimum basis set which is used for the construction of the potential of the continuum orbital and the radial matrix elements. Hence the HF method is particularly suited for the study of the effect of the bound-orbital basis set, but so far there have been only a few attempts in this direction.[4, 46, 51–53]

The neon $K-LL$ transitions have been most thoroughly studied in this respect.[4, 46, 52, 53] KELLY[4] displayed several values for the Ne $K-L_1L_1$ rate calculated by assuming different bound-state orbitals. The continuum orbital was calculated in the field of the $(1s)^2(2p)^6$ ion and orthogonalized to $1s$ and $2s$ using the Silverstone-Huzinaga method described in Sect. 19 β. Different choices of $1s$ orbitals as either the initial- or final-state RHF function and $2s$ orbitals as a RHF or spin-polarised function for the initial state lead to a scatter of rates ranging between 7.60×10^{-4} a.u. to 11.4×10^{-4} a.u. CALLAN et al.[52] and HOWAT et al.[53] have shown using a variety of orbital bases that this scatter prevails also for the other $K-LL$ rates. It also appears that the value of the $G^1(2s\,2p)$ Slater integral, employed in the CI between $(2s)^{-2}$ and $(2p)^{-2}\,{}^1S$, is critical. Following the procedure of CALLAN et al., Fig. 11 attempts to illustrate the manner in which the R^k integrals involved in the Ne $K-LL$ rates

[52] Callan, E.J., Krueger, T.K., McDavid, W.L. (1969): Bull. Amer. Phys. Soc. *14*, 830

[53] Howat, G., Åberg, T., Goscinski, O. (1976): Extended Abstracts of Int. Conf. Phys. X-ray Spectra, NBS Gaithersburg, Maryland, Deslattes, R.D. (ed.), p. 35

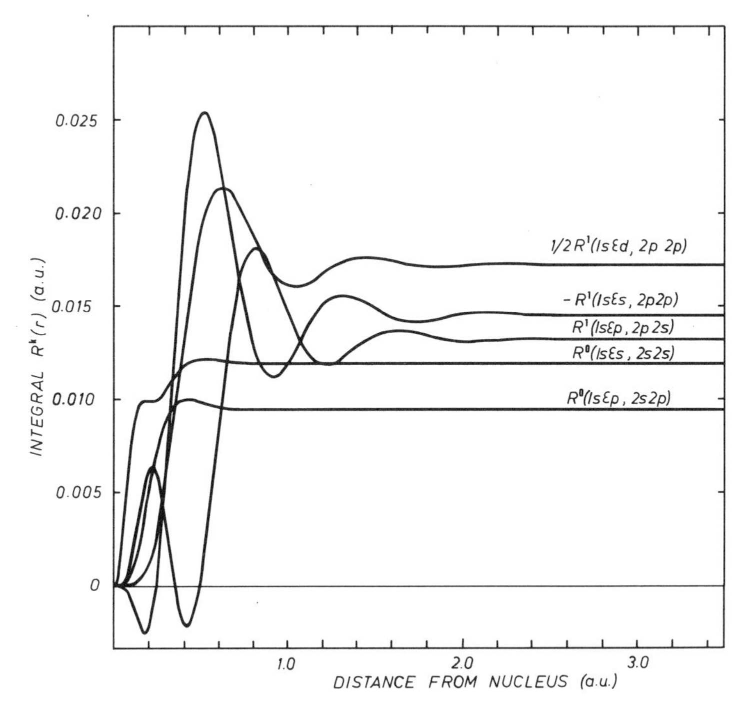

Fig. 11. The integral (20.1) as a function of the distance r from nucleus for various final L holes in neon. The expectation values of r are 0.16 a.u. (1s), 0.82 a.u. (2s), and 0.81 a.u. (2p) for the basis set used

build up as a function of r. The ordinate in Fig. 11 is given by

$$R^k(r) = \int_0^r dr' \, P_{\varepsilon l}(r') \, P_{n'l'}(r') \, Y^k(1s, n''l''; r')/r'. \tag{20.1}$$

The bound orbitals were the $(2s)^{-1}(2p)^{-1}$ transition operator set employed by Howat et al.[46] It is seen that the R^0 functions build up in the region of the $2s$ node and the R^1 functions, which have a pronounced oscillatory character, build up almost over the size of the whole atom. Hence R^0 and R^1 integrals are expected to be rather sensitive to details in the $2s$ and $2p$ wave functions in accordance with the calculations of $K-LL$ amplitudes with various L-shell wave functions. The $1s$ dependence of the amplitude is less significant but not negligible. This complex behavior of the R^k functions makes the dependence on orbital basis set difficult to systematize. It is also doubtful whether the simplified treatment of the Auger effect as a localized phenomenon that occurs very close to the nucleus[54] would lead to an understanding of this dependence.

This sensitivity to discrete orbital basis is of some concern when the effects of orbital relaxation are considered. We have already stressed in Sect. 18 that two optimal calculations on initial and final states are necessary to account for

[54] Fano, U., Theodosiou, C.E., Dehmer, J.L. (1976): Rev. Mod. Phys. *48*, 49

the energetics of the process. Hence one might also suppose that the "best" single particle basis for the calculation of the amplitudes would consist of both initial- *and* final-bound orbitals with the continuum function calculated in the field of the residual ion. As a consequence there is a nonorthogonality problem, unless it is possible to retain some link with initial and final configurations through either a modification of the potential[4] or a choice of an orbital basis set that characterizes the transition rather than the initial and final configurations separately.[55]

In order to explore the relaxation effect on the amplitudes using non-orthogonal orbitals, it is necessary, according to Sect. 10α, to replace the WENTZEL ansatz in the customary many-electron form[56]

$$W=2\pi\left|\langle\Psi_\varepsilon|\sum_{\nu<\mu}\frac{1}{r_{\nu\mu}}|\Phi\rangle\right|^2$$

by

$$W=2\pi|\langle\Psi_\varepsilon|H-E|\Phi\rangle|^2, \tag{20.2}$$

which involves the total Hamiltonian and the energy of the initial hole state. It has been found that the one-electron operator and the E term in (20.2) give nontrivial contributions to the rate.[57] Explicitly this is shown by (10.2), which applies when the initial- and final-state wave functions are single determinants of UHF spin orbitals.

Returning now to the usual RHF case, where the wave fuctions Ψ_ε and Φ are generally linear combinations of determinants, (10.2) may be used determinant by determinant to give the relaxed orbital amplitude formula. For $K-L_1L_1$ transitions in neon or neon-like ions this technique yields, for example, the formula

$$\begin{aligned}\langle\Psi_\varepsilon|H-E|\Phi\rangle=&-\langle 1s|1s\rangle\langle 2p|2p\rangle^6\Big[R^0(\varepsilon s\,1s\,2s\,2s)\\&+\langle\varepsilon s|2s\rangle\Big(\langle 1s|f|2s\rangle+\frac{R^0(1s\,1s\,1s\,2s)}{\langle 1s|1s\rangle}\\&+\frac{6R^0(1s\,2p\,2s\,2p)-R^1(1s\,2p\,2p\,2s)}{\langle 2p|2p\rangle}\Big)\\&+\langle 1s|2s\rangle\Big(\langle\varepsilon s|f|2s\rangle+\frac{R^0(1s\,\varepsilon s\,1s\,2s)-2R^0(1s\,\varepsilon s\,2s\,1s)}{\langle 1s|1s\rangle}\\&+\frac{6R^0(\varepsilon s\,2p\,2s\,2p)-R^1(\varepsilon s\,2p\,2p\,2s)}{\langle 2p|2p\rangle}\Big)\\&-\frac{\langle\varepsilon s|1s\rangle R^0(1s\,1s\,2s\,2s)}{\langle 1s|1s\rangle}-\mathscr{E}\langle\varepsilon s|2s\rangle\langle 1s|2s\rangle\Big],\end{aligned} \tag{20.3}$$

[55] See V.76

[56] See, e.g., the review of BURHOP and ASAAD in the Introduction; this is also the form employed by FRANCESCHETTI and MILLER (Proc. Int. Conf. Phys. X-ray Spectra (NBS, Gaithersburg 1976) p. 43; see also Miller, D.L., Franceschetti, D.R. (1976): J. Phys. B 9, 1471) in their study of nonradiative rates in light atoms using nonorthogonal orbitals

[57] See II.48 and Howat, G., Åberg, T., Goscinski, O. (1976): Proc. Sec. Int. Conf. Inn. Shell Ioniz. Phenom. (Freiburg) p. 126

where we use the approximation $\mathscr{E}=2\varepsilon_{2s}-F^0(2s\,2s)$. In (20.3) the two first orbitals in the Slater integrals (8.10) are final-state orbitals. If we use the approximations $\langle 1s|1s\rangle=\langle 2p|2p\rangle=1$ and $R^k(ffii)=R^k(fiii)$ inside the square brackets, we can eliminate the one-electron integrals by using the RHF equations for the initial $2s$ orbital. The final result using this procedure corresponds effectively to (10.6) in which the $\mathscr{E}$ term is missing.

In this way, the effect of nonorthogonality has been investigated for Ne, Mg, and Ar $K-LL$ and Mg $K-LM_1(M_1M_1)$ processes in LS coupling.[46] For the $K-LL$ case it is observed that the overlap correction to the lowest-order amplitude is at most a few percent for $K-L_1L_{23}\,{}^1P$, $K-L_{23}L_{23}\,{}^1S$, and ${}^1D$, while the $K-L_1L_1\,{}^1S$ and $K-L_1L_{23}\,{}^3P$ overlap corrections lower these rates by approximately 10%, 5% (Ne), 10%, 6% (Mg) and 8%, 8% (Ar), respectively. In particular, for neon, it was observed that the neglect of E in (20.2) would lead to an anomalously large overlap correction which would reduce the $K-L_1L_{23}\,{}^3P$ lowest-order rate by about 30%. Thus for consistency the *full* ΔE_{SCF} expression must be employed.

When one considers the $K-LM_1$ and $K-M_1M_1$ transitions in magnesium, the overlap correction becomes much more important than was apparent in the $K-LL$ case: the nonorthogonality corrections were estimated to increase the lowest-order rates of $K-LM_1\,{}^1S$ by 6%, of ${}^3S$ by 1%, of ${}^1P$ by 10%, of ${}^3P$ by 30%, and of $K-M_1M_1\,{}^1S$ by 78%. To see the reason for this, we make the same approximations as above, i.e., $R^k(ffii)\cong R^k(fiii)$, to remove the one-electron integrals from the overlap correction. For $K-M_1M_1$ transitions we obtain

$$\begin{aligned}\langle\Psi_\varepsilon|H-E|\Phi\rangle\cong -\{&R^0(\varepsilon s\,1s\,3s\,3s)\\&-\langle 1s|3s\rangle\,[R^0(\varepsilon s\,3s\,3s\,3s)+\tfrac{3}{2}R^0(\varepsilon s\,1s\,1s\,3s)]\\&-\langle\varepsilon s|3s\rangle\,R^0(1s\,3s\,3s\,3s)\\&-\langle 2s|3s\rangle\,[R^0(\varepsilon s\,1s\,2s\,3s)+R^0(\varepsilon s\,1s\,3s\,2s)]\\&-\langle\varepsilon s|1s\rangle\,R^0(1s\,1s\,3s\,3s)-\langle\varepsilon s|2s\rangle\,R^0(1s\,2s\,3s\,3s)\\&-\langle 1s|2s\rangle\,R^0(\varepsilon s\,2s\,3s\,3s)-\mathscr{E}\langle\varepsilon s|3s\rangle\langle 1s|3s\rangle\},\end{aligned}\tag{20.4}$$

where $\mathscr{E}=2\varepsilon_{3s}-F^0(3s\,3s)$. The dominant term in the correction is

$$\langle 2s|3s\rangle\,[R^0(\varepsilon s\,1s\,2s\,3s)+R^0(\varepsilon s\,1s\,3s\,2s)],\tag{20.5}$$

which can be seen to involve an expression reminiscent of the $K-L_1M_1\,{}^1S$ rate formulae. The coupling to the weak M_1M_1 transition results in a substantial modification in rate, as we saw above. Indeed, for all these transitions, it was found that the dominant correction could be ascribed to linking weak with (relatively) strong transitions. In particular, the $K-L_{23}M_1\,{}^{1,3}P$ correction links with the much larger $K-L_1L_{23}$ rates. Thus, it is implied that, when two final states differ by effectively a single excitation, e.g., LM and LL, it is possible that the relaxation correction to the smaller of the two rates may be quite large. This is a provisional statement since the overlap matrix element which appears as a coefficient, like $\langle 2s|3s\rangle$ in (20.5), must also be large enough to make the "link" effective. The work of CHEN et al.[51] has also revealed a significant influence of relaxation on CK and Super CK transitions, whereas FAEGRI and KELLY[58] find similar relaxation effects in the $K-LL$ spectrum of the HF molecule as in neon.

δ) Correlation effects. Electron correlation strongly perturbs the relative intensities of nonradiative transitions in the same manner as does intermediate coupling discussed in Sect. 20α. A most striking and familiar example is the substantial redistribution of intensity betwen $K-L_1L_1\,{}^1S$ and $K-L_{23}L_{23}\,{}^1S$ when configuration mixing between the $(2s)^{-2}$ and $(2p)^{-2}$ hole states[59] is incorporated into the expression for the amplitudes. Indeed, using $X_{\alpha\beta}$ wave functions, BHALLA[60] found good agreement between theory and experiment for Ne $K-LL$ transitions after introducing this interaction. This type of limited CI is now recognized to be an important feature of electron correlation for the calculation of nonradiative transition matrix elements. Like spin-orbit coupling, electron correlation can be a major source of intensity in formally forbidden channels due to this mixing between $N-2$ residual ion wave functions of the same symmetry. These features are well illustrated by the $M_{45}-N_1N_{23}\,{}^1P_1$ transitions in krypton, as discussed by MEHLHORN.[61] In this particular case, it is found that a mixed-coupling (see Sect. 9β) calculation without CI overestimates the intensity relative to the $M_{45}-N_{23}N_{23}\,{}^1S_0$ intensity by more than a factor of two. The discrepancy has been attributed to configuration mixing between ${}^1P_1$ states arising from configurations

$$(1)=4s\,4p^5; \qquad (2)=4s^2\,4p^3(^2P)\,5s;$$
$$(3)=4s^2\,4p^3(^3P)\,4d; \qquad (4)=4s^2\,4p^3(^2D)\,4d.$$

Diagonalization of the $N-2$ electron Hamiltonian with spin-orbit coupling mixes these four states. For the root associated with configuration $4s\,4p^5$, the intensity calculated before mixing is to be scaled by a factor $|c_1|^2$ since in the frozen-core model employed there is no contribution from the other states. Agreement between theory and experiment was enhanced considerably. Moreover, one can see that the intensity lost by the "parent" $4s\,4p^5$ is distributed among the erstwhile forbidden "discrete double nonradiative" transition by mixing in a contribution from the allowed transition. Indeed, these transitions are observed with substantial intensities, especially that to state (4), which acquires an intensity of 0.78 relative to the parent.

It is apparent that this CI model is of the same type as that described in Sect. 18β for the energetics of the transition and will operate mainly in the valence region of the ion. Most investigations have been confined to this

---

[58] Faegri, Jr., K., Kelly, H.P. (1979): Phys. Rev. A *19*, 1649

[59] See V.52

[60] Bhalla, C.P. (1973): Phys. Lett. *44* A, 103

[61] Mehlhorn, W. (1976) in: Photoionization and other probes of many-electron interactions, Wuilleumier, F.J. (ed.), New York: Plenum, p. 309

model and the effects of such configuration mixing.[62] However, it is well known that for outer-shell autoionizing states, final-state interchannel interactions strongly modify their spectra.[63] The first work to demonstrate their importance for Auger processes was the MBPT analysis of Ne $K-LL$ transitions by KELLY.[3] It is now understood that the inclusion of electron correlation beyond the simple CI scheme is vital for a satisfactory acount of the $K-LL$ Auger emission spectra of light atoms.[46] It has also been shown that interchannel interactions cause deviations from statistical transition probability ratios for $2p$ shake-off following the creation of a $1s$ hole in neon.[64]

In the work of HOWAT et al.[46] the scattering approach of Sect. 10β to incorporate these higher-order effects into the nonradiative transition amplitude has been employed. While the MBPT method includes *all* correlation effects implicitly, the scattering approach, at its simplest level, is more suited to the description of final-state interchannel coupling in a transparent manner. According to (10.8), the conventional WENTZEL formula is to be replaced by an expression revealing that the final-state continuum wave functions interact with each other as well as with the discrete state. In lowest order the partial rate is given by (10.9), where E is the initial hole-state energy. This formula prompts several observations. The form of (10.9) dictates that if the residual ion states of channels α and β have the same symmetry, they must be diagonal in the $N-2$ electron Hamiltonian. This is the interaction introduced by ASAAD[59] for $K-LL$ tranitions. Once this interaction is included, it can "link" channels for which no previous interchannel term existed e.g. $(2s)^{-2}\,{}^1S\varepsilon s\,{}^2S$ and $(2p)^{-2}\,{}^1D\varepsilon d\,{}^2S$. These terms are not small for neon.[46] Finally, as pointed out in Sect. 10β, only for a continuum orbital calculated with the correct nonlocal HF potential as in (19.5) is the intrachannel interaction $V_{\alpha\alpha}$ identically zero.

The effect of the final-state interchannel interaction can be seen in Fig. 12. Several calculations for Ne $K-LL$ rates in LS coupling are displayed for a variety of discrete orbital basis sets: $X_{\alpha\beta}$ and HF initial state $(1s)^{-1}$ orbitals,[65] transition operator orbitals for the $(1s)^{-1}\to(2s)^{-1}(2p)^{-1}$ transition,[46] and Kelly's basis, which consists of $1s$ and $\mathcal{E}l$ calculated for the final state and $2s$, $2p$ for the initial state[3]. The alteration in the calculated rates is dramatic and shows that the interchannel terms are just as important as the residual ion CI. The transitions $K-L_1L_1$ and $K-L_{23}L_{23}\,{}^1S$ lose and gain intensity by factors of about two over the lowest-order estimate. Also indicated in the figure is a strong dependence on the orbital basis set which seems to be mainly due to the dependence of the lowest-order amplitude on the orbitals employed. The $X_{\alpha\beta}$ results, which do not include the intrachannel interaction, seem to be in good agreement with the experimental observations. However, HOWAT et al.[65]

[62] Aksela, H., Aksela, S. (1974): Rep. of Dep. Phys. (Oulu University) No 41, unpublished; Aksela, S., Väyrynen, J., Aksela, H. (1974): Phys. Rev. Lett. *33*, 999; Aksela, H., Aksela, S. (1974): J. Phys. B 7, 1262; McGuire, E.J. (1977): Phys. Rev. A *16*, 2365

[63] Fano, U., Cooper, J.W. (1968): Rev. Mod. Phys. *40*, 441; Fano, U. (1975): J. Opt. Soc. Am. *65*, 979

[64] Chattarji, D., Mehlhorn, W., Schmidt, V. (1978): J. Electron. Spectosc. *13*, 97

[65] Howat, G., Åberg, T., Goscinski, O., Soong, S.C., Bhalla C.P., Ahmed, M. (1977): Phys. Lett. *60* A, 404

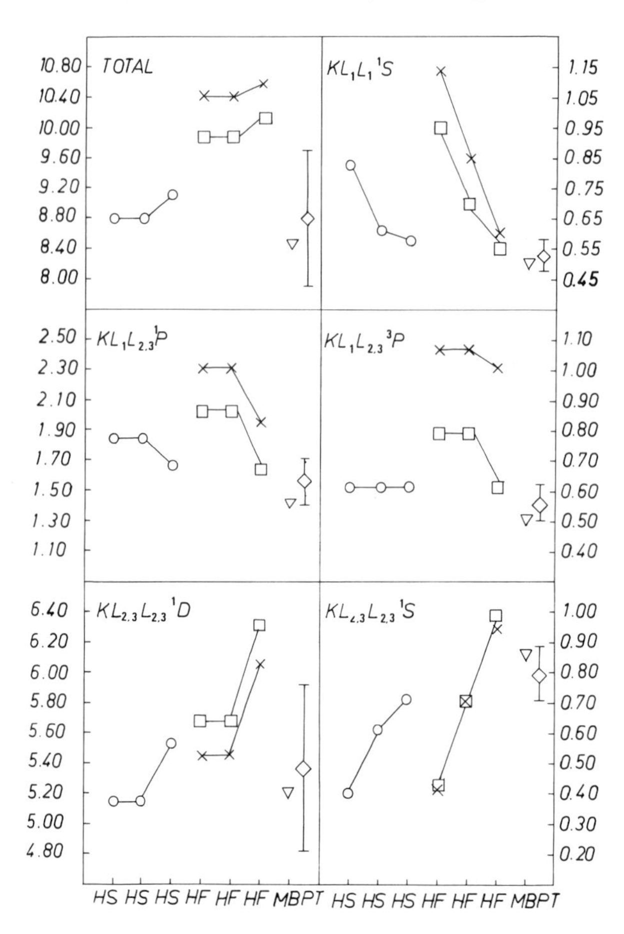

Fig. 12. Neon $K-LL$ Auger rates in units of 10^{-3} a.u. Crosses - the initial-state HF basis. Circles - the initial-state HS basis. Squares - KELLY's mixed HF basis. For each basis there are three rates which, from left to right, correspond to no correlation, the inclusion of $2s^{-2}\,^1S_0-2p^{-2}\,^1S_0$ core mixing only, and the inclusion of interchannel mixing in the final state. Triangles - KELLY's CORR I MBPT calculation (II.51). Diamonds - the most probable experimental values with approximately a 10% uncertainty in the determination of the absolute width of the neon K state (see the analysis of Krause, M.O., Oliver, J.H. (1979): J. Phys. Chem. Ref. Data 8, 329)

suggested that the neglected intrachannel terms will have the effect of shifting the $X_{\alpha\beta}$ rates away from the experimental region towards the corresponding HF values. Surprisingly, despite the considerable redistribution of individual rates, the total decay rate changes little. No sum rule seems to be operative, although there appears to be extensive cancellation of terms. Similar behavior to the Ne case has also been found for Mg $K-LL$ transitions.[46] The application to Ar $K-LL$ processes[46] reveals the interesting feature that the intercontinuum coupling terms are smaller in magnitude than in the Ne and Mg cases. In conjunction with the growth of the lowest-order amplitude with atomic number, it is inferred that such final-state interactions are likely to be of less importance at medium and large Z for $K-LL$ transitions.

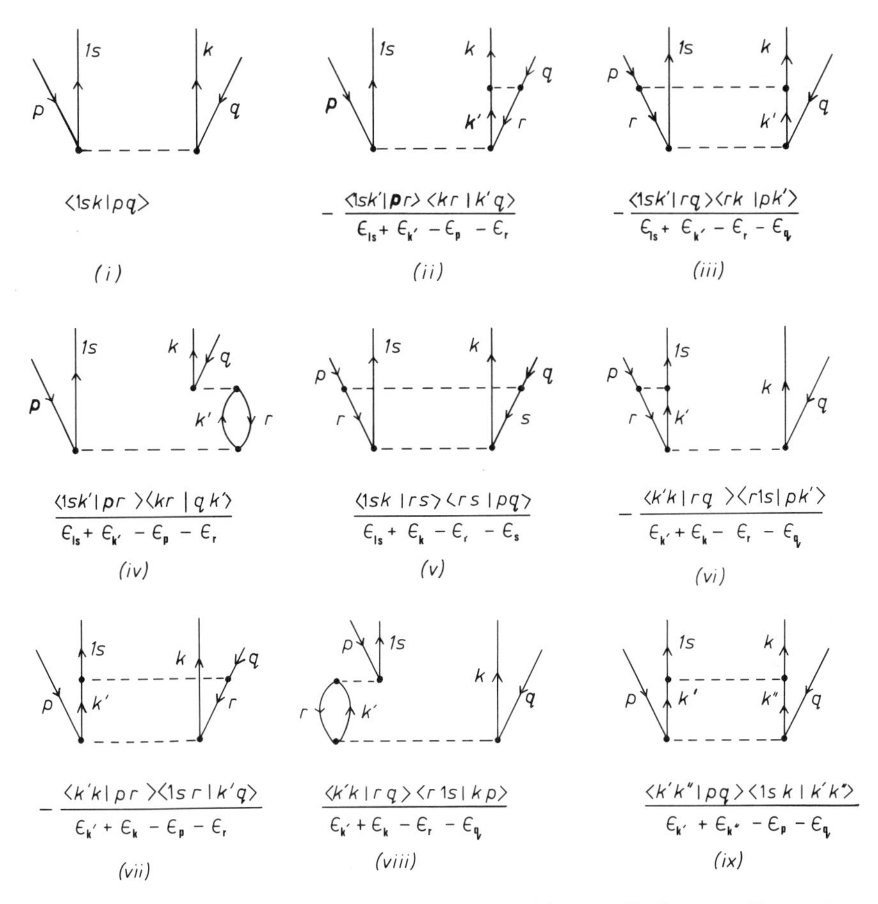

Fig. 13. Diagrams contributing to the neon $K-LL$ transition amplitudes according to KELLY (II.51) and their interpretation. In the diagrams, p and q denote the final L holes, r and s alternative core holes, and k the ejected electron. Note that the discrete and continuous summation over k' and k'' is not indicated explicitly

The major interactions which are missing from the calculations described above are those pertaining to initial-state correlation.[3] Figure 13 shows the diagrams which are involved in KELLY's MBPT calculation. The interpretation of diagrams (ii)–(iv) shows that they correspond to the interchannel interaction, whereas diagram (v) corresponds with $r=2s(2p)$, $s=2s(2p)$ to $2s^{-2}-2p^{-2}$ CI. Diagrams (vi)–(ix) correspond to CI in the initial state (see also Sect. 10γ). It is apparent from Fig. 12 that when these correlations are added to the results with CI in the final state only, the calculations are in good agreement with the experimental observations. In particular, $2p^6$ intrashell correlations are vital to the $K-L_{23}L_{23}\,{}^1S$, ${}^1D$ transitions in Ne. The reduction of the latter rate, the dominant mode for $K-LL$, is of major importance to the K width. For the other transitions, the initial-state correlations seem to exhibit a certain degree of cancellation among themselves. If this type of correlation declines in importance in the same manner as interchannel terms seem to do as Z increases, a lowest-order relativistic treatment including the $2s^{-2}-2p^{-2}$ CI and IC (determined by the MCDF method) may be sufficient for $K-LL$ transitions at intermediate and large Z values.

Appendix A. Some formulae for the treatment of the continuous spectrum

In his paper[1] on the configuration interaction in the continuum, FANO used DIRAC's procedure[2] involving the delta function for the diagonalization of the Hamiltonian matrix with respect to a continuous basis. This procedure and the use of the delta and related functions in general for describing the continuum have found their rigorous justification in the theory of generalized functions.[3] We briefly review, without any attempt of rigor, the background of these methods, which are also used to obtain solutions of the scattering equations.

Consider an *ordinary* one-dimensional function $y=f(x)$ which is defined almost everywhere for $-\infty<x<\infty$ and which is Lebesque integrable in each finite interval $a\leqq x\leqq b$. For such a function there exists a functional

$$(f,\varphi)=\int_{-\infty}^{+\infty} f(x)\,\varphi(x)\,dx \tag{A.1}$$

for every $\varphi(x)$ which is an infinitely differentiable "test" function with compact support (i.e., it vanishes outside some finite interval of x). The functional (A.1) is clearly a continuous linear functional on the space R of the "test" functions $\varphi(x)$. Now this is taken as a *definition* of a generalized function so that it is only defined through values of functionals. To be more precise, a generalized function is any functional f satisfying the conditions

(a) $(f,\alpha_1\varphi_1+\alpha_2\varphi_2)=\alpha_1(f,\varphi_1)+\alpha_2(f,\varphi_2)$ for all φ_1 and φ_2 in R and any numbers α_1 and α_2

(b) if $\varphi_1\ldots,\varphi_\nu\ldots\to 0$ in R, then $(f,\varphi_1),(f,\varphi_2)\ldots(f,\varphi_\nu)\to 0$.

Note that this definition does not necessarily imply that f can be represented by the formula (A.1). For example, Dirac's delta function $\delta(x)$, which fulfils the relation $(\delta,\varphi)=\varphi(0)$, satisfies (a) and (b) but (δ,φ) can apparently not be obtained from an ordinary function by calculating the integral in (A.1). However, if there is a sequence $f_1,f_2\ldots$ of ordinary functions such that

$$\lim_{\nu\to\infty}(f_\nu,\varphi)=(f,\varphi) \tag{A.2}$$

---

[1] See I.3

[2] Dirac, P. (1927): Z. Phys. *44*, 585

[3] This appendix is based on the following monographs: Shilov, G.E. (1968): Generalized functions and partial differential equations, New York: Gordon and Breach; Vladimirov, V.S. (1971): Equations of mathematical physics, New York: Dekker

for any φ, then (f, φ) defines a generalized function f. Actually, (A.2) still defines a generalized function if each (f_ν, φ) is a generalized function. This result can be written symbolically $f = \lim_{\nu \to \infty} f_\nu$. For example $(\varepsilon = \nu^{-1})$,

$$\delta(x) = \lim_{\varepsilon \to 0} \frac{1}{\pi} \frac{\varepsilon}{x^2 + \varepsilon^2}. \tag{A.3}$$

A frequently used generalized function is the "principal value" function $P(1/x)$, which is defined by

$$\left(P\frac{1}{x}, \varphi\right) = \mathscr{P} \int \frac{\varphi(x)}{x} dx = \lim_{\varepsilon \to 0} \left(\int_{-\infty}^{-\varepsilon} \frac{\varphi(x)}{x} dx + \int_{+\varepsilon}^{\infty} \frac{\varphi(x)}{x} dx \right). \tag{A.4}$$

Since a generalized function f multiplied by an infinitely differentiable function $\alpha(x)$ is a generalized function, it follows that

$$(f\alpha, \varphi) = (f, \alpha\varphi). \tag{A.5}$$

Hence

$$\left(xP\frac{1}{x}, \varphi\right) = \left(P\frac{1}{x}, x\varphi\right) = \int_{-\infty}^{+\infty} \varphi\, dx = (1, \varphi),$$

so that

$$xP\left(\frac{1}{x}\right) = 1 \tag{A.6}$$

for *all* x in the space of the generalized functions. According to (A.3) and (A.5),

$$(x\delta(x), \varphi) = (\delta(x), x\varphi) = 0. \tag{A.7}$$

Hence the general solution of the equation $fx = 0$ is $f(x) = c\delta(x)$, where c is an arbitrary number. By combining this result with (A.6) it can be shown that the general solution of the equation $fx = g$, where g is an ordinary function, is

$$f = P\frac{g(x)}{x} + c\delta(x). \tag{A.8}$$

This is how we obtain the solutions of (2.9) and (3.4) by the transformation (2.10).

If $\alpha(x)$ is a differentiable function, then

$$\lim_{\varepsilon \to 0} \frac{\alpha(x)}{x \pm i\varepsilon} = P\frac{\alpha(x)}{x} \mp i\pi\alpha(0)\,\delta(x). \tag{A.9}$$

The proof is based on the fact that

$$\lim_{\varepsilon \to 0} \int \frac{\alpha(x)\,\varphi(x)\,dx}{x \pm i\varepsilon} = \lim_{\varepsilon \to 0} \alpha(0)\,\varphi(0) \int_{-R}^{+R} \frac{x \mp i\varepsilon}{x^2 + \varepsilon^2} dx + \int_{-R}^{+R} \frac{\varphi(x)\,\alpha(x) - \varphi(0)\,\alpha(0)}{x} dx$$

$$= \mp i\pi\alpha(0)\,[\delta(x), \varphi(x)] + \left[P\frac{\alpha(x)}{x}, \varphi(x)\right]$$

for every $\varphi(x)$ which is zero for $|x|>R$. The result (A.9) is often written in the form

$$\lim_{\nu\to 0}\int_{-\infty}^{+\infty}\frac{\alpha(x)\,dx}{x-x_0\mp i\nu}=\mathscr{P}\int_{-\infty}^{+\infty}\frac{\alpha(x)\,dx}{x-x_0}\pm i\pi\alpha(x_0). \tag{A.10}$$

This limiting procedure is used in connection with the Lippmann-Schwinger equations and a number of other equations.

The derivation of boundary conditions (2.12), (2.13), and (3.5) is based on the validity of

$$\lim_{r\to\infty}\mathscr{P}\int_0^{\infty}\frac{e^{\pm ik(\tau)r}\,\alpha(\tau)\,d\tau}{E-\tau}=\mp i\pi\alpha(E)\,e^{\pm ik(E)r}, \tag{A.11}$$

where $\tau=k^2/2$. This result can be obtained from (A.10) by the complex integration of the integral in the left-hand side, which gives

$$\int_{-\infty}^{\infty}\frac{e^{\pm ikr}\,\alpha(k)\,dk}{k-k_0-i\nu}=\begin{cases}2\pi i e^{-\nu}\,\alpha(k_0)\,e^{ik_0r}\\ 0.\end{cases}$$

Another consequence of this result, namely, the formula

$$\lim_{\nu\to 0}\int_0^{\infty}\frac{k\sin kr_{12}\,dk}{(k-i\nu)^2-k_1^2}=+\frac{\pi}{2}e^{ik_1r_{12}}, \tag{A.12}$$

is needed for the derivation of (11.18).

Appendix B. Relationship between the configuration interaction and the projection-operator approach in the theory of resonances

Following FESHBACH,[4] two Hermitian projection operators, P and Q, such that

$$P+Q=1,\qquad P^2=P,\qquad Q^2=Q, \tag{B.1}$$

are introduced. From (B.1) it follows that $PQ=QP=0$. These projection operators are used to partition the Schrödinger equation

$$(H-E)\,\Psi=0 \tag{B.2}$$

into an equivalent pair of coupled equations

$$(H_P-E)\,P\Psi=-H_{PQ}\,Q\Psi \tag{B.3}$$

$$(H_Q-E)\,Q\Psi=-H_{QP}P\Psi, \tag{B.4}$$

[4] See I.16

where $H_P = PHP$, $H_Q = QHQ$, $H_{PQ} = PHQ$, and $H_{QP} = QHP$. Suppose that the inverse $(H_Q - E)^{-1}$ exists. Then Q can be solved from (B.4) and the solution can be substituted in (B.3). The result is

$$(H_P + V - E)\,P\Psi = 0, \tag{B.5}$$

where

$$V = H_{PQ}(E - H_Q)^{-1}\,H_{QP}. \tag{B.6}$$

From (B.5) it is seen that the component $P\Psi$ is an eigenfunction of an effective Hamiltonian $H_P + V$. The fundamental idea in FESHBACH's approach is the use of P as a projection operator which selects so-called open channels. This implies that if Ψ is a stationary scattering state which fulfils the Schrödinger equation (B.2), then

$$\lim_{r\to\infty} P\Psi = \lim_{r\to\infty} \Psi \tag{B.7}$$

for any r. According to (B.7), the wave function $P\Psi$ yields the same scattering properties as Ψ itself. Therefore, instead of using the Schrödinger equation (B.2) to study the scattering problem, it is possible to use (B.5), where P and Q can be chosen in the best way. Suppose that P has been chosen so that it includes all open channels. Then $Q = 1 - P$ is such that all eigenfunctions of QHQ are bound and belong to the so-called closed channels. If $H_Q\chi_\nu = E_\nu\chi_\nu$, then $Q = 1 - P = \sum_\nu |\chi_\nu\rangle\,\langle\chi_\nu|$. This shows that for any value of E different from E_ν the potential V can be written as

$$V = \sum_\nu H_{PQ}\frac{|\chi_\nu\rangle\,\langle\chi_\nu|}{E - E_\nu}H_{QP}, \tag{B.8}$$

which reflects the resonance behavior through the denominators. We assume that one of the eigenvalues E_μ is separated from the rest, and determine the solutions of (B.5) in its vicinity. Equation (B.5) is rewritten in the following form:

$$(H' - E)\,P\Psi = -\frac{H_{PQ}\chi_\mu\langle\chi_\mu|H_{QP}|P\Psi\rangle}{E - E_\mu}, \tag{B.9}$$

where all terms of V except the one on left has been absorbed in H'. The solutions of the homogenous equation

$$(H' - E)\,P\Psi = 0 \tag{B.10}$$

are denoted by $P\chi^+$ or $P\chi^-$ depending on whether they fulfil the outgoing or ingoing wave boundary conditions. Using the Green's function we obtain the solution of the inhomogeneous equation as

$$P\Psi^\pm = P\chi^\pm + \lim_{\nu\to 0}\,[E - H' \pm \mathrm{i}\nu]^{-1}\frac{H_{PQ}\chi_\mu\langle\chi_\mu|H_{QP}|P\Psi^\pm\rangle}{E - E_\mu}. \tag{B.11}$$

The same result could be obtained using transformation (2.10) and applying the appropriate boundary conditions. From (B.11) it follows that

$$\langle\chi_\mu|H_{QP}|P\Psi^\pm\rangle=\langle\chi_\mu|H_{QP}|P\chi^\pm\rangle+\Lambda_\mu\Delta_\mu, \tag{B.12}$$

where

$$\Lambda_\mu=\frac{\langle\chi_\mu|H_{QP}|P\Psi^\pm\rangle}{E-E_\mu},$$

$$\Delta_\mu=\lim_{\nu\to 0}\langle\chi_\mu|H_{QP}[E-H'\pm i\nu]^{-1}H_{PQ}|\chi_\mu\rangle. \tag{B.13}$$

As a consequence, result (B.11) can also be written

$$P\Psi^\pm=P\chi^\pm+\lim_{\nu\to 0}[E-H'\pm i\nu]^{-1}H_{PQ}\chi_\mu\frac{\langle\chi_\mu|H_{QP}|P\chi^\pm\rangle}{E-E_\mu-\Delta_\mu}, \tag{B.14}$$

from which the resonance properties can be deduced in the projection operator approach.[5]

In order to establish the relationship between the wave functions (B.14) and (3.6), we identify χ_μ as Φ and assume that $Q=|\Phi\rangle\langle\Phi|$. Then H' reduces to PHP and the projection operator P is given by

$$P=\sum_{\beta=1}^{N_c}\int_0^\infty|\chi_{\beta\tau}^\pm\rangle\langle\chi_{\beta\tau}^\pm|\,d\tau, \tag{B.15}$$

where the wave functions $\chi_{\beta\tau}^\pm$ are solutions of the equation

$$[PHP-(\tau+E_\beta)]\,\chi_{\beta\tau}^\pm=0. \tag{B.16}$$

Solving this equation is equivalent to the diagonalization of the continuum matrix (2.8).

The introduction of the operator $Q=|\Phi\rangle\langle\Phi|$ in (B.14) gives

$$P\Psi^\pm=\chi_{\alpha\varepsilon}^\pm+\frac{\langle\Phi|H|\chi_{\alpha\varepsilon}^\pm\rangle}{E-E_\Phi-\Delta}\lim_{\nu\to 0}[E-PHP\pm i\nu]^{-1}PH\Phi, \tag{B.17}$$

which, by virtue of (B.15), takes the form

$$P\Psi^\pm=\chi_{\alpha\varepsilon}^\pm+\frac{\langle\Phi|H|\chi_{\alpha\varepsilon}^\pm\rangle}{E-E_\Phi-\Delta}\cdot\lim_{\nu\to 0}\sum_{\beta=1}^{N_c}\int_0^\infty\frac{\langle\chi_{\beta\tau}^\pm|H|\Phi\rangle\,\chi_{\beta\tau}^\pm\,d\tau}{E-E_\beta-\tau\pm i\nu}, \tag{B.18}$$

where

$$\Delta=\lim_{\nu\to 0}\sum_{\beta=1}^{N_c}\int_0^\infty\frac{|\langle\chi_{\alpha\tau}^\pm|H|\Phi\rangle|^2\,d\tau}{E-E_\beta-\tau\pm i\nu}. \tag{B.19}$$

---

[5] See I.17 and Chung, K.T. (1972): Phys. Rev. A 6, 1809

Taking into account the orthogonality $\langle\Phi|\chi^{\pm}_{\alpha\tau}\rangle=0$ and using notations of Sect. 2 gives finally

$$P\Psi^{\pm}=\chi^{\pm}_{\alpha\mathcal{E}}+\lim_{\nu\to 0}\frac{M^{\pm}_{\alpha}(\mathcal{E},\mathcal{E})}{E-E_{\mathrm{r}}\pm\mathrm{i}\frac{\Gamma}{2}}\sum_{\beta=1}^{N_{\mathrm{c}}}\int_{0}^{\infty}\frac{M^{\pm}_{\beta}(\tau,E)^{*}\,\chi^{\pm}_{\beta\tau}\,d\tau}{E-E_{\beta}-\tau\pm\mathrm{i}\nu}. \tag{B.20}$$

In order to get the total wave function, $Q\Psi^{\pm}$ must be added to $P\Psi^{\pm}$. According to (B.4) and (B.12),

$$Q\Psi^{\pm}=\Lambda_{\mu}\chi_{\mu}, \tag{B.21}$$

where Λ_{μ} is given by (B.12) and (B.13). Again using the notations of Sect. 2, (B.21) can be written in the form

$$Q\Psi^{\pm}=\frac{M^{\pm}_{\alpha}(\mathcal{E},\mathcal{E})}{E-E_{\mathrm{r}}\pm\mathrm{i}\frac{\Gamma}{2}}\Phi. \tag{B.22}$$

A comparison of (B.20) and (B.22) with (3.6) shows that the projection-operator approach of FESHBACH[4] and the configuration-interaction approach of FANO[1] are identical provided the projection operators Q and P are identified according to (3.10).

Appendix C. A derivation of the Fano profile from scattering theory

We consider photoabsorption in the case of two channels, i.e., $N_{\mathrm{c}}=2$. Channel 1, which we shall call the elastic channel, is associated with a one-photon continuum $|\omega\rangle$ attached to the atomic ground state $|i\rangle$. Channel 2, which we shall call the inelastic channel, is associated with an electron continuum $|f,\varepsilon\rangle$ and the photon vacuum state $|0\rangle$. These continua interact with each other and with a discrete electronic state $|\Phi\rangle$ attached to $|0\rangle$. The situation is analogous to that of Fig. 5, except that the N-electron bound state $|\Phi,0\rangle$ replaces the $(N-1)$-electron state $|i,0\rangle$ and the outgoing wave channel $|i,\omega\rangle$ the ingoing-wave channel $|f,\omega\rangle$. The channels $|f,\varepsilon,0\rangle$ and $|\lambda,\varepsilon,0\rangle$, which describe the outgoing electrons, correspond both to ingoing waves.

Hamiltonian (4.3) induces a coupling between $|\Phi,0\rangle$ and $|i,\omega\rangle$ through H_{int} and between $|\Phi,0\rangle$ and $|f,\varepsilon,0\rangle$ through H_{el}. In addition, the elastic and inelastic channels interact through H_{int}. This interaction gives rise to the *direct* ionization. Interactions within the channels are neglected and only the $\boldsymbol{p}\cdot\boldsymbol{A}$ term is considered in H_{int}. The cross section for the scattering from the elastic channel into the inelastic channel is proportional to $|T'^{+}_{21}|^{2}$, where

$$T'^{+}_{21}=T^{+}_{21}+\frac{\langle\chi^{-}_{2\varepsilon}|H-E|\Phi,0\rangle\langle\Phi,0|H-E|\chi^{+}_{1\varepsilon}\rangle}{E-E_{\mathrm{r}}+\mathrm{i}\frac{\Gamma}{2}} \tag{C.1}$$

according to (4.1). In (C.1) Γ, which is assumed to be weakly dependent of E in the vicinity of $E_r \cong E_\Phi$, is given according to (4.2) by

$$\Gamma(E) = 2\pi\left(|\langle\chi_{1\mathcal{E}}^{-}|H-E|\Phi,0\rangle|^2 + |\langle\chi_{2\mathcal{E}}^{-}|H-E|\Phi,0\rangle|^2\right). \tag{C.2}$$

According to (2.20), the scattering wave functions $\chi_{1\mathcal{E}}^{+}$ and $\chi_{2\mathcal{E}}^{-}$ are given by

$$\begin{aligned}\chi_{1\mathcal{E}}^{+} &= |i\rangle|\omega\rangle + \lim_{\nu\to 0}\int_0^\infty \frac{|f,\tau\rangle|0\rangle\langle f,\tau,0|H_{\text{int}}|i,\omega\rangle\, d\tau}{E-E_f-\tau+i\nu}\\ \chi_{2\mathcal{E}}^{-} &= |f,\mathcal{E}\rangle|0\rangle + \lim_{\nu\to 0}\int_0^\infty \frac{|i\rangle|\omega\rangle\langle i,\omega|H_{\text{int}}|f,\mathcal{E},0\rangle\, d\omega}{E-E_i-\omega-i\nu}\end{aligned} \tag{C.3}$$

to first order in H_{int}. Here both $|f,\mathcal{E}\rangle$ and $|\omega\rangle$ are assumed for simplicity to be normalized per unit energy range. In accordance with (C.3) T_{21}^{+} is put equal to

$$D_f(\mathcal{E},\omega) = \langle f,\mathcal{E},0|H_{\text{int}}|i,\omega\rangle \tag{C.4}$$

in (C.1). It also follows that

$$\langle\chi_{2\mathcal{E}}^{-}|H-E|\Phi\rangle = M(\mathcal{E}) + \lim_{\nu\to 0}\int_0^\infty \frac{D_\Phi(\omega)\, D_f(\mathcal{E},\omega)\, d\omega}{E-E_i-\omega+i\nu} \tag{C.5}$$

and

$$\langle\Phi,0|H-E|\chi_{1\mathcal{E}}^{+}\rangle = D_\Phi(\omega) + \lim_{\nu\to 0}\int_0^\infty \frac{M(\mathcal{E})\, D_f(\tau,\omega)\, d\tau}{E-E_f-\tau+i\nu}. \tag{C.6}$$

Note that since $|f,\mathcal{E}\rangle$ fulfil the standing wave boundary condition (2.2), $M(\mathcal{E})$ and $D_f(\mathcal{E},\omega)$ can be assumed real. In (C.5) and (C.6) $D_\Phi(\omega)$ stands for the notation

$$D_\Phi(\omega) = \langle\Phi,0|H_{\text{int}}|i,\omega\rangle. \tag{C.7}$$

From (C.5) and (C.6) it follows using (A.10) that

$$\begin{aligned}T_{21}^{\prime +} = D_f(\mathcal{E},\omega) &+ \frac{1}{E-E_r+i\frac{\Gamma}{2}}\\ &\cdot[M'(\mathcal{E}) - i\pi D_\Phi(\omega)\, D_f(\mathcal{E},\omega)]\,[D'_\Phi(\omega) - i\pi M(\mathcal{E})\, D_f(\mathcal{E},\omega)],\end{aligned} \tag{C.8}$$

where the matrix elements

$$D'_\Phi(\omega) = \langle\Phi',0|H_{\text{int}}|i,\omega\rangle \quad\text{and}\quad M'(\mathcal{E}) = \langle\Phi'',0|H-E|f,\mathcal{E}\rangle \tag{C.9}$$

use the wave functions

$$\Phi' = \Phi + P\int_0^\infty \frac{M(\tau)|f,\tau\rangle\, d\tau}{E-E_f-\tau} \quad\text{and}\quad \Phi'' = \Phi + P\int_0^\infty \frac{D_f(\mathcal{E},\omega)|i,\omega\rangle\, d\omega}{E-E_i-\omega}. \tag{C.10}$$

Consequently (C.8) takes as a function of $\lambda=2(E-E_r)/\Gamma$ the form

$$T'^{+}_{21}=D_f(\mathcal{E},\omega)\left(1+\frac{q'-i}{\lambda+i}\right) \tag{C.11}$$

provided the principal parts of (C.10) are neglected in the cross terms involving $i\pi$. The parameter q' becomes after a similar approximation

$$q'=\frac{\Gamma_{nr}}{\Gamma}\left(1-\frac{\Gamma_r}{q^2\Gamma_{nr}}\right)q, \tag{C.12}$$

where $q=D'_\Phi(\omega)/\pi M(\mathcal{E})D_f(\mathcal{E},\omega)$ is the Fano parameter. In (C.12) Γ_{nr} represents the nonradiative "width" $\Gamma_{nr}=2\pi M(\mathcal{E})^2$ which together with the radiative "width" $\Gamma_r=2\pi D_\Phi(\omega)^2$, gives according to (C.2) the sum rule $\Gamma=\Gamma_r+\Gamma_{nr}$.

The direct ionization cross section σ^d is proportional to $D_f(\mathcal{E},\omega)^2$. Hence the total cross section σ is given according to (C.11) by

$$\sigma=\sigma^d\frac{(\lambda+q')^2}{\lambda^2+1}, \tag{C.13}$$

which is a generalization of the Fano profile. This result, related to that of NITZAN[6] and ARMSTRONG et al.,[7] accounts for the radiative width Γ_r that in a more general case could also include contributions from other radiative transitions than those from $|\Phi\rangle$ to $|i\rangle$. If we put $\Gamma_r=0$, q' reduces to q in (C.8) and Γ to Γ_{nr} in λ so that (C.13) becomes the original Fano profile.[1] In Fig. 14 we have plotted a few Fano profiles that clearly show the resonance behavior of σ in the vicinity of E_r. If q is very large, then (C.15) represents a Lorentz curve, in which case Γ is the halfwidth and can be interpreted as the total nonradiative transition probability.

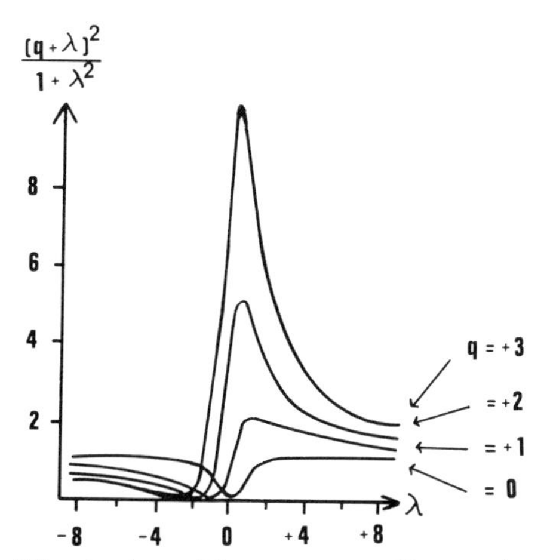

Fig. 14. Fano profiles for different values of the parameter q. For negative q values the scale of the abscissa should be reversed

---

[6] Nitzan, A. (1974): Mol. Phys. 27, 65

[7] See I.2

Appendix D. Equivalence of electron and hole wave functions with respect to symmetry

In this appendix we wish to show that if the wave function (8.52), which is given by

$$|\mathscr{E}, SLM_S M_L\rangle = \sum_{\alpha=1}^{Q} C_\alpha(\mathscr{E})\, D_\alpha(\dots,(nlm_l m_s),\dots), \tag{D.1}$$

is an eigenfunction of L^2 and S^2, which belongs to the eigenvalues $L(L+1)$ and $S(S+1)$, respectively, then the conjugate wave function (8.55), which is given by

$$|\mathscr{H}, SLM_S M_L\rangle = (-1)^{M_L} \sum_{\alpha=1}^{Q} C_\alpha(\mathscr{E})\, D_\alpha(\dots,(nl-m_l-m_s)^{-1},\dots), \tag{D.2}$$

is also an eigenfunction of L^2 and S^2, which belongs to the same eigenvalues.

The proof is based on the following formulae (e.g., Ref. [11])

$$L^2 = (L_x + \mathrm{i}L_y)(L_x - \mathrm{i}L_y) + L_z^2 - L_z \tag{D.3}$$

$$(L_x \pm \mathrm{i}L_y)\,|SLM_S M_L\rangle = \sqrt{(L \mp M_L)(L \pm M_L + 1)}\,|SLM_S M_L \pm 1\rangle \tag{D.4}$$

$$(L_x \pm \mathrm{i}L_y)\, D = \sum_v \sqrt{(l_v \mp m_{l_v})(l_v \pm m_{l_v} + 1)}\, D(\dots,(nlm_{l_v} \pm 1 m_{s_v}),\dots), \tag{D.5}$$

which are also valid for $\boldsymbol{S}$ in the analogous form. Hence we shall only consider $\boldsymbol{L}$ since the proof is similar for $\boldsymbol{S}$.

Suppose

$$|\mathscr{E}, SLSL\rangle = \sum_{\alpha=1}^{Q'} C'_\alpha(\mathscr{E})\, D'_\alpha(\dots,(nlm_l m_s),\dots),$$

then it is possible to construct the wave function

$$\Psi = \sum_{\alpha=1}^{Q'} C'_\alpha(\mathscr{E})\, \bar{D}'_\alpha(\dots,(nlm_l m_s)^{-1},\dots)$$

where each determinant $\bar{D}'_\alpha$ has been obtained from the corresponding D'_α in $|\mathscr{E}, SLSL\rangle$ by including for each incomplete shell those spin orbitals which are missing in D'_α. According to (D.5), $(L_x + \mathrm{i}L_y)\, D'_\alpha = 0$ for each α so that $(L_x - \mathrm{i}L_y)\, \bar{D}'_\alpha = 0$. Hence $L^2\psi = (L_z^2 - L_z)\,\psi = L(L+1)\,\psi$ according to (D.3), which shows that ψ is an eigenfunction of L^2 belonging to $L(L+1)$.

Suppose now that $|\mathscr{E}, SLSL-1\rangle$ is given by

$$|\mathscr{E}, SLSL-1\rangle = \sum_{\alpha=1}^{Q} a_\alpha D_\alpha .$$

Since

$$(L_x - \mathrm{i}L_y)|\mathscr{E}, SLSL\rangle = \sqrt{2L} \cdot |\mathscr{E}, SLSL-1\rangle$$

according to (D.4), it follows that

$$a_\alpha = \frac{1}{\sqrt{2L}} \sum_{\nu=1}^{Q} C'_\nu(\mathscr{E}) \langle D_\alpha | L_x - \mathrm{i}L_y | D'_\nu \rangle .$$

On the other hand, since $\psi = |\mathscr{H}, SL-S-L\rangle$, it follows from (D.4) that

$$(L_x + \mathrm{i}L_y)\psi = \sqrt{2L} \cdot |\mathscr{H}, SL-S-L+1\rangle .$$

Hence the coefficients in

$$|\mathscr{H}, SL-S-L+1\rangle = \sum_{\alpha=1}^{Q} b_\alpha \bar{D}_\alpha$$

can be obtained from

$$b_\alpha = \frac{1}{\sqrt{2L}} \sum_{\nu=1}^{Q} C'_\nu(\mathscr{E}) \langle \bar{D}_\alpha | L_x + \mathrm{i}L_y | \bar{D}'_\nu \rangle .$$

Evidently $\langle D_\alpha | L_x - \mathrm{i}L_y | D'_\nu \rangle$ and $\langle \bar{D}_\alpha | L_x + \mathrm{i}L_y | \bar{D}'_\nu \rangle$ must be identical except for a common phase factor. That this is indeed true follows from (D.5) since $\langle D_\alpha | L_x - \mathrm{i}L_y | D'_\nu \rangle$ can only be different from zero if D'_ν has at least one spin orbital u for which $m_l = m'_l + 1$, where m'_l pertains to a spin orbital v in $\bar{D}'_\nu$. In addition, D_α must contain that spin orbital v, in which case $\langle D_\alpha | L_x - iL_y | D'_\nu \rangle$ produces a nonzero contribution equal to

$$\sqrt{[l + (m'_l + 1)][l - (m'_l + 1) + 1]} = \sqrt{(l - m'_l)(l + m'_l + 1)} .$$

It is clear that this coefficient also arises from $\langle \bar{D}_\alpha | L_x + \mathrm{i}L_y | \bar{D}'_\nu \rangle$ provided $\bar{D}_\alpha$ has the spin orbital u pertaining to $m'_l + 1$ which would then not be present in D_α. Since $L_x - \mathrm{i}L_y$ is a sum of one-electron operators, D_α and D'_ν can only differ by v and u, so that a u in D_α would imply two u's in D'_ν. Since this is impossible, D_α does not contain u and the coefficients a_α and b_α are identical except for a phase factor. By continuing the step-up and step-down procedure for both $\boldsymbol{L}$ and $\boldsymbol{S}$ and by using the same arguments as above for each step, the required relationship between the wave functions (D.1) and (D.2) is established.

Appendix E. Conversion and normalization factors

Usually the atomic matrix elements are given in atomic units. The unit for distance is the radius of the first Bohr orbit, $a_0 = 4\pi\varepsilon_0 \hbar^2 / me^2 = 5.29177$

$\times 10^{-11}$ m; for energy, 1 hartree $= \mathrm{H} = e^2/4\pi\varepsilon_0 a_0 = 43.598 \times 10^{-19}$ J $= 27.21$ eV; and for time, $\tau = 4\pi\varepsilon_0 a_0 \hbar/e^2 = 2.419 \times 10^{-17}$ s.[8]

In order to display the units we write the transition probability corresponding to (9.1) in a simplified form and use SI units:

$$w = \frac{2\pi}{\hbar} \left| \langle f | \frac{e^2}{4\pi\varepsilon_0 r_{12}} | i \rangle \right|^2 \rho(E_\mathrm{f}), \tag{E.1}$$

where the value of the final-state density $\rho(E_\mathrm{f})$ depends on the choice of the normalization of the continuum wave function $|f\rangle$. The quantity

$$M^2 = \left| \langle f | \frac{e^2}{4\pi\varepsilon_0 r_{12}} | i \rangle \right|^2 \rho(E_\mathrm{f}) \tag{E.2}$$

has the unit of energy. If atomic units are used in M^2, then

$$w = \frac{2\pi e^2}{4\pi\varepsilon_0 \hbar a_0} M^2 = \frac{2\pi}{\tau} M^2, \tag{E.3}$$

where τ is the atomic unit of time. The width Γ is related to the probability w by $\Gamma = \hbar w$ so that

$$\Gamma = 2\pi H M^2, \tag{E.4}$$

where H is the atomic unit of energy. Hence in atomic units both w and Γ are expressed in the form $2\pi |\langle f | r_{12}^{-1} | i \rangle|^2 \rho(E_\mathrm{f})$. Usually w is required in units of eV/$\hbar$ and Γ in units of eV, in which case the atomic-unit value must be multiplied by 27.21 eV.

The continuum orbital is normalized in many ways in the literature. The simplest and most common choice is the normalization per unit energy range, in which case $\rho(E_\mathrm{f}) = 1$. The corresponding normalization factor C in the asymptotic form

$$R_{\varepsilon l}(r) \sim \frac{C}{r} \sin\theta(k, r) \tag{E5}$$

of the radial part of the continuum wave function is given by $C = \sqrt{2/\pi k}$. MCGUIRE[9] sometimes uses $C = \sqrt{4/k}$, i.e., a normalization to 2π per hartree so that $w = M^2$ with $\rho(E_\mathrm{f}) = 1$ in some of his works. KELLY[10] chooses $C = 1$ so that $w = (4/k) M^2$ and $\rho(E_\mathrm{f}) = 1$. Especially in the older literature, the continuum wave function is adjusted to yield one electron ejected per unit time. This corresponds to $C = \sqrt{4/k}$ so that $w = 2\pi M^2$ provided $\rho(E_\mathrm{f}) = 1/2\pi$.

The difference in the normalization per unit energy range and that corresponding to one electron ejected per unit time becomes clearer than above if we use SI units. In the former case $C = \sqrt{2/\pi\hbar v}$ and $\rho(E_\mathrm{f}) = 1$, in the latter case

[8] Cohen, E.R., Taylor, B.N. (1973): J. Phys. Chem. Ref. Data *2*, 663

[9] See II.44

[10] See II.51

$C=\sqrt{4/v}$ and $\rho(E_f)=h^{-1}$. Here $v=\hbar k/m$ and w is expressed as $2\pi\hbar^{-1}M^2$ in both cases. Sometimes there is also a normalization per unit wave number, in which case $C=\sqrt{2/\pi}$ and $\rho(E_f)=(\hbar v)^{-1}$.

Acknowledgments. Our work has profited from the happy collaboration with Osvaldo Goscinski. We would also like to thank him for many helpful and stimulating discussions. We are most grateful to Werner Mehlhorn for his adivce, encouragement and critical reading of the manuscript. Many colleagues have helped us with comments and information about their research. We would like especially to thank L. Armstrong, Jr., W.N. Asaad, C.P. Bhalla, B. Crasemann, J.P. Desclaux, H. Klar, M.O. Krause, F.P. Larkins, E.J. McGuire, V. Radojević, V. Schmidt, B.W. Shore, N. Stolterfoht, S. Stenholm, J. Tulkki, and U. Wille for their assistance. Our sincere thanks are due to Eeva-Kaarina Viinikka for her collaboration on the phase problem. We wish to thank especially Eva Lindström for her skilful typing of the manuscript. We are also indebted to Maija Peltonen for her help with the bibliography and Mirja Rautiainen for preparing the figures. This work has been supported by Imperical Chemical Industries and Science Research Council (UK) (GH) and Finnish Academy of Sciences (TÅ). The travel support by Helsinki University of Technology is also gratefully acknowledged.

General references

[*1*] Mott, N.F., Massey, H.S.W. (1965): The theory of atomic collisions, 3rd edition, Oxford: Clarendon Press
[*2*] Goldberger, M.L., Watson, K.M. (1964): Collision theory, New York: Wiley
[*3*] Rodberg, L.S., Thaler, R.M. (1967): Introduction to the quantum theory of scattering, New York: Academic Press
[*4*] Taylor, J.R. (1972): Scattering theory, New York: Wiley
[*5*] Joachain, C.J. (1975): Quantum collision theory, Amsterdam: North-Holland
[*6*] Landau, L.D., Lifshitz, E.M. (1962): Quantum mechanics, Oxford: Pergamon Press
[*7*] Messiah, A. (1961/1962): Quantum mechanics, Vols. I and II, Amsterdam: North-Holland
[*8*] Edmonds, A.R. (1974): Angular momentum in quantum mechanics, Princeton: Princeton University Press
[*9*] Condon, E.U., Shortley, G.H. (1964): The theory of atomic spectra, Cambridge: University Press
[*10*] Bethe, H., Salpeter, E.E. (1957) in: Handbuch der Physik, Vol. XXXV, Flügge, S. (ed.), Berlin, Göttingen, Heidelberg: Springer
[*11*] Sobel'man, I.I. (1972): Introduction to the theory of atomic spectra, Oxford: Pergamon Press
[*12*] Slater, J.C. (1960): Quantum theory of atomic structure, Vols. I and II, New York: McGraw-Hill
[*13*] Froese Fischer, Charlotte (1977): The Hartree-Fock method for atoms, New York: Wiley
[*14*] Shore, B.W., Menzel, D.H. (1968): Principles of atomic spectra, New York: Wiley
[*15*] Mizushima, M. (1970): Quantum mechanics of atomic spectra and atomic structure, New York: Benjamin
[*16*] de-Shalit, A., Talmi, I. (1963): Nuclear shell theory, New York: Academic Press
[*17*] Akhiezer, A.I., Berestetskii, V.B. (1965): Quantum electrodynamics, New York: Wiley

Subject Index

Aberration coefficient 246
– – for spherical sector analyzer 246
Absorption instruments 145
– –, double beam 145
– –, double ion chamber 147
– –, split beam 146
– of radiation 129, 143
– – – by excited atoms 178
– – – by ions 178
Acetaldehyde 417
Acetylacetons 439
Acetylenes 414, 434
Additive shifts 433
Adsorbate core line shapes 361
Ag, PES of metal valence bands 459
–, – – outer valence shell 309
Air, photoelectron spectrum 233
Alkali halides 442, 444
– metals 127
– –, PES of outer valence shell 309
– –, photoionization cross sections 169ff.
Al Kα radiation 228
Allenes 409
Analyzer
–, dispersive 244
–, undispersive 250
Angle of twist 438
Angular distributions of Auger electrons, coincidence experiments 554, 555
– – – –, high-energy electron impact 548, 549
– – – –, noncoincidence experiments 549ff.
– – – –, photoionization 547, 548
– – of photoelectrons (*see also* asymmetry parameter *β*) 31, 101ff., 102, 104, 127, 128, 155, 199, 424
– – – –, angular momentum transfer formulation 31ff.
– – – –, approximation of vanishing final state interactions 39
– – – –, central potential approximation 40
– – – –, Cooper-Zare formulation 40
– – – – from *p*-subshells 101ff.
– – – – from *s*-subshells 106ff.
– – – –, interchannel interaction effects 103
– – – – of adsorbates 455
– – – –, relativistic effects 108ff.
– – – – within a resonance 112
– – – –, Ar 201, 202
– – – –, C 107
– – – –, Cl 105, 106
– – – –, Cs 111, 205
– – – –, Kr 201, 203
– – – –, Ne 202
– – – –, Xe 104, 110, 201, 203, 204, 207
– momentum transfer 31
– – – formulation for *β* 31ff.
– – – theory 426
Anharmonicity 414
Anisotropy coefficients 551, 552
Approximate self-consistent molecular orbital theory, CNDO 435
– – – –, CNDO/S 421, 436
– – – –, INDO 435
– – – –, INDO/S 436
– – – –, LNDO/S 436
– – – –, MINDO 435
– – – –, RSPT 421
– – – –, SPINDO 436
Ar, autoionizing structure in absorption spectrum 169
–, *β*-parameter 102, 201, 202
–, double photoionization cross section 119, 195
–, ionization threshold 126
–, KLM Auger spectrum 308
–, partial photoionization cross section 53, 56, 62, 67, 71, 79, 81, 99
–, total photoionization cross section 164
–, 3*s* photoelectron spectrum 305, 306
Arsabenzene 430
AsH_3 407, 413, 415
Asymmetry parameter *β* (*see also* angular distribution of photoelectrons) 40, 101ff., 155, 200ff., 209, 289, 311
– – for the rare gases 291
– – of the $n=2$ satellite in He 313
– –, rotational state dependence 428
– –, vibrational state dependence 427
Atomic binding energies 315
Au, PES of metal valence bands 459
–, – – outer valence shell 309
Auger kinetic energies for calibration 272
– process 196
– transition, amplitudes 518, 538, 581ff.
– – component rate 519
– –, definition 501

Auger transition energy 501, 566ff.
- - group rate 519
- - intensity 523, 524
- - multiplet rate 519
- - term rate 519
Autodetachment resonances 446
Autoionization 132, 143, 159, 166, 399, 484ff.
-, Fano theory of 63
- state 428
Average energy, many-configuration jj 575
- -, single-configuration jj 541
- transition energy 576
Azomethane 441

Ba 174, 175, 309
-, photoelectron spectra 314, 315
-, satellite photoelectron lines 314
Band structure of Cu 456
- - of GaSe 453
- -, three-dimensional 452
- -, two-dimensional 450
Be, double photoionization cross section 119
-, photoionization cross section 91, 116
Beer's law 130
Benzene 359, 388, 389, 390, 411, 425, 428, 430
-, monosubstituted 430
-, shake up spectrum 359
-, valence photoelectron spectrum 387
-, vibrational energies 412
Benzoyl chloride 395
B_2H_6 348
BF_3 348, 397
Br_2^+ 412
BrO 442
Binding energies for free atoms 316
- - for krypton 298
- - for xenon 298
- energy, adiabatic 254
- -, atomic 293
- -, correlation effects 565, 566
- -, ΔSCF 285
- - for gases 272
- - for neon and H_2O 343
- - for solids 272
- -, nuclear size effects 564, 565, 588
- -, perturbation analysis 560
- -, quantum-electrodynamical effects 563, 564
- -, relativistic effects 562, 563
- -, vertical 284, 295
Biochemical process 436
Biological molecules 436
Body frame 476
Born-Haber cycle 354
Born-Oppenheimer approximation 294
Boundary conditions
- - for radial wave functions 497, 583, 588
- -, ingoing wave 475, 477, 486
- -, outgoing wave 472, 477
- -, standing wave 478
Bound state intensity method 397
Branching ratio 133, 152
Breit interaction 536, 538, 563
Breit-Wigner formula 483, 487
Brillouin's theorem 511
Butadiene 438

C1*s* shifts in fluoromethane 340
C_2 447
Calculation of vibrational structure 418
Calibration of photoelectron spectrum 268
Capillary discharge for uv-radiation 237
Carbonyl 434
Cd 152, 177
-, photoionization cross section 178
Ce, photoionization cross section of 4*d*-shell 51
Central field model, nonrelativistic 496ff.
- - -, relativistic 533
- potential approximation 39
- - - results for β 101
- - - results for σ 40
CH_2 442
CH_3 441
CH_4 305, 391, 393, 407, 426, 432
C_2H_6 407
C_3H_6 407
C_4H_6 437
Channel electron multiplier 255
CH_3Br 404, 405
CH_3Br^+ 422
CH_3Cl 404
CH_3Cl^+ 422
CH_3Ga 444
CH_3I 404, 414
CH_3I^+ 414, 422
CH_3In 444
CH_3OH 432
$(CH_3)_2Cd$ 444
$(C_2H_5)_2Cd$ 444
$(C_2H_5)_4Pb$ 444
$(C_6H_5)_4Sn$ 444
Chemical reaction 438
- shift 221
- - between surface and bulk metal atoms 355
- - of core electron lines 337
- -, model of 339
- -, thermodynamic model 351
Chemisorbed CO on tungsten 381
Chloroethenes 395
C_4F_6 437
$C_4H_2F_4$ 437
C_5H_5X(X=N, P, As, Sb, Bi) 434

$(CH_2OH_2)_2$ 433
Cl, β-parameter of 3s-shell 106
–, – of 3p-shell 105
ClO 442
Close-coupling method 66, 91
Cloud chamber 127, 128
C_2N_2 420, 421
CNDO, complete neglect of differential overlap method 435
CNDO/S, screened CNDO method 421, 436
CO 126, 368, 373, 396, 397, 426, 427, 432, 456, 457
–, adsorbed on nickel 457
–, adsorbed on tungsten 380
–, shake-up spectrum 357
CO^+ 400
CO_2 126, 374, 397, 401, 426, 427
–, photoelectron spectrum 266, 268
CO_2^+ 411, 422
C_3O_2 412
Coincident experiments 162
Composite molecule approach 433
Configuration interaction 287, 401, 506, 520, 579, 604
– –, form of the electric dipole transition matrix element 27, 28
– –, form of the final-state wave function 60ff.
– – in the final continuum state 60ff., 288, 301, 310
– – in the final ionic state 288, 302, 304, 307, 308, 379, 400
– – in the final state 288, 302
– – in the initial state 288, 302, 308, 310, 400
– –, interchannel interaction 61
– –, intrachannel interaction 61, 62
Continuum wave functions, nonrelativistic 497, 498, 582
– – –, normalization 498, 583, 588, 591, 618
– – –, numerical procedures 589ff.
– – –, radial 497, 498, 536, 537
– – –, relativistic 536, 537, 587ff.
Convolution integral 279
Cooper Minimum 55
Cooper-Zare formula for β 40
– – –, numerical results 101ff.
Core hole localization 346
– line multiplet structure 376
– – – – in molecules 376
– – – – in solids 376
– shift model including relaxation 345
Correlation effects 123, 169 (*see also* Electron correlation)
– – on binding energies 565, 566
– – on nonradiative transition amplitudes 604ff.
– – – transition energies 578, 581
COS 401, 426, 427
Coster-Kronig transition amplitudes 518, 538, 581ff.
– – component rate 519
– – definition 501
– – energy 501, 566ff.
– – group rate 519, 520
– – intensity 523, 524
– – multiplet rate 519
– – term rate 519
Coulomb approximation 89
– gauge 6, 490
– matrix element 502, 505
– phase shift 476, 497, 583
Cr 309
$Cr(CO)_6$ 397
CS 401
CS_2 401, 426, 427
Cs 127
–, β-parameter of 6s-shell 111, 205
–, photoionization cross section 57, 170, 171, 173
Cu 309, 459
CuO, Cu_2O 363
CuPd alloys 366, 464
Cylindrical capacitor analyzer 155, 246
– mirror analyzer 155, 247, 248
CX_4-compounds 404
1,4-cyclohexadiene 430

Decay probability, nonradiative 487, 488, 494
– –, partial 487
– –, radiative 492, 494
– –, relativistic 536
– rate of electron emission 283
Deconvolution method 279
ΔE_{SCF} method 512, 558, 565, 575, 576
Density of states structure 458
Deuteration 417
DF^+ 415, 424
Diarsine 438
Diazocyclo-hexadienone 439
Dibromamine 440
Dichloramine 440
Differential scattering cross section 160
Diphosphine 438
Dirac-Fock method, general 543, 563, 571, 575
– –, radial equation 587
Dirac-Slater calculation 406
Direct transitions in photoelectron spectra 448
Discrete basis set theories 100
Dispersion compensation 231
Dispersive analyzer 244
Doppler broadening 235
– – of HeI resonance radiation 235
Double electron excitation 159
– ion chamber 147

Double photoionization 113, 193
– –, Ar 119, 195
– –, Be 119
– –, He 113, 193
– –, Ne 116, 194
– –, Xe 199

Eigenchannel method 97
Electric-dipole approximation 6, 289
Electron affinities 180
– – of small molecules 445, 446
– binding energy 389
– correlation 123, 169, 196
– – theories for photoionization processes 59ff.
(*see also* names of individual theoretical methods: close-coupling, configuration interaction, discrete basis set method, eigenchannel method, hyperspherical coordinate method, many-body perturbation theory, multiconfiguration Hartree-Fock theory, quantum defect theory, random phase approximation, *R*-matrix method.)
– detector 255
– energy analyzer 153, 243
– – –, cylindrical mirror analyzer 246
– – –, ellipsoidal mirror 251
– – –, parallel plate mirror 248
– – –, spherical capacitor 245
– – –, spherical grid retarding potential 250
– – –, toroidal condenser 250
– impact ionization 399
– spectrometer 215
– –, iron-free double focusing 247
– –, time-of-flight 252
Electron spectroscopy, history of 218
Empirical rules of intensities of photoelectron bands 395
Energy level 507
– term 503
Equilibrium system 438
Error analysis 134
ESCA diffraction 380, 383
– – for adsorption systems 385
– –, single scattering model 385
– spectrometer for studies of solids and surfaces 264
Ether 438
Ethylene 401, 402, 417, 433
– glycol 438
Eu 309, 461
–, photoelectron spectrum 337

F_2^+ 415
Fano line profiles 168, 485, 615
Fermi's Golden Rule 527
Feshbach projection operators 483, 556, 610ff.
Final state density 490, 516, 517, 617
– – photon-electron interaction 492ff.
– – wavefunction 7ff.
– – –, incoming-wave boundary condition 8ff., 13, 20ff.
– – –, in-states 10
– – – normalization 7, 12, 14, 17
– – –, multichannel 15ff.
– – –, out-states 10, 13, 20ff.
– – –, partial wave expansion 11, 14
– – –, single channel 8ff.
Fluctuating dipoles 402
Fluorescence 159
Fock operator 509, 525
Formic acid 417
Four-spinor 532, 537
Franck-Condon principle 295
– – factors 418
Frozen-core approximation 501
Furan 430

Ga 309
GaSe 451
Gas-liquid chromatography 440
Gas-phase reactions 439, 442
Gauge invariance of the Schrödinger equation 26
– transformation 531
Ge 309
GeH_4 407
GeO 441
Gd 461
Generalized function 608
– oscillator strength 161
Gradient method 421
Grating monochromator 237
Greens function method 294, 421
Ground state potential model 343, 345, 348
Group electronegatives 434
– shift analysis 350
– – for carbon 351
– – model 350

H_2 422, 426, 429
–, values of asymmetry parameter β 430
H_2^+ 429
H_2O 286, 370, 391, 393, 413, 415, 427, 431, 432, 441
H_2O^+ 422, 424
H_2S 413, 415, 427
H_2S^+ 422, 424, 417
H_2Se 413, 415
H_2Te 413, 415
Hallucinogenic activity 435
Halogene atoms, PES of outer valence shell 309
– diatomic molecule 403, 414

Ham effect 422
HAM/3, hydrogenic atoms in molecules method 436
Hamiltonian, atomic 6
–, interaction 6, 26
Hammett parameter 434
Hartree-Fock method 68
– –, equations 284
– –, multiconfiguration 80
– –, radial equation 582
– –, restricted 511
– –, term averaged 511
– –, unrestricted 509, 510, 560
Hartree-Slater method 286
HBr^+ 415
HCN^+ 411
He^- 447
He, autoionizing structure 169
–, β-parameter 311
–, double photoionization cross section 113, 193
–, ionization threshold 126
–, photoelectron spectrum 309ff.
–, photoionization cross section 115, 164
HeI resonance radiation 235
– excitation 388
HeII resonance radiation 237
Heat pipes 149
Hexafluorobutadiene 438
HF 431
HF^+ 415, 424
Hg 152, 177
–, photoelectron spectrum 186
–, photoelectron spectrum of 4f-shell 266
–, photoelectron spectrum of valence shells 187
–, photoionization cross section 179
Hole configuration and wave functions 513ff.
Holographic grating 143
Hopfield continuum 128, 140
Hot band 412
Hückel method 435
– –, extended 435
Hydrazine 438
Hydrogen cyamide 417
Hydrogenic approximation 42
Hyperconjugation 435
Hyperspherical coordinate method 115

I_2^+ 412
ICl, photoelectron spectrum 266
Incoming wave boundary condition 8ff., 13, 20ff.
INDO, intermediate neglect of differential overlap method 435
INDO/S, screened INDO model 436
Inelastic mean free path, universal curve 275
Initial density of states (IDOS) structure 459
Inner-shell vacancy state energies 556ff.
– – –, notation 499
In-states 10
Intensity model for x-ray excited spectra 395
– of Auger transition 523, 524
– of main band 393
– of multielectron transition 289
– of multiplets for open shell atoms 328
– of one-electron transition 289
– of photoelectron bands, empirical rules 395
– of photoelectron line 273
Interchannel interaction 61, 526, 527, 546, 605, 606
– – effects on β 103
– – effects on σ 81
– –, limit of vanishing 36ff.
Interhalogen molecules 414
Intermediate coupling, nonrelativistic 506, 507, 520, 521
– –, relativistic 540ff., 594
Intrachannel interaction 61
Intramolecular field strength 444
Inversion barrier 415
Ion chambers 159
– –, double 147
Ionization efficiency 132
– yield 148
Iron tetracarbonylolefin complexes 395

Jahn-Teller interaction 389, 422
– –, active mode 410
– – coupling parameter 407
– –, distortions for point groups 408
– – effect, pseudo 409
– – theorem 406
jj-coupling 508, 517
jj to LS transformation 509, 516

K 127, 128, 152, 171
–, oxidation of films 382
Ketene 417
Keto-enol tautomerism 439
K-matrix 16ff., 479, 480
–, Hamiltonian eigenstates in terms of the 16
– integral equation 16
– on-the-energy-shell 17
–, relation to S-matrix 21
Kohn variational method 582
Koopmans' theorem 285, 510, 512, 560
– –, breakdown of 393
Kr, autoionizing structure 143, 166
–, β-parameter 201, 203
–, ionization potential 126
–, partial cross section 190
–, photoionization cross section 164, 170

La, photoionization cross section of 4*d*-shell 53
Lab frame 476
Lagrange-Helmholtz-relation 252
Lamb shift 491, 564, 574ff.
Latter tail correction 585
Length form for the transition matrix element 23ff.
Level shift
- -, nonradiative 482, 556, 557
- -, radiative 491
Li 171
Ligand-field theory 444
Light sources 136
- -, continuum 137, 140
- -, dc glow 137
- -, discrete 137
- -, duoplasmatron 137
- -, Hopfield continuum 140
- -, spark 138
- -, synchrotron 140
Lippmann-Schwinger equation 479, 493
Liquid sample handling in photoelectron spectroscopy 260
Liquid ESCA 261
LNDO/S, local neglect of differential overlap method/screened 436
Local exchange potential, nonrelativistic 584ff., 598
- - -, relativistic 589
Lone-pair orbital splitting 438
LS-coupling 403, 503ff., 517
Luminosity 157, 244
2,6-lutidine 430

Magic angle 54° 44′ 156, 424
Many-body perturbation theory 68ff., 294, 528, 529, 581, 605ff.
- - - -, choice of zero order basis 69, 72–74
- - - - for double photoionization 116ff.
- - - -, virtual double excitations in photoionization processes 74
Many-configuration jj-average energy 575
Many-electron transitions 400
- - wave function 505, 506, 508, 537, 538
Mass spectrometers 152
MCl_4 395
McLeod gauges 136
Mercury streaming 136
Metallocenes 393
Metal valence bands 459
- vapour oven 259
Methane 313, 367, 407, 427
Methanol 417
-, photoelectron spectrum of liquid 261
Methyl halides 405, 406
- cyanide 417
- radical 417
Mg Kα radiation 228
- - -, monochromatized 259
MgO, photoelectron spectrum 220
Microwave discharge to produce transient species 259
MINDO, modified INDO method 435
Mixed coupling 521
Mn 309
MnF_2 378, 379
Möller's formula 535, 536
Molecular subunit 432
- conformation 438
Molecules on metal surfaces 380
-, ionically bonded 442
MO methods for characterization of spectra 431
Monochromatized Al Kα 230
Monochromators 143
Monopole transition 293
Morse potential 418
Multichannel electron detection 255
- plates 255
- quantum defect theory 90ff.
- *R*-matrix theory 98
- scattering
- -, nonresonant 476–480
- -, resonant 480–483
- wavefunction 15ff
Multiconfiguration Hartree-Fock theory 80
- interaction method, Dirac-Fock 543, 563, 575, 577, 578
- - -, Hartree-Fock 511, 574, 575, 578
Multiple ionization 159, 193
- scattering calculation 397, 426, 436
- - MS-Xα approach 436
Multiplet 524

N, 1*s* spectrum of NH_4NO_3 374
-, - - of N_2 368
N_2 126, 368, 369, 373, 374, 390, 391, 393, 396, 397, 401, 427, 432
-, shake-up spectrum 357
-, vibrational energies 413
N_2^+ 400
Na 171, 172
NaCl 384, 385
Natural excitation 524
- line width 235
- - - of He I resonance radiation 235
- - - of photoelectron line 281
Ne, autoionizing structure 169
-, β-parameter of 2*p*-shell 202
-, binding energies 286
-, double photoionization cross section 116, 194
-, ionization threshold 126

-, $p \rightarrow d$ transitions 56
-, photoionization cross section 164, 184
-, 1s photoelectron line 297
-, 1s photoelectron spectrum 296ff.
-, 1s satellite photoelectron spectrum 299
Negative ions 180, 444
NH_3 126, 370, 407, 413, 415, 416, 432
N_2H_2 442
NO 126, 368, 377, 378, 397, 426, 427
NO^+ 414
NO_2 126
NS 442
Nondispersive analyzer 250
Nonlocal potentials 24ff.
Nonradiative transition amplitudes, effect of bound orbital basis on 600, 601, 605, 606
- - -, effect of local potential and kinetic energy on 598, 599
- - energies, ab initio calculations 574ff.
- - -, semiempirical treatments 566ff.
- - rates
- - -, closed-shell case 519, 520
- - -, nonrelativistic 519ff.
- - -, open-shell case 521ff.
- - -, relativistic 536ff.
Normalization screening theory 46
-, wavefunction 7, 12, 14, 17
Nuclear potential 588
- size effects on binding energies 564, 565
- - - on nonradiative transition energies 575

O_2 126, 368, 377, 378, 396, 397, 404, 426
O_2^+ 414, 415, 424
O_3 126, 442, 447
O_3^- 447
OH 442
Open-shell system 389
Orbital energy 284, 369
- - eigenvalues for Ar 298
- - shift 341, 342
- - shifts in second row elements 341
- - sum rule 435
Os 461
Oscillator strength, atomic 33ff.
- -, generalized 35, 161
- -, optical 133, 161
- -, relations of continuum and discrete spectra 89
- -, relation to photoionization cross section 34
Out-states 10, 13, 20
Overlap matrix element 512, 525
Oxidation of potassium films 382

P_2, vibrational energies 413, 447
Pair correlation energy 288
Parallel plate electron analyzer 155, 248
Paramagnetic molecules 376
Parity 496, 504, 517, 537
Partial photoionization cross sections 133, 152, 182
- - - -, Ar 190, 195
- - - -, Be 91, 116
- - - -, Ce 51
- - - -, Cs 57
- - - -, He 192
- - - -, Hg 188
- - - -, Kr 190
- - - -, Ne 184, 189, 194
- - - -, Xe 188, 198, 199
Pauli matrices 532
- Principle, random phase approximation violations of the 76
Pentafluoropyridine 430
Perfluoro effect 432
PH_3 407, 413, 415
Phase conventions, Condon and Shortley 497
- -, Fano and Racah 506
- factors 514, 515, 518, 521, 537
Phase-sensitive detection 332
Phonon broadening 374
Phosphabenzene 430
Phosphoryl chloride 395
Photodetachment 180, 444, 446
Photoelectric cross section 275
- - -, differential 289
- effect 219
- yield, Al 154
Photoelectron line, intensity 273
- -, natural width 281
- -, width 278
- spectrometer for gases and condensed matter 228
- spectroscopy 129, 154, 215
- -, basic processes 215
- -, experimental arrangement 215
- -, gas cell arrangement 257
- -, liquid sample handling 260
- -, sample preparation 256
- -, solid sample handling 262
- spectrum, calibration of 268
- -, deconvolution method 279
- -, density of states structure 459
- -, general features 265
- - of paraaminofluorobenzene 338
- -, temperature dependence of 438
- -, theoretical models 283
Photon source for photoelectron excitation 224
- -, Mζ 234
- -, uv 234
- -, x-ray 227
Photon-electron interaction 485, 490, 532
Photoionization 123

Photoionization cells 145
- cross section 7, 29, 30, 132, 164
- - -, central potential approximation 39ff.
- - -, delayed maximum at threshold 47ff.
- - -, high energy behaviour 42ff.
- - -, hydrogenic approximation 42ff.
- - -, measurement of 396
- - -, minimum above threshold 47, 55ff.
- - -, near threshold features 46ff.
- - -, vanishing final state interaction approximation 39ff.
- - -, Al 174
- - -, Ar 73, 74, 80, 82, 98, 164
- - -, Ba 175
- - -, Cd 178
- - -, Cs 171, 173
- - -, He 164
- - -, Hg 179
- - -, K 171
- - -, Kr 164, 166, 170
- - -, Li 171
- - -, Na 171, 172
- - -, Ne 164, 184
- - -, Rb 171
- - -, Xe 164, 167
- - -, Zn 176
- efficiency 132
- from solids, selection rules 448
- measurements 163
π-orbital splitting 438
$\pi \rightarrow \pi$ transitions 359
pKα activity 434
Plasmon excitation in photoelectron spectra 363
-, extrinsic 364
-, intrinsic 364
PN, vibrational energies 413
Polarization 144
- effects, long-range 67
Polarized radiation 238
Polyacetylenes 414, 434
Polymers 338, 339
Potassium 127, 128, 152, 171
- films, oxidation of 382
- halides 376
Potential barriers 47ff.
-, central atomic 39ff.
- curve gradients 370
-, Herman-Skillman 41
-, non-local 24ff.
-, perturbation theory for screened Coulomb 45
-, vector 6
Predissociative state 415
Preredardation electron lens 216, 252
- of photoelectrons 254
Pseudopotential 584
- approaches 436
Pullman's index 437
Purine 436
Pyridazines 438
Pyridine 393, 395, 430, 434
Pyrrol 430

Quantum defect theory 86ff.
- - -, multichannel 90ff.
- - -, single channel 86ff.
Quantum-electrodynamical effects on binding energies 563ff.
- - on nonradiative transition energies 575, 576

Radial Schrödinger equation 497
- - -, nonrelativistic 582, 583
- - -, relativistic 587
Random phase approximation 75ff., 294
- - -, configuration-interaction interpretations 77ff.
- - -, extensions 83ff.
- - - interchannel interaction predictions 81ff.
- - -, Pauli principle violations 76, 77
- - - with exchange 296, 303, 529, 530
Rare earth elements 309, 461
Rb 127, 171
Re 461
Reaction rate 440
Reduced zone scheme 448
Relativistic effects 393
- - on asymmetry parameter β 108ff.
- - on binding energies 562, 563
- - on nonradiative transition amplitudes 593ff.
- - - - energies 571ff., 575
Relaxation 524, 560, 569, 602
- energies for s-subshells 287
- energy 285, 342, 347, 370
- -, division of 347
- - for boron 1s ionization 348
Renner active mode 410
- effect 416, 422
Reorganization energy 285
Resolution of spherical sector electron spectrometer 245
Resonance, autoionization profile of a 63ff., 73, 91ff., 99
-, β parameter within a 93, 107, 112
- lines 139
- scattering, Auger electron emission 488
- -, autoionization 486
- -, general 480ff.
- -, photon emission 489
R-matrix theory 94ff.
- - -, multichannel 98ff.

- - - relation to eigenchannel method 97
- - -, single channel 95ff.
Rotating anode 228
Rotational excitations 422
Rowland circle 230, 231
RSPT, Rayleigh-Schrödinger perturbation theory method 421
Russell-Saunders coupling 403, 503ff., 517
Rydberg-Klein-Rees potential 419

S_2^+ 415
Sample preparation for photoelectron spectroscopy 256
Satellite lines in the photoelectron spectrum 310
- - in adsorbate core photoelectron spectra 361
- - in benzene 360
- - in core electron spectra of solids 360
- - in transition metal core spectra 362
- structure of valence bands 400
- transitions in molecular spectra 357
SbH_3 407, 413, 415
SbH_3^+ 409
Selection rules in photoionization from solids 448
Self-absorption of HeI resonance radiation 235, 236
Self-consistent-field method 286, 510, 574ff.
Semi-classical harmonic oscillar description 418
SF_6 397
Shake off 292
Shake up 266, 292
- - spectrum of N_2 357
- - - benzene 359
- - - of CO 359
Shape of metal core electron line 365
- resonance 397, 428
Shell 498, 499
Short-lived species 440
Signal to background ratio in electron spectra 232
SiH_4 407
SiO 441
SiF_4 349
Single-configuration jj-average energy 541
Single scattering model for ESCA diffraction 385
Singularity index 365
Slater determinant 284, 501
- integral 503
Sm 309, 461
S matrix, nonresonant 474–479
-, resonant 481, 482
Sn 309
SnH_4 407
SO 442
SO^+ 415
SO_2 447
Soft X-ray emission spectra 371
- - - - of Co and N_2 373
Solid sample handling in photoelectron spectroscopy 262
Spectra, resonance lines 139
Spectrograph 143
Spherical analyzer (180°) 155, 245
- harmonics 496, 497
SPINDO, spectroscopic-potentials-adjusted INDO method 436
Spin orbital 498, 501
Spin-orbit interaction 389, 444, 498, 499, 506, 562, 572
- coupling parameter 404
- splitting 403
Stationary scattering wave function 472–475
Statistical weights 184, 185
- -, Xe 185
Strength of a transition 524, 548
Subshell 499
Substituent effect on the vibrational structure 417
Sudden approximation 292
Sulphide 438
Sulphoxides 438
Sulphur-containing molecules 342
Sum rule for satellite intensities 293
Super Coster-Kronig transition 288, 501
- - - amplitudes 518, 538, 581ff.
- - - component rate 519
- - - energy 501, 566ff.
- - - group rate 519, 520
- - - in Xe 302
- - - intensity 523, 524
- - - multiplet rate 519
- - - term rate 519
Supersonic jet technique 258
Symmetry coordinates 419
- species of vibronic levels in linear molecules 411
Synchrotron radiation 136, 240
- - in excitation of electron spectra 240
- -, universal curve 241

t-butyl 441
TeH_2 393
Te_2^+ 404
Teller-Redlich product rule 418
Temperature variation technique 412
Thallium halides 444, 445
Theoretical models for photoelectron spectra 283
Thermodynamic model 353
- - of chemical shift 351, 354

Thionanisole 438
Thiophene 430
Thomas-Reiche-Kuhn sum rule 34
Three-step model 449
Threshold law for multiple ionization 120
- maximum in σ 47ff.
- photoelectron spectroscopy 415
Through-bond interaction 435
Through-space interaction 435
Time-of-flight electron spectrometer 252
T matrix, nonresonant 479, 480
-, resonant 482, 485, 486
Toroidal condenser electron analyzer 250
Transient species 440
Transition metal, photoionization cross section of $3p$-shell 54
- method for calculation of SCF binding energies 286
- operator method 346, 512, 575
- potential model 348, 349
Trithiapentalene 370, 371
Two-dimensional band structures 450
Two-electron wave function 508

UF_6 393, 394
UF_6^+ 403
UV line source 236
- polarizer 239
- radiation source 234, 238, 240

Va, photoionization cross section of $3p$-shell 53
Valence bands, metal 459
- -, alloy 461
- - in hydrocarbon 465
- - of Au, Ag and Cu 460
- photoelectron spectrum 387
- - - of alkanes 466
- - - of benzene 387
Variational collapse 557ff.
Velocity form for the transition matrix element 23ff.
Vibrational coupling constants 421
- energies of benzene 412
- excitations 294, 338
- - in core electron spectra 366
- overlap integral 419
- selection rules 409
- - -, calculation of 418
- structure and orbital shape 412
Vibronic interaction, type a 410, 428
- -, type b 410
Virtual bound states in alloys 464
Voigt function 279

Weisskopf-Wigner approximation 489
Wentzel ansatz 516, 527, 528
Width of photoelectron line 278
-, nonradiative 494
-, radiative 494

X_α method 286
- -, nonrelativistic 575
- -, relativistic 575
Xe 126
-, absorption coefficient 167
-, β-parameter 94, 103, 109ff., 201ff.
-, double ionization cross section 199
-, partial cross section of $5s$-shell 83, 188
-, photoionization cross section 164
-, super Coster-Kronig transitions 302
-, $4s-4p$ photoelectron spectrum 304
XeF_2 444
XeF_4 444
X-ray source 229, 230
- monochromatization 230, 232
- monochromator 231

Yb 309

Zn 152, 176

A monthly journal

Applied Physics A
Solids and Surfaces

Applied Physics A "Solids and Surfaces" is devoted to concise accounts of experimental and theoretical investigations that contribute new knowledge or understanding of phenomena, principles or methods of applied research.
Emphasis is placed on the following fields (giving the names of the responsible co-editors in parentheses):

Solid-State Physics
Semiconductor Physics (**H. J. Queisser,** MPI Stuttgart)
Amorphous Semiconductors (**M. H. Brodsky,** IBM Yorktown Heights)
Magnetism (Materials, Phenomena) (**H. P. J. Wijn,** Philips Eindhoven)
Metals and Alloys, Solid-State Electron Microscopy (**S. Amelinckx,** Mol)
Positron Annihilation (**P. Hautojärvi,** Espoo)
Solid-State Ionics (**W. Weppner,** MPI Stuttgart)

Surface Physics
Surface Analysis (**H. Ibach,** KFA Jülich)
Surface Physics (**D. Mills,** UC Irvine)
Chemisorption (**R. Gomer,** U. Chicago)

Surface Engineering
Ion Implantation and Sputtering (**H. H. Andersen,** U. Aarhus)
Laser Annealing (**G. Eckhardt,** Hughes Malibu)
Integrated Optics, Fiber Optics, Acoustic Surface-Waves (**R. Ulrich,** TU Hamburg)

Special Features:
Rapid publication (3–4 months)
No page charges for concise reports
50 complimentary offprints
Microform edition available

Articles:
Original reports and short communications.
Review and/or tutorial papers
To be submitted to:
Dr. H. K. V. Lotsch, Springer-Verlag, P. O. Box 105280, D-6900 Heidelberg, FRG

Springer-Verlag
Berlin
Heidelberg
New York

W. Demtröder

Laser Spectroscopy

Basic Concepts and Instrumentation
2nd corrected printing. 1982. 431 figures.
Approx. 715 pages. ISBN 3-540-10343-0

This textbook is an introduction to modern techniques and intrumentation in laser spectroscopy. After an elementary discussion of basic subjects such as absorption and dispersion of light, coherence, line broadening effects and saturation phenomena a detailed outline of spectroscopic instrumentation, such as spectrographs, interferometers and photodetectors is given.
The main part of the book deals with recently developed techniques in laser spectroscopy and the increased sensitivity and spectral resolution they have made possible. Doppler-free methods are particularly emphasized.
This book helps close the gap between classical works on optics and spectroscopy and more specialized publications on modern research in this field. It is addressed to graduate students in physics and chemistry as well as scientists just starting out in this field.

Electron Spectroscopy for Surface Analysis

Editor: **H. Ibach**
With contributions by numerous experts
1977. 123 figures, 5 tables. XI, 255 pages. (Topics in Current Physics, Volume 4). ISBN 3-540-08078-3

Contents: Introduction. - Design of Electron Spectrometers for Surface Analysis. - Electron-Excited Core Level Spectroscopies. - Electron Diffraction and Surface Defect Structure. - Photoemission Spectroscopy. - Electron Energy Loss Spectroscopy.

H. Haken

Synergetics

An Introduction
Nonequilibrium Phase Transitions and Self-Organization in Physics, Chemistry and Biology
2nd enlarged edition. 1978. 152 figures, 4 tables.
XII, 355 pages. (Springer Series in Synergetics, Volume 1). ISBN 3-540-08866-0

"Synergetics, according to Professor Haken, is the study of how component subsystems can interact to produce structure and coherent motion on a macroscopic scale. In fact it is a theory of selforganisation with applications not only in physics and chemistry but also - biology and sociology. In this book an introduction is given to the basic physical ideas and mathematical methods to be used. The text is imaginatively written and well illustrated by an amazing variety of examples drawn from such diverse fields as laser physics, fluid dynamics, mechanical engineering, chemical reactions, ecology and morphogenesis. ... Professor Haken is to be congratulated in producing such a readable introduction to a subject still in its infancy."
Physics Bulletin

J. Kessler

Polarized Electrons

1976. 104 figures. IX 223 pages. (Texts and Monographs in Physics). ISBN 3-540-07678-6

"... a self-contained, intelligible introduction to the physics of spinpolarized free electrons. This monograph is an excellent guide for graduate students and research scientists interested in understanding either the growing field of polarized electrons or the insights this field can provide for other areas of physics."
Physics Today

Laser Spectroscopy

of Atoms and Molecules
Editor: **H. Walther**
With contributions by numerous experts
1976. 137 figures, 22 tables. XVI, 383 pages.
ISBN 3-540-07324-8

Contents: Atomic and Molecular Spectroscopy with Lasers. - Infrared Spectroscopy with Tunable Lasers. - Double-Resonance Spectroscopy of Molecules by Means of Lasers. - Laser Raman Spectroscopy of Gases. - Linear and Nonlinear Phenomena in Laser Optical Pumping. - Laser Frequency Measurements, the Speed of Light, and the Meter.

Raman Spectroscopy of Gases and Liquids

Editor: **A. Weber**
With contributions by numerous experts
1979. 103 figures, 25 tables. XI, 318 pages
(Topics in Current Physics, Volume 11)
ISBN 3-540-09036-3

"... An attractive feature of this volume is the candid perspective used by all of the authors in treating their respective areas of specialty..."
IEEE J. of Quantum Electronics

Springer-Verlag
Berlin
Heidelberg
New York